Microbial Interventions in Agriculture and Environment

Dhananjaya Pratap Singh
Vijai Kumar Gupta • Ratna Prabha
Editors

Microbial Interventions in Agriculture and Environment

Volume 1 : Research Trends, Priorities and Prospects

Editors
Dhananjaya Pratap Singh
Department of Biotechnology
ICAR – NBAIM
Maunath Bhanjan, Uttar Pradesh, India

Ratna Prabha
Department of Biotechnology
ICAR – NBAIM
Maunath Bhanjan, Uttar Pradesh, India

Vijai Kumar Gupta
Department of Chemistry & Biotechnology,
School of Science
Tallinn University of Technology
Tallinn, Estonia

ISBN 978-981-13-8393-9 ISBN 978-981-13-8391-5 (eBook)
https://doi.org/10.1007/978-981-13-8391-5

This Springer imprint is published by the registered company Springer Nature Singapore Pte Ltd.
The registered company address is: 152 Beach Road, #21-01/04 Gateway East, Singapore 189721, Singapore

Foreword

Despite the simplicity of the way of life, the microorganisms present an abundant genetic diversity in the nature coming from their chromosomal DNAs as well as plasmid DNAs, obtained through the evolution over time so that they could adapt to the conditions of all marine and terrestrial ecosystems.

There were billions of years of ecological adaptation to fulfill their function on planet Earth, which is to provide a balance of nutrients essential for the maintenance of global ecosystems. This high biological diversity of the world's microbiota has provided human beings with practical, economical, and/or ecological opportunities for most of the known microorganisms by exploiting their metabolic capacities at the biochemical and molecular levels and defining the multiple functions associated with the simplest cell. Their long history of evolutionary diversification has provided them with a variety of genes, proteins, enzymes, and metabolites (primary and secondary) that can be of great help for the sustainable use of the environment, agriculture, and bioindustry, as well as the health of living things in general.

In recent decades, research on microorganisms has been intensified, with substantial advances in the biotechnology area, reflecting new ideas, thoughts, approaches, methods, media, tools, techniques, and, in particular, biotechnological products from the global microbiota. Genetic sequencing techniques with the aid of computational methods have placed the research with microorganisms at a much higher level.

Furthermore, application-oriented research encompassing microbe-based products (live cells and metabolic bioformulations) has exclusively gone in favor of agriculture and the environment.

The book *Microbial Interventions in Agriculture and Environment* in its first volume *Research Trends, Priorities and Prospects* is presenting a consolidated account of the authenticated work from the renowned authors working worldwide to explore the myth tagged with the tiny, often unseen but functionally sound, living species. Structurally arranged chapters cover principles and mechanisms, methods and tools, approaches and illustrations, and practices and applications in the most updated but lively manner to make the volume workable for the existing as well as new readership. The subject is discussed in the simplistically narrative form, and the topics are covered in a precise manner to reflect an open account of the research area before the readers. My optimistic view takes me to state that the research community will surely be benefitted from this compilation as it explores many of the hidden and undiscovered facts about microbial life to which the world is looking for. I am sure that this volume will certainly enlighten researchers, faculties, scholars, students, and professionals because of its collectively composed literature on a vast subject.

Luiz Antonio de Oliveira
Former Vice Director
National Institute for Amazonian
Research (INPA/MCTI),
Manaus, Amazonas, Brazil

Preface

Our environment encompasses microbial life-forms at a very higher level in any agroecosystem. Microbial communities act as the most powerful and vital principles of biology to regulate almost all kinds of life processes. Be it the maintenance of biogeochemical and carbon cycle, degradation and decomposition of undesired biological or anthropogenic resources, management of nutrients and minerals in agroecosystems, causal or suppression of biotic stresses (diseases) in organisms, or regulation of overall functionalities of the associative organisms in the soil, air, and water, microorganisms find their implicit and intertwined role. This is why microbial communities comprise the real unseen wealth on the Earth. Microbial life-forms possess tremendous genetic diversity always clubbed with the functional potentialities. Diverse range of functions in the microbial communities arise due to their adaptation capabilities in the continuously changing environments, even under the extreme ones. Therefore, because of an array in chemical diversity, it is believed these unseen organisms not only silently intervene but explicitly regulate carbon sequestration, gaseous exchange, nutrient acquisition, and mineralization processes. Such processes ultimately benefit soils and crops due to improved soil fertility and plant health.

Microbial research has witnessed tremendous developments in the past few decades. Advancements in newly evolved principles, methodologies, protocols, instruments, computation tools, and techniques have led to establish this field to a new height. The outcomes of genomics, proteomics, and metabolomics studies have resulted in classifying newer functions for the isolated and characterized genes, proteins, and metabolites in holistic way. The work on the genomics of prokaryotic and eukaryotic microbial species has made taxonomic as well as functional revelations possible. Since microbial life-forms are integrated components of both the agricultural and environmental systems, majority of work was devoted in isolating, culturing, identifying, and characterizing functions of the genes, proteins, metabolites, or even the organisms. By knowing the functions, the microbial species or their products (genes, proteins, or metabolites) have been assigned specific attributes for agricultural and environmental benefits. Although we are exposed to very little volume of overall microorganisms due to the lack of culturability and speculate to know only 1% of the total communities, work on the rest of the unknown microbial world has remained attractive and interesting but challenging too. Advancements in sequencing technologies have helped in the emergence of microbial meta-omics research era which not only addresses holistic manner of taxonomic diversity in any ecosystem but also counts on the functional attributes of the communities to decipher their role in varied habitats. We are now blessed with the knowledge on multifarious aspects of microbial life and functions and look into the priorities and prospects of this area for future generations.

The book *Microbial Interventions in Agriculture and Environment* in its volume *Research Trends, Priorities and Prospects* addresses current research leads and emerging trends to identify and fix the prioritized areas and escalate prospective future projections in the application of microbial research for the benefit of agriculture and environment. The volume covers discussions on the prospective journey of one of the most important beneficial fungi, *Trichoderma*, in combating global warming and hothouse conditions. Mitigation strategies of crops under microbial interaction systems have also been discussed. Applications of prospective tools like metagenomics and metabolomics for deciphering taxonomic and functional aspects of habitat-specific microbiomes are covered to a greater extent for generating insights on the use of emerging out-of-box approaches, tools, and techniques. Research trends on microbial applications in agriculture and environment included application-based impact assessment in crops for plant growth promotion, biological control, bioremediation, biodegradation, decomposition, and bioconversion. The volume covers the work on wide applications of microbial inoculants on crops to signalize potential benefits for crop production across the world. Microbial species have remained the best source of enzymes, proteins, metabolites, biofuels, biorefineries, foods, and feed resources. The content of this volume also discusses these aspects reflecting potential and prospective applications of microbial life in such emerging areas. Understanding microbial interactions within their own communities and environment and with the associated plants (plant-microbe interactions) has remained a challenging task. Research encompassing microbial interactions has been narrated in terms of modulating physiological, biochemical, and molecular changes in the interacting species and, thereby, mitigating biotic and abiotic challenges by the plant and microbial species. These areas, which are discussed in detail in this volume, hold valuable prospects for future agriculture which is facing pressure from the emergence of newer disease-causing agents, continuous climate change, and global warming.

We believe that the work presented in this volume has potential prospective values for the readers who are in search of meaningful collection of literature source on microbial interventions in agriculture and environment. We are sure that this volume will invite wide attention of the targeted readership worldwide in pursuance of the structurally sound, thoughtfully compiled, and well-cited subject content on all-time microbial research trends, priorities and prospects.

Maunath Bhanjan, Uttar Pradesh, India — Dhananjaya Pratap Singh
Tallinn, Estonia — Vijai Kumar Gupta
Maunath Bhanjan, Uttar Pradesh, India — Ratna Prabha

Contents

About the Editors and Contributors

Editors

Dhananjaya Pratap Singh is presently Principal Scientist in Biotechnology at the ICAR-National Bureau of Agriculturally Important Microorganisms, Maunath Bhanjan, India. He completed his master's degree from G. B. Pant University of Agriculture and Technology, Pantnagar, and his Ph.D. in Biotechnology from Banaras Hindu University, Varanasi. His research interests include plant-microbe interactions, bioprospecting of metabolites of microbial and plant origin, microbe-mediated stress management in plants, metabolomics-driven search for small molecules, and bioinformatics in microbial research. He has been working on the societal implications of microbial biotechnology pertaining to microbe-mediated crop production practices and rapid composting of agro-wastes at farm and farmer's levels. He has successfully performed outreach of such technologies to farming community for adoption at field scale. He has been associated with the development of supercomputing infrastructure for agricultural bioinformatics in microbial domain in India under the National Agricultural Bioinformatics Grid (NABG) program of the ICAR. He is an Associate of the National Academy of Agricultural Sciences (NAAS), India, and has been awarded with several prestigious awards including Dr. A. P. J. Abdul Kalam Award for Scientific Excellence. With many publications in the journals of national and international repute, he has also edited five books on microbial research with Springer Nature and other publishers.

Vijai Kumar Gupta is the Senior Scientist, ERA Chair of Green Chemistry, Tallinn University of Technology, Estonia. His area of research interests include bioactive natural products, microbial biotechnology and applied mycology, bioprocess technology, biofuel and biorefinery research, and glycobiotechnology of plant-microbe interaction. He is the Secretary of the European Mycological Association and Country Ambassador of the American Society for Microbiology. He is a Fellow of Linnaean Society and Mycological Society of India and Associate Fellow of the National Academy of Biological Sciences, India, and Indian Mycological Association. He has been the Editor of reputed journals of international recognition and edited 28 books with publishers like Elsevier, Wiley-Blackwell, Frontiers Media SA, Taylor & Francis, Springer Nature, CABI, and De Gruyter. To his credit, he also has a vast number of research publications and review papers in internationally reputed high-impact factor journals. He also holds two IPs in the area of microbial biotechnology for sustainable product development.

Ratna Prabha is currently working as DST Women Scientist at the ICAR-National Bureau of Agriculturally Important Microorganisms, India. With doctorate in Biotechnology and master's degree in Bioinformatics, she has been actively involved in different research activities. Her research interests lie in microbe-mediated stress management in plants, database development, comparative microbial analysis, phylogenomics and pangenome analysis, metagenomics data analysis, and microbe-mediated composting technology development and dissemination. She has been engaged in developing various digital databases on plants and microbes and has various edited and authored books, many book chapters, and different research papers and review articles in journals of international repute.

Contributors

Sabiha Aaysha Department of Biotechnology, Bundelkhand University, Jhansi, Uttar Pradesh, India

Jayanthi Abraham Microbial Biotechnology Laboratory, School of Biosciences and Technology, VIT, Vellore, Tamil Nadu, India

Temoor Ahmed Department of Bioinformatics & Biotechnology, Government College University, Faisalabad, Pakistan

Roshan Ara Department of Biotechnology, Bundelkhand University, Jhansi, Uttar Pradesh, India

Hassan Azaizeh Institute of Applied Research (Affiliated with University of Haifa), The Galilee Society, Shefa-Amr, Israel
Department of Environmental Science, Tel Hai College, Qiryat Shemona, Israel

Ivelina Babrikova Department of Microbiology and Environmental Biotechnologies, Agricultural University – Plovdiv, Plovdiv, Bulgaria

Rhaissa R. Barbosa Metabolomics Research Group, Departamento de Química Orgânica, Instituto de Química, Universidade Federal da Bahia, Salvador, Brazil

Hemant P. Borase School of Life Sciences, Kavayitri Bahinabai Chaudhari North Maharashtra University, Jalgaon, Maharashtra, India
C. G. Bhakta Institute of Biotechnology, Uka Tarsadia University, Surat, Gujarat, India

Ambalal Babulal Chaudhari School of Life Sciences, Kavayitri Bahinabai Chaudhari North Maharashtra University, Jalgaon, India

Navin Dharmaji Dandi School of Life Sciences, Kavayitri Bahinabai Chaudhari North Maharashtra University, Jalgaon, India

Suvendu Das Institute of Agriculture and Life Science, Gyeongsang National University, Jinju, Republic of Korea

Catherine P. de Almeida Metabolomics Research Group, Departamento de Química Orgânica, Instituto de Química, Universidade Federal da Bahia, Salvador, Brazil

Prajkata Deshmukh Department of Biotechnology, National Institute of Pharmaceutical Education and Research-Guwahati, Guwahati, Assam, India

Sonali Dua Amity Institute of Microbial Biotechnology, Amity University, Noida, Uttar Pradesh, India

Mahek Fatima Amity Institute of Microbial Biotechnology, Amity University, Noida, Uttar Pradesh, India

Prerna Gautam Amity Institute of Microbial Biotechnology, Amity University, Noida, Uttar Pradesh, India

Pavankumar M. Gavit School of Life Sciences, Kavayitri Bahinabai Chaudhari North Maharashtra University, Jalgaon, India

Danail Georgiev Faculty of Biology, Department of Microbiology, University of Plovdiv, Plovdiv, Bulgaria

Vijai Kumar Gupta Department of Chemistry and Biotechnology, School of Science, Tallinn University of Technology, Tallinn, Estonia

Gary E. Harman Cornell University, Geneva, NY, USA

Sabir Hussain Department of Environmental Sciences & Engineering, Government College University, Faisalabad, Pakistan

Syed Sarfraz Hussain Department of Biological Sciences, Forman Christian College (A Chartered University), Lahore, Pakistan
School of Agriculture, Food & Wine, Waite Campus, University of Adelaide, Adelaide, SA, Australia

Prasad Jape School of Life Sciences, Kavayitri Bahinabai Chaudhari North Maharashtra University, Jalgaon, India

Abiya Johnson Department of Biotechnology, National Institute of Pharmaceutical Education and Research-Guwahati, Guwahati, Assam, India

Raksha Ajay Kankariya School of Life Sciences, Kavayitri Bahinabai Chaudhari North Maharashtra University, Jalgaon, India

Namera C. Karun Department of Biosciences, Mangalore University, Mangalore, Karnataka, India

Brijendra K. Kashyap Department of Biotechnology, Bundelkhand University, Jhansi, Uttar Pradesh, India

Megha Kastwar Department of Biotechnology, Bundelkhand University, Jhansi, Uttar Pradesh, India

Shubhangi Kaushik Department of Biotechnology, National Institute of Pharmaceutical Education and Research-Guwahati, Guwahati, Assam, India

Jae-Yean Kim Division of Applied Life Science (BK21 Plus program), Plant Molecular Biology and Biotechnology Research Center, Gyeongsang National University, Jinju, South, Korea

Sunil H. Koli School of Life Sciences, Kavayitri Bahinabai Chaudhari North Maharashtra University, Jalgaon, Maharashtra, India

Gagan Kumar Department of Mycology and Plant Pathology, Institute of Agricultural Sciences, Banaras Hindu University, Varanasi, India

Lallawmsangi Molecular Microbiology and Systematic Laboratory, Department of Biotechnology, Mizoram University, Aizawl, Mizoram, India

Lalrokimi Molecular Microbiology and Systematic Laboratory, Department of Biotechnology, Mizoram University, Aizawl, Mizoram, India

Éva Laslo Faculty of Economics, Socio-Human Sciences and Engineering, Department of Bioengineering, Sapientia Hungarian University of Transylvania, Miercurea Ciuc, Romania

Vincent Vineeth Leo Molecular Microbiology and Systematic Laboratory, Department of Biotechnology, Mizoram University, Aizawl, Mizoram, India

Yang-Rui Li Key Laboratory of Sugarcane Biotechnology and Genetic Improvement (Guangxi), Ministry of Agriculture, Sugarcane Research Center, Chinese Academy of Agricultural Sciences, Guangxi Key Laboratory of Sugarcane Genetic Improvement, Sugarcane Research Institute, Guangxi Academy of Agricultural Sciences, Nanning, China
College of Agriculture, Guangxi University, Nanning, China

Anupam Maharshi Department of Mycology and Plant Pathology, Institute of Agricultural Sciences, Banaras Hindu University, Varanasi, India

Vijay Maheshwari School of Life Sciences, Kavayitri Bahinabai Chaudhari North Maharashtra University, Jalgaon, India

Faisal Mahmood Department of Environmental Sciences & Engineering, Government College University, Faisalabad, Pakistan

Natasha Manzoor Department of Soil and Water Sciences, China Agricultural University, Beijing, China

Gyöngyvér Mara Faculty of Economics, Socio-Human Sciences and Engineering, Department of Bioengineering, Sapientia Hungarian University of Transylvania, Miercurea Ciuc, Romania

Jose Mathew Department of Biotechnology, Bundelkhand University, Jhansi, Uttar Pradesh, India

Jyoti Misri Division of Animal Science, Indian Council of Agricultural Research, New Delhi, India

Bhavana V. Mohite School of Life Sciences, Kavayitri Bahinabai Chaudhari North Maharashtra University, Jalgaon, Maharashtra, India

Sher Muhammad Department of Bioinformatics & Biotechnology, Government College University, Faisalabad, Pakistan

Arpan Mukherjee Department of Botany, Institute of Science, Banaras Hindu University, Varanasi, India

Soumya Nair Microbial Biotechnology Laboratory, School of Biosciences and Technology, VIT, Vellore, Tamil Nadu, India

Chandrakant P. Narkhede School of Life Sciences, Kavayitri Bahinabai Chaudhari North Maharashtra University, Jalgaon, Maharashtra, India

Muhammad Noman Department of Bioinformatics & Biotechnology, Government College University, Faisalabad, Pakistan

Satish V. Patil School of Life Sciences, Kavayitri Bahinabai Chaudhari North Maharashtra University, Jalgaon, Maharashtra, India
North Maharashtra Microbial Culture Collection Centre (NMCC), Kavayitri Bahinabai Chaudhari North Maharashtra University, Jalgaon, Maharashtra, India

Vikas S. Patil University Institute of Chemical Technology, Kavayitri Bahinabai Chaudhari North Maharashtra University, Jalgaon, Maharashtra, India

Ratna Prabha Department of Biotechnology, ICAR – NBAIM, Maunath Bhanjan, Uttar Pradesh, India

Richa Raghuwanshi Department of Botany, Institute of Science, Banaras Hindu University, Varanasi, India

Jamatsing D. Rajput School of Life Sciences, Kavayitri Bahinabai Chaudhari North Maharashtra University, Jalgaon, Maharashtra, India

Paulo R. Ribeiro Metabolomics Research Group, Departamento de Química Orgânica, Instituto de Química, Universidade Federal da Bahia, Salvador, Brazil

Birinchi Kumar Sarma Department of Mycology and Plant Pathology, Institute of Agricultural Sciences, Banaras Hindu University, Varanasi, India

Muhammad Shahid Department of Bioinformatics & Biotechnology, Government College University, Faisalabad, Pakistan

Amita Sharma Department of Agriculture, Shaheed Udham Singh College of Research and Technology, Tangori, Punjab, India

Parul Sharma Department of Basic Sciences, Dr. Y.S. Parmar University of Horticulture and Forestry, Solan, Himachal Pradesh, India

V. Sharma Department of Biotechnology, National Institute of Pharmaceutical Education and Research-Guwahati, Guwahati, Assam, India

Vimal Sharma Department of Biochemistry, Royal Global University, Guwahati, Assam, India

Rahul Mahadev Shelake Division of Applied Life Science (BK21 Plus program), Plant Molecular Biology and Biotechnology Research Center, Gyeongsang National University, Jinju, South Korea

Stefan Shilev Department of Microbiology and Environmental Biotechnologies, Agricultural University – Plovdiv, Plovdiv, Bulgaria

Akanksha Singh Department of Biotechnology, Bundelkhand University, Jhansi, Uttar Pradesh, India

Bhim Pratap Singh Molecular Microbiology and Systematic Laboratory, Department of Biotechnology, Mizoram University, Aizawl, Mizoram, India

Dhananjaya Pratap Singh Department of Biotechnology, ICAR – NBAIM, Maunath Bhanjan, Uttar Pradesh, India

Harikesh Bahadur Singh Department of Mycology and Plant Pathology, Institute of Agricultural Sciences, Banaras Hindu University, Varanasi, India

Mohini Prabha Singh Punjab Agriculture University, Ludhiana, India

Pratiksha Singh Key Laboratory of Sugarcane Biotechnology and Genetic Improvement (Guangxi), Ministry of Agriculture, Sugarcane Research Center, Chinese Academy of Agricultural Sciences, Guangxi Key Laboratory of Sugarcane Genetic Improvement, Sugarcane Research Institute, Guangxi Academy of Agricultural Sciences, Nanning, China
College of Agriculture, Guangxi University, Nanning, China
Guangxi Key Laboratory of Crop Genetic Improvement and Biotechnology, Nanning, China

Rajesh Kumar Singh Key Laboratory of Sugarcane Biotechnology and Genetic Improvement (Guangxi), Ministry of Agriculture, Sugarcane Research Center, Chinese Academy of Agricultural Sciences, Guangxi Key Laboratory of Sugarcane Genetic Improvement, Sugarcane Research Institute, Guangxi Academy of Agricultural Sciences, Nanning, China
College of Agriculture, Guangxi University, Nanning, China
Guangxi Key Laboratory of Crop Genetic Improvement and Biotechnology, Nanning, China

Rajni Singh Amity Institute of Microbial Biotechnology, Amity University, Noida, Uttar Pradesh, India

Sunita Singh Division of Food Science and Post-harvest Technology, ICAR-Indian Agricultural Research Institute, New Delhi, India

Manoj K. Solanki Key Laboratory of Sugarcane Biotechnology and Genetic Improvement (Guangxi), Ministry of Agriculture, Sugarcane Research Center, Chinese Academy of Agricultural Sciences, Guangxi Key Laboratory of Sugarcane Genetic Improvement, Sugarcane Research Institute, Guangxi Academy of Agricultural Sciences, Nanning, China

Manoj Kumar Solanki Department of Food Quality & Safety, Institute for Post-harvest and Food Sciences, The Volcani Center, Agricultural Research Organization, Rishon LeZion, Israel

Qi Qi Song Key Laboratory of Sugarcane Biotechnology and Genetic Improvement (Guangxi), Ministry of Agriculture, Sugarcane Research Center, Chinese Academy of Agricultural Sciences, Guangxi Key Laboratory of Sugarcane Genetic Improvement, Sugarcane Research Institute, Guangxi Academy of Agricultural Sciences, Nanning, China
College of Agriculture, Guangxi University, Nanning, China

Kandikere R. Sridhar Department of Biosciences, Mangalore University, Mangalore, Karnataka, India

Nikolay Vassilev Faculty of Sciences, Department of Chemical Engineering, University of Granada, Granada, Spain

Pankaj Prakash Verma Institute of Agriculture and Life Science, Gyeongsang National University, Jinju, Republic of Korea
Department of Basic Sciences, Dr. Y.S. Parmar University of Horticulture and Forestry, Solan, Himachal Pradesh, India

Rajnish Kumar Verma Department of Botany, Dolphin PG College of Science and Agriculture, Chunni Kalan, Punjab, India

Li-Tao Yang College of Agriculture, Guangxi University, Nanning, China
Guangxi Key Laboratory of Crop Genetic Improvement and Biotechnology, Nanning, China

50 Years of Development of Beneficial Microbes for Sustainable Agriculture and Society: Progress and Challenges Still to Be Met—Part of the Solution to Global Warming and "Hothouse Earth"

1

Gary E. Harman

1.1 Introduction

1.1.1 The Challenge

About 80 years ago, the first papers were published on *Trichoderma* described its potential uses for the control of plant diseases (Weindling 1932, 1934). Mycorrhizal fungi in plants were discovered even earlier (Berch et al. 1882) and N-fixing *Rhizobia* even earlier than that (https://www.mcdb.ucla.edu/Research/Hirsch/imagesb/HistoryDiscoveryN2fixingOrganisms.pdf). All three genera are widely used in agriculture and colonize plant roots. While each has distinct morphology and genetic bases, they also have significant similarities in terms of their effects on plants (Harman and Uphoff 2018; Shoresh et al. 2010).

All three, as well as other organisms, colonize roots extensively and create season-long benefits, but only a few *Trichoderma* strains were realized to have this ability in the 1980s (Ahmad and Baker 1987; Ahmad and Baker 1988). These organisms are true plant symbionts (Harman et al. 2004). Since they live in and exist in plant roots, they must acquire sugars and other nutrients from plants. This root colonization typically or frequently results in increased growth of plants, of both their shoots and roots, as well as inducing increased resistance to biotic and abiotic stresses and improved nutrient use efficiency. Both the plant and the microbe benefit from the association, which meets the classical definition of a symbiotic relationship, for *Trichoderma* (Harman et al. 2004).

In fact, the long-term internal colonization of the microbe into the root results in what can be called an enhanced holobiont (EH) as distinguished from a plant that lacks

G. E. Harman (✉)
Cornell University, Geneva, NY, USA
e-mail: geh3@cornell.edu

D. P. Singh et al. (eds.), *Microbial Interventions in Agriculture and Environment*,
https://doi.org/10.1007/978-981-13-8391-5_1

these integrated colonists. Holobionts are assemblages of different species that form ecological units (Margulis and Fester 1991). In this concept, plant and animal organisms do not exist as genetic units in isolation, but instead are associated with an entire ecosystem of other organisms, including fungi, bacteria, and many other microbes.

As will be described, specific *Trichoderma* species and other endophytic bacteria and fungi colonize roots and become symbiotic with plants. Effective strains, added as seed treatments, soil drenches, or other methods where they come into contact with roots colonize the roots and therefore become self-assembling with plants. If certain strains are sufficiently effective, the result is a plant that performs more efficiently because of its root colonists.

With EH, plants in the field can have many advantages without requiring extensive investments in plant breeding and genetics. The delivery of highly effective symbionts allows improvements in plant productivity that are similar to the goals of plant breeders. Such improvements are discussed in the next section. So, one challenge is to discover and deliver to plants in the field effective microbes that (a) colonize plants and (b) are very effective in producing EHs.

The advantages of EHs include:

1. Bigger and more rapid growing plants with larger shoots and roots
2. More efficient acquisition/efficiency of use of nutrients, either acquired from soil or as fixed N
3. Upregulation of plant genes and pathways that provide:
   (a) Greater resistance to diseases or pests through induced resistance, thus resulting in less disease or pest damage in both shoots and roots
   (b) Resistance to drought, salt, and other environmental stresses
   (c) Ability to overcome the toxic and damaging effects of reactive oxygen species (ROS) produced by biotic and abiotic stresses or overexcitement of the photosynthetic apparatus (Nath et al. 2013)
   (d) Enhanced levels of photosynthesis
4. Better seed germination, frequently
5. Enhanced rooting and establishment of cuttings and transplants

Once these benefits are acknowledged, a second large challenge emerges—for scientists and technologists to develop further knowledge and technology so that the advantages of EH can be realized to meet societal needs.

1.1.2 Societal Needs

According to the United Nations (Anonymous 2012), the world needs an Ever-Green Revolution, akin but expanded in scope from the original Green Revolution that has enabled the world to feed itself. This revolution would:

- Be sustainable with fewer pesticides and less fertilizer
- Contribute to economic justice
- Double crop productivity

- Ensure that cutting-edge research is rapidly moved from laboratory to field

This Ever-Green Revolution is required because the world faces continually growing demands for food and fiber given that its human population will probably reach ten billion persons by 2050. Global food production will thus need to increase by about 50%, while the arable land and the water available for such production are declining. At the same time, stresses posed by both biotic and biotic are increasing due to climate change. Limitations on arable land and water demonstrate that "we must … create higher-yielding crops, producing more usable product with lower inputs" (Anonymous 2017).

Typically, achieving higher crop yields has relied upon the enhancement of crops' genetic potentials through plant breeding, coupled with ever-higher increments of agronomic inputs. However, such strategies are producing more limited gains in plant yield as the Green Revolution approaches its limits (Adlas and Achoth 2006; Janaiah et al. 2005; Long et al. 2015). Agrochemical inputs such as fertilizers provide diminishing returns (Janaiah et al. 2005). In recent decades, increases in yield have decelerated as input use rates are already very high in the most productive areas, and remaining cultivable areas are less likely to be similarly responsive, the best areas having already been exploited.

One of the greatest limitations on plant breeding methods to enhance crops' yield has been that they have not been able to achieve increases in plants' photosynthetic efficiency (Long et al. 2015). The conversion of light energy into organic biomass is the fundamental basis of life on Earth, and thus it is of critical and fundamental importance. There have been hopes raised that in the case of rice, for example, a C4 pathway for photosynthesis could be engineered into a plant that relies, like most plants, on the C3 pathway for converting light into stored energy (Mitchell and Sheehy 2006). However, these and other goals in increasing photosynthetic efficiency have not been successful. Complexity is inordinate, so success if ever achievable lies decades in the future (University of Oxford 2017).

In addition, the world faces serious problems because of enhanced levels of greenhouse gases, nitrate pollution of waterways, and other environmental issues (Committee on Geoengineering Climate 2015). The holobionts described here have demonstrated capabilities to alleviate some of these problems, but this has not been much explored or exploited.

The author believes that major societal needs can be met, at least in part with EHs. It is the overall challenge, then, for the scientific of the present and the future, to realize this potential.

The sections that follow are the author's approaches to both discover and apply the technology as it has progressed over the decades.

1.1.3 The Early Years 1978–1980

My research career began in 1965 while an undergraduate at Colorado State University. I had mostly paid my college expenses working in gas stations and the like. As a junior at CSU, I sought a job in a research lab and met Dr. Ralph (Tex)

Baker. He was a highly energetic and imaginative faculty member, and he gave me a job as a technician working with *Fusarium* (*Hyphomces*) on the initiation of the sexual cycle of this fungus. Tex allowed me to develop this project on my own. This project gave me my first taste of scientific problem-solving, and I found this very satisfying—I am still doing this and still find it interesting and exciting. As a result of this work, I published a paper in *Phytopathology* as an undergraduate and with sole authorship (Harman 1967). Having a sole-authored publication as an undergraduate is, of course, highly unusual, but at that time I was unaware of this. The mentoring and help by Tex was a critical event in my development as a scientist.

After graduating from CSU with a Bachelor of Science degree, I began work directly on a PhD at Oregon State University with Malcom Corden. I chose a research project to purify polygalacturonase, an enzyme that was involved in pathogenesis by fungi in many plants. It was a highly challenging project that two previous students had not been able to solve. In retrospect, it was a poor choice as a PhD project since it did not involve setting a hypothesis but instead was an exercise in purely technological development. If I had been unable to separate the enzyme from other proteins, I would not have had a thesis that was acceptable. However, at the end of my time and funding at OSU, I was able to purify the enzyme and published the result (Harman and Corden 1972). The PhD was awarded in 1970, although the work was completed in October, 1969. Thus, from a BS to a PhD was only about 3 years—this was too short a time to develop as a professional scientist and person, but it allowed me to take the next step.

That next step was a postdoctoral associate at NC State University with Guy Gooding and Teddy Hebert. The project was quite unusual—at that time (1969) tobacco was being recognized as a public health threat. Essentially all tobacco was infected with tobacco mosaic virus. It was hypothesized that the health threat was due to the presence/effects of TMV, and I was asked to investigate this possibility. I found nothing to support that hypothesis, but it was a great learning experience. I learned about setting hypotheses and conducting hypothesis-based research.

Towards the end of my appointment there, I was expecting to have a job in the NCSU system in Wilmington in research and extension of blueberry and gladiolus diseases. However, the Vietnam War was raging then, and there were students killed at Kent State University in protests against the war. All the campuses of the North Carolina system, and elsewhere, were shaken by protests. The students of the various campuses in North Carolina agreed to march in Raleigh to that state's capital building. I joined that demonstration and was appalled to see that the thousands of students on the grounds of the capital building were completely ringed by state police with lethal armaments. I left, and, fortunately, no one was injured or harmed. However, the next day, I no longer had any prospects for the Wilmington position—the NCSU faculties were conservative, and I had crossed over a line.

1.2 Cornell: 1970–1980

However, I was fortunate enough to secure a faculty position at Cornell University and started work on July 1, 1970 at the age of 25. The position was unique—I had a 100% research position to work on the physiology of parasitism of seed pathogens and seed-associated microbes. This was a very broad and unique description, and I could see several good opportunities. One area was dealing with seed-storage diseases such as *Aspergillus* spp., and I pursued this avenue for a time with several publications. Another, reported in an article published in *Nature* on the role of free radicals in seed aging, was of high impact (Harman and Mattick 1976).

Another area with wide significance which I began to explore was the potential use of changes in the seed's microflora and its potential for biocontrol. Working with Charles Eckenrode, an entomologist, we discovered that the microflora on seeds produces volatile metabolites that Dipterans (seed corn maggots and others) use to detect germinating seeds. The adult insects use this microbial cue to lay eggs near seeds that are just beginning to germinate (Eckenrode et al. 1975). If we altered the composition of the spermosphere (the area on and just around the germinating seed), we could change these volatile cues. Alteration of the microbial community in the spermosphere changed the volatiles emanating from the germinating seed, lessening damage by the insect. To modify the spermosphere, we treated seeds with *Chaetomium globosum*, which produces its own volatile metabolites and limits the growth of the bacteria that produce the original metabolites that stimulate oviposition (Harman et al. 1978). This was my first foray into altering the composition of the microbiome around plants to accomplish biocontrol.

1.2.1 Beginnings of *Trichoderma* Research, 1980–1990

In 1990, I returned to Colorado State University to work with Tex Baker on a sabbatic leave. It was my great good fortunate that Ilan Chet, Professor at the Hebrew University of Jerusalem, was also at CSU at the same time, also on a sabbatic leave. Tex, working in Colombia, had discovered that some soils become suppressive to diseases, and they jointly began trying to discover the active principle for this. They discovered that a strain of *T. hamatum* found in that soil was able to suppress the pathogen *Rhizoctonia solani* also residing there (Chet and Baker 1981).

This was a fertile new area of research, and the three of us, with help from graduate students, began to investigate seed treatments using that strain to control *R. solani* and *Pythium* spp. This was successful, and the results were soon published (Harman et al. 1980). We also examined the in vitro interactions of *T. hamatum* with the pathogens and described their mycoparasitic interactions (Chet et al. 1981). I and many other scientists in the field considered mycoparasitism to be the principal mechanism of biocontrol; however, we know now that it is, at best, an incomplete description of events.

The year of interaction with Chet and Baker was extremely fruitful. It began my research with *Trichoderma*, which still is continuing. Moreover, it established a connection with Ilan Chet that continued until his retirement in about 2010. This relationship was essential. It enabled Harman and Chet to obtain numerous grants from the USA-Israel Binational Agricultural Research Development Fund (BARD). This was one of the few granting agencies that gave equal weight to conducting applied and basic research concurrently. Without BARD, my program could not have succeeded over the years because inadequate funds would have been available for working on the application of specific strains or other technologies. Close relationships and friendships forged are frequently critical to scientific careers. In addition, the Cornell Biotechnology program provided a number of small grants over the years. This program was designed specifically for small company research funding a required a match from the small companies.

The strains discovered from the Colombian soils were highly promising, and I sent them back to my lab at Cornell, where a postdoctoral fellow, Jonathan Hubbard, attempted to repeat the work that we were doing in Colorado. Unfortunately these efforts failed. Strains and methods successful in Colorado were ineffective in New York. Upon my return to Cornell, we investigated why this occurred. We hypothesized that strains from New York soils might be better adapted to local conditions and isolated several strains that were promising. We found that the soils in NY were poorer in iron than those in Colorado, and the presence of spermosphere bacteria produced siderophores that prevented growth of the strains from Colombia (Hubbard et al. 1983). The strains from New York produced their own siderophores which helped them compete for iron with the bacteria (Hubbard et al. 1983).

It was about this time that I attended a meeting of a group focused on *Trichoderma* and *Gliocladium* primarily for plant and biocontrol applications. At that meeting, protoplast fusion was described by Douglas Everleigh of Rutgers as a method of obtaining asexual hybrids of these nonsexual fungi. It was at this time that I made a very important decision which has guided my career ever since. I concluded that biocontrol organisms had a high probability of improving plant agriculture and benefiting human society. This implied that it was inadequate simply to publish papers in journals, but ***there had to be some application of the technology if it was ever to be more than a laboratory curiosity***. This required commercial production of the technology by whatever means necessary to accomplish this goal.

The investigation of protoplast fusion, which was funded by BARD, was productive. We made inter-strain fusions between numerous different strains and species, and then proceeded with a monumental screening effort to sort through thousands of different progeny strains. This occurred both through tests in soils infested with several different pathogens and with the limited genetic tools available to us at the time (Stasz and Harman 1990; Stasz et al. 1988a, b, 1989). Thomas Stasz, a postdoctoral student and former graduate student with me, was instrumental in this development. A few strains were highly improved over their original parental strains. With this genetic improvement, not only did we obtain strains that were less affected by the pathogens, but we also saw significant improvements in plants' root development. The result of one of the early screenings is shown in Fig. 1.1. One of the strains produced and evaluated was *T. harzianum* 1295-22, now widely known as T22.

Fig. 1.1 Results of one of the first assays done with protoplast fusion progeny in the laboratory in the mid-1980s. Seeds of cucumbers were treated with conidia of various strains suspended in Methocel (a methyl cellulose material used as a sticker). These treated seeds were planted in a natural field soil amended with the pathogen *Pythium ultimum*, which causes seed and root rot. After several days of growth, the seedlings were removed from the soil and the roots carefully washed and placed in test tubes with water for viewing. In the tubes are representative seedlings from seeds treated with Methocel (M), the parental strain T12 (12), with T95 (95), and strains 1295-7 (7) and 1295-22 (22). The hypothesis was that if we asexually combined T12, which was isolated from NY soils and was able to compete with bacteria in the New York soils probably by production of siderophores (Hubbard et al. 1983) with T95, which and which was derived from the strains derived from Columbia that had been modified by mutation to be rhizosphere competent (Ahmad and Baker 1987), we would be able to obtain progeny that combined the useful properties of both parental strains. In reality, a few of the progeny that we obtained were considerably superior in capabilities to either strain, as evidenced by increased root growth in this illustration. *T. afroharzianum* (formerly *T. harzianum*) strain 1295-22 (T22) is now used around the world as a biocontrol agent that also promotes plant growth. (Harman 2000)

The designation 1295-22 derived from the fact that it was a fusion between T95, which was derived from the strain discovered from Colombian soils by Baker and Chet, and T12, which was a local NY strain. It was the 22nd protoplast fusion progeny that we selected from this fusion, hence 1295-22. We had a very long development phase since the progeny were not stable when first isolated, and in fact, protoplast fusion is a tremendous system for production of diversity within *Trichoderma* (Harman et al. 1998; Stasz and Harman 1990; Stasz et al. 1989). The genetics are complex, involving initial multikaryon formation and recombination events between nuclei of the two parental strains that give rise to novel phenotypes (Harman et al. 1998). A few of these grow faster, are strongly rhizosphere-competent, and have proven over the years to provide season-long benefits to plants as will be described later.

We patented *T. harzianum* strain T22 together with sister strains (Harman et al. 1993b) through Cornell and published the results (Harman et al. 1989; Stasz et al. 1988a). The rights to the patent was licensed from Cornell to the Eastman Kodak Company, which was planning to make this strain and another native strain (Smith et al. 1990, 1991, 1992) the flagship products of their new agricultural biotechnology division. Kodak planned to use a new fermentation facility in Rochester NY to produce the organism using liquid fermentation. High densities of conidia were produced, but they had very poor shelf-life after drying. We were able to develop methods to increase stability of these conidia (Jin et al. 1991, 1992, 1996). However, and very importantly, Kodak obtained one of the very first US Environmental Protection Agency registrations for a strain of *Trichoderma* for biological control. Thus, from the beginning, T22 could legally be sold for control of plant diseases in the USA.

Unfortunately, Kodak had little expertise or understanding of microbial agricultural technologies, nor did anyone else. At the time they began this commercialization effort, there was no knowledge of technologies required for large-scale production of products with ***adequate shelf-life*** for agriculture distribution systems.

After a few years of development (1990), Kodak was faced with a huge settlement with the Polaroid Corporation for patent infringement. This settlement effectively ended the Company's fledgling efforts at commercialization of its diverse biological investments, including the production facility in Rochester. Kodak attempted to sell the technology related to T22 but was unsuccessful.

Much of the success of any scientific (and probably other fields as well) depends upon response to challenges that arise. Challenges and roadblocks can be career stoppers, but if the responses are successful, they can also lead to important accomplishments. It may be that in my dual roles of academic scientist and entrepreneur, the difficulties encountered may have been more prominent than other researchers' experience. The inability of Kodak to proceed with the development of the T22 technology was such a roadblock.

1.3 Roadblocks and Opportunities

1.3.1 Market Introduction and Acceptance of T22 and Other Early Strains

As described above, we had developed what appeared to be very useful strains in T22 as well as other strains. And progress was initially encouraging with the licensing and development that was begun with Kodak. However, after Kodak was unable to proceed with development of this technology, another direction had to be found. This occurred with the development of TGT Inc./BioWorks.

Our search for possible partners found no established companies that were willing to license and develop T22 and related technologies. It became apparent that if this technology was ever to be used, it had to be accomplished by me and my co-workers. Therefore, co-inventor Stasz and Russell Howard started a company, TGT Inc., which was later renamed BioWorks. I was its Chief Scientific Officer (CSO), Howard was the CEO, and Stasz was the Vice-President. I remained a full-time Professor at Cornell and became the Department Chair for a newly merged Department of Horticultural Sciences. As the company's CSO, I would be developing the technology, which was feasible since my primary efforts at Cornell were being directed toward advancing this technology. We requested and received the rights from the Cornell Research Foundation to practice the inventions, i.e., use the strains (Harman et al. 1993b; Smith et al. 1991, 1992). We formed TGT as a corporation, but I received no salary. We knew that the production systems developed by Kodak were not adequate to produce a viable product with good shelf-life. Therefore, at Cornell we developed a semisolid production system that was effective; the basic process is published as mentioned earlier (Harman et al. 1996).

Thus, as seen from the rest of this chapter, the development of commercial systems for biocontrol and other advantages of *Trichoderma* was woven together with more basic development of systems analysis and biology at Cornell. This combination worked very well. Quite frequently we obtained information on the commercial side that was useful for work at Cornell and vice versa. The two sides of this development partnership were synergistic, and progress was more rapid with this hybrid approach than it would have been otherwise.

Largely as a consequence of having both research at Cornell and commercialization at TGT/BioWorks, T22 is now used around the world, especially as a greenhouse soil amendment. I believe that without the coordinated development at TGT and Cornell, T22 would never have been used commercially. It was the first, or at least among the first, *Trichoderma* strains to be so commercialized. This pioneering strain has supported the development of these fungi for agricultural use around the world.

At TGT, we had started a company but lacked business experience, which was a huge limitation that we had to overcome. Almost immediately, we received a large order from an agricultural distributor. We had to scramble to produce sufficient product and did so in a rather primitive way. But this was later refined and has become a highly efficient process. We shipped the product and then waited for repeat orders that never came. Almost a year later, one of the co-founders visited a warehouse of the distributor and found the product that we had shipped sitting on the floor, never used. We asked why, and the people in the warehouse told us that they had no idea what it was for, or how to use it. We assumed that the agricultural distributor would tell farmers what it was and how to use it. But this did not happen, and will not happen. The people who develop and provide the product or strain have to tell the farmers or other customers what it is and what its advantages are. No one else will do this.

At the same time, we began to raise funds from investors. This had to be done on the basis of the personal credibility of the company founders and inventors, and we managed to raise $1 million, which was a tremendous achievement. In order to do this, however, each of the co-founders had to invest as well, and we did.

We developed a production system with quality control and began to sell the product. Soon, it became obvious that our production system had to be upgraded and expanded, and we raised additional funds to do this. That we had a very good Board of Directors was very helpful. However, in the mid-1990s, the Board determined that two of the co-founders needed to be replaced. I remained and became the interim CEO of the company, now named BioWorks. This was a fulfilling experience that I cherished.

Cornell allowed me to change my appointment at Cornell to 60% time with the remainder devoted to BioWorks. It should be noted that this was the only time that I ever received direct salary from this or any other company that I was involved with while employed at Cornell. It is, in my opinion, an unacceptable conflict of interest to receive salary compensation from both the university and corporate entities while remaining a full-time university faculty member (conflicts of interest are described later in brief).

At the same time, I was co-editing the two-volume book on *Trichoderma* that has become a standard monograph (Harman and Kubicek 1998; Kubicek and Harman 1998). I remained as CEO for about 9 months and then had to make a decision: would I remain as the CEO of BioWorks and resign my professorship at Cornell or vice versa? I could not really do both. It was a very difficult decision, but it was agreed that William Foster would become the CEO, and I remained working with BioWorks on a part-time basis for a year or two. Eventually, I decided to return to Cornell on a full-time basis as it was not a workable position for the largest stockholder and former CEO of a company to be occupying a role under the CEO. Foster is still the CEO of BioWorks, and it has done well. We expected at BioWorks that T22 would not be its only product, but also that genes from *Trichoderma* and other organisms could produce highly valuable enzymes and serve as transgenes in plants as commercial products. As noted later, this was a good idea based on sound technology, but it was unsuccessful in the commercial marketplace.

1.3.2 Acceptance of Concepts of Abilities of Strains That We Developed Relative to the Dogmas of the Day

Throughout my career, there have been times when those around me in the scientific and academic communities have disbelieved or not accepted my concepts and direction of thinking. In most cases, through persistence and having publications or grants at critical times, I was able to overcome at least some of these. For new students, I suggest: be skeptical of pronouncements like "everybody knows that xxx" because this shuts off further and critical thinking. And do not reflexively accept during scientific discussions a dismissal of arguments by the statement like "Prof. YY says zzz." Neither of these assertions should be acceptable to new or established scientists. Science proceeds by superseding common dogmas and by proceeding along paths of inquiry that innovators believe or wish to develop, regardless of scientific assumptions or fashions of the day.

The first of these I experienced in the late 1980s. We were finding that *Trichoderma* strains, properly formulated and delivered, could fulfill many useful functions, not just biocontrol. Moreover, the advantages that they conferred lasted not just for a short time, but could provide advantages for an entire growing season, or more. These and other concepts for further investigation were not accepted by my peers, and both papers and grant proposals were being rejected. To counter this, I proposed to the journal *Plant Disease* that I write a review/synthesis paper to be entitled "Myths and dogmas for biocontrol: Changes in perspective derived from research with *Trichoderma harzianum* T-22." This proposal was accepted and the paper published (Harman 2000).

The concepts expressed in that paper were well received and widely accepted. Getting this paper published was absolutely essential for me to proceed down both the academic and commercial pathways that I have followed. It is important to note that this paper was heavy on concepts and compiled data from many different sources. It probably would have been impossible to put all of this into a single research paper. But this was a review paper. Such conceptually formulated papers are critical, in my view, to making scientific progress of any field. This paper was crucial to gaining both commercial and academic acceptance for the technology that I was developing.

At the same time, the laboratory at Cornell made considerable advances in applied aspects of *Trichoderma* application, including seed treatment (Taylor and Harman 1990; Taylor et al. 1988, 1991) and fermentation/production of *Trichoderma* (Jin et al. 1991, 1992, 1996). These were enabling technologies that permit the commercial production of *Trichoderma*.

1.3.3 Chitinolytic and Glucanolytic Enzymes and Genes

Not every technology we developed was commercially successful. Some projects provided excellent scientific discoveries, but could not be successfully developed due to circumstances beyond my control. The mid-1990s were very productive in my lab at Cornell. I had a long series of visiting scientists in my laboratory, and most of our effort focused on identifying and working with chitinolytic and glucanolytic enzymes and genes from *Trichoderma*. We also developed improved methods of fermentation and production of *Trichoderma* spores and laid the groundwork for understanding the changes in gene regulation of plants that were induced by *Trichoderma*. An outstanding group of visiting scientists took an interest in this work including Arne Tronsmo (Agricultural University of Norway), Matteo Lorito (University of Naples), Jie Cheng (Jiao Tong University, China), Lodovica Gullino (University of Torino, Italy), Clemens Peterbauer (Technical University of Vienna, Austria), Claudio Altomare (CNR, Bari, Italy), and Antonio Llobell (University of Seville, Spain). These were accompanied by some very good post-doctoral fellows, including Christopher Hayes and Xixuan Jin. Roxanne Broadway, a faculty colleague, was essential to the discovery of the enzymes that would function at high pH levels. Earlier students included Eric Nelson, Alex Sivan, Izhak Hadar, and Wei-Lang Chao.

The chitinolytic and glucanolytic enzymes from *Trichoderma* were purified and were found to be very strongly antifungal and to be strongly synergistic with each other and with other compounds (Di Pietro et al. 1993; Harman et al. 1993a; Lorito et al. 1993a, b, 1994a, b, c, 1996). The genes were isolated (Hayes et al. 1994; Lorito 1998) and found to markedly induce resistance in plants when introduced as transgenes (Bolar et al. 2000, 2001; Lorito et al. 1998). Later, chitinolytic enzymes and genes that act at alkaline pH level from *Streptomyces* were found to be effective against insects both as purified enzymes and as transgenes (Broadway et al. 1995, 1998).

All of this research was highly significant scientifically, but was not translated into useful products because of market forces. In large part, this was because of the enhanced regulations and public pushback to transgenic technologies in all forms. No small company could compete in this market space, since many millions of dollars and years of development were required for commercialization. In essence, the public outcry against transgenic technologies made commercial development only the province of large companies with very deep pockets. Since this was my specific area at BioWorks after I resigned as CEO, and there was no market for our transgenic technologies, BioWorks could no longer afford me. We sought large commercial partners to no avail; at that time, mid-size and large companies were merging. We obtained initial interest from several companies, but invariably mergers and acquisitions occurred, and the priorities and personnel at the potential acquiring and partner companies changed.

In my opinion, without this translation, even very good research in our general field will not be long remembered if only scientific papers result, and there is no application. These technologies were well accepted scientifically, and I owe a debt of gratitude to the excellent scientists who worked on these projects. Their science was excellent, and several of them remain good friends. However, I expect that these achievements will be less remembered than will those that are commercial successes as well as academically significant.

1.3.4 Selected Strains of *Trichoderma* Are Plant Symbionts

As we developed *Trichoderma* technologies, it became apparent that some of the strains were plant symbionts, as mentioned earlier. Late in the 1980s, we began to observe in commercial field trials that seed treatments with T22 frequently, but not always, resulted in bigger and greener plants than when lacking the organism. We knew by this time that T22 was a very efficient root colonist, but the effects that we were observing were on the shoots of plants as well as in the roots, and T22 did not colonize shoots. Not only were inoculated plants bigger and greener, but also diseases were controlled systemically. Clearly, there was not just a single effect, but there were evidently multiple aspects of the changes induced in the plants.

The scientific methods of proteomic analysis and determining differential gene expression were becoming available, and it seemed essential to me that we catalog and identify the multiple genetic changes that must be occurring to produce the phenotypic characteristic observed. To finance this effort, we wrote grant proposal after grant proposal, and all were turned down. The reviews were all similar with the assessment that we were just on a "fishing expedition" and that we needed to be focusing on single hypothesis. Finally, however, BARD did fund this research for corn, and through the efforts initially of Jie Chen and especially from Michal Shoresh who joined by lab with this funding, we completed a study that showed we could identify 91 upregulated proteins in the shoots and 30 downregulated genes. Some of them were definitely involved in photosynthesis as increased starch was deposited in the shoots, while fewer genes were affected in the roots (Shoresh and Harman 2008). At about this same time, other groups also reported similar findings (Alfano et al. 2007; Marra et al. 2006; Segarra et al. 2007). From these studies it became evident that these so-called fishing expeditions were revealing that these organisms were indeed inducing fundamental changes within plants. These wholesale and fundamental changes would never have been revealed by the very specific and narrow approaches advocated by the reviewers. This taught me something about the limitations of how much contemporary scientific investigation is conducted.

1.4 Lessons from Commercialization

As stated earlier, in about 1984, I decided that if I was going to make a career out of studying beneficial microbes, then they had to be used on a commercial-scale for the public good. It was not going to be enough for me to just do good research and publish it. No matter how good the journals were, there was only a short time period when anyone, including me, would remember what was published. However, a product that is used commercially and that is used for the public good is a better legacy.

Further, if anyone was going to use the technologies that I was developing, then I needed to do also the work to make it useful to the public. Ideally, a company would pick up university technology like that which I was developing and would commercialize it, but I have had little success along this avenue. The early work at Kodak was a start in that direction, but it was unsuccessful due to that company's financial issues. Moreover, it (nor we at the time) had any idea how to commercialize microbial technologies.

If any company other than one that is very experienced and already in the field decides to develop a microbial technology, there is a very steep slope that must be climbed. The following sections describe this uphill climb. The barriers to entry are numerous, and most apply equally to most products or processes that are new.

1.4.1 University Technology Is Very Far from a Commercial Product

Earlier I described how essential it is that there be proper strain selection; only six strains of the thousands that I have tested have proved to be commercially useful. And even when a useful strain is identified, there are numerous questions that have to be answered. One of these is the basic question of what application is expected? What will the market pay for it? How much must be applied for it to be effective? How can a high quality product be manufactured, and what quality control processes need to be put in place to ensure that the product is reproducible and that it will perform as advertised? There are many chasms that exist (valleys of death) between initial discovery and even initial product introduction. These are presented in Fig. 1.2, taken from Moore (1991). Answering the questions above, along with the adaptive research needed to produce products based on the answers, is time-consuming and expensive.

Unfortunately, there is almost no public or grant funding that is available for product development. There is significant funding for basic research and, paradoxically, to test products and develop uses for existing products. However, there is nothing at all for the all-important product development phase that is required to cross the first chasm. If products like T22 are to be developed, then, in my experience, it is up to the discoverer of the technology to develop and find funding for the all-critical product development phase. This is particularly true for a product that is novel, such as a strain of *Trichoderma* where few, if any, have been developed and commercialized before. This type of product, which is new and novel, is particularly hard to create and provide to the marketplace since it is a disruptive technology to the existing market relationships, with projects and their manufacturers resisting displacement. This differs fundamentally from other products, such as a new variety of wheat, for which markets exist and where marketing channels are in place.

Particularly problematic is the infrastructure designed to manufacture the product (in this case for growing and formulating the fungus). Effective quality control (freedom from contamination) is essential, as is formulated into a product that has sufficient shelf-life for the marketplace. In many cases there need to be liquid and powder/granular forms of the product to meet different market niches, necessitating development of different formulations of the product.

Then, once a product is in place, customers are needed. This is the second valley of death (Fig. 1.2). Who will pay for the product once it is introduced? If they will buy it, how much will they pay? How is product performance evaluated? In the case of our organisms this has required hundreds of field trials all carefully done and statistically evaluated. Is the product cost-effective to the end user? How good are the data that support this conclusion? Then, a marketing team must be put into place to convey all this information to end users. These are large tasks requiring specialized teams.

Then, a very big question is getting the necessary regulatory approvals. At the least there is one agency per country, so if multiple countries are involved, then

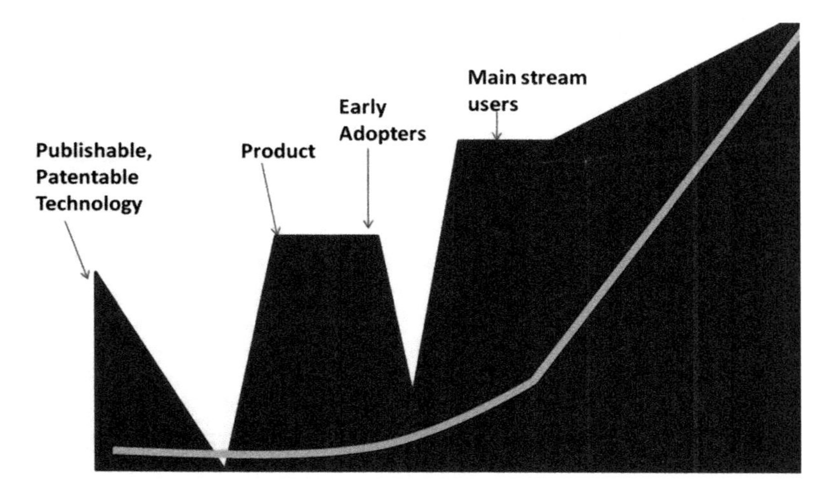

Fig. 1.2 A diagram that represents the issues affecting commercialization of microbial agents or any other disruptive technology. The blue shapes represent the stages along the way. The development of a microbial agent that is sufficiently advanced to enable publication or patent application is only the beginning. Once a promising agent is in hand, then a product must be developed that has the characteristics of stable shelf-life and efficacy that is sufficient for the marketplace. The first gap in the blue shapes is the first "valley of death" (Moore 1991) where sufficient resources must be allocated to create prototypes of a useful product. There are almost no public sources of such funds. At about this time, resources may be necessary to register the product with regulatory agencies. Altogether, several million dollars and 1–2 years are required for this step. Then, once a product is available, it must be tested and provided to potential users/customers. Especially for a new type of product, such as T22, most people will be skeptical, particularly if it is designed to be an alternative to reliable chemical pesticides, as T22 was. Efficacy and some measurable improvements to existing products must be shown. Attributes such as environmental friendliness are useful and compelling to only a limited set of customers. The first buyers of the product are the "early adopters," and they are notoriously fickle. They buy a new product, try it, and then are likely to abandon it for the next new product. This gap between the product and adoption by mainstream customers is the second valley of death, where products and companies frequently do not survive. It is only after the mainstream buyers make the new product a standard for their operations is survival of the product assured. The difficulties in crossing the chasms or "valleys of death" are made more acute by the need to build up the revenues derived from sales. The yellow line provides an estimate of the timing of sales revenues. They only become significant after mainstream customers make the product part of their standard operations. For T22, this passage took at least 10 years and many millions of investor dollars

there are multiple regulatory agencies that must be satisfied. If toxicology evaluations must be satisfied, this requires hiring outside contractors because most regulatory authorities have very specialized requirements that are almost impossible for universities or companies to meet unless they have special departments devoted only to this. It probably would have been impossible for TGT/BioWorks to successfully deploy T22 into the markets if Kodak had not already registered it as a "microbial pesticide" (the US regulatory term for a biological product that controls plant pests or pathogens regardless of mechanism).

1.4.2 Steps to Commercialization Are Very Expensive

A minimum cost to accomplish the steps to bring a product to market is, at the least, about $8 million and several years of development (Harman et al. 2010). Even with the Kodak registration, several rounds of fund-raising, totaling several million dollars, had to be accomplished by TGT/BioWorks. Answers to the questions of technology and market development emerged as we went along, but the experience was, at best, bruising and difficult. At the time, no one had the answers to the questions noted here and above, so there was no option. Having said this, it is possible to bring *Trichoderma* and similar products to *local* markets where the product is grown and distributed in its original form, with no processing or registration (Harman et al. 2010).

Two examples of this are production of the microbial agent on a grain or other solid substrate and delivering the product on this substrate for application at relatively high rates to grower fields. This requires no consideration of shelf-life and delivery beyond, at most, a 1-day's car drive away. Other village-level production systems also are possible (Selvamukilan et al. 2006). A different system for localized delivery was production of the agents in liquid fermentation systems and application through fertigation systems directly to the crop. Both of these systems were successful, but unsuited to large-scale production and no regulatory approvals were required (Harman et al. 2010).

1.4.3 Conflicts of Interest with Universities

If, as in my case, the technology was developed within a university with company ties to the inventor formed, then there is significant potential for conflicts of interest between the company, the university inventor, and the university. When we started TGT/BioWorks, there were no rules, so I was very careful to tell the university what I planned to do, and to make sure that there were no surprises. Of course, the inventions were made at Cornell University and the patents were filed by the University. Then they were licensed back to the company, which paid royalties to the university. As a consequence, Cornell University has received about $1.8 million from royalties and fees paid for patents that were licensed by companies that I started and for which I was usually the primary science officer. I was very careful to make sure that my employment at Cornell received first priority and that everything I was doing was fully disclosed both to the university and to the public. Almost nothing was proprietary beyond the usual time required for publication. Almost everything had to be patented, but the timing of publication was never an issue. It is possible to file provisional applications to protect intellectual property rights for a year. During that year, the filing is never examined and at the end of the year, a decision to file or not had to be made by the University. The inventor has conflicts in such decisions, especially when a for-profit company is involved, and cannot enter into the discussion.

Even with these precautions, and even with the funds provided as royalties to Cornell, eventually the ever-increasing burden of conflict of interests required me to retire from Cornell. Universities face increasing scrutiny relative to Conflicts of

Interest. In the case of Cornell, increasing levels of barriers were put into place, all administered through an anonymous committee. It indicated to me what it thought would be required for me and my program to be in compliance with regulations, and I met their goals. Unfortunately, due to the layers of bureaucracy between me and the committee, communication was very poor. While I was able to meet their requirements, I could not overcome the barriers of communication between us. Resolution of any conflict of interest barriers could easily have been overcome, I believe, had I been able to meet and to talk directly with the anonymous committee, but this was not allowed.

1.5 The ABM/Cornell Years 1990–2012

1.5.1 Development of Concepts of Enhanced Holobionts

Earlier I described the necessity, for gaining acceptance, of having proteomic and related discoveries which described the fundamental changes occurring in plants as a consequence of root colonization by symbiotic strains of *Trichoderma*. A similar change of direction was occurring on the commercial side. BioWorks had begun what became a very successful marketing approach that focused on greenhouse applications. It made great progress in developing products and market approaches that were and are effective (see https://www.bioworksinc.com/). T22 still was and is a mainstay of their product line, but it has added other products as well.

However, as I was winding up my employment at BioWorks, I wished to pursue what I viewed as exciting developments with corn and other crops. As a consequence of commercial field trials, we began to see that T22 added as a seed treatment could sometimes dramatically improve both shoot and root growth of corn and other row crops (Fig. 1.3). I wished to pursue this exciting new prospect, but the new CEO wished only to focus on row crops.

In 2000, a new company, Advanced Biological Marketing (ABM), was formed. This company was conceptualized in part to proceed with the *Trichoderma* technologies I had developed. Leon Bird and Daniel Custis were the early leaders of this company, and they visualized an important role for commercial R&D. This is highly unusual for a small company, so I had an immediate role in doing contract research from Cornell. ABM licensed the use of T22 for row-crop seed treatment with a focus on *Trichoderma* seed treatments on corn, wheat, and other row crops. It also focused on better seed treatments with *Rhizobia* for soybeans and other legumes. *Rhizobium*, and especially *Bradyrhizobium*, is particularly difficult to develop as a commercial product since, unless properly formulated, it has very poor shelf-life on seeds. Working with Dan Custis in cooperative contract research, we developed an encapsulated *Bradyrhizobium* product that had much improved shelf-life on seeds (Harman and Custis 2006). A similar product and process are used today.

ABM conducted hundreds of field trials with T22 on cereal crops, especially wheat and corn. Unfortunately, these trials, plus research that we did at Cornell, indicated that on some genotypes of corn, the use of T22 did not result in improvements

Fig. 1.3 An example of many comparisons that demonstrated that seed treatments with selected *Trichoderma* spp. improved the growth of corn and other crops even when no disease was present. This was first observed in commercial field trials prior to 1990. In the example above, seeds were treated only with standard chemical pesticides (left) or with a mixture of *T. afroharzianum* and *T. atroviride* applied over the standard chemical pesticide mix. This mixture of *Trichoderma* spp. was the active ingredients of the commercial product SabrEx™ (www.abm1st.com) that was discovered by the author (Harman 2014). There have been hundreds of commercial trials conducted by companies that evaluated the efficacy of different strains and formulations, and these have been essential both to commercial development and to provide the underpinnings of the basic technologies leading to EHs. The plants produced have larger roots and are greener, with higher total levels of C (presumably from higher photosynthetic levels) on a per hectare basis. This phenomenon has been noted very many times, and this is from a field in Wisconsin. (Photograph by Harman, and used with the approval of ABM and also shown in Harman and Uphoff, 2019). NOTE: This photograph was recently included in Harman and Uphoff, 2019, which is an open-source publication. Is this a problem?

in the yield of corn. In some cases, yields with T22 inoculation were actually reduced (Harman 2006). ABM contracted with Cornell and my lab to produce new strains of *Trichoderma* that gave consistently good yield results with corn and other crops. We went back to the protoplast fusion progeny collection from which we obtained T22 and found progeny from other fusions that seemed improved over T22.

Cornell applied for patents on these strains, and the patents were granted (Harman 2014). ABM licensed these strains, and we delivered three of them to ABM in 2006. These were certain strains of *T. afrohazianum* and T. *atroviride*. ABM began to test combinations of these strains in field trials between 2008 and 2010. These particular strains were selected because they (a) increased plant growth in the greenhouse, (b) induced systemic resistance to disease, and (c) increased nitrogen use efficiency in the greenhouse (Harman et al. 2018; Harman and Mastouri 2010). Very recently, many studies from labs around the world have demonstrated that root colonization by endophytic *Trichoderma* strains enhance photosynthesis in various plants by induction of expression of a large number of genes involved in photosynthetic pathways (Harman et al. 2019).

In 2010, ABM began to market different mixtures of these for treatment of wheat, corn, and cotton, and in combination with *Rhizobia* on legumes, they have become widely used in the USA and around the world. These strains provide both enhanced yields in several crops and also, and very importantly, induce resistance to abiotic stresses such as drought and salt. We were invited to write a review/knowledge synthesis paper on this by *Nature Reviews Microbiology*. This paper showed convincingly that some strains of *Trichoderma* are plant symbionts and that they induce numerous beneficial effects on plants (Harman et al. 2004).

This paper, plus documented changes in gene and protein expression in crop plants, began to demonstrate that most of the beneficial effects on plants were due to systemic changes in the plants' gene expression. Several other papers were published at about the same time that demonstrated this to be the case (Alfano et al. 2007; Marra et al. 2006; Shoresh and Harman 2008). As this research and development in both the academic and commercial arenas demonstrated, selected strains of *Trichoderma* induced all of the beneficial changes in plant performance that are listed at the outset of this chapter.

So, back at Cornell, we were faced with a dilemma that our strains, as well as those of other researchers, induced all of the effects noted as advantages of holobionts listed above. Did all of these result from upregulation of different genes, or was there some underlying principle underlying all of these effects? It was now apparent that there was an underlying principle which appeared to be the upregulation of genetic systems that ameliorate the negative effects of reactive oxygen species (ROS) which are produced by/in plants by stress (Mittler 2002) and even by overexcitement of the photosynthetic systems in plants by bright sunlight (Nath et al. 2013).

Thanks to the work of graduate student Fatemeh (Aileen) Mastouri, we found that *Trichoderma* strains systematically upregulate the entire network of genes that produce enzymes which catalyze the reduction of antioxidants from oxidized back to reduced forms. Materials such as glutathione and ascorbic acid effectively detoxify ROS, but once this reaction occurs, then the antioxidants are in an oxidized form and must be reduced for another reaction to occur. Thus, there is a cycle of oxidization and reduction of ROS that is catalyzed by the upregulated enzymes. In addition, enzymes like superoxide dismutase directly interact with ROS to reduce them to less toxic forms (Mastouri 2010; Mastouri et al. 2010, 2012).

1.5.2 Extension of EH Concepts to Other Endophytic Root Symbionts

From these and other papers (Tyagi et al. 2017; Zhang et al. 2016), it is clear that regulation and modulation of ROS to a physiologically optimal level is essential for good plant growth, especially under stress. Moreover, this seems to be a general phenomenon underlying root endophytes' improvement of plant growth and resistance to diseases and other stresses. Other endophytic root colonists that have similar mechanisms include arbuscular mycorrhizal fungi (AMF) (Mo et al. 2016), *Piriformospora indica* (Sun et al., 2010; Tyagi et al., 2017; Vadassery et al., 2009), and *Rhizobia* (Defez et al. 2017; Fukami et al. 2018).

They all also have the capability of increasing plant photosynthetic capabilities (Chi et al. 2010; Dani et al. 2017; Mastouri 2010; Mo et al. 2016; Porcel et al. 2015; Vargas et al. 2009), as summarized and discussed in Harman and Uphoff (2018).

Important also is the fact that root colonization by these fungi can result in much greater root volume and depth (Harman and Uphoff 2018). This is clearly important for crop plants' resistance to drought, but it is also an important component of soil processes and functions, including the necessary increase in soil organic matter (SOM) that is essential to agricultural productivity (Lal 2004).

All of this recent development, including the concept of EHs and the realization that this concept of the value of endophytic symbiotic root colonizers, occurred after my retirement from Cornell in 2012. ABM offered me to position of Chief Scientific Officer, and we built a state-of-the-art research facility here in Geneva, New York. This was staffed with competent persons, including some who had been in my lab at Cornell. At that time, I also recruited a person to be trained to serve as a successor CSO. This was successful, and Dr. Molly Cadle-Davidson is now the CSO, following my retirement from ABM in 2017.

1.6 Challenges for the Research Community

In this chapter, I have outlined the development and concepts of *Trichoderma* strains and other beneficial microbes with a particular focus on their abilities of EHs to make broad improvements in agriculture. It is also becoming apparent that they have also large potential to contribute to alleviating certain global ills.

As mentioned at the outset, the world needs to increase its production of food and fiber to meet a growing population, and to do so in the face of increased environmental stress and a dwindling and degraded land base, with less adequate and reliable water supply. The purposeful development of systems to produce EHs that are relatively simple for farmers to utilize and that are self-assembling in the agricultural environment, i.e., after reliable root colonization from seed treatments the roots are rapidly colonized and better plant health and vigor will ensue, can provide a powerful means to meet certain global needs, such as adapting to and mitigating climate change.

There is urgent need to alleviate the accumulation of greenhouse gases such as CO_2 and NO_x in the atmosphere as these are driving global warming. The accumulating effects of these gases are clearly endangering our civilization. Some of the NO_x derives from N fertilizers applied to agricultural lands. The buildup of NO_x in the atmosphere and the leaching of NO_3 into waterways and water bodies can be curbed and lessened through better agricultural practices and by deeper and greater root development so that greater amounts of atmospheric N are incorporated into plants and especially their roots rather than being released into the atmosphere and waterways. Farm profits could also be increased by reducing the need for chemical fertilizers and through their more efficient use.

If the amplification of endophytic root microbes produces crop EHs with (a) greater photosynthetic capabilities, and (b) deeper roots, then more CO_2 will be sequestered from the air and, eventually, transported into the roots. While the above-ground plant parts when harvested will by various routes release any sequestered C into the atmosphere as CO_2, the C that is sequestered in the plant roots, especially in absence of deep tillage, will remain in the soil. As the roots decompose, much of that C will be retained in soil as SOM. This SOM can become part of a storage system for excess C, and the amounts are not trivial. Because of plants' C/N ratio of 10:1, both and N will be incorporated into soil. If SOM is increased, the amount of N that can be evolved into the atmosphere as NO_x or into waterways as NO3 will be reduced.

Worldwide, it is estimated that an increase of 25–50% in root C together with moderate increases in deeper rooting could increase the stores of C in the soil by some 35–100 Mt./year (Paustian et al. 2016). This is equal to 80.5–230 Mt. of atmospheric CO_2. Withdrawing this amount of carbon from the carbon cycle would contribute significantly to reducing the greenhouse gases that are contributing to global climate change, since the total amount of annual CO_2 increase in the atmosphere is presently about 16 GT (Committee on Geoengineering Climate 2015). Such reductions can continue year after year as plant growth continues to be stimulated (Harman and Uphoff 2018). We estimate that the promotion of endophytes to establish EHs could increase the amount of C sequestered by about 12.5 t/ha, with an amount of 46 t/Ha of C removed from the atmosphere (Harman and Uphoff 2019; Harman et al. 2019). If multiplied across the world, this could be a potent mitigator and help to actually reverse some of the accumulation of especially CO_2 at the same time that production food supplies are being enhanced or sustained.

If this is the case, then this removal of C from the atmosphere ought to be recognized and rewarded, and farmers ought to be given incentives and compensated for this effort. If systems such as carbon-cap trading or carbon farming (a system whereby tax credits are provided to farmers for demonstrated increases in SOM) were implemented, farm income would increase while environmental and agronomic objectives are furthered.

In short, I am suggesting that EHs can make farming not only a method of producing food and fiber, but this could become part of the efforts to make the world a more livable planet. Such "carbon farming" could become both an agricultural ethic and a source of farm economic sustainability.

For this to occur, the C sequestered would have to be verified. Fortunately, there are satellite remote-sensing systems that can measure photosynthetic activity. These are already being used for the US Corn Belt which has the highest photosynthetic rates in the world at the height of the growing season (Guanter et al. 2014). Increases in photosynthetic rates can thus be measured relatively easily and reliably.

The other effect of EH promotion, root growth and development, is more difficult to measure since roots may be 2–3 m deep in the soil, and, in fact, they need to be if the soil sequestration of C is to be effective. We need to develop nondestructive methods for measuring root biomass in soil without digging, such as ground-penetrating radar (Delgado et al. 2017) or electrode resistivity imaging (Amato et al. 2008). Direct soil coring with quantification of roots (Frasier et al. 2018) may be an option. Further quantification can be obtained by using reporter genes of both microbes and plants and/or by C isotope tagging (Killham and Yeomans 2001). But it should be possible to ascertain how (and how much) using EH technology is contributing to withdrawing C (and N) from the atmosphere. Neither enhanced photosynthetic efficiency nor deep rooting has yet been the focus of any studies on endophytic root colonists of EHs.

However, the value of these capabilities is probably greater than the advantages of symbiotic endophytes that are already known and exploited, such as biocontrol. It is my challenge to research community, then, to develop EHs for the benefit of agriculture and of society at large. It is probable that other agriculturally based technologies can be envisioned and implemented so that farming can be a sustainable contributor to mitigate global climate change and other environmental goals.

There is an imperative to accomplish these tasks quickly as this is a promising way to meet multiple objectives – producing multiple benefits – at low cost with easily accessible technology that capitalizes on productive potentials already existing in nature. The levels of global warming that can currently be anticipated would result in geo- and biophysical feedback loops that are irreversible and would result in a "hothouse earth" scenario that is likely to be cataclysmic to human society. Examples include feedbacks such as disruption of ocean currents due to melting of ice at the poles that would forever alter present weather patterns and dynamics; still-greater increases in the levels of CO_2 in the atmosphere because of more and fiercer forest fires; and melting of the permafrost (Steffan et al. 2018). If these events should occur, even if the C emission goals of the Paris accord are met, further acceleration of global warming would result which disastrously alter possibilities for our lives on Earth. Dramatic efforts are needed to reorient our societies, policies, and behavior, to reduce C emissions and to increase plants' photosynthetic activity sustainably so as to reduce the dynamics of unbearable warming, and to maintain an acceptably stable natural environment on Earth.

Under the "hothouse earth" scenario, sea levels could rise dramatically (Strauss et al. 2015), and significant portions of the Earth may be rendered unhospitable to human life (Steffan et al. 2018). Avoiding this future probably requires not only limitations on C emissions but also actual removal of greenhouse gases from the atmosphere. The extension of EHs and the biological systems that we describe here are practical and attainable systems to accomplish this goal as well as to make sustainable increases in agricultural productivity in the face of increasingly inhospitable conditions on Earth.

We know that there is no single solution which will abate and possibly reverse current adverse trends. Promoting EH agriculture will not in itself prevent the climate disasters that may occur, but it does hold out the prospect of multiple agronomic, economic, and environmental benefits at reasonable cost. As we understand better the potency of many kinds of microbiomes, we should not forgo these opportunities because we do not recognize and utilize our human interdependency with the microbial realm. So, this is a dire challenge to the readers of this chapter. It is almost certain that other tools can be developed based on the increases of the impacts of intentionally managed plant-microbe interactions. It is critical that we develop these as rapidly as possible. Use the tools that have been developed to improve plants and ward off at least the worst effects of global climate change!

Acknowledgments Finally, this chapter has recognized many who have contributed to my research and to myself personally. Cornell Professor Norman Uphoff generously contributed his time and expertise in the review of this chapter.

I would be remiss if I did not recognize the love of my life, my wife Barbara Jean Harman, who has put up with me and with all the companies and developments for the last 53 years.

References

Adlas J, Achoth L (2006) Is the Green Revolution vanishing? Empirical evidence from TFP analysis for rice. Poster paper for International Association of Agricultural Economists conference, Australia, August 12–18. https://ageconsearch.umn.edu/bitstream/25561/1/pp060473.pdf

Ahmad JS, Baker R (1987) Rhizosphere competence of *Trichoderma harzianum*. Phytopathology 77:182–189

Ahmad JS, Baker R (1988) Implications of rhizosphere competence of *Trichoderma harzianum*. Can J Microbiol 34:229–234

Alfano G, Lewis Ivey ML, Cakir C, Bos JIB, Miller SA, Madden LV et al (2007) Systemic modulation of gene expression in tomato by *Trichoderma harzianum* 382. Phytopathology 97:429–437

Amato M, Basso B, Celano G, Bitella G, Morelli G, Rossi R (2008) In situ detection of tree root distribution and biomass by multi-electrode resistivity imaging. Tree Physiol 28:1441–1448

Anonymous (2012) United Nations Secretary-General's panel on global sustainability. Resilient people, resilient planet: a future worth choosing. United Nations, New York

Anonymous (2017) Strategies for survival. Nat Plants 3:907. https://doi.org/10.1038/s41477-017-0081-x

Berch SM, Massicotte HB, Tackaberry LE (2005) Republication of a translation of "The vegetative organs of *Monotropa hypopitys* L." published by F. Kamienski in 1882, with an update on Monotropa mycorrhizas. Mycorrhiza 15(5):323–332.

Bolar JP, Norelli JL, Wong K-W, Hayes CK, Harman GE, Aldwinckle HS (2000) Expression of endochitinase from *Trichoderma harzianum* in transgenic apple increases resistance to apple scab and reduces vigor. Phytopathology 90:72–77

Bolar JP, Norelli JL, Harman GE, Brown SK, Aldwinckle HS (2001) Synergistic activity of endochitinase and exochitinase from *Trichoderma atroviride* (*T. harzianum*) against the pathogenic fungus (*Venturia inaequalis*) in transgenic apple plants. Transl Res 10:533–543

Broadway RM, Williams DL, Kain WC, Harman GE, Lorito M, Labeda DP (1995) Partial characterization of chitinolytic enzymes from *Streptomyces albidoflavus*. Lett Appl Microbiol 20:271–276

Broadway R, Gongora C, Kain WC, Sanderson JP, Monroy JA, Bennett KC et al (1998) Novel chitinolytic enzymes with biological activity against herbivorous insects. J Chem Ecol 24:985–998

Chet I, Baker R (1981) Isolation and biocontrol potential of *Trichoderma harzianum* from soil naturally suppressive of *Rhizoctonia solani*. Phytopathology 71:286–290

Chet I, Harman GE, Baker R (1981) *Trichoderma hamatum*: its hyphal interactions with *Rhizoctonia solani* and *Pythium* spp. Microb Ecol 7:29–38

Chi F, Yang P, Han F, Jing Y, Shen S (2010) Proteomic analysis of rice seedlings infected by *Sinorhizobium meliloti* 1021. Proteomics 10:1861–1874. https://doi.org/10.1002/pmic.200900694

Committee on Geoengineering Climate, B.o.A.S.a.C., Ocean Studies Board, National Research Council (2015) Climate intervention: carbon dioxide removal and reliable sesquestration. National Academies Press, Washington DC

Dani F, Zain CRCM, Isahak A, Fathurrahman F, Sulaiman N, Uphoff N et al (2017) Relationships observed between *Trichoderma* inoculation and characteristics of rice grown under System of Rice Intensification (SRI) vs. conventional methods of cultivation. Symbiosis 72:45–59. https://doi.org/10.1007/s13199-016-0438-3

Defez R, Andreozzi A, Dickinson M, Charlton A, Tadini L, Pesaresi P et al (2017) Improved drought stress response in alfalfa plants nodulated by an IAA over-producing rhizobium strain. Front Microbiol 8:2466. https://doi.org/10.3389/fmicb.2017.02466

Delgado A, Hays DB, Bruton RK, Ceballos H, Novo A, Boi E et al (2017) Ground penetrating radar: a case study for estimating root bulking rate in cassava (*Manihot esculenta* Crantz). Plant Methods 13:65. https://doi.org/10.1186/s13007-017-0216-0

Di Pietro A, Lorito M, Hayes CK, Broadway RM, Harman GE (1993) Endochitinase from *Gliocladiumirens*: isolation, characterization, and synergistic antifungal activity in combination with gliotoxin. Phytopathology 83:308–313

Eckenrode CJ, Harman GEGE, Webb DRDR (1975) Seed-borne microorganisms stimulate seedcorn maggot egg laying. Nature 256:487–488

Frasier I, E N, Fernandesz R, Quiroga A (2018) Direct field method for root biomass quantification in agrosystems. MethodsX 3:513–519

Fukami J, de la Osa C, Javier Ollero F, Megias M, Hungria M (2018) Co-inoculation of maize with *Azospirillum brasilense* and *Rhizobium tropici* as a strategy to mitigate salinity stress. Funct Plant Biol 45:328–339. https://doi.org/10.1071/fp17167

Guanter L, Zhang Y, Jung M, Joiner J, Voigt M, Berry JA et al (2014) Global and time-resolved monitoring of crop photosynthesis with chlorophyll fluorescence. Proc Natl Acad Sci U S A 111:E1327–E1333. https://doi.org/10.1073/pnas.1320008111

Harman GE (1967) Physiology of sexual reproduction in *Hypomyces solani* f. *cucurbitae*. III. Perithecium formation on media containing compounds involved in shikimic acid pathway. Phytopathology 57:1138–1139

Harman GE (2000) Myths and dogmas of biocontrol. Changes in perceptions derived from research on *Trichoderma harzianum* T-22. Plant Dis 84:377–393

Harman GE (2006) Overview of mechanisms and uses of *Trichoderma* spp. Phytopathology 96:190–194

Harman GE (2014) US Patents 8,716,001, 8,877,480, 8,877,480 *Trichoderma* strains that induce resistance to plant diseases and/or increase plant growth

Harman GE, Corden ME (1972) Purification and partial characterization of the polygalacturonases produced by *Fusarium oxysporum* f. sp. *lycopersici*. Biochim Biophys Acta 264:328–338

Harman GE, Custis D (2006) US Patent 9,090,884. Formulations of viable microorganisms and their method of use

Harman GE, Kubicek CP (1998) *Trichoderma* and *Gliocladium*, vol. 2. Enzymes, biological control and commercial applications. Taylor and Francis, London

Harman GE, Mastouri F (2010) Enhancing nitrogen use efficiency in wheat using *Trichoderma* seed inoculants. In: Antoun H, Avis T, Brisson L, Prevost D, Trepanier M (eds) Biology of plant-microbe Ineractions. International Society for Plant-Microbe Intereactons, St. Paul, p 4

Harman GE, Mattick LR (1976) Association of lipid oxidation with seed aging and death. Nature 260:323–324

Harman GE, Eckenrode CJ, Webb DR (1978) Alteration of spermosphere ecosystems affecting oviposition by the bean seed fly and attack by soilborne fungi on germinating seeds. Ann Appl Biol 90:1–6

Harman GE, Chet I, Baker R (1980) *Trichoderma hamatum* effects on seed and seedling disease induced in radish and pea by *Pythium* spp. on *Rhizoctonia solani*. Phytopathology 70:1167–1172

Harman GE, Taylor AG, Stasz TE (1989) Combining effective strains of *Trichoderma harzianum* and solid matrix priming to improve biological seed treatments. Plant Dis 73:631–637

Harman GE, Hayes CK, Lorito M, Broadway RM, Di Pietro A, Peterbauer C et al (1993a) Chitinolytic enzymes of *Trichoderma harzianum*: purification of chitobiosidase and endochitinase. Phytopathology 83:313–318

Harman GE, Stasz TE, Weeden NF (1993b) Fused Biocontrol Agents. 5,260,213

Harman GE, Lattore B, Agosin A, San Martin R, Riegel DG, Nielsen PA et al (1996) Biological and integrated control of Botrytis bunch rot of grapes using *Trichoderma* spp. Biol Control 7:259–266

Harman GE, Hayes CK, Ondik KL (1998) Asexual genetics in *Trichoderma* and *Gliocladium*: mechanisms and implications. In: Kubicek CP, Harman GE (eds) *Trichoderma* and *Gliocladium*, vol 1. Taylor and Francis, London, pp 243–270

Harman GE, Howell CR, Viterbo A, Chet I, Lorito M (2004) *Trichoderma* species – opportunistic, avirulent plant symbionts. Nat Rev Microbiol 2:43–56

Harman GE, Obregón MA, Samuels GJ, Lorito M (2010) Changing models of biocontrol in the developing and developed world. Plant Dis 94:928–939

Harman GE, Cadle-Davidson M, Nosir W (2018) Patent application WO2017192117A1. Highly effective and multifunctional microbial compositions and uses

Harman GE, Doni F, Khadka RB, Uphoff N (2019) Endophytic strains of increase plants' photosynthetic capability. J Appl Microbiol

Harman GE, Uphoff N (2019) Symbiotic root-endophytic soil microbes improve crop productivity and provide environmental benefits. Scientifica 2019:1–25

Hayes CK, Klemsdal S, Lorito M, Di Pietro A, Peterbauer C, Nakas JP et al (1994) Isolation and sequence of an endochitinase-encoding gene from a cDNA library of *Trichoderma harzianum*. Gene 138:143–148

Hubbard JP, Harman GE, Hadar Y (1983) Effect of soilborne *Pseudomonas* sp. on the biological control agent, *Trichoderma hamatum*, on pea seeds. Phytopathology 73:655–659

Janaiah A, Otsuka K, Hossain M (2005) Is the productivity impact of the Green Revolution in rice vanishing? Empirical evidence from TFP analysis. Econ Polit Wkly 40:5596–5600

Jin X, Hayes CK, Harman GE (1991) Principles in the development of biological control systems employing *Trichoderma* species against soil-borne plant pathogenic fungi. In: Letham GC (ed) Symposium on industrial mycology, Mycological Society of America, Brock/ Springer series in contemporary biosciences

Jin X, Harman GE, Taylor AG (1992) Conidial biomass and desiccation tolerance in *Trichoderma harzianum*. Biol Control 1:237–243

Jin X, Taylor AG, Harman GE (1996) Development of media and automated liquid fermentation methods to produce desiccation-tolerant propagules of *Trichoderma harzianum*. Biol Control 7:267–274

Killham K, Yeomans C (2001) Rhizosphere carbon flow measurement and implications: from isotopes to reporter genes. Plant Soil 232:91–96. https://doi.org/10.1023/a:1010386019912

Kubicek CP, Harman GE (1998) *Trichoderma* and *Gliocladium*, vol. 1. Basic biology, taxonomy and genetics. Taylor and Francis, London

Lal R (2004) Soil carbon sequestration impacts on global climate change and food security. Science 304:1623–1627

Long SP, Marshall-Colon A, Zhu X-G (2015) Meeting the global food demand of the future by engineering crop photosynthesis and yield potential. Cell 161:56–66

Lorito M (1998) Chitinolytic enzymes and their genes. In: Harman GE, Kubicek CP (eds) *Trichoderma* and *Gliocladium*, vol 2. Taylor and Francis, London, pp 73–99

Lorito M, Di Pietro A, Hayes CK, Woo SL, Harman GE (1993a) Antifungal, synergistic interaction between chitinolytic enzymes from *Trichoderma harzianum* and *Enterobacter cloacae*. Phytopathology 83:721–728
Lorito M, Harman G, Hayes C, Broadway R, Tronsmo A, Woo SL et al (1993b) Chitinolytic enzymes produced by *Trichoderma harzianum*: antifungal activity of endochitinase and chitobiosidase. Phytopathology 83:302–307
Lorito M, Di Pietro A, Hayes CK, Harman GE (1994a) Purification and characterization of a glucan 1,3-ß-glucosidase and a N-acetyl-ß-glucosaminidase from *Trichoderma harzianum*. Phytopathology 84:398–405
Lorito M, Hayes CK, Zoina A, Scala F, Del Sorbo G, Woo SL et al (1994b) Potential of genes and gene products from *Trichoderma* sp. and *Gliocladium* sp. for the development of biological pesticides. Mol Biotechnol 2:209–217
Lorito M, Peterbauer C, Hayes CK, Harman GE (1994c) Synergistic interaction between fungal cell wall degrading enzymes and different antifungal compounds enhances inhibition of spore germination. Microbiology (Reading) 140:623–629
Lorito M, Woo SL, D'Ambrosio M, Harman GE, Hayes CK, Kubicek CP et al (1996) Synergistic interaction between cell wall degrading enzymes and membrane affecting compounds. Molec. Plant-Microbe Interact 9:206–213
Lorito M, Woo SL, Garcia Fernandez I, Colucci G, Harman GE, Pintor-Toro JA et al (1998) Genes from mycoparasitic fungi as a source for improving plant resistance to fungal pathogens. Proc Natl Acad Sci U S A 95:7860–7865
Margulis L, Fester R (1991) Symbiosis as a source of evolutionary innovation. MIT Press, Cambridge, MA
Marra R, Ambrosino P, Carbone V, Vinale F, Woo SL, Ruocco M et al (2006) Study of the three-way interaction between *Trichoderma atroviride*, plant and fungal pathogens using a proteome approach. Curr Genet 50:307–321
Mastouri F (2010) Use of *Trichoderma* spp. to improve plant performance under abiotic stress. PhD. Cornell University, Ithaca
Mastouri F, Bjorkman T, Harman GE (2010) Seed treatments with *Trichoderma harzianum* alleviate biotic, abiotic and physiological stresses in germinating seeds and seedlings. Phytopathology 100:1213–1221
Mastouri F, Bjorkman T, Harman GE (2012) *Trichoderma harzianum* strain T22 enhances antioxidant defense of tomato seedlings and resistance to water deficit. Mol Plant-Microbe Interact 25:1264–1271
Mitchell PL, Sheehy JE (2006) Supercharging rice photosynthesis to increase yield. New Phytol 171:688–693. https://doi.org/10.1111/j.1469-8137.2006.01855.x
Mittler R (2002) Oxidative stress, antioxidants and stress tolerance. Trends Plant Sci 7:405–410
Mo Y, Wang Y, Yang R, Zheng J, Liu C, Li H, et al (2016) Regulation of plant growth, photosynthesis, antioxidation and osmosis by an arbuscular mycorrhizal fungus in watermelon seedlings under well-watered and drought conditions. Front Plant Sci 7. https://doi.org/10.3389/fpls.2016.00644
Moore GA (1991) Crossing the chasm. HarperCollins, New York
Nath K, Jajoo A, Poudyal RS, Timilsina R, Park YS, Aro E-M et al (2013) Towards a critical understanding of the photosystem II repair mechanism and it regulation under stress conditions. FEBS Lett 587:3372–3381
Paustian K, Campell N, Dorich C, Marx E, Swan A (2016) Assessment of potential greenhouse gas mitigation from changes to crop root management and architecture. Booz Allen Hamilton, Washington DC
Porcel R, Redondo-Gomez S, Mateos-Naranjo E, Aroca R, Garcia R, Manuel Ruiz-Lozano J (2015) Arbuscular mycorrhizal symbiosis ameliorates the optimum quantum yield of photosystem II and reduces non-photochemical quenching in rice plants subjected to salt stress. J Plant Physiol 185:75–83. https://doi.org/10.1016/j.jplph.2015.07.006
Rifai MA (1969) A revision of the genus *Trichoderma*. Mycol Pap 116:1–56

Samuels GJ, Hebbar PK (2015) *Trichoderma*. Identification and Agricultural Properties. The American Phytopathological Society, St. Paul

Segarra G, Casanova E, Bellido D, Odena MA, Oliveira E, Trillas I (2007) Proteome, salicylic acid and jasmonic acid changes in cucumber plants inoculated with *Trichoderma asperellum* strain T34. Proteomics 7:3943–3952

Selvamukilan B, Rengalakshmi S, Tamizoli P, Nair S (2006) Village-level production and use of biocontrol agents and biofertilizers. In: Uphoff N et al (eds) Biological approaches to sustainable soil systems. CRC Press, Baco Raton

Shoresh M, Harman GE (2008) The molecular basis of maize responses to *Trichoderma harzianum* T22 inoculation: a proteomic approach. Plant Physiol 147:2147–2163

Shoresh M, Mastouri F, Harman GE (2010) Induced systemic resistance and plant responses to fungal biocontrol agents. Annu Rev Phytopathol 48:21–43

Smith VL, Wilcox WF, Harman GE (1990) Potential for biological control of Phytophthora root and crown rots of apple by *Trichoderma* and *Gliocladium* spp. Phytopathology 80:880–885

Smith VL, Wilcox WW, Harman GE (1991) US Patent 4,996,157. Biological control of *Phytophthora* by *Gliocladium*. 4,996,157

Smith VL, Wilcox W, Harman GE (1992) US Patent 5,165,928. Biological control of *Phytophthora* by *Gliocladium*

Stasz TE, Harman GE (1990) Nonparental progeny resulting from protoplast fusion in *Trichoderma* in the absence of parasexuality. Exp Mycol 14:145–159

Stasz TE, Harman GE, Weeden NF (1988a) Protoplast preparation and fusion in two biocontrol strains of *Trichoderma harzianum*. Mycologia 80:141–150

Stasz TE, Weeden NF, Harman GE (1988b) Methods of isozyme electrophoresis for *Trichoderma* and *Gliocladium* species. Mycologia 80:870–874

Stasz TE, Harman GE, Gullino ML (1989) Limited vegetative compatibility following intra- and interspecifc protoplast fusion in *Trichoderma*. Exp Mycol 13:364–371

Steffan W, Rockstriom J, Richardson K, Lenton TM, Folke C, Liverman D et al. (2018) Trajectories of the earth system in the Anthropocene. PNAS http://www.pnas.org/content/pnas/early/2018/07/31/1810141115.full.pdf: 8

Strauss BH, Kulp S, Leverman A (2015) Carbon choices determine US cities committed to future below sea level. PNAS 112:13508–13513

Sun C, Johnson J, Cai D, Sherameti I, Oelmueller R, Lou B (2010) Piriformospora indica confers drought tolerance in Chinese cabbage leaves by stimulating antioxidant enzymes, the expression of drought-related genes and the plastid-localized CAS protein. J Plant Physiol 167:1009–1017. https://doi.org/10.1016/j.jplph.2010.02.013

Taylor AG, Harman GE (1990) Concepts and technologies of selected seed treatments. Annu Rev Phytopathol 28:321–339

Taylor AG, Klein DE, Whitlow TH (1988) Smp: solid matrix priming of seeds. Sci Hortic 37:1–11

Taylor AG, Min T-G, Harman GE, Jin X (1991) Liquid coating formulation for the application of biological seed treatments of *Trichoderma harzianum*. Biol Control 1:16–22

Tyagi J, Varma A, Pudake RN (2017) Evaluation of comparative effects of arbuscular mycorrhiza (*Rhizophagus intraradices*) and endophyte (*Piriformospora indica*) association with finger millet (*Eleusine coracana*) under drought stress. Eur J Soil Biol 81:1–10. https://doi.org/10.1016/j.ejsobi.2017.05.007

University of Oxford (2017) Breakthrough in efforts to 'supercharge' rice and reduce world hunger. http://www.ox.ac.uk/news/2017-10-19-breakthrough-efforts-supercharge-rice-and-reduce-world-hunger

Vadassery J, Tripathi S, Prasad R, Varma A, Oelmueller R (2009) Monodehydroascorbate reductase 2 and dehydroascorbate reductase 5 are crucial for a mutualistic interaction between Piriformospora indica and Arabidopsis. J Plant Physiol 166:1263–1274. https://doi.org/10.1016/j.jplph.2008.12.016

Vargas WA, Mandawe JC, Kenerley CM (2009) Plant-derived sucrose is a key element in the symbiotic association between *Trichoderma virens* and maize plants. Plant Physiol 151:792–808

Weindling R (1932) *Trichoderma lignorum* as a parasite of other soil fungi. Phytopathology 22:837–845

Weindling R (1934) Studies on a lethal principle effective in the parasitic action of *Trichoderma lignorum* on *Rhizoctonia solani* and other soil fungi. Phytopathology 24:1153–1179

Zhang S, Gan Y, Xu B (2016) Application of plant-growth-promoting fungi *Trichoderma longibrachiatum* T6 enhances tolerance of wheat to salt stress through improvement of antioxidative defense system and gene expression. Front Plant Sci 7:1405. https://doi.org/10.3389/fpls.2016.01405

Metabolomics Approaches in Microbial Research: Current Knowledge and Perspective Toward the Understanding of Microbe Plasticity

2

Paulo R. Ribeiro, Rhaissa R. Barbosa, and Catherine P. de Almeida

2.1 Introduction

2.1.1 Microbial Metabolomics Research in the Post-genomic Era

The central dogma of molecular biology describes how genetic information stored within genes of a living organism flows into proteins: DNA → RNA → protein (Jafari et al. 2017). The advances conquered during the current post-genomic era have brought to light a paradigm shift in which new genetic research has enabled simultaneous analyses at the level of transcripts, proteins and metabolites (Illig and Illig 2018; Marcone et al. 2018; Singh et al. 2018; Van Der Heul et al. 2018). Enzymes are proteins, except for some catalytic RNA molecules, and their activity depends on several factors, including native protein integrity, conformation, pH, and temperature. Thousands of different enzymes work coordinately to ensure that all required chemical reactions occurs flawlessly within each individual cell (Raveendran et al. 2018; Scrutton 2017). The metabolic signature of an organism can be characterized by identifying the pathways (sets of enzymes) encoded in its genome. However, this metabolic signature is better assessed by applying advanced metabolomics approaches. Metabolomics encompasses the qualitative and quantitative analysis of the complete set of metabolites (metabolome) of an organism (Ribeiro et al. 2018). Therefore, by using a metabolomics approach it is possible to study key molecules of the metabolism of a given organism (Alcalde and Fraser 2016; Barkal et al. 2016; Bean et al. 2016; Beloborodova et al. 2018; D'Sousa Costa et al. 2015; Ribeiro et al. 2015). Metabolomics approaches in microbial research have allowed the identification and characterization of thousands of distinct chemical reactions that occur as the microorganism grow and divide. In general, microbial metabolomics researches

P. R. Ribeiro (✉) · R. R. Barbosa · C. P. de Almeida
Metabolomics Research Group, Departamento de Química Orgânica, Instituto de Química, Universidade Federal da Bahia, Salvador, Brazil
e-mail: pauloribeiro@ufba.br

D. P. Singh et al. (eds.), *Microbial Interventions in Agriculture and Environment*,
https://doi.org/10.1007/978-981-13-8391-5_2

apply two different approaches: untargeted and targeted. The untargeted approach encompasses the qualitative or semiquantitative characterization of the entire metabolome without prior knowledge of the metabolites to be analysed, whereas the targeted approach focuses on specific metabolites. These two approaches may focus on intracellular (fingerprinting) or extracellular (footprinting) metabolites (Baptista et al. 2018; Götz et al. 2018; López-Gresa et al. 2017; Sander et al. 2017).

A general workflow of metabolomics in microbial researches encompass four consecutive steps: sampling, extraction, data acquisition, and data processing (Azzollini et al. 2018; Chatzimitakos and Stalikas 2016; Maansson et al. 2016). The sampling step requires the separation, mainly by centrifugation, and quenching of the intra- and extracellular metabolites produced by the microorganism in the biological material. The extraction step includes mechanical cell disruption and the use of buffers or organic solvents to obtain the intra- and extracellular metabolites. The data acquisition step requires the use of advanced chromatographic and spectroscopic tools to determine metabolite content within a sample as well as its molecular structure. In some cases, prior to the data acquisition step it is necessary to perform metabolite derivatization. The last step requires the use of sophisticated softwares and chemometric tools to perform data processing and extraction. At this stage, statistical analysis are applied to identify discriminant group of metabolites responsible for the overall alterations in the studied system (Fig. 2.1).

An important aspect of any metabolomics study is to fully characterize the metabolome of an organism. Therefore, in order to obtain a broad picture of the microbial metabolome, researches usually apply several chromatography and metabolite detection techniques. Gas and liquid chromatography (GC and LC, respectively) are frequently used for metabolite separation, whereas nuclear magnetic resonance (NMR) and mass spectrometry (MS) are the most used techniques for metabolite detection (Fernand et al. 2017; Ortiz-Villanueva et al. 2017; Schelli et al. 2017; Vinci et al. 2018; Wang et al. 2016).

2.1.2 Initial Analysis of the Microbial Metabolomics Studies

In the following section, we will present a detailed systematic literature review of the current knowledge, main findings, and perspective toward the understanding of

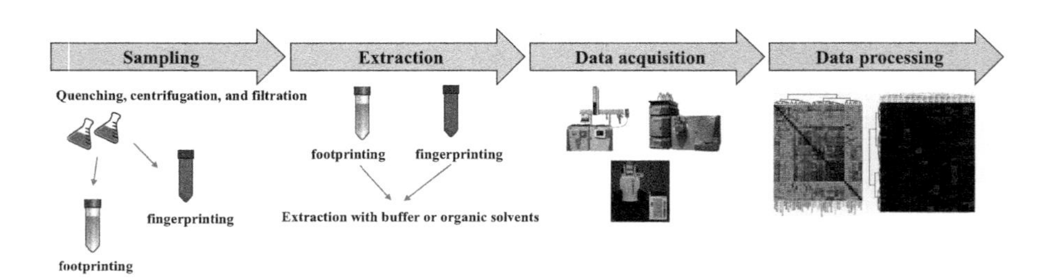

Fig. 2.1 Schematic representation of a general microbial metabolomics workflow

microbe plasticity in microbial metabolomics research. This book chapter covers manuscripts published between January 2014 and August 2018 that were available from scientific databases such as "Google Scholar," "PubMed," "ScienceDirect," "SpringerLink," and "Web of Science – Clarivate Analytics." The search through the scientific databases was performed using the keywords "metabolomics" or "metabolite profiling," along with "microorganism," "microbe," "fungi," "bacteria," "plasticity," and "biofilm." The chemical structures in this paper were drawn using ChemDraw Ultra 12.0.

Bacteria was, by far, the most well-studied microorganism with 44% of the published manuscripts, followed by fungi (30%), and lichen (7%) (Fig. 2.2a). Intracellular metabolome (fingerprinting) was assessed by 62% of the published manuscripts, whereas extracellular metabolome (footprinting) was assessed by 38% (Fig. 2.2b). Liquid chromatography–mass spectrometry (LC-MS) was used by 45% of the published manuscripts, followed by gas chromatography–mass spectrometry (GC-MS) (36%) and nuclear magnetic resonance (NMR) (19%) (Fig. 2.2c).

2.1.3 Metabolomics for Microbial Bioactive Metabolites

The worrisome diffusion of antibiotic-resistant microorganisms has encouraged new and advanced research to development new and more efficient antimicrobial metabolites (Bosso et al. 2018; Santos et al. 2018; Tracanna et al. 2017). Microorganisms constitute an important source of new bioactive metabolites that aid the development of new drugs and chemicals used for industrial and agricultural purposes (Honoré et al. 2016; Kildgaard et al. 2014; Romoli et al. 2014; Yogabaanu et al. 2017). Potentially new bioactive metabolites can be obtained by associating the discovery of new natural products with semisynthetic remodeling (Mgbeahuruike et al. 2017; Pintilie et al. 2018; Yang et al. 2018). Metabolomics is an important ally in drug discovery since it provides new strategies and methods to perform reliable and fast identification of new bioactive metabolites from different organisms

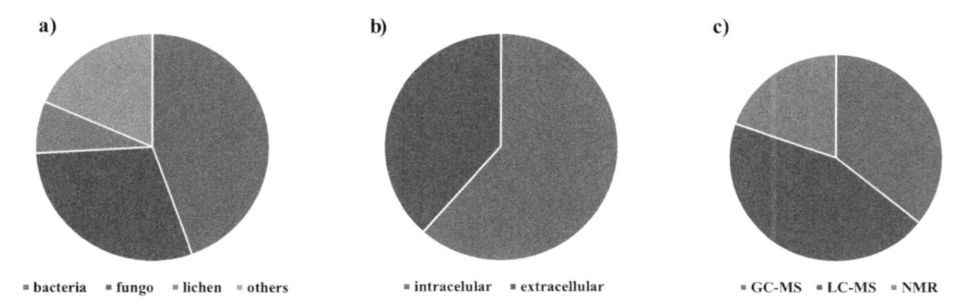

Fig. 2.2 (**a**) Type of microorganisms, (**b**) portion of the metabolome, and (**c**) techniques used for data acquisition of microbial metabolomics studies

(Bittencourt et al. 2015; Hakeem Said et al. 2017; Koistinen et al. 2018; Liao et al. 2018; Maansson et al. 2016; Santos et al. 2018).

An innovative approach using accurate dereplication by ultra-high performance liquid chromatography (UHPLC) and a high-resolution mass spectroscopy (HRMS) was used to find new bioactive metabolites from species of Aspergillus, Penicillium, and Emericellopsis from marine origin (Kildgaard et al. 2014). Dereplication techniques are key components of natural product screening and discovery since they allows rapidly and efficiently discrimination between previously known compounds and potential new bioactive metabolites within a crude extract (Hubert et al. 2017). Several metabolites were identified including small polyketides, non-ribosomal peptides, terpenes, and meroterpenoids. Four new metabolites related to asperphenamate were identified from *Penicillium bialowiezense* (Fig. 2.3). Asperphenamate is a natural anticancer phenylalanine dipeptide analog derivative with an *N*, *N′*-substituted phenylalanine–phenylalaninol ester framework (Liu et al. 2016b). Asperphenamate was initially isolated from *Aspergillus flavus* and later on from raw malt, which was used to treat hyperplasia of mammary glands (Clark et al. 1977), and it exhibits antitumor activity toward a number of cell lines (Li et al. 2012; Yuan et al. 2012). Therefore, dereplication by UHPLC-HRMS allowed the identification of potentially new bioactive metabolites.

Additionally, helvolic acid was identified in the culture of *Emericellopsis* sp. strain (IBT 28361). Helvolic acid is a nortriterpenoid first isolated from an endophyte fungal, *Xylaria* sp. (Fig. 2.4) (Ratnaweera et al. 2014). Helvolic acid showed antibacterial activity against *Bacillus subtilis*, *Enterococus faecalis*, methicillin-resistant *Staphylococcus aureus*, *Mycobacterium tuberculosis*, *Pseudomonas aeruginosa*, *Ralstonia solanacearum*, *Streptococcus pneumonia*,

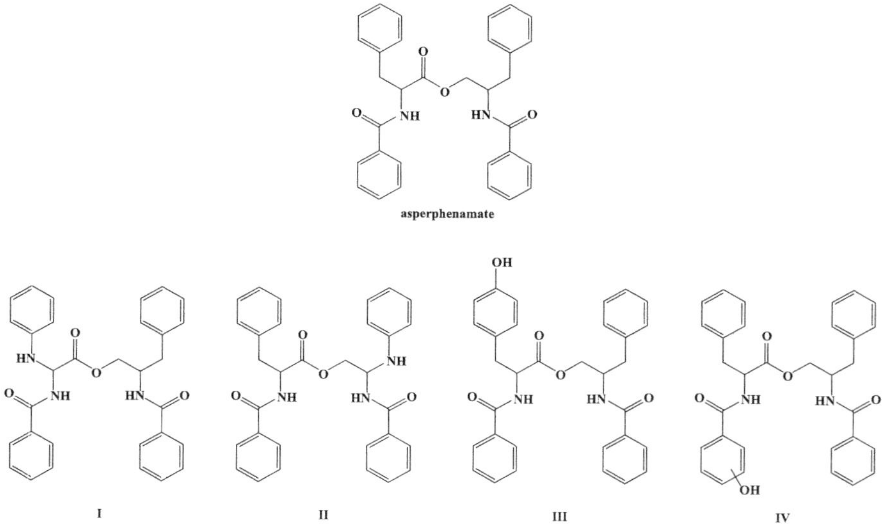

Fig. 2.3 Chemical structure of asperphenamate and its four new related metabolites (I–IV)

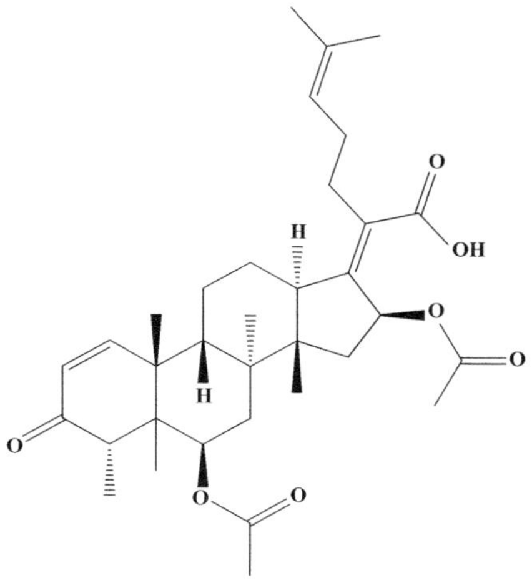

Fig. 2.4 Chemical structure of the antibacterial metabolite helvolic acid

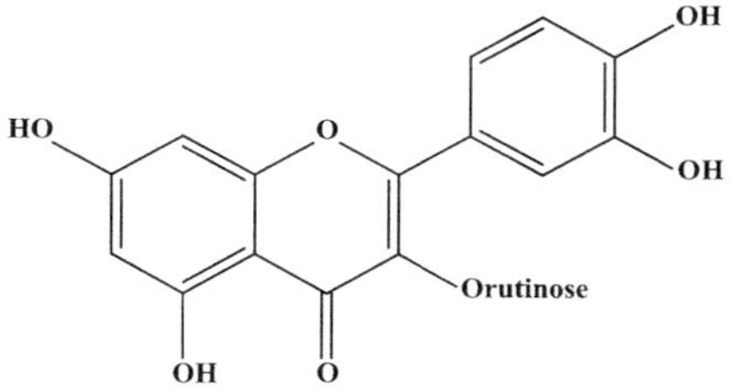

Fig. 2.5 Chemical structure of rutin

and *Xanthomonas campestris* (Luo et al. 2017; Ratnaweera et al. 2014; Sanmanoch et al. 2016; Yang et al. 2017).

Aspergillus flavus was isolated as an endophytic fungus from the Indian medicinal plant *Aegle marmelos* (Patil et al. 2015). Extracts produced from this fungus culture showed great antibacterial activity against *Escherichia coli*, *Pseudomonas aeruginosa*, *Salmonella abony*, *S. typhi*, *Bacillus subtilis*, and *Staphylococcus aureus*. Additionally, extracts showed DPPH scavenging activity, and membrane-stabilizing activity. Targeted high performance liquid chromatography (HPLC) identified rutin (Fig. 2.5) was the main compound produced in the extracts and most likely responsible for the observed activities. The chemical structure of rutin encompass the flavonol quercetin attached to a disaccharide rutinose moiety. This flavonoid is mainly found in plants, with rare occurrence in fungi (Patil et al. 2015). Therefore, these results shows that endophytic fungi are potential sources of bioactive metabolites and the combination with advanced metabolomics analysis allows their exploitation for medicinal, agricultural, and industrial uses.

Some bioactive metabolites presents biopreservation properties, especially of food related products. Honoré et al. (2016) used reversed-phase liquid chromatography-mass spectrometry (LC-MS) in an untargeted footprinting approach to assess the metabolome of *Lactobacillus paracasei*. Bioassay-guided fractionation and comprehensive screening was applied to identify potential antifungal metabolites. The antifungal property was measured by the capacity to inhibit the relative growth of the two *Penicillium* strains. The untargeted footprinting approach allowed the identification of glucose, amino acids such as leucine, isoleucine, phenylalanine, methionine, tryptophane, proline, and tyrosine, along with adenosine, and adenine. Additionally, a series of 2-hydroxy acids were identified: lactic acid, 2-hydroxy-4-methylpentanoic acid, 2-hydroxy-3-phenylpropanoic acid, 2-hydroxy-3-(4-hydroxyphenyl) propanoic acid, 2-hydroxy-3-phenylpropanoic acid, 2-hydroxy-(4-hydroxyphenyl) propanoic acid, and 2-hydroxy-4-methylpropanoic acid. These metabolites showed minimal inhibitory concentration for 50% inhibition (MIC_{50}) varying from 5 to 10 $mg.mL^{-1}$ against two *Penicillium* strains. Three undescribed antifungal metabolites, along with three known were detected from *Lb. paracasei* (Honoré et al. 2016).

Several studies supports that microorganisms isolated from marine samples are a promising source of new bioactive metabolites. Kim et al. (2016b) used the intestine of the golden sea squirt to isolate the wild-type bacterial strain. The strain was identified as Pseudoalteromonas sp. by 16S rDNA sequence analysis. LC-MS analysis of the ethyl acetate extract revealed that presence of nine metabolites belonging to the 4-hydroxy-2-alkylquinoline class and were identified as pseudane III, IV, V, VI, VII, VIII, IX, X, and XI (Fig. 2.6).

Additionally, two new metabolites from marine bacteria were identified: 2-isopentylqunoline-4-one and 2-(2,3-dimetylbutyl)qunoline-4-(1H)-one (Fig. 2.6). Pseudane VI and VII possess anti-melanogenic and antiinflamatory activities (Kim et al. 2016b, 2017), whereas pseudane IX showed strong anti-Hepatitis C virus activities (Wahyuni et al. 2014).

Betancur et al. (2017) isolated actinobacteria strains from sediment, invertebrate and algae samples collected from coral reefs in the Colombian Caribbean Sea. Species belonging to the genera *Streptomyces*, *Micromonospora*, and *Gordonia* were identified within the isolated bacteria by 16S rRNA gene sequencing. They used LC-MS analysis to identify new antimicrobial and quorum quenching metabolites against pathogens from the isolated actinobacteria strains. Six out of the 24 isolates showed promising results regarding the antimicrobial activities. Dereplication was applied to identify new bioactive metabolites by excluding well-known active metabolites or inactive natural products.

Twenty-eight entities did not present any hits and may represent new compounds. Dereplication indicates the presence of possible antibacterial and anthelmintic activity pyridine derivatives from *Streptomyces tendae* and *S. piericidicus*, δ-lactones inducer of anthracycline production from *S. viridochromogenes*, antibacterial fatty acid derivatives from *S. globisporus*, antifungal and antibacterial anthraquinone derivatives from Streptomyces sp., antitumor and antifungal alkaloids from *S. thioluteus*, and antibacterial macrolides from *S. griseus* (Betancur et al. 2017). Some of the well-known active metabolites included streptomycin-D, youlenmycin, inostamycin-b, pterulamide III, bistheonellic acid B, and mechercharmycin A (Fig. 2.7) (Betancur et al. 2017). Streptomycin is an antibiotic used to treat several types of infection (Schatz et al. 1944) and it shows quorum sensing inhibitory

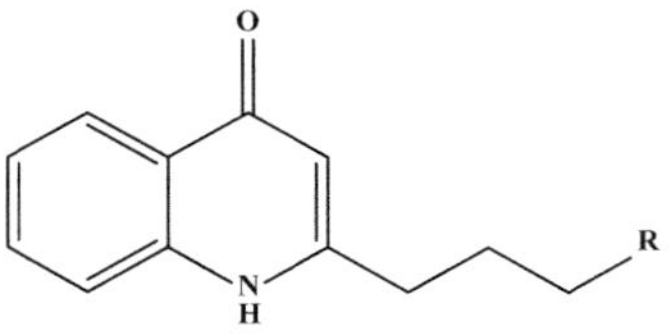

pseudane III, (R = H)
pseudane IV, (R = CH_3)
pseudane V, (R = C_2H_5)
pseudane VI, (R = C_3H_7)
pseudane VII, (R = C_4H_9)
pseudane VIII, (R = C_5H_{11})
pseudane IX, (R = C_6H_{13})
pseudane X, (R = C_7H_{15})
pseudane XI, (R = C_8H_{17})

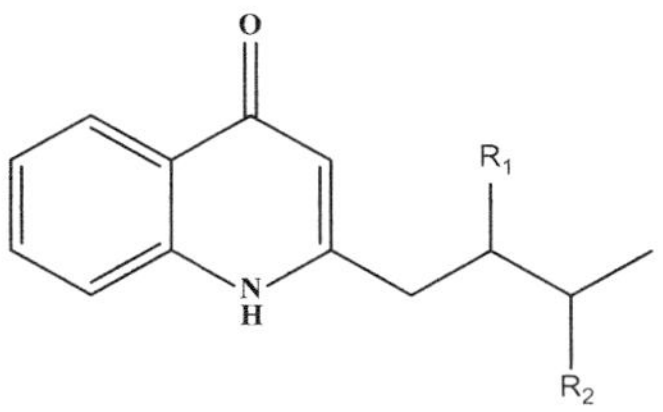

2-isopentylqunoline-4-one, (R_1 = H; R_2 = CH_3)
2-(2,3-dimetylbutyl)qunoline-4-(1H)-one, (R_1 = CH_3; R_2 = CH_3)

Fig. 2.6 Chemical structure of bioactive metabolites produced by a wild-type bacterial strain isolated from the intestine of the golden sea squirt (*Halocynthia aurantium*)

activity in *Acinetobacter baumannii* (Saroj and Rather, 2013). Inostamycin-b showed antimicrobial activities against *Staphylococcus aureus*, *Micrococcus luteus*, *Bacillus anthracis*, *B. subtilis*, *Corynebacterium bovis*, and *Mycobacterium smegmatis* (Odai et al. 1994). Pterulamide III is a cytotoxic linear peptide isolated from Pterula species (Lang et al. 2006), whereas mechercharmycin A is a antitumor cyclic peptide-like isolated from *Thermoactinomyces* sp. obtained from marine source (Kanoh et al. 2005).

Gnavi et al. (2016) also worked with marine-derived microorganism, but instead of bacteria they isolated sterile mycelia from *Flabellia petiolata* collected in the Mediterranean Sea. Species belonging to the genera *Biatriospora*, *Beauveria*, *Massarina*, *Microascacea*, *Roussoellacea*, and *Knufia* were identified within the isolated bacteria by sequencing the nrDNA internal transcribed spacer (ITS) and large ribosomal subunit (LSU) partial regions. Antibacterial activity was assessed against the multidrug-resistant (MDR) bacteria *Burkholderia metallica*, *Pseudomonas aeruginosa*, *Klebsiella pneumoniae* and *Staphylococcus aureus*. These bacteria are involved in cystic fibrosis and nosocomial infections. LC-MS analysis was applied to identify intra- and extracellular metabolites produced from each fungal strain and revealed the presence of 2-aminodocosa-6,17-dien-1,3-diol, 2-aminooctadecan-1,3,4-triol, 2-aminooctadecan-1,3-diol, phytoceramide C2, aphidicolin, scopularide A, bis(2-ethylhexyl) hexanedioic acid, fusoxysporone, and ergostane derivatives ergosta-5,7,22-trien-3-β-ol and ergosta-3,5,7,9(11),22-pentaene (Fig. 2.8). These metabolites might be responsible for the antibacterial activity of the extracts (Gnavi et al. 2016).

The actinomycete *Streptomyces sparsus* VSM-30 was isolated from deep sea sediments of Bay of Bengal. LC- and GC-MS analyses of the extracellular metabolites from the ethyl acetate extract revealed that presence of tryptophan dehydrobutyrine diketopiperazine, maculosin, 7-o-demethyl albocycline,

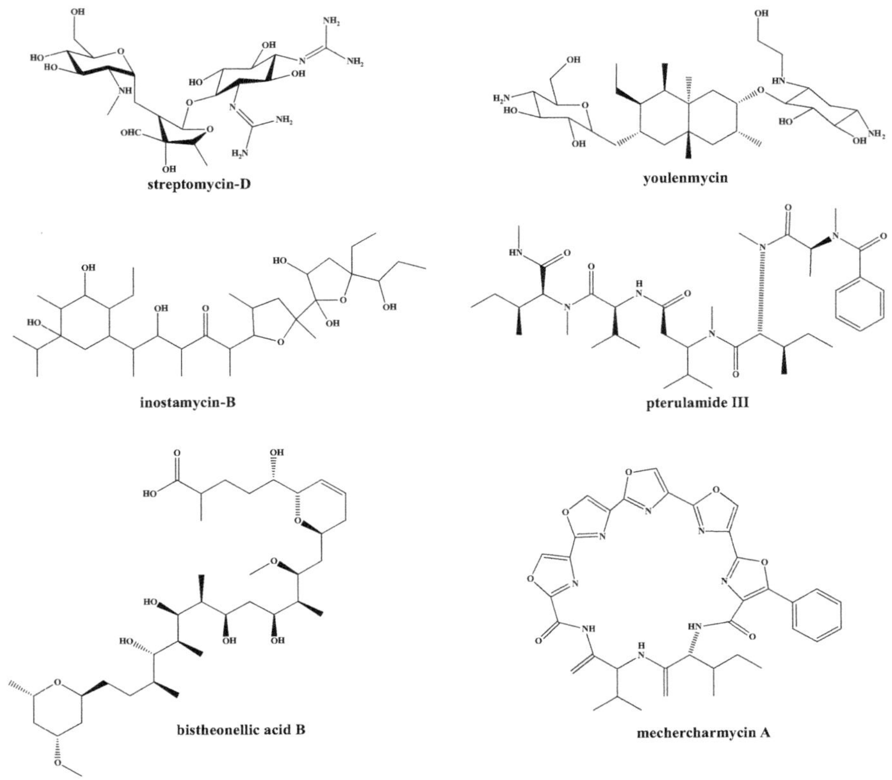

Fig. 2.7 Chemical structure of antimicrobial produced by actinobacteria strains from sediment, invertebrate and algae samples collected from coral reefs in the Colombian Caribbean Sea

albocycline M-2, 7-o-demethoxy-7-oxo albocycline, dotriacontane, 11-decyl-tetracosane, diheptyl phthalate, 1-hexadecanesulfonyl chloride, L-alanyl-L-tryptophan, phthalic acid ethyl pentyl ester, 4-trifluoroacetoxyhexadecane, and 1H-imidazole 4,5-dihydro-2,4-dimethyl. These metabolites contribute for the biological activities of the extract (Managamuri et al. 2017). Yogabaanu et al. (2017) used high performance liquid chromatography (HPLC) to identify variations in the extracellular metabolome of soil fungi in response to temperature variation and to screen for antimicrobial metabolites. These fungi are found at the Arctic and Antarctic regions and were obtained from the National Antarctic Research Centre Fungal Collection, from the University of Malaya, Kuala Lumpur. The antimicrobial activity of these fungi was assessed against *B. subtilis*, *B. cereus*, *E. coli*, *E. faecalis*, and *P. aeruginosa* by disk diffusion assay through the inhibition zone produced. They showed that culture temperature influenced the metabolome of the fungal strains. However, they failed to identify the metabolites detected in the crude extracts, limiting their results to the presentation of the retention times of the metabolites (Yogabaanu et al. 2017). Romoli et al. (2014) also worked with

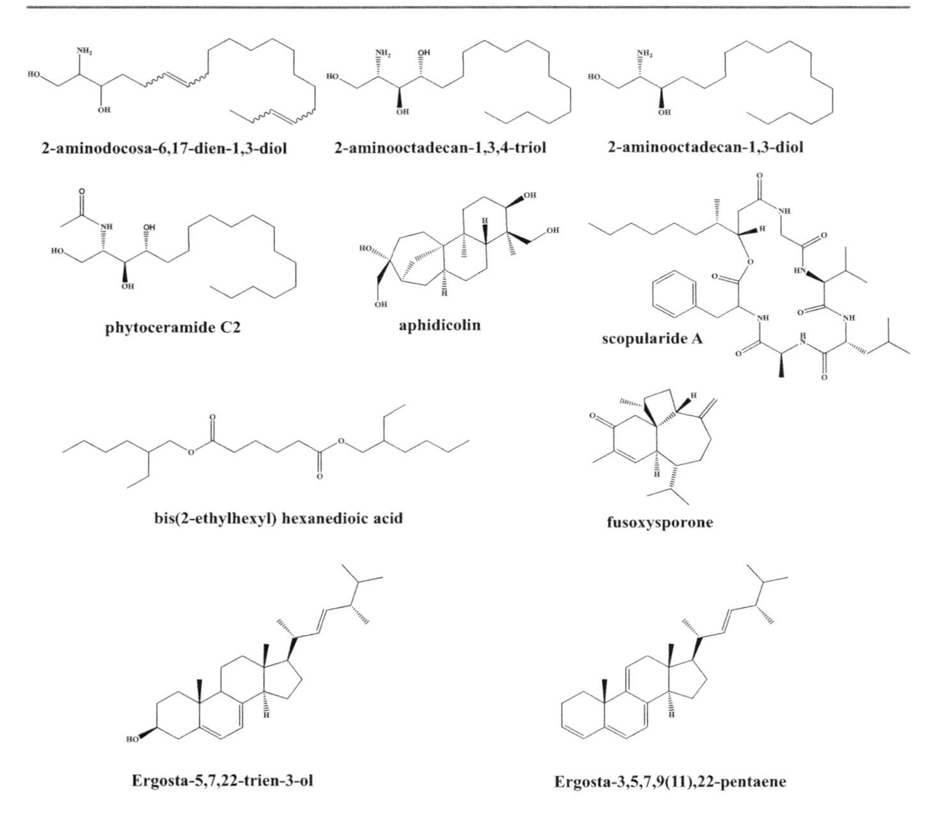

Fig. 2.8 Chemical structure of the metabolites produced by *Flabellia petiolata* sterile mycelia

microorganisms that inhabits the Antarctic region, more specifically the *Pseudoalteromonas* TB41 bacteria strain. These bacteria produced a wide range of volatile compounds (VOCs) that inhibit the growth of *Burkholderia cepacia* complex (Bcc) strains. The Bcc strains are opportunistic pathogens of cystic fibrosis patients (Sfeir 2018; Van Dalem et al. 2018). Solid phase micro extraction (SPME) gas chromatography–mass spectrometry (GC-MS) analysis allowed the identification of 30 VOCs, including some alcohols, along with some nitrogen- and sulfur-rich compounds (Romoli et al. 2014). Microbes produce a diverse range of volatile compounds that may function as key components in the cross-microbial relationships within a microbiota. However, the authors did to correlate the concentration of the identified VOCs with the possible inhibitory effects on the growth of Bcc strains.

2.1.4 Metabolomics for Microbial Biofilms

Microbial biofilms consist of a multicellular microbial conglomerate embedded in complex extracellular matrix adhered on a solid surface. In contrast, planktonic cells are single-cell organisms that may drift or disperse in a liquid medium (Azeredo et al. 2017). Microbial biofilms are ubiquitous in nature and may constitute a microorganism survival strategy to unfavorable environmental conditions (Landini et al. 2010). The formation of a microbial biofilm formation involves microbial adhesion, and accumulation of an extracellular matrix. This extracellular matrix is composed of macromolecules such as proteins, polysaccharides, humic substances, and extracellular DNA. Microbial biofilm formation is tightly regulated by a combination of environmental and physiological cues, such as nutrient availability, and cellular stress (Assaidi et al. 2018; Favre et al. 2018; Jin et al. 2018; Landini et al. 2010). They can be beneficial or have a negative impact providing tolerance to antibiotic treatment, and enhancing virulence of many pathogenic bacteria. This is especially concerning when biofilms are formed on industrial settings or on medical devices (Azeredo et al. 2017; Jin et al. 2018; Landini et al. 2010). Due to its versatility, metabolomics has been applied to biofilm research in order to identify the biochemical changes between the planktonic and biofilm phenotypes, to assess the chemical composition of biofilms, and to monitor in vivo biofilm formation and development.

Microbe cells within biofilms and their planktonic equals are morphologically and physiologically distinct from each other. There is a strong correlation between pathogenic biofilms and diseases, since microbe biofilms are more resistant to unfavorable environmental conditions. Therefore, it is crucial to develop strategies to assess in vivo biofilm formation to understand the mechanism underlying the dynamic biochemical changes occurring during the transition between the planktonic and biofilm phenotypes. Bacterial membranes encompass a diverse panel of amphiphilic phospholipids, such as phosphatidylglycerol, phosphatidylethanolamine, cardiolipin, phosphatidylcholine, and phosphatidylinositol, as well as ornithine lipids, glycolipids, and sphingolipids (Sohlenkamp and Geiger 2015). *Escherichia coli* has been used for a long time as the perfect organism to study membrane lipids. However, bacterial membrane lipid composition may vary among different bacteria species and environmental growth conditions (Sohlenkamp and Geiger 2015). Benamara et al. (2014) assessed the dynamics of the phospholipid composition of the *P. aeruginosa* membranes (fingerprint) throughout biofilm development on glass wool. They applied gas chromatography-mass spectrometry (GC-MS) to assess the lipidome dynamics in response to the biofilm age (i.e., from 1-, 2-, to 6-day-old biofilm). Phosphatidylethanolamines (PE 30:1, 31:0, 38:1, 38:2, 39:1, and 39:2) and phosphatidylglycerols (PG 31:0, 38:0, 38:1, 38:2, 39:1, and 39:2) were the predominant lipids on *P. aeruginosa* inner and outer membrane. Lipidome changes was more significant for the biofilm phenotype than for the planktonic counterpart. Heavier and branched-chains phospholipids decreased in the outer membrane, whereas cyclopropylated phospholipids increased in both membranes with the biofilm age. Curiously, the lipidome of the oldest biofilms were more similar to the metabolome of the planktonic phenotype (Benamara et al. 2014). Accumulation of phosphatidylethanolamine derivatives were observed in the biofilm phenotype of *Pseudoalteromonas lipolytica*, whereas ornithine lipids were preferably produced by the planktonic phenotype (Favre et al. 2018). Analysis of the intracellular metabolome of *Desulfovibrio vulgaris* planktonic and biofilm phenotypes by GC- and LC-MS showed that metabolites related to fatty acid biosynthesis such as lauric, mysistic, palmitoleic, and stearic acids were up-regulated in the biofilm as compared to the planktonic phenotype (Zhang et al. 2016). These results support the hypothesis that membrane related metabolites are important for the formation, maintenance and function of microbial biofilms as well as their differentiation from the planktonic phenotype. Membrane composition re-modelling supported *Streptococcus intermedius* growth and adaptation to anaerobic conditions. *S. intermedius* plasticity under oxygen depletion allows it to coexist in biofilms, both as a commensal and a pathogen (Fei et al. 2016).

Microbial biofilm development on biotic and abiotic surfaces act as a continual source of contamination. Microbial biofilm formation by Salmonella spp. has profound consequences in many industries, since they are notoriously difficult to eradicate (Corcoran et al. 2013; Keelara et al. 2016; Patel et al. 2013). Wong et al. (2015) applied GC-MS to detect biochemical change between intracellular and extracellular metabolites produced by the biofilm and planktonic phenotypes of Salmonella spp. cells and Salmonella biofilms of different ages. Alanine, glutamic acid, glycine, and ornithine showed the major contribution to discriminate between the extracellular metabolome of planktonic and biofilm phenotypes, whereas succinic acid, putrescine, pyroglutamic acid, and N-acetylglutamic acid acted as major contributors to discriminate between the intracellular metabolome of planktonic and biofilm phenotypes. Similarly, amino acids were responsible for the main discrimination among the samples of different days of biofilm growth. However, the intracellular showed no significant differences in response to age (Wong et al. 2015). Undoubtedly, central carbon and nitrogen metabolism plays a crucial role on the differentiation of the intracellular and extracellular metabolome of planktonic and biofilm phenotypes. For example, Stipetic et al. (2016) reported that *Staphylococcus aureus* planktonic and biofilm phenotypes showed differences in in arginine biosynthesis. Ząbek et al. (2017) applied quantitative NMR to assess the metabolome of the planktonic and biofilm phenotypes (1 and 2-day-old biofilm) of *Aspergillus pallidofulvus* and reported that the levels of the extracellular leucine, arginine, choline, betaine, *N*-acetylglucosamine, and phenylalanine were upregulated after 24 h of growth. Additionally, organic acids such as threoic, aspartic, docosanoic, malonic, hydrobenzoic and keto-gluconic, as well as the carbohydrates fructose, mannose, cellobiose, and maltose are important intracellular metabolites produced by the planktonic and biofilm phenotypes of *Vibrio fischeri* ETJB1H (Chavez-Dozal et al. 2015).

Borgos et al. (2015) used high-resolution liquid chromatography-mass spectrometry fingerprinting as a rapid, sensitive and noninvasive technique to assess the formation and development of *P. aeruginosa* biofilm between 0 and 196 h after inoculation. Despite the fact that these authors applied an untargeted approach, they identified, in the positive ESI mode, a compound with m/z 211.0867 ($M+H^+$ ion) that changed in response to both strain and sampling time. The unknown compound was unambiguously annotated as pyocyanine (Borgos et al. 2015). Pyocyanin is a virulence factor produced by *P. aeruginosa,* which shows antimicrobial activity against Gram-positive bacteria (Gharieb et al. 2013). The authors, however, failed to present a time-course analysis of the pyocyanin content, limiting their discussion to the ANOVA results. We can only infer that as an antimicrobial compound, pyocyanin is produce in order to ensure proper formation of the microbe conglomerate preventing the incorporation of unwanted microorganisms.

Ammons et al. (2014) applied quantitative NMR to assess dynamic biochemical changes between the planktonic and biofilm phenotypes of methicillin-resistant and methicillin-susceptible *Staphylococcus aureus*. *S. aureus* is considered a wound bioburden since it forms a sort of colonizing biofilm as major contributor to nonhealing wounds (DeWitt et al. 2018; Kim et al. 2018). Principal component analysis based on both intracellular and extracellular metabolites differentiated the phenotypes. Amino acid uptake, lipid catabolism, and butanediol fermentation are key features distinguishing the phenotypes (Ammons et al. 2014). Additionally, they claimed that a shift in metabolism from energy production to assembly of cell-wall components and matrix deposition may also play a role in distinguishing between the planktonic and biofilm phenotypes. This is a farfetched hypothesis since they did not identified any cell-wall metabolite components, with the exception of some pyrimidine nucleotides that may serve as precursors for synthesis of teichoic acids and peptidoglycan in *S. aureus*. Schelli et al. (2017) used HPLC-MS/MS to assess the metabolome of two *S. aureus* strains in response to methicillin exposure. As expected, methicillin exposure disturbed the metabolome of the methicillin susceptible *S. aureus* in a greater extent than of the methicillin resistant strain (Schelli et al. 2017).

2.1.5 Metabolomics for Microbial Biomarkers

2.1.5.1 Abiotic Stresses: Light and Oxygen Availability and Salinity

Microbial biomarkers can be classified in three types: exposure, effect and susceptibility. The identification of reliable biomarkers is important for a wide number of purposes, and may provide information related to microbial metabolism concerning exposure, growth and adaptation under biotic and abiotic stress conditions. Abiotic stress conditions encompass light, oxygen, and water availability along with salinity and low and high temperature.

Ultra performance liquid chromatography was used to compare the extracellular metabolome of *Aspergillus nidulans* during fungal development in the dark or light (Bayram et al. 2016). Light not only accelerated asexual development of *A. nidulans*, but also induced the production of the antitumoral metabolites terrequinone A and emericellamide (Fig. 2.9). Terrequinone A is a bisindolylquinone derivative with tumor growth inhibitory activity (He et al. 2004), whereas emericellamide is an antibiotic compound of mixed origins with polyketide and amino acid building blocks (Chiang et al. 2008; Newman 2016). Dark conditions, however, led to the preference of the sexual development and the accumulation of the polyketide mycotoxin sterigmatocystin, along with the antraquinones asperthecin, and emodin (Fig. 2.9) (Bayram et al. 2016). These metabolites could be used as potential

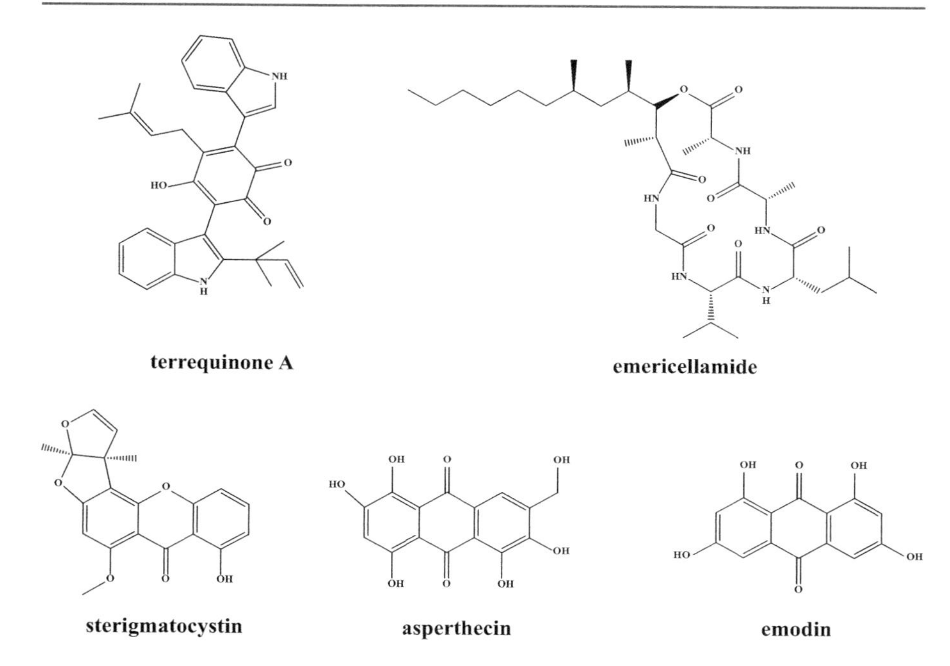

Fig. 2.9 Chemical structure of biomarkers produced from *A. nidulans* during fungal development in the dark or light

biomarkers for the genus *Aspergillus* during fungal development in the dark or light conditions.

Oxygen availability may also alter the metabolome composition of a microorganism and for that reason Fei et al. (2016) used hydrophilic interaction chromatography (HILIC–TOF–MS) to assess the effect of oxygen availability on the intra and extracellular metabolome of *Streptococcus intermedius* strain B196. Oxygen depletion enhanced pyrimidine and purine metabolism and the central carbon metabolism. This microorganism showed high plasticity, especially under anaerobic growth conditions, in which *S. intermedius* adaptation to oxygen depletion involved the re-modelling of the cellular membrane composition. Alterations in cellular membrane composition, especially fatty acids and phospholipids, are an important adaptation mechanism present several microorganisms (Suutari and Laakso 1994) and may also be targeted for biomarker discovery in response to oxygen availability.

Nuclear magnetic resonance spectroscopy (^{1}H NMR) was applied to evaluate the effect of salt stress on the intracellular metabolome of four halophilic bacterial isolates. *Halomonas hydrothermalis*, *Bacillus aquimaris*, *Planococcus maritimus* and *Virgibacillus dokdonensis* were isolated from a saltern region in India. Metabolites involved in the glycolytic pathway, pentose phosphate pathway and citric acid cycle showed a salt-dependent increase. These are main energy-generating pathways, which may explain the fact that cellular homeostasis was favored over

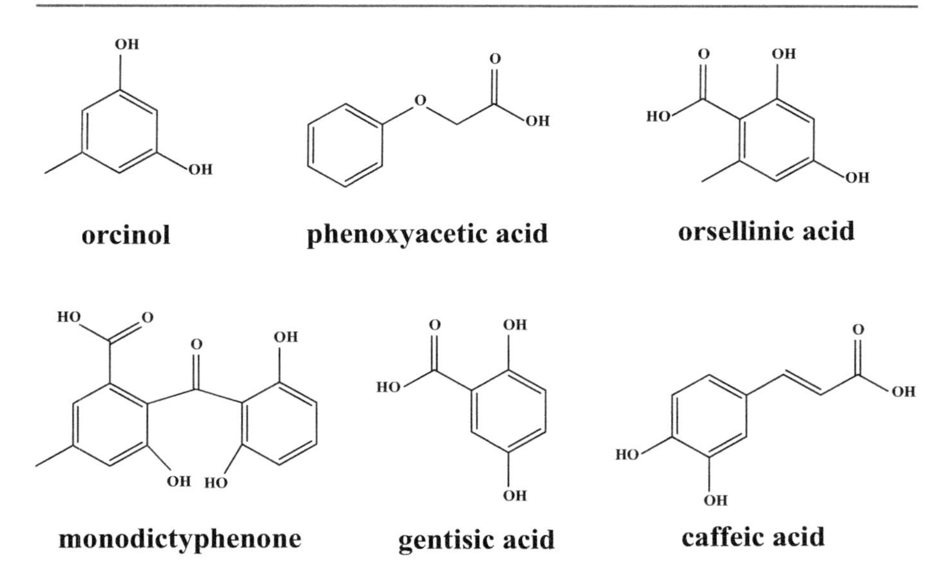

Fig. 2.10 Chemical structure of possible biomarkers produced from *A. nidulans* during fungal development in the presence of ionic liquids' stimulants cholinium chloride and 1-ethyl-3-methylimidazolium chloride

growth (Joghee and Jayaraman 2014). Alves et al. (2016) evaluated the effect of ionic liquids' stimuli (cholinium chloride or 1-ethyl-3-methylimidazolium chloride) on the extracellular metabolome of *Aspergillus nidulans*. ^{1}H NMR analyses revealed that both ionic liquids stimulated production of acetyl-CoA. Acetyl-CoA is an important precursor for secondary metabolites and non proteinogenic amino acids. Ionic liquids stimulated the production of orcinol, phenoxyacetic acid, orsellinic acid, monodictyphenone, gentisic acid, and caffeic acid (Fig. 2.10) (Alves et al. 2016). Taken together, these studies highlights the importance of metabolomics for strain enhancement and phenotypic analysis of microorganisms unfavorable conditions.

2.1.5.2 Metabolic Engineering

Metabolic engineering approaches aim at developing microbial cell factories to modulate metabolic pathways for metabolite over production or to improve cellular properties optimizing genetic and regulatory processes within cells (Chae et al. 2017; Lian et al. 2018). Metabolomics along with systems biology, synthetic biology and evolutionary engineering methodologies have allowed a fast and impressive advancing of metabolic engineering to rewire cellular metabolism.

Saccharomyces cerevisiae is a single-celled eukaryote commonly used for metabolic engineering research. The fact that the genome of this microorganism has been fully sequenced and is easily manipulated makes it an attractive model species for metabolic engineering approaches. Kim et al. (2016a) used GC-TOF-MS to investigate the intracellular metabolome of *S. cerevisiae* in order to identify possible ethanol tolerance mechanisms. For that, an ethanol-tolerant mutant yeast iETS3 was constructed and its metabolome was compared to the wild-type S. *cerevisiae* BY4741. Several metabolites from the central carbon and nitrogen metabolism were identified including amines, amino acids, fatty acids, organic acids, phosphates, sugars and sugar alcohols. Principal component and hierarchical clustering analyses showed a clear separation of the metabolite profiles of iETS3 and BY4741. Metabolites involved in cell membrane composition, glutamate and trehalose metabolism were identified as possible biomarkers for ethanol tolerance in *S. cerevisiae* (Kim et al. 2016a).

Pichia pastoris is used for process-scale production of recombinant secreted proteins, but often shows low productivity. Tredwell et al. (2017) used recombinant *P. pastoris* strains with varying levels of unfolded protein response (UPR) induction to study cellular stress responses and to identify potential biomarkers of UPR induction. NMR metabolic profiling of the intra- and extracellular metabolome allowed the quantification of 32 metabolites including 15 amino acids (alanine, arginine, asparagine, aspartate, glutamate, glutamine, histidine, isoleucine, leucine, lysine, methionine, phenylalanine, serine, tyrosine, and valine), 9 organic acids (benzoate, citrate, formate, fumarate, lactate, malate, succinate, 2-oxoisocaproate, and 3-methyl-2-oxovalerate), glucose, and some other metabolites. The authors suggested that the metabolites identified from both cell extracts and supernatants could be used as potential biomarkers for future high-throughput screening of large numbers of *P. pastoris* clones.

Campylobacter jejuni is a human bacterial pathogen described as one of the most common causes of food poisoning worldwide leading to self-limited diarrheal illness. In some cases, macrolide antibiotics are required for its efficient treatment (Chen et al. 2018; Ranjbar et al. 2017). The resistance gene erm(B) and mutations in rplD and rplV (23S rRNA) has led to the development of macrolide-resistant Campylobacter strains (Wang et al. 2014). Fu et al. (2018) have applied UHPLC-MS/MS to investigate the intracellular metabolome of a susceptible (NCTC 11168) and a resistant (NCTC 11168 with ermB) strain of *C. jejuni*. The resistance gene erm(B) had a deep impact on membrane integrity and stability, what was confirmed by the reduced biofilm formation capability of resistant strain as compared to the susceptible strain. Thirty-six metabolites were identified as potential biomarkers to differentiate the susceptible and resistant *C. jejuni* stain. These metabolites are involved in cell signaling, membrane integrity and stability, and energy-generating pathways. These results highlight important metabolic regulatory pathways associated with resistant of *C. jejuni*.

Microbes often produce metabolites as a survival strategy, especially when growing in harsh environments. Scarce nutrient and water availability may impose competition among microorganisms and, therefore, compelling them to produce metabolites to protect themselves, to control the proliferation of other microorganisms, or to acquire certain advantage when challenged by other microorganisms within a given microbiota. This process is called quorum quenching, which is usually proceeded by the quorum sensing. Quorum sensing is a transcriptional regulatory mechanism by which microorganisms regulate population density through chemical signaling. Molecules secreted by microorganisms are a form of intra- and interspecies communication that helps bacteria coordinate their behavior (Gökalsın and Sesal 2016; Liu et al. 2016a; Padder et al. 2018). For example, *P. aeruginosa* uses quorum sensing to regulate the production of secreted virulence factors through the *N*-acyl-homoserine lactone (AHL)-dependent quorum sensing system (Davenport et al. 2015). Davenport et al. (2015) used a mutant unable to produce quorum sensing signaling molecules and its wild-type progenitor to investigate their extracellular metabolome throughout the growth curve by GC-MS and UHPLC-MS analyses. Metabolites involved on the central primary metabolism were affected suggesting that the AHL-dependent quorum sensing induces a general reprogramming of the metabolome. The content of metabolites involved in energy-generating pathways, such as citrate, malate, succinate and fumarate (tricarboxylic acid cycle intermediates) was higher in the mutant unable to produce quorum sensing signaling molecules as compared to its wild-type progenitor. Depletion of TCA intermediates might be responsible for the consistent lower growth yield of the wild-type as compared to mutant (Davenport et al. 2015).

2.2 Conclusion and Perspectives

Compared to transcriptomics approaches, metabolomics has the advantage of the relatively smaller number of metabolites, which makes it easier to extract relevant information of a given biological system. Additionally, metabolite composition rapidly responds to cellular activity and external stimuli. Metabolomics is becoming increasingly widespread toward understanding of microbe plasticity due to its unique ability to generate fast and robust functional data. We presented a detailed systematic literature review of the current knowledge, main findings, and perspective toward the understanding of microbe plasticity in microbial metabolomics research. Taken together, these studies demonstrated that metabolomics can be a useful tool for the identification of microbial bioactive metabolites, for the characterization of biofilms and for biomarkers discovery through strain improvement and phenotypic analysis of microorganisms.

References

Alcalde E, Fraser PD (2016) Metabolite profiling of Phycomyces blakesleeanus carotene mutants reveals global changes across intermediary metabolism. Microbiology (United Kingdom) 162:1963–1971

Alves PC, Hartmann DO, Núñez O, Martins I, Gomes TL, Garcia H, Galceran MT, Hampson R, Becker JD, Pereira CS (2016) Transcriptomic and metabolomic profiling of ionic liquid stimuli unveils enhanced secondary metabolism in *Aspergillus nidulans*. BMC Genomics 17:284

Ammons MCB, Tripet BP, Carlson RP, Kirker KR, Gross MA, Stanisich JJ, Copié V (2014) Quantitative NMR metabolite profiling of methicillin-resistant and methicillin-susceptible

Staphylococcus aureus discriminates between biofilm and planktonic phenotypes. J Proteome Res 13:2973–2985
Assaidi A, Ellouali M, Latrache H, Mabrouki M, Hamadi F, Timinouni M, Zahir H, El Mdaghri N, Barguigua A, Mliji EM (2018) Effect of temperature and plumbing materials on biofilm formation by *Legionella pneumophila* serogroup 1 and 2–15. J Adhes Sci Technol 32:1471–1484
Azeredo J, Azevedo NF, Briandet R, Cerca N, Coenye T, Costa AR, Desvaux M, Di Bonaventura G, Hébraud M, Jaglic Z, Kačániová M, Knøchel S, Lourenço A, Mergulhão F, Meyer RL, Nychas G, Simões M, Tresse O, Sternberg C (2017) Critical review on biofilm methods. Crit Rev Microbiol 43:313–351
Azzollini A, Boggia L, Boccard J, Sgorbini B, Lecoultre N, Allard PM, Rubiolo P, Rudaz S, Gindro K, Bicchi C, Wolfender JL (2018) Dynamics of metabolite induction in fungal co-cultures by metabolomics at both volatile and non-volatile levels. Front Microbiol 9:72
Baptista R, Fazakerley DM, Beckmann M, Baillie L, Mur LAJ (2018) Untargeted metabolomics reveals a new mode of action of pretomanid (PA-824). Sci Rep 8:5084
Barkal LJ, Theberge AB, Guo CJ, Spraker J, Rappert L, Berthier J, Brakke KA, Wang CCC, Beebe DJ, Keller NP, Berthier E (2016) Microbial metabolomics in open microscale platforms. Nat Commun 7:10610
Bayram O, Feussner K, Dumkow M, Herrfurth C, Feussner I, Braus GH (2016) Changes of global gene expression and secondary metabolite accumulation during light-dependent Aspergillus nidulans development. Fungal Genet Biol 87:30–53
Bean HD, Rees CA, Hill JE (2016) Comparative analysis of the volatile metabolomes of *Pseudomonas aeruginosa* clinical isolates. J Breath Res 10:047102
Beloborodova NV, Olenin AY, Pautova AK (2018) Metabolomic findings in sepsis as a damage of host-microbial metabolism integration. J Crit Care 43:246–255
Benamara H, Rihouey C, Abbes I, Ben Mlouka MA, Hardouin J, Jouenne T, Alexandre S (2014) Characterization of membrane lipidome changes in *Pseudomonas aeruginosa* during biofilm growth on glass wool. PLoS ONE 9:e108478
Betancur LA, Naranjo-Gaybor SJ, Vinchira-Villarraga DM, Moreno-Sarmiento NC, Maldonado LA, Suarez-Moreno ZR, Acosta-González A, Padilla-Gonzalez GF, Puyana M, Castellanos L, Ramos FA (2017) Marine Actinobacteria as a source of compounds for phytopathogen control: an integrative metabolic-profiling/bioactivity and taxonomical approach. PLoS ONE 12:e0170148
Bittencourt MLF, Ribeiro PR, Franco RLP, Hilhorst HWM, de Castro RD, Fernandez LG (2015) Metabolite profiling, antioxidant and antibacterial activities of Brazilian propolis: use of correlation and multivariate analyses to identify potential bioactive compounds. Food Res Int 76:449–457
Borgos SEF, Skjåstad R, Tøndervik A, Aas M, Aasen IM, Brunsvik A, Holten T, Iversen OJ, Ahlen C, Zahlsen K (2015) Rapid metabolic profiling of developing *Pseudomonas aeruginosa* biofilms by high-resolution mass spectrometry fingerprinting. Ann Microbiol 65:891–898
Bosso M, Ständker L, Kirchhoff F, Münch J (2018) Exploiting the human peptidome for novel antimicrobial and anticancer agents. Bioorg Med Chem 26:2719–2726
Chae TU, Choi SY, Kim JW, Ko YS, Lee SY (2017) Recent advances in systems metabolic engineering tools and strategies. Curr Opin Biotechnol 47:67–82
Chatzimitakos TG, Stalikas CD (2016) Qualitative alterations of bacterial metabolome after exposure to metal nanoparticles with bactericidal properties: a comprehensive workflow based on1H NMR, UHPLC-HRMS, and metabolic databases. J Proteome Res 15:3322–3330
Chavez-Dozal A, Gorman C, Nishiguchi MK (2015) Proteomic and metabolomic profiles demonstrate variation among free-living and symbiotic vibrio fischeri biofilms microbial genetics, genomics and proteomics. BMC Microbiol 15:226
Chen JC, Tagg KA, Joung YJ, Bennett C, Watkins LF, Eikmeier D, Folster JP (2018) Report of erm(B) *Campylobacter jejuni* in the United States. Antimicrob Agents Chemother 62
Chiang YM, Szewczyk E, Nayak T, Davidson AD, Sanchez JF, Lo HC, Ho WY, Simityan H, Kuo E, Praseuth A, Watanabe K, Oakley BR, Wang CCC (2008) Molecular genetic mining of the

Aspergillus secondary metabolome: discovery of the emericellamide biosynthetic pathway. Chem Biol 15:527–532

Clark AM, Hufford CD, Robertson LW (1977) Two metabolites from *Aspergillus flavipes*. Lloydia 40:146–151

Corcoran M, Morris D, De Lappe N, O'Connor J, Lalor P, Dockery P, Cormican M (2013) *Salmonella enterica* biofilm formation and density in the centers for disease control and prevention's biofilm reactor model is related to serovar and substratum. J Food Prot 76:662–667

Davenport PW, Griffin JL, Welch M (2015) Quorum sensing is accompanied by global metabolic changes in the opportunistic human pathogen *Pseudomonas aeruginosa*. J Bacteriol 197:2072–2082

DeWitt JP, Stetson CL, Thomas KL, Carroll BJ (2018) Extensive cutaneous botryomycosis with subsequent development of Nocardia-positive wound cultures. J Cutan Med Surg 22(3):344–346

D'Sousa Costa CO, Ribeiro PR, Loureiro MB, Simões RC, De Castro RD, Fernandez LG (2015) Phytochemical screening, antioxidant and antibacterial activities of extracts prepared from different tissues of *Schinus terebinthifolius* Raddi that occurs in the coast of Bahia, Brazil. Pharmacogn Mag 11:607–614

Favre L, Ortalo-Magné A, Pichereaux C, Gargaros A, Burlet-Schiltz O, Cotelle V, Culioli G (2018) Metabolome and proteome changes between biofilm and planktonic phenotypes of the marine bacterium *Pseudoalteromonas lipolytica* TC8. Biofouling 34:132–148

Fei F, Mendonca ML, McCarry BE, Bowdish DME, Surette MG (2016) Metabolic and transcriptomic profiling of *Streptococcus intermedius* during aerobic and anaerobic growth. Metabolomics 12:1–13

Fernand MG, Roullier C, Guitton Y, Lalande J, Lacoste S, Dupont J, Ruiz N, Pouchus YF, Raheriniaina C, Ranaivoson E (2017) Fungi isolated from Madagascar shrimps- investigation of the *Aspergillus niger* metabolism by combined LC-MS and NMR metabolomics studies. Aquaculture 479:750–758

Fu Q, Liu D, Wang Y, Li X, Wang L, Yu F, Shen J, Xia X (2018) Metabolomic profiling of *Campylobacter jejuni* with resistance gene ermB by ultra-high performance liquid chromatography-quadrupole time-of-flight mass spectrometry and tandem quadrupole mass spectrometry. J Chromatogr B Anal Technol Biomed Life Sci 1079:62–68

Gharieb MM, El-Sheekh MM, El-Sabbagh SM, Hamza WT (2013) Efficacy of pyocyanin produced by *Pseudomonas aeruginosa* as a topical treatment of infected skin of rabbits. Biotechnol Indian J 7:184–193

Gnavi G, Palma Esposito F, Festa C, Poli A, Tedesco P, Fani R, Monti MC, de Pascale D, D'Auria MV, Varese GC (2016) The antimicrobial potential of algicolous marine fungi for counteracting multidrug-resistant bacteria: phylogenetic diversity and chemical profiling. Res Microbiol 167:492–500

Gökalsın B, Sesal NC (2016) Lichen secondary metabolite evernic acid as potential quorum sensing inhibitor against *Pseudomonas aeruginosa*. World J Microbiol Biotechnol 32:150

Götz F, Longnecker K, Kido Soule MC, Becker KW, McNichol J, Kujawinski EB, Sievert SM (2018) Targeted metabolomics reveals proline as a major osmolyte in the chemolithoautotroph *Sulfurimonas denitrificans*. MicrobiologyOpen 7:e00586

Hakeem Said I, Rezk A, Hussain I, Grimbs A, Shrestha A, Schepker H, Brix K, Ullrich MS, Kuhnert N (2017) Metabolome comparison of bioactive and inactive *Rhododendron* extracts and identification of an antibacterial cannabinoid(s) from *Rhododendron collettianum*. Phytochem Anal 28:454–464

He J, Wijeratne EMK, Bashyal BP, Zhan J, Seliga CJ, Liu MX, Pierson EE, Pierson Iii LS, VanEtten HD, Gunatilaka AAL (2004) Cytotoxic and other metabolites of *Aspergillus* inhabiting the rhizosphere of sonoran desert plants. J Nat Prod 67:1985–1991

Honoré AH, Aunsbjerg SD, Ebrahimi P, Thorsen M, Benfeldt C, Knøchel S, Skov T (2016) Metabolic footprinting for investigation of antifungal properties of *Lactobacillus paracasei*. Anal Bioanal Chem 408:83–96

Hubert J, Nuzillard JM, Renault JH (2017) Dereplication strategies in natural product research: how many tools and methodologies behind the same concept? Phytochem Rev 16:55–95

Illig L, Illig T (2018) Metabolomics and molecular imaging in the post-genomic era. In: P5 medicine and justice: innovation, unitariness and evidence. Springer, Cham, pp 12–21
Jafari M, Ansari-Pour N, Azimzadeh S, Mirzaie M (2017) A logic-based dynamic modeling approach to explicate the evolution of the central dogma of molecular biology. PLoS ONE 12:e0189922
Jin C, Yu Z, Peng S, Feng K, Zhang L, Zhou X (2018) The characterization and comparison of exopolysaccharides from two benthic diatoms with different biofilm formation abilities. An Acad Bras Cienc 90:1503–1519
Joghee NN, Jayaraman G (2014) Metabolomic characterization of halophilic bacterial isolates reveals strains synthesizing rare diaminoacids under salt stress. Biochimie 102:102–111
Kanoh K, Matsuo Y, Adachi K, Imagawa H, Nishizawa M, Shizuri Y (2005) Mechercharmycins A and B, cytotoxic substances from marine-derived *Thermoactinomyces* sp. YM3-251. J Antibiot 58:289–292
Keelara S, Thakur S, Patel J (2016) Biofilm formation by environmental isolates of *Salmonella* and their sensitivity to natural antimicrobials. Foodborne Pathog Dis 13:509–516
Kildgaard S, Mansson M, Dosen I, Klitgaard A, Frisvad JC, Larsen TO, Nielsen KF (2014) Accurate dereplication of bioactive secondary metabolites from marine-derived fungi by UHPLC-DAD-QTOFMS and a MS/HRMS library. Mar Drugs 12:3681–3705
Kim S, Kim J, Song JH, Jung YH, Choi IS, Choi W, Park YC, Seo JH, Kim KH (2016a) Elucidation of ethanol tolerance mechanisms in *Saccharomyces cerevisiae* by global metabolite profiling. Biotechnol J 11:1221–1229
Kim WJ, Kim YO, Kim JH, Nam BH, Kim DG, An CM, Lee JS, Kim PS, Lee HM, Oh JS, Lee JS (2016b) Liquid chromatography-mass spectrometry-based rapid secondary-metabolite profiling of marine *Pseudoalteromonas* sp. M2. Mar Drugs 14:24
Kim ME, Jung I, Lee JS, Na JY, Kim WJ, Kim YO, Park YD, Lee JS (2017) Pseudane-VII isolated from *Pseudoalteromonas* sp. M2 ameliorates LPS-induced inflammatory response in vitro and in vivo. Mar Drugs 15:336
Kim BE, Goleva E, Hall CF, Park SH, Lee UH, Brauweiler AM, Streib JE, Richers BN, Kim G, Leung DYM (2018) Skin wound healing is accelerated by a lipid mixture representing major lipid components of *Chamaecyparis obtusa* plant extract. J Investig Dermatol 138:1176–1186
Koistinen VM, da Silva AB, Abrankó L, Low D, Villalba RG, Barberán FT, Landberg R, Savolainen O, Alvarez-Acero I, de Pascual-Teresa S, Van Poucke C, Almeida C, Petrásková L, Valentová K, Durand S, Wiczkowski W, Szawara-Nowak D, González-Domínguez R, Llorach R, Andrés-Lacueva C, Aura AM, Seppänen-Laakso T, Hanhineva K, Manach C, Bronze MR (2018) Interlaboratory coverage test on plant food bioactive compounds and their metabolites by mass spectrometry-based untargeted metabolomics. Metabolites 8
Landini P, Antoniani D, Burgess JG, Nijland R (2010) Molecular mechanisms of compounds affecting bacterial biofilm formation and dispersal. Appl Microbiol Biotechnol 86:813–823
Lang G, Mitova MI, Cole ALJ, Din LB, Vikineswary S, Abdullah N, Blunt JW, Munro MHG (2006) Pterulamides I-VI, linear peptides from a Malaysian *Pterula* sp. J Nat Prod 69:1389–1393
Li Y, Luo Q, Yuan L, Miao C, Mu X, Xiao W, Li J, Sun T, Ma E (2012) JNK-dependent Atg4 upregulation mediates asperphenamate derivative BBP-induced autophagy in MCF-7 cells. Toxicol Appl Pharmacol 263:21–31
Lian J, Mishra S, Zhao H (2018) Recent advances in metabolic engineering of *Saccharomyces cerevisiae*: new tools and their applications. Metab Eng 50:85–108
Liao X, Hu F, Chen Z (2018) Identification and quantitation of the bioactive components in *Osmanthus fragrans* fruits by HPLC-ESI-MS/MS. J Agric Food Chem 66:359–367
Liu L, Gui M, Wu R, Li P (2016a) Progress in research on biofilm formation regulated by LuxS/AI-2 quorum sensing. Shipin Kexue Food Sci 37:254–262
Liu Q, Li W, Sheng L, Zou C, Sun H, Zhang C, Liu Y, Shi J, Ma E, Yuan L (2016b) Design, synthesis and biological evaluation of novel asperphenamate derivatives. Eur J Med Chem 110:76–86

López-Gresa MP, Lisón P, Campos L, Rodrigo I, Rambla JL, Granell A, Conejero V, Bellés JM (2017) A non-targeted metabolomics approach unravels the VOCs associated with the tomato immune response against *Pseudomonas syringae*. Front Plant Sci 8:1118

Luo X, Zhou X, Lin X, Qin X, Zhang T, Wang J, Tu Z, Yang B, Liao S, Tian Y, Pang X, Kaliyaperumal K, Li JL, Tao H, Liu Y (2017) Antituberculosis compounds from a deep-sea-derived fungus *Aspergillus* sp. SCSIO Ind09F01. Nat Prod Res 31:1958–1962

Maansson M, Vynne NG, Klitgaard A, Nybo JL, Melchiorsen J, Nguyen DD, Sanchez LM, Ziemert N, Dorrestein PC, Andersen MR, Gram L (2016) An integrated metabolomic and genomic mining workflow to uncover the biosynthetic potential of bacteria. mSystems 1

Managamuri U, Vijayalakshmi M, Ganduri VSRK, Rajulapati SB, Bonigala B, Kalyani BS, Poda S (2017) Isolation, identification, optimization, and metabolite profiling of *Streptomyces sparsus* VSM-30. 3 Biotech 7:217

Marcone GL, Binda E, Berini F, Marinelli F (2018) Old and new glycopeptide antibiotics: from product to gene and back in the post-genomic era. Biotechnol Adv 36:534–554

Mgbeahuruike EE, Yrjönen T, Vuorela H, Holm Y (2017) Bioactive compounds from medicinal plants: focus on Piper species. S Afr J Bot 112:54–69

Newman DJ (2016) Predominately uncultured microbes as sources of bioactive agents. Front Microbiol 7:1832. https://doi.org/10.3389/fmicb.2016.01832

Odai H, Shindo K, Odagawa A, Mochizuki J, Hamada M, Takeuchi T (1994) Inostamycins B and C, new polyether antibiotics. J Antibiot 47:939–941

Ortiz-Villanueva E, Benavente F, Piña B, Sanz-Nebot V, Tauler R, Jaumot J (2017) Knowledge integration strategies for untargeted metabolomics based on MCR-ALS analysis of CE-MS and LC-MS data. Anal Chim Acta 978:10–23

Padder SA, Prasad R, Shah AH (2018) Quorum sensing: a less known mode of communication among fungi. Microbiol Res 210:51–58

Patel J, Singh M, Macarisin D, Sharma M, Shelton D (2013) Differences in biofilm formation of produce and poultry *Salmonella enterica* isolates and their persistence on spinach plants. Food Microbiol 36:388–394

Patil MP, Patil RH, Maheshwari VL (2015) Biological activities and identification of bioactive metabolite from endophytic *Aspergillus flavus* L7 isolated from *Aegle marmelos*. Curr Microbiol 71:39–48

Pintilie L, Stefaniu A, Ioana Nicu A, Maganu M, Caproiu MT (2018) Design, synthesis and docking studies of some novel fluoroquinolone compounds with antibacterial activity. Rev Chim 69:815–822

Ranjbar R, Babazadeh D, Jonaidi-Jafari N (2017) Prevalence of *Campylobacter jejuni* in adult patients with inflammatory bacterial diarrhea, East Azerbaijan, Iran. Acta Med Mediterr 33:901–908

Ratnaweera PB, Williams DE, de Silva ED, Wijesundera RLC, Dalisay DS, Andersen RJ (2014) Helvolic acid, an antibacterial nortriterpenoid from a fungal endophyte, *Xylaria* sp. of orchid *Anoectochilus setaceus* endemic to Sri Lanka. Mycology 5:23–28

Raveendran S, Parameswaran B, Ummalyma SB, Abraham A, Mathew AK, Madhavan A, Rebello S, Pandey A (2018) Applications of microbial enzymes in food industry. Food Technol Biotechnol 56:16–30

Ribeiro PR, Ligterink W, Hilhorst HWM (2015) Expression profiles of genes related to carbohydrate metabolism provide new insights into carbohydrate accumulation in seeds and seedlings of *Ricinus communis* in response to temperature. Plant Physiol Biochem 95:103–112

Ribeiro PR, Canuto GAB, Brito VC, Batista DLJ, de Brito CD, Loureiro MB, Takahashi D, de Castro RD, Fernandez LG, Hilhorst HWM, Ligterink W (2018) Castor bean metabolomics: current knowledge and perspectives toward understanding of plant plasticity under stress condition. In: Disaster risk reduction, 1st edn. Springer Singapore, pp 237–253.

Romoli R, Papaleo MC, De Pascale D, Tutino ML, Michaud L, LoGiudice A, Fani R, Bartolucci G (2014) GC-MS volatolomic approach to study the antimicrobial activity of the antarctic bacterium *Pseudoalteromonas* sp. TB41. Metabolomics 10:42–51

Sander K, Asano KG, Bhandari D, Van Berkel GJ, Brown SD, Davison B, Tschaplinski TJ (2017) Targeted redox and energy cofactor metabolomics in *Clostridium thermocellum* and *Thermoanaerobacterium saccharolyticum* Mike Himmel. Biotechnol Biofuels 10:285

Sanmanoch W, Mongkolthanaruk W, Kanokmedhakul S, Aimi T, Boonlue S (2016) Helvolic acid, a secondary metabolite produced by neosartorya spinosa KKU-1NK1 and its biological activities. Chiang Mai J Sci 43:483–493

Santos PM, Batista DLJ, Ribeiro LAF, Boffo EF, de Cerqueira MD, Martins D, de Castro RD, de Souza-Neta LC, Pinto E, Zambotti-Villela L, Colepicolo P, Fernandez LG, Canuto GAB, Ribeiro PR (2018) Identification of antioxidant and antimicrobial compounds from the oilseed crop *Ricinus communis* using a multiplatform metabolite profiling approach. Ind Crop Prod 124:834–844

Saroj SD, Rather PN (2013) Streptomycin inhibits quorum sensing in *Acinetobacter baumannii*. Antimicrob Agents Chemother 57:1926–1929

Schatz A, Bugle E, Waksman SA (1944) Streptomycin, a substance exhibiting antibiotic activity against Gram-positive and Gram-negative bacteria. Proc Soc Exp Biol Med 55:66–69

Schelli K, Rutowski J, Roubidoux J, Zhu J (2017) *Staphylococcus aureus* methicillin resistance detected by HPLC-MS/MS targeted metabolic profiling. J Chromatogr B Anal Technol Biomed Life Sci 1047:124–130

Scrutton NS (2017) Enzymes make light work of hydrocarbon production. Science 357:872–873

Sfeir MM (2018) *Burkholderia cepacia* complex infections: more complex than the bacterium name suggest. J Infect 77:166–170

Singh RK, Lee JK, Selvaraj C, Singh R, Li J, Kim SY, Kalia VC (2018) Protein engineering approaches in the post-genomic era. Curr Protein Pept Sci 19:5–15

Sohlenkamp C, Geiger O (2015) Bacterial membrane lipids: diversity in structures and pathways. FEMS Microbiol Rev 40:133–159

Stipetic LH, Dalby MJ, Davies RL, Morton FR, Ramage G, Burgess KEV (2016) A novel metabolomic approach used for the comparison of *Staphylococcus aureus* planktonic cells and biofilm samples. Metabolomics 12:75

Suutari M, Laakso S (1994) Microbial fatty acids and thermal adaptation. Crit Rev Microbiol 20:285–328

Tracanna V, de Jong A, Medema MH, Kuipers OP (2017) Mining prokaryotes for antimicrobial compounds: from diversity to function. FEMS Microbiol Rev 41:417–429

Tredwell GD, Aw R, Edwards-Jones B, Leak DJ, Bundy JG (2017) Rapid screening of cellular stress responses in recombinant Pichia pastoris strains using metabolite profiling. J Ind Microbiol Biotechnol 44:413–417

Van Dalem A, Herpol M, Echahidi F, Peeters C, Wybo I, De Wachter E, Vandamme P, Piérard D (2018) In vitro susceptibility of *Burkholderia cepacia* complex isolated from cystic fibrosis patients to ceftazidime-avibactam and ceftolozane-tazobactam. Antimicrob Agents Chemother 62

Van Der Heul HU, Bilyk BL, McDowall KJ, Seipke RF, Van Wezel GP (2018) Regulation of antibiotic production in Actinobacteria: new perspectives from the post-genomic era. Nat Prod Rep 35:575–604

Vinci G, Cozzolino V, Mazzei P, Monda H, Savy D, Drosos M, Piccolo A (2018) Effects of *Bacillus amyloliquefaciens* and different phosphorus sources on maize plants as revealed by NMR and GC-MS based metabolomics. Plant Soil 429:437–450

Wahyuni TS, Widyawaruyanti A, Lusida MI, Fuad A, Soetjipto, Fuchino H, Kawahara N, Hayashi Y, Aoki C, Hotta H (2014) Inhibition of hepatitis C virus replication by chalepin and pseudane IX isolated from *Ruta angustifolia* leaves. Fitoterapia 99:276–283

Wang Y, Zhang M, Deng F, Shen Z, Wu C, Zhang J, Zhang Q, Shen J (2014) Emergence of multidrug-resistant *Campylobacter* species isolates with a horizontally acquired rRNA methylase. Antimicrob Agents Chemother 58:5405–5412

Wang Z, Li MY, Peng B, Cheng ZX, Li H, Peng XX (2016) GC-MS-based metabolome and metabolite regulation in serum-resistant *Streptococcus agalactiae*. J Proteome Res 15:2246–2253

Wong HS, Maker GL, Trengove RD, O'Handley RM (2015) Gas chromatography-mass spectrometry-based metabolite profiling of *Salmonella enterica* serovar typhimurium differentiates between biofilm and planktonic phenotypes. Appl Environ Microbiol 81:2660–2666

Yang MH, Li TX, Wang Y, Liu RH, Luo J, Kong LY (2017) Antimicrobial metabolites from the plant endophytic fungus *Penicillium* sp. Fitoterapia 116:72–76

Yang YT, Zhu JF, Liao G, Xu HJ, Yu B (2018) The development of biologically important spirooxindoles as new antimicrobial agents. Curr Med Chem 25:2233–2244

Yogabaanu U, Weber JFF, Convey P, Rizman-Idid M, Alias SA (2017) Antimicrobial properties and the influence of temperature on secondary metabolite production in cold environment soil fungi. Pol Sci 14:60–67

Yuan L, Li Y, Zou C, Wang C, Gao J, Miao C, Ma E, Sun T (2012) Synthesis and in vitro antitumor activity of asperphenamate derivatives as autophagy inducer. Bioorg Med Chem Lett 22:2216–2220

Ząbek A, Junka A, Szymczyk P, Wojtowicz W, Klimek-Ochab M, Młynarz P (2017) Metabolomics analysis of fungal biofilm development and of arachidonic acid-based quorum sensing mechanism. J Basic Microbiol 57:428–439

Zhang Y, Pei G, Chen L, Zhang W (2016) Metabolic dynamics of *Desulfovibrio vulgaris* biofilm grown on a steel surface. Biofouling 32:725–736

Is PGPR an Alternative for NPK Fertilizers in Sustainable Agriculture? 3

Éva Laslo and Gyöngyvér Mara

3.1 Sustainable Agriculture and Environmental Problems of Current Fertilizing Methods

Agricultural production as a user of natural resources has a significant influence on the state of the environment. In agricultural practice, focus has shifted to its environmental impact and effect on the population's wellbeing and living standards. This concept of sustainable agriculture was formulated as the main challenge globally. The rapid population growth of the earth has given rise to major concerns about the food supply. It is expected that the global population will increase from 7.2 to 9.6 billion by 2050. If the consumption habits remain unchanged, the lands used for crop production and production efficiency have to be increased. This phenomenon gives rise to concern about maintaining the world ecosystem functions and services. The solution relies on the development and innovation of sustainable agriculture, which achieves crop production without polluting the environment and causing damage. The origin of the word "sustain," is derived from the Latin word *sustinere*, having the meaning of maintain, long-term support or permanence.

Considering agriculture, sustainable farming systems describe the management systems that are able to maintain their productivity and their benefits to society for an indefinite period of time. This agricultural system must be a resource preserver, socially encouraging, economically competitive, and environmentally friendly (Valkó 2017).

The phrase "Sustainable agriculture" became known in literature in the 1980s, when the Worldwatch Institute published a work on sustainable societies. In 1990, the Senate of the United States Congress introduced the Sustainable Agriculture Research and Education Act, which dealt with developing technical guides for

É. Laslo · G. Mara (✉)
Faculty of Economics, Socio-Human Sciences and Engineering,
Department of Bioengineering, Sapientia Hungarian University of Transylvania,
Miercurea Ciuc, Romania
e-mail: maragyongyver@uni.sapientia.ro

D. P. Singh et al. (eds.), *Microbial Interventions in Agriculture and Environment*,
https://doi.org/10.1007/978-981-13-8391-5_3

low-input sustainable agricultural production methods and initiation of a national training program in sustainable agriculture. It was defined that sustainable agriculture comprises crop and livestock production in an integrated system with site-specific application and durability. Regarding the definition, this system provides humanity with food. It contributes to the enhancement of environmental quality, natural resources, and society. The nonrenewable materials are used as effectively as possible, combining the natural biological cycles and controls. It also maintains the economic viability of agricultural operations (Gold 2016).

One of the concerns of modern agricultural practice is ecological worry. This includes the deterioration of soil productivity, desertification, water pollutants, such as fertilizers, eutrophication, etc. In sustainable agroecosystems, it is emphasized to keep the natural resource base and to depend on the minimum use of artificial inputs outside the agricultural system (Itelima et al. 2018).

The supply of necessary nutrients is one of the major challenges of agricultural production. The traditional chemical forms of fertilizers used in plant production result in significant growth of the crops. In general, farmers use an overdose of fertilizers to maximize crop production. Approximately 50–70% of conventional fertilizers used are lost in the environment, and the consequences are the negative impact on the environment (eutrophication, water and soil contamination) and health. For example, the nitrite with other pollutants can disturb the nervous system, cause heart diseases, and different types of cancer. The fertilizer industry uses a very high amount of energy for the production of these compounds (Singh et al. 2017).

Common practice for enhancement of cultivated crop production is the use of different forms of fertilizer. For the N supply, urea, ammonium nitrate, diammonium phosphate, etc., are used (Hermary 2007). The exaggerated treatment of plants with N fertilizers contributes to the increase of root biomass. Owing to this fact, a high absorption of the other nutrients can occur, resulting in a lack of micronutrients in the soil. Because nitrate is absorbed by plants in the fast growing stage, the soil may release significant amounts of it. This results in nitrogen loss. Another negative impact of N fertilizer is on global warming, due to the ammonia and NOx gases. It was shown that the use of various fertilizers (P_2O_5, K_2O, urea) for cereal crops resulted in leaching NO_3^-, loses of phosphorus, nitrogen, and ammonia volatilization.

Phosphorus is an essential macronutrient and also one of the major limiting factors in crop production. A large amount of phosphorus exists in soils in an immobilized form that is unavailable for plants, and therefore chemical fertilization is used. P fertilizer is taken out from P-rich rock in the form of phosphate, which is a finite resource (Karamesouti and Gasparatos 2017). It was evaluated that 5.7 billion hectares of land throughout the world are deficient in P, which underlines the importance of phosphorus as a limiting factor (Granada et al. 2018). Owing to the high rate of added phosphorus immobilization in soil from fertilization in agricultural, routinely, twice as much or more P fertilizers are used than needed. It was also estimated that annual P utilization will increase yearly by 2.5% (Sattari et al. 2012). Besides the fact that fertilizers are expensive and need finite resources, they are also harmful to the environment, soil structure, properties, composition, and microbiota (White 2008). As an alternative solution, natural phosphate rocks used in combination with phosphate solubilization bacteria (PSB) under field conditions can be used as P fertilizers (Kaur and Reddy 2015).

Globally, potassium represents the seventh most abundant element that occurs in the earth's crust. The different forms of potassium in the soil are: mineral K, exchangeable K, non-exchangeable K, and dissolved K^+ ions. From this, plants can reach only 1–2% in the form of solution and exchangeable K. This mineral is essential for plants because it is involved in different growth and development mechanisms and takes part in cell membrane function. The forms of potassium found as minerals are potassium sulfate or chloride. In agricultural production, it is used as potassium sulfate, in most cases under the name of potash or arcanite. The negative impact of the use of the mined form is that it can easily leach. The consequence of K leaching is its accumulation in different aquatic ecosystems harboring the vegetation (Meena et al. 2016).

In agricultural practice, the expanded use of chemical fertilizers has contributed to the deterioration of water and soil and caused irreversible impacts on the biosphere too. Many researchers emphasize that the solution lies in sustainable resource management. One possible measure is the use of biofertilizers to reduce the negative impact of synthetic manures. These microbial products can contribute to plant nutrient acquisition without the depletion of natural resources (Verma et al. 2018).

3.2 Role of Bacteria in Nutrient Management of Plants

3.2.1 Plant Nutrition Requirements

There are 13 essential mineral elements divided into major elements and micronutrients based on the concentration needed by the plant. The majority of elements are primarily taken up by the root transport system in ionic form, other elements, such as C, H, and O from water and air.

Nitrogen is found in both organic and inorganic form in plants, with predominantly organic prevalence, comprising amino-acids, enzymes, nucleic acids, chlorophyll, and alkaloids. Nitrogen in inorganic form (NO_3^-) can accumulate in plant tissues. The nitrogen content of plants varies between 0.5% and 5% of the dry weight. N is available for root absorption either as NO_3^- or NH_4^+. NO_3^- moves in the soil basically by mass flow, while NH_4^+ by diffusion, and they are absorbed at the root surface. Uptake of NO_3^- stimulates the uptake of cations, while uptake of NH_4^+ restricts cations.

Phosphorus is the component of phospholipids, proteins, nucleic acids, adenosine triphosphate (energy providing molecule), and phytin. The phosphorus content of plants varies between 0.1% and 0.5% of the dry weight, and it is present in soil in organic (50–70% of total P content, in the form of phytin) and inorganic (30–50% of total P content, in the form of Al, Fe, and Ca phosphates) form. It is available for root absorption in $H_2PO_4^-$ and HPO_4^{2-} anionic forms, moves in the soil primarily by diffusion and root hair abundance increases the opportunity of P uptake (Lambers et al. 2006).

Potassium has as major function in the plant water status and cell turgor pressure maintenance and is involved in stomatal functioning. It is also required for carbohydrate accumulation and translocation as well as for enzyme activation. The potassium content of plants varies between 0.5% and 5% of dry weight. Potassium moves in the soil mostly by diffusion and partially by mass flow, and it is absorbed as K^+ cation. The root density and soil oxygen has a notable effect on its uptake.

3.2.2 Plant Main Mechanisms of Nutrient Acquisition

The uptake of soil nutrients is affected by several factors, such as soil properties and nutrient content, plant root properties (size, architecture, morphology, substance release), and rhizosphere microorganisms. Plant roots forage for nutrients. Transport from soil to root is realized through mass flow, diffusion or root interception. The uptake of nutrients occurs through membrane transporter proteins on the root surface. Owing to the continuous uptake, nutrient concentration on the root surface is decreased, generating a concentration gradient from soil to root surface. Plants differ in nutrient uptake capacity, but there is a clear correlation between root hair development and plant nutritional level in the case of nitrate, phosphate, and potassium; the uptake being facilitated by root length and volume. Different mechanisms play a role in N, P, and K uptake (Jungk 2001).

Nitrogen from soil is available for plants in organic (urea, amino acids, and small peptides) and inorganic (nitrate and ammonium) form, but the organic forms contribute to plant N nutrition only in special environments, and therefore the inorganic forms are considered universal. The acquisition of nitrogen depends on the root architecture and uptake activity through plasma membrane. High affinity transporters and low affinity transporters are located in the plasma membrane, serving nutrient uptake. The two types of transporters were developed because of the large variation in nitrate concentration, low affinity when external nitrate concentration is high and high affinity when nitrate concentration is low in the cell external environment. Nitrate uptake is realized through NPF (nitrate transporter 1/peptide transporter family) and NRT2 (nitrate transporter 2) transporter proteins. NPF transporter proteins have low affinity for nitrate, whereas NRT2 transporter proteins are high affinity transporters (Pii et al. 2015). The members of the latter protein family for nitrate transport require another NAR2 protein. Experimental data shows that, in the case of Arabidopsis plants, NRT2 transporters (consisting of seven different genes) accounted for 95% of high-affinity nitrate influx; some of the proteins being involved in nitrate uptake from soil (*AtNRT2.4* and *AtNRT2.5* genes were expressed), whereas others in apoplastic transport (*AtNRT2.1* gene was expressed) (Kiba and Krapp 2016).

Ammonium transport is mediated by the AMT transporter superfamily encoding high affinity ammonium transporters. In Arabidopsis, six AMT genes exist, three encoding transporters that absorb ammonium by a direct route from soil and one encoding apoplastic transporter (Kiba and Krapp 2016).

The nutrient uptake can also be modulated by root growth and development, when under mild nitrogen limitation the increased absorptive surface and scavenging root system make it possible for the plants to adapt to nutrient availability.

Phosphorus is obtained in the form of inorganic phosphate (Pi) in the form of several cations (PO_3^{-4}, HPO_2^{-4}, H_2PO^{-4}) depending on the pH. The most easily accessed form of P for plants is H_2PO^{-4}. Inorganic phosphate uptake is an energy mediated process realized through phosphate/H^+ symporter. These membrane proteins are included in phosphate transporter (PT), among which the PHT1 family (phosphate transporter 1) is the most studied (Nussaume et al. 2011). PHT1 are highly expressed in roots and comprise nine members in *Arabidopsis thaliana*. Phosphate transporter genes are transcriptionally induced under Pi starvation condition (Gu et al. 2016).

Potassium is essential for many physiological processes in the plants; therefore, the concentration in the cytosol is maintained within the 100–200 mM range. Potassium is absorbed as K^+ anion through high and low affinity mechanisms depending on external concentration. In the case of high external potassium concentration, K^+ uptake is passive and is realized through membrane channels, whereas in the case of low external potassium concentration, the high affinity uptake is mediated by H^+/K^+ symporter (Nieves-Cordones et al. 2014). These two systems were described in *Arabidopsis thaliana*, where the passive membrane transport is realized through the inward-rectifier K^+ channel (AKT1), whereas in the case of low K^+ concentration the high affinity K^+ transporter (AtHAK5) is involved in potassium uptake (Ródenas et al. 2017). In the case of high external K^+ concentration, it was observed that non-selective cation channels sensitive to Ca^{2+}can also contribute to potassium uptake.

3.2.3 Role of PGP Bacteria in Plant Nutrient Management

Nitrogen, phosphorus, and potassium uptake by roots is strictly dependent on their availability in soil. Plant roots in addition to water and nutrient uptake also synthetize and secrete diverse compounds called root exudates that act as chemical attractants for soil microbes. These chemical compounds regulate the rhizosphere microbial community. The biogeochemical cycles of major nutrients are mainly managed by microbial processes. The rhizosphere bacteria therefore affect the nutritional and physiological status of plants (Ahemad and Kibret 2014; Sahu et al. 2018).

Rhizobacteria can alter nitrogen availability in soil through several processes, such as soil organic matter decomposition, atmospheric N_2 fixation, nitrification, and denitrification. Owing to the fact that the total nitrogen in soil is present mainly in organic form (90%), which is unavailable to the plants, the role of the rhizosphere bacteria in soil organic matter mineralization is important. Proteins, nucleic acids, and other organic compounds containing N are decomposed and transformed into the plant available form as ammonia through the process called ammonification.

Another microbial process that plays a role in plant nitrogen management is the biological nitrogen fixation by diazotrophs, when the atmospheric N_2 is turned into plant-utilizable forms. The biological nitrogen fixation can be realized by free-living diazotrophs (*Cyanobacteria, Proteobacteria, Archaea,* and *Firmicutes*) not associated with plants and by symbiotic diazotrophs (*Rhizobium* and *Bradyrhizobium* in the case of legumes, *Frankia, Nostoc, Azolla* in the case of non-legumes). In the biological nitrogen fixation process, the atmospheric N_2 is transformed into ammonia by the microorganisms using the nitrogenase enzyme system, found in both free-living and symbiotic systems. Nitrogenase genes (*nif*) are found in a cluster of seven operons, including structural genes, regulatory genes for the synthesis of enzymes, and others important for functioning, encoding 20 different proteins (Saha et al. 2017).

Nitrification processes are realized by *Nitrosococcus* and *Nitrobacter* bacteria that transform soil ammonia into a plant available form in nitrite (NO_2^-) and nitrate (NO_3^-). Through denitrification, nitrites and nitrates are converted by denitrifying bacteria (for example *Pseudomonas, Paracoccus, Alcaligenes, Bacillus*) back to gaseous form (NO_x). The presence of NO_x in the soil can trigger plant growth and development and also has a positive impact on root acquisition processes (Takahasi and Morikawa 2014).

Rhizobacteria, besides nutrient mobilization, can also enhance the nutrient uptake of the plant. In the case of maize, a single inoculation with *Bacillus sp.*, *Acinetobacter sp.*, and *Klebsiella sp.* notably increased the N uptake of the plant; in early growth, the majority of N was assimilated from soil urea source, while in later growth through N fixation (Kuan et al. 2016). *Achromobacter sp.* were also reported as enhancers of NO_3^- uptake in *Brassica napus*. In *Arabidopsis thaliana* seedlings, the inoculation with *Phyllobacterium brassicacearum* increased the NO_3^- uptake in the first period, but decreased after 7 days. Data regarding the role of PGP rhizobacteria in altering NO_3^- uptake across the root plasma membrane are still contradictory. Information about the role of plant growth promoting rhizobacteria on the plant acquisition of NH_4^+ and urea is scarce. In the case of *Cucurbita moschata,* it was observed that the plants supplemented with NO_3^+, NH_4^+, and NO_3NH_4 and inoculated with bio-inoculant (Bionutrients AG 8-1-9, containing the mixture of *Bacillus subtilis, B. amyloliquefaciens, B. pumilus, B. licheniformis,* and *Saccharomyces cerevisiae*) showed an increase in biomass and N, P, K, and Mn concentration in leaves (Tchiaze et al. 2016).

Phosphorus is present in soil as phosphates in organic form as phytic acid and inorganic form bound to Fe, Al, and Ca that reduces its solubility. In addition, the application of fertilizers applied as inorganic phosphates are 75% immobilized in soil, and therefore they cannot solve the plant nutritional problem (Tóth et al. 2014). Less than 5% of soil P is taken up by plants in the form of HPO_4^{2-} and H_2PO_4. Phosphate solubilizing bacteria (PSB) provide soluble phosphate for plants mainly due to the presence of low molecular weight organic acids (gluconic acid, citric acid), whereas plants supply bacteria with carbon compounds for their growth. Low molecular weight organic acids through ligand exchange desorb phosphate, and once released it is available for plants. Besides the increased phosphate availability, PGP rhizobacteria can enhance the phosphate uptake of the plants by stimulating the plasma membrane H^+-ATP-ase in plant roots (Pii et al. 2015). Soil microbes beside organic acids can also produce enzymes, such as phosphatases and phytases, in soil releasing phosphates. Soil bacteria belonging to *Aerobacter, Acinetobacter, Acromobacter, Agrobacterium, Azospirillum, Azotobacter, Bacillus, Burkholderia, Enterococcus, Enterobacter, Erwinia, Flavobacterium, Micrococcus, Pantoea, Pseudomonas, Rhizobium,* and *Serratia* genera were described as having phosphate solubilizing activity (Anzuay et al. 2015; Pii et al. 2015).

Potassium uptake can be modified by K-solubilizing microbes that excrete low molecular organic acids, mainly citric, oxalic, tartaric, succinic acids, but production of ferulic, coumaric, syringic, and malic acid was also reported. Organic acids dissolute K+ from minerals by lowering pH (acidolysis) and forming metal-organic complexes with Si^{4+} ion and bringing the K into solution. Biofilms, capsular polysaccharides, polymers, and low molecular weight ligands produced by soil microbiota are able to mobilize potassium through the weathering process (Ahmad et al. 2016).

Since molecular fingerprinting is used in microbial community analysis, the ability of plants to select species specific microbiome was demonstrated (Pii et al. 2015). The composition of root exudates (low and high molecular weight organic compounds) varies among plant species and with environmental factors. These exudates (mainly low molecular weight) are an accessible C source for microbes and act as chemoattractants, and therefore the microbes are more abundant in the root proximity. Plant and bacteria communicate in the rhizosphere through complex signals, and as a result of this communication, the type of relationship is settled (detrimental, neutral or beneficial). In this context, rhizosphere bacteria that play an important role in plant nutrient acquisition processes depend on plant species and genotype, plant-microbe communication, and environmental conditions (Miransari 2014; Rosier et al. 2018).

3.3 PGPR as Bio Inoculants in Practical Use

In integrated nutrient management, the use of bio-inoculants is spreading. Bio-inoculants are based on selected bacterial strains that increase access to the inaccessible nutrient for plant growth and development. They also contribute to the improvement of soil sustainability and productivity and are, therefore, considered a tool for green agriculture. It was reported that the market of microbial inoculants worldwide will increase from $440 million in 2012 to $1295 million by 2020 (Owen et al. 2015). Microbial inoculants are applied to host plant surface, seed or soil. After colonizing the environment, they can exert their effect. Depending on their mechanism – contributing to the availability of nutrients – they can be grouped as nitrogen fixers (N-fixer), potassium and phosphorus solubilizers. In agricultural systems, different bacterial formulations are applied as bio-inoculants based on nitrogen fixing and phosphorus and potassium solubilizing microorganisms. It was revealed that single strains also exert beneficial effects, but mixed inoculants are more productive and effective.

Through biological nitrogen fixation, different microorganisms use their complex enzyme systems to transform atmospheric N into an assimilable N form, such as ammonia. The efficiency of this process is affected by different factors, such as climatic, soil or host genotype or the complex host bacteria interaction. It was revealed that the efficiency of legume–rhizobia symbiosis with approximately 13–360 kg N/ha is higher than the non-symbiotic systems, where the measured values range between 10 and 160 kg N/ha. Many experiments focused on measuring the amount of fixed nitrogen in different plant species, for example, in groundnut the fixed N varied between 126 and 319 kg N/ha, in soybean 3–643 kg N/ha, in pigeon pea 77–92 kg N/ha, in cowpea between 25 and 100 kg N/ha, in green gram 71–74 kg N/ha, and in black gram 125–143 kg N/ha (Gopalakrishnan et al. 2015).

Nitrogen fixing biofertilizers are grouped as free-living bacteria, for example *Azotobacter*, *Bejerinkia*, *Clostridium*, *Klebsiella*, *Anabaena*, and *Nostoc*. Bacteria from *Rhizobium*, *Frankia*, *Anabaena*,and *Azollae* genera belong to the symbiotic group, whereas bacteria from *Azospirillum* genera belong to the associative symbiotic species. Atmospheric nitrogen fixation is one of their direct plant promotion effects. These bacterial formulations in most cases are crop specific (Bhat et al. 2015).

The genus of *Azospirillum* belong to the family Spirilaceae. Their contact with plants is based on associative symbiosis. Host plants are those that possess the C4-dicarboxyliac pathway of photosynthesis. They are proposed for the inoculation of maize, sugarcane, sorghum, and pearl millet. These bacterial species were also detected in the rhizosphere of different plants, such as rice, maize, sugarcane, pearl millet, vegetables, and plantation crops. There are reports of applying them as biofertilizer for diverse crops, such as barley, castor, cotton, coffee, coconut, jute, linseeds, maize, mustard, oat, rice, rubber, sesame, sorghum, sugar beets, sunflower, tobacco, tea, and wheat (Bhat et al. 2015).

It was detected that these bacteria are able to fix nitrogen to 20–40 kg/ha. A worldwide improved inoculation effect was determined in the case of *A.lipoferum* and *A.brasilense. Azospirillium brasilense* with *Rhizobium meliloti* plus 2,4D exerted beneficial effects on wheat, improving the harvested grain's N, P, and K content (Askary et al. 2009). In the case of maize, *A. lipoferum* CRT1, a commercial isolate, showed a positive effect on sugar metabolism (Rozier et al. 2017). It was reported that *A. brasilense* Ab-V5, besides influencing the photosynthesis metabolism in maize, also positively influenced the nitrogen supply under nitrogen limiting conditions (Calzavara et al. 2018).

From the family *Azotobacteriaceae: A. vinelandii, A. beijerinckii, A. insignis,* and *A. macrocytogenes* are the most known species. These bacterial species take part in the global nitrogen cycle due to their role in atmospheric nitrogen fixation. *Azotobacter sp.* are able to fix atmospheric nitrogen in the rhizospheric relationship with maize and wheat. It was shown that the application of *Azotobacter sp.* strains in mustard and wheat increased the plant growth rate, yield, and nitrogen level. In the case of *Brassica juncea,* the inoculation with *Azotobacter chroococcum* contributed to the stimulation of plant growth, whereas in *Fagopyrum esculentum* the inoculation with *Azotobacter aceae* contributed to nitrogen assimilation (Gouda et al. 2018).

It was also reported that *Azotobacter vinelandii* has a synergistic effect with *Rhizobium sp.*, promoting the formation of nodules on the roots of different leguminous plants, such as soybean, pea, and clover (Gopalakrishnan et al. 2015). Bioinoculants based on *Pseudomonas* species were also reported as having an effect on nitrogen assimilation. Rice seedlings inoculated with *Pseudomonas stutzeri* A15 showed 1.5- and threefold higher shoot length and root dry weight contrary to the control plants. It was proposed that this bacterial strain contributed to the nitrogen fixation (Pham et al. 2017).

Another form of nitrogen supply to plants is based on *Rhizobium*–legume symbiosis. That is a host dependent complex biochemical relationship. It was remarked on the global market that in 2012 the prevalent biofertilizers were rhizobium-based formulations (Bhardwaj et al. 2014).

Biofertilizers can also be used for the phosphorus supply of crop plants. The result of the phosphate mobilizing and solubilizing biofertilizers is the increase of the P mobilization in soil, where the soluble form of this nutrient is low. Different microorganisms were reported to have the ability to solubilize phosphorus. These include bacterial species belonging to genera *Arthrobacter, Bacillus, Beijerinckia, Burkholderia, Enterobacter, Erwinia, Flavobacterium, Mesorhizobium, Microbacterium, Rhizobium, Rhodococcus, Pseudomonas,* and *Serratia*. The above mentioned phosphorus solubilizing bacteria used as bio-inoculants improved the plant growth and yield in agricultural soils. Beyond bacterial strains, there are also microscopic fungi with phosphate mobilization capacity belonging to the *Aspergillus, Fusarium, Penicillium,* and *Sclerotium* genera. This type of biofertilizer is defined as a broad spectrum biofertilizer (Alori et al. 2017). It was revealed that the plant growth promoting effect was associated with phosphate solubilization in *Triticum aestivum* treated with *Azotobacter chroococcum*, in *Camellia sinensis* inoculated with *Bacillus megaterium*, and in *Cucumis sativus* treated with *Bacillus megaterium var. phosphaticum*. *Enterobacter agglomerans* used as an inoculant for *Solanum lycopersicum* showed phosphate solubilization effect. Co-inoculation of *Bradyrhizobium japonicum* with different phosphate mobilizing bacteria, such as *Pseudomonas chlororaphis* and *Pseudomonas putida,* resulted in phosphate solubilization in *Glycine max* (Gouda et al. 2018).

In many studies, it was shown that the applied phosphate solubilizing bacteria, beyond increasing phosphorus uptake of plants, contributed to the improvement of plant yield. Plant growth was detected in the case of wheat inoculated with *Serratia sp.*, in sweetleaf inoculated with *Burkholderia gladioli,* and in maize treated with *Burkholderia cepacia* (Alori et al. 2017).

Total weight and length of Chinese cabbage was increased by *Pseudomonas aeruginosa*. Rice shoot length was increased with the application of *Bacillus thuringiensis*. Productivity of wheat was achieved by the application of *Azotobacter chroococcum* and *Bacillus subtilis* (Singh et al. 2017). The phosphorus mobilizing *Rhizobium tropici* CIAT899 in beans contributed to the enhancement of nodule number and mass, and it also increased the shoot dry weight and the root growth.

The existing form of potassium in soil is insoluble rock or silicate. Numerous plant growth promoting bacteria, due to organic acid production, are able to release potassium in an accessible form to plants. The potassium solubilizer bacteria include, for example, *Bacillus edaphicus, B. ferrooxidans, B. mucilaginosus, B. megaterium var. phosphaticum, B. subtilis, Burkholderia sp., Enterobacter hormaechei, Paenibacillus sp.,* and *Pseudomonas sp* (Meena et al. 2016). As part of the soil bacterial community, they have a key role in the potassium cycle. These bacteria are used in potassium solubilizing or mobilizing biofertilizers. It is revealed that the result of potassium solubilizing and mobilizing biofertilizer consists of the weathering reaction of potassium bearing minerals from natural available sources. The efficiency of bio-inoculants is influenced by different factors, such as the potassium solubilization mechanism, applied strains, nutritional status of soil, minerals, and other environmental conditions (Etesami et al. 2017).

The use of these microorganisms in greenhouse or in field conditions as seed or seedling inoculants resulted in the increase of germination percentage, plant growth, and yield. The enhancement of K uptake by plants was also shown. In different plants, such as cotton, rape, eggplant, peanut, maize, sorghum, wheat, Sudan grass, potato, tomato, and tea, the growth promotion was detected due to the beneficial effect of microorganisms (Etesami et al. 2017).

A beneficial effect of *Bacillus mucilaginosus* strain RCBC13 on tomato plant was observed, resulting in an increase of 125% in biomass. The potassium and phosphorus uptake was more than 150% compared to uninoculated plants (Etesami et al. 2017). In two field experiments, the potassium-solubilizing *Bacillus cereus* and *Pseudomonas sp.* contributed to the improvement of potassium uptake, and this nutrient use efficiency also enhanced the tomato yield (Etesami et al. 2017). In wheat, *Bacillus sp.* significantly increased the N, P, and K content and the yield compared to the uninoculated control. Field experiments in hot pepper inoculated with the phosphate solubilizers *Bacillus megaterium* and *Bacillus mucilaginosus* resulted in beneficial effects on photosynthesis, biomass harvest, and fruit yield (Sindhu et al. 2016). In the case of rice plants, the grain yield resulted from a sample inoculated with a potassium solubilizer microorganism increased from 4419 to 5218 kg/ha.

It was reported that the efficiency of bacterial strains with potassium mobilizing or solubilizing capacity as bio-inoculants was higher when they were used in combination with soil minerals, such as mica, feldspar, or rock phosphate (Meena et al. 2016).

Numerous bacterial strains were reported as having beneficial effects on plants due to the improvement of nutrient uptake of plants. By using the potential of these bacterial strains, either in single or in complex formulations, a decrease in chemical fertilizer utilization can be achieved, suiting the requirements of an environmentally friendly and sustainable agricultural production.

References

Ahemad M, Kibret M (2014) Mechanisms and applications of plant growth promoting rhizobacteria: current perspective. J King Saud Univ – Sci 26:1–20. https://doi.org/10.1016/j.jksus.2013.05.001

Ahmad M, Nadeem SM, Naveed M, Zahir ZA (2016) Potassium-solubilizing bacteria and their application in agriculture. In: Meena V, Maurya B, Verma J, Meena R (eds) Potassium solubilizing microorganisms for sustainable agriculture. Springer, New Delhi. https://doi.org/10.1007/978-81-322-2776-2_21

Alori ET, Glick BR, Babalola OO (2017) Microbial phosphorus solubilization and its potential for use in sustainable agriculture. Front Microbiol 8:971–979. https://doi.org/10.3389/fmicb.2017.00971

Anzuay MS, Ludueña LM, Angelini JG, Fabra A, Taurian T (2015) Beneficial effects of native phosphate solubilizing bacteria on peanut (*Arachis hypogaea* L) growth and phosphorus acquisition. Symbiosis 66:89–97. https://doi.org/10.1007/s13199-015-0337-z

Askary M, Mostajeran A, Amooaghaei R, Mostajeran M (2009) Influence of the co-inoculation Azospirillum brasilense and rhizobium meliloti plus 2,4-D on grain yield and N, P, K content of *Triticum aestivum* (Cv. Baccros and Mahdavi). Am-Eurasian J Agric Environ Sci 5(3):296–307

Bhardwaj D, Ansari MW, Sahoo RK, Tuteja NV (2014) Biofertilizers function as key player in sustainable agriculture by improving soil fertility, plant tolerance and crop productivity. Microb Cell Factories 13:66. https://doi.org/10.1186/1475-2859-13-66

Bhat T, Ahmad L, Ganai MA, Shams-Ul-Haq KOA (2015) Nitrogen fixing biofertilizers; mechanism and growth promotion: a review. J Pure Appl Microbiol 9:1675–1690

Calzavara AK, Paiva PHG, Gabriel LC, Oliveira ALM, Milani K, Oliveira HC, Bianchini E, Pimenta JA, de Oliveira MCN, Dias-Pereira J, Stolf-Moreira R (2018) Associative bacteria influence maize (*Zea mays* L.) growth, physiology and root anatomy under different nitrogen levels. Plant Biol 20:870–878. https://doi.org/10.1111/plb.12841

Etesami H, Emami SE, Alikhani HA (2017) Potassium solubilizing bacteria (KSB): mechanisms, promotion of plant growth, and future prospects – a review. J Soil Sci Plant Nutr 17:897–911
Gold MV (2016) Sustainable agriculture: the basics. In: Etingoff K (ed) Sustainable agriculture and food supply scientific, economic and policy enhancements. Apple Academic Press, Oakville
Gopalakrishnan S, Sathya A, Vijayabharathi R, Varshney RK, Gowda CLL, Krishnamurthy L (2015) Plant growth promoting rhizobia: challenges and opportunities. 3 Biotech 5:355–377. https://doi.org/10.1007/s13205-014-0241-x
Gouda S, Kerry RG, Das G, Paramithiotis S, Shin HS, Patra JK (2018) Revitalization of plant growth promoting rhizobacteria for sustainable development in agriculture. Microbiol Res 206:131–140. https://doi.org/10.1016/j.micres.2017.08.016
Granada CE, Passaglia LMP, de Souza EM, Sperotto RA (2018) Is phosphate solubilization the forgotten child of plant growth-promoting rhizobacteria? Front Microbiol 9:2054. https://doi.org/10.3389/fmicb.2018.02054
Gu M, Chen A, Sun S, Xu G (2016) Complex regulation of plant phosphate transporters and the gap between molecular mechanisms and practical application: what is missing? Mol Plant 9:396–416. https://doi.org/10.1016/j.molp.2015.12.012
Hermary H (2007) Effects of some synthetic fertilizers on the soil ecosystem
Itelima JU, Bang WJ, Sila MD, Onyimba IA, Egbere OJ (2018) A review: biofertilizer – a key player in enhancing soil fertility and crop productivity. Microbiol Biotechnol Rep 2:73–83. https://doi.org/10.26765/DRJAFS.2018.4815
Jungk A (2001) Root hairs and the acquisition of plant nutrients from soil. J Plant Nutr Soil Sci 164:121–129. https://doi.org/10.1002/1522-2624(200104)
Karamesouti M, Gasparatos D (2017) Sustainable management of soil phosphorus in a changing world. In: Rakshit A, Abhilash P, Singh H, Ghosh S (eds) Adaptive soil management: from theory to practices. Springer, Singapore. https://doi.org/10.1007/978-981-10-3638-5_9
Kaur G, Reddy MS (2015) Effects of phosphate-solubilizing bacteria, rock phosphate and chemical fertilizers on maize-wheat cropping cycle and economics. Pedosphere 25:428–437. https://doi.org/10.1016/s1002-0160(15)30010-2
Kiba T, Krapp A (2016) Plant nitrogen acquisition under low availability: regulation of uptake and root architecture. Plant Cell Physiol 57:707–714. https://doi.org/10.1093/pcp/pcw052
Kuan KB, Othman R, Abdul Rahim K, Shamsuddin ZH (2016) Plant growth-promoting rhizobacteria inoculation to enhance vegetative growth, nitrogen fixation and nitrogen remobilisation of maize under greenhouse conditions. PLoS One 11:e0152478. https://doi.org/10.1371/journal.pone.0152478
Lambers H, Shane MW, Cramer MD, Pearse SJ, Veneklaas EJ (2006) Root structure and functioning for efficient acquisition of phosphorus: matching morphological and physiological traits. Ann Bot 98(4):693–713. https://doi.org/10.1093/aob/mcl114
Meena VS, Bahadur I, Ram B, Ashok M, Meena RK, Meena SK (2016) Potassium-solubilizing microorganism in evergreen agriculture: an overview. In: Meena V, Maurya B, Verma J, Meena R (eds) Potassium solubilizing microorganisms for sustainable agriculture. Springer, New Delhi. https://doi.org/10.1007/978-81-322-2776-2_1
Miransari M (2014) Plant growth promoting rhizobacteria. J Plant Nutr 37:2227–2235. https://doi.org/10.1080/01904167.2014.920384
Nieves-Cordones M, Alemán F, Martínez V, Rubio F (2014) K^+ uptake in plant roots. The systems involved, their regulation and parallels in other organisms. J Plant Physiol 171:688–695. https://doi.org/10.1016/j.jplph.2013.09.021
Nussaume L, Kanno S, Javot H, Marin E, Pochon N, Ayadi A, Nakanishi TM, Thibaud M-C (2011) Phosphate import in plants: focus on the PHT1 transporters. Front Plant Sci 2:83. https://doi.org/10.3389/fpls.2011.00083
Owen D, Williams AP, Griffith GW, Withers PJA (2015) Use of commercial bio-inoculants to increase agricultural production through improved phosphorus acquisition. Appl Soil Ecol 86:41–54. https://doi.org/10.1016/j.apsoil.2014.09.012

Pham VT, Rediers H, Ghequire MG, Nguyen HH, De Mot R, Vanderleyden J, Spaepen S (2017) The plant growth-promoting effect of the nitrogen-fixing endophyte *Pseudomonas stutzeri* A15. Arch Microbiol 199:513–517. https://doi.org/10.1007/s00203-016-1332-3

Pii Y, Mimmo T, Tomasi N, Roberto T, Cesco S, Crecchio C (2015) Microbial interactions in the rhizosphere: beneficial influences of plant growth-promoting rhizobacteria on nutrient acquisition process. A review. Biol Fertil Soils 51:403–415. https://doi.org/10.1007/s00374-015-0996-1

Ródenas R, García-Legaz MF, López-Gómez E, Martínez V, Rubio F, Ángeles Botella M (2017) $NO_3^{\,-}$, $PO_4^{\,3-}$ and $SO_4^{\,2-}$ deprivation reduced LKT1-mediated low-affinity K^+ uptake and SKOR-mediated K^+ translocation in tomato and Arabidopsis plants. Physiol Plant 160(4):410–424. https://doi.org/10.1111/ppl.12558

Rosier A, Medeiros FHV, Bais HP (2018) Defining plant growth promoting rhizobacteria molecular and biochemical networks in beneficial plant-microbe interactions. Plant Soil 428:35–55. https://doi.org/10.1007/s11104-018-3679-5

Rozier C, Hamzaoui J, Lemoine D, Czarnes S, Legendre L (2017) Field-based assessment of the mechanism of maize yield enhancement by Azospirillum lipoferum CRT1. Sci Rep 7:7416–7427. https://doi.org/10.1038/s41598-017-07929-8

Saha B, Saha S, Das A, Bhattacharyya PK, Basak N, Sinha AK, Poddar P (2017) Biological nitrogen fixation for sustainable agriculture. In: Meena V, Mishra P, Bisht J, Pattanayak A (eds) Agriculturally important microbes for sustainable agriculture. Springer, Singapore. https://doi.org/10.1007/978-981-10-5343-6_4

Sahu A, Bhattacharjya S, Mandal A, Thakur JK, Atoliya N, Sahu N, Manna MC, Patra AK (2018) Microbes: a sustainable approach for enhancing nutrient availability in agricultural soils. In: Meena V (ed) Role of rhizospheric microbes in soil. Springer, Singapore. https://doi.org/10.1007/978-981-13-0044-8_2

Sattari SZ, Bouwman AF, Giller KE, van Ittersum MK (2012) Residual soil phosphorus as the missing piece in the global phosphorus crisis puzzle. Proc Natl Acad Sci 109:6348–6353. https://doi.org/10.1073/pnas.1113675109

Sindhu SS, Parmar P, Phour M, Sehrawat A (2016) Potassium-solubilizing microorganisms (KSMs) and its effect on plant growth improvement. In: Meena V, Maurya B, Verma J, Meena R (eds) Potassium solubilizing microorganisms for sustainable agriculture. Springer, New Delhi. https://doi.org/10.1007/978-81-322-2776-2_13

Singh RP, Kumar S, Sainger M, Sainger PA, Barnawal D (2017) Eco-friendly nitrogen fertilizers for sustainable agriculture. In: Rakshit A, Abhilash P, Singh H, Ghosh S (eds) Adaptive soil management: from theory to practices. Springer, Singapore. https://doi.org/10.1007/978-981-10-3638-5_11

Takahashi M, Morikawa H (2014) Nitrogen dioxide is a positive regulator of plant growth. Plant Signal Behav 9:e28033. https://doi.org/10.4161/psb.28033

Tchiaze AI, Taffouo VD, Fankem H, Kenne M, Baziramakenga R, Ekodeck GE, Antoun H (2016) Influence of nitrogen sources and plant growth-promoting rhizobacteria inoculation on growth, crude fiber and nutrient uptake in squash (*Cucurbita moschata* Duchesne ex Poir.). Not Bot Horti Agrobot 44:53–59. https://doi.org/10.15835/nbha44110169

Tóth G, Guicharnaud R-A, Tóth B, Hermann T (2014) Phosphorus levels in croplands of the European Union with implications for P fertilizer use. Eur J Agron 55:42–52. https://doi.org/10.1016/j.eja.2013.12.008

Valkó G (2017) A fenntartható mezőgazdaság indikátorrendszerének kialakítása az Európai Unió tagországaira vonatkozóan. Központi Statisztikai Hivatal, Budapest, pp 19–31

Verma RK, Sachan M, Vishwakarma K, Upadhyay N, Mishra, RK, Tripathi DK, Sharma S (2018) Role of PGPR in sustainable agriculture: molecular approach toward disease suppression and growth promotion. In: Meena V (ed) Role of rhizospheric microbes in soil. Springer, Singapore. https://doi.org/10.1007/978-981-13-0044-8_9

White PJ (2008) Efficiency of soil and fertilizer phosphorus use: reconciling changing concepts of soil phosphorus behaviour with agronomic information. In: Syers JK, Johnston AE, Curtin D (eds) Experimental agriculture 45(01). Food and Agricultural Organization of the United Nations, Rome, p 108. https://doi.org/10.1017/S0014479708007138

Soil: Microbial Cell Factory for Assortment with Beneficial Role in Agriculture

4

Pratiksha Singh, Rajesh Kumar Singh, Mohini Prabha Singh, Qi Qi Song, Manoj K. Solanki, Li-Tao Yang, and Yang-Rui Li

4.1 Introduction

Soil is a multifaceted physiochemical and living substrate (Berendsen et al. 2012). It offers the medium of life for humans, animals, plants, micro-organisms, etc. The soil is the most common intermediate in plants for growth and progress. If the conditions of the soil are good for the plant then it affects the crop yield, and the

P. Singh · R. K. Singh (✉)
Key Laboratory of Sugarcane Biotechnology and Genetic Improvement (Guangxi), Ministry of Agriculture, Sugarcane Research Center, Chinese Academy of Agricultural Sciences, Guangxi Key Laboratory of Sugarcane Genetic Improvement, Sugarcane Research Institute, Guangxi Academy of Agricultural Sciences, Nanning, China

College of Agriculture, Guangxi University, Nanning, China

Guangxi Key Laboratory of Crop Genetic Improvement and Biotechnology, Nanning, China

M. P. Singh
Punjab Agriculture University, Ludhiana, India

Q. Q. Song · Y.-R. Li
Key Laboratory of Sugarcane Biotechnology and Genetic Improvement (Guangxi), Ministry of Agriculture, Sugarcane Research Center, Chinese Academy of Agricultural Sciences, Guangxi Key Laboratory of Sugarcane Genetic Improvement, Sugarcane Research Institute, Guangxi Academy of Agricultural Sciences, Nanning, China

College of Agriculture, Guangxi University, Nanning, China

M. K. Solanki
Key Laboratory of Sugarcane Biotechnology and Genetic Improvement (Guangxi), Ministry of Agriculture, Sugarcane Research Center, Chinese Academy of Agricultural Sciences, Guangxi Key Laboratory of Sugarcane Genetic Improvement, Sugarcane Research Institute, Guangxi Academy of Agricultural Sciences, Nanning, China

L.-T. Yang
College of Agriculture, Guangxi University, Nanning, China

Guangxi Key Laboratory of Crop Genetic Improvement and Biotechnology, Nanning, China

D. P. Singh et al. (eds.), *Microbial Interventions in Agriculture and Environment*,
https://doi.org/10.1007/978-981-13-8391-5_4

properties (soil pH, texture, and structure) of the soil function as the ability to maintain the plant life (Wong et al. 2015). Water retention and the nutrients existing in the soil give benefits to the plant's health and are influenced by soil texture (Bronick and Lal 2005). The texture of the soil refers to the total amount of sand, silt, clay, and organic matter present, which depends on the topographical places and seasons. Organic soil and clay have improved nutrients and water holding capacity over sandy soil, but soil that has more clay and organic matter than water capacity, for an elongated time period, leads to waterlogged soil (Wong et al. 2015). Therefore, soil containing the optimal percentage of sand, silt, clay, and organic matter is used as an ideal condition for agricultural and farming. Apart from this, soil structure is also a crucial feature that regulates the effects of the soil and permits it to maintain plant life, soil carbon sequestration, and water quality (Bronick and Lal 2005). Therefore, favorable soil structure supports progress in soil fertility, agronomic output, and improves soil permeability (Bronick and Lal 2005). The pore spaces present between the soil particles affect the water and air movement within the rhizospheres and nutrient accessibility for plant root growth and microbial actions (Wong et al. 2015). The pH of the soil measures the alkalinity or acidity of the soil, which is another significant property that directly affects the accessibility of nutrients uptake for the plant. Macronutrients have a tendency to be less accessible at low pH, and in the case of high pH, micronutrients tend to be less accessible; therefore, the perfect soil pH range is 6.0–6.5 (Wong et al. 2015).

Soil and micro-organisms are allied with each other because the function of soil is to protect and provide the medium for existence of the different micro-organisms, and these micro-organisms help to maintain the soil fertility and reduce the toxic material from industrial waste and diverse forms of chemical fertilizers, herbicides, fungicides, insecticides that are externally applied by the farmers to increase plant development and crop production. The communities of soil look for a maximum pool of biological microbial diversity with regular interaction with the plant (Berendsen et al. 2012). The maximum and different kinds of microbial populations are one of the major resources of healthy soils; in addition, they have balanced nutrients, pH value, water holding capacity, and battle unfavorable weather conditions hazards (Meena and Jha 2018). An approximate number of microbial species varies from a few to millions/billions in the soil, and 99% of microbes cannot be cultured in medium, classified, or identified, and even to date we are discovering their important roles, but it is particularly difficult for such micro-organisms. There are some known microbial floras, i.e., bacteria, fungi, actinomycetes, protozoa, soil nematodes, and algae present everywhere on the earth. They are present in all soil types, which are cultivated and noncultivated areas, including rhizosphere soils, sediments soil, rocks and rock crevices soils, deserts and sandy soils, thermal and volcanic ash soils, beaches and floodplains soil, organic soils, grassland soils, mountain soils, marsh and moorlands, glaciers and ice sheets, temperate forest soils, etc. These soils' microbial flora play a key role in nutrient cycling and are a good indicator of soil health and fertility. Several isolation techniques have been discovered by scientists and co-workers, such as dilution plate techniques (Waksman 1911, 1944), baiting method (Harvey 1925), uncoated glass slide (Rossi and Ricardo 1927; Cholodny 1930), agar film method (Jones and Mollison 1948), immersion tube method (Chester 1948), soil plate method and modified soil plate method (Warcup 1950), immersion plate technique (Thornton 1952), root maceration (Stover and Waite 1953), dilution frequency method (Allen 1957), direct microscopic examination (Conn 1981), and many other techniques used for isolating more and novel microbes in different soil types by specific and nonspecific media. The current chapter pays attention to the diversity of soil and rhizospheric microbes, especially to their beneficial role in agriculture with the method of PGPR and biocontrol agents that are beneficial for plant growth and the environment. Over the past 200 years several soil/plant bacteria and fungi have been continuously thought of and confirmed to be beneficial microbes promoting plant growth and suppressing pathogens, but this information has yet to be broadly exploited in the agricultural microbiology and biotechnology sector (Berg 2009). Additionally, an ongoing increase of the human population continues to demand more food products worldwide; therefore, we have to find alternatives methods to increase the food production without adding harmful chemicals while maintaining the soil health by using agriculturally important soils microbial flora. In the past 15–20 years, many researchers and scientists have considerably increased our awareness of the mechanisms in use by PGPR (Glick 2012). Agriculturally important micro-organisms, such as nitrogen-fixing bacteria, antagonistic/biological control agents (BCAs), plant growth promoting rhizobacteria (PGPR), and plant growth promoting fungi (PGPF), play an important part in these important challenges, and they fulfill a major role for plants and soil in changed environmental situations (Raaijmakers et al. 2009; Hermosa et al. 2011). However, the practical use of PGPR and PGPF efforts to increase the plant growth promotion is a necessity that includes acquiring an enhanced understanding of the mechanisms and how they works in plants with different conditions.

Soil and rhizospheric bacteria have the capability to grow a broad range of nutrient substances (both organic and inorganic). The concentration of microbial colonies established in the plant's roots is probably high because different types of nutrients existing in the plant roots, such as sugars, vitamins, enzymes, amino acids, and organic acids, are then used by rhizospheric bacteria to support their growth and assimilation (Whipps 1990). In rhizosphere, the density of bacterial populations is approx 10–1000 times more than in soil, whereas the average in a laboratory medium is still 100-fold lower than that (Lugtenberg and Kamilova 2009). Also, plant growth promoting bacteria enhance the growth of a plant by two modes, i.e., directly and indirectly (Glick 2012) (Fig. 4.1). In general, direct plant growth promotion mechanisms are phosphate solubilization, phytohormones (cytokinin, ethylene, gibberellic acid), ammonia, siderophore production, and nitrogen-fixation activity, while the indirect mechanisms are the production of antibiotics, HCN, hydrolytic enzymes (chitinase, endoglucanase, protease, cellulose), and induced systemic resistance (Fig. 4.2). In Table 4.1, we report some recent PGP strains isolated from different rhizosphere and soil. The production of 1-aminocyclopropane-1-carboxylate (ACC)

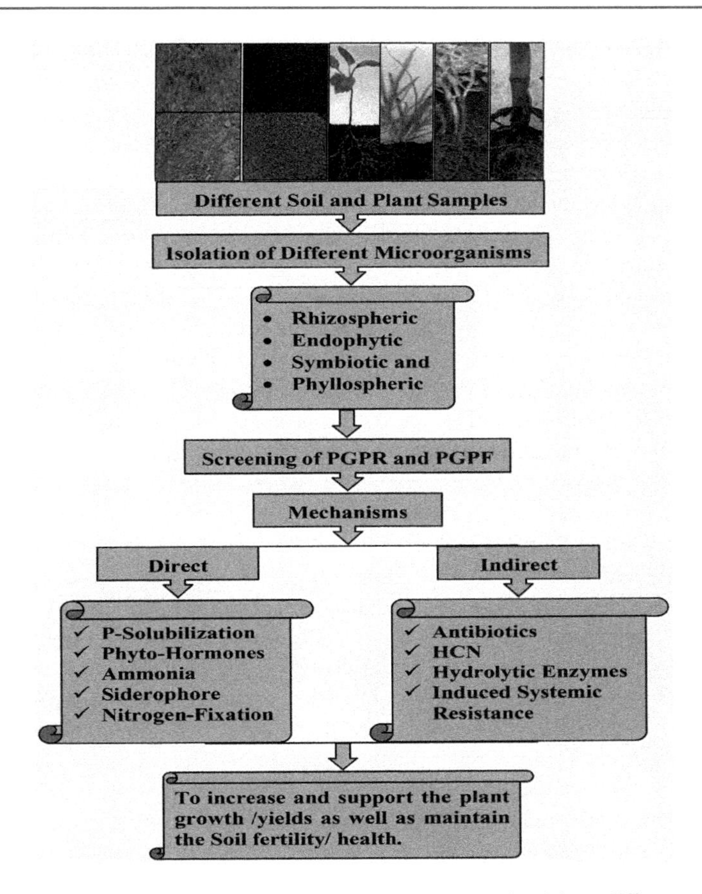

Fig. 4.1 Schematic representation of PGPR/ PGPF microbes isolated from different sources and their important mechanisms are categorized into two groups for different plant growth promoting traits

deaminase activity by PGPR also improves the plant defense activity under stress conditions by decreasing the ethylene levels (Zahir et al. 2009).

4.2 Mechanisms of Direct PGPR

4.2.1 Phosphate

Phosphorus (P) is one of the most essential macronutrients necessary for plant growth development after nitrogen (N). It is the most essential plant nutrient, which affects almost all of the plant's growth and development (Wang et al. 2009). It plays a key role in nearly all the metabolic processes, including signal transduction, energy transfer, crop quality, photosynthesis, disease resistance, and respiration macromolecular biosynthesis (Khan et al. 2010). In soils, P is available in both

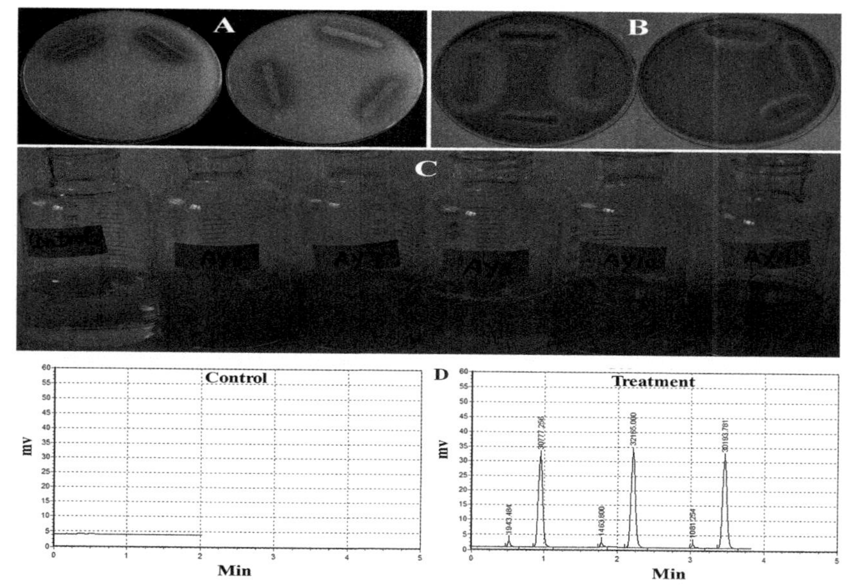

Direct Mechanisms

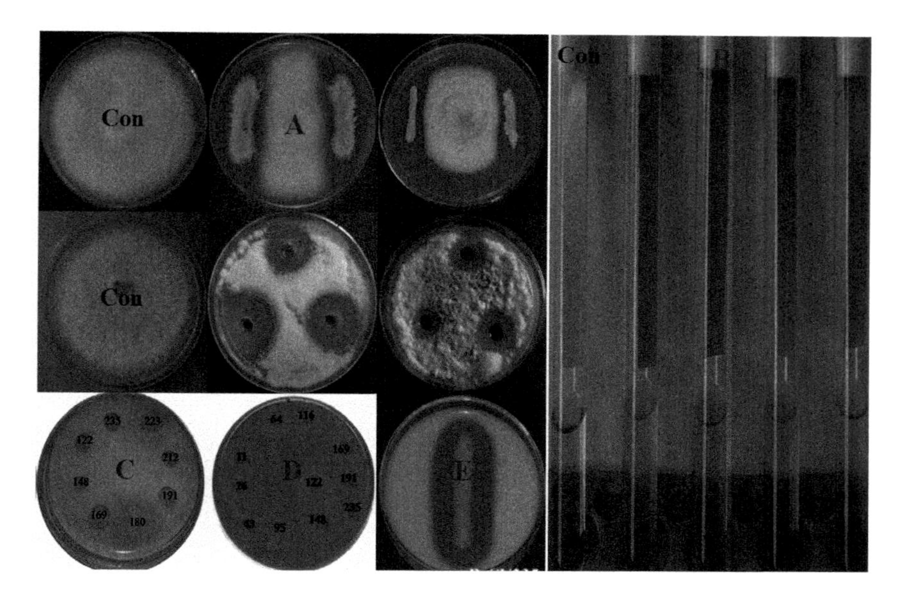

Indirect Mechanisms

Fig. 4.2 Different plant growth promoting mechanisms are categories in two groups: Direct mechanisms; (**a**) Phosphate solubilization (**b**) Siderophore production (**c**) Ammonia production and (**d**) Acetylene reduction assay for nitrogenase activity. Indirect mechanisms; (**a**) Antibiotic production (**b**) Hydrogen cyanide (**c**) Hydrolytic enzyme like chitinase (**d**) Endoglucanase and (**e**) Protease activity

Table 4.1 List of plant growth promotion rhizobacteria isolated from different plants

PGPR strains	PGPR mechanisms	Sources/host plant	References
Pseudomonas plecoglossicida, Stenotrophomonas maltophilia, Bacillus cereus, Enterobacter sp.	P-solubilisation, Siderophore, IAA, HCN	Turmeric	Vinayarani and Prakash (2018)
Acinetobacter sp. *Pseudomonas putida, Pseudomonas aeruginosa*			
Agrobacterium tumefaciens, Achromobacterinsolitus, Stenotrophomonas sp., *Stenotrophomonas maltophilia, Chryseobacterium* sp., *Flavobacterium* sp., *Pseudoxanthomonas Mexicana, Cupriavidus* sp., *Bordetella petrii*	P-solubilisation, siderophore, IAA, ammonia,	Maize and rice rhizosphere, wheat root	Youseif (2018)
Bacillus pumilus, B. cereus, B. licheniformis, B. subtilis	P-solubilisation, siderophore, IAA, ACC, ammonia, HCN	Chickpea	Sharma et al. (2018)
Pseudomonas koreensis, Pseudomonasentomophila	P-solubilisation, Siderophore, IAA, nitrogen fixation	Sugarcane	Li et al. (2017)
Stenotrophomonas maltophilia	P-solubilisation, Siderophore, IAA, ACC, HCN, ammonia	Sorghum bicolor	Singh and Jha (2017)
Pseudomonas aeruginosa	P-solubilisation, siderophore, IAA, ACC, ammonia, Nitrogenase	Achyranthes aspera	Devi et al. (2017)
Bacillus megaterium, Pseudomonasnitroreducens	P-solubilisation, Siderophore, IAA, nitrogen fixation	Sugarcane	Solanki et al. (2016)
Stenotrophomonas maltophilia, B. amyloliquefaciens, B. subtilis subsp. *Spizizenii, B. subtilis* subsp. *subtilis, Bacillus subtilis,*	IAA, ARA, antifungal activity	Cucumber	Islam et al. (2016)
Agrobacierium tumefaciens, Klebsiella sp., *Ochrobactrum anthropic, Pseudomonas stutzeri, Pseudomonas sp.*	IAA, ACC, ARA, P-solubilisation,	Arthrocnemum indicum	Sharma et al. (2016)
StenotrophomonasMaltophilia, Agrobacterium tumefaciens	P-solubilisation, Siderophore, IAA, nitrogen fixation	Sugarcane	Li and Glick (2005)

(continued)

Table 4.1 (continued)

PGPR strains	PGPR mechanisms	Sources/host plant	References
Stenotrophomonas rhizophila, Bacillus sp., *Acetobactor pasteurianus, Stenotrophomonas*	P-solubilisation, IAA, ARA	Wheat	Majeed et al. (2015)
Rhodococcus sp., *Pseudomonas* sp., *Arthrobacter nicotinovorans, Arthrobacter* sp., *Burkholderia*	IAA, ACC, Siderophores, HCN, ammonia,	Contaminated soil	Sofia and Paula (2014)
Bacillus sp.	IAA, siderophores, antifungal activity, ammonia, P-solubilization	Santalum album	Pradhan et al. (2014)
Pseudomonas stutzeri, Bacillus sp., *Bacillus subtilis, Enterobacter* sp., *Enterobacter cloacae, Pseudomonas putida, Bacillus subtilis* subsp. *inaquosorum*	IAA, siderophores, antifungal activity, HCN, P- solubilization, nitrogen fixation	A number of different plant species	El-Sayed et al. (2014)
B. subtilis	P-solubilisation, Siderophore, IAA,	Chickpea	Singh et al. (2013)
Lysinibacillus fusiformis	P-solubilisation, Siderophore, IAA	Chickpea	Singh et al. (2012)
Klebsiella, Enterobacter	P-solubilisation, Siderophore, IAA	Sugarcane	Lin et al. (2012)
Pseudomonas sp., *Pseudomonas fluorescens, Burkholderiaglumae*	ACC deaminase, IAA, HCN, Siderophore, ammonia,		Rashid et al. (2012)
Bacillus	ACC deaminase, IAA, HCN siderophore,P-solubilization, lytic enzyme,		Kumar et al. (2012)
Bacillus pumilus	P-solubilisation, siderophore, IAA,	Wheat	Tiwari et al. (2011)
*Acinetobacter*spp.	IAA, P-solubilization, Siderophores	Pennisetumglaucum	Rokhbakhsh-Zamin et al. (2011)
Pseudomonas putida	IAA, siderophores, ammonia, HCN, Exo-polysaccharides, P-solubilization	Mustard	Ahemad and Khan (2011)
Pseudomonas sp.	IAA, siderophores	Alyssum serpyllifolium	Ma et al. (2011)

(continued)

Table 4.1 (continued)

PGPR strains	PGPR mechanisms	Sources/host plant	References
Enterobacter asburiae	IAA, siderophores, HCN, ammonia, P-solubilization	Mustard	Ahemad and Khan (2010)
Rahnellaaquatilis	P-solubilization, IAA, ACCdeaminase	Sugarcane	Mehnaz et al. (2010)
Stenotrophomonas maltophilia	Nitrogenase activity, IAA, P-solubilization, ACC deaminase	Sugarcane	Mehnaz et al. (2010)
Paenibacilluspolymyxa	IAA, siderophores	Peppers	Phi et al. (2010)
Rhizobium phaseoli	IAA	Mung bean	Zahir et al. (2010)

forms, i.e., organic and inorganic, but various limiting factors are available for plant root uptake for growth because it is present in an unavailable form (Sharma et al. 2013). In soil, on average, the level of phosphorus content present is about 0.05% (w/w); and only 0.1% of this phosphorus is used for plants (Zhu et al. 2011). Of the P available in soil, 95–99% is in the insoluble form and cannot be taken up by the plants; therefore, a majority of the chemical fertilizer applied by farmers on a normal basis is to raise the phosphorus uptake in plants. However, it is the nonstop application of chemical fertilizers in excess amounts that leads to soil and environmental contamination. Micro-organisms play an important task in natural phosphorus, and the solubilization of insoluble P by micro-organisms was reported by Pikovskaya (1948). Several reports are available that examine the ability of different micro-organisms bacteria, fungi, actinomycetes, algae, and VAM for P solubilizing capabilities and improvement of the plant nutrient acquisition (Table 4.2). There are a large group of different microbial genera present in soil and plant rhizospheres that are able to transform insoluble P to a soluble form of P, which is available for plants, i.e., called phosphate solubilizing micro-organisms (Sperberg 1958; Khan et al. 2014). The insoluble forms of tri-calcium, aluminum, iron phosphate [$(Ca_3PO_4)_2$, (Al_3PO_4), (Fe_3PO_4)], etc., might be changed to soluble P by using those microbes to solubilize the P, and they are exhibited in diverse soil ecosystems (Sharma et al. 2013). A significantly higher amount of P solubilizing microorganisms is available in the rhizosphere as compared with non-rhizosphere soil (Katznelson et al. 1962). Visually detecting the ability of P solubilizing microorganisms is possible by using Pikovskaya medium plates for isolation and screening, which shows a clear zone of inhibitions about the microbial colonies (Fig. 4.3). Therefore, it is of great interest to study the application of microbial inoculants activities possessing P solubilizing properties in agricultural soils for management strategies that are capable of improving crop growth, increasing yields, diminishing

Table 4.2 Biodiversity of phosphorus solubilizing microorganisms (PSM)

S. No	Microorganisms	Species	References
1	Bacteria	**Gram positive:** *Bacillus* sp.,*Bacillus brevis*, *B. cereus* var. *albolactis*, *B. circulans*, *B. coagulans*, *B. firmus*, *B. megaterium*, *B. megaterium* var. *phosphaticum*, *B. mesentricum*, *B. mycoides*, *B. polymyxa*, *B. pumilus*, *B. pulvifaciens*, *B. sphaericus*, *B. subtilis*, *B. licheniformis*, *B. amyloliquefaciens*, *B. atrophaeus*, *B. fusiformis*, *B. coagulans*, *B. chitinolyticus*, *B. subtilis*, *Clostridium* sp.,	Alori et al. (2017), Li et al. (2017), Kishore et al. (2015), and Sharma et al. (2013)
		Gram negative: *Acetobacter diazotrophicus*, *Achromobacter* sp., *Aerobacter aerogenes*, *Agrobacterium radiobacter*, *Agrobacterium* sp., *Alcaligenes* sp., *Arthrobacter mysorens*, *Bradyrhizobium* sp., *Brevibacterium* sp., *Burkholderia cepacia*, *Citrobacter freundii*, *Enterobacter aerogenes*, *Enterobacter agglomerans*, *Enterobacter asburiae*, *Enterobacter cloacae*, *Escherichia freundii*, *Escherichia intermedia*, *Erwinia herbicola*, *Flavobacterium* sp., *Gluconobacter diazotrophicus*, *Micrococcus* sp., *Mycobacterium* sp., *Nitrosomonas* sp., *Pseudomonas calcis*, *P. cepacia*, *P. fl uorescens*, *P. putida*, *P. rathonia*, *P. striata*, *P. syringae*, *Serratia marcescens*, *S. phosphaticum*, *Thiobacillus ferrooxidans*, *T. thiooxidans*, *Rahnella aquatilis*, *Rhizobium meliloti*, *Xanthomonas* sp., *Azotobacter chroococcum*, *Kluyvera ascorbata*, *Azospirillum brasilense*, *A. lipoferum*, *Acinetobacter calcoaceticus*, *Paenibacillus Ralstonia*, *Citrobacter* sp., pseudomonas sp., *P. koreensis*, *P. entomophila*, *P. monteilii*, *P. mosselii*, *P. putida*, *P. striata*, *Erwinia* sp., *Nitrobacter* sp., *Thiobacillus ferroxidans*, *T. thioxidans*, *Rhizobium meliloti*, *Xanthomonas* sp.	

(continued)

Table 4.2 (continued)

S. No	Microorganisms	Species	References
2	Fungi	*Achrothecium* sp., *Alternaria tenuis*, *Arthrobotrys*, *Aspergillus aculeatus*, *A. awamori*, *A. carbonum*, *A. flavus*, *A. foetidus*, *A. fumigatus*, *A. japonicus*, *A. nidulans*, *A. nidulans* var. *acristatus*, *A. niger*, *A. rugulosus*, *A. terreus*, *A. wentii*, *Cephalosporium* sp., *Chaetomium globosum*, *Cladosporium herbarum*, *Cunninghamella* sp., *C. elegans*, *Curvularia lunata*, *Fusarium oxysporum*, *Helminthosporium* sp., *Humicola lanuginosa*, *H. inslens*, *Mortierella* sp., *Micromonospora* sp., *Mucor* sp., *Myrothecium roridum*, *Oidiodendron* sp., *Paecilomyces lilacinus*, *P. fusisporus*, *Penicillium aurantiogriseum*, *P. bilaji*, *P. digitatum*, *P. funiculosum*, *P. lilacinum*, *P. oxalicum*, *P. pinophilum*, *P. rubrum*, *P. rugulosum*, *P. simplicissimum*, *P. variabile*, *Phoma* sp., *Populospora mytilina*, *Pythium* sp., *Rhizoctonia solani*, *Rhizopus* sp., *Sclerotium rolfsii*, *Torulaspora globosa*, *Torula thermophila*, *Trichoderma harzianum*, *T. viridae*, *Schwanniomyces occidentalis*, *Emericella rugulosa*, *Penicillium camemberti*, *Colletotrichum* sp.	Alori et al. (2017), Kishore et al. (2015), Sharma et al. (2013)
		Yeast: *Yarrowia lipolytica*, *Schizosaccharomyces pombe*, *Pichia fermentas*	
3	Actinomycetes	*Actinomyces* sp., *Actinomyces coelicolor*, *Streptomyces* sp., *Streptomyces violascens*, *S. noboritoensis*, *S. cinereorectus*, *S. cinnabarinus*, *Microbacterium aurantiacum*, *M. kitamiense*, *Angustibacter luteus*, *Kocuria flava*, *Isoptericola hypogeus*, *Agromyces soli*, *Kocuria palustris*, *Microbacterium yannicii,Isoptericola variabilis*, *Nocardia* sp., *Streptoverticillium* sp., *Thermoactinomycetes* sp., *Micromonospora* sp., *Streptomyces californicus,* S. *exfoliates,* S. *rimosus,* S. *fulvissimus,* S. *lydicus,* S. *chromogenus,* S. *fulvissimus,* S. *filipinensis,* S. *purpureus,* S. *griseoviridis,* S. *longisporoflavus,* S. *xanthochromogenes, Streptoverticillium olivoverticillatum,* S. *nogalater, Streptomyces nogalater,* S. *pactum,* S. *aureofaciens,* S. *chattanoogensis,* S. *cellulosae*	Nandimath et al. (2017), Alori et al. (2017), and Kishore et al. (2015)
4	Cyanobacteria	*Anabena* sp., *Calothrix braunii*, *Hapalosiphon fontinalis*, *Nostoc* sp., *Phormidium* sp., *Scytonema* sp.,*Scytonema cincinnatom*, *Tolypothrix tenuis*, *Tolypothrix ceylonica*, *Westiellopsis prolifica*	Alori et al. (2017), Kishore et al. (2015), and Sharma et al. (2013)
5	VAM	*Glomus fasciculatum*, *Glomus etunicatum*	Sharma et al. (2013) and Saxena et al. (2014)

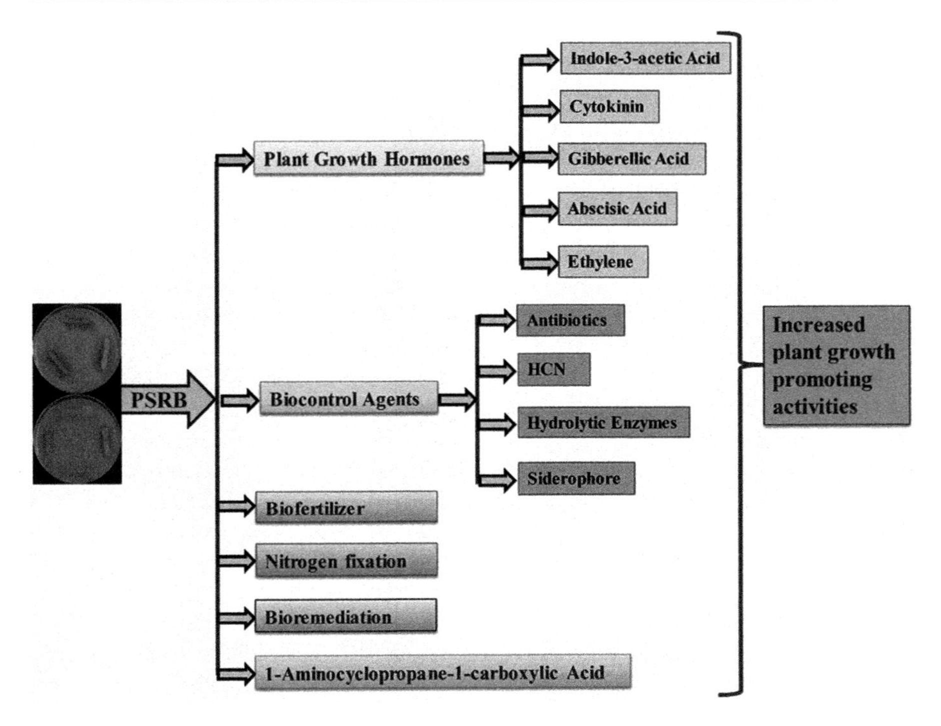

Fig. 4.3 Drawing arrangements of selected phosphate solubilizing rhizospheric bacteria and their important mechanisms identified for plant growth promotion with different PGPR activity

the environmental pollution, and acting as a substitute to added chemical-based P fertilizers applications.

4.2.2 Plant Growth Regulators

Plant growth regulators are organic constituents, which are available in very low amounts (<1 mM) to encourage, prevent, or adapt growth and progress of the plants (Damam et al. 2016), and they are not synthesized naturally. These are also called exogenous plant hormones and can be induced by the PGPR group of microbes, such as bacteria, fungi, actinomycetes, etc. These microbes have the capability to generate plant growth hormones, such as auxins, gibberellins, cytokinins, ethylene, and abscisic acid (Vejan et al. 2016). They play a key role in controlling almost all features for plants growth, i.e., physiological, morphological, and biochemical processes. All hormones, such as auxins, gibberellins, cytokinins, ethylene, and abscisic acid, are synthesized directly or indirectly by PGPR (Vejan et al. 2016).

4.2.2.1 Indole-3-Acetic Acid (IAA) (Auxins)

Auxin is one of the most significant molecules that is regulated by a maximum number of plant developments directly or indirectly (Tanimoto 2005). The greatest lively and well-known auxins in plants is indole-3-acetic acid (Hayat et al. 2010), and it is produced by several PGPR microbes. It is also implicated in all features of root expansion in the plant, especially to control plant augmentation and progress, such as cell elongation, cell division, tissue differentiation, and aid apical dominance (Goswami et al. 2016). Plants treated with IAA long-term have shown high development of the different root parts of the plant that allow contact with soil to access a greater amount of nutrients present for the expansion of root surface area, eventually adding to the whole plant growth system (Aeron et al. 2011). Approximately 80% of the rhizospheric microbial flora produce IAA; therefore, when we screened and applied IAA producing PGPR in the field to improve the endogenous IAA levels in plants, it consequently had a significant result on the root system by increasing the size, weight, and number of branches in plants (Goswami et al. 2016). Therefore, these changes guide to boost the capability to explore the soil for nutrient replacements, to improve plant's nutrition puddle and growth ability (Ramos-Solano et al. 2008). In this chapter, we reported that a number of different PGPRs from different species isolated from soil and rhizosphere produce IAA for supporting plant growth and progress (Table 4.1). Another hormone is also regulated by auxin for synthesis, i.e., strigolactones (Al-Babili and Bouwmeester 2015).

4.2.2.2 Gibberellins

Gibberellin (GA) is a large group of hormones that represents as many as 136 diverse structured molecules (Goswami et al. 2016). The developmental processes in higher plants, including the germination of seeds, stem elongation, flowering and fruit development, and leaf and stem growth are prejudiced by gibberellins (Hedden and Phillips 2000; Bottini et al. 2004), and GA governs the majority of physiological consequences, i.e., shoot elongation (Spaepen and Vanderleyden 2011). About 128 plant species from a total of 136 GAs are known to date, 7 fungal species form 28 GAs, and 7 bacterial species from only 4 GAs (GA1, GA3, GA4, and GA20) have been recognized (MacMillan 2001). This class of phytohormones is a common structure of skeleton for 19–20 carbon atoms, and therefore these hormones are capable of translocating from the roots to the aerial parts of the plant (Goswami et al. 2016).

4.2.2.3 Cytokinin

To control the plant growth from the cellular to tissue level and from organ to the whole plant, the cytokinin hormone plays a major role for activating cell separation (Letham and Palni 1983; Francis and Sorrell 2001; Sakakibara 2006). It occurs in two forms, i.e., free and tRNA-bound, and it is crucial for plants in the regulation of various physiological developments (Letham and Palni 1983; Stirk and Van Staden 2010). Cytokinins excite the plant cell division, vascular cambium, differentiation and encourage root hairs formation, reduce lateral root formation, and primary root elongation (Aloni et al. 2006; Riefler et al. 2006). During the plant growing progressions, such as the construction of embryo vasculature, nutritional signaling, leaf expansion, branching, chlorophyll production, and root growth, promotion of seed germination and delay of senescence are also very much influenced by cytokinins

(Letham and Palni 1983; Howell et al. 2003). The initiation and extension of axillary buds, out from shoot apical, were described and well connected with the levels of cytokinins (Bangerth et al. 2000; Shimizu-Sato and Mori 2001; Yong et al. 2014). The plants with reduction of endogenous cytokinins have dissimilar morphological and evolved changes, such as shorter shoot internodes, delayed flowering, fewer flowers, less leaf surface area, smaller shoot apical meristems, improved root growth, and a longer root meristem (Schmulling 2002). The concentrations of phytohormones are directly regulated by the rates of biosynthesis, metabolism, inactivation, and degradation with homeostasis below the control of both internal and external factors (Sakakibara 2006). It also occurs in the form of free base riboside or ribotide, where riboside is a biologically active form and riboside is a form of transportation via the xylem system (Wang et al. 2017). The riboside form of cytokinins is afterward transformed into the energetic form by a different enzyme at the shoot (Sakakibara 2006). Today, several cytokinins producing microbes associated with plant rhizosphere and soil have been well characterized and identified.

4.2.2.4 Ethylene

Ethylene (C_2H_4) is a small and simple gaseous molecule, typically linked with the ripening and senescence measures in the plant (De Martinis et al. 2015). It is another plant hormone known to control numerous processes, such as the ripening of fruits and abscission of leaves (Reid 1981). A high level of ethylene induces defoliation and the cellular processes that direct stem and root growth inhibition (Li and Glick 2005), and under stress conditions, they can stop certain progress, such as elongation, nitrogen fixation, etc. Many PGPRs help to facilitate the growth and development of a plant by reducing stress responses within plants by decreasing the concentration of ethylene levels. Ethylene phytohormone regulates metabolism on the molecular, cellular, and even whole plant level in various roles (Schaller 2012; Khan and Khan 2014). It is also mediated by plant floral organ abscission, which is one of the vital processes for plant progress (Khan and Khan 2014), and is prejudiced by the act of plants below the ideal and demanding environments by intermingling through other signaling molecules (Muller and Munne-Bosch 2015; Thao et al. 2015). The concentration in the cell and sensitivity of plant hormone depend on the activity of ethylene (Arraes et al. 2015; Sun et al. 2016).

4.2.2.5 Abscisic Acid

Abscisic acid (ABA) is generally known as a "stress hormone", which responds to different environmental stresses, i.e., both biotic and abiotic (Zhang 2014). It plays an essential function for a diverse array of stresses, such as heavy metals, drought, salinity, temperature, radiation, thermal or heat stress, etc., and it also improves a diversity of processes, such as seed germination, seed dormancy, and closing of stomata (Vishwakarma et al. 2017). ABA performs a number of functions at the cell stage, such as regulating the enzyme production essential for cell defense from lack of moisture (Li et al. 2014), and it is constant in high temperatures (Zhang 2014). It also controls a variety of exterior and interior parameters for the growth and progress of plants. In plants, ABA is considered to be a possible applicant for the messaging among the roots and shoots system through stress by some other plant consequent signals and also cooperates with other plant signals disturbed by organ-to-organ communication (Vishwakarma et al. 2017). A direct application of PGPR microbes is recognized as a bio-protector that may improve the plant growth hormone and crop fortification.

4.3 Siderophore Production

Siderophores are the metal-chelating agents whose main purpose is to imprison the unsolvable ferric iron from diverse habitats (Nagoba and Vedpathak 2011), and research in this field initiated around five years ago. The livelihood of all micro-organisms requires Iron (Fe) for growth and expansion because it is involved in numerous metabolic processes, such as electron transport chain, oxidative phosphorylation, tricarboxylic acid cycle, oxygen metabolism, DNA and RNA syntheses, and photosynthesis (Fardeau et al. 2011; Aguado-Santacruz et al. 2012), and several biosyntheses, such as antibiotics, vitamins, porphyrins, toxins, cytochromes, siderophores, pigments, aromatic compounds, and nucleic acid synthesis are required (Messenger and Ratledge 1985). Recently, Fe was observed to perform a crucial function in the microbial biofilm formation and it regulates the surface motility and stabilizes the polysaccharide matrix of the micro-organism (Weinberg 2004; Glick et al. 2010; Chhibber et al. 2013). To live in a low availability or an iron-depleted environment, micro-organisms make certain organic compounds with low molecular masses called siderophores (Ahmed and Holmstrom 2014). According to Schwyn and Neilands (1987), in Fe-limiting conditions, micro-organisms and plants produce siderophore metal chelating agents with low molecular weight (200–2000 Da), and it is not only causal in plant and microbe's nourishment but also in accumulation of other environmental functions. Available literature shows both gram (−) ve and (+) ve bacteria are synthesized in siderophores for the iron-deprived situation (Tian et al. 2009; Saharan and Nehra 2011). In general, the majority of aerobic and facultative anaerobic bacteria were established to produce siderophores in an iron anxiety situation (Neilands 1995). Depending on the functional characteristics, siderophores are classified into three main categories, i.e., hydroxamates, catecholates, and carboxylates, additionally 500 different types of siderophores are identified, and 270 have been structurally characterized (Boukhalfa et al. 2003).

4.3.1 Biotechnological Applications of Siderophores

Siderophore is a small biological organic particle produced by a variety of micro-organisms having a large application in a diversity of fields, such as agriculture, microbial ecology, and medicine (Saha et al. 2015; Ahmed and Holmstrom 2014):

- *Agriculture*
  Enhances growth and pathogen biocontrol of plants
  Improves soil fertility
- *Microbial Ecology and Taxonomy*
  Enhances the growth of unculturable micro-organisms in artificial media
  Alters the microbial community
  Enhances bio-remediation of heavy metals
- *Siderophore as Medicine*
  Trojan horse antibiotics
  Iron overload therapy
  Removal of transuranic elements
  Antimalarial activity
  Cancer therapy
- *Bioremediation of Environmental Pollutants*
  Petroleum hydrocarbons
  Nuclear fuel reprocessing
  Bio-bleaching of pulps
- *Siderophore as an Optical Biosensor*

4.4 Nitrogen Fixation

Nitrogen (N) is the most biologically essential element in the environment and is a fundamental macro-nutrient for plant development, after water. It is also an important constituent of proteins, nucleic acids, and an essential nutrient for all organisms and reinforced for all forms of life (Dobereiner 1997). Micro-organisms that perform biological nitrogen fixation have countless important characters in both the soil and plants. Plant growth is directly dependent upon a satisfactory amount of fixed nitrogen, and nitrogen available in the soil is absorbed by roots in the form of ammonium and nitrates. Only prokaryote organisms have the capability to use atmospheric nitrogen through a process known as biological nitrogen fixation (BNF), which changes the atmospheric nitrogen to ammonia, and this form can be used by different plants (Lam et al. 1996; Franche et al. 2009). These microbes not only supply nitrogen to the plant for growth and progress but also to some other PGPR characters as an excellent bio-fertilizer agent used in agriculture; they include N fixation, phytohormone production (auxins, cytokinins, and gibberellins), improved nutrient uptake, stress resistance, P-solubilization, siderophores, IAA production, exopolysaccharides, ACC (1-Aminocyclopropane-1-Carboxylate acid) deaminase activity, bioremediation, biofertilizers, biopesticides, and biocontrol (Singh et al. 2017a, b; Li et al. 2017). Diazotrophs bacterium are found in an extensive diversity of environments: free-living in the soil, water, associative symbioses with grasses, termite guts, actinorhizal relationship with woody plants, cyanobacterial symbioses with plants, and root-nodule symbiosis with legumes (Dixon and Kahn 2004). Diazotrophs are responsible for BNF of the complex nitrogenase

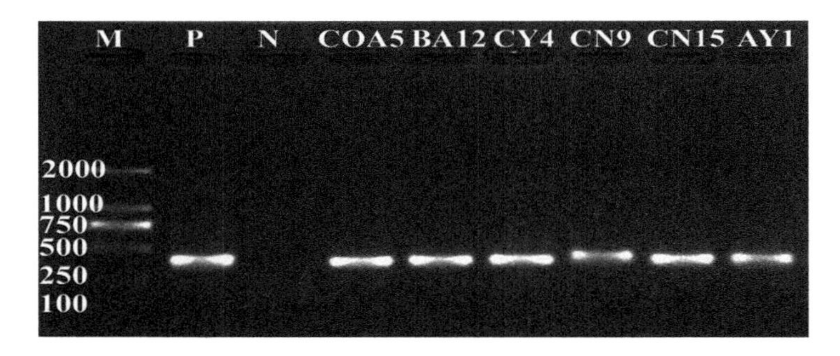

Fig. 4.4 PCR-amplification of *nifH* gene from genomic DNA of different *Pseudomonas* species. M, molecular size marker from 100 to 2000 bp. P is a positive control and N is negative

enzyme, and this enzyme is composed of three subunits that are controlled by a multifaceted system with multiple genes (Figueiredo et al. 2013). Nitrogenase I is dependent on iron and molybdenum and is encoded with the nif gene, nitrogenase II is dependent on vanadium and is encoded in the vnf gene, and nitrogenase III is dependent on iron and is encoded in the anf gene (Franche et al. 2009; Canfield et al. 2010). Out of these, nif genes are accountable for encoding highly conserved subunits in free-living and associative microorganisms (Zehr et al. 2003; Franche et al. 2009), and the nif gene has been used to describe the genomic diversity of diazotrophs (Zehr et al. 2003). The nif genes comprise nifD, nifH, and nifK, which completely encode proteins of the nitrogenase enzyme complex (Figueiredo et al. 2013). nifH is a functional gene, which is encoded with the Fe-protein of nitrogenase, and it is well conserved and studied as compared to other nif genes, which have been used for phylogenetic investigation of the diazotrophic bacterial community (Zehr et al. 2003; Franche et al. 2009). Li et al. (2017) also studied and amplified the nifH gene and PCR products of accurate size from the total extracted genomic DNA of strains, and positive nifH gene amplification produces an amplified fragment of about 360 bp (Fig. 4.4).

4.4.1 Benefits of Using Biological Nitrogen Fixation (BNF)

- *Reduced Production Cost*: Farmers apply higher doses of fertilizers, so applying BNF then reduces the production costs.
- *Solve Environment Problem*: The use of BNF inoculants as substitutes for chemical fertilizers reduces the environmental contamination.
- *Effectiveness*: BNF-fertilizer is more effective than chemical fertilizers for the long term.
- *Improved Yields*: BNF inoculants regularly increase the crop yields and protection in any crops.
- *Maintained Soil Fertility and Structure*: Nitrogen-fixing micro-organisms play a significant role in both soil and plants and also maintain the soil fertility with progress for plants.

- *Sustainability*: BNF is part of the integrated pest management of sustainable agricultural systems.

4.5 Mechanisms of Indirect PGPR

4.5.1 Antibiotics

Antibiotics are a varied group of organic compounds with low molecular weight (Sahu et al. 2018), and a number of PGPR microbes produce secondary metabolites, which are very harmful or have toxic effects that leads to growth inhibition of the selected phytopathogenic micro-organisms. Antibiotics production is one of the most powerful tools to study and understand the mechanisms of biocontrol for selected PGPR against several plant pathogenic fungi (Shilev 2013). According to Glick et al. (2007), antibiotics production is a general biocontrol mechanism of PGPR, which acts as an antagonistic action against several phytopathogens. PGPR micro-organisms, such as *Bacillus*, *Pseudomonas*, actinomycetes, and fungal species, play a major role in inhibiting the phytopathogenic microbes by producing one or more antibiotics. The antagonistic PGPR contrive the destruction of plant phytopathogens by the excretion of extracellular secondary metabolites that inhibit even at very low amounts (Goswami et al. 2016). The primary six main categories, phloroglucinol, phenazines, pyrrolnitrin, pyoluteorin, cyclic lipopeptides, and hydrogen cyanide, are antibiotic compounds associated with biocontrol for the inhibition of root diseases (Haas and Defago 2005). Recently, *Pseudomonas* and *Bacillus* species produced lipopeptide biosurfactants that have shown indirect biocontrol potential because of their positive results on modest interactions with bacteria, fungi, oomycetes, protozoa, nematodes, and plants organisms (de Bruijn et al. 2007; Raaijmakers et al. 2010). Several antibiotics were isolated from the different fungal and bacterial strains, and this variety comprises mechanisms that prevent the synthesis of pathogen cell walls, membrane structures of cells, and also inhibit the development of the small ribosome subunit of beginning complexes (Maksimov et al. 2011). PGPR is produced by antibiotics, such as DAPG, phenazine-1-carboxylic acid, phenazine-1-carboxamide, pyoluteorin, pyrrolnitrin, oomycinA, viscosinamide, butyrolactones, kanosamine, zwittermicin A, aerugine, rhamnolipids, cepaciamide, ecomycins, pseudomonic acid, azomycin, cepafungins, and karalicin (Fernando et al. 2005), and these antibiotics are known to contain antiviral, antimicrobial, antihelminthic, phytotoxic, antioxidant, cytotoxic, antitumor, and PGP activities (Kim 2012).

4.5.2 Hydrogen Cyanide (HCN) Production

A number of PGPR have biocontrol activity, but some mechanisms can act in a different way and synthesize a volatile organic compound that is identified as hydrogen cyanide (HCN). In most of the cases reported of PGPR strains producing HCN, it is used as a biocontrol mechanism, and if only a lower amount of HCN is produced it would not be active mainly in defense and spread to most of the fungal pathogens (Olanrewaju et al. 2017). The role of disease suppression through HCN production is a long known study (Keel et al. 1989). A secondary metabolite commonly produces a hydrogen cyanide gas through PGPR strains, and this gas is known to harmfully affect the root metabolism and growth (Schippers et al. 1990). The production rate of HCN by PGPR microbes can also differ depending upon the crop types, possibly owing to change in the amino acid composition of root exudates (Sahu et al. 2018), and it has been exposed to have an advantageous result on the plants (Voisard et al. 1989). A maximum cyanogenesis occurs in the transition from exponential to stationary phase (Askeland and Morrison 1983), and it is also influenced by numerous environmental aspects, such as iron, phosphate, and oxygen concentrations (Knowles and Bunch 1986). HCN production and disease control are both very important for iron sufficiency (Keel et al. 1989; Voisard et al. 1989); therefore, HCN production is closely related to siderophore metabolism and genes (Blumera and Haas 2000). HCN toxicity is affected in its capacity to prevent cytochrome c oxidase along with other significant metallo-enzymes (Nandi et al. 2017). Several PGPR bacterial genera, such as *Rhizobium, Pseudomonas, Alcaligenes, Bacillus,* and *Aeromonas,* have revealed HCN production (Li et al. 2017; Singh et al. 2013; Ahmad et al. 2008; Das et al. 2017). The suppression of chickpea pathogens (*Fusarium oxysporum f.* sp. *ciceri* race 1, *F. solani* and *Macrophomina phaseolina*) causing wilt and root rot disease has been attributed to HCN (Singh et al. 2013).

4.5.3 Hydrolytic Enzymes

Hydrolytic enzymes are one of the key mechanisms used for biocontrol agents to control various phytopathogens (Kobayashi et al. 2002; Shaikh and Sayyed 2015). Hydrolytic enzymes, such as chitinases, cellulase, glucanases, proteases, phosphatase, dehydrogenase, lipases, etc., are involved in the lysis of the pathogen cell wall (Neeraja et al. 2010), and several micro-organisms produced various polymeric compounds, such as cellulose, hemicellulose, chitin, and protein, which can be hydrolyzed by the lytic enzymes and inhibit the growth and activities of pathogens by secreting lytic enzymes (Tariq et al. 2017). Plant growth promoting rhizobacteria produce various hydrolytic enzymes, which are used as biocontrol agents that can directly suppress the growth and activities of phytopathogens via lysis of cell wall, thus improving the plant growth and progress. First, biocontrol microbes secrete the extra cellular enzymes for the opening of high molecular weight substrates, such as cellulose, chitin, pectin, and lignin, to mineralize organic compounds to minerals (N, P, S) and other elements (Mankau 1962). Hydrolytic enzymes are proficient at the breaking down of glycosidic bonds in chitin, and therefore they perform a dynamic function in the biological control action of several plant diseases by degrading the cell wall of phytopathogens (Jadhav et al. 2017). Hydrolytic enzymes, such as chitinase, glucanase, protease, and cellulase are of foremost curiosity owing to their capability to damage and lysis the fungal cell wall; therefore, hydrolytic enzymes are engaged as biocontrol agents of various fungal pathogens (Mabood et al. 2014). It is one of the important processes for environment-friendly control of various soil-borne pathogens (Aeron et al. 2011). The structural integrity of the cell wall for targeted pathogens is affected by these hydrolytic enzymes (Budi et al. 2000), and they have important potential for inhibiting the phytopathogens in biocontrol methods (Mabood et al. 2014). Singh et al. (2012) studied the optimization of media components for chitinase production and antifungal activities of a potent biocontrol strain of *Lysinibacillus fusiformis* B-CM18 against the soil-borne pathogens *Fusarium oxysporum f.* sp. *ciceri, F. solani, F. oxysporum f.* sp. *Lycopersici*, and *Macrophomina phaseolina.*

4.5.4 Induced Systemic Resistance (ISR)

Several nonpathogenic rhizobacteria have been exposed to contain diseases during plant defense mechanisms by stimulating an inducible that renders the host further resistant to phytopathogen entrance, a phenomenon termed induced systemic resistance (ISR) (Ongena and Thonart 2006; Van Loon and Bakker 2006). Many PGPR induce the systemic resistance in various plants against numerous environmental stressors (Prathap and Ranjitha, 2015), and an ISR in plants is very complicated, which has been moderately elucidated in some plant model systems, i.e., Arabidopsis (Devendra et al. 2007). According to Van Loon et al. (1998), induced resistance is a condition of an improved self-protective capacity developed by plants after properly stimulated. Previously, ISR was described in plant carnation to be systemically protected by strain *P. fluorescens* (WCS417r) against *F. oxysporum f.* sp. *Dianthi* (Van Peer et al. 1991) and in cucumber plants, by Wei et al. (1991). Various diseases caused by fungal, bacterial, viral, insects, and nematodes are able to be reduced by the application of PGPR (Naznin et al. 2012); besides, ISR involves the signaling of jasmonate and ethylene inside the plant, and these phytohormones stimulate the plant's host defense responses against a diversity of plant pathogens (Glick 2012). Several individual PGPR bacterial workings stimulate ISR, such as lipopolysaccharides (LPS), flagella, siderophores, cyclic lipopeptides, 2, 4- di-acetylphloroglucinol, homoserine lactones, and volatiles, such as acetoin and 2, 3-butanediol (Doornbos et al. 2012; Berendsen et al. 2015). A plant defense system is activated by the vascular system through pathogenic attack, and a signal is produced that results in the activation of an enormous quantity of defense enzymes, such as chitinase, ß-1, 3-glucanase, phenylalanine ammonia lyase, polyphenol oxidase, peroxidase, lipoxygenase, SOD, CAT, and APX along with some proteinase inhibitors (Gouda et al. 2018). ISR helps to control several plant diseases; nevertheless, it is not precise against a particular pathogen (Kamal et al. 2014). The method of ISR in disease suppression given by bacterium confirmed on a specific pathosystem is found through proving the spatial separation of the pathogen and the resistance-inducing agent to ignore any direct antagonistic interaction (Ongena and Jacques 2008).

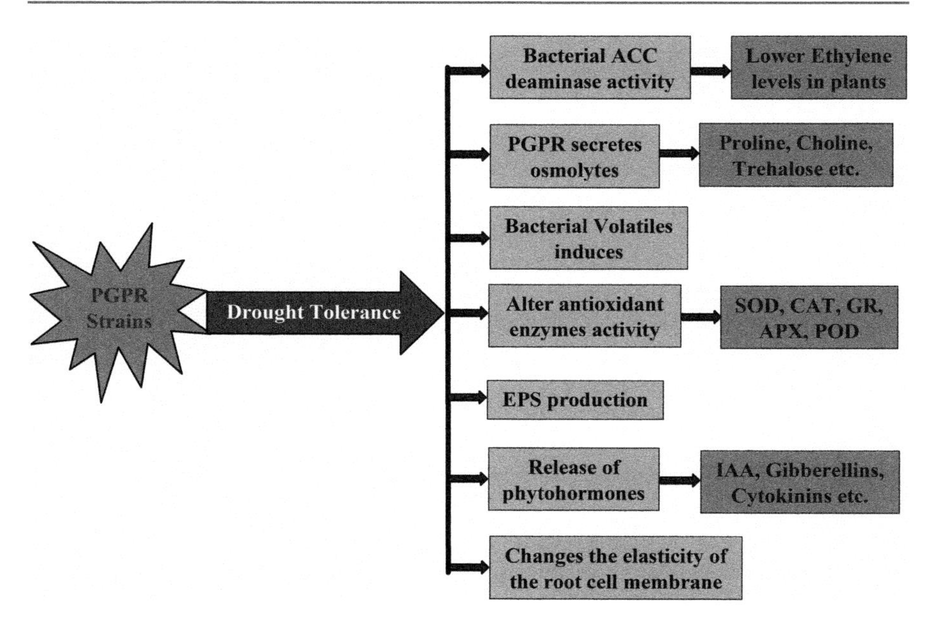

Fig. 4.5 Mechanism of drought tolerance induced by plant growth promoting rhizobacteria in the plant

Figure 4.5 shows systematized ISR can be universally observed as a three-step procedure, and it also decreases the destruction from phytopathogens, which are vigorously completed from foliage, flowers or fruits, and thus is effective against root-infecting pathogens (Ongena and Jacques 2008).

4.6 Mechanisms of PGPR Facilitated Drought Stress Tolerance

Anything that causes a negative consequence on plant growth and progress is known as a stress (Foyer et al. 2016). The beneficial functions of PGPR micro-organisms in plant growth and progress, biocontrol activity, nutrient management in addition to various direct and indirect mechanisms are well known, but the function of these microbes in the management of various stresses both biotic and abiotic is very important to know. Plants are normally subjected to diverse environmental stresses and they have to develop exact response mechanisms (Ramegowda and Senthil-Kumarb 2015).Various studies are available to understand the molecular mechanisms concerned in both biotic and abiotic stress tolerance management (Pontigo et al. 2017; Singh et al. 2017a, b). This present chapter focused on the role of PGPR in helping plants to survive drought stress, and induced PGPR bacteria promising clarification for the management of plant drought tolerance mechanisms contain: (1) phytohormones production, such as abscisic acid, gibberellic acid, cytokinin, and indole-3-acetic acid, (2) ACC deaminase enzyme to reduce the ethylene level in roots, (3) induced systemic tolerance by bacterial compounds, and (4) bacterial exopolysaccharides (Yang et al. 2009; Kim et al. 2013; Timmusk et al. 2014) (Fig. 4.5). Drought tolerance is probably likely to cause various serious problems for plant growth and progress throughout 50% of the arable lands by 2050 (Ashraf 1994; Vinocur and Altman 2005; Kasim et al. 2013). Farooq et al. (2009) also mentioned the special effects of drought stress on plant systems:

- Crop growth, progress, and yield
- Water relatives
- Nutrient relatives
- Photosynthesis
  Stomatal oscillations
  Photosynthetic enzymes
  Adenosine triphosphate synthesis
- Assimilate segregating
- Respiration
- Oxidative impairment

According to Vurukonda et al. (2016), numerous mechanisms have been reported for PGPR mediated drought stress tolerance in plants:

- Variation of phytohormonal activity in imparting drought tolerance in plants
- Role of volatile compounds in inducing drought tolerance
- PGPR change root morphology under drought stress
- Role of ACC deaminase activity produced by rhizobacteria in drought stress tolerance
- Osmolytes in informing drought tolerance in plants
- Exopolysaccharide (EPS) production by PGPR and improvement of drought stress
- Changing antioxidant enzyme in the defense system
- Molecular studies in the improvement of drought stress by PGPR
- Co-inoculation of PGPR for improvement of drought stress

4.7 1-Aminocyclopropane-1-Carboxylic Acid Deaminase

The enzyme 1-aminocyclopropane-1-carboxylic acid (ACC) deaminase was first characterized by Honma and Shimomura (1978), and later it was exposed to be closely involved in the rise of plant growth by plant growth promoting bacteria (Glick et al. 1998). A number of plant growth promoting bacteria contain enzyme ACC deaminase, and this enzyme can cleave ACC, which is an instantaneous precursor of ethylene in crops, to a-ketobutyrate and ammonia; therefore, a lower level of ethylene is produced (Glick 1995, 2005, Glick et al. 1998), and the role of ACC deaminase in plant growth-promoting rhizobacteria was suggested by Glick et al.

(1998). A procedure was developed by Glick on a minimal media containing ACC as a nitrogen source and isolated pseudomonas, which helps as a plant growth promotion activity (Glick 1995). ACC deaminase (EC: 4.1.99.4) was noticed in *Penicillium citrinum* (Honma 1993), and, in 1978, *Pseudomonas* sp. strain ACP from yeast, *Hansenula saturnus* (Honma and Shimomura 1978). A number of bacterial strains containing ACC deaminase have been isolated in laboratory conditions worldwide, and PGPR activity has been significantly reported. The ACC deaminase enzyme exists in the cytoplasm of bacteria species at a very low level until it is encouraged by ACC, and this enzymatic activity is a comparatively slow process (Jacobson et al. 1994). Furthermore, regarding the mechanisms of plant growth promotion producing microbes, many PGPR bacteria have excited plant growth through the ACC deaminase enzyme that promotes plant growth activity by reducing plant ethylene levels. Bacterial species that produce the ACC deaminase enzyme can inoculate plants for resistance to the injurious effect of ethylene synthesized as an indicator of stress conditions, thus producing tolerance to diverse environmental conditions and harmful pathogens, i.e., abiotic and biotic stresses (Zahir et al. 2009).

4.8 Conclusion

Worldwide, the majority of countries and their economies depend on the agriculture system. Therefore, the interaction between different PGPR and crops is a significant move toward improving food production for the increasing global population, in addition to maintaining the quality of soil, plant tolerance, crop productivity, lowering the chemical fertilizers costs that actively contribute to unstable climate conditions, water supply, etc., in the accessible conditions of worldwide variation. Application of PGPR strains with modern tools and techniques can meet all the above criteria and serve as a satisfactory explanation in sustainable agriculture. Plant diseases by various pathogens plus unstable environmental conditions are the main problem for yield losses; therefore, the use of chosen PGPR strains to protect the crops along with increasing the agricultural products over the past few decades were reported by many researchers and scientists. However, accurate information regarding how indigenous PGPR accomplishes the benefits for different plant rhizospheric is not fully understood in the environment and must be improved. To select an appropriate PGPR soil rhizosphere microbe as a biofertilizers or biocontrol agent, a multi-disciplinary exploration with the endeavor of combining applications in biotechnology, microbiology, nanotechnology, and chemical engineering collectively offering novel formulations technology with huge possible opportunities must be explored for future sustainable agricultural developments.

References

Aeron A, Kumar S, Pandey P et al (2011) Emerging role of plant growth promoting rhizobacteria in agrobiology. In: Maheshwari DK (ed) Bacteria in agrobiology: crop ecosystems. Springer, Berlin/Heidelberg, pp 1–36. https://doi.org/10.1007/978-3-642-18357-7-1

Aguado-Santacruz GAA, Moreno-Gómez BA, Jiménez-Francisco BB et al (2012) Impact of the microbial siderophores and phyto siderophores on the iron assimilation by plants: a synthesis. Rev Fitotec Mex 35:9–21

Ahemad M, Khan MS (2010) Influence of selective herbicides on plant growth promoting traits of phosphate solubilizing *Enterobacter asburiae* strain PS2. Res J Microbiol 5:849–857

Ahemad M, Khan MS (2011) Assessment of plant growth promoting activities of rhizobacterium *Pseudomonas putida* under insecticide-stress. Microbiol J 1:54–64

Ahmad F, Ahmad I, Khan MS et al (2008) Screening of free-living rhizospheric bacteria for their multiple plant growth promoting activities. Microbiol Res 163:173–181

Ahmed E, Holmstrom SJM (2014) Siderophores in environmental research: roles and applications. Microb Biotechnol 7:196–208

Al-Babili S, Bouwmeester HJ (2015) Strigolactones, a novel carotenoid-derived plant hormone. Annu Rev Plant Biol 66:161–186

Allen LE (1957) Experiments in soil bacteriology. Burgess, Minneapolis

Aloni R, Aloni E, Langhans M et al (2006) Role of cytokinin and auxin in shaping root architecture: regulating vascular differentiation, lateral root initiation, root apical dominance and root gravitropism. Ann Bot 97:883–893

Alori ET, Glick BR, Babalola OO et al (2017) Microbial phosphorus solubilization and its potential for use in sustainable agriculture. Front Microbiol 8:971. https://doi.org/10.3389/fmicb.2017.00971

Arraes FBM, Beneventi MA, de Sa MEL et al (2015) Implications of ethylene biosynthesis and signaling in soybean drought stress tolerance. BMC Plant Biol 15:213. https://doi.org/10.1186/s12870-015-0597-z

Ashraf M (1994) Breeding for salinity tolerance in plants. Crit Rev Plant Sci 13:17–42

Askeland RA, Morrison SM (1983) Cyanide production by *Pseudomonas fluorescens* and *Pseudomonas aeruginosa*. Appl Environ Microbiol 45:1802–1807

Bangerth F, Li CJ, Gruber J et al (2000) Mutual interaction of auxin and cytokinins in regulating correlative dominance. Plant Growth Regul 32:205–217

Berendsen RL, Pieterse CMJ, Bakker PAHM et al (2012) The rhizosphere microbiome and plant health. Trends Plant Sci 17:478–486

Berendsen RL, Verk MCV, Stringlis IA et al (2015) Unearthing the genomes of plant-beneficial *Pseudomonas* model strains WCS358, WCS374 and WCS417. BMC Genomics 16:539

Berg G (2009) Plant–microbe interactions promoting plant growth and health: perspectives for controlled use of microorganisms in agriculture. Appl Microbiol Biotechnol 84:11–18

Blumera C, Haas D (2000) Iron regulation of the *hcn ABC* genes encoding hydrogen cyanide synthase depends on the anaerobic regulator ANR rather than on the global activator GacA in *Pseudomonas fluorescens* CHA0. Microbiology 146(10):2417–2424

Bottini R, Cassan F, Piccoli P (2004) Gibberellin production by bacteria and its involvement in plant growth promotion and yield increase. Appl Microbiol Biotechnol 65:497–503

Boukhalfa H, Lack J, Reilly SD et al (2003) Siderophore production and facilitated uptake of iron and plutonium in *P. putida*. AIP Conf Proc 673:343–344

Bronick CJ, Lal R (2005) Soil structure and management: a review. Geoderma 124:3–22

de Bruijn DI, de Kock MJD, Yang M et al (2007) Genome-based discovery, structure prediction and functional analysis of cyclic lipopeptide antibiotics in *Pseudomonas* species. Mol Microbiol 63:417–428

Budi SW, Van Tuinen D, Arnould C et al (2000) Hydrolytic enzyme activity of *Paenibacillus* sp. strain B2 and effects of the antagonistic bacterium on cell integrity of two soil borne pathogenic fungi. Appl Soil Ecol 15:191–199

Canfield D, Glazer AN, Falkowski PD (2010) The evolution and future of earth's nitrogen cycle. Science 330:192–196

Chester CGC (1948) A contribution to the study of fungi in the soil. Trans Brit Mycol Soc 30:100–117

Chhibber S, Nag D, Bansal S et al (2013) Inhibiting biofilm formation by *Klebsiella pneumoniae* B5055 using an iron antagonizing molecule and a bacteriophage. BMC Microbiol 13:174–183

Cholodny N (1930) Uber eine neue methode zur untersuchung der Boden microflora. Arch Microbiol 1:620–652

Conn HJ (1981) The microscopic study of bacteria and fungi in the soil. NY Agric Exp St Tech Bull 64:3–20

Damam M, Kaloori K, Gaddam B et al (2016) Plant growth promoting substances (phytohormones) produced by rhizobacterial strains isolated from the rhizosphere of medicinal plants. Int J Pharm Sci Rev 37:130–136

Das K, Prasanna R, Saxena AK et al (2017) Rhizobia: a potential biocontrol agent for soilborne fungal pathogens. Folia Microbiol. https://doi.org/10.1007/s12223-017-0513-z

De Martinis D, Tomotsugu K, Chang C et al (2015) Ethylene is all around. Front Plant Sci 6:76. https://doi.org/10.3389/fpls.2015.00076

Devendra KC, Prakash A, Johri BN et al (2007) Induced systemic resistance (ISR) in plants: mechanism of action. Indian J Microbiol 47:289–297

Devi KA, Pandey G, Rawat AKS et al (2017) The endophytic symbiont—*Pseudomonas aeruginosa* stimulates the antioxidant activity and growth of *Achyranthes aspera* L. Front Microbiol 8:1897. https://doi.org/10.3389/fmicb.2017.01897

Dixon R, Kahn D (2004) Genetic regulation of biological nitrogen fixation. Nat Rev Microbiol 2:621–631

Döbereiner J (1997) Importância da fi xaçãobiológica de nitrogênio para a agricultura sustentável. Biotecnol Ciênc Desenvolv 1:2–3. Encarte Especial

Doornbos RF, van Loon LC, Peter AHM et al (2012) Impact of root exudates and plant defense signaling on bacterial communities in the rhizosphere. Rev Sustain Dev 32:227–243

El-Sayed WS, Akhkha A, El-Naggar MY, Elbadry M (2014) In vitro antagonistic activity, plant growth promoting traits and phylogenetic affiliation of rhizobacteria associated with wild plants grown in arid soil. Front Microbiol 5:651. https://doi.org/10.3389/fmicb.2014.00651

Fardeau S, Mullie C, Dassonville-Klimpt A et al (2011) Bacterial iron uptake: a promising solution against multidrug resistant bacteria. In: Méndez-Vilas A (ed) Science against microbial pathogens: communicating current research and technological advances. Formatex, Badajoz, pp 695–705

Farooq M, Wahid A, Kobayashi N et al (2009) Plant drought stress: effects, mechanisms and management. Agron Sustain Dev 29:185–212

Fernando DWG, Nakkeeran S, Zhang Y et al (2005) Biosynthesis of antibiotics by PGPR and its relation in biocontrol of plant diseases. In: Siddiqui ZA (ed) PGPR: biocontrol and biofertilization. Springer, Dordrecht, pp 67–109

Figueiredo MVB, Mergulhão ACES, Sobral JK et al (2013) Biological nitrogen fixation: importance, associated diversity, and estimates. In: Arora NK (ed) Plant microbe Symbiosis: fundamentals and advances. © Springer, India. https://doi.org/10.1007/978-81-322-1287-4_10

Foyer CH, Rasool B, Davey JW et al (2016) Cross-tolerance to biotic and abiotic stresses in plants: a focus on resistance to aphid infestation. J Exp Bot 7:2025–2037

Franche C, Lindstrom K, Elmerich C et al (2009) Nitrogen-fixing bacteria associated with leguminous and non-leguminous plants. Plant Soil 321:35–59

Francis D, Sorrell DA (2001) The interface between the cell cycle and plant growth regulators: a mini review. Plant Growth Regul 33:1–12

Glick BR (1995) The enhancement of plant growth by free-living bacteria. Can J Microbiol 41:109–117

Glick BR (2005) Modulation of plant ethylene levels by the bacterial enzyme ACC deaminase. FEMS Microbiol Lett 251:1–7

Glick BR (2012) Plant growth-promoting bacteria: mechanisms and applications. Scientifca 2012:963401. https://doi.org/10.6064/2012/963401
Glick BR, Penrose DM, Li J et al (1998) A model for the lowering of plant ethylene concentrations by plant growth-promoting bacteria. JTB 190:63–68
Glick BR, Todorovic B, Czarny J et al (2007) Promotion of plant growth by bacterial ACC deaminase. Crit Rev Plant Sci 26:227–242
Glick R, Gilmour C, Tremblay J et al (2010) Increase in rhamnolipid synthesis under iron-limiting conditions influences surface motility and biofilm formation in *Pseudomonas aeruginosa*. J Bacteriol 192(12):2973–2980
Goswami D, Thakker JN, Dhandhukia PC et al (2016) Portraying mechanics of plant growth promoting rhizobacteria (PGPR): a review. Cogent Food Agric 2:1127500
Gouda S, Kerry RG, Das G et al (2018) Revitalization of plant growth promoting rhizobacteria for sustainable development in agriculture. Microbiol Res 206:131–140
Haas D, Défago G (2005) Biological control of soil-borne pathogens by *fluorescent pseudomonads*. Nat Rev Microbiol 3:307–319
Harvey TV (1925) Asurvey of the watermolds and *Pythiums* occurring in the soils of Chapel Hill. J Eisha Mitchell Sci Soc 41:151–164
Hayat R, Ali S, Amara U et al (2010) Soil beneficial bacteria and their role in plant growth promotion: a review. Ann Microbiol 60:579–598
Hedden P, Phillips AL (2000) Gibberellin metabolism: new insights revealed by the genes. Trends Plant Sci 5:523–530
Hermosa R, Botella L, Alonso-Ramírez A et al (2011) Biotechnological applications of the gene transfer from the beneficial fungus *Trichoderma harzianum*spp. to plants. Plant Signal Behav 6(8):1235–1236
Honma M (1993) Stereospecific reaction of 1-aminocyclopropane-1-carboxylate deaminase. In: Pech JC, Latche A, Balague (eds) Cellular and molecular aspects of the plant hormone ethylene. Kluwer Academic Publishers, Dordrecht, pp 111–116
Honma M, Shimomura T (1978) Metabolism of 1-aminocyclopropane-1-carboxylic acid. Agri Biol Chem 42:1825–1831
Howell SH, Lall S, Che P et al (2003) Cytokinins and shoot development. Trends Plant Sci 8:453–459
Islam S, Akanda AM, Prova A et al (2016) Isolation and identification of plant growth promoting rhizobacteria from cucumber rhizosphere and their effect on plant growth promotion and disease suppression. Front Microbiol 6:1360. https://doi.org/10.3389/fmicb.2015.01360
Jacobson CB, Pasternak JJ, Glick BR et al (1994) Partial purification and characterization of 1-aminocyclopropane-1-carboxylate deaminase from the plant growth promoting rhizobacterium Pseudomonas putida GR12-2. Can J Microbiol 40:1019–1025
Jadhav HP, Shaikh SS, Sayyed RZ et al (2017) Role of hydrolytic enzymes of Rhizoflora in biocontrol of fungal Phytopathogens: an overview. In: Mehnaz S (ed) Rhizotrophs: plant growth promotion to bioremediation, microorganisms for sustainability, vol 2. Springer Nature, Singapore. https://doi.org/10.1007/978-981-10-4862-3_9
Jones PCT, Mollison JE (1948) A technique for quantitative estimation of soil microorganisms. J Gen Microbiol 2:54–69
Kamal R, Gusain YS, Kumar V (2014) Interaction and symbiosis of fungi, Actinomycetes and plant growth promoting rhizobacteria with plants: strategies for the improvement of plants health and defense system. Int J Curr Microbiol Appl Sci 3:564–585
Kasim WA, Osman ME, Omar MN et al (2013) Control of drought stress in wheat using plant growth promoting bacteria. J Plant Growth Regul 32:122–130
Katznelson H, Peterson EA, Rovatt JW et al (1962) Phosphate dissolving microorganisms on seed and in the root zone of plants. Can J Bot 40:1181–1186
Keel C, Voisard C, Berling CH et al (1989) Iron sufficiency, a prerequisite for suppression of tobacco black root rot in *Pseudomonas fluorescens* strain CHA0 under gnotobiotic conditions. Phytopathology 79:584–589

Khan MIR, Khan NA (2014) Ethylene reverses photosynthetic inhibition by nickel and zinc in mustard through changes in PS II activity, photosynthetic nitrogen use efficiency, and antioxidant metabolism. Protoplasma 251:1007–1019. https://doi.org/10.1007/s00709-014-0610-7

Khan MS, Zaidi A, Ahemad M et al (2010) Plant growth promotion by phosphate solubilizing fungi – current perspective. Arch Agron Soil Sci 56:73–98

Khan MS, Zaidi A, Ahmad E et al (2014) Mechanism of phosphate solubilization and physiological functions of phosphate-solubilizing microorganisms. In: Khan MS et al (eds) Phosphate solubilizing microorganisms. Springer, Cham. https://doi.org/10.1007/978-3-319-08216-5_2

Kim SD (2012) Colonizing ability of *Pseudomonas fluorescens*2112, among collections of 2,4-diacetylphloroglucinol-producing *Pseudomonas fluorescens* spp. in pea rhizosphere. J Microbiol Biotechnol 22:763–770

Kim YC, Glick B, Bashan Y et al (2013) Enhancement of plant drought tolerance by microbes. In: Aroca R (ed) Plant responses to drought stress. Springer, Berlin

Kishore N, Pindi PK, Ram RS (2015) Phosphate-solubilizing microorganisms: a critical review. In: Bahadur B et al (eds) Plant biology and biotechnology: volume I: plant diversity, organization, function and improvement. Springer, India. https://doi.org/10.1007/978-81-322-2286-6_12

Knowles CJ, Bunch AW (1986) Microbial cyanide metabolism. Adv Microb Physiol 27:73–111

Kobayashi DY, Reedy RM, Bick JA et al (2002) Characterization of chitinase gene from *Stenotrophomonas maltophilia* strain 34S1 and its involvement in biological control. Appl Environ Microbiol 68:1047–1054

Kumar P, Dubey RC, Maheshwari DK et al (2012) Bacillus strains isolated from rhizosphere showed plant growth promoting and antagonistic activity against phytopathogens. Microbiol Res 167:493–499

Lam HM, Coschigano KT, Oliveira IC et al (1996) The molecular-genetics of nitrogen assimilation into amino acids in higher plants. Annu Rev Plant Physiol Plant Mol Biol 47:569–593

Letham DS, Palni LMS (1983) The biosynthesis and metabolism of cytokinins. Ann Rev Plant Physiol 34:163–197

Li QSL, Glick SBR (2005) The effect of native and ACC deaminase containing Azospirillum brasilense Cdl843 on the rooting of carnation cuttings. Can J Microbiol 51:511–514

Li C, Yue J, Wu X et al (2014) An ABA-responsive DRE-binding protein gene from Setariaitalica, SiARDP, the target gene of SiAREB, plays a critical role under drought stress. J Exp Bot 65:5415–5427. https://doi.org/10.1093/jxb/eru302

Li HB, Singh RK, Singh P, Song QQ, Xing YX, Yang LT, Li YR et al (2017) Genetic diversity of nitrogen-fixing and plant growth promoting *pseudomonas* species isolated from sugarcane rhizosphere. Front Microbiol 8:1268. https://doi.org/10.3389/fmicb.2017.01268

Lin L, Li Z, Hu C, Zhang X et al (2012) Plant growth-promoting nitrogen-fixing *Enterobacteria* are in association with sugarcane plants growing in Guangxi, China. Microbes Environ 27(4):391–398

Lugtenberg B, Kamilova F (2009) Plant-growth-promoting rhizobacteria. Ann Rev Microbiol 63:541–556

Ma Y, Rajkumar M, Luo Y et al (2011) Inoculation of endophytic bacteria on host and non-host plants-effects on plant growth and Ni uptake. J Hazard Mater 195:230–237

Mabood F, Zhou X, Smith DL et al (2014) Microbial signaling and plant growth promotion. Can J Plant Sci 94:1051–1063

MacMillan J (2001) Occurrence of gibberellins in vascular plants, fungi, and bacteria. J Plant Growth Regul 20:387–442

Majeed A, Abbasi MK, Hameed S et al (2015) Isolation and characterization of plant growth-promoting rhizobacteria from wheat rhizosphere and their effect on plant growth promotion. Front Microbiol 6:198. https://doi.org/10.3389/fmicb.2015.00198

Maksimov IV, Abizgil'dina RR, Pusenkova LI et al (2011) Plant growth promoting rhizobacteria as alternative to chemical crop protectors from pathogens (review). Appl Biochem Microbiol 47:333–345

Mankau R (1962) Soil fungistasis and nematophagous fungi. Phytopathology 52:611–615

Meena RP, Jha A (2018) Conservation agriculture for climate change resilience: a microbiological perspective. In: Kashyap PL, Srivastava AK, Tiwari SP, Kumar S (eds) Microbes for climate resilient agriculture, 1st edn. Wiley. Published 2018 by Wiley
Mehnaz S, Baig DN, Lazarovits G et al (2010) Genetic and phenotypic diversity of plant growth promoting rhizobacteria isolated from sugarcane plants growing in Pakistan. J Microbiol Biotechnol 20:1614–1623
Messenger AJM, Ratledge C (1985) Siderophores. In: Young MM (ed) Comprehensive biotechnology, 3rd edn. Pergamon press, New York, pp 275–295
Muller M, Munne-Bosch S (2015) Ethylene response factors: a key regulatory hub in hormone and stress signaling. Plant Physiol 169:32–41. https://doi.org/10.1104/pp.15.00677
Nagoba B, Vedpathak D (2011) Medical applications of siderophores. Eur J Gen Med 8:229–235
Nandi M, Selin C, Brawerman G et al (2017) Hydrogen cyanide, which contributes to *Pseudomonas chlororaphis* strain PA23 biocontrol, is upregulated in the presence of glycine. Biol Control 108:47–54
Nandimath AP, Karad DD, Gupta SG et al (2017) Consortium inoculum of five thermo-tolerant phosphate solubilizing Actinomycetes for multipurpose biofertilizer preparation. Iran J Microbiol 9:295–304
Naznin HA, Kimura M, Miyazawa M et al (2012) Analysis of volatile organic compounds emitted by plant growth promoting fungus phoma sp. GS8- 3 for growth promotion effects on tobacco. Microbe Environ 28:42–49
Neeraja C, Anil K, Purushotham P et al (2010) Biotechnological approaches to develop bacterial chitinases as a bioshield against fungal diseases. Crit Rev Biotechnol 30:231–241
Neilands JB (1995) Siderophores: structure and function of microbial iron transport compounds. J Biol Chem 270:26723–26726
Olanrewaju OS, Glick BR, Babalola OO et al (2017) Mechanisms of action of plant growth promoting bacteria. World J Microbiol Biotechnol 33:197
Ongena M, Jacques P (2008) Bacillus lipopeptides: versatile weapons for plant disease biocontrol. Rev Trends Microbiol 16:115–125. https://doi.org/10.1016/j.tim.2007.12.009
Ongena M, Thonart P (2006) Resistance induced in plants by non-pathogenic microorganisms: elicitation and defense responses. In: Teixeira da Silva JA (ed) Floriculture, ornamental and plant biotechnology: advances and topical issues, vol 3. Global Science Books, London, pp 447–463
Phi QT, Yu-Mi P, Keyung-Jo S et al (2010) Assessment of root-associated *Paenibacillus polymyxa* groups on growth promotion and induced systemic resistance in pepper. J Microbiol Biotechnol 20:1605–1613
Pikovskaya RI (1948) Mobilization of phosphorus in soil in connection with vital activity of some microbial species. Microbiol 17:362–370
Pontigo S, Godoy K, Jiménez H et al (2017) Silicon-mediated alleviation of aluminum toxicity by modulation of Al/Si uptake and antioxidant performance in ryegrass plants. Front Plant Sci 8:642
Pradhan A, Pinheiro JP, Seena S et al (2014) Polyhydroxyfullerene binds cadmium ions and alleviates metal-induced oxidative stress in *Saccharomyces cerevisiae*. Appl Environ Microbiol 80(18):5874–5881. https://doi.org/10.1128/AEM.01329-14
Prathap M, Ranjitha KBD (2015) A critical review on plant growth promoting rhizobacteria. J Plant Pathol Microbiol 6:1–4
Raaijmakers JM, Paulitz TC, Steinberg C et al (2009) The rhizosphere: a playground and battlefield for soilborne pathogens and beneficial microorganisms. Plant Soil 321:341–361
Raaijmakers JM, de Bruijn I, Nybroe O et al (2010) Natural functions of lipopeptides from *Bacillus* and *Pseudomonas*: more than surfactants and antibiotics. FEMS Microbiol Rev 34:1037–1062
Ramegowda V, Senthil-Kumarb M (2015) The interactive effects of simultaneous biotic and abiotic stresses on plants: mechanistic understanding from drought and pathogen combination. J Plant Physiol 176:47–54
Ramos-Solano B, Barriuso J, Gutiérrez-Mañero FJ et al (2008) Physiological and molecular mechanisms of plant growth promoting rhizobacteria (PGPR). In: Ahmad I, Pichtel J, Hayat S (eds)

Plant–bacteria interactions: strategies and techniques to promote plant growth. Wiley VCH, Weinheim, pp 41–54. https://doi.org/10.1002/9783527621989.ch3
Rashid S, Charles TC, Glick BR et al (2012) Isolation and characterization of new plant growth promoting bacterial endophytes. Appl Soil Ecol 61:217–224
Reid MS (1981) The role of ethylene in flower senescene. Acta Hortic 261:157–169
Riefler M, Novak O, Strnad M et al (2006) Arabidopsis cytokinin receptor mutants reveal functions in shoot growth, leaf senescence, seed size, germination, root development, and cytokinin metabolism. Plant Cell 18:40–54
Rokhbakhsh-Zamin F, Sachdev D, Kazemi-Pour N et al (2011) Characterization of plant-growth-promoting traits of Acinetobacter species isolated from rhizosphere of *Pennisetum glaucum*. J Microbiol Biotechnol 21:556–566
Rossi G, Ricardo S (1927) L'seae microscopico ebacteriologicoiretto del ferrenoagrario. Nuovi Ann Minist Agric 7:457–470
Saha M, Sarkar S, Sarkar B et al (2015) Microbial siderophores and their potential applications: a review. Environ Sci Pollut Res. https://doi.org/10.1007/s11356-015-4294-0
Saharan BS, Nehra V (2011) Plant growth promoting rhizobacteria: a critical review. Life Sci Med Res 21:1–30
Sahu B, Singh J, Shankar G et al (2018) *Pseudomonas fluorescens* PGPR bacteria as well as biocontrol agent: a review. IJCS 6:01–07
Sakakibara H (2006) Cytokinins: activity, biosynthesis, and translocation. Annu Rev Plant Biol 57:431–449
Saxena J, Minaxi, Jha A et al (2014) Impact of a phosphate solubilizing bacterium and an arbuscular mycorrhizal fungus (*Glomus etunicatum*) on growth, yield and P concentration in wheat plants. Clean Soil Air Water. https://doi.org/10.1002/clen.201300492
Schaller GE (2012) Ethylene and the regulation of plant development. BMC Biol 10:9. https://doi.org/10.1186/1741-7007-10-9
Schippers B, Bakker A, Bakker P et al (1990) Beneficial and deleterious effects of HCN-producing pseudomonads on rhizosphere interactions. Plant Soil 129(1):75–83
Schmülling T (2002) New insights into the functions of cytokinins in plant development. J Plant Growth Regul 21:40–49
Schwyn B, Neilands JB (1987) Universal chemical assay for the detection and determination of siderophores. Anal Biochem 160:47–56
Shaikh SS, Sayyed RZ (2015) Role of plant growth–promoting rhizobacteria and their formulation in biocontrol of plant diseases. In: Arora NK (ed) Plant microbes Symbiosis: applied facets. Springer, New Delhi, pp 337–351
Sharma SB, Sayyed RZ, Trivedi MH, Gobi TA et al (2013) Phosphate solubilizing microbes: sustainable approach for managing phosphorus deficiency in agricultural soils. Springer Plus 2:587
Sharma S, Kulkarni J, Jha B (2016) Halotolerant rhizobacteria promote growth and enhance salinity tolerance in peanut. Front Microbiol 7:1600. https://doi.org/10.3389/fmicb.2016.01600
Sharma A, Kashyap PL, Srivastava AK et al (2018) Isolation and characterization of halotolerant *bacilli* from chickpea (*Cicer arietinum* L.) rhizosphere for plant growth promotion and biocontrol traits. Eur J Plant Pathol. https://doi.org/10.1007/s10658-018-1592-7
Shilev S (2013) Soil Rhizobacteria regulating the uptake of nutrients and undesirable elements by plants. In: Arora NK (ed) Plant microbe Symbiosis: fundamentals and advances. Springer, India, pp 147–150
Shimizu-Sato S, Mori H (2001) Control of outgrowth and dormancy in axillary buds. Plant Physiol 127:1405–1413
Singh RK, Kumar DP, Solanki MK et al (2012) Optimization of media components for chitinase production by chickpea rhizosphere associated *Lysinibacillus fusiformis* B-CM18. JBM 52:1–10
Singh RK, Kumar DP, Solanki MK et al (2013) Multifarious plant growth promoting characteristics of chickpea rhizosphere associated Bacilli help to suppress soil-borne pathogens. Plant Grow Regul 73:91–101

Singh RK, Singh P, Li HB et al (2017a) Soil–plant–microbe interactions: use of nitrogen-fixing bacteria for plant growth and development in sugarcane. In: Singh DP et al (eds) Plant-microbe interactions in agro-ecological perspectives. Springer Nature, Singapore. https://doi.org/10.1007/978-981-10-5813-4_3

Singh S, Tripathi DK, Singh S et al (2017b) Toxicity of aluminium on various levels of plant cells and organism: a review. Environ Exp Bot 137:177–193

Sofia IAP, Paula MLC (2014) Phosphate-solubilizing rhizobacteria enhance *Zea mays* growth in agricultural P-deficient soils. Ecol Eng 73:526–535

Solanki MK, Wang Z, Wang FY et al (2016) Intercropping in sugarcane cultivation influenced the soil properties and enhanced the diversity of vital diazotrophic bacteria. Sugar Tech. https://doi.org/10.1007/s12355-016-0445-y

Spaepen S, Vanderleyden J (2011) Auxin and plant-microbe interactions. Cold Spring Harb Perspect Biol 3:a001438. https://doi.org/10.1101/cshperspect.a001438

Sperberg JI (1958) The incidence of apatite-solubilizing organisms in the rhizosphere and soil. Aust J Agric Res 9:778

Stirk WA, Van Staden J (2010) Flow of cytokinins through the environment. Plant Growth Regul 62:101–116

Stover RH, Waite BH (1953) An improved method of isolating *Fusarium* sp. from plant tissues. Phytopathology 43:700–701

Sun X, Zhao T, Gan S et al (2016) Ethylene positively regulates cold tolerance in grapevine by modulating the expression of ethylene response factor 057. Sci Rep 6:24066. https://doi.org/10.1038/srep24066

Tanimoto E (2005) Regulation and root growth by plant hormones-roles for auxins and gibberellins. Crit Rev Plant Sci 24:249–265

Tariq M, Noman M, Ahmed T et al (2017) Antagonistic features displayed by plant growth promoting Rhizobacteria (PGPR): a review. J Plant Sci Phytopathol 1:038–043

Thao NP, Khan MIR, Thu NBA et al (2015) Role of ethylene and its cross talk with other signaling molecules in plant responses to heavy metal stress. Plant Physiol 169:73–84. https://doi.org/10.1104/pp.15.00663

Thornton RH (1952) The screened immersion plate. A method of isolating soil microorganisms. Research 5:190–191

Tian F, Ding Y, Zhu H et al (2009) Genetic diversity of siderophore-producing bacteria of tobacco rhizosphere. Braz J Microbiol 40:276–284

Timmusk S, Islam A, Abd El D et al (2014) Drought-tolerance of wheat improved by rhizosphere bacteria from harsh environments: enhanced biomass production and reduced emissions of stressvolatiles. PLoS One 9:1–13

Tiwari S, Singh P, Tiwari R et al (2011) Salt-tolerant rhizobacteria-mediated induced tolerance in wheat (*Triticum aestivum*) and chemical diversity in rhizosphere enhance plant growth. Biol Fertil Soils 47:907–916

Van Loon LC, Bakker PAHM (2006) Induced systemic resistance as a mechanism of disease suppression by rhizobacteria. In: Siddiqui ZA (ed) PGPR: biocontrol and biofertilization. Springer, Dordrecht, pp 39–66

Van Loon LC, Bakker PAHM, Pieterse CMJ et al (1998) Systemic resistance induced by rhizosphere bacteria. Annu Rev Phytopathol 36:453–483

Van Peer R, Niemann GJ, Schippers B et al (1991) Induced resistance and phytoalexin accumulation in biological control of fusarium wilt of carnation by *Pseudomonas* sp. strain WCS417r. Phytopathology 91:728–734

Vejan P, Abdullah R, Khadiran T et al (2016) Role of plant growth promoting rhizobacteria in agricultural sustainability—a review. Molecules 21:573. https://doi.org/10.3390/molecules21050573

Vinayarani G, Prakash HS (2018) Growth promoting rhizospheric and endophytic bacteria from *Curcuma longa* L. as biocontrol agents against rhizome rot and leaf blight diseases. Plant Pathol J 34(3):218–235

Vinocur B, Altman A (2005) Recent advances in engineering plant tolerance to abiotic stress: achievements and limitations. Curr Opin Biotechnol 16:123–132
Vishwakarma K, Upadhyay N, Kumar N et al (2017) Abscisic acid signaling and abiotic stress tolerance in plants: a review on current knowledge and future prospects. Front Plant Sci 8:161. https://doi.org/10.3389/fpls.2017.00161
Voisard C, Keel C, Haas D, Dèfago G et al (1989) Cyanide production by *Pseudomonas fluorescens* helps suppress black root rot of tobacco under gnotobiotic conditions. EMBO J 8:351–358
Vurukonda SSKP, Vardharajula S, Shrivastava M et al (2016) Enhancement of drought stress tolerance in crops by plant growth promoting rhizobacteria. Microbiol Res 184:13–24
Waksman SA (1911) Do fungi live and produce mycelium in the soil? Soil Sci NS 44:320–322
Waksman SA (1944) Three decades with soil fungi. Soil Sci 58:89–114
Wang X, Wang Y, Tian J et al (2009) Over expressing AtPAP15 enhances phosphorus efficiency in soybean. Plant Physiol 151:233–240
Wang J, Tian C, Zhang C et al (2017) Cytokinin signaling activates WUSCHEL expression during axillary meristem initiation. Plant Cell 29:1373–1387
Warcup JH (1950) The soil plate method for isolation of soil fungi. Nature (London) 166:117–118
Wei G, Kloepper JW, Tuzun S et al (1991) Induction of systemic resistance of cucumber to *Colletotrichum orbiculare* by select strains of plant growth-promoting rhizobacteria. Phytopathology 81:1508–1512
Weinberg ED (2004) Suppression of bacterial biofilm formation by iron limitation. Med Hypotheses 63:863–865
Whipps J (1990) Carbon utilization. In: Lynch JM (ed) The rhizosphere. Wiley-Interscience, Chichester, pp 59–97
Wong WS, Tan SN, Ge L et al (2015) The importance of phytohormones and microbes in biofertilizers. In: Maheshwari DK (ed) Bacterial metabolites in sustainable agroecosystem, sustainable development and biodiversity, vol 12. Springer, Cham. https://doi.org/10.1007/978-3-319-24654-3_6
Yang J, Kloepper JW, Ryu CM et al (2009) Rhizosphere bacteria help plants tolerate abiotic stress. Trends Plant Sci 14:1–4
Yong JWH, Letham DS, Wong SC et al (2014) *Rhizobium*-induced elevation in xylem cytokinin delivery in pigeonpea induces changes in shoot development and leaf physiology. Funct Plant Biol 41:1323–1335
Youseif SH (2018) Genetic diversity of plant growth promoting rhizobacteria and their effects on the growth of maize plants under greenhouse conditions. AOAS. https://doi.org/10.1016/j.aoas.2018.04.002
Zahir ZA, Ghani U, Naveed M et al (2009) Comparative effectiveness of pseudomonas and *Serratia* sp. containing ACC-deaminase for improving growth and yield of wheat (*Triticum aestivum* L.) under salt-stressed conditions. Arch Microbiol 191:415–424
Zahir ZA, Shah MK, Naveed M et al (2010) Substrate dependent auxin production by *Rhizobium phaseoli* improves the growth and yield of *Vigna radiata* L. under salt stress conditions. J Microbiol Biotechnol 20:1288–1294
Zehr JP, Jenkins BD, Short SM et al (2003) Nitrogenase gene diversity and microbial community structure: a cross-system comparison. Appl Environ Microbiol 5:539–554
Zhang D (2014) Abscisic acid: metabolism, transport and signaling. Springer, New York
Zhu F, Qu L, Hong X, Sun X et al (2011) Isolation and characterization of a phosphate solubilizing halophilic bacterium *Kushneria* sp. YCWA18 from Daqiao Saltern on the coast of yellow sea of China. Evid Based Complement Alternat Med 2011:615032. https://doi.org/10.1155/2011/615032

Insights into the Unidentified Microbiome: Current Approaches and Implications

5

Ratna Prabha, Dhananjaya Pratap Singh, and Vijai Kumar Gupta

5.1 Introduction

The Earth is dominated by microorganisms, which are phylogenetically divergent and functionally prominent to provide an array of biogeochemical multifunctionalities for the environmental and agricultural sustainability. Microbial communities play critical and real-time role in maintaining multiple ecological functions like mineral recycling, primary production, carbon sequestration, decomposition, biodegradation and bioremediation and regulation of climatic changes in the soils and water bodies (van der Heijden et al. 2008; Wagg et al. 2014; Bardgett and van der Putten 2014). Their genomic diversity and metabolic functions are currently among the most significant areas of research (Whitman et al. 1998; Ghazanfar et al. 2010; Sloan et al. 2006) due to a rich pool of valuable genes, proteins, enzymes and metabolites (Smith and Chapman 2017) that help microorganisms to offer ecological services (Delgado-Baquerizo et al. 2016). However, despite the advancements in the newer tools, techniques and methodologies, the overall microbial diversity of any particular ecosystem and ecological functions linked to various genera/species still remains unidentified, and we hardly know as less as only a single percent of the microbial communities (Bell et al. 2005; Peter et al. 2011).

The pivotal role played by microbial communities in regulating biogeochemical cycles on the Earth during the early ages of its evolution and making the atmosphere oxygenic made it suitable for all living being. Still, microorganisms are the silent

R. Prabha · D. P. Singh (✉)
Department of Biotechnology, ICAR – NBAIM,
Maunath Bhanjan, Uttar Pradesh, India
e-mail: dhananjaya.singh@icar.gov.in

V. K. Gupta
Department of Chemistry and Biotechnology, School of Science, Tallinn University of Technology, Tallinn, Estonia

D. P. Singh et al. (eds.), *Microbial Interventions in Agriculture and Environment*,
https://doi.org/10.1007/978-981-13-8391-5_5

ecological workers for recycling mineral nutrients and organic compounds that facilitate soil health by improving soil structure and fertility, contribute to plant nutrition and health and maintain ecosystem functions (Sathya et al. 2016). Evolutionary diversification of microorganisms parallel to the evolution of the Earth made their inhabitation possible in all kinds of habitats including those of extreme environments too (Li et al. 2014). Since wide diversification enabled these organisms biologically, genetically, metabolically and functionally diversified, it becomes more pertinent, dynamic and practically viable to uncover hidden diversity of microbial communities in any particular environment, establish a link between genetic and functional metabolic diversity and identify microorganisms with potential functions for utilization for agricultural and environmental benefits.

Prokaryotes were the first forms and thus the ancestors of all living beings on the Earth with cells having nucleus-independent hereditary information, following which complex eukaryotic life evolved from a unique prokaryotic endosymbiosis (Lane 2011). The prokaryotic abundance within the ocean and soil is estimated to be 1.2×10^{29} and 2.6×10^{29}, while that on the ocean and terrestrial subsurfaces is 3.5×10^{30} and $0.25–2.5 \times 10^{30}$, respectively (Whitman et al. 1998; Skilbeck 2012). Estimated unseen majority of prokaryotes represent the vast living material on the Earth surface and constitute an interface for the living (biotic) and nonliving (abiotic) niche. This is why they are among the most important reasons of significance in regulating biogeochemical cycles and transforming organic matter into farm inputs (Zeglin 2015). The microbial types, community composition and qualitative processes may substantially differ in the freshwater, sea and terrestrial environment, and the functional behaviour of communities along with their interactions with other species may form complex networks (Zak et al. 2003; Olff et al. 2009; Monard et al. 2016). The present scenario of global changes continuum in ecosystem functions, especially those linked with changing climate, down-regulated trophic levels and invasion of species, has generated a surge of focussed interest in understanding microbial identities and ecological services in terms of regulation of nutrient recycling, decomposition and mineralization processes, fixation of minerals, production of antibiotics and metabolites, interaction patterns among species and with plants, climate resilient crop productivity, suppression of pests and diseases, mitigation of biotic changes, pollutant remediation and reclamation of soil health (Beasley et al. 2012; Burkepile et al. 2006; Cline and Zak 2015).

Majority of microbes inhabit plants as well as animals, not only as pathogens but also as beneficial mutuality associate, that reflect symbioses (Torto-Alalibo et al. 2010). Such associations represent multiple microbial mechanisms for plant benefits like effector protein interaction, host defence mitigation and nutrient acquisition and involve complex, highly specialized molecular processes regulated by a number of associated genes; therefore, it becomes important to characterize microbial identity and understand multitrophic interactions they undergo for their survival and performance with plants and animals under any ecological niche (Chibucos and Tyler 2009; Farrar et al. 2014; Meena et al. 2017). Knowing communities of microorganisms and the kind of interactions they undergo within them (Salazar and de los Reyes-Gavilan et al. 2016; Proal et al. 2017), with their hosts and non-hosts (Badri et al. 2009; Bonfante and Anca 2009) and with the abiotic environments (Li et al.

2014) opens avenues to link classical ecological succession of communities to the whole evolutionary processes. Functional and constitutive interactions among the species and the environment have led to the generation of enormous genetic, structural and functional diversity among the species. During the interactions, the gene transfer mechanisms (lateral or horizontal) and processes of conjugation, transduction and transformation led to the exchange of genetic information easily within prokaryotic organisms which probably enabled microbial population to adapt and evolve in due course of time following distinct diversified patterns (Popa and Dagan 2011; Ku and Martin 2016). While the adaptation to the environmental conditions leads to the survival, evolutionary diversification due to the evolution and transfer of new genes allows organisms to adopt and give more wide complexity, diversity and fast evolution of new species (Elena and Lenski 2003; Andersson et al. 2015). Therefore, a better knowledge of the associative interactions of microbial communities with their hosts and other habitats is critical for understanding the issues closely connected with the ecological success of agricultural practices, loss or gain in agricultural productivity and food security, remediation of environmental contaminants, mitigation of global climate abruption, greenhouse gas emissions, and most importantly the quality of human, plant and animal health.

5.2 Habitat Diversity Made Microbial Communities Diverse

The tagline 'everything is everywhere, but the environment selects' (de Wit and Bouvier 2006) stands very practical for microbial communities and their habitation. Ubiquitous microorganisms find their home in diverse environmental conditions. They are known to thrive well in most of the natural habitats but, at the same time, can live happily in lone or composite extreme environmental conditions including highest cold and frost (Kirchman et al. 2010; Hamdan et al. 2013; De Maayer et al. 2014; Glaring et al. 2015; Li et al. 2017); high temperature (Haizhou Li et al. 2015); thermal vents and hot springs (Benson et al. 2011; Bizzoco and Kelley 2013); saline, alkaline, acidic and arid soils (Keshri et al. 2013; Steven et al. 2013; Kalwasińska et al. 2017); volcanoes (Henneberger et al. 2006); heavy metal-contaminated soils; polyaromatic hydrocarbons (PAHs); and pesticides (Thavamani et al. 2012; Bell et al. 2016; de Souza et al. 2017). Within the specific habitats, these organisms were evolved under the changing environments and thus, have undergone mutations and adaptation strategies for billions of years. Habitat pressure not only strengthened the genetic composition of the organisms, which were further facilitated from gene transfers events but also equipped them with the metabolic capabilities linked with their survival and adaptation. Therefore, while studying microbial communities in any natural or extreme habitats, it becomes pertinent to examine the genetics and dynamics of evolutionary adaptation, impact of environmental pressure in adaptation and evolutionary consequences (Elena and Lenski 2003; Delgado-Baquerizo et al. 2016).

In any natural habitat, including that of agricultural soils which are mostly intruded by human interventions, microbial communities are worth maintaining multiple ecosystem functions and services ('multifunctionality') as key biological constituents (Wagg et al. 2014; Bardgett et al. 2014). Agricultural soils are among the most prominent habitats for the microbial communities and studies under controlled conditions indicated improved soil functions from multifunctional diversity of microorganisms but, this has not been addressed at a global scale in wider perspective taking different soil types and crop plants (Bodelier 2011; Miki et al. 2014). Soil associated factors, belowground roots, rhizosphere and the microenvironment of the plants along with the plant genotypes help maintain microbial diversity and community composition for their mutual benefits (Wardle et al. 2004), but our understanding is limited by the knowledge available on the relationship between microbial community structure and taxa, genera or species linked multifunctionalities and major supporting soil-benefiting processes such as decomposition of litter and mineralization of organic matter bound-nutrients that facilitates matter and energy transfer in the tripartite interactions among the soil matter, plant roots and microorganisms (Wardle et al. 2004; Wagg et al. 2014). Knowing realistic communities, mechanisms of interactions and functions in the habitats, especially in the agricultural soils again becomes important in the time of abrupt climate change, which may invite loss of specific microbial communities and associated functions (Philippot et al. 2013) or invasion of pathogenic species in the soils (van Elsas et al. 2012) in a limited time frame or forever (Delgado-Baquerizo et al. 2016).

In comparison to belowground rhizosphere microbial diversity, the information on the communities inhabiting the aboveground plants are scarce and thus the questions like what type of microorganisms? how they are established? and what functions they are performing? remain least answered (Vorholt 2012). The findings that plant-like biosynthetic pathways were present in bacteria (Moore et al. 2002) and fungus endophytes biologically synthesize natural products in plants (Cook et al. 2004) encouraged the surge to explore the possibility of the origin of many metabolites that were isolated from the plants as the products of microbial origin (Newman and Cragg 2015). Plant-associated microbes act as a great source of secondary metabolites (Gunatilaka 2006) including antibiotics (Kasuri et al. 2014b) and anticancer agents (Mohana Kumara et al. 2012). Plants themselves are a prominent habitat of microbial communities as they host both epiphytic and endophytic microorganisms (Kasuri et al. 2014a) and in many cases the fungal endophytes that were vertically transmitted (Hodgson et al. 2014) or even found hosting other bacteria that produce metabolites with definite functions (Partida-Martinez and Hertweck 2005). Endophytic microorganisms as natural symbionts in the plants contribute enormous beneficial impacts on plant development, health and production. Therefore, exploring beneficial fungal or bacterial endophytic communities isolated, identified and functionally characterized from the agricultural crops and understanding their interaction mechanisms, products and services within the plants could be of prospective interest. Potshangbam et al. (2017) reported identification of non-tissue-specific dominant fungal endophyte genera *Fusarium*, *Sarocladium*, *Aspergillus* and *Penicillium*. Further, *Acremonium* sp. in maize and *Penicillium simplicissimum* in rice were identified following fungal DNA isolation and amplification of ITS1 and ITS4 regions. These endophytes promoted plant growth due to

IAA production ability, acted as antagonists by producing ß-1,3-glucanase and cellulase activity and tolerated temperature stress with high temperature withstanding capacity of 50 °C (Potshangbam et al. 2017). Likewise, pink-pigmented methylotrophic bacteria isolated from various crops like sugarcane, pigeonpea, mustard, potato and radish produced phytohormones and thereby, enhanced seed germination and growth of wheat (Meena et al. 2012). However, besides being of paramount importance, especially for agricultural productivity, extensive and integrative work on phyllosphere microbial communities focusing on identification, functional trait characterization and possibilities of potential applications of endophytic or epiphytic microorganisms for crop growth limits their potentials to become applicable for crop health management. We are also lagging behind over the studies on predominating bacterial phyla, plant and environmental factors shaping such communities on the phyllosphere (leaves, stem, flowers, fruits), community chemistry, periodical community growth on developing plant parts, adaptation behaviour and multipartite interactions with host and among other inhabiting microbial partners (Vorholt 2012). Therefore, the use of advanced methods and tools for deciphering structural and functional indigenous phyllosphere communities will facilitate to open newer avenues for the development of scalable plant protection and growth-promoting microbial agents for multi-field and multi-crop applications.

5.3 Microbial Diversity: Unanswered Questions

Because of their evolution in every possible niche over billions of years, microorganisms were supposed to evolve strategies towards overcoming and adapting multiple stressing environmental conditions on this biosphere. They acted as bridging link for lowering down the hyper-stressing environmental conditions and thus, facilitated the multicellular organisms to evolve under conducive evolutionary atmosphere. However, despite the amazing potential of microbial cosmos and its own ecosystems, which may behave by and large like large-scale ecosystems with a few exceptions (Gibbons and Gilbert 2015), our understanding about the microbial biogeography of a particular niche is limited and still, we know little about the microbial ecosystem function. The distribution of microbial diversity across any large-scale ecosystems on the biosphere, whether it be soil, aquatic environment or above and below-ground plant-surroundings, and their role in the functioning and making the system sustainable is also less understood (Gibbons and Gilbert 2015). What is the taxonomic numbers of microbes on the Earth? How many bacterial species can live in a micro-niche or in unit volume of marine or fresh-water or of fertile soil or sediment? Is there a vertical and longitudinal divergence of prokaryotic diversity? Is the microbial diversity, coupled with other organisms living together in the ecosystem, display idiosyncratic relationship? What is the root cause of so many diverse microbial genera and species? From how many diverse species, how many total interactions exist? All these and so many other questions remained unanswered and made the studies on microbial community ecology more challenging. Microbial community-drive activities in the soil and rhizosphere are responsible for various ecosystem functions including acquisition, mobilization and distribution of nutrients, decomposition of organic matter, cycling of soil minerals, aeration and aggregation of soils, filtration and bioremediation of pollutants, plant growth promotion, suppression of diseases and causal agents and production of release of greenhouse gases (Singh and Singh 2014). Valuable functions directly correlated with plant and soil health made microbial communities indispensable for rhizosphere and plant surroundings. However, for crop plants, we have yet to decipher what are the specific types of microbial communities associated with the plant genotypes, how periodical communities change with the plant growth and development and what exact role they dispense. Plants host epiphytic and endophytic microbial communities (Hardoim et al. 2015) with the aboveground phyllosphere (Lindow and Brandl 2003; Whipps et al. 2008). Due to their genotypic constitution and environmental inhabitancy, plant roots have the capabilities to attract and recruit microbial communities of their own choice and need (Angel et al. 2016; Lareen et al. 2016). Answers to the questions like how many and what kind of microorganisms, especially those that can be cultivated live in association with crop plants? how plants engage microbial communities for their benefits? how microbial associations immunize plants against pathogenic attacks? and to what extent microbial interaction, both belowground and aboveground, can help plants tolerate abiotic stresses? help us devise new and novel strategies based on microbial inoculation of crop plants to support sustainable agricultural productivity (O'Callaghan 2016).

The development of accurate, rapid and universally adopted methods for the determination of microbial diversity and their functions is therefore, essential and highly desirable to construct solid, reproducible and consistent data sets enabling large spatial and temporal studies possible. The parameters for studying microbial diversity need to include multiple methods taking into account greater integration of the available taxonomic background and datasets. Also the holistic approaches for community profiling targeting structural or functional prospects shall also be taken into account. Even large-scale efforts based on cultivable approaches have yielded less than 1% bacteria from the environmental samples (Amann et al. 1995), which, depending upon the variable cultivable enrichment techniques, can be extended to as much as 10% in freshwater lake (Bruns et al. 2003) or 23% marine sediment sample (Köpke et al. 2005). Therefore, the scalability of the culture conditions and the media remains questioned because of the facts that microorganisms live in communities and rarely act alone and they depend on the interactive activities of other organisms to grow successfully and express their potential functions (Boon et al. 2014). A deeper understanding about the shifts in microbial communities and associated in any specific niche cannot be developed unless we record the diversity composition and functions and explore how communities are responding to the changing niche conditions in real time. For targeting all these questions, we need to have appropriate integrated methods, approaches and interactive pipelines in hands to address taxonomic, phylogenetic and trait-based functional parameters that may include biochemical and metabolic modules, gene functions and genomic properties. Developing effective and efficient methods for deciphering microbial diversity in aquatic, soil and rhizosphere ecosystem remains a long-standing challenge (Thies

2007). The shift from isolate (culture)-based methods to genetic assessment-based rapid techniques have provided evidence for taxonomic and functional microbial diversity in recent years and have become popular in examining taxonomic abundance and geographical distribution of communities in recent years (Zarraonaindia et al. 2013). The community-centric approach, therefore, is becoming a powerful tool to address challenges and opportunities (Teeling and Glockner 2012) and help understand niche-specific microbial diversity.

5.4 Addressing Community Complexity Challenges

Growing with the evolutionary challenges, microbial diversity expanded in a big way which has led to the development of cellular, biophysical, physiological, metabolic and genomic complexities in the organisms. Within a given habitat of dynamic and interactive nature, microbial communities are exposed to behave and perform complex functions of global implications that often cumulatively drive the habitability of the biosphere. Both the metabolic and genomic complexity of microbial communities has made it difficult to understand or even predict their responses towards changes in the environment under short-term or long-term conditions (Gronstal 2016). Microbial interactions within their own communities, higher organisms (plants, microflora and fauna) and their living micro-environment make the community behaviour so intricate to decipher the engagement of interacting communities, types of multipartite interactions and intrusive mechanisms at cellular, metabolic and/or molecular level (Zuñiga et al. 2017). Large size of microbial diversity, multifunctional communities and manifold environmental interactions gives rise to multiphasic complexities and thus, limiting our ability to decipher whole microbiome composition and functions to the most appropriate extent.

The best examples are the hot desert terrestrial edaphic systems representing soils, cryptic and refuge niches and rhizosphere associated microorganisms growing under critical harsh conditions that yielded magnificent microbial diversity all over the world (Makhalanyane et al. 2015). Limitations of the traditional microbiological, metabolic or molecular methods in both scalability and precision or our inability to decipher and detect patterns of the enormous data of sequences generated due to high-throughput genomics technologies restricted our understanding of complex microbial communities which are poorly identified, understood and characterized (Hill et al. 2002). Comparative methodological analysis of various techniques used real as well as simulated 16S rRNA pyrosequencing datasets to characterize microbial communities in diverse environment (Kuczynski et al. 2010) or comparison of various DNA extraction methods for recovery of soil protists can improve our understanding to apply appropriate methods for community analysis. Presently, parallel to the cultivation-based methods, omics-based strategies including 16S rDNA or ITS sequencing, whole genome sequencing of identified microorganisms, metagenomics and metatranscriptomics-based microbial community analysis, proteomics and metaproteomics-based functional protein characterization and global metabolic profiling through metabolomics being applied to decipher taxonomic, metabolic and functional diversity of microbes in different habitats (Zarraonaindia et al. 2013; Ravin et al. 2015; Jansson and Baker 2016) have gained importance. Besides, approaches for modelling microbial diversity in ecosystem at large or small scales while considering possible environmental interactions (Larsen et al. 2012), targeting microbial metabolic networks in natural and engineered environment (Perez-Garcia et al. 2016), understanding competitive and cooperative metabolic interactions in organisms (Freilich et al. 2011), contextualizing complex data types in microbial networks (Faust and Raes 2012), predicting responses of communities to environmental perturbations using computational modelling (Zuñiga et al. 2017), defining microbial community dynamics based on the studies on energetic and metabolic interaction networks (Embree et al. 2015) and addressing systems biology of complex microbial ecology (Bordbar et al. 2014), community functioning (Röling et al. 2010) and host-microbe interactions using metabolomics (Heinken and Thiele 2015) have emerged as potential strategies to target complex community dynamics at genomic and metabolic level and interpret possible interactive functions with the environment.

The estimates that a single gram of the soils harbour almost ten billion microbial population with vast different diversity reflect huge unchartered microbial dynamics (Roselló-Mora and Amann 2001). Likewise, marine aquatic microorganisms that are potential source of commercial secondary metabolites, bioactive molecules and possess bioremediation, decomposition and biodegradation capabilities hold enormous challenges for community analysis (Das et al. 2006). Although defining complete microbial diversity remain challenging, the complexity, divergence and variability can be addressed at various biological parameters, especially the genetic variability within taxonomic groups (genera and species), number (species richness in confined region), species evenness (relative taxon-based abundance with diverse functional groups-guilds) of the communities (Thies 2008). Spatial and temporal patterns of microbial diversity are also obscure and therefore, estimating prokaryote diversity in natural normal and sub-normal ecosystems is a challenging priority in microbial ecology (Dimitriu et al. 2008; Swirglmaier et al. 2015). Other challenging and important aspects to be addressed in microbial ecology are the range of processes, complexities of multitrophic interactions and ultimate benefits (functional outcomes) of whole community-level characterization for the plant, soil and other organisms living together (Guttman et al. 2014).

5.5 Identifying the Unidentified: Culturable vs Nonculturable Approaches

The interconnecting realities that microorganisms rarely act alone NCBI database, accessed on Nov 2018, they depend on metabolic activities of other organisms to grow, reproduce and function (Stolyar et al. 2007) and majority of microbial life forms has eluded cultivation in appropriate culture conditions restricted the expansion of deeper understanding about microbial communities. With any environmental sample that need to undergo investigations from microbial communities perspectives, it becomes important to (1) isolate microorganisms and establish identity; (2) characterize functional traits of the isolated organisms; (3) assess the overall microbial communities in the sample; and (4) establish taxa- or genera-linked metabolic functions of the communities. Culturable approaches based on the microscopic observations and standard microbiological protocols that include methods of isolation and enrichment, culture media and growth conditions (Jett et al. 1997; Sanders 2012) are accurate, reliable and perfect for the identification of microorganisms and their trait characterization (Janssen 2008). However, stringent quality control measures for culture media and growth conditions should be practical enough to ensure recovery of as much organisms as possible from any environmental sample (Cantarelli et al. 2003).

Abundance and diversity of culturable microbiome assumed success in isolating, identifying and characterizing bacterial (bacteria, actinobacteria, methylotrophs, cyanobacteria) and fungal isolates in rhizosphere (Ahmad et al. 2008), endophytic and epiphytic conditions (Yandigeri et al. 2012; Meena et al. 2012), rice phyllosphere (Venkatachalam et al. 2016) and water bodies (Lee et al. 2014). Polyphasic analysis (both microscopic and DNA-based) (Prakash et al. 2007) further revealed taxonomic and phylogenetic assessment comprising isolation of total DNA, 16S rRNA amplification and ITS (internal transcribed spacer) RNA gene for fungus followed by sequencing and analysis (Ellis et al. 2003; Štursa et al. 2009; Zappelini et al. 2015). The procedure is followed by trait-based characterization of the isolates, which is based on metabolic capabilities, gene functions and genomic properties. Routine procedures like plate counts for determining colony-forming units (CFU) per millilitre, total counts based on microscopic determination of cells stained with 4,6-diamidino-2-phenylindole (DAPI) and counts that considers microscopic determination of microbial cells stained with 5′-cyano-2,3-ditoly tetrazolium chloride (CTC) are assessed (Ultee et al. 2004).

Molecular microbial diversity analysis of 16S rRNA genes revealed the non-culturability of the bacterial divisions OP10, OP11, BRC, SC3, WS2, WS3 and TM7 but only their sequences are known (Schloss and Handelsman 2004). These are considered as "candidate divisions" while the culturable microorganisms are "weeds" of the microbes representing a little fraction of entire microbial domain (Hugenholtz 2002). Molecular techniques based on 16S rRNA alone for establishing diversity of prokaryotic organisms (Griffiths et al. 2000) coupled with multiple statistical tools that address modelling, prediction and analysis of diversity indices estimate species richness and perform analysis of rarefaction curve (Hughes et al. 2001) largely facilitated microbial taxonomy. Besides such methods that target genomic DNA from the isolates, approaches based on highly conservative properties such as MALDI-TOF mass spectroscopy or the analysis of fatty acid methyl ester (FAME) composition can also be used as powerful, accurate and reliable tool for the identification of organisms to species level.

Microbial diversity represents a vast assemblage of communities inhabiting a particular environment. Various methods involve microbial culture on multiple media (use of selective media Nutrient Agar, Luria–Bertani medium for identification purpose), Biolog-based method and molecular tools (Garbeva et al. 2004;

Hugenholtz 2002; Kirka et al. 2004). Physiological, morphological and colonial differences are among the most used parameters for differentiation of bacterial population, although it is not necessary that they provide accurate and exact microbial identification (Ghazanfar et al. 2010). Identified culture of any microorganism is essentially a need for many obvious reasons (Alain and Querellou 2009). Firstly, the cultures with their fully explored physiological, molecular and metabolic potential can be conserved and utilized for various usages as per the agricultural and environmental needs. Secondly, microbial cells are required for deciphering cellular characteristics using high-end microscopy engaging electron or confocal laser microscopes. Thirdly, whole genomic DNA of an organism is required for high-throughput DNA sequencing to develop deeper understanding on physiological and metabolic characteristics of the microbes. Essentially, an identified and taxonomically named culture is required for full taxonomic characterization and submission to the biodiversity repositories. Over and above, easy to culture microorganisms with their specific metabolic and functional traits can help in assessment of microbial diversity of any specific niche. However, low cultivability of microorganisms in communities, lack of appropriate growth media and growth conditions and metabolic complexity of the organisms restrict the transition from "non-culturable" to "culturable" state under laboratory conditions. For microbial species identification, various DNA-based methods are in practice (DeLong 2005). In this age of high-throughput sequencing, a huge number of 16S sequences are consistently being reported from various habitats and deposited to public databases, which could now be used to generate more precise knowledge about taxonomy and phylogenetic lineage of the culturable isolates. Similarly, only 9352 complete genome sequences are available for prokaryotic organisms (NCBI database, accessed on Nov. 2018). These studies explored novel genes, proteins and metabolic pathways in the organisms that inhabited different habitats, grown in varying media conditions, deposited in various repositories (Table 5.1). Therefore, besides the existing limitations of cultivability, the approach has potential, prominent and practical applications to grab powerful microbial representatives that could be utilized for their multifunctional potentials in food, agriculture and environment.

Evidences for the facts that only a handful microbial population can be cultured come from the microscopic observations that show enormous organisms under microscope than those which appear on the culture plates as colonies (Stewart 2012). It was proposed that the organisms under the microscopic field may be dead cells to become active on culture media, but many of these cells that did not grow on plates were shown to be metabolically active (Roszak and Colwell 1987). In fact, microorganisms rarely grow alone outside natural habitat conditions, where they live in communities and depend on the activities of other organisms for their growth and performance. These communities differ in their genetics, energetics, metabolic potential and the way they interact with the inhabited microenvironment, pose difficulties for their isolation in lone on media conditions. Another strong evidence that support non-cultivability of the majority of microbial species arise from DNA sequence data from the environmental DNA after PCR amplification followed by cloning or high-throughput sequencing and characterizing 16S rRNA phylogeny of

Table 5.1 Identification of microbial communities from various extreme environmental conditions, habitats and their functions

Environment/ condition(s)	Organisms/communities and functions	Methodology	Habitat(s)	References
Extreme cold and alkaline	Dominant bacterial community and less dominant archaea; *Cyanobacteria* and phototrophic *Proteobacteria;* putative anaerobic *Firmicutes* and *Bacteroidetes*	High-throughput pyrosequencing of 16S rRNA genes and analysis of ikaite community structure	Ikka Fjord in Southern Greenland. Permanently cold (less than 6 °C) and alkaline (above pH 10) environment	Glaring et al. (2015)
Extreme cold	Relative abundance of *Betaproteobacteria, Chloroflexi, Bacteroidetes, Deltaproteobacteria, Gammaproteobacteria* and *Firmicutes* were higher, *Actinobacteria, Alphaproteobacteria, Acidobacteria, Betaproteobacteria, Chloroflexi, Bacteroidetes* and *Deltaproteobacteria,* accounting for >80% bacterial sequences	FLX 454 pyrosequencing and sequence data analysis	Qinghai Province, China, elevation ranging from 2900 to 4000 m, annual temperature of −1.7 °C, mean air temperature approximately −3.1 °C with a maximum of 5.7 °C	Li et al. (2017)
Extreme cold water bodies	*Seohaeicola, Loktanella,* and *Halomonas, Leptolyngbya, Oscillatoria,* and *Nodularia, Hyphomicrobium, Rhodobacter,* and *Pedomicrobium* as dominant taxa	DNA extraction; 454 pyrosequencing of the 16S rRNS gene	The Salar de Huasco, Chile, mean air temp 5.0 °C	Anguilar et al. (2016)
Sub-arctic glacier	Dominant *Proteobacteria* at 5 °C and 10 °C and members of *Chloroflexi, Acidobacteria* and *Verrucomicrobia* in addition to *Proteobacteria at* 22 °C	Next-generation sequencing of the 16S rRNA gene	Glacier forefield Styggedalsbreen Norway	Mateos-Rivera et al. (2016)

(continued)

Table 5.1 (continued)

Environment/ condition(s)	Organisms/communities and functions	Methodology	Habitat(s)	References
Antarctic snow pack	Proteobacteria predominant phylum (57%) comprising 12 genera: *Brevundimonas*, *Afipia*, *Rhizobium*, *Sphingomonas*, *Pigmentiphaga*, *Polaromonas*, *Variovorax*, *Acinetobacter*, *Moraxella*, *Pseudomonas*, *Rhodopseudomonas* and *Stenotrophomonas*. 82 isolates phylogenetically related to Proteobacteria (including α-, β-, and γ-proteobacteria), Actinobacteria, Firmicutes and Bacteroidetes, and 1 lineage of domain eukaryota: Basidiomycota	Enrichment and isolation on tryptone soy broth, nutrient broth, Zobell marine broth, Antarctic media R2A and R3A, DNA isolation, amplification, cloning and sequencing of 16S/18S rDNA from snow	Princess Elizabeth Land region of East Antarctica, ambient air temperature −1 to −32 °C, mean annual temperature −16 °C	Antony et al. (2016)
High temperature	Dominant bacterial genera *Caldisericum*, *Thermotoga* and *Thermoanaerobacter,* archaeal genera *Vulcanisaeta* and *Hyperthermus,* genera *Vulcanisaeta*, *Thermofilum*, *Hyperthermus*, *Methanocaldococcus* and *Methanosaeta;* ammonia-oxidizing activity	Culture-independent method that combines CARD-FISH, qPCR and abundance of 16S rRNA and amoA genes	Steep thermal gradient (50–90 °C); Tengchong Geothermal Field, China	Haizhou Li et al. (2015)
Hot spring	Proteobacteria (50% abundance) followed by Bacteroidetes (13% abundance) and Firmicutes (significant abundance); *Deinococcus*, Verrucomicrobia, Planctomycetes, and Chloroflexi (low abundance)	Metagenome extraction and analysis with next generation technology (bTEFAP)	Water samples of Ma'in and Afra hot springs, Jordan	Hussein et al. (2017)
High temperature	Archaeal dominance by the sequences related to *Methanobacterium formicicum* and *Methanothermobacter thermautotrophicus*, and bacterial dominance by sequences related to *Hydrogenophilus* and *Deferribacter*	16S rRNA gene libraries and culture-based methods	Dissolved-in-water type gas field; Japan (46 and 53 °C tem from a depth of 700–800 m	Mochimaru et al. (2007)

High temperature heavy oil reservoir	Methanogens (*Methanomethylovorans, Methanoculleus, Methanolinea, Methanothrix*, and *Methanocalculus*) aerobic organotrophic bacteria (*Tepidimonas, Pseudomonas, Acinetobacter*), as well as of denitrifying (*Azoarcus, Tepidiphilus, Calditerrivibrio*), fermenting (*Bellilinea*), iron-reducing (*Geobacter*) sulphur-reducing bacteria (*Desulfomicrobium, Desulfuromonas*)	Isolation of microbial DNA and RNA from back-flushed water and preparation of clone libraries for 16S rRNA gene and cDNA of 16S rRNA	Dagang high-temperature oilfield, Hebei Province, China depths of 1206–1435 m, temperature 59 °C	Nazina et al. (2017)
Geothermal steam	Halophilic archaea	Culturable and unculturable, media enrichment technique	Hydrothermal vents or fumaroles, steam waters in Russia (Kamchatka) and the USA (Hawaii, New Mexico, California and Wyoming)	Ellis et al. (2008)
Hydrothermally modified volcanic soils, acidic conditions	Microbial communities	Extremophiles	DNA techniques	Henneberger et al. (2006)
High sodium carbonate (soda) conditions; stable elevated pH	Alphaproteobacteria (mostly *Rhodobacteraceae*) and Gammaproteobacteria (including *Halomonas* and *Thioalkalivibrio*), Firmicutes (aerobic *Bacillus*, anaerobic *Clostridia)*, Bacteroidetes (*Cytophaga*, *Flexibacter*, *Flavobacterium*, *Bacteroides*, *Salinibacter*), Cyanobacteria *Arthrospira* and *Anabaenopsis*, and purple phototrophic bacteria from families of *Ectothiorhodospiraceae*, *Chromatiaceae* and *Rhodobacteraceae*	Culture-dependent and independent methods	Saturated alkaline brines in Lake Magadi (Kenya, Africa); hypersaline soda lake brines (total salinity >250 g/L); North American and Central Asian soda lakes	Dadheech et al. (2013), Krienitz et al. (2013), and Sorokin et al. (2014)

(continued)

Table 5.1 (continued)

Environment/ condition(s)	Organisms/communities and functions	Methodology	Habitat(s)	References
Saline soda lime	Bacterial community Proteobacteria, Firmicutes, Bacteroidetes and Actinobacteria Archaeal community *Candidatus Halobonum* and *Halorubrum and members of Phenylobacterium, Skermanella, Bryobacter, Simkania, Salinibacter, Psychromonas, Halomonas, Synechococcus, Haloferula* and *Euhalothece*	PCR amplification of 16S rRNA genes, pyrosequencing for total bacterial community	Janikowo, Kuyavia, Central Poland (with chlorides concentration around 30 g kg^{-1})	Kalwasińska et al. (2017)
Saline-alkaline soil	Actinobacteria, Fermicutes, Proteobacteria, Bacteroidetes, Chloroflexi, Acidobacteria, Planctomycetes, Nitrospira, cyanobacteria Archaea-*Euryarchaeota* dominated by family *Halobacteriaceae, Methanobacterium, Methanocella,*	Clone libraries construction using 16S rRNA and key functional gene(s) of carbon fixation (*cbbL*), nitrogen fixation (*nifH*), ammonia oxidation (*amoA*) and sulphur metabolism (*apsA*), q-PCR	Una soil, coastal region of Gujarat, India	Keshri et al. (2013)
Soda lake	Dominated communities *Alpha*- and *Gammaproteobacteria* (30% and 18% of the sequences), gram-positive group and *Chloroflexi*-related sequences (both 13% of the sequences), *Epsilonproteobacteria*-like sequences found and *Deltaproteobacteria*, *Cyanobacteria* (*Synechococcus*) detected	Culture-dependent and independent methods, Denaturing gradient gel electrophoresis (DGGE) of bacterial and archaeal 16S rRNA genes	Meromictic lake in lower Grand Coulee, Eastern Washington State	Dimitriu et al. (2008)

Dryland with soil depth and parental material (biological soil crusts)	*Cyanobacteria* and *Proteobacteria* demonstrated, high relative abundance in the biocrusts with *Actinobacteria*, *Chloroflexi* and *Archaea*	Quantitative PCR, amplification, and sequencing of bacterial 16S rRNA genes	Island in the Sky district of Canyonlands National Park, UT	Steven et al. (2013)
Hot desert edaphic systems				Makhalanyane et al. (2015)
Eroded soils due to industrial activities	Core microbiome comprised 64.4% bacterial and 62.4% fungal genera; *Gammaproteobacteria* (24% dominance) with *Pseudomonas* genera (72%), dominance of fungal taxa *Hebeloma* and *Geopora*	Sequencing of 16S rRNA and internal transcribed spacer (ITS) fungal RNA gene amplicons from chlor-alkali residue	Chlor-alkali tailings dump	Zappelini et al. (2015)
Mangrove ecosystem	*Proteobacteria* (65.7%), *Bacteriodes* (11.83%), *Firmicutes* (5.56%) and *Actinobacteria* (3.61%)	Environmental DNA isolation and sequencing on Illumina HiSeq platform 2500, annotation using MG-RAST	Pyannur, Panangod, Vallarpadam and Madakal, Kerala	Imchen et al. (2017)

the communities (Keller and Zengler 2004; Stewart 2012). Although, a volume of work on the isolation and identification of microbial species from various habitats using cultivable approach exist, it reflects very small proportion of natural diversity in a habitat of microbial communities (Joint et al. 2010) and, thus, fails to represent almost 90–99% of the estimated microorganisms on the Earth (Štursa et al. 2009). Technology-driven advancements that enabled direct sequencing of environmental DNA and RNA have now opened avenues for the generation of metagenomic and metatranscriptomic data and their analysis for structural and functional assessment of microbial communities in diverse various habitats (Prosser 2015).

1. Amplified Ribosomal DNA Restriction Analysis
2. FAME (Fatty acid methyl esters) analysis
3. Fluorescence In Situ Hybridization (FISH)
4. Multilocus sequence typing
5. Physiological profiling/ Carbon substrate utilization
6. Plate count method
7. Random Amplified Polymorphic DNA (RAPD)/DNA Amplification Fingerprinting (DAF)
8. Ribosomal Intergenic Spacer Analysis

5.5.1 Amplified Ribosomal DNA Restriction Analysis (ARDRA)

ARDRA utilizes nucleotide sequence alterations available in the PCR product of 16S rRNA genes. RFLP using 16S rRNA and 16S-23S rRNA genes has been developed as a technique for molecular identification, characterization and differentiation of bacterial species within genera and species level. Cook and Meyers (2003) developed a rapid method for the identification of most of the microorganisms based on RFLP techniques of 16S rRNA gene. Several other authors also reported that RFLP technique is very efficient for discriminating microorganisms up to inter- as well as intra-generic level (Steingrube et al. 1997; Wilson et al. 1998; Laurent et al. 1999). The technique is considered to be a useful tool for screening environmental bacterial, actinomycetes, fungal isolates and/or clone libraries (Sjoling and Cowan 2003). It also contributes to an exploration of the diversity of the microbial communities of the different samples analysed. Molecular diversity study of these isolates showed a good variability with 16S rDNA-RFLP of all the isolates with respect to their morphological and chemotaxonomic characteristics (Alves et al. 2002; Laurent et al. 1999). Cluster analysis approach was used to test the reliability of RFLP to preselect different organisms and to obtain a rough overview of the diversity before sequencing. ARDRA analysis reveals fragments of rRNA genes that are specific for communities. The technique seems to be most informative if used with sequence information for quantitative tracking of environmental samples. The amplified product of ribosomal gene from the environment genomic material (DNA) is usually digested with restriction endonucleases (e.g. tetracutter *AluI* and *HaeIII*). The restriction fragments were run on agarose/polyacrylamide gels. The method seems to be efficient and useful for time to time and fast monitoring of microbial communities and in making comparative analysis of diversity in response to environmental changes. This technique shows usefulness in identifying unique clones and estimating OTUs in environmental cloned libraries developed on the basis of restriction profile of the clones (Smith et al. 1997). Whole community ARDRA reflected diversity of soils contaminated with copper (Smith et al. 1997; Rastogi and Sani 2011). Restriction digestion of 16S rRNA gene using tetracutter or hexacutter endonucleases yields variable distinct restriction patterns. In each of the restriction patterns about two to five restricted fragments of varying sizes were found that is based on Jaccard's similarity index. Major and minor clusters were formed after combined restriction patterns. It is a type of co-dominant molecular marker in which digestion of a particular gene part or loci is allowed at one time. However, ARDRA has limitations in generating restriction profiles from complex microbial population.

5.5.2 Phospholipid Fatty Acid (PLFA) or FAME Analysis

More than 300 fatty acids ranging from C2 to C24 are an integral component of a diverse range of bacteria which usually harbour qualitative (at genus level) and quantitative (at species level) compositional differences in these compounds (de Carvalho and Caramujo 2014). As a useful phenotypic biochemical tool for bacterial classification and characterization, the composition of fatty acid methyl esters (FAME) is usually considered as stable marker and independent of plasmids, mutations or cell damage (Banoweltz et al. 2006), but culture conditions may influence fatty acid profiles (Scherer et al. 2003). Since the type and compositional abundance of phospholipid fatty acids (PLFA) are genotype-driven and differ from organism to organism inhabiting different habitats, it acts as a tool to characterize and differentiate diverse microbial communities and changes in the overall community composition with time (Tunlid and White 1992; Vandamme et al. 1996). Fatty acid signatures can differentiate taxonomic diversity of microbial population (Welch 1991). For example, FAME analysis was used to analyse microbial communities and their population dynamics in chemically contaminated soils (Siciliano and Germida 1998; Kelly et al. 1999), characterization of *Bacillus mycoides* (Von Wintzingerode et al. 1997), microbial communities in groundwater (Green and Scow 2000), foodborne bacterial pathogens and aerobic endospores of bacilli (Whittaker et al. 2005) and spores of *Bacillus cereus* T-Strains on different culture media (Ehrhardt et al. 2010, 2015). FAME is notably applied for the analysis of bacterial community differentiation but it was applicable to differentiate plant parasitic nematodes also and separate profiles were developed for *Rotylenchulus reniformis* and *Meloidogyne incognita*, species and races in *Meloidogyne* genera and various stages of life of *Heterodera glycines* which usually do not segregate using canonical analysis (Sekora et al. 2009). Overall, the FAME analysis method has wide applicability, mostly because it is a culture-independent technique, but it has limitations of fast degradation of fatty acids from the dead cells, preciseness in fatty acid extraction from environmental samples and appropriate database for comparative profiling and differentiation.

5.5.3 Fluorescence In Situ Hybridization (FISH)

The technique facilitates in situ phylogenetic identification and estimation of microbial cells in an individual by cell hybridization with the probes of set size oligonucleotide. At various taxonomic level, 16S rRNA genes targeting molecular probes were reported (Amann et al. 1995). Usually, the probes of 18–30 bp long nucleotides are used. They have a fluorescent dye at the 5′ end to allow the detection of cellular rRNA bound probe using epifluorescence microscopic method. The fluorescent signal intensity has correlation with the content of cellular rRNA vs growth rates to reflect cell metabolic state. For a high-resolution automated analysis, FISH can be clubbed with flow cytometry to explore fixed microbial population. The method was used to monitor bacterial population dynamics in different agricultural soils contaminated with s-triazine herbicides (Caracciolo et al. 2010). Different molecular probes have been used for targeting specific bacterial phylogenetic groups from subdivisions of Proteobacteria and Planctomycetes.

A modified method using catalysed reporter deposition (CARD) utilizes hybridization signal that is increased using tyramide-labelled fluorochromes (Pernthaler et al. 2002). The method permits accumulation of fluorescent probes at targeted sites at which ultimately signal intensity and sensitivity is increased. A more advanced imaging technique clubbed with FISH and secondary-ion mass spectrometry (SIMS) was introduced by Li et al. (2008). The method was used for detection and quantitative analysis of ammonia-oxidizing bacteria (AOB) (Mobarry et al. 1996; Briones et al. 2002). Most FISH is applied reliably to study microbial population in the nutrient-poor environments such as soil and roots where the bacterial cell numbers are too low to enumerate.

5.5.4 Multilocus Sequence Typing (MLST)

MLST is a reliable, reproducible and efficient DNA sequence method with prominent examples of deciphering microbial population. It uses DNA sequences from different multiple regions of the genome for making discrimination of strains in the populations (Maiden et al. 1998) to reveal microbial ecological and evolutionary analyses. The method is also named by synonyms such as multiple gene genealogical analysis (MGGA) or comparative genealogical analysis (CGA) (Xu 2006). There have been more advantages of using multiple loci for the analysis over single locus-based analysis as it generates more robust information leading to conclusion, reflects higher representation of multiple regions in the genome and deals more efficiently with the horizontal gene transfers which are common in prokaryotic populations (Xu 2006). Existence of public databases based on MLST information makes data sharing more feasible among the among researchers. The technique has been used to explore ecological genetics of environmental population. It can describe fine-scale gene and genotypic diversity of microbial communities. MLST also allows the characterization of strains and clones of medical importance (Feil and Enright 2004; Urwin and Maiden 2003). However, environmentally diverse but agriculturally more relevant group of microorganisms are less explored using MLST (Xu 2006).

5.5.5 Physiological Profiling (BIOLOG)

BIOLOG is a technique that utilizes 96-well microtitre plates to characterize active functional diversity and bacterial identification via carbon source utilization profile (Garland and Mills 1991). Gram-negative (GN) and gram-positive (GP) microtitre plates with 95 different carbon sources and one control well without a substrate are available from BIOLOG. GN and GP plates were devised for bacterial characterization and not for community analysis. Since some of the fungal communities are not able to reduce tetrazolium dye after substrate utilization in GN and GP plates, fungus-specific plates are also available for their identification. Community-level physiological profiling (CLPP) is very common in soil microbial community identification and analysis using BIOLOG (Derry et al. 1999).

5.5.6 Plate Count Method

Plate count or direct viable count method was used in the beginning of microbe's identification and discrimination for the diversity analysis based on colony spreading. Microbes having high growth rate and fungi produce large amount of spores when allowed to grow on minimal agar plates (Dix and Webster 1995). Traditional methods can give information on the active but heterotrophic population on nutrient agar media. Viable count method has some lacuna and problems in enumerating organisms from soils on growth media (Tabacchioni et al. 2000), maintain growth parameters conditions like temperature, pH and light and do not prompt to culture greater bacterial and fungal population.

5.5.7 Random Amplified Polymorphic DNA (RAPD) and DNA Amplification Fingerprinting (DAF)

RAPD and DAF are the methods that utilize DNA amplification with a short length primer (approx. ten nucleotides). The short sequences randomly anneal in the genomic DNA at multiple sites under low annealing temperature conditions (Franklin et al. 1999). PCR amplicons of different sizes are generated in the single reaction and are separated on agarose or polyacrylamide gel on the basis of their genetic diversity in the microbial communities. It is a type of dominant molecular marker that allows many loci or gene part amplification at one time in one sample of DNA with single PCR reaction. However, usually co-dominant markers are more informative than dominant markers. RAPD/DAF is an extensively used technique for fingerprinting of whole microbial population as it serves to identify closely related bacterial species (Franklin et al. 1999). The technique is sensitive to the experimental conditions, quality and quantity of template DNA and the primers being used. Thus, the use of several primers and reaction conditions for preparing a comparative profile of relatedness in communities and obtaining the most differentiating pattern among species and strains is always appreciable. The RAPD study with 14 random primers have revealed changes in the microbial diversity of soils contaminated with the chemical inputs (triazolone pesticide) and fertilizers (ammonium bicarbonate) (Yang et al. 2000). RAPD analysis demonstrated that pesticide-treated soils maintained identical diversity profile at the DNA level as the control soils having no contamination.

5.5.8 Ribosomal Intergenic Spacer Analysis (RISA)

RISA is a PCR amplification-based technique that involves a part of intergenic spacer region (ISR) found between the small and large (16S; 23S) ribosomal subunits (Fisher and Triplett 1999) which are having significant heterogeneity in length and sequence of nucleotides. The profiles of RISA of the environmental samples can be generated from the primers annealed to conserved regions in 16S/23S rRNA genes to reveal most of the dominant bacteria. It gives community-specific profile in which different bands correspond to one specific organism belonging to the original population. The automated RISA, also called ARISA, involves the use of a fluorescence-labelled forward primer. This helps in the detection of ISR fragments automatically using a laser detector. ARISA profiles generated from these soils were distinct and contained several diagnostic peaks with respect to size and intensity. Characterization of bacterial profile from different soils with distinct vegetation cover and physicochemical properties has been shown using ARISA (Ranjard et al. 2001). These results reflected that the technique is effective and sensitive for creating distinction in complex bacterial communities. ARISA allows simultaneous analysis of multiple environmental samples but the technique may provide overestimated diversity richness of microbial population (Fisher and Triplett 1999).

5.6 Methods for Identification of Unculturable Microbes

Various molecular techniques have been devised to provide an insight on a diversity of those microbes which are yet not cultured in the laboratory (Table 5.1). Many of these techniques rely on the use of rRNA genes that facilitated cloning and sequencing to help in the characterization of marine bacterial and archaeal collections (Díez et al. 2001). Yet, these techniques are not worthy when many different samples are under consideration as analysis of clone libraries is time-consuming.

We have described some of the techniques most commonly used for the analysis of uncultured microorganisms.

1. SSCP (Single strand confirmation polymorphism)
2. Terminal Restriction Fragment Length Polymorphism
3. DNA Micro-arrays
4. Denaturing gradient gel electrophoresis
5. Cloning
6. Whole Microbial Genome Sequencing
7. Quantitative real time PCR

Some of these culture-independent methods also focus on putting a connection between microbial community function with the genetic identity of key organisms. One of these methods is stable-isotope probing (SIP) which is able of identifying microorganisms responsible for particular biogeochemical processes (Manefield et al. 2002).

5.6.1 Single-Strand Conformation Polymorphism (SSCP)

SSCP involves denaturation of environmental PCR products and separation of single-stranded DNA fragments on nondenaturing polyacrylamide gel (Schwieger and Tebbe 1998). Separation usually involves sequence differences of most commonly a single base pair. This further results in different secondary structure to create measurable mobility in the gel based on differences. SSCP does not require GC clamping of primers, gels of gradient nature and specific electrophoretic apparatus. In this way, the method is more simplified and easily adaptable in comparison to DGGE. In this technique, the DNA bands may get excised out of the gel, reamplified further and then sequenced finally to provide sequences preferably for small fragments (between 150 and 400 bp) only (Muyzer 1999). The technique limits in being its higher rate of DNA strand reannealing after initial denaturation. This may be corrected at the time of PCR by the use of phosphorylated primer which is specifically digested with lambda exonuclease. SSCP has been worked out to discriminate between cultures of *Bacillus subtilis*, *Pseudomonas fluorescens* and *Sinorhizobium meliloti* from the soils (Schwieger and Tebbe 1998). It was used for the analysis of bacterial communities in the rhizosphere of *Medicago sativa* and common weed *Chenopodium album*. Based on the analysis, it was affirmed that every plant harbours its own distinct rhizosphere community despite the same growing conditions and soils of the plants.

5.6.2 Terminal Restriction Fragment Length Polymorphism (TRFLP)

Having similarities with the ARDRA, TRFLP has a major difference of using 5¢ fluorescently labelled primer at the time of the PCR reaction. PCR products thus obtained are digested using restriction enzyme and terminal restriction fragments (T-RFs) are separated on DNA sequencers (Thies 2007). It uses the detection of terminally fluorescent labelled restriction fragments only. This leads to the simplification of banding patterns that further allow analysis of complex microbial communities in the environment. This has the advantage to have community diversity assessment based on the analysis of size, numbers and height of peaks of T-RFs. Each T-RF is considered as a representative of single OTU. With the use of bioinformatics tools, web-based T-RFLP analysis programs are now helping researchers to assign identities of fragment sequences putatively while comparing that with the databases of known 16S rDNA sequences. T-RFLP generates characteristic pattern for restriction enzymes which need application of two or more enzymes in general. However, the method underestimates community diversity due to its limitation of resolving a limited number of bands per gel (<100) and sharing of similar T-RF length called OTU overlap or OTU homoplasy for bacterial species. Overall, the method constitutes a robust community diversity assessment index. The method has provided evidences that the T-RFLP results are usually well correlated with those of cloned libraries. The technique was used to understand the biogeographical profile of soil bacterial communities and the impact of environmental factors shaping community composition and functional diversity. Ninety-eight soil samples representing a wide environmental diversity (temperature, pH and geography) were collected from North and South America by Fierer and Jackson (2006). Differentially higher bacterial diversity was recorded in neutral soil samples than in acidic soils using TRFLP method. In this way, the method was proved to be authenticated predictors of animal and plant diversity.

5.6.3 DNA Microarrays

Microarrays are representing a powerful tool to recognize, identify, depict and characterize microbial species living in the natural habitats. DNA–DNA hybridization is most commonly being used with DNA microarrays for detecting and identifying bacteria (Cho and Tiedje 2001). Instead of the genomes or genes, the DNA fragments provide the advantage of rejecting the need to keep microbial cultures. This is because the genes can be cloned into plasmids or else amplified by continuous PCR. DNA microarray is able to identify a large number of genetic determinants in a rapid and simultaneous manner and thus assists in various aspects of clinical, environmental and industrial microbiology. The technique assists in assessing genomic information in pathogenic species like as identification of their virulence-associated factors and antibiotic resistance genes. Various microarrays were developed for the detection of bacteria and the assessment of microbial community communities (Zhou 2003). Loy et al. (2002) created a microarray with probes specific to known groups of sulphate-reducing prokaryotes and used it for identification and characterization of most of the reference strains and their diversity analysis in various environmental habitats. This approach was also applied on other group of organisms also such as Rhodocyclales (Loy et al. 2005) and *Enterococcus* species (Lehner et al. 2005). DNA microarrays usually face problems of specificity, sensitivity, reproducibility and quantifiable level of diversity analysis in natural habitats due to their skewed distribution which facilitates cross-hybridization among closely related species. This has other related problems like genetic variation of strains and varying efficiency of DNA isolation too (Xu 2006).

5.6.4 Denaturing Gradient Gel Electrophoresis (DGGE)/ Temperature Gradient Gel Electrophoresis (TGGE)

DGGE and TGGE are the almost identical techniques to explore complex microbial community in any ecosystem. Both techniques have the similarity in basic principles besides there is a chemical denaturants gradient in DGGE while a temperature gradient is created in TGGE. A complex functional microbial community can be identified from different samples at a time by using this technique. These are fingerprinting techniques which provide a similar pattern of the samples processing and the information obtained. DGGE specifically gives fast comparative results and data across different communities and the specific phylogenetic information can be gained from the eliminated bands (Díez et al. 2001). DGGE based on the PCR amplification of 16S rRNA gene is a method to study the dynamic behaviour of complex microbial assemblages along with isolation of microorganisms in pure culture (Casamayor et al. 2000). DGGE has been widely used for investigation of distribution patterns of marine bacterial assemblages (Riemann et al. 1999; Schauer et al. 2000) and marine pico-eukaryotic organisms (Liu et al. 1997; Marsh 1999; Moeseneder et al. 1999; Pace 1997). DGGE is capable of detecting multiple species simultaneously even on a large scale (Zijnge et al. 2006). A major benefit of DGGE is that specific bands can be further proceeded for sequencing and thus in the environmental samples presence of any particular phylotype can be scrutinized (Casamayor et al. 2000; Muyzer et al. 1997). Though DGGE profiles are a representative fingerprint of the particular community under study (Zijnge et al. 2006).

5.6.5 Cloning

Cloning is one of the most widely used methods for analysis of uncultured microbes. It involves cloning of PCR product of natural sample and then sequencing of desired fragment of genes (DeSantis et al. 2007) and then further comparison with nucleotide sequence databases like the GenBank (NCBI) or Ribosomal Database Project (RDP) and then identification of clone on the basis of these results (DeSantis et al. 2007). Cloning of 16S rRNA genes allows to identify diversity of any region along with identification of novel taxa (if any). However, it is a cumbersome work, e.g. for describing 50% richness in any soil sample, more than 40,000 clones are required (Dunbar et al. 2002). On an average, clone libraries of 16S rRNA genes with <1000 sequences represent a very small fraction of diversity present in that area. With cloning, there exist many problems such as insufficient clone sequencing, labour-intensiveness and requiring more time and cost. However, despite these limitations, clone libraries still hold recognition of "gold standards" as far as initial microbial diversity survey is considered (DeSantis et al. 2007). As there is progress in sequencing methods and costs also reducing day by day, immense advancement is likely to occur in this approach of microbial diversity analysis.

5.6.6 Whole Microbial Genome Sequencing

As there is a boom in the number of whole genomes sequenced and with increasing technology, sequencing of entire microbial genome is a widespread and cumulative approach for better understanding of microbial ecology and functions. The techniques of short-read sequencing, e.g. pyrosequencing, have reduced the time and cost of microbial whole genome sequencing projects (Metzker 2010). Whole genome sequencing possesses the potential to unravel functions of microbes at a molecular level and can be directly employed for various applications like community ecology, bioremediation, bioenergy and many more (Ikeda et al. 2003; Kirka et al. 2004). Microbial Genomic Resources at NCBI (National Center for Biotechnology Information), a public repository of sequenced genomes of prokaryotes, helped researchers for generating comparative analysis of genomes and functions.

5.6.7 Quantitative Real-Time PCR (qRT-PCR)

Quantitative PCR methods like real-time PCR, TaqMan PCR and competitive PCR (cPCR) aim to quantify the number of gene copies/transcript level in environmental samples. The benefit of real-time PCR in comparison to other PCR-based quantification methods is that it spotlights on the logarithmic phase of product accumulation instead of end-product abundance. Therefore, in comparison to other techniques, it is more accurate as it is less affected by amplification efficiency or diminution of a reagent. This method is also free from the risk of contamination, as there is not any requirement of processing after PCR. A major drawback of real-time PCR is that it requires quite expensive instruments like thermocycler and reagents. Q-PCR or real-time PCR (RT-PCR) was used extensively in exploring microorganisms to evaluate their abundance using taxonomic and functional gene markers (Bustin et al. 2005; Smith and Osborn 2009). The technique has also been effectively employed for detection of physiological bacterial groups such as ammonia oxidizers, methane oxidizers, and sulphate reducers in the environmental samples quantitatively (Foti et al. 2007). The gene abundance of any functional or structural gene in terms of copy number in an ecological system can be estimated by quantitative real-time PCR method. This analysis requires the standard curve preparation from serially diluted copies of that gene product. In conclusion, the technique gives the real data of the functionality of an ecosystem.

5.7 Common Microbial Identification Methods

A few methods also exist which are used for identification of culturable as well as unculturable microbes. These include:

1. Repetitive DNA PCR
2. Ribotyping
3. DNA Sequencing

5.7.1 Repetitive DNA PCR

Repetitive DNA PCR also known as Rep PCR was developed by Versalovic et al. (1991) for the fingerprinting bacterial genomes through investigation of strain-specific patterns while using repetitive DNA elements for PCR amplification. In Rep PCR, repetitive extragenic palindromic (REP) and enterobacterial repetitive intergenic consensus (ERIC) sequence elements are used for typing purposes. REP fragments used were of palindromic nature and readily create stem-loop structures that facilitated multiple functions for highly conserved but dispersed elements (Newbury et al. 1987; Yang and Ames 1988). ERIC sequences are highly conserved central inverted repeat located in extragenic regions with 126 bp element (Sharples and Lloyd 1990; Hulton et al. 1991). BOX sequence (154 bp) is another repetitive element which is used to design the PCR primers (Versalovic et al. 1994). REP sequences, enterobacterial repetitive intergenic consensus (ERIC) sequences and BOX elements are frequently distributed bacteria genomes (Versalovic et al. 1991, 1994). Thus, usually three primer sets corresponding to REP, ERIC and BOX sequences, respectively, were commonly used for rep-PCR-based genomic fingerprinting. In general, the protocol is known as rep PCR, but specifically they are referred as REP-PCR, ERIC-PCR and BOX-PCR (Gillings and Holley 1997). Rep PCR was used to differentiate strains among the genetic diversity of plant pathogens. Each primer set is useful to fingerprint diverse bacteria and plant-associated actinomycetes (Clark et al. 1998; de Bruijn 1992; Louws et al. 1994, 1998; Rademaker et al. 1998). The technique has successfully classified and differentiated strains of *E. coli* (Lipman et al. 1995), *Rhizobium meliloti* (de Bruijn 1992), *Bradyrhizobium japonicum* (Judd et al. 1993), *Streptomyces* spp. and *Xanthomonas* spp. The rep PCR can also be applied in medical and environmental microbiology (Louws et al. 1997; Rademaker and de Bruijn 1997; Rademaker et al. 1998; Versalovic et al. 1998).

5.7.2 Ribotyping

Ribotyping also referred to as 'molecular fingerprinting' focus on genes encoding 16S rRNA for identifying microorganisms (Farber 1996; Hartel et al. 2002; Samadpour 2002). Ribotyping gives DNA fingerprints of genes coding for rRNA which are highly conserved in microorganisms. The genetic fingerprints of the bacterial isolates from various samples can also be compared. Ribotyping is one of the most rapid and specific method of bacterial identification, used worldwide (Reysenbach et al. 1992; Farag et al. 2001). The sequence of small-subunit rRNA differs in orderly manner crossways phylogenetic lines and encloses segments that are conserved at the species, genus, or kingdom level. By using oligonucleotide primers at sequences conserved throughout the eubacterial kingdom, Wilson et al. (1990) amplified bacterial 16S rDNA sequences with the PCR. Ribotyping also helps in upholding the amounts of bacterial 16S ribosomal DNA sequences from the point of sequencing and probing and also for detection and identification of known pathogens which are difficult to grow in laboratory (Table 5.1).

5.7.3 DNA Sequencing

Molecular methods are usually based on differences in the DNA sequence of distinguishing organism subtypes. Previously, DNA sequencing was based on the detection of radioactive labelling reaction products. Modern DNA sequencing methods utilize fluorescent nucleotides to label the DNA and reading the sequence with the instrument. In detail, the process involves PCR amplification of DNA followed by sequencing reactions of the products. Sometimes, RNA is also used as the starting material (Boettger 1989; Simpson et al. 2002). Compared to the techniques like PFGE, Rep-PCR or RAPD analysis that focus on whole chromosome, sequencing of the DNA considers only small part of sites that vary potentially among bacterial strains. The variability within the selected sequences should be enough for differentiating various strains of particular species. 16S rRNA genes have been constantly used for identification purposes of new organisms that have variations between strains (Lane et al. 1985; Woese 1987; Ward and Fraser 2005). The intergenic regions (16S-23S rRNA genes) have also been used to identify variability among organisms (Houpikian and Raoult 2001).

A comparative advantages and disadvantages of the methods in identification of microbial community is given in Table 5.2.

5.8 Conclusion

Microbial diversification parallel to the evolution of the Earth's history has blessed these organisms with inhabitation and adaptation capabilities in all kinds of habitats including those of extreme environments too (Li et al. 2014). Since wide diversification enabled these organisms biologically, genetically, metabolically and functionally diversified, it becomes more pertinent, dynamic and practically viable to uncover hidden diversity of microbial communities in any particular environment, establish a link between genetic and functional metabolic diversity and identify microorganisms with potential functions for utilization for agricultural and environmental benefits. Although there exists a huge volume of work on the identification and characterization of microorganisms from various habitats, it also faces plenty of limitations too in terms of techniques, methods, protocols and culture media conditions. Metagenomics have facilitated characterization of microbial taxonomy in a given habitat, and linking the same with the functions, this has to go a long way to prove its worth. New technologies are the array of hope in this area to decipher community taxonomy and functions while understanding ecological role assigned to the microorganisms by the nature in the complex habitats.

Table 5.2 Comparison of various molecular methods for their technical competence in the analysis of microbial processes and their identification

Techniques	Applications	Disadvantages
1. PCR	Sensitive, easy working, rapid	Semi-quantitative due to the kinetics of PCR product accumulation, dynamics only 1000-fold
2. RT-PCR	Sensitive and rapid	Semi-quantitative due to the kinetics of PCR product accumulation, dynamics only 1000-fold
3. c-PCR (integrated culture-polymerase chain reaction (C-PCR))	Most commonly used for revealing enteric viruses in environmental samples with high concentrations and provides sensitive, precise results in 2–5 days	Didn't work well for samples which have either low viral concentrations or possess toxic materials or inhibitors
4. Fingerprinting	Rapid means to screen large number of samples for comparative microbial community analysis	Major contribution by dominant microbes
5. Real-time PCR	Extremely sensitive, mostly used for quantitative measurements of gene transcription i.e. change in expression of any gene over a time period	Liable to inhibition by contaminants, if any in biological samples; incorrect data analysis, or unwarranted conclusions are also common
6. DNA microarray	The technique is used to identify bacteria species or to assess structural as well as functional bacterial diversity. Method is not confined to PCR biases and can contain thousands of target genes	It can only detect the most abundant species. Either enriched cultures or minimal diversity culture can be used otherwise the cross-hybridization can be a problem
7. Fluorescence in-situ hybridization (FISH)	FISH can be done at cellular level or in-situ, successfully used to study spatial distribution of bacteria in biofilms	Lack of sensitive when used with complex environmental samples
8. SIP	Able to identify microbes involved in any specific geochemical processes	Susceptible to biases due to the incubation with the isotope and further cycling of the stable isotope
9. Metagenomics	Useful for whole community analysis	Analysis and interpretation is difficult and cumbersome

Acknowledgement RP is thankful to DST for financial support under DST-Women Scientist Scheme-B (KIRAN Program) (Grant No. DST/WOS-B/2017/67-AAS).

References

Ahmad F, Ahmad I, Khan MS (2008) Screening of free-living rhizospheric bacteria for their multiple plant growth promoting activities. Microbiol Res 163:173–181

Alain K, Querellou J (2009) Cultivating the uncultured: limits, advances and future challenges. Extremophiles 13:583–594

Alves MH, Campos-Takaki GM, Porto ALF, Milanez AI (2002) Screening of *Mucor* spp. for the production of amylase, lipase, polygalacturonase and protease. Braz J Microbiol 33:325–330
Amann RI, Ludwig W, Schleifer KH (1995) Phylogenetic identification and in situ detection of individual microbial cells without cultivation. Microbiol Rev 59(1):143–169
Andersson DI, Jerlström-Hultqvist J, Näsvall J (2015) Evolution of new functions de novo and from preexisting genes. Cold Spring Harb Perspect Biol 7(6):a017996. https://doi.org/10.1101/cshperspect.a017996
Angel R, Conrad R, Dvorsky M, Kopecky M, Kotilínek M, Hiiesalu I, Schweingruber F, Doležal J (2016) The root-associated microbial community of the world's highest growing vascular plants. Microb Ecol 72:394–406
Anguilar P, Acosta E, Dorador C, Sommaruga R (2016) Large differences in bacterial community composition among three nearby extreme waterbodies of the high Andean plateau. Front Microbiol. https://doi.org/10.3389/fmicb.2016.00976
Antony R, Sanyal A, Kapse N, Dhakephalkar PK, Thamban M, Nair S (2016) Microbial communities associated with Antarctic snow pack and their biogeochemical implications. Microbiol Res 192:192–202. https://doi.org/10.1016/j.micres.2016.07.004
Badri DV, Weir TL, Van Der Lelie D, Vivanco JM (2009) Rhizosphere chemical dialogues: plant–microbe interactions. Curr Opin Biotechnol 20:642–650
Banoweltz GM, Whittaker GW, Dierksen KP, Azevedo MD, Kennedy AC, Griffith SM, Steiner JJ (2006) Fatty acid methyl ester analysis to identify source of soil in surface water. J Environ Qual 3:133–140
Bardgett RD, van der Putten WH (2014) Belowground biodiversity and ecosystem functioning. Nature 515:505–511
Beasley JC, Olson ZH, Devault TL (2012) Carrion cycling in food webs: comparisons among terrestrial and marine ecosystems. Oikos 121:1021–1026. https://doi.org/10.1111/j.1600-0706.2012.20353.x
Bell T et al (2005) The contribution of species richness and composition to bacterial services. Nature 436:1157–1160
Bell TH, Stefani FOP, Abram C, Champagne J, Yergeau E, Hijri M, St-Arnaud M (2016) A diverse soil microbiome degrades more crude oil than specialized bacterial assemblages obtained in culture. Appl Environ Microbiol 82:5530–5541. https://doi.org/10.1128/AEM.01327-16
Benson CA, Bizzoco RW, Lipson DA, Kelley ST (2011) Microbial diversity in nonsulfur, sulfur and iron geothermal steam vents. FEMS Microbiol Ecol 76:74–88
Bizzoco RLW, Kelley ST (2013) Microbial diversity in acidic high-temperature steam vents. In: Polyextremophiles. pp 315–332. https://doi.org/10.1007/978-94-007-6488-0_13
Bodelier PLE (2011) Toward understanding, managing, and protecting microbial ecosystems. Front Microbiol 2:80
Boettger EC (1989) Rapid determination of bacterial ribosomal RNA sequences by direct sequencing of enzymatically amplified DNA. FEMS Microbiol Lett 65:171–176
Bonfante P, Anca IA (2009) Plants, mycorrhizal fungi, and bacteria: a network of interactions. Annu Rev Microbiol 63:363–383
Boon E, Meehan CJ, Whidden C, Wong DH-J, Langille MGI, Beiko RG (2014) Interactions in the microbiome: communities of organisms and communities of genes. FEMS Microbiol Rev 38(1):90–118. https://doi.org/10.1111/1574-6976.12035
Bordbar A, Monk JM, King ZA, Palsson BO (2014) Constraint-based models predict metabolic and associated cellular functions. Nat Rev Genet 15:107–120
Briones AM, Okabe S, Umemiya Y, Ramsing NB, Reichardt W, Okuyama H (2002) Influence of different cultivars on populations of ammonia-oxidizing bacteria in the root environment of rice. Appl Environ Microbiol 68:3067–3075
Bruns A, Nübel U, Cypionka H, Overmann J (2003) Effect of signal compounds and incubation conditions on the culturability of freshwater bacterioplankton. J Appl Environ Microbiol 69:1980–1989
Burkepile DE et al (2006) Chemically mediated competition between microbes and animals: microbes as consumers in food webs. Ecology 87:2821–2831

Bustin SA, Benes V, Nolan T, Pfaffl MW (2005) Quantitative real-time RT-PCR – a perspective. J Mol Endocrinol 34:597–601
Cantarelli VC, Inamine E, Brodt TCZ, Secchi C, de Souza PF, Amaro MC, Batalha AA, Ligiero SD (2003) Quality control for microbiological culture media. Is it enough to follow the NCCLS M22-A2 procedures? Braz J Microbiol 34(Suppl.1):8–10. ISSN 1517-8382
Caracciolo AB, Bottoni P, Grenni P (2010) Fluorescence in situ hybridization in soil and water ecosystems: a useful method for studying the effect of xenobiotics on bacterial community structure. Toxicol Environ Chem 92:567–579
Casamayor EO, Schäfer H, Bañeras L, Pedrós-Alió C, Muyzer G (2000) Identification of and spatio-temporal differences between microbial assemblages from two neighboring sulfurous lakes: a comparison of microscopy and denaturing gradient gel electrophoresis. Appl Environ Microbiol 66:499–508
Chibucos MC, Tyler BM (2009) Common themes in nutrient acquisition by plant symbiotic microbes, described by the Gene Ontology. BMC Microbiol 9(Suppl 1):S6. https://doi.org/10.1186/1471-2180-9-S1-S6
Cho J-C, Tiedje JM (2001) Bacterial species determination from DNA–DNA hybridization by using genome fragments and DNA microarrays. Appl Environ Microbiol 67:3677–3682
Clark CA, Chen C, Ward-Rainey N, Pettis GS (1998) Diversity within Streptomyces ipomoeae based on inhibitory interactions, rep-PCR, and plasmid profiles. Phytopathology 88(11):1179–1186. https://doi.org/10.1094/PHYTO.1998.88.11.1179
Cline LC, Zak DR (2015) Soil microbial communities are shaped by plant-driven changes in resource availability during secondary succession. Ecology 96:3374–3385
Cook AE, Meyers PR (2003) Rapid identification of filamentous actinomycetes to the genus level using genus-specific 16S rRNA gene restriction fragment patterns. Int J Syst Evol Microbiol 53:1907
Cook D, Gardner DR, Pfister JA, Grum D (2004) Biosynthesis of natural products in plants by fungal endophytes with an emphasis on Swainsonine. In: Jetter R (ed) Phytochemicals – biosynthesis, functions and applications, Recent advances in phytochemistry. Springer, Switzerland, pp 23–34
Dadheech PK, Glöckner G, Casper P, Kotut K, Mazzoni CJ, Mbedi S, Krienitz L (2013) Cyanobacterial diversity in the hot spring, pelagic and benthic habitats of a tropical soda lake. FEMS Microbiol Ecol 85:389–401
Das S, Lyla PS, Khan SA (2006) Marine microbial diversity and ecology: importance and future perspectives
De Bruijn FJ (1992) Use of repetitive (repetitive extragenic palindromic and enterobacterial repetitive intergenic consensus) sequences and the polymerase chain reaction to fingerprint the genomes of Rhizobium meliloti isolates and other soil bacteria. Appl Environ Microbiol 58:2180–2187
de Carvalho CCCR, Caramujo M-J (2014) Fatty acids as a tool to understand microbial diversity and their role in food webs of Mediterranean temporary ponds. Molecules 19:5570–5598
De Maayer P, Anderson D, Cary C, Cowan DA (2014) Some like it cold: understanding the survival strategies of psychrophiles. EMBO Rep 15:508–517. https://doi.org/10.1002/embr.201338170
de Souza AJ, de Andrade PAM, Pereira AP d A, Andreote FD, Tornisielo VL, Regitano JB (2017) The depleted mineralization of the fungicide chlorothalonil derived from loss in soil microbial diversity. Sci Rep 7:14646. https://doi.org/10.1038/s41598-017-14803-0
de Wit R, Bouvier T (2006) 'Everything is everywhere, but, the environment selects'; what did Baas Becking and Beijerinck really say? Environ Microbiol 8:755–758
Delgado-Baquerizo M, Maestre FT, Reich PB, Jeffries TC, Gaitan JC, Encinar D, Berdugo M, Campbell CD, Singh BK (2016) Microbial diversity drives multifunctionality in terrestrial ecosystems. Nat Commun 7:10541. https://doi.org/10.1038/ncomms10541
DeLong EF (2005) Microbial community genomics in the ocean. Nat Rev Microbiol 3:459–469

Derry AM, Staddon WJ, Kevan PG, Trevors JT (1999) Functional diversity and community structure of micro-organisms in three arctic soils as determined by sole-carbon-source-utilization. Biodivers Conserv 8:205–221

DeSantis TZ, Brodie EL, Moberg JP, Zubieta IX, Piceno YM, Andersen GL (2007) High-density universal 16S rRNA microarray analysis reveals broader diversity than typical clone library when sampling the environment. Microb Ecol 53:371–383

Díez B, Pedrós-Alio C, Marsh TL, Massana R (2001) Application of denaturing gradient gel electrophoresis (DGGE) to study the diversity of marine picoeukaryotic assemblages and comparison of DGGE with other molecular techniques. Appl Environ Microbiol 67(7):2942

Dimitriu PA, Pinkart HC, Peyton BM, Mormile MR (2008) Spatial and temporal patterns in the microbial diversity of a meromictic soda lake in Washington State. Appl Environ Microbiol 74(15):4877–4888. https://doi.org/10.1128/AEM.00455-08

Dix NJ, Webster J (1995) Fungal ecology. Chapman & Hall, London

Dunbar J, Barns SM, Ticknor LO, Kuske CR (2002) Empirical and theoretical bacterial diversity in four Arizona soils. Appl Environ Microbiol 68:3035–3045

Ehrhardt CJ, Chu V, Brown TC, Simmons TL, Swan BK, Bannan J, Robertson JM (2010) Use of fatty acid methyl ester profiles for discrimination of *Bacillus cereus* T-strain spores grown on different media. Appl Environ Microbiol 76:1902–1912. https://doi.org/10.1128/AEM.02443-09

Ehrhardt CJ, Murphy DL, Robertson JM, Bannan JD (2015) Fatty acid profiles for differentiating growth medium formulations used to culture *Bacillus cereus* T-strain spores. J Forensic Sci 60(4):1022–1029. https://doi.org/10.1111/1556-4029.12771. Epub 2015 Apr 9

Elena SF, Lenski RE (2003) Microbial genetics: evolution experiments with microorganisms: the dynamics and genetic bases of adaptation. Nat Rev Genet 4:457–469. https://doi.org/10.1038/nrg1088

Ellis RJ, Morgan P, Weightman AJ, Fry JC (2003) Cultivation-dependent and -independent approaches for determining bacterial diversity in heavy-metal-contaminated soil. Appl Environ Microbiol 69:3223–3230. https://doi.org/10.1128/AEM.69.6.3223-3230

Ellis DG, Bizzoco RW, Kelley ST (2008) Halophilic Archaea determined from geothermal steam vent aerosols. Environ Microbiol 10(6):1582–1590. https://doi.org/10.1111/j.1462-2920.2008.01574.x

Embree M, Liu JK, Al-Bassam MM, Zengler K (2015) Networks of energetic and metabolic interactions define dynamics in microbial communities. Proc Natl Acad Sci U S A 112:15450–15455

Farag AM, Goldstein JN, Woodward DF, Samadpour M (2001) Water quality in three creeks in the backcountry of Grand Teton National Park, USA. J Freshw Ecol 16:135–143

Farber JM (1996) An introduction to the hows and whys of molecular typing. J Food Protect 59:1091–1101

Farrar K, Bryant D, Cope-Selby N (2014) Understanding and engineering beneficial plant–microbe interactions: plant growth promotion in energy crops. Plant Biotechnol J 12:1193–1206. https://doi.org/10.1111/pbi.12279

Faust K, Raes J (2012) Microbial interactions: from networks to models. Nat Rev Microbiol 10:538–550

Feil EJ, Enright MC (2004) Analyses of clonality and the evolution of bacterial pathogens. Curr Opin Microbiol 7:308–313

Fierer N, Jackson RB (2006) The diversity and biogeography of soil bacterial communities. Proc Natl Acad Sci U S A 103:626–631

Fisher MM, Triplett EW (1999) Automated approach for ribosomal intergenic spacer analysis of microbial diversity and its application to freshwater bacterial communities. Appl Environ Microbiol 65:4630–4636

Foti M, Sorokin DY, Lomans B, Mussman M, Zacharova EE, Pimenov NV, Kuenen JG, Muyzer G (2007) Diversity, activity, and abundance of sulfate-reducing bacteria in saline and hypersaline soda lakes. Appl Environ Microbiol 73:2093–3000

Franklin RB, Taylor DR, Mills AL (1999) Characterization of microbial communities using randomly amplified polymorphic DNA (RAPD). J Microbiol Methods 35:225–235

Freilich S, Zarecki R, Eilam O, Segal ES, Henry CS, Kupiec M et al (2011) Competitive and cooperative metabolic interactions in bacterial communities. Nat Commun 2:589
Garbeva P, van Veen JA, van Elsas JD (2004) Microbial diversity in soil: selection microbial populations by plant and soil type and implications for disease suppressiveness. Annu Rev Phytopathol 42:243–270
Garland JL, Mills AL (1991) Classification and characterization of heterotrophic microbial communities on the basis of patterns of community-level sole-carbon-source utilization. Appl Environ Microbiol 57(8):2351–2359
Ghazanfar S, Azim A, Ghazanfar MA, Anjum MI, Begum I (2010) Metagenomics and its application in soil microbial community studies: biotechnological prospects. J Anim Plant Sci 6(2):611–622
Gibbons SM, Gilbert JA (2015) Microbial diversity—exploration of natural ecosystems and microbiomes. Curr Opin Genet Dev 35:66–72. https://doi.org/10.1016/j.gde.2015.10.003
Gillings M, Holley M (1997) Repetitive element PCR fingerprinting (rep-PCR) using enterobacterial repetitive intergenic consensus (ERIC) primers is not necessarily directed at ERIC elements. Lett Appl Microbiol 25(1):17–21
Glaring MA, Vester JK, Lylloff JE, Abu Al-Soud W, Sørensen SJ, Stougaard P (2015) Microbial diversity in a permanently cold and alkaline environment in Greenland. PLoS One 10(4):e0124863. https://doi.org/10.1371/journal.pone.0124863
Green CT, Scow KM (2000) Analysis of phospholipid fatty acids (PLFA) to characterize microbial communities in aquifers. Hydrogeol J 8:126–141
Griffiths RI, Whiteley ADS, O'Donnell AG, Bailey MJ (2000) Rapid method for coextraction of DNA and RNA from natural environments for analysis of ribosomal DNA- and rRNA-based microbial community composition. Appl Environ Microbiol 66(12):5488
Gronstal A (2016) A view into the complexity of microbial communities. Nat Sci Rep. https://astrobiology.nasa.gov/news/a-view-into-the-complexity-of-microbial-communities/
Gunatilaka AAL (2006) Natural products from plant-associated microorganisms: distribution, structural diversity, bioactivity, and implications of their occurrence. J Nat Prod 69:509–526. https://doi.org/10.1021/np058128n
Guttman DS, McHardy AC, Schulze-Lefert P (2014) Microbial genome-enabled insights into plant–microorganism interactions. Nat Rev Genet 15:797–813. https://doi.org/10.1038/nrg3748
Haizhou Li H, Yang Q, Li J, Gao H, Ping Li P, Zhou H (2015) The impact of temperature on microbial diversity and AOA activity in the Tengchong Geothermal Field, China. Sci Rep 5:17056. https://doi.org/10.1038/srep17056
Hamdan LJ, Coffin RB, Sikaroodi M, Greinert J, Treude T, Gillevet PM (2013) Ocean currents shape the microbiome of Arctic marine sediments. ISME J 7(4):685–696. pmid:23190727
Hardoim PR, van Overbeek LS, Berg G, Pirttilä AM, Compant S, Campisano A, Döring M, Sessitsch A (2015) The hidden world within plants: ecological and evolutionary considerations for defining functioning of microbial endophytes. Microbiol Mol Biol Rev 79(3):293–320
Hartel PG, Summer JD, Hill JL, Collins JV, Entry JA, Segars WI (2002) Geographic variability of *Escherichia coli* ribotypes from animals in Idaho and Georgia. J Environ Qual 31(4):1273–1278
Heinken A, Thiele I (2015) Systems biology of host-microbe metabolomics. Wiley Interdiscip Rev Syst Biol Med 7:195–219
Henneberger RM, Walter MR, Anitori RP (2006) Extraction of DNA from acidic, hydrothermally modified volcanic soils. Environ Chem 3(2):100–104. https://doi.org/10.1071/EN06013
Hill JE, Seipp RP, Betts M, Hawkins L, Van Kessel AG, Crosby WL, Hemmingsen SM (2002) Extensive profiling of a complex microbial community using high throughput sequencing. Appl Environ Microbiol 68(6):3055–3066
Hodgson S, Cates C, Hodgson J, Morley NJ, Sutton BC, Gange AC (2014) Vertical transmission of fungal endophytes is widespread in forbs. Ecol Evol 4:1199–1208. https://doi.org/10.1002/ece3.953
Houpikian P, Raoult D (2001) 16S/23S rRNA intergenic spacer regions for phylogenetic analysis, identification, and subtyping of Bartonella species. J Clin Microbiol 39(8):2768–2778

Hugenholtz P (2002) Exploring prokaryotic diversity in the genomic era. Genome Biol 3:Reviews 0003
Hughes JB, Hellmann JJ, Ricketts TH, Bohannan BJM (2001) Counting the uncountable: statistical approaches to estimating microbial diversity. Appl Environ Microbiol 67(10):4399–4406
Hulton CSJ, Higgins CF, Sharp PM (1991) ERIC sequences: a novel family of repetitive elements in the genomes of *Escherichia coli*, Salmonella typhimurium and other enterobacteria. Mol Microbiol 5:825–834
Hussein EI, Jacob JH, Shakhatreh MAK, Al-razaq MAA, Juhmani AS-F, Cornelison CT (2017) Exploring the microbial diversity in Jordanian hot springs by comparative metagenomic analysis. MicrobiologyOpen 6:e00521. https://doi.org/10.1002/mbo3.521
Ikeda H, Ishikawa J, Hanamoto A, Shinose M, Kikuchi H, Shiba T, Sakaki Y, Hattori M, Omura S (2003) Complete genome sequence and comparative analysis of the industrial micro-organism Streptomyces avermitilis. Nat Biotechnol 21:526–531
Imchen M, Kumavath R, Barh D, Azevedo V, Ghosh P, Viana M, Wattam AR (2017) Searching for signatures across microbial communities: metagenomic analysis of soil samples from mangrove and other ecosystems. Sci Rep 7:8859. https://doi.org/10.1038/s41598-017-09254-6
Janssen P (2008) New cultivation strategies for terrestrial microorganisms. In: Zengler K (ed) Accessing uncultivated microorganisms. ASM Press, Washington, DC, pp 173–192. https://doi.org/10.1128/9781555815509.ch10
Jansson JK, Baker ES (2016) A multi-omic future for microbiome studies. Nat Microbiol 1:16049
Jett BD, Hatter KL, Huycke MM, Gilmore MS (1997) Simplified agar plate method for quantifying viable bacteria. BioTechniques 23:648–650
Joint I, Mühling M, Querellou J (2010) Culturing marine bacteria – an essential prerequisite for biodiscovery. Microb Biotechnol 3(5):564–575. https://doi.org/10.1111/j.1751-7915.2010.00188.x
Judd AK, Schneider M, Sadowsky MJ, de Bruijn FJ (1993) Use of repetitive sequences and the polymerase chain reaction technique to classify genetically related Bradyrhizobiumjaponicum serocluster 123 strains. Appl Environ Microbiol 59:1702–1708
Kalwasińska A, Felföldi T, Szabó A, Deja-Sikora E, Kosobucki P, Walczak M (2017) Microbial communities associated with the anthropogenic, highly alkaline environment of a saline soda lime, Poland. Antonie Van Leeuwenhoek 110(7):945–962. https://doi.org/10.1007/s10482-017-0866-y
Keller M, Zengler K (2004) Tapping into microbial diversity. Nat Rev Microbiol 2:141–150
Kelly JJ, Haggblom M, Tate RL III (1999) Changes in soil microbial communities over time resulting from one time application of zinc: a laboratory microcosm study. Soil Biol Biochem 31:1455–1465
Keshri J, Mishra A, Jha B (2013) Microbial population index and community structure in saline–alkaline soil using gene targeted metagenomics. Microbiol Res 168:165–173. https://doi.org/10.1016/j.micres.2012.09.005
Kirchman DL, Cottrell MT, Lovejoy C (2010) The structure of bacterial communities in the western Arctic Ocean as revealed by pyrosequencing of 16S rRNA genes. Environ Microbiol 12(5):1132–1143
Kirka JL, Beaudettea LA, Hartb M, Moutoglisc P, Klironomosb JN, Lee H, Trevors JT (2004) Methods of studying soil microbial diversity. J Microbiol Methods 58:169–188
Köpke B, Wilms R, Engelen B, Cypionka H, Sass H (2005) Microbial diversity in coastal subsurface sediments: a cultivation approach using various electron acceptors and substrate gradients. Appl Environ Microbiol 71:7819–7830
Krienitz L, Dadheech PK, Kotut K (2013) Mass developments of the cyanobacteria Anabaenopsis and Cyanospira (Nostocales) in the soda lakes of Kenya: ecological and systematic implications. Hydrobiologia 703:79–93
Ku C, Martin WF (2016) A natural barrier to lateral gene transfer from prokaryotes to eukaryotes revealed from genomes: the 70 % rule. BMC Biol 14:89. https://doi.org/10.1186/s12915-016-0315-9

Kuczynski J, Liu Z, Lozupone C, McDonald D, Fierer N, Knight R (2010) Microbial community resemblance methods differ in their ability to detect biologically relevant patterns. Nat Methods 7:813–819

Kusari S, Singh S, Jayabaskaran C (2014a) Biotechnological potential of plant-associated endophytic fungi: hope versus hype. Trends Biotechnol 32:297–303. https://doi.org/10.1016/j.tibtech.2014.03.009

Kusari S, Lamsho M, Kusari P, Gottfried S, Zuhlke S, Louven K et al (2014b) Endophytes are hidden producers of maytansine in Putterlickia roots. J Nat Prod 77:2577–2584. https://doi.org/10.1021/np500219a

Lane N (2011) Energetics and genetics across the prokaryote-eukaryote divide. Biol Direct 6:35. https://doi.org/10.1186/1745-6150-6-35

Lane DJ, Pace B, Olsen GJ, Stahl DA, Sogin ML, Pace NR (1985) Rapid determination of 16S ribosomal RNA sequences for phylogenetic analyses. Proc Natl Acad Sci U S A 82:6955–6959

Lareen A, Burton F, Schäfer P (2016) Plant root-microbe communication in shaping root microbiomes. Plant Mol Biol 90:575–587

Larsen PE, Gibbons SM, Gilbert JA (2012) Modeling microbial community structure and functional diversity across time and space. FEMS Microbiol Lett 332:91–98. https://doi.org/10.1111/j.1574-6968.2012.02588.x

Laurent F, Provost F, Boiron P (1999) Rapid identification of clinically relevant *Nocardia* species to genus level by 16S rRNA gene PCR. J Clin Microbiol 37:99–102

Lee E, Ryan UM, Monis P, McGregor GB, Bath A, Gordon C, Paparini A (2014) Polyphasic identification of cyanobacterial isolates from Australia. Water Res 59:248–261. https://doi.org/10.1016/j.watres.2014.04.023

Lehner A, Loy A, Behr T, Gaenge H, Ludwig W, Wagner M, Schleifer K-H (2005) Oligonucleotide microarray for identification of Enterococcus species. FEMS Microbiol Lett 246:133–142

Li T, Wu T-D, Mazeas L, Toffin L, Guerquin-Kern J-L, Leblon G, Bouchez T (2008) Simultaneous analysis of microbial identity and function using NanoSIMS. Environ Microbiol 10: 580–588.

Li S-J, Hua Z-S, Huang L-N, Li J, Shi S-H, Chen L-X, Kuang J-L, Liu J, Hu M, Shu W-S (2014) Microbial communities evolve faster in extreme environments. Sci Rep 4:6. https://doi.org/10.1038/srep06205

Li Y, Adams J, Yu Shi Y, Wang H, He J-S, Chu H (2017) Distinct soil microbial communities in habitats of differing soil water balance on the Tibetan Plateau. Sci Rep. Article Number 46407. https://doi.org/10.1038/srep46407

Lindow SE, Brandl MT (2003) Microbiology of the phyllosphere. Appl Environ Microbiol 69:1875–1883

Lipman LJ, de Nijs A, Gaastra W. Isolation and identification of fimbriae and toxin production by Escherichia coli strains from cows with clinical mastitis. Vet Microbiol. 1995;47(1-2):1–7.

Liu WL, Marsh TL, Cheng H, Forney LJ (1997) Characterization of microbial diversity by determining terminal restriction fragment length polymorphisms of genes encoding 16S rRNA. Appl Environ Microbiol 63:4516–4522

Louws FJ, Fulbright DW, Stephens CT, de Bruijn FJ (1994) Specific genomic fingerprints of phytopathogenic *Xanthomonas* and *Pseudomonas* pathovars and strains generated with repetitive sequences and PCR. Appl Environ Microbiol 60:2286–2295

Louws FJ, Schneider M, de Bruijn FJ (1997) Assessing genetic diversity of microbes using repetitive sequence-based PCR (rep-PCR). In: Toranzos G (ed) Nucleic acid amplification methods for the analysis of environmental microbes. Technomic Publishing, Lancaster, pp 63–94

Louws FJ, Bell J, Medina-Mora CM, Smart CD, Opgenorth D et al (1998) Rep-PCR-mediated genomic fingerprinting: a rapid and effective method to identify *Clavibacter michiganensis*. Phytopathology 88:862–868

Loy A, Lehner A, Lee N, Adamczyk J, Meier H, Ernst J, Schleifer KH, Wagner M. Oligonucleotide microarray for 16S rRNA gene-based detection of all recognized lineages of sulfate-reducing prokaryotes in the environment. Appl Environ Microbiol. 2002;68(10):5064–81.

Loy A et al (2005) 16S rRNA gene-based oligonucleotide microarray for environmental monitoring of the betaproteobacterial order "Rhodocyclales". Appl Environ Microbiol 71:1373–1386

Maiden MC, Bygraves JA, Feil E, Morelli G, Russell JE, Urwin R, Zhang Q, Zhou J, Zurth K, Caugant DA, Feavers IM, Achtman M, Spratt BG (1998) Multilocus sequence typing: a portable approach to the identification of clones within populations of pathogenic microorganisms. Proc Natl Acad Sci U S A 95(6):3140–3145

Makhalanyane TP, Valverde A, Gunnigle E, Frossard A, Ramond J-B, Cowan DA (2015) Microbial ecology of hot desert edaphic systems. FEMS Microbiol Rev 39:203–221. https://doi.org/10.1093/femsre/fuu011

Manefield M, Whiteley AS, Griffiths RI, Bailey MJ (2002) RNA stable isotope probing, a novel means of linking microbial community function to phylogeny. Appl Environ Microbiol 68(11):5367–5373

Marsh TL (1999) Terminal restriction fragment length polymorphism (T-RFLP): an emerging method for characterizing diversity among homologous populations of amplification products. Curr Opin Microbiol 2:323–327

Mateos-Rivera A, Yde JC, Wilson B, Finster KW, Reigstad LJ, Øvreås L (2016) The effect of temperature change on the microbial diversity and community structure along the chronosequence of the sub-arctic glacier forefield of Styggedalsbreen (Norway). FEMS Microbiol Ecol 92(4):fnw038. https://doi.org/10.1093/femsec/fiw038

Meena KK, Kumar M, Kalyuzhnaya MA, Yandigeri MS, Singh DP, Saxena AK, Arora DK (2012) Epiphytic pink-pigmented methylotrophic bacteria enhance germination and seedling growth of wheat (*Triticum aestivum*) by producing phytohormone. Antonie Van Leeuwenhoek 101:777–786. https://doi.org/10.1007/s10482-011-9692

Meena et al (2017) Abiotic stress responses and microbe-mediated mitigation in plants: the omics strategies. Front Plant Sci. https://doi.org/10.3389/fpls.2017.00172

Metzker ML (2010) Sequencing technologies – the next generation. Nat Rev Genet 11:31–46

Miki T et al (2014) Biodiversity and multifunctionality in a microbial community: a novel theoretical approach to quantify functional redundancy. Proc R Soc Lond B 281:20132498

Mobarry BK, Wagner M, Urbain V, Rittmann BE, Stahl DA (1996) Phylogenetic probes for analyzing abundance and spatial organization of nitrifying bacteria. Appl Environ Microbiol 62:2156–2162

Mochimaru H, Yoshioka H, Tamaki H, Nakamura K, Kaneko N, Sakata S, Imachi H, Sekiguchi Y, Uchiyama H, Kamagata Y (2007) Microbial diversity and methanogenic potential in a high temperature natural gas field in Japan. Extremophiles 11:453–461

Moeseneder MM, Arrieta JM, Muyzer G, Winter C, Herndl GJ (1999) Optimization of terminal-restriction fragment length polymorphism analysis for complex marine bacterioplankton communities and comparison with denaturing gradient gel electrophoresis. Appl Environ Microbiol 65:3518–3525

Mohana Kumara P, Zuehlke S, Priti V, Ramesha BT, Shweta S, Ravikanth G et al (2012) Fusarium proliferatum an endophytic fungus from Dysoxylum binectariferum Hook.f, produces rohutikine, a chromane alkaloid possessing anti-cancer activity. Antonie Van Leeuwenhoek 101:323–329. https://doi.org/10.1007/s10482-011-9638-2

Monard C, Gantner S, Bertilsson S, Hallin S, Stenlid J (2016) Habitat generalists and specialists in microbial communities across a terrestrial-freshwater gradient. Sci Rep 6:37719. https://doi.org/10.1038/srep37719

Moore BS, Hertweck C, Hopke JN, Izumikawa M, Kalaitzis JA, Nilsen G et al (2002) Plant-like biosynthetic pathways in bacteria: from benzoic acid to chalcone. J Nat Prod 65:1956–1962. https://doi.org/10.1021/np020230m

Muyzer G (1999) DGGE/TGGE a method for identifying genes from natural ecosystems. Curr Opin Microbiol 2:317–322

Muyzer G, Brinkhoff T, Nübel U, Santegoeds C, Schäfer H, Wawer C (1997) Denaturing gradient gel electrophoresis (DGGE) in microbial ecology. In: Akkermans ADL, van Elsas JD, de Bruijn FJ (eds) Molecular microbial ecology manual, vol 3.4.4. Kluwer Academic Publishers, Dordrecht, pp 1–27

Nazina TN, Shestakova NM, Semenova EM, Korshunova AV, Kostrukova NK, Tourova TP, Min L, Feng Q, Poltaraus AB (2017) Diversity of metabolically active bacteria in water-flooded high-temperature heavy oil reservoir. Front Microbiol. https://doi.org/10.3389/fmicb.2017.00707

Newbury SF, Smith NH, Robinson EC, Hiles ID, Higgins CF (1987) Stabilization of translationally active mRNA by prokaryotic REP sequences. Cell 48:297–310
Newman DJ, Cragg GM (2015) Endophytic and epiphytic microbes as "sources" of bioactive agents. Front Chem. https://doi.org/10.3389/fchem.2015.00034
O'Callaghan M (2016) Microbial inoculation of seed for improved crop performance: issues and opportunities. Appl Microbiol Biotechnol 100:5729–5746
Olff H, Alonso D, Berg MP, Eriksson BK, Loreau M, Piersma T, Rooney N (2009) Parallel ecological networks in ecosystems. Philos Trans R Soc Lond Ser B Biol Sci 364:1755–1779. https://doi.org/10.1098/rstb.2008.0222
Pace NR (1997) A molecular view of microbial diversity and the biosphere. Science 276:734–740
Partida-Martinez LP, Hertweck C (2005) Pathogenic fungus harbours endosymbiotic bacteria for toxin production. Nature 437:884–888. https://doi.org/10.1038/nature03997
Perez-Garcia O, Lear G, Singhal N (2016) Metabolic network modeling of microbial interactions in natural and engineered environmental systems. Front Microbiol 7:673–703
Pernthaler A, Preston CM, Pernthaler J, DeLong EF, Amann R (2002) A comparison of fluorescently labeled oligonucleotide and polynucleotide probes for the detection of pelagic marine bacteria and archaea. Appl Environ Microbiol 68:661–667
Peter H et al (2011) Function-specific response to depletion of microbial diversity. ISME J 5:351–361
Philippot L et al (2013) Loss in microbial diversity affects nitrogen cycling in soil. ISME J 7:1609–1619
Popa O, Dagan T (2011) Trends and barriers to lateral gene transfer in prokaryotes. Curr Opin Microbiol 14:615–623. https://doi.org/10.1016/j.mib.2011.07.027
Potshangbam M, Devi SI, Sahoo D, Strobel GA (2017) Functional characterization of Endophytic fungal community associated with *Oryza sativa* L. and *Zea mays* L. Front Microbiol 8
Prakash O, Verma M, Sharma P, Kumar M, Kumari K, Singh A, Kumari H, Jit S, Gupta SK, Khanna M, Lal R (2007) Polyphasic approach of bacterial classification—an overview of recent advances. Indian J Microbiol 47(2):98–108. https://doi.org/10.1007/s12088-007-0022-x
Proal AD, Lindseth IA, Marshall TG (2017) Microbe-microbe and host-microbe interactions drive microbiome dysbiosis and inflammatory processes. Discov Med 23(124):51–60
Prosser JI (2015) Dispersing misconceptions and identifying opportunities for the use of 'omics' in soil microbial ecology. Nat Rev Microbiol 13:439–446. https://doi.org/10.1038/nrmicro3468
Rademaker JLW, de Bruijn FJ (1997) Characterization and classification of microbes by rep-PCR genomic finger-printing and computer assisted pattern analysis. In: Caetano-Anolles G, Gresshoff PM (eds) DNA markers: protocols, applications and overviews. Wiley, New York, pp 151–171
Rademaker JLW, Louws FJ, Versalovic J, De Bruijn FJ (1998) Characterization of the diversity of ecologically important microbes by rep-PCR genomic fingerprinting. In: Molecular microbial ecology manual, supplement. Kluwer Academic Publishers, Dordrecht, pp 1–16
Ranjard L et al (2001) Characterization of bacterial and fungal soil communities by automated ribosomal intergenic spacer analysis fingerprints: biological and methodological variability. Appl Environ Microbiol 67:4479–4487
Rastogi G, Sani RK (2011) Molecular techniques to assess microbial community structure, function, and dynamics in the environment. In: Ahmad I et al (eds) Microbes and microbial technology: agricultural and environmental applications. Springer, New York, pp 29–58
Ravin NV, Mardanova AV, Skryabin KG (2015) Metagenomics as a tool for the investigation of uncultured microorganisms. Genetika 51:519–528
Reysenbach A-L, Giver LJ, Wickham GS, Pace NR (1992) Differential amplification of rRNA genes by polymerase chain reaction. Appl Environ Microbiol 58:3417–3418
Riemann L, Steward GF, Fandino LB, Campbell L, Landry MR, Azam F (1999) Bacterial community composition during two consecutive NE Monsoon periods in the Arabian Sea studied by denaturing gradient gel electrophoresis (DGGE) of rRNA genes. Deep-Sea Res 46:1791–1811
Röling WF, Ferrer M, Golyshin PN (2010) Systems approaches to microbial communities and their functioning. Curr Opin Biotechnol 21:532–538
Roselló-Mora R, Amann R (2001) The species concept for prokaryotes. FEMS Microbiol Rev 25:39–67

Roszak DB, Colwell RR (1987) Survival strategies of bacteria in the natural environment. Microbiol Rev 51:365–379

Salazar N, de los Reyes-Gavilan CG (2016) Editorial: insights into microbe–microbe interactions in human microbial ecosystems: strategies to be competitive. Front Microbiol. https://doi.org/10.3389/fmicb.2016.01508

Samadpour M (2002) Microbial source tracking: principles and practice. In: Microbiological source tracking workshop-abstracts. February 5, 2002. Irvine, CA. NWRI Abstract Report NWRI-02-01. National Water Research Institute, Fountain Valley, pp 5–10

Sanders ER (2012) Aseptic laboratory techniques: plating methods. J Vis Exp 63:3064

Sathya A, Vijayabharathi R, Gopalakrishnan S (2016) Soil microbes: the invisible managers of soil fertility. In: Microbial inoculants in sustainable agricultural productivity. Springer, New York, pp 1–16

Schauer M, Massana R, Pedros-Alio C (2000) Spatial differences in bacterioplankton composition along the Catalan coast (NW Mediterranean) assessed by molecular fingerprinting. FEMS Microbiol Ecol 33:51–59

Scherer C, Muller K-D, Rath P-M, Ansorg RAM (2003) Influence of culture conditions on the fatty acid profiles of laboratory-adapted and freshly isolated strains of Helicobacter pylori. J Clin Microbiol 41:1114–1117

Schloss PD, Handelsman J (2004) Status of the microbial census. Microbiol Mol Biol Rev 68(4):686–691

Schwieger F, Tebbe CC (1998) A new approach to utilize PCR-single-strand conformation polymorphism for 16S rRNA gene-based microbial community analysis. Appl Environ Microbiol 64:4870–4876

Sekora NS, Lawrence KS, Agudelo P, van Santen E, McInroy JA (2009) Using FAME analysis to compare, differentiate, and identify multiple nematode species. J Nematol 41:163–173

Sharples GJ, Lloyd RG (1990) A novel repeated DNA sequence located in the intergenic regions of bacterial chromosomes. Nucleic Acids Res 18(22):6503–6508

Siciliano SD, Germida JJ (1998) Biolog analysis and fatty acid methyl ester profiles indicate that Pseudomonad inoculants that promote phytoremediation alter the root-associated microbial community of *Bromus biebersteinii*. Soil Biol Biochem 30:1717–1723

Simpson JM, Santo Domingo JW, Reasoner DJ (2002) Microbial source tracking: state of the science. Environ Sci Tech 36(24):5279–5288

Singh DP, Singh HB (2014) Trends in soil microbial ecology. Studium Press LLC, Houston

Sjöling S, Cowan DA (2003) High 16S rDNA bacterial diversity in glacial meltwater lake sediment, Bratina Island, Antarctica. Extremophiles 7(4):275–282

Skilbeck G (2012) Drastic measures: a revised estimate of Earth's microbes. http://theconversation.com/drastic-measures-a-revised-estimate-of-earths-microbes-9101

Sloan WT, Lunn M, Woodcock S, Head IM, Nee S, Curtis TP (2006) Quantifying the roles of immigration and chance in shaping prokaryote community structure. Environ Microbiol 8:732–740

Smith DR, Chapman MR (2017) Economical evolution: microbes reduce the synthetic cost of extracellular proteins. mBio 1(3):e00131–e00110. https://doi.org/10.1128/mBio.00131-10

Smith CJ, Osborn AM (2009) Advantages and limitations of quantitative PCR (Q-PCR)-based approaches in microbial ecology. FEMS Microbiol Ecol 67(1):6–20

Smith E, Leeflang P, Wernars K (1997) Detection of shifts in microbial community structure and diversity in soil caused by copper contamination using amplified ribosomal DNA restriction analysis. FEMS Microbiol Ecol 23:249–261

Sorokin DY, Berben T, Melton ED, Overmars L, Vavourakis CD, Muyzer G (2014) Microbial diversity and biogeochemical cycling in soda lakes. Extremophiles 18:791–809

Steingrube VA, Wilson RW, Brown BA, Jost KC Jr, Blacklock Z, Gibson JL, Wallace RJ Jr (1997) Rapid identification of the clinically significant species and taxa of aerobic actinomycetes, including Actinomadura, Gordona, Nocardia, Rhodococcus, Streptomyces, and Tsukamurella isolates, by DNA amplification and restriction endonuclease analysis. J Clin Microbiol 35:817–822

Steven B, Gallegos-Graves LV, Belnap J, Kuske CR (2013) Dryland soil microbial communities display spatial biogeographic patterns associated with soil depth and soil parent material. FEMS Microbiol Ecol 86:101–113. https://doi.org/10.1111/1574-6941.12143

Stewart EJ (2012) Growing unculturable bacteria. J Bacteriol 194:4151–4160

Stolyar S, Van Dien S, Hillesland KL, Pinel N, Lie TJ, Leigh JA, Stahl DA (2007) Metabolic modeling of a mutualistic microbial community. Mol Syst Biol 3:92

Štursa P, Uhlík O, Kurzawová V, Koubek J, Ionescu M, Strohalm M, Lovecká P, Macek T, Macková M (2009) Approaches for diversity analysis of cultivable and non-cultivable bacteria in real soil. Plant Soil Environ 55:389–396

Swirglmaier K, Keiz K, Engel M, Geist J, Raeder U (2015) Seasonal and spatial microbial diversity along a trophic gradient in the interconnected lakes of the Osterseen Lake District, Bavaria. Front Microbiol 6:1168

Tabacchioni S, Chiarini L, Bevivino A, Cantale C, Dalmastri C (2000) Bias caused by using different isolation media for assessing the genetic diversity of a natural microbial population. Microb Ecol 40:169–176

Teeling H, Glockner FO (2012) Current opportunities and challenges in microbial metagenome analysis—a bioinformatic perspective. Brief Bioinform 13(6):728–742. https://doi.org/10.1093/bib/bbs039

Thavamani P, Malik S, Beer M, Megharaj M, Naidu R (2012) Microbial activity and diversity in long-term mixed contaminated soils with respect to polyaromatic hydrocarbons and heavy metals. J Environ Manag 99:10–17

Thies JE (2007) Soil microbial community analysis using terminal restriction fragment length polymorphisms. Soil Sci Soc Am J 71:579–591

Thies JE (2008) Molecular methods for studying microbial ecology in the soil and rhizosphere. In: Nautiyal CS, Dion P (eds) Molecular mechanisms of plant and microbe coexistence. Soil biology, vol 15. Springer-Verlag, Berlin. https://doi.org/10.1007/978-3-540-75575-3

Torto-Alalibo T, Collmer CW, Gwinn-Giglio M, Lindeberg M, Meng S, Chibucos MC, Tseng T-T, Lomax J, Biehl B, Ireland A, Bird D, Dean RA, Glasner JD, Perna N, Setubal JC, Collmer A, Tyler BM (2010) Unifying themes in microbial associations with animal and plant hosts described using the gene ontology. Microbiol Mol Biol Rev 74:479–503. https://doi.org/10.1128/MMBR.00017-10

Tunlid A, White DC (1992) Biochemical analysis biomass, community structure, nutritional status and metabolic activity of microbial communities on soil. In: Stotzky G, Bollag JM (eds) Soil biochemistry, vol 7. Marcel Dekker, New York, pp 229–262

Ultee A, Souvatzi N, Maniadi K, Konig H (2004) Identification of the culturable and nonculturable bacterial population in ground water of a municipal water supply in Germany. J Appl Microbiol 96:560–568. https://doi.org/10.1111/j.1365-2672.2004.02174.x

Urwin R, Maiden MCJ (2003) Multi-locus sequence typing: a tool for global epidemiology. Trends Microbiol 11:479–487

van der Heijden MGA et al (2008) The unseen majority: soil microbes as drivers of plant diversity and productivity in terrestrial ecosystems. Ecol Lett 11:296–310

van Elsas JD et al (2012) Microbial diversity determines the invasion of soil by a bacterial pathogen. Proc Natl Acad Sci U S A 24:1159–1164

Vandamme P, Pot B, Gillis M, de Vos P, Kersters K, Swings J (1996) Polyphasic taxonomy, a consensus approach to bacterial systematics. Microbiol Rev 60(2):407–438

Venkatachalam S, Ranjan K, Prasanna R, Ramakrishnan B, Thapa S, Kanchan A (2016) Diversity and functional traits of culturable microbiome members, including cyanobacteria in the rice phyllosphere. Plant Biol 18:627–637

Versalovic J, Koeuth T, Lupski JR (1991) Distribution of repetitive DNA sequences in eubacteria and application to fingerprinting of bacterial genomes. Nucleic Acids Res 19:6823–6831

Versalovic J, Schneider M, de Bruijn FJ, Lupski JR (1994) Genomic fingerprinting of bacteria using repetitive sequence based PCR (rep-PCR). Methods Mol Cell Biol 5:25–40

Versalovic J, de Bruijn FJ, Lupski JR (1998) Repetitive sequence-based PCR (rep-PCR) DNA fingerprinting of bacterial genomes. In: de Bruijn FJ, Lupski JR, Weinstock GM (eds) Bacterial genomes: physical structure and analysis. Chapman & Hall, New York, pp 437–454

Von Wintzingerode F, Rainey FA, Kroppenstedt RM, Stackebrandt E (1997) Identification of environmental strains of *Bacillus mycoides* by fatty acid analysis and species-specific rDNA oligonucleotide probe. FEMS Microbiol Ecol 24:201–209

Vorholt JA (2012) Microbial life in the phyllosphere. Nat Rev Microbiol 10:828–840. https://doi.org/10.1038/nrmicro2910

Wagg C et al (2014) Soil biodiversity and soil community composition determine ecosystem multifunctionality. Proc Natl Acad Sci U S A 111:5266–5270

Ward N, Fraser CM (2005) How genomics has affected the concept of microbiology. Curr Opin Microbiol 8(5):564–571

Wardle DA et al (2004) Ecological linkages between aboveground and belowground biota. Science 304:1629–1633

Welch DF (1991) Applications of cellular fatty acid analysis. Clin Microbiol Rev 4(4):422–438

Whipps JM, Hand P, Pink D, Bending GD (2008) Phyllosphere microbiology with special reference to diversity and plant genotype. J Appl Microbiol 105:1744–1755

Whitman WB, Coleman DC, Wiebe WJ (1998) Prokaryotes: the unseen majority. PNAS 95(12):6578–6583

Whittaker P, Fry FS, Curtis SK, Al-Khaldi SF, Mossoba MM, Yurawecz MP, Dunkel VC (2005) Use of fatty acid profiles to identify food-borne bacterial pathogens and aerobic endospore-forming bacilli. J Agric Food Chem 53:3735–3742

Wilson KH, Blitchington RB, Green RC (1990) Amplification of bacterial 16S ribosomal DNA with polymerase chain reaction. J Clin Microbiol 28:1942–1946

Wilson WH, Turner S, Mann NH (1998) Population dynamics of phytoplankton and viruses in a phosphate limited mesocosm and their effect on DMSP and DMS production. Estuar Coast Shelf Sci 46(Suppl A):49–59

Woese CR (1987) Bacterial evolution. Microbiol Rev 51(2):221–271

Xu J (2006) Microbial ecology in the age of genomics and metagenomics: concepts, tools, and recent advances. Mol Ecol 15:1713–1731

Yandigeri MS, Meena KK, Singh D, Malviya N, Singh DP, Solanki MK, Yadav AK, Arora DK (2012) Drought-tolerant endophytic actinobacteria promote growth of wheat (*Triticum aestivum*) under water stress conditions. Plant Growth Regul 68:411–420. https://doi.org/10.1007/s10725-012-9730-2

Yang Y, Ames GF (1988) DNA gyrase binds to the family of prokaryotic repetitive extragenic palindromic sequences. Proc Natl Acad Sci U S A 85(23):8850–8854

Yang Y, Yao J, Hu S, Qi Y. 2000. Effects of agricultural chemicals on DNA sequence diversity of soil microbial community: a study with RAPD marker. Microb. Ecol. 39, 72–79

Zak DR, Holmes WE, White DC, Peacock AD, Tilman D (2003) Plant diversity, soil microbial communities, and ecosystem function: are there any links? Ecology 84:2042–2050. https://doi.org/10.1890/02-0433

Zappelini C, Karimi B, Foulon J, Lacercat-Didier L, Maillard F, Valot B, Blaudez D, Cazaux D, Gilbert D, Yergeau E, Greer C, Chalot M (2015) Diversity and complexity of microbial communities from a chlor-alkali tailings dump. Soil Biol Biochem 90:101–110

Zarraonaindia I, Smith DP, Gilbert JA (2013) Beyond the genome: community-level analysis of the microbial world. Biol Philos 28(2):261–282

Zeglin LH (2015) Stream microbial diversity in response to environmental changes: review and synthesis of existing research. Front Microbiol 6:454

Zhou J (2003) Microarrays for bacterial detection and microbial community analysis. Curr Opin Microbiol 6:288–294

Zijnge V, Welling GW, Degener JE, van Winkelhoff AJ, Abbas F, Harmsen HJ (2006) Denaturing gradient gel electrophoresis as a diagnostic tool in periodontal microbiology. J Clin Microbiol 44:3628–3633

Zuñiga C, Zaramela L, Zengler K (2017) Elucidation of complexity and prediction of interactions in microbial communities. Microb Biotechnol 10:1500–1522. https://doi.org/10.1111/1751-7915.12855

Interactions in Soil-Microbe-Plant System: Adaptation to Stressed Agriculture

6

Stefan Shilev, Hassan Azaizeh, Nikolay Vassilev, Danail Georgiev, and Ivelina Babrikova

6.1 Introduction

In the last decades, the intensive agriculture faces serious problems originated by the necessity of higher yield and quality of the agricultural products. Expecting the number of world population to reach almost nine billion by 2050, the necessity of more quantity and more secure food is evident. As a result, in all industrialized countries an intensive fertilization is applied as a method for yield increase. All these applications, incl. pesticide application, conducted to changes in the agroecosystem having as a result accumulation of chemicals in soil, water and plant production, and decreasing crop productivity. On the other hand, the climate change conducted to aggravation of unfavorable environmental conditions, such as drought, nutrient scarcity, high temperature, between others. In that situation, the agricultural

Authors' Contributions Study concepts, design, and editing: S. Shilev. Abstract and Introduction: S. Shilev. Section 6.2: H. Azaizeh. Section 6.3: S. Shilev, D. Georgiev, and I. Babrikova. Section 6.4: N. Vassilev. Concluding remarks: H. Azaizeh and S. Shilev. Manuscript final version approval: All authors read and approved the manuscript.

S. Shilev (✉) · I. Babrikova
Department of Microbiology and Environmental Biotechnologies, Agricultural University – Plovdiv, Plovdiv, Bulgaria
e-mail: stefan.shilev@au-plovdiv.bg

H. Azaizeh
Institute of Applied Research (Affiliated with University of Haifa), The Galilee Society, Shefa-Amr, Israel

Department of Environmental Science, Tel Hai College, Qiryat Shemona, Israel

N. Vassilev
Faculty of Sciences, Department of Chemical Engineering, University of Granada, Granada, Spain

D. Georgiev
Faculty of Biology, Department of Microbiology, University of Plovdiv, Plovdiv, Bulgaria

D. P. Singh et al. (eds.), *Microbial Interventions in Agriculture and Environment*,
https://doi.org/10.1007/978-981-13-8391-5_6

sector is trying to find other approaches, more suitable and more acceptable, in order to reach the objectives.

One unique alternative of the conventional method is the improvement of naturally occurring interactions between plant roots and soil microbial communities, such as mycorrhizal fungus and PGPR. Stimulation of the development of these communities could help the plants to battle the environmental stresses that lead to reduced plant quality and productivity. It is known that diverse microbial genera are vital for soil fertility, also because they are involved in different biotic activities and occupy plenty of environmental niches related to nutrient cycles. Their activity could be defined as very important for sustainable crop production (Kidd et al. 2009).

This very interesting, natural, and applicable strategy for improving agricultural production is more visible and more investigated during the last decade (Fig. 6.1). More than 830 papers indexed in Scopus were published for the last 10 years, while their citations for that time are almost 8000.

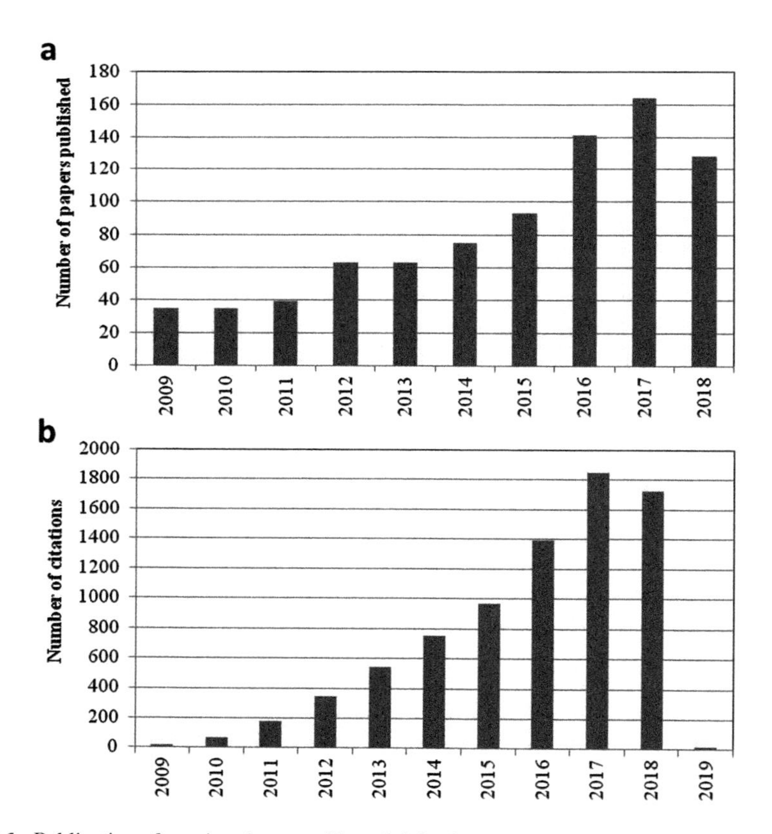

Fig. 6.1 Publications featuring the use of beneficial microorganisms in agriculture (**a**) and their citation (**b**) during the last 10 years. Source: *Scopus*; search parameters: ("plant growth-promoting" or "beneficial microorganisms") and (agriculture). The search was performed in the end of September 2018

On the other hand, a number of review papers collected plenty of information concerning the beneficial microorganisms and their investigation, abilities, habitats, and ecology (Vejan et al. 2016; Gupta et al. 2015; de Souza et al. 2015; Kong and Glick 2017; Gouda et al. 2018; Kidd et al. 2009). In the present review paper, we aimed to give an additional and recent focus on beneficial microorganisms from the point of view of the important mechanisms and agricultural applications.

6.2 Interactions of Root Exudates and Rhizosphere and the Role of Beneficial Microorganisms for Healthy Plants

It is well known that plant roots release different compounds in order to attract and select microbial consortium in their rhizosphere environment, where these plants associated microorganisms, use different mechanisms to influence plant health and growth. The rhizosphere microbes help the plant roots to uptake soil nutrients from the soil matrix. Therefore, the processes of root growth in rhizosphere have an enormous impact on soil nutrient transformations and mobilizations and the efficient usage by the different plant species. Plant roots regulate their morphologies to suite environmental conditions in the soils. In addition, they significantly modify rhizospheric processes through their physiological activities, specially by the exudation of different small molecules, namely, organic acids, sugars, phosphatases, signaling compounds, proteins, and redox compounds (Hinsinger et al. 2009; Zhang et al. 2010; Marschner 2012). Root exudates are divided into two main classes: low molecular weight compounds, such as amino acids, organic acids, sugars, phenolics, hormones, and other secondary metabolites, and high molecular weight compounds, such as carbohydrates, proteins, and others (Bais et al. 2006; Badri and Vivanco 2009). In addition, root exudation includes the secretion of various ions, free oxygen and water, enzymes, mucilage, and diverse organic compounds to attract special rhizosphere microbes (Bais et al. 2006). The microbes live in the rhizosphere and interact with living entities with the diverse metabolites released by plant roots. These interactions influence the plant growth and development and change nutrient dynamics, which may change the plants' susceptibility toward diseases and abiotic stresses which affect plant health as well as plant growth (Yadav et al. 2015). The root produces different chemical signal molecules that attract different microbes toward it. Positive interactions include growth regulation mimicking molecules that support plant growth, development, and cross-species signaling with other rhizospheric microorganisms. The plant roots excrete almost 10–40% of their photosynthetically fixed carbon in the form of exudates and certain signaling molecules or antimicrobials for soil microorganisms. Such exudates contribute to the selection of the specific microbial consortium adapted to the specific rhizosphere of certain plant species (Guttman et al. 2014). The qualitative and quantitative composition of the root exudates is specified by the plant species and cultivars, their developmental stages, different environmental conditions and the presence of different microbial communities associated with the roots (Badri and

Vivanco 2009). These differences specify microbial community structure in the rhizosphere microenvironment that generates certain degree of specificity for plant species to gain mutual interactive benefits.

The root-induced rhizosphere mechanisms help plants in mobilization and acquisition of soil nutrients and at the same time regulate nutrient use efficiency by different species. The phenomenon is supposed to significantly contribute to crop production and plant health and sustainability (Zhang et al. 2010). Plant roots respond to environmental stimuli through the secretion of a diverse range of compounds depending upon their nutritional status and soil conditions (Cai et al. 2012; Carvalhais et al. 2013). These exudates interfere with the interacting plant-microbial species and constitute a significant proportion of efficiency of the microbial consortium in the rhizosphere (Cai et al. 2009, 2012; Carvalhais et al. 2013).

6.2.1 How Plants Are Able to Shape Their Associated Microbial Communities for Their Benefit?

The consortium of the microbial community in the rhizosphere is affected by the physicochemical conditions in the rhizospheric soil due to plant exudation; therefore, it differs considerably from those microbes present in the bulk soil because of root activities which mainly involve exudation which substantially affects their types. The efficiency of root/rhizosphere is mainly managed through (1) manipulating plant root growth, (2) regulating rhizosphere processes of the interaction between plant and microbes, and (3) optimizing root zone management of the different cropping systems (Shen et al. 2013).

It is well known that roots of soil plants are colonized by a diverse consortium of microbes that collectively function for the benefits of the plants and the microbes. Various studies have pointed out the influence of the microbiome on the host plants. The mechanisms by which plants select and shape associated microbial communities have been worked out but with little pace and attention. What are the drivers for the composition of the root-associated microbial communities? Different studies have shown that soil type was identified as the major factor affecting the composition of microbial communities in the rhizosphere (Schreiter et al. 2014), as the soil is a diverse reservoir for microorganisms that can be a potential source to colonize roots. However, under identical environmental conditions and soils, the plant genotype is the main factor which affects the structural and the functional diversity of root communities, where the plant is controlling and selecting its own microbial consortium (Reinhold-Hurek and Hurek 2011). When grown side by side in the same soil, different plant species harbor partially different microbiome consortiums despite they are grown under the same conditions and the same soil type. Several observations showed that root communities varied in different plant genera and species. However, root microbiome composition can diversify at the subspecies level, as was documented for cultivars of potato (Andreote et al. 2010) and rice, *Oryza sativa* (Hardoim et al. 2011). Functional diversity is also affected by variety: Remarkable varietal differences in root-associated nitrogenase gene fragment (nifH)-expressing communities were detected in rice; even cultivars representing sister lineages from the same crossing differed in their active diazotrophic microbiome (Knauth et al. 2005). Therefore, plant influence appeared to be heritable, as an interspecies rice hybrid showed an intermediate profile of the parental species. Differences in patterns of plant root exudation, which can have positive or negative effects on microbial population (Bais et al. 2006), are likely to play an important role in the development of plant type- and developmental of stage-specific microbiomes (Berg et al. 2014). For example, in a comparison of wheat, maize, rape, and barrel clover plants, root exudates significantly shaped the microbial consortium structure in the rhizospheric soil which is controlled by the plant genotype (Haichar et al. 2008). Root exudates of rice plants collected under sterile conditions induced a global transcriptomic response in the endophytic bacterium *Azoarcus* sp. strain BH72, and expression of genes required for endophytic colonization were elevated, suggesting that the bacterium was primed for the endophytic lifestyle by exudates (Shidore et al. 2012).

Brachypodium species are important in investigation of grasses due to the growing availability of genetic resources including a fully sequenced genome and the availability of a large collection of accessions. The *Brachypodium* rhizospheric microbial community and the root exudation profiles showed similar profile to those reported for wheat rhizospheres and different to *Arabidopsis* type; therefore, it was proposed that *Brachypodium* is a good model to investigate the microbiome of wheat (Kawasaki et al. 2016b).

6.2.2 Factors Affecting Plant Exudation and Beneficial Microbes

Several factors are influencing the production of root exudates including plant type, age, light type and intensity, soil microflora, soil fertilizer, soil pH, and other environmental factors and their interactions. Composition of root-associated microbial communities is controlled by factors arising from interactions with other microbes as well as regulated at an environmental condition (e.g., pH, temperature) or host level (plant species) (Wagner et al. 2016; Widder et al. 2016; Wemheuer et al. 2017). In addition to the rhizosphere pH changes induced by cation-anion imbalance, other processes such as root organic acid release, root and microbial respiration, and redox-coupled pH changes are involved in the change of the pH level (Hinsinger et al. 2003). Although many carboxylic acids are released from roots, the primary acids contributing to pH shifts are mainly citric and malic acid (Jones et al. 2003), which are mainly investigated under hydroponic conditions; however, there is lack of information under field conditions.

Despite there are many exudates that can potentially indirectly enhance nutrient acquisition through activation of the rhizosphere microbial biomass, there are few cases where these mechanisms have been proven to be of direct significance under field conditions. This is the case due to the lack of available techniques, and the difficulties in performing rhizosphere experimentation under field conditions. The released organic acids as an example can directly affect the behavior of inorganic P in the soil in several ways which cause in the end for the release of P into solution (Jones 1998). Changes of rhizospheric pH and/or exudation by using complexing agents allow to stimulate desorption of nutrients (e.g., Fe, P) from the soil growth matrix and increase their solubility in soil solution and subsequently their uptake and translocation into the plant (Duffner et al. 2012; Römheld 1987; Vance et al. 2003; Briat 2008); in addition, plant roots react to different environmental conditions through the secretion of various compounds which interfere in the plant-microbial interaction, being considered as an important factor in the efficiency of the inoculants to stimulate plant growth through different mechanisms (Bais et al. 2006; Cai et al. 2009, 2012; Carvalhais et al. 2013).

Endophytic bacteria consisting of different genera have been detected in a wide range of plant species, which can promote plant growth and/or resistance to diseases as well as environmental stresses by a variety of mechanisms including the fixation of atmospheric nitrogen for the benefits of the plants (Stoltzfus et al. 1997; Reinhold-Hurek and Hurek 1998) or the production of antibiotics and phytohormones required for protection against diseases and for better plant growth (Lodewyckx et al. 2002; Lugtenberg et al. 2002; Sturz et al. 2000); therefore, nowadays, many endophytes are used in agricultural cropping systems as biofertilizers and/or biological control agents for sustainable agriculture (Sturz et al. 2000; Lugtenberg et al. 2002). Analyzing the influence of fertilizer application and mowing frequency on bacterial endophytes in several grass species showed that management regimes influenced endophytic communities structure, and the observed responses were grass species-specific (Wemheuer et al. 2017). This might be attributed to several microbial specifically associated with a single grass species, and the structural and functional community patterns showed no correlation to each other, indicating that plant species-specific selection of endophytes is controlled by functional rather than phylogenetic traits (Wemheuer et al. 2017). Based on the comparison of microbiome data for the different root-soil zones and on knowledge of bacterial functions, a three-step enrichment model for shifts in community structure from bulk soil toward roots was suggested (Reinhold-Hurek et al. 2015).

Several studies have shown that exudates can select the microbial communities, so they are specific to certain plant species or even plant genotype stimulating or inhibiting particular microbial populations associated with the plant roots of particular species (Chaparro et al. 2014; Alegria Terrazas et al. 2016; Kawasaki et al. 2016b; Martin et al. 2017). In seagrass species, it was found that bacteria isolated from the roots of *Zostera marina* and *Halodule wrightii* showed positive chemotactic responses and preferential substrate utilization to root exudates and root extracts (Kilminster and Garland 2009). Other studies using 13C or 14C labeling have directly followed the flow of carbon source from the specific seagrasses into certain sediment bacteria (Holmer et al. 2001; Kaldy et al. 2006). Since there is very importance of the root microbiomes to host plant health, therefore, there is a need to better understand the controls and drivers of microbial compositions in seagrass systems, where the light availability controls primary productivity, reduced light may impact root exudation amount and type and consequently the composition of the root microbiome, where using 16S rDNA sequencing revealed that microbial diversity and composition strongly influenced by the presence of the specific seagrass roots, and the root microbiomes and were unique to each seagrass species under investigation (Martin et al. 2018).

Seagrasses uptake inorganic and organic nutrients through leaves and roots, where fixation of atmospheric N into ammonia by diazotrophic bacteria is considered as an important additional source of N covering the nutrient requirements of these plant species (Garcias-Bonet et al. 2016). Seagrass roots are also colonized by a diverse microbial community that are important for N fixation (Bagwell et al. 2002; Garcias-Bonet et al. 2016), sulfate reduction and oxidation (Küsel et al. 2008), phosphate solubilization (Ghosh et al. 2012), and nutrient processing (Trevathan-Tackett et al. 2017). Rhizospheric microbial community and their interaction with the plant can influence the productivity of the crops, where the microbial consortium can benefit plant growth by increasing nutrient supply to plants, suppressing pathogens, and by carrying out other roles. Plant growth-promoting (PGP) strains of *Azospirillum* and *Herbaspirillum* colonize *Brachypodium* roots and enhance the growth of some *Brachypodium* genotypes under low or no N conditions (Amaral et al. 2016). Inoculations with the PGP strain *Bacillus subtilis* B26 increased *Brachypodium* biomass and also enhanced plant drought resistance.

Plants release exudates into the rhizosphere which can alter the rhizosphere microbial community structure and diversity compared to the bulk soil where each plant species harbors specific rhizospheric microbial consortium depending on plant species as well as plant-microbial interaction (Berg and Smalla 2009). Root exudation is also influenced by various biotic and abiotic factors in the surrounding environment, which may lead to a significant shift in the rhizosphere microbiota composition (Lakshmanan et al. 2012; Kawasaki et al. 2012, 2016a). There is a requirement to understand the plant-soil interface sufficiently well to allow the rhizosphere to be engineered and adapted to benefit plant fitness in cereals (Zhang et al. 2015; Ryan et al. 2009). Characterizing the core microbial communities in the rhizosphere and identifying the major root exudates are critical inputs to such models. This information was collected in model plants such as *Arabidopsis* (Lundberg et al. 2012) and in crop species such as wheat (Ai et al. 2015; Donn et al. 2015), rice (Edwards et al. 2015), and maize (Peiffer et al. 2013). An interesting study using *Arabidopsis*, *Brachypodium*, and *Medicago* to investigate the shifts in the microbial populations in the soil over successive plantings, which lead to suggest three models, modified the soil microbiomes differently (Tkacz et al. 2015).

Plant age affects the composition of rhizosphere microorganism consortium and the stage of plant maturity controls the significance of rhizosphere effect and the degree of response to specific microorganisms (Buée et al. 2009). The flowering stage of plants is the most active period of plant metabolism and growth, where the mycorrhizosphere microorganism level increases during this stage and leads to increase of exudates content and composition (Walker et al. 2003; Tahat et al. 2008). Some microbes were found to be more effective at the flowering stage than in the seedling stage or at the full maturity stage (Bais et al. 2006). The effect of light type and intensity on the production of pectin and polygalacturonase (PG) in the root exudates of *Trifolium alexandrinum* showed that the pectin methyl esterase and PG increased with an increase in the duration of light to which plants were exposed during the experiment which indicates the importance of light and intensity (Chhonkar 1978).

Phosphorus (P) is a major yield-determining nutrient in legume, where the major problem with P nutrition is not the P content present in soil but its bioavailability to plants, as inorganic P gets immobilized in acid soils with Fe^{3+} and Al^{3+}, whereas in calcareous soils, P is fixed with Ca^{2+} (Liao et al. 2006). It was shown that low molecular weight exudates like carboxylic acids, sugars, phenolics, and amino acids have a major role in enhancing P acquisition (Carvalhais et al. 2011; Vengavasi et al. 2016, 2017). Two soybean genotypes with contrasting root exudation potential and P uptake efficiency (P-efficient) and (P-inefficient) were grown under natural environment with low and sufficient P availability to assess growth and photosynthetic efficiency and to establish relationship between photo-biochemical processes and root exudation showed that different exudates by roots revealed significant genotypic variation in soybean responses to sufficient and low P availability which indicate the importance of the plant genotype in the plant/microbial interaction for the mutual benefits (Vengavasi and Pandey 2018).

6.2.3 The Contribution of Mycorrhizae and Endophytic Bacteria on Nutrient Acquisition

Vesicular-arbuscular mycorrhizal (VAM) fungi symbioses are an association between obligate biotrophic fungi and more than 80% of land plants and depend on living plant roots for the supply of organic carbon, and they represent the largest component of the soil fungal community (Gosling et al. 2006). VAM and endophytes promote the growth of plants in various ways similar to rhizosphere bacteria (Etesami et al. 2014). The presence of VAM in the rhizosphere or plant roots may change root exudation by the colonized plants, where mycorrhizal plants often grow better than non-VAM plants, in most instances due to higher mineral uptake where colonization has been shown to change the amount and quality of host root exudates (Azaizeh et al. 1995; Marschner 1995). They also play an important role in plant resistance to water and salt stress (Miransari et al. 2008) and acidity and phytotoxic levels of Al in the soil environment (Seguel et al. 2013) and in improving soil structure through the exudation of various compounds (Wu et al. 2008). Some plants colonized with VAM can be more appropriate to uptake heavy metals such as As-contaminated water than soils (Caporale et al. 2014).

Dormant arbuscular mycorrhizal (AM) fungal spores are not only adapted to adverse environments but are also the most effective means of colonization, where the colonization ratio of AM fungi is largely correlated with spore germination which is the precondition of symbiosis with the plants. During the pre-symbiotic phase, many factors (such as a rhizosphere environment, high flavonoid content, presence of soil microorganisms, and plant cell suspension culture) can induce spore germination and promote hyphal growth without the presence of a host plant (Gianinazzi-Pearson et al. 1996; Graham 1982). In addition, root exudates can increase the length and degree of branching of AM fungi hyphae (Tamasloukht et al. 2003) and play an important role in plant-microbe interactions in the rhizosphere zone (Karin et al. 2013). Some studies have shown that root exudates or host extracts can stimulate spore germination; however, others have indicated negative or inconsequential effects (Hepper and Jakobsen 1983; Bécard and Fortin 1988).

In addition, it was found that AM fungal spores can uptake glucose as a carbon source from the environment (Bücking et al. 2008), where the glucose, N sources, and root exudates have great effect on amino acids metabolism in vitro and on spore germination of various AM fungi (Gachomo et al. 2009; Jin and Jiang 2011). The availability of exogenous inorganic N and organic N to the AM fungal spores using only CO_2 for germination generated more than five times more internal free amino acids than those in the absence of exogenous N (Wang et al. 2015b), where the supply of exogenous nitrate to spores with only CO_2 resulted in rise to more than ten times more asparagine than that found without exogenous N supply. The most interesting result was that root exudates were better than glucose at promoting AM spore germination, and exhibited interactions with certain forms of N such as urea and nitrate in the presence of root exudates to increase the spore germination rate and the hyphal length of certain AM fungi (Wang et al. 2015b).

Mineral nutrients such as P or Fe are very reactive and strongly bound to soil particles, where its availability is generally low, especially in calcareous soils, where plant species differ greatly in their capacity to acquire nutrients from soil such as Fe, P or other minerals from calcareous soils, whereas others cannot extract enough nutrients to persist on such soils (Lambers et al. 2008b). Nutrient acquisition from calcareous soils involves rhizosphere processes, such as the exudation of phosphate mobilizing carboxylates (Hinsinger et al. 2001) or the release of Fe-chelating phytosiderophores (Ma et al. 2003; Robin et al. 2008). In order that P is assimilated by plants, the organic P should be converted into inorganic or low molecular weight of organic acids. Phosphatases are enzymes that can hydrolyze phosphate esters and anhydrides including phosphoprotein phosphatases, phosphodiesterases, diadenosine, acid phosphatases, and other types (Zimmermann 2003). Phosphate acquisition from soils with low P concentrations in solution was shown to be enhanced by mycorrhizal symbioses (Richardson et al. 2009). However, even when P acquisition or plant growth are not enhanced in the presence of mycorrhizal fungi, the P taken up by the fungus may represent a major fraction of the total amount of P acquired by the mycorrhizal plants (Smith et al. 2003). Approximately 80% of all higher plant species can form a mycorrhizal symbiosis; of these, the AM association is the most common (Brundrett 2009), especially on relatively young soils (Lambers et al. 2008a). Plants benefit from the fungi because these acquire different nutrients, which are inaccessible because of distance from the plant roots or occurrence as forms that are unavailable and the AM assist the plants to mobilize them from the soil, and the fungi obtain organic compounds produced by the plant (Smith and Read 2008).

The "hyphosphere" represents the soil influenced by the external phase of the mycorrhizal fungus where the release of various compounds and mycorrhizal hyphae can influence microbial activity and nutrient dynamics in the hyphosphere soil, particularly ectomycorrhizal mycelium, which is capable of releasing various hydrolytic enzymes to mobilize nutrients from organic sources (Chalot and Brun 1998), in addition to other compounds (Sun et al. 1999). AM fungi can secrete large amounts of glycoproteins into the soil environment (Rillig et al. 2002, 2003), which may represent a recalcitrant pool of the carbon source in some soils (Rillig et al. 2001). Some of these exuded compounds may subsequently be reabsorbed by the mycorrhizal hyphae just as roots can reabsorb exuded compounds. Microbial activity and composition has been shown to be affected in the hyphosphere of AM fungi (Andrade et al. 1997; Filion et al. 1999; Staddon et al. 2003). Mycorrhizal colonization can influence exudation process in other ways where the ectomycorrhizal colonization increases root longevity (King et al. 2002) while both increased and decreased root longevity has been also reported following AMF colonization (Hodge 2001; Atkinson et al. 2003). The decomposition of the mycorrhizal root is also likely to differ from that of the nonmycorrhizal root because of the different chemistry as a result of the fungus being present in various plant tissues (Langley and Hungate 2003). Thus, rhizodeposition processes from mycorrhizal roots markedly differ from nonmycorrhizal roots (Azaizeh et al. 1995).

The microbes colonizing the rhizosphere also influence plant root exudation process where many studies have shown that the colonization of AM fungi has changed the plant root exudation qualitatively and quantitatively, e.g., increasing secretions of N, phenolics, and gibberellins and reducing secretions of total sugars, potassium ions, phosphorus, and other compounds (Jones et al. 2004). Several studies have shown that different ectomycorrhizal fungi have distinct effects on the amount and the composition of plant root exudates (Fransson and Johansson 2010). The inoculation with ectomycorrhizal fungus and (or) rhizobacteria can alter root exudation quantitatively and qualitatively, where an interesting study has shown that both the abundance and the type of root-associated fungi have influenced plant root exudation rates (Meier et al. 2013).

Nitrogen acquisition can be enhanced greatly by symbiotic N_2 fixation process, which is common in legumes (Vessey et al. 2005), where the symbiotic microorganisms can play a key role in accessing complex organic N; however, in some mycorrhizal systems, saprotrophs play a pivotal role in making N available to the plants. The AM fungi also increase N nutrition by extending the absorption "mycorrhizosphere" zone due to hyphal extensions (Jonsson et al. 2001; Lerat et al. 2003), where the increase in N uptake was related to the stimulation of bacteria growing in the rhizosphere.

Root exudates are considered as one of the mechanisms that explain the ability of AM to suppress or increase different soilborne diseases (Mukerji et al. 2002), where in response to pathogen attack, plants release root exudates, such as oxalic acids, phytoalexins, proteins, and other unknown organic compounds that affect beneficial as well as pathogenic microbes (Steinkellner et al. 2007). The composition of root exudates varies among different plant species and affected by various environmental conditions (Marschner 1995; Tahat et al. 2011). Although it is believed that root exudates play a major role in the infection and colonization of hosts by AM, the actual role or mode of action of exudates was elucidated only in the last few years (Smith and Read 2008). The germination of *Fusarium oxysporum f.* sp. *Lycopersici* as an example was inhibited in the presence of root exudates from the tomato plants (Scheffknecht et al. 2006). Root exudates can have direct defensive traits against various pathogens, where pathogen-activated plant defenses can result in root secretion of various antimicrobial compounds, where it was shown that root-derived antimicrobial metabolites from *Arabidopsis* confer resistance to a variety of *Pseudomonas syringae* pathovars (Bais et al. 2005). In another work, it was shown that transgenic plants which produce antimicrobial proteins can influence rhizosphere microbial communities (Glandorf et al. 1997). The hyphal length of *Glomus mosseae* was greatly affected by the exudates of mycorrhizal plant species, and the growth of *Ralstonia solanacearum* was suppressed due to *G. mosseae* spores germination (Tahat et al. 2010), and exudates from mycorrhizal strawberry plants suppressed the sporulation of *P. fragariae* in in vitro study (Norman and Hooker 2000).

The microbiomes colonizing the roots are critical for plant growth and health due to their influence on biogeochemical cycling and nutrient acquisition, induction of host defense to various pathogens due to the production of plant growth regulators such as hormones and antibiotics (Reinhold-Hurek et al. 2015; Alegria Terrazas et al. 2016). The plant growth-promoting bacteria (PGPB) found in the rhizosphere are capable of enhancing the growth of plants and protecting them from different diseases and abiotic stresses (Grover et al. 2011; Glick 2012). PGPB are good microbes because they colonize roots and supply favorable environmental conditions for growth development and function of the different plant species. It was shown that non-symbiotic endophytic relationships occur within the intercellular spaces of plant tissues, which contain high levels of compounds and inorganic nutrients available for the growth of these microbes (Bacon and Hinton 2006). The success and efficiency of PGPB for agricultural crops are influenced by various factors, and their efficiency in root colonization is closely associated with microbial competition and survival in the soil, as well as cell-to-cell communication via quorum sensing which is considered nowadays as the main factor in this process (Meneses et al. 2011; Alquéres et al. 2013; Beauregard et al. 2013).

6.3 Microbial Tools for Plant Stress Alleviation

Plants are often exposed to different unfavorable influences such as nutrient scarcity, drought, high temperatures, toxic element, etc. In these conditions, they reduce their growth and quality of agricultural products. It is known that more than 80% of soil fertility is due to the microorganisms. The interactions between plants and bacteria can be generalized into three types: positive, negative, and neutral (Whipps 2001). Most of the autochthonous plant-associated rhizobacteria benefit from the interactions, despite they are neutral or positive for the plant. Many rhizospheric bacteria in some conditions could negatively influence the plant development due to pathogenic or parasitic activity or secretion of phytotoxic compounds (Beattie 2006). In opposite, plant growth-promoting rhizobacteria (PGPR) possess tools that help in plant growth and development, even in stress conditions. The bacteria should possess several abilities in order to be characterized as PGPR (Kloepper 1994): (a) they must have the ability to colonize plant root surface; (b) they must be able to grow up, multiply, and compete with other microbial populations; and (c) they must be able to promote plant growth. So, they must be beneficial to the plant. PGPR are often classified according to the place in plant that they occupy as intracellular (iPGPR) or extracellular (ePGPR), depending on the level of association with the root cells. The iPGPR live in the root cells, such as nodules, while the ePGPR are allocated on the surface (rhizoplane) (Gray and Smith 2005; Shilev et al. 2012a). The following bacterial genera, among many others, can be associated to the ePGPR: *Agrobacterium*, *Flavobacterium*, *Micrococcus*, *Azotobacter*, *Cyanobacteria*, *Azospirillum*, *Bacillus*, *Burkholderia*, *Caulobacter*, *Pseudomonas*, *Arthrobacter*, and *Chromobacterium* (Bhattacharyya and Jha 2012; Tilak et al. 2005).

On the other hand, strains such as *Azorhizobium*, *Bradyrhizobium*, *Mesorhizobium*, and *Rhizobium*, part of the family *Rhizobiaceae*, are the iPGPR. Most of rhizobacteria are Gram-negative rods, while Gram-positive ones are less presented. These and other authors reported that numerous communities of actinomycetes are also in the rhizosphere, where display beneficial traits (Bhattacharyya and Jha 2012; Merzaeva and Shirokikh 2006). Plants react to the environmental conditions through the secretion of a number of compounds which influence the plant-microbe interactions, being considered an important instrument for the efficiency of beneficial microorganisms. In addition, soil health is a very important and influences population growth due to several soil characteristics: soil type, nutrients accession, existence of toxic compounds, etc. The results of bacterial promotion on plant growth and development generally are more visible in case of negative conditions to the plants – abiotic (salinity, drought, toxic elements, etc.) or biotic (pathogens) stresses (Glick 2015; Shilev 2013). PGPR may act also as biocontrol agents and indirectly may improve plant development through their activity against phytopathogens. Also, PGPR can directly improve plant growth by facilitating the availability of nutrients or changing the levels of phytohormone (Glick 2005). Consortium of three rhizobacteria significantly increased germination, root and shoot length and fresh and dry weight of wheat plant compared to single inoculation of any rhizobacteria and uninoculated control. It has been suggested that this consortium could be used for the production of an effective bioinoculant for eco-friendly and sustainable production of wheat (Kumar et al. 2018).

6.3.1 Mechanisms that Directly Promote Plant Growth

The direct mechanisms of microbial actions that support plant growth are diverse. They are related to nutrient uptake; nitrogen, phosphorus, potassium, or iron accession to the plants; or the production of phytohormones, siderophores, and exopolysaccharides.

6.3.1.1 Phytohormone Production

Plants produce phytohormones that regulate their own processes in a different way (biochemical, physiological, or morphological) and are important in order to boost the agricultural production (Lugtenberg et al. 2002; Somers et al. 2004). As organic substances their production is strictly regulated by the plant but could be synthesized exogenously by microorganisms or synthetically for use as plant growth regulators. Soil microorganisms are known to produce several compounds that are characterized as phytohormones: auxins, cytokinins, and gibberellins.

Auxin

The auxins are important compounds that regulate several plant processes directly or indirectly (Tanimoto 2005). One of the most studied and important auxin is the indole-3-acetic acid (IAA) that is involved in a plenty of physiological plant processes: induction of plant response (Navarro et al. 2006) and plant development (Gravel et al. 2007), especially root elongation, root hair formation, or lateral root formation, also depending on the IAA levels (Kidd et al. 2009). On the other hand, the production of these phytohormones is widely distributed among rhizospheric bacteria, thus playing an important role in plant-bacterial interactions (Glick 2015). Other researchers informed that majority of isolated rhizobacteria of rice are IAA producers (Souza et al. 2013). Matsukawa and coauthors (2007) suggested that IAA produced by plants and bacteria in rhizosphere acts as a start for *Streptomyces* to increase antibiotic production.

According to Wang and collaborators (2011), bacterial IAA is an important instrument for plant growth promotion, while it directly stimulates plant cell elongation and cell division. In pot experiment with sand/peat substrate under salt-produced stress (100 mM NaCl) was found that *Pseudomonas* inoculants increase the fresh weight of sunflower with more than 10% and accumulate less Na^+ and more K^+, while the strain *Pseudomonas fluorescence* CECT 378 supported up to 66% increment in leaves, 34% in stems, and 16% in roots, and the effect of wild-type strain was more pronounced in shoots with almost 30%. Both strains were found to be IAA and siderophore producers in in vitro experiments (Shilev et al. 2012b). Furthermore, the endogenous IAA in plant tissues and the sensitivity of the plant to IAA are also key factors to determine if the effect of bacterial IAA in plant growth is positive or negative. In plant roots, the level of endogenous IAA may be optimal or suboptimal for supporting plant growth (Pilet and Saugy 1987); therefore, the IAA produced by bacteria could modify the IAA level to optimal or almost optimal, resulting in either PGP or suppression (Kong and Glick 2017). There are different pathways that use L-tryptophan as a precursor of IAA production. Most of the beneficial bacteria synthesize it via indole-3-pyruvate pathway (Lambrecht et al. 2000), while the biosynthesis in plant-beneficial bacteria is inducible (Patten and Glick 1996).

Rhizobacteria are known to possess dual ability – synthesizing and catabolizing IAA (Duca et al. 2014). The same authors suggested that some bacteria stimulate plant growth by metabolizing IAA synthesized by plants when it is detrimentally higher than normal levels. In this way, the degradation of IAA in case of alteration of endogenous plant production could be also a plant growth promotion mechanism. The capacity of catabolizing IAA has been characterized in *Pseudomonas putida* 1290. This strain uses IAA as in the same time a unique source of carbon, nitrogen, and energy. Moreover, the strain 1290 produces IAA in medium with added L-tryptophan. In co-inoculation in radish (*Raphanus sativus* L.) roots, this strain lowered the negative effects of high IAA concentrations produced by the pathogen *P. syringae*. So, this strain can prevent pathogen attack to radish root, but also stimulate their growth (Leveau and Lindow 2005). In addition, the plant-derived IAA is also an attractant that bacteria can use to have a competitive advantage over the bacteria that lack chemotactic capability (Scott et al. 2013).

Cytokinins and Gibberellins

Cytokinins and gibberellins, like IAA, play a crucial role in the regulation of plant growth and development. The cytokinins are involved in protein synthesis, seed germination, cell division, and metabolite transport among others (Salamone et al. 2005; Frugier et al. 2008; Hussain and Hasnain 2011), while the gibberellins participated in cell division, activation of membranes, stimulation of fluorescence, etc. (Tanimoto 2005). Plant growth promotion by bacteria-producing cytokinins (*Rhizobium* spp., *Pantoea agglomerans*, *Paenibacillus polymyxa*, *Rhodospirillum rubrum*, *Bacillus subtilis*, *Azotobacter* spp., *Pseudomonas fluorescens*) and gibberellins (*Bacillus licheniformis*, *Bacillus pumilus*, *Azospirillum* sp.) has been reported by diverse authors (Gutiérrez-Mañero et al. 2001; Pertry and Vereecke 2009). Arkhipova and coauthors (2007) suggested that cytokinin-producing bacteria improve plant growth in moderate drought conditions. On the other hand, Glick (2012) reveals that cytokinin levels produced by PGPR are lower than those from phytopathogens, so that the effect of the cytokinins from the PGPR on plant growth is stimulatory, while the effect of the pathogens is inhibitory.

6.3.1.2 Nitrogen Fixation

Nitrogen is a very important nutrient for the whole living beings, because of its key role in the organic molecules. It is part of DNA, proteins, etc., but its crucial role in physiological and biochemical processes is well known (Krapp 2015). Nitrogen could be a very important obstacle for yield production in deficient soils. That is the reason for the excessive use of agrochemicals in agricultural practices in the last decades. However, most of the fertilization load is not utilized by the plants but conducted to increasing contamination (eutrophication and acidification) and increased financial expenses (Vimal et al. 2017). The magnificent ability of the beneficial bacteria to fix atmospheric N_2 is known, but not very well understood by non-specialists. No other living beings are capable to perform this extraordinary action taking a gaseous molecule, converting it in mineral compound, and finally releasing it to the others (plants, microorganisms, etc.).

The soil nitrogen fixation is due to two kinds of microorganisms – symbiotic and free-living – which contribution to the global nitrogen load is of about 180×10^6 tons per year, divided into 80–20% between both groups (Graham 1988). As was discussed earlier, organic inputs in rhizosphere from the roots alter microbial biodiversity, thus increasing also N uptake. In any case, N_2 fixation is very "expensive" from the point of view of energy consumption, because to reduce 1 mole of elemental N_2 microorganisms spend 16 moles of adenosine triphosphate (ATP). Symbiotic N_2 fixation is a process exclusively driven by bacteria, as they are the unique organisms capable to take the elemental nitrogen possessing the enzyme nitrogenase and reducing it to ammonia in the root nodules (Kidd et al. 2009). Genes that encode N_2 fixation ability are present in both free-living and symbiotic bacteria. They are involved in activation of iron-proteins and in the biosynthesis of cofactor of iron and molybdenum and donation of electrons. These genes are found in clusters of 20–25 kb with 7 operons encoding of about 20 proteins. The symbiotic N_2-fixing bacteria are considered as iPGPR spread in genera *Rhizobium*, *Mesorhizobium*, *Azorhizobium*, *Bradyrhizobium*, etc., belonging to the *Rhizobiaceae* family, although some authors do not recognize them as PGPR except when the association is with non-legumes (Dobbelaere et al. 2003). The most famous association is with Fabaceae plant species (pea, alfalfa, garden peas, soybeans, etc.). Also, *Frankia* species and some endophytes are considered iPGPR too. On the other hand, the non-symbiotic N_2-fixing bacteria include *Azotobacter*, *Azospirillum*, *Pseudomonas*, *Agrobacterium*, *Erwinia*, *Bacillus*, and *Burkholderia*, among others (Gray and Smith 2005). Because of high energy requirements and relatively low metabolic activity, the productivity of ePGPR in N_2 fixation is limited. According to good agricultural practices, N mineral fertilization is between 150 and 250 kg/ha/year, depending on the crop, state of development, etc. Compared to the productivity of ePGPR, which is around 5-15-20 kg/ha/year (Dobbelaere et al. 2003), it is evident that a combination of tools (more growth-promoting capabilities) is needed for the characterization of that strain as beneficial and continues further with the exploration of possibilities for formulation as biofertilizer.

6.3.1.3 Phosphate Solubilization

Phosphorus is a very important nutrient for the plant and is also required by plants for normal development, in appropriate amounts for optimal growth. Generally, in soil, it exists in two forms, as organic and inorganic phosphates. Microorganisms are capable to convert the insoluble phosphates (organic or inorganic) into accessible to the plant forms, thus increasing the crop yield (Igual et al. 2001; Rodriguez et al. 2006). According to Goldstein (1994), the amount of soluble phosphorus in the soil is commonly quite low, usually at levels of 1 ppm or less. Plants can absorb different forms of phosphorus but the major part is absorbed in the forms of HPO_4^{-2} or $H_2PO_4^{-1}$. The fixation or precipitation of phosphorus in the soil is strongly dependent on the pH and the soil types. Some authors describe the release of soluble phosphorus by microorganisms (Ohtake et al. 1996; McGrath et al. 1998; Rodriguez and Fraga 1999).

Phosphorus plays an important role in almost all metabolic processes, including energy conversion, signal transduction, respiration, molecular biosynthesis, and photosynthesis (Anand et al. 2016). However, 95–99% of the phosphorus is present in insoluble, immobilized, or precipitated forms; therefore, it is difficult for plants to absorb it. Organic acids of low molecular weight synthesized by different soil bacteria solubilize inorganic phosphorus (Sharma et al. 2013b). Phosphate-solubilizing PGPR involves the genera *Arthrobacter*, *Bacillus*, *Beijerinckia*, *Burkholderia*, *Enterobacter*, *Microbacterium*, *Pseudomonas*, *Erwinia*, *Rhizobium*, *Mesorhizobium*, *Flavobacterium*, *Rhodococcus*, and *Serratia*; they have been found to enhance plant growth and yield (Oteino et al. 2015). These data are presented by Gouda et al. (2018).

A large number of microorganisms, such as *Pseudomonas* spp., *Agrobacterium* spp. and *Bacillus circulans*, exhibit the ability to assist in the absorption of inorganic phosphorus by solubilization and mineralization (Babalola and Glick 2012). Others involve strains like *Azotobacter*, *Bacillus*, *Burkholderia*, *Erwinia*, *Rhizobium*, *Bradyrhizobium*, *Sinomonas* and *Thiobacillus*, even *Salmonella* and *Serratia* (Postma et al., 2010; Kumar et al., 2014; Zhao et al., 2014). Various types of molds and yeasts, which function in a similar way, include the strains *Alternaria*, *Arthrobotrys*, *Aspergillus*, *Cephalosporium*, *Cladosporium*, *Cunninghamella*, *Fusarium*, *Glomus*, *Micromonospora*, *Myrothecium*, *Oidiodendron*, *Paecilomyces*, *Penicillium*, *Phoma*, *Pichia fermentans*, *Populospora*, *Pythium*, *Rhizoctonia*, *Rhizopus*, *Saccharomyces*, *Schizosaccharomyces*, *Sclerotium*, *Torula*, and *Trichoderma* between many others (Srinivasan et al. 2012; Alori et al. 2017; Sharma et al. 2013a). Different bacterial strains have the capability to dissolve bioinavailable phosphate (mineral phosphate) compounds (Rodriguez and Fraga 1999; Rodriguez et al. 2006). Strains of genera *Rhizobium*, *Pseudomonas*, *Bacillus*, among others, are very effective solving phosphates (Illmer and Schinner 1992; Halder and Chakrabarty 1993; Rodriguez and Fraga 1999; Banerjee et al. 2006). Biosynthesis of different organic acids is involved in phosphate solubilization by bacteria (Rodriguez and Fraga 1999). Also, organic acid biosynthesized with microbial origin plays a role in phosphate-dissolving (2-ketogluconic acid). This compound is found to be produced by *Rhizobium leguminosarum*, *Rh. meliloti*, *Bacillus firmus*, and other soil bacteria (Kidd et al. 2009). Other microorganisms (*Bacillus licheniformis* and *B. amyloliquefaciens*) were proven to excrete lactic, isovaleric, isobutene, and acetic acids (Hayat et al. 2010).

The activity of different phosphatases in rhizosphere indicates that phosphatase activity is significant in the rhizosphere mainly at pH below 7. Many acidic phosphatases are synthesized by bacteria of genera *Rhizobium*, *Pseudomonas*, *Bacillus*, etc. (Chen et al. 2006). There is other information regarding the dissolution of phosphates by *Rhizobium* (Halder et al. 1990) and by the non-symbiotic nitrogen-fixing *Azotobacter*. The efficacy of the *Mesorhizobium* strain has been shown to improve the growth and absorption of phosphorus in chickpea and barley plants without the addition of phosphates. The most common mechanism used by microorganisms to dissolve tricalcium phosphates is acidification of the near environment releasing organic acids (Rodríguez and Fraga 1999). Gene manipulations of these bacteria have been used to improve plant yield (Rodríguez et al. 2006). Although some of the bacterial (such as *Pseudomonads* and *Bacillus*) and fungal strains (*Aspergillus* and *Penicillium*) have been identified as PSMs, their relative performance under in-situ conditions is not reliable, so there is a need for genetically modified strains which have more pronounced qualities (Ingle and Padole 2017).

A significant number of microorganisms – phytase producers of various taxonomic groups – bacteria, yeasts, and molds, have been found to synthesize enzymes with certain biochemical properties and catalytic capacity, which depend primarily on the producer and the medium conditions. Plants and the other autotrophic organisms are always the first link in the food chain (primary producers), after which the various species of the animal kingdom may continue it. Finally, the microorganisms end the food chain with demineralization of the final products. Phosphate groups give this molecule a high-negative charge and therefore a strong binding ability that reduces the nutrient bioavailability of amino acids and minerals such as Ca^{2+}, Zn^{2+}, and $Fe^{2}+$ (Haros et al. 2001). Many essential metal ions ($Ca^{2}+$, $Zn^{2}+$, $Fe^{2}+$) are associated with IP_6 and form precipitates under neutral or slightly alkaline conditions. The stability of the complexes formed between IP_6 and the metal ions at low pH values is in the following order $Zn^{2}+ > Cu^{2}+ > Co^{2}+ > Mn^{2}+ > Ca^{2}+$, whereas at pH 7.4 the order is $Cu^{2}+ > Zn^{2}+ > Ni^{2}+ > Co^{2}+ > Mn^{2}+ > Fe^{2}+ > Ca^{2}+$. These complexes are insoluble and this is the main reason why the bioavailability of minerals in high phytic acid diets is reduced. The simultaneous presence of two different types of cations increases the amount of IP_6-metal complex precipitates (Simpson and Wise 1990).

Microorganisms of various taxonomic groups – bacteria, yeasts, and molds – produce phytases (Dvorakova 1998; Vohra and Satyanarayana 2003; Vats and Banerjee 2004). Phytase-synthesizing microorganisms were isolated from a significant number of sources, including soil, fermented food/raw materials, contaminated water, gastrointestinal fluids of ruminants, and plant roots. In almost all mold producers, enzymes are excreted in the culture medium, and for this reason, they most often affect the absorption of phosphorus from plants. Both intracellular and extracellular production has been reported for the bacteria. Recently, data on yeast phytase producers indicated exclusively intracellular activity, but lately enzyme secreting strains were also cited (Lambrechts et al. 1992; Nakamura et al. 2000). Volfova et al. (1994) isolated several *Aspergillus niger* strains that produced phytases, the most active being *A. niger 89* and *A. niger 92*. Both strains synthesize the enzyme during active cell growth and simultaneously produce organic acids that lower the pH of the medium and thus contribute to the chemical degradation of phytates. In the case of solid-phase cultivation of producers from the *Aspergillus*, *Mucor*, and *Rhizopus* genera, phytases were also synthesized, the cultivation of *Aspergillus ficuum* in wheat bran medium being the most effective (Fujita et al. 2000). Many enzymes are released in the soil, such as cellulases, hemicellulases, amylases, pectinases, and fungal protein, which increase the absorption and the biologic value of nutrients absorbed by plants from the soil (Bogar and Srakers 2003).

Pandey et al. (2001) investigate strains of *Schwanniomyces castellii*, *Schw. occidentalis*, *Hansenula polymorpha*, *Arxula adeninivorans*, *Rhodotorula gracilis*, and others. An increase in the amount of phytase is often observed in the study of soils with low phosphate content. This is reported for *Candida tropicalis* and *Yarrowia lipolytica* (Hirimuthugoda et al. 2007). It has been found that certain yeast species secrete the enzyme in the soil: *Schwanniomyces castellii* (Segueilha et al. 1992), *Arxula adeninivorans* (Sano et al. 1999), *Pichia spartinae*, and *P. rhodanensis* (Nakamura et al. 2000). Lambrechts et al. (1992) examined 21 yeast strains of 10 species and selected 5 of them – *Candida tropicalis CBS 5696*, *Torulopsis candida CBS 940*, *Debaryomyces castelli CBS 2923*, *Kluyveromyces fragilis U1*, and *Schwanniomyces castellii CBA 2863* – which grow well in a medium with sodium phytate as the sole carbon source. *Schwanniomyces castellii* has a higher phytase potential than other phytase-producing yeasts. Its ability to degrade phytate in some natural raw materials – wheat bran and cottonseed meal has been studied by Segueilha et al. (1993).

Candida krusei WZ-001 was isolated from soil from Dalian Province in China (Quan 2002). The phytase isolated from *Pichia anomala* is characterized by high pH and thermostability and broad substrate specificity, indicating that this strain can develop in different soil types (Vohra and Satyanarayama 2002). Sano and co-workers (1999) reported a very high extracellular activity is characteristic of strains of the species *Arxula adeninivorans*.

In *Schw. castellii* the phytase production decreases when the content of organic or inorganic phosphate increases (Pandey et al. 2001).

Pavlova et al. (2008) isolated yeasts from samples of soils, roots, mosses from the Bulgarian base on the Livingston Peninsula in Antarctica for the first time. They identified them as representatives of different genera and species and examined them for the production of extracellular and cell-associated phytases in environments containing calcium phytate. They cite the strain *Cryptococcus laurentii* AL27 as the most promising one.

Several types of bacteria, such as *Lactobacillus amylovorus*, *Escherichia coli*, *Bacillus subtilis*, *B. amyloliquefaciens*, *Klebsiella* spp., and others, have been studied for phytase biosynthesis (Pandey et al. 2001). The ability to produce indolylacetic acid and to mineralize organic phosphorus by phytase are characteristic of some rhizobacteria. These properties were recorded in *Bacillus sp.* and *Paenibacillus sp.* (Acuca et al. 2011). Phytase activity was also detected in *B. amyloliquefaciens* DS11 (Kim et al. 1999). Several researchers (Shimizu 1992; Griener et al. 1993; Kim et al. 1998) investigated bacterial strains of *Bacillus* spp. and *E. coli*, isolated from soil near the roots of legumes. Yoon et al. (1996) consider that with the exception of strains *Enterobacter* spp. and *B. subtilis*, the phytases of the other bacteria are intracellular. *B. subtilis* strains grow very well on scalded soybeans that are rich in phytates, without other nutritional supplements, indicating that the strains can be beneficial for the uptake of organic phosphorus by plants. During their cultivation, the phytase activity reached a maximum on the 5th day (Shimizu 1992).

6.3.1.4 Siderophore Production

Bacterial activities could conduct to an improvement of plant nutrient uptake, which also results in higher growth and development even in stressful conditions. One of the very important elements is the iron. It takes part in various microbial enzymes, so its importance is proven. In any case, the iron in the aerobic environment exists mainly as Fe^{3+} forming insoluble complexes hydroxides and oxyhydroxides, unavailable to microorganisms and plants. To "solve" this problem bacteria have developed an efficient strategy to make the complexes available. In an iron-deficient environment, they synthesize low molecular weight compounds (<1000 Da) named siderophores (Neilands 1983). These molecules have affinity to metal ions forming complexes, although the siderophores act as solubilizing agents for much more ions from minerals or less soluble organic compounds, such as Al, Cd, Pb, Cu, Zn, etc. (Schalk et al. 2011). According to Boukhalfa and Crumbliss (2002), more than 500 different siderophores are identified. Despite this, metal binding side of the siderophores are α-hydroxycarboxylic acid, catechol, or hydroxamic acid moieties sites and thus can be classified as hydroxycarboxylate-, catecholate-, or hydroxamate-type siderophores (Raymond and Denz 2004). Many siderophores are polypeptides and are synthesized by the non-ribosomal peptide synthetase multienzyme family, which is also responsible for the synthesis of most of microbial peptide antibiotics.

In addition, many of the hydroxamate and α-hydroxy acid-containing siderophores are not polypeptides. They are produced by dicarboxylic acid and either diamine or amino alcohol building molecules linked by amide or ester. Such siderophores are constructed by the non-ribosomal peptide synthetase – independent siderophore pathway, which is widely utilized in bacteria (Rajkumar et al. 2010). The structure and biosynthesis of siderophores are studied in the last years by different authors (Miethke and Marahiel (2007); Barry and Challis (2009). According to Jalal and van der Helm (1991) and Madigan and coauthors (1997), the siderophores form complexes with Fe^{3+} 1:1, which is taken up by the bacterial plasma membrane, reducing Fe^{3+} to Fe^{2+} after liberating the ion in cell plasma. This mechanism of iron uptake in bacteria is described by Krewulak and Vogel (2008). Although the siderophores are produced by pathogens and free-living and symbiotic nitrogen-fixing bacteria, they are most common in PGPR. The beneficial bacteria possess many abilities that improve plant development even in unfavorable conditions where advantage given by siderophores is more evident. However, the function of siderophores is bound to the metal ion uptake improving the Fe nutrition, especially in an extreme environment as scarcity of nutrition or metal contamination. On the other hand, the siderophore production may alter positively the synthesis of IAA, thus increasing overall effect of beneficial bacteria (Dimkpa et al. 2008).

Costa and collaborators (2014) analyzing PGPR data found that 64% of the isolates and 100% of all bacterial genus presented siderophore production. Plants often capture Fe^{3+}-siderophore bacterial complexes utilizing them and do not suffer depletion mediation by bacterial siderophores (Dimkpa et al. 2009). In addition, Pahari and Mishra (2017) reported that siderophore producing isolates significantly increase the growth parameters like root length, shoot length, and biomass of okra (*Abelmoschus esculentus* L.) but also showed antagonistic effect against different phytopathogens including *Rhizoctonia solani* (ITCC-186) and *Fusarium oxysporum* (ITCC-578). According to Berendsen and coauthors (2015), siderophores are one of the key factors stimulating induced systematic resistance in plants against phytopathogens. *Azospirillum brasilense* produces siderophores that expressed in vitro activity against *Colletotrichum acutatum* (anthracnose producing microbe). Inoculated plants of strawberry with the same bacterial population were able to decrease their disease symptoms (Tortora et al. 2011). Pattan et al. (2017) discussed that isolated siderophore showed the antagonists effects against human pathogenic *Pseudomonas aeruginosa* and on phytopathogenic fungi. In maize research,

Szilagyi-Zecchin and collaborators (2014) found that endophytic strains from *Bacillus* sp. express various PGP characteristics, including siderophore production, and these were efficient against the growth of *Fusarium verticillioides*, *Colletotrichum graminicola*, *Bipolaris maydis*, and *Cercospora zeae-maydis* fungi.

6.3.1.5 Exopolysaccharide Production

Exopolysaccharide (EPS) production is very important for certain abilities of beneficial microorganisms. They could be defined as high molecular weight compounds of intracellular, structural, and extracellular EPSs found in bacteria, algae, and plants. They display a wide spectrum of variety and are from importance in biofilm formation, root colonization, formation of shielding from desiccation, and stress protection, among others (Gupta et al. 2015; Qurashi and Sabri 2012; Tewari and Arora 2014). EPSs produced by *P. putida* strain GAP-p45 alleviate salt produce stress to sunflower seedlings (Sandhya et al. 2009). According to Parada and co-workers (2006), EPSs are very important for the beneficial bacteria in their interactions with the plant using them as signal molecules and providing defense response of infection. Many of the EPS-producing beneficial microorganisms play a vital role in soil fertility and agricultural sustainability (*Rhizobium* sp., *Azotobacter vinelandii*, *Bacillus drentensis*, *Enterobacter cloacae*, *Agrobacterium* sp., *Xanthomonas* sp.).

6.3.2 Indirect Mechanisms

Indirect mechanisms are those through which the PGPR suppress or prevent negative effects on plants provoked mostly by abiotic or biotic stresses.

6.3.2.1 Antibiosis

The use of microbial antagonism against phytopathogens in agriculture is not a new approach. Biocontrol uses beneficial (non-pathogenic) microorganisms that suppress the development of unwanted, harmful microorganisms and thus is one of the most studied biocontrol issues in the last years (Ulloa-Ogaz et al. 2015). According to Ramadan and co-workers (2016), most of the *Pseudomonas* strains produce antifungal antibiotics (phenazines, phenazine-1-carboxylic acid, phenazine-1-carboxamide, pyrrolnitrin, pyoluteorin, 2,4diacetylphloroglucinol, rhamnolipids, oomycin A, cepaciamide A, ecomycins, viscosinamide, etc.), bactericines (andazomycin), and so on (Ramadan et al. 2016). In addition, *Bacillus* sp. also produces a wide range of antagonistic substances with ribosomal or non-ribosomal origin (subtilosin A, subtilintas A, sublancin; chlorotetain bacilysin, mycobacillin, rhizocticins, difficidin, etc.) (Wang et al. 2015a). Bacterial antibiosis is recorded in different investigations with *Bacillus* inoculating alfalfa seedlings, *Pseudomonas* in wheat, etc. (Vejan et al. 2016; Gupta et al. 2015).

6.3.2.2 Induced Systematic Resistance

Induced systematic resistance (ISR) is a specific physiological state of enhanced defensive capability as a response to determined stressors. Beneficial microbes could induce such resistance in different ways activating the mechanisms through several signals (bacterial components), such as cyclic lipopeptides; siderophores; lipopolysaccharides; 2,4-diacetylphloroglucinol; volatiles, like 2,3-butanediol and acetoin; and homoserine lactones (Berendsen et al. 2015). In this way, plants are "immunized" against a broad spectrum of pathogens; thus, future attacks are repelled. *Bacillus* and *Pseudomonas* species were found to improve plant defense against phytopathogens in many plants through ISR. On the other hand, *Xanthomonas campestris* (black rot) in cabbage was suppressed by biocontrol agent *Paenibacillus*, inducing systematic resistance (Ghazalibigla et al. 2016). Although the ecological niches of ePGPM and iPGPM are different, they use similar mechanisms to suppress phytopathogens and promote plant growth (Shilev 2013). The effect of combined population of PGPR was studied in chilli, showing ISR and growth promotion in greenhouse condition. The authors (Audipudi et al. 2016) concluded that the combined application is more appropriate to be used because of the combination of several different mechanisms presented in distinct microbial populations. Studying the potential of *Pseudomonas aeruginosa* PM12 in the induction of ISR against *Fusarium* in tomato plants, Fatima and Anjum (2017) found strong antifungal effect of 3-hydroxy-5-methoxy benzene methanol of bacterial extracts after GC-MS analysis. Thus, the compound showed intensive remodulation in defense-related pathways against *Fusarium oxysporum*. In conclusion, the application of beneficial microorganisms as biocontrol agents against soilborne pathogens could be an advantage in integrated pest management.

6.3.2.3 Enzyme Production

Ethylene is a very important regulator of plant growth and development, especially in the case of stresses (Gamalero and Glick 2012; Hao et al. 2007). According to Swain (1974), ethylene is associated with the environmental stress, and then plant increases the internal concentration of the phytohormone. These factors are mainly of abiotic origin, such as water stress, salinity, toxic metals, extreme temperature, etc. As a response to the abiotic stress, plants inhibit their growth increasing root endogenous ethylene production. Thus, plant roots have limited growth that reflects to whole plant in reduction of plant biomass. In such conditions, different mechanisms are known that reduce the concentration of ethylene in plants. One of them involves the activity of the bacterial enzyme 1-aminocyclopropane-1-carboxylate (ACC) deaminase (Glick et al. 1998). A model of lowering ethylene concentration in plants by beneficial soil bacteria that possess the enzyme ACC deaminase is proposed. ACC deaminase-containing beneficial bacteria can facilitate plant growth and development through the conversion of the immediate ethylene precursor ACC into α-ketobutyrate and ammonia, thus reducing the levels of plant ethylene and improving plant growth in an unfavorable abiotic environment (Glick 2012; Gamalero and Glick 2012; Nascimento et al. 2018). In this case, beneficial bacteria utilize ACC as a sole nitrogen source. Stress induces ACC oxidase in the plant so that there is an increased flux resulting in a first increase of ethylene that induces the transcription of protective genes in the plant. In that moment, bacterial ACC deaminase is activated by the increased concentration of ACC as a result from the function of ACC plant synthesis so that the level of next ethylene peak is decreased tremendously till 90%. Because oxidase has a greater affinity for ACC than does ACC deaminase, when ACC deaminase-producing bacteria are present, plant ethylene levels are dependent upon the ratio of ACC oxidase to ACC deaminase (Glick et al. 1998). This approach is very often applied in phytoremediation strategies dealing with toxic metals. Plant-beneficial bacteria that possess ACC deaminase association improve the growth of the plant, as well as their metal tolerance (Rodriguez et al. 2008).

Pseudomonas sp. and *Acinetobacter* sp. possess ACC deaminase and produce IAA in salt stress environment in the rhizosphere of barley and oats, thus promoting plant growth (Chang et al. 2014). Iqbal and collaborators (2012) observed improved growth characteristics of lentil, such as number of nodules, weights, etc., but also nitrogen content in grains. All these were related to the lowered ethylene production through the plant growth-promoting *Pseudomonas* sp. strains possessing ACC deaminase activity. In other study, Ahmad and collaborators (2013) reported about growth enhancement and quality improvement of mung beans when *Rhizobium* and *Pseudomonas* strains were inoculated under salt stress conditions. In addition, Shaharoona and co-authors (2006) reported that in the same crop, the co-inoculation of *Bradyrhizobium* and of ACC deaminase presenting strain resulted in the stimulation of nodulation. Similarly, Ali and co-workers (2014) found that tomato plants treated previously with the endophytic *P. migulae* and *P. fluorescens* showed ACC deaminase activity and presented improved growth under high salinity stress compared with plants treated previously with an ACC deaminase-deficient mutant and control.

Finally, bacterial ACC deaminase activity can be divided into two sections, based on the level of enzymatic activity (Glick 2010): high ACC deaminase-expressing microorganisms and low ACC deaminase-expressing microorganisms. The first ones are situated near to the plant surfaces and include plenty of microorganisms from rhizosphere, phyllosphere and also endophytes. In contrary, low ACC deaminase-expressing microbes only adhere to specific plants or are only present in determinate tissues. These microorganisms do not lower the whole level of ethylene produced by the plant, but they could prevent a certain increase in ethylene levels. This kind of beneficial microorganisms includes most of the *Rhizobium* sp. (Glick 2005). Also, genus diversity of beneficial bacteria exhibiting ACC deaminase activity had been identified in a wide spectrum of genera such as *Pseudomonas*, *Achromobacter*, *Azospirillum*, *Bacillus*, *Burkholderia*, *Acinetobacter*, *Ralstonia*, *Agrobacterium*, *Enterobacter*, *Alcaligenes*, *Serratia*, *Rhizobium*, etc. (Kang et al. 2010; Onofre-Lemus et al. 2009).

6.3.2.4 VOCs

Biocontrol strains may produce volatile organic compounds (VOCs) that possess antagonistic activity against phytopathogenic fungi, bacteria, or nematodes. In a biocontrol study, the VOCs of *P. fluorescens* WR-1 not only showed a concentration-dependent bacteriostatic effect on the growth of *R. solanacearum* but also could inhibit its virulence habilities. The VOCs can spread over a long distance and bacteriostatic environment persists around the plant rhizosphere compared to the antibiotics, which can be effective only if biocontrol agents colonize plant roots effectively (Raza et al. 2016). On the other hand, the interaction between VOCs of *Bacillus subtilis* and *Ralstonia solanacearum* and plant results in growth promotion and induced systemic resistance against the bacterial wild pathogen *R. solanacearum* (Tahir et al. 2017).

6.4 Production and Formulation of Bacterial Biofertilizers

A biofertilizer could be defined as the formulated product containing one or more microorganisms that enhance the nutrient status (and the growth and yield) of plants by either replacing soil nutrients or by making nutrients more available to plants or by increasing plant access to nutrients (Malusa and Vassilev 2014). The oldest, officially recognized invention on plant-beneficial microorganisms appeared in 1896 (Nobbe and Hiltner 1896). However, the application of plant-beneficial microorganisms started in the 1950s when seeds were coated with bacterial cultures to promote plant growth and development (Brown 1974). Now, the production and commercialization of plant-beneficial microorganisms is one of the most active fields of the biotech industry. The market of plant growth promoters is estimated at 946.6 million of dollars in 2015, but this value will increase with 14.08% till the end of 2022. Due to the increasing human population the need for agro-chemical products will raise, which will simultaneously increase the need of biofertilizers (Markets and markets 2016).

In general, the main steps of a biostimulant production follow a scheme, which includes up to eight key experimental groups of studies (Fig. 6.2).

Selection of plant-beneficial bacterial strains is normally carried out using criteria-specific properties including plant growth-promoting or antimicrobial metabolites, competing with other (local) soil microorganisms for nutrients, and demonstrating sufficient level of genetic stability and resistance toward various abiotic and biotic factors (Herrmann and Lesueur 2013). Selection is generally oriented to two main groups of plant-beneficial activities: fixation of nitrogen by mutualistic endosymbionts, such as *Rhizobium,* and mutualistic, rhizospheric plant growth promoters. It should be noted, however, that biofertilizers are normally characterized by multifunctional properties, which affect all aspects of nutrition and growth, various stresses, and interactions with other organisms in the soil-plant systems (Berg et al. 2014; Vacheron et al. 2013; Vassileva et al. 2010). For example, nitrogen-fixing bacteria, in addition to their main function, may manifest other properties typical for bacteria stimulating plant growth and development such as biosynthesis of phytohormones, siderophores, amino acids, polysaccharides, etc., thus increasing the overall benefits to plants (Pathak and Kumar 2016).

The selected bacteria should be easily cultivated, preserving their metabolic functions. Soil microorganisms, including bacteria, are living in the soil – a complex medium creating specific environment for each living organism. Many bacteria were isolated from soil and characterized and their plant-beneficial properties described (Jacoby et al. 2017). However, all these studies are carried out in an

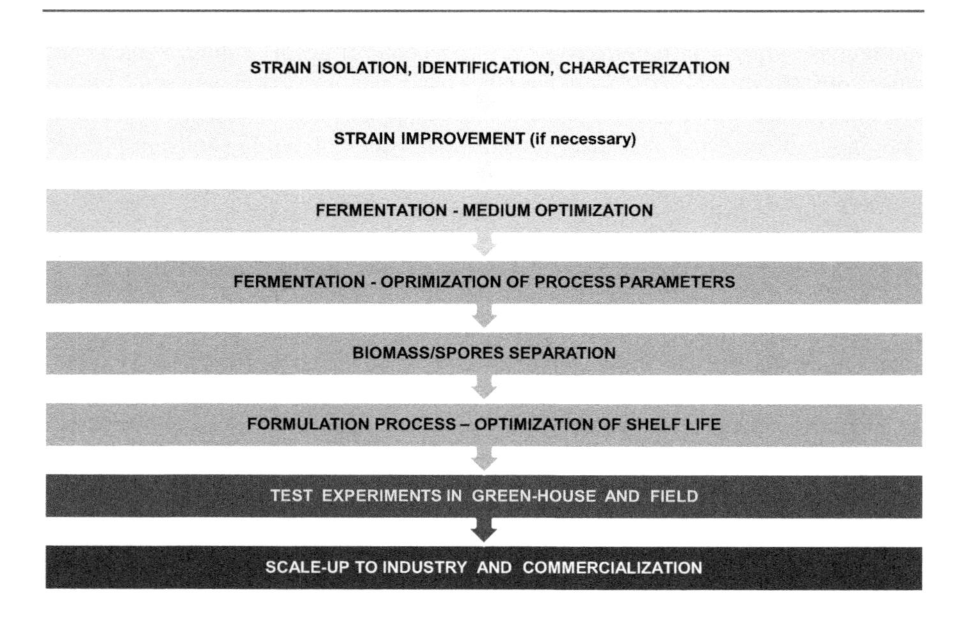

Fig. 6.2 Main steps in biofertilizer/biocontrol development and production. (Modified from Vassilev and Mendes 2018)

artificial, synthetic medium incapable of mimicking the endogenous abiotic and biotic conditions required for microbial growth, and many bacteria are categorized as uncultivable in such conditions (Pham and Kim 2012).

In the production of biofertilizers for commercial use, a high-quality biomass and/or spores of the target inoculum is required that further must retain high viability during the formulation process, storage, transportation, and after application to seed or in soil. The biomass or spore production is carried out in fermenters in conditions of liquid submerged or solid-state fermentation (SSF) processes based on inexpensive media (Malusa et al. 2012; Vassilev et al. 2015; Vassilev and Mendes 2018). Extensive studies are needed to optimize medium composition, process parameters, and transfer the laboratory technology to large-scale production. According to the type of the process, fermentations depend on the inoculum age and size, medium composition, concentration of the medium constituents and their ratio, water activity, the level of dissolved oxygen/aeration, addition of precursors, humidity, temperature, initial pH and its maintenance (if necessary), and time of harvest. The liquid submerged bacterial fermentation is well studied and described. In submerged cultures, bacteria and fungi may produce biomass and spores. This process depends on the microorganism, its nutritional medium or fermentation process (McCoy et al. 1988). During the last 20 years, the SSF has attracted more attention in the biotechnology industry although in general, the biofertilizer production process in SSF is more suitable in small-scale conditions. SSF has been defined as the bioprocess carried out in the absence, or near-absence, of free water; however, the substrate must possess enough moisture to support the growth and metabolic activity of the microorganism (Costa et al. 2018). SSF is a process, which includes a unique solid, liquid, and gaseous phase interactions, thus ensuring advantageous microbial growth and metabolic activity. This eco-friendly process, which is normally based on solid agro-industrial wastes used as a substrate or cell-carrier, is particularly characterized by lower energy and water needs, lack of contaminants, and high metabolic target activity (Thomas et al. 2013). To use submerged fermentation or SSF is a question of economic choice and also depends on the bacterial properties and specificity of the formulation procedure.

While the cultivation processes and optimization of fermentation parameters for bacterial growth and biomass/spore production are well studied, formulation procedures are far from the market and farmers' requirements (Lesueur et al. 2016). Biofertilizer, produced as a result of solid-state fermentation, in fact is a ready-to-use commercial product, thus avoiding the formulation step of the overall production process (Mendes et al. 2015). The final product of the fermentation, containing mineralized agro-industrial waste, bacterial biomass, and all released metabolites are simply dried ground and introduced into sterile bags. The problem with the SSF-based formulation is that the products are bulky, thus requiring extra space for processing and storage. For this reason, the liquid-state fermentation is preferred to obtain large quantities of biomass often in shorter time (Jambhulkar et al. 2016). In conditions of submerged fermentation, bacteria can be separated from fermentation broth and further concentrated and formulated. Alternatively, both biomass and medium can be formulated to form granules, pellets, wettable powders, or liquids. It is widely accepted that the fermentation process for the production of abundant and dense biomass/spores is the most decisive part of the overall production of a final biofertilizer commercial product (Bashan et al. 2014).

Four groups of biofertilizer formulates are commercialized depending on the carrier material: soil-related materials, plant derivates, inert carriers, lyophilized and oil dried bacteria, liquid carriers, and capsule-based carriers (Bashan 1998).

The carrier is the inoculant portion that ensures the effective release of the bacterial cells. Carrier diversity is enormous including water, vermiculite, perlite, calcium sulfate, calcium phosphate, coal, biochar, mineral soil, vegetable oil, corn cob, natural and artificial polysaccharides, etc. (Bashan 1998). The carrier characteristics include to be easily available and inexpensive and chemically stable; to be non-toxic, thus ensuring a friendly environment for the microorganism; to maintain sufficient humidity; to be capable of delivering of metabolically active cells in the soil; to be easy to process; and to ensure cell viability after determined periods of storage. Here, we will describe the most widespread and the most innovative carriers and formulations.

Peat is the solid carrier of choice for biofertilizer formulation, but it is not easily available and is expensive (Stephens and Rask 2000). However, this carrier material is well known, and farmers are familiar with its application. It is important to note that peat, with its high surface area and high water-holding capacity ensures bacterial metabolic activity and cell multiplication continues during the storage period.

Another solid carrier is biochar produced by pyrolysis of biomass under limited oxygen availability. Biochar can derive from plant biomass or animal bones and due to the specific porous structure and properties is an excellent carrier for soil microorganisms (Hale et al. 2014; Warren et al. 2009). Biochar enhances the soil physical and biochemical properties, and particularly animal biochar serves as a high-quality P source (Vassilev et al. 2013).

Talc, a metamorphic mineral composed of hydrated magnesium silicate, is frequently used as a formulation agent. It easily permits bio-preparations of more than one plant-beneficial microorganism (Shanmugam et al. 2011; Sahu et al. 2013). It is interesting to mention that talc formulations can be used directly and as a suspension to the seeds or as a spray. Talc is also used as filler, adding bulkiness to formulations based on costly polysaccharides used in immobilized-cell-based technologies (Sahu and Brahmaprakash 2016).

Liquid formulations, based on aqueous medium containing all components necessary for microbial growth, are now gaining popularity for different types of plant-beneficial microorganisms. Current liquid biofertilizer formulations are easy to handle and adapted for existing seeding equipment or directly in soil (Herrmann and Lesueur 2013). They can be produced in conditions of liquid submerged fermentations, processed aseptically, and maintained metabolically active before use (Mahanty et al. 2017). This kind of formulations is cost-effective, as they do not need solid carrier material and normally contain high bacterial concentrations thus allowing the application of a lower quantity compared to carrier-based biofertilizers. Liquid biofertilizers can be enriched with cell protectant and additives to improve inoculant performance during storage and in soil (Sahu and Brahmaprakash 2016). However, liquid formulated biofertilizers are very sensitive when applied on seeds and need addition of high molecular weight polymers to improve their survival (Singleton et al. 2002; Vassilev et al. 2017a). Liquids with biofertilizing properties could be produced without cells and/or using immobilized cells – a technique, which allows for more effective, multifunctional products (Mendes et al. 2017; Vassilev et al. 2017b).

During the last 20 years, application of immobilization methods in the field of biofertilizer and biocontrol production was observed (Bashan 1998; Vassilev et al. 2001, 2005; Malusa et al. 2012). Immobilization methods make use of non-toxic polymeric natural compounds such as alginate, agar, agarose, polyacrylamide gel, pectin, chitosan, etc. (Bashan et al. 2014; Vassilev et al. 2015). Compared to free cell systems, immobilization of plant-beneficial microorganisms offers advantages including enhanced metabolic activity and stability, better fermentation control, and low risk of contamination (Vassilev et al. 2007). In case of applications in disturbed soils or soil-plant systems, formulated plant-beneficial microorganisms encapsulated in natural gel carriers ensure very efficient barrier against biotic and abiotic stresses (Cassidy et al. 1996; Vassileva et al. 1999; Vassilev et al. 2012). In addition, results show a strong positive effect of such kind of formulations on both maintenance of viability/metabolic activity during storage and upon introduction into soil and delivery of these active cells, thus ensuring higher rhizosphere microbial enrichment compared to the direct introduction of the respective free microbial forms (Vassileva et al. 1999). To immobilize bacterial cells, processes such as spray drying, interfacial polymerization, and gelation are widely studied. However, novel techniques are continuously developed to improve the viability and resistance of cells during drying procedures and storage (Vassilev et al. 2015).

The tendencies in the field of production and formulation of biofertilizers for the nearest future are well determined (Vassilev et al. 2015; Bashan et al. 2014). Briefly, co-cultivation (fermentation) processes, addition of medium constituents with both nutrient and protective properties, development of large-scale SSF processes, liquid cell-free biofertilizers, economically acceptable immobilization techniques, and inclusion of fillers and additives in the immobilization matrixes are among the most acceptable and easy to develop scientific procedures in biofertilizer production and formulation. The most important is to create smart systems based on a multifaceted technological approach gathering achievements from various scientific fields.

6.5 Conclusion

The interactions between plants and microbes in the rhizosphere are complex where the root exudation is the key point in this interaction. There are too many papers demonstrating that interactions in rhizosphere are mediated directly or indirectly by root exudates. However, recently the sequencing technology allows investigating the interactions at the community level. Furthermore, studies analyze the interactions at functional level identifying the signals involved in interactions among different species which is the key point in the utilization of these processes for the benefits of the crops and sustainability of the plant species. Root exudates are crucial and fundamental signals in plant, fungal, and microbe communications in the soil. They are some kind of messengers that intermediate communication between all partners in the rhizosphere. Thus, the rhizosphere with all beauty presented above, is extremely complex, with continuously changing characteristics. What is important is that *we know that we don't know sufficient*, but we have to reach more deeply in the research of interactions between the microbes, the plants, the other soil organisms, and the abiotic environment.

Acknowledgements This review paper was partially supported by project 09-18 of Agricultural university-Plovdiv.

N. Vassilev has received Project Grant (CTM2014-53186-R) from the Spanish Ministerio de Ciencia e Innovaci6n and EC FEDER Fund.

References

Acuca J, Jorquera M, Martínez O, Menezes-Blackburn D et al (2011) Indole acetic acid and phytase activity produced by rhizosphere bacilli as affected by pH and metals. J Soil Sci Plant Nutr 11(3):1–12

Ahmad M, Zahir ZA, Khalid M (2013) Efficacy of *Rhizobium* and *Pseudomonas* strains to improve physiology, ionic balance and quality of mung bean under salt-affected conditions on farmer's fields. Plant Physiol Biochem 63:170–176

Ai C, Liang GQ, Sun JW, Wang XB, He P, Zhou W (2015) Reduced dependence of rhizosphere microbiome on plant-derived carbon in 32-year long-term inorganic and organic fertilized soils. Soil Biol Biochem 80:70–78

Alegria Terrazas R, Giles C, Paterson E, Robertson-Albertyn S et al (2016) Plant-microbiota interactions as a driver of the mineral turnover in the rhizosphere. Adv Appl Microbiol 10:1–67

Ali S, Charles TC, Glick BR (2014) Amelioration of high salinity stress damage by plant growth-promoting bacterial endophytes that contain ACC deaminase. Plant Physiol Biochem 80:160–167. https://doi.org/10.1016/j.plaphy.2014.04.003

Alori ET, Glick BR, Babalola OO (2017) Microbial phosphorus solubilization and its potential for use in sustainable agric. Front Microbiol 8:971. https://doi.org/10.3389/fmicb.2017.00971

Alquéres S, Meneses C, Rouws L, Rothballer M et al (2013) The bacterial superoxide dismutase and glutathione reductase are crucial for endophytic colonization of rice roots by *Gluconacetobacter diazotrophicus* PAL5. Mol Plant-Microbe Interact 26:937–945

Amaral FP, Pankievicz VCS, Arisi ACM, Souza EM, Pedrosa F, Stacey G (2016) Differential growth responses of *Brachypodium distachyon* genotypes to inoculation with plant growth promoting rhizobacteria. Plant Mol Biol 90(6):689–697. https://doi.org/10.1007/s11103-016-0449-8

Anand K, Kumari B, Mallick MA (2016) Phosphate solubilizing microbes: an effective and alternative approach as bio-fertilizers. Int J Pharm Sci 8(2):37–40

Andrade G, Mihara KL, Linderman RG, Bethlenfalvay GJ (1997) Bacteria from rhizosphere and hyphosphere soils of different arbuscular–mycorrhizal fungi. Plant Soil 192:71–79

Andreote FD, Rocha UN, Araujo WL, Azevedo JL, van Overbeek LS (2010) Effect of bacterial inoculation, plant genotype and developmental stage on root-associated and endophytic bacterial communities in potato (*Solanum tuberosum*). Anton Leeuw 97:389–399

Arkhipova TN, Prinsen E, Veselov SU et al (2007) Cytokinin producing bacteria enhance plant growth in drying soil. Plant Soil 292(1):305–315

Atkinson D, Black KE, Forbes PJ, Hooker JE, Baddeley JA, Watson CA (2003) The influence of arbuscular mycorrhizal colonization and environment on root development in soil. Eur J Soil Sci 54:751–757

Audipudi AV et al. (2016) Effect of mixed inoculations of plant growth promoting rhizobacteria of chilli on growth and induced systemic resistance of *Capsicum fruitescence* L. 4th Asian PGPR Conference Recent trends in PGPR research for sustainable crop productivity 9-20 ref 27

Azaizeh H, Marschner H, Roemheld V, Wittenmayer R (1995) Effects of a vesicular-arbuscular mycorrhizal fungus and other soil microorganisms on growth, mineral acquisition and root exudation of soil-grown maize plants. Mycorrhiza 5:321–327

Babalola OO, Glick BR (2012) The use of microbial inoculants in African agriculture: current practice and future prospects. J Food Agric Environ 10:540–549

Bacon CW, Hinton DM (2006) Bacterial endophytes: the endophytic niche, its occupants, and its utility. In: Gnanamanickam SS (ed) Plant-associated bacteria. Springer, Dordrecht, pp 155–194

Badri DV, Vivanco JM (2009) Regulation and function of root exudates. Plant Cell Environ 32(6):666–681. https://doi.org/10.1111/j.1365-3040.2009.01926.x

Bagwell CE, La Rocque JR, Smith GW et al (2002) Molecular diversity of diazotrophs in oligotrophic tropical seagrass bed communities. FEMS Microbiol Ecol 39:113–119. https://doi.org/10.1016/S0168-6496(01)00204-5

Bais HP, Prithiviraj B, Jha AK, Ausubel FM, Vivanco JM (2005) Mediation of pathogen resistance by exudation of antimicrobials from roots. Nature 434:217

Bais HP, Weir TL, Perry LG, Gilroy S, Vivanco JM (2006) The role of root exudates in rhizosphere interactions with plants and other organisms. Annu Rev Plant Biol 57:233–266

Banerjee MR, Yesmin L, Vessey JK (2006) Plant growth promoting rhizobacteria as biofertilizers and biopesticides. In: Rai MK (ed) Handbook of microbial biofertilizers. Haworth Press, New York

Barry SM, Challis GL (2009) Recent advances in siderophore biosynthesis. Curr Opin Chem Biol 13:1–11

Bashan Y (1998) Inoculants of plant growth-promoting bacteria for use in agriculture. Biotechnol Adv 16:729–770

Bashan Y, de-Bashan LE, Prabhu SR, Hernandez JP (2014) Advances in plant growth-promoting bacterial inoculant technology: formulations and practical perspectives (1998–2013). Plant Soil 378:1–33
Beattie GA (2006) Plant-associated bacteria: survey, molecular phylogeny, genomics and recent advances. In: Gnanamanickam SS (ed) Plant-associated bacteria. Springer, Dordrecht, pp 1–56
Beauregard PB, Chai Y, Vlamakis H, Losick R, Kolter R (2013) *Bacillus subtilis* biofilm induction by plant polysaccharides. Proc Natl Acad Sci U S A 110:E1621–E1630
Bécard G, Fortin JA (1988) Early events of vesicular-arbuscuLar mycorrhiza formation on Ri T-DNA transformed roots. New Phytol 108:211–218
Berendsen RL, Verk MCV, Stringlis IA et al (2015) Unearthing the genomes of plant-beneficial *Pseudomonas* model strains WCS358, WCS374 and WCS417. BMC Genomics 16:539
Berg G, Smalla K (2009) Plant species and soil type cooperatively shape the structure and function of microbial communities in the rhizosphere. FEMS Microbiol Ecol 68:1–13
Berg G, Grube M, Schloter M, Smalla K (2014) Unraveling the plant microbiome: looking back and future perspectives. Front Microbiol 5:148
Bhattacharyya PN, Jha DK (2012) Plant growth-promoting rhizobacteria (PGPR): emergence in agriculture. World J Microbiol Biotechol 28:1327–1350
Bogar B, Srakers G (2003) Optimization of phytase production by solid substrate fermentation. J Ind Microbiol Biotechnol 30:183–189
Boukhalfa H, Crumbliss AL (2002) Chemical aspects of siderophore mediated iron transport. Biometals 15:325–339
Briat JF (2008) Iron dynamics in plants. Adv Bot Res 46:137–180
Brown ME (1974) Seed and root bacterization. Annu Rev Phytopathol 12:181–197
Brundrett MC (2009) Mycorrhizal associations and other means of nutrition of vascular plants: understanding the global diversity of host plants by resolving conflicting information and developing reliable means of diagnosis. Plant Soil 320:37–77. https://doi.org/10.1007/s11104-008-9877-9
Bücking H, Abubaker J, Govindarajulu M, Tala M et al (2008) Root exudates stimulate the uptake and metabolism of organic carbon in germinating spores of *Glomus intraradices*. New Phytol 180:684–695
Buée M, De Boer W, Martin F, van Overbeek L, Jurkevitch E (2009) The rhizosphere zoo: an overview of plant-associated communities of microorganisms, including phages, bacteria, archaea, and fungi, and of some of their structuring factors. Plant Soil 321:189–212
Cai T, Cai W, Zhang J, Zheng H, Tsou AM et al (2009) Host legume-exuded antimetabolites optimize the symbiotic rhizosphere. Mol Microbiol 73:507–517
Cai Z, Kastell A, Knorr D, Smetanska I (2012) Exudation: an expanding technique for continuous production and release of secondary metabolites from plant cell suspension and hairy root cultures. Plant Cell Rep 31:461–477
Caporale AG, Sarkar D, Datta R, Punamiya P, Violante A (2014) Effect of arbuscular mycorrhizal fungi (*Glomus* spp.) on growth and arsenic uptake of vetiver grass (*Chrysopogon zizanioides* L.) from contaminated soil and water systems. J Soil Sci Plant Nutr 14:955–972
Carvalhais LC, Dennis PG, Fedoseyenko D (2011) Root exudation of sugars, amino acids, and organic acids by maize as affected by nitrogen, phosphorus, potassium, and iron deficiency. J Plant Nutr Soil Sci 174:3–11. https://doi.org/10.1002/jpln.201000085
Carvalhais LC, Dennis PG, Fan B, Fedoseyenko D et al (2013) Linking plant nutritional status to plant-microbe interactions. PLoS One 8:e68555
Cassidy MB, Lee H, Trevors JT (1996) Environmental applications of immobilized microbial cells: a review. J Ind Microbiol 16:79–101
Chalot M, Brun A (1998) Physiology of organic nitrogen acquisition by ectomycorrhizal fungi and ectomycorrhizas. FEMS Microbiol Rev 22:21–44
Chang P, Gerhardt KE, Huang X-D, Yu X-M, Glick BR, Gerwing PD, Greenberg BM (2014) Plant growth-promoting bacteria facilitate the growth of barley and oats in salt-impacted soil: implications for phytoremediation of saline soils. Int J Phytoremediation 16:1133–1147. https://doi.org/10.1080/15226514.2013.821447

Chaparro JM, Badri DV, Vivanco JM (2014) Rhizosphere microbiome assemblage is affected by plant development. ISME J 8:790–803. https://doi.org/10.1038/ismej.2013.196

Chen YP, Rekha PD, Arun AB, Shen FT, Lai WA, Young CC (2006) Phosphate solubilizing bacteria from subtropical soil and their tricalcium phosphate solubilizing abilities. Appl Soil Ecol 34:33–41

Chhonkar PK (1978) Influence of light on pectic enzymes in root exudates of *Trifolium alexandrinum* inoculated with *Rhizobium trifolii*. Zentralbl Bakteriol Naturwiss 133(1):50–53. Co., New York, NY

Costa PB, Granada CE, Ambrosini A, Moreira F, de Souza R et al (2014) A model to explain plant growth promotion traits: a multivariate analysis of 2,211 bacterial isolates. PLoS One 9(12):e116020. https://doi.org/10.1371/journal.pone.0116020

Costa JAV, Treichel H, Kumar V, Pandey A (2018) Advances in solid-state fermentation. In: Pandey A, Larroche CH, Soccol C (eds) Current developments in biotechnology and bioengineering. Curr Adv Solid-State Ferm. Elsevier B.V. 1–17

Dimkpa CO, Svatos A, Dabrowska P, Schmidt A, Boland W, Kothe E (2008) Involvement of siderophores in the reduction of metal-induced inhibition of auxin synthesis in *Streptomyces* spp. Chemosphere 74:19–25

Dimkpa CO, Merten D, Svatos A, Büchel G, Kothe E (2009) Siderophores mediate reduced and increased uptake of cadmium by *Streptomyces tendae* F4 and sunflower (*Helianthus annuus*), respectively. J Appl Microbiol 5:687–1696

Dobbelaere S, Vanderleyden J, Okon Y (2003) Plant growth-promoting effects of diazotrophs in the rhizosphere. Crit Rev Plant Sci 22:107–149

Donn S, Kirkegaard JA, Perera G, Richardson AE, Watt M (2015) Evolution of bacterial communities in the wheat crop rhizosphere. Environ Microbiol 17:610–621

Duca D, Lorv J, Patten CL, Rose D, Glick BR (2014) Indole-3-acetic acid in plant–microbe interactions. Antonie Van Leeuwenhoek 106:85–125

Duffner A, Hoffland E, Temminghoff EJM (2012) Bioavailability of zinc and phosphorus in calcareous soils as affected by citrate exudation. Plant Soil 361:165–175

Dvorakova J (1998) Phytase: sources, preparation and exploitation. Folia Microbiol 43:323–338

Edwards J, Johnson C, Santos-Medellõn C, Lurie E, Podishetty NK, Bhatnagar S et al (2015) Structure, variation, and assembly of the root-associated microbiomes of rice. Proc Natl Acad Sci U S A 112(8):E911–EE20. https://doi.org/10.1073/pnas.1414592112. WOS:000349911700014

Etesami H, Mirseyed H, Alikhani H (2014) In planta selection of plant growth promoting endophytic bacteria for rice (*Oryza sativa* L.). J Soil Sci Plant Nutr 14:491–503

Fatima S, Anjum T (2017) Identification of a potential ISR determinant from *Pseudomonas aeruginosa* PM12 against *Fusarium* wilt in tomato. Front Plant Sci 8:848

Filion M, St-Arnaud M, Fortin JA (1999) Direct interaction between the arbuscular mycorrhizal fungus *Glomus intraradices* and different rhizosphere micro-organisms. New Phytol 141:525–533

Fransson PMA, Johansson EM (2010) Elevated CO_2 and nitrogen influence exudation of soluble organic compounds by ectomycorrhizal root systems. FEMS Microbiol Ecol 71:186–196. https://doi.org/10.1111/j.1574-6941.2009.00795.x

Frugier F, Kosuta S, Murray JD, Crespi M, Szczyglowski K (2008) Cytokinin: secret agent of symbiosis. Trends Plant Sci 13:115–120

Fujita J, Budda N, Tujimoto M, Yamane Y, Fukuda H, Mikami S, Kizaki Y (2000) Isolation and characterization of phytase isozymes produced by *Aspergillus oryzae*. Biotechnol Lett 22:1797–1802

Gachomo E, Allen JW, Pfeffer PE, Govindarajulu M, Douds DD, Jin HR, Nagahashi G, Lammers PJ, Shachar-Hill Y, Bücking H (2009) Germinating spores of *Glomus intraradices* can use internal and exogenous nitrogen sources for *de novo* biosynthesis of amino acids. New Phytol 184:399–411

Gamalero E, Glick BR (2012) Ethylene and abiotic stress tolerance in plants. In: Ahmad P, Prasad MNV (eds) Environmental adaptations and stress tolerance of plants in the era of climate change. Springer, New York, pp 395–412

Garcias-Bonet N, Arrieta JM, Duarte CM, Marbà N (2016) Nitrogen fixing bacteria in Mediterranean seagrass (*Posidonia oceanica*) roots. Aquat Bot 131:57–60. https://doi.org/10.1016/j.aquabot.2016.03.002

Ghazalibigla H et al (2016) Is induced systemic resistance the mechanism for control of black rot in *Brassica oleracea* by a *Paenibacillus* sp.? Biol Control 92:195–201

Ghosh U, Subhashini P, Dilipan E et al (2012) Isolation and characterization of phosphate-solubilizing bacteria from seagrass rhizosphere soil. J Ocean Univ China 11:86–92. https://doi.org/10.1007/s11802-012-1844-7

Gianinazzi-Pearson V, Dumas-Gaudot E, Gollotte A, Tahiri-Alaoui A, Gianinazzi S (1996) Cellular and molecular defence-related root responses to invasion by arbuscular mycorrhizal fungi. New Phytol 133:45–57

Gianinazzi-Pearson V, Branzanti B, Gianinazzi S (1989) In vitro enhancement of spore germination and early hyphal growth of a vesicular-arbuscular mycorrhizal fungus by host root exudates and plant flavonoids. Symbiosis 7:243–255

Glandorf DC, Bakker PA, Van Loon LC (1997) Influence of the production of antibacterial and antifungal proteins by transgenic plants on the saprophytic soil microflora. Acta Bot Neerl 46:85–104

Glick BR (2005) Modulation of plant ethylene levels by the bacterial enzyme ACC deaminase. FEMS Microbiol Lett 251:1–7

Glick BR (2010) Using soil bacteria to facilitate phytoremediation. Biotechnol Adv 28:367–374

Glick B (2012) Plant growth-promoting bacteria: mechanisms and applications. Scientifica 2012:1–15

Glick BR (2015) Beneficial plant–bacterial interactions. Springer, Cham

Glick BR, Penrose DM, Li J (1998) A model for the lowering of plant ethylene concentrations by plant growth promoting bacteria. J Theor Biol 190:63–68

Goldstein AH (1994) Involvement of the quinoprotein glucose dehydrogenase in the solubilization of exogenous mineral phosphates by Gram negative bacteria. In: Torriani-Gorni A, Yagil E, Silver S (eds) Phosphate in microorganisms: cellular and molecular biology. ASM Press, Washington, DC, pp 197–203

Gosling P, Hodge A, Goodlass G, Bending GD (2006) Arbuscular mycorrhizal fungi and organic farming. Agric Ecosyst Environ 113:17–35

Gouda S, Kerry R, Dasc G, Paramithiotisd S, Shine H, Patra J (2018) Revitalization of plant growth promoting rhizobacteria for sustainable development in agriculture. Microbiol Res 206:131–140

Graham JH (1982) Effect of citrus root exudates on germination of chlamydospores of the vesicular-arbuscular mycorrhizal fungus *Glomus epigaeum*. Mycologia 74:831–835

Graham PH (1988) Principles and application of soil microbiology. Prentice Hall, Upper Saddle River

Gravel V, Antoun H, Tweddell RJ (2007) Growth stimulation and fruit yield improvement of greenhouse tomato plants by inoculation with *Pseudomonas putida* or *Trichoderma atroviride*: possible role of indole acetic acid (IAA). Soil Biol Biochem 39(8):1968–1977

Gray EJ, Smith DL (2005) Intracellular and extracellular PGPR: commonalities and distinctions in the plant-bacterium signaling processes. Soil Biol Biochem 37:395–412

Grover M, Ali SKZ, Sandhya V, Rasul A, Venkateswarlu B (2011) Role of microorganisms in adaptation of agriculture crops to abiotic stresses. World J Microbiol Biotechnol 27:1231–1240

Gupta G, Parihar SS, Ahirwar NK, Snehi SK, Singh V (2015) Plant growth promoting rhizobacteria (pgpr): current and future prospects for development of sustainable agriculture. J Microbiol Biochem Technol 7:96–102. https://doi.org/10.4172/1948-5948.1000188

Gutiérrez-Mañero FJ, Ramos B, Probanza A, Mehouachi J, Talon M (2001) The plant growth-promoting rhizobacteria *Bacillus pumilus* and *Bacillus licheniformis* producehigh amounts of physiologically active gibberellins. Physiol Plant 111:206–211

Guttman D, McHardy AC, Schulze-Lefert P (2014) Microbial genome-enabled insights into plant-microorganism interactions. Nat Rev Genet 15:797–813

Haichar FZ, Marol C, Berge O, Rangel-Castro JI, Prosser JI et al (2008) Plant host habitat and root exudates shape soil bacterial community structure. ISME J 2:1221–1230
Halder AK, Chakrabarty PK (1993) Solubilization of inorganic phosphate by *Rhizobium*. Folia Microbiol 38:325–330
Halder AK, Mishra AK, Bhattacharya P, Chakrabarthy PK (1990) Solubilization of rock phosphate by Rhizobium and Bradyrhizobium. J Gen Appl Microbiol 36:1–92
Hale L, Luth M, Kenney R et al (2014) Evaluation of pinewood biochar as a carrier of bacterial strain *Enterobacter cloacae* UW5 for soil inoculation. Appl Soil Ecol 84:192–199
Hao Y, Charles TC, Glick BR (2007) ACC deaminase from plant growth promoting bacteria affects crown gall development. Can J Microbiol 53:1291–1299
Hardoim PR, Andreote FD, Reinhold-Hurek B, Sessitsch A, van Overbeek LS, van Elsas JD (2011) Rice root-associated bacteria: insights into community structures across 10 cultivars. FEMS Microbiol Ecol 77:154–164
Haros M, Rosell M, Benedito C (2001) Fungal phytase as a potential breadmaking additive. Eur Food Res Technol 213:317–322
Hayat R, Ali S, Amara U, Khalid R, Ahmed I (2010) Soil beneficial bacteria and their role in plant growth promotion: a review. Ann Microbiol 60:579–598. https://doi.org/10.1007/s13213-010-0117-1
Hepper CM, Jakobsen I (1983) Hyphal growth from spores of the mycorrhizal fungus *Glomus caledonius*: effect of amino acids. Soil Biol Biochem 15:55–58
Herrmann L, Lesueur D (2013) Challenges of formulation and quality of biofertilizers for successful inoculation. Appl Microbiol Biotechnol 97:8859–8873
Hinsinger P, Fernandes Barros ON, Benedetti MF, Noack Y, Callot G (2001) Plant-induced weathering of a basaltic rock: experimental evidence. Geochim Cosmochim Acta 65:137–152
Hinsinger P, Plassard C, Jaillard B, Tang CX (2003) Origins of root-mediated pH changes in the rhizosphere and their responses to environmental constraints: a review. Plant Soil 248:43–59
Hinsinger P, Bengough AG, Vetterlein D, Young IM (2009) Rhizosphere: biophysics, biogeochemistry and ecological relevance. Plant Soil 321:117–152
Hirimuthugoda N, Zhenming C, Longfei W (2007) Probiotic yeasts with phytase activity identified from the gastrointestinal tract of sea cucumbers. SPC Beche de Mer Inform Bull 26:31–32
Hodge A (2001) Arbuscular mycorrhizal fungi influence decomposition of, but not plant nutrient capture from, glycine patches in soil. New Phytol 151:725–734
Holmer M, Andersen FO, Nielsen SL, Boschker HTS (2001) The importance of mineralization based on sulfate reduction for nutrient regeneration in tropical seagrass sediments. Aquat Bot 71:1–17. https://doi.org/10.1016/S0304-3770(01)00170-X
Hussain A, Hasnain S (2011) Interactions of bacterial cytokinins and IAA in the rhizosphere may alter phytostimulatory efficiency of rhizobacteria. World J Microbiol Biotechnol 27:2645–2654
Igual JM, Valverde A, Cervantes E, Velázquez E (2001) Phosphatesolubilizing bacteria as inoculants for agriculture: use of updated molecular techniques in their study. Agronomie 21:561–568
Illmer P, Schinner F (1992) Solubilization of inorganic phosphates by microorganisms isolated from forest soil. Soil Biol Biochem 24:389–395
Ingle K, Padole D (2017) Phosphate solubilizing microbes: an overview. Int J Curr Microbiol Appl Sci 1:844–852
Iqbal MA, Khalid M, Shahzad SM, Ahmad M, Soleman N et al (2012) Integrated use of *Rhizobium leguminosarum*, plant growth promoting rhizobacteria and enriched compost for improving growth, nodulation and yield of lentil (*Lens culinaris* Medik). Chilean J Agric Res 72:104–110
Jacoby R, Peukert M, Succurro A, Koprivova A, Kopriva S (2017) The role of soil microorganisms in plant mineral nutrition-current knowledge and future directions. Front Plant Sci 8:1617
Jalal MAF, van der Helm D (1991) Isolation and spectroscopic identification of fungal siderophores. In: Winkelmann G (ed) CRC handbook of microbial iron chelates. CRC Press, p 235–269
Jambhulkar PP, Sharma P, Yadav R (2016) Systems for introduction of microbial inoculants in the field. In: Singh DP, Singh HB, Prabha R (eds) Microbial inoculants in sustainable agricultural productivity. Vol 2. Functional applications. Springer, New Delhi, pp 199–218

Jin HR, Jiang DH, Zhang PH (2011) Effect of carbon and nitrogen availability on the metabolism of amino acids in the germinating spores of arbuscular mycorrhizal fungi. Pedosphere 21:432–442
Jones DL (1998) Organic acids in the rhizosphere – a critical review. Plant Soil 205:25–44
Jones DL, Farrar J, Giller KE (2003) Associative nitrogen fixation and root exudation – what is theoretically possible in the rhizosphere? Symbiosis 35:19–38
Jones DL, Hodge A, Kuzyakov Y (2004) Plant and mycorrhizal regulation of rhizodeposition. New Phytol 163:459–480. https://doi.org/10.1111/j.1469-8137.2004.01130.x
Jonsson LM, Nilsson LC, Wardle DA, Zackrisson O (2001) Context dependent effects of ectomycorrhizal species richness on tree seedling productivity. Oikos 93:353–364
Kaldy JE, Eldridge PM, Cifuentes LA, Jones WB (2006) Utilization of DOC from seagrass rhizomes by sediment bacteria: 13C-tracer experiments and modeling. Mar Ecol Prog Ser 317:41–55. https://doi.org/10.3354/meps317041
Kang BG, Kim WT, Yun HS, Chang SC (2010) Use of plant growth-promoting rhizobacteria to control stress responses of plant roots. Plant Biotechnol Rep 4:179–183
Karin H, Anna M, Andreas V, Franz H, Siegrid S (2013) Alterations in root exudation of intercropped tomato mediated by the arbuscular mycorrhizal fungus *Glomus mosseae* and the soilborne pathogen *Fusarium oxysporum* f. sp. *Lycopersici*. J Phytopathol 161:763–773
Kawasaki A, Watson ER, Kertesz MA (2012) Indirect effects of polycyclic aromatic hydrocarbon contamination on microbial communities in legume and grass rhizospheres. Plant Soil 358:162–175. https://doi.org/10.1007/s11104-011-1089-z. WOS:000308190400015
Kawasaki A, Warren CR, Kertesz MA (2016a) Specific influence of white clover on the rhizosphere microbial community in response to polycyclic aromatic hydrocarbon (PAH) contamination. Plant Soil 401:365–379. https://doi.org/10.1007/s11104-015-2756-2. WOS:000372947800026
Kawasaki A, Donn S, Ryan PR, Mathesius U, Devilla R, Jones A et al (2016b) Microbiome and exudates of the root and rhizosphere of *Brachypodium distachyon*, a model for wheat. PLoS One 11:e0164533. https://doi.org/10.1371/journal.pone.0164533
Kidd P, Barceló J, Bernal MP, Navari-Izzo F, Poschenrieder C, Shilev S, Clemente R, Monteroso C (2009) Trace element behavior at the root-soil interface: implications in phytoremediation. J Environ Exp Bot 67:243–259
Kilminster K, Garland J (2009) Aerobic heterotrophic microbial activity associated with seagrass roots: effects of plant type and nutrient amendment. Aquat Microb Ecol 57:57–68. https://doi.org/10.3354/ame01332
Kim YO, Kim HK, Bae KS, Yu JH, Oh TK (1998) Purification and properties of a thermostable phytase from Bacillus sp. DS11. Enzym Microb Technol 22:2–7
Kim YO, Lee JK, Oh BC, Oh TK (1999) High-level of a recombinant thermostable phytase in *Bacillus subtilis*. Biosci Biotechnol Biochem 63:2205–2207
King JS, Albaugh TJ, Allen HL, Buford M, Strain BR, Dougherty P (2002) Below-ground carbon input to soil is controlled by nutrient availability and fine root dynamics in loblolly pine. New Phytol 154:389–398
Kloepper JW (1994) Plant growth-promoting rhizobacteria (other systems). In: Okon Y (ed) *Azospirillum*/plant associations. CRC Press, Boca Raton, pp 111–118
Knauth S, HurekT BD, Reinhold-Hurek B (2005) Influence of different *Oryza* cultivars on expression of *nifH* gene pools in roots of rice. Environ Microbiol 7:1725–1733
Kong Z, Glick BR (2017) The role of plant growth-promoting bacteria in metal phytoremediation. Adv Microb Physiol 71:97–132
Krapp A (2015) Plant nitrogen assimilation and its regulation: a complex puzzle with missing pieces. Curr Opin Plant Biol 25:115–122
Krewulak KD, Vogel HJ (2008) Structural biology of bacterial iron uptake. Biochim Biophys Acta 1778:1781–1804
Kumar S, Bauddh K, Barman SC, Singh RP (2014) Amendments of microbial bio fertilizers and organic substances reduces requirement of urea and DAP with enhanced nutrient availability and productivity of wheat (*Triticum aestivum* L.). Ecol Eng 71:432–437. https://doi.org/10.1016/j.ecoleng.2014.07.007

Kumar P, Thakur S, Dhingra GK, Singh A, Pal MK et al (2018) Inoculation of siderophore producing rhizobacteria and their consortium for growth enhancement of wheat plant. Biocatal Agric Biotechnol 15:264–269
Küsel K, Blöthe M, Schulz D, Reiche M, Drake HL (2008) Microbial reduction of iron and porewater biogeochemistry in acidic peatlands. Biogeosciences 5(6):1537–1549
Lakshmanan V, Kitto SL, Caplan JL, Hsueh YH, Kearns DB, Wu YS (2012) Microbe-associated molecular patterns-triggered root responses mediate beneficial rhizobacterial recruitment in Arabidopsis. Plant Physiol 160(3):1642–1661. https://doi.org/10.1104/pp.112.200386. WOS:000310584200037
Lambers H, Chapin FS, Pons TL (2008a) Plant physiologica ecology, 2nd edn. Springer, New York
Lambers H, Shaver G, Raven JA, Smith SE (2008b) N and Pacquisition change as soils age. Trends Ecol Evol 23:95–103
Lambrechts C, Boze H, Molin G, Galzy P (1992) Utilization of phytate by some yeasts. Biotech Lett 14:61–66
Lambrecht M, Okon Y, Vande Broek A, Vanderleyden J (2000) Indole-3-acetic acid: a reciprocal signalling molecule in bacteria–plant interactions. Trends Microbiol 8(7):298–300
Langley JA, Hungate BA (2003) Mycorrhizal controls on belowground litter quality. Ecology 84:2302–2312
Lerat S, Lapointe L, Gutjahr S, Piché Y, Vierheilig H (2003) Carbon partitioning in a split-root system of arbuscular mycorrhizal plants is fungal and plant species dependent. New Phytol 157:589–595
Lesueur D, Deaker R, Herrmann L, Bräu L, Jansa J (2016) The production and potential of biofertilizers to improve crop yields. In: Arora NK et al (eds) Bioformulations: for sustainable agriculture. Springer, New Delhi
Leveau JHJ, Lindow SE (2005) Utilization of the plant hormone indole-3-acetic acid for growth by *Pseudomonas putida* strain 1290. Appl Environ Microbiol 71:2365–2371
Liao H, Wan H, Shaff J, Wang X, Yan X, Kochian LV (2006) Phosphorus and aluminum interactions in soybean in relation to aluminum tolerance. Exudation of specific organic acids from different regions of the intact system. Plant Physiol 141:674–684. https://doi.org/10.1104/pp.105.076497
Lodewyckx C, Vangronsveld J, Porteous F, Moore ERB et al (2002) Endophytic bacteria and their potential applications. Crit Rev Plant Sci 21:583–606. https://doi.org/10.1080/0735-260291044377
Lugtenberg BJ, Chin AWTF, Bloemberg GV (2002) Microbe-plant interactions: principles and mechanisms. Antonie Van Leeuwenhoek 81:373–383
Lundberg DS, Lebeis SL, Paredes SH, Yourstone S, Gehring J, Malfatti S et al (2012) Defining the core *Arabidopsis thaliana* root microbiome. Nature 488(7409):86–90. https://doi.org/10.1038/nature11237. WOS:000307010700038
Ma JF, Ueno H, Ueno D, Rombolà AD, Iwashita T (2003) Characterization of phytosiderophore secretion under Fe deficiency stress in *Festuca rubra*. Plant Soil 256:131–137
Madigan MT et al. (1997) Brock's biology of microorganisms. Prentice-Hall and Southern Illinois University
Mahanty T, Bhattacharjee S, Goswami M, Bhattacharyya P, Das B, Ghosh A, Tribedi P (2017) Biofertilizers: a potential approach for sustainable agriculture development. Environ Sci Pollut Res 24:3315–3335
Malusa E, Vassilev N (2014) A contribution to set a legal framework for biofertilisers. Appl Microbiol Biotechnol 98:6599–6607
Malusa E, Sas–Paszt L, Ciesielska J (2012) Technologies for beneficial microorganisms inocula used as biofertilizers. Sci World J 49:1206
Markets and markets (2016) Biofertilizers market by type (nitrogen-fixing, phosphate-solubilizing, potash-mobilizing), microorganism (*Rhizobium*, *Azotobacter*, *Azospirillum*, *Cyanobacteria*, P-solubilizer), mode of application, crop type, form, and region – global forecast to 2022
Marschner H (1995) Mineral nutrition of higher plants, 2nd edn. Academic, London, p 889
Marschner P (2012) Mineral nutrition of higher plants, 3rd edn. Academic, London

Martin BC, Statton J, Siebers AR, Grierson PF, Ryan MH, Kendrick GA (2017) Colonizing tropical seagrasses increase root exudation under fluctuating and continuous low light. Limnol Oceanogr. https://doi.org/10.1002/lno.10746
Martin BC, Gleeson D, Statton J, Siebers AR, Grierson P, Ryan MH, Kendrick GA (2018) Low light availability alters root exudation and reduces putative beneficial microorganisms in seagrass roots. Front Microbiol 8:2667
Matsukawa E, Nakagawa Y, Iimura Y, Hayakawa M (2007) Stimulatory effect of indole-3-acetic acid on aerial mycelium formation and antibiotic production in *Streptomyces* sp. Actinomycetologica 21:32–39
McCoy CW, Samson RA, Boucias DG (1988) In: Ignoffo CM, Mandava NB (eds) Handbook of natural pesticides: microbial pesticides, part A. Entomogenous protozoa and fungi. CRC Press, Boca Raton, pp 151–236
McGrath JW, Hammerschmidt F, Quinn JP (1998) Biodegradation of phosphonomycin by Rhizobium huakuii PMY1. Appl Environ Microbiol 64:356–358
Meier IC, Avis PG, Phillips RP (2013) Fungal communities influence root exudation rates in pine seedlings. FEMS Microbiol Ecol 83:585–595. https://doi.org/10.1111/1574-6941.12016
Mendes GO, Silva NMRM, Anastacio TC, Vassilev NB, Ribeiro JI, Silva IR, Costa MD (2015) Optimization of *Aspergillus niger* rock phosphate solubilization in solid-state fermentation and use of the resulting product as a P fertilizer. Microb Biotechnol 8:930–939
Mendes G, Galvez A, Vassileva M, Vassilev N (2017) Fermentation liquid containing microbially solubilized P significantly improved plant growth and P uptake in both soil and soilless experiments. Appl Soil Ecol 117–118:208–211
Meneses CH, Rouws LF, Simões-Araújo JL, Vidal MS, Baldani JI (2011) Exopolysaccharide production is required for biofilm formation and plant colonization by the nitrogenfixing endophyte *Gluconacetobacter diazotrophicus*. Mol Plant-Microbe Interact 24:1448–1458
Merzaeva OV, Shirokikh IG (2006) Colonization of plant rhizosphere by actinomycetes of different genera. Microbiology 75:226–230
Miethke M, Marahiel MA (2007) Siderophore-based iron acquisition and pathogen control. Microbiol Mol Biol Rev 71:413–451
Miransari M, Bahrami HA, Rejali F, Malakouti MJ (2008) Using arbuscular mycorrhiza to reduce the stressful effects of soil compaction on wheat (*Triticum aestivum* L.) growth. Soil Biol Biochem 40:1197–1206
Mukerji KG, Manoharachary C, Chamola B (2002) Techniques in mycorrhizal studies. Kluwer Academic Publishers, London/Dordrecht, pp 285–296
Nakamura Y, Fukuhara H, Sano K (2000) Secreted phytase activities of yeasts. Biosci Biotechnol Biochem 64:841–844
Nascimento FX, Rossi MJ, Glick BR (2018) Ethylene and 1-aminocyclopropane-1-carboxylate (ACC) in plant–bacterial interactions. Front Plant Sci 9:114. https://doi.org/10.3389/fpls.2018.00114
Navarro L, Dunoyer P, Jay F, Arnold B, Dharmasiri N, Estelle M et al (2006) A plant miRNA contributes to antibacterial resistance by repressing auxin signaling. Science 312:436–439
Neilands JB (1983) Siderophores. In: Eichhorn L, Marzilla LG (eds) Advances in inorganic biochemistry. Elsevier, p 137–166
Nobbe F, Hiltner L (1896) Inoculation of the soil for cultivating leguminous plants. US Patent n. 570813
Norman JR, Hooker JE (2000) Sporulation of *Phytophthora fragariae* shows greater stimulation by exudates of non-mycorrhizal than by mycorrhizal strawberry roots. Mycol Res 104:1069–1073
Ohtake H, Wu H, Imazu K, Ambe Y, Kato J, Kuroda A (1996) Bacterial phosphonate degradation, phosphite oxidation and polyphosphate accumulation. Resour Conserv Recycl 18:125–134
Onofre-Lemus J, Hernández-Lucas I, Girard L, Caballero-Mellado J (2009) ACC (1-aminocyclo propane-1-carboxylate) deaminase activity, a widespread trait in *Burkholderia* species, and its growth-promoting effect on tomato plants. Appl Environ Microbiol 75:6581–6590

Oteino N, Lally RD, Kiwanuka S, Lloyd A, Ryan D, Germaine KJ, Dowling DN (2015) Plant growth promotion induced by phosphate solubilizing endophytic *Pseudomonas* isolates. Front Microbiol 6:745

Pahari A, Mishra BB (2017) Antibiosis of siderophore producing bacterial isolates against phytopathogens and their effect on growth of okra. Int J Curr Microbiol App Sci 6:1925–1929

Pandey A, Szakacs G, Soccol CR, Rodriguez–Leon JA, Soccol VT (2001) Production, purification and properties of microbial phytases. Bioresour Technol 77:203–214

Parada M, Vinardell J, Ollero F, Hidalgo A, Gutiérrez R et al (2006) *Sinorhizobium fredii* HH103 mutants affected in capsular polysaccharide (KPS) are impaired for nodulation with soybean and *Cajanus cajan*. Mol Plant-Microbe Interact 19:43–52

Pathak DV, Kumar M (2016) Microbial inoculants as biofertilizers and biopesticides. In: Singh DP et al (eds) Microbial inoculants in sustainable agricultural productivity: research perspectives, vol 1. Springer, New Delhi, pp 197–209

Pattan J, Kajale S, Pattan S (2017) Isolation, production and optimization of siderophores (iron chilators) from *Pseudomonas fluorescence* NCIM 5096 and *Pseudomonas* from soil rhizosphere and marine water. Int J Curr Microbiol App Sci 6:919–928

Patten CL, Glick BR (1996) Bacterial biosynthesis of indole-3-acetic acid. Can J Microbiol 42:207–220

Pavlova K, Gargova S, Tankova Z (2008) Phytase from Antarctic yeast strain *Cryptococcus laurentii* AL27. Folia Microbiol 53:29–34

Peiffer JA, Spor A, Koren O, Jin Z, Tringe SG, Dangl JL (2013) Diversity and heritability of the maize rhizosphere microbiome under field conditions. Proc Natl Acad Sci U S A 110:6548–6553. https://doi.org/10.1073/pnas.1302837110. WOS:000318041500067

Pertry I, Vereecke D (2009) Identification of *Rhodococcus fascians* cytokinins and their modus operandi to reshape the plant. Proc Natl Acad Sci 106:929–934

Pham VHT, Kim J (2012) Cultivation of unculturable soil bacteria. Trends Biotechnol 30:475–484

Pilet PE, Saugy M (1987) Effect of root growth on endogenous and applied IAA and ABA. Plant Physiol 83:33–38

Postma J, Nijhuis EH, Someus E (2010) Selection of phosphorus solubilizing bacteria with biocontrol potential for growth in phosphorus rich animal bone charcoal. Appl Soil Ecol 46:464–469

Quan C (2002) Purification and properties of a phytase from *Candida krusei* WZ-001. J Biosci Bioeng 94:419–425

Qurashi AW, Sabri AN (2012) Bacterial exopolysaccharide and biofilm formation stimulate chickpea growth and soil aggregation under salt stress. Braz J Microbiol 43:1183–1191

Rajkumar M, Ae N, Narasimha M, Prasad V, Freitas H (2010) Potential of siderophore-producing bacteria for improving heavy metal. Phytoextraction 28:142–149

Ramadan EM, Abdel Hafez AA, Hassan EA, Saber FM (2016) Plant growth promoting rhizobacteria and their potential for biocontrol of phytopathogens. Afr J Microbiol Res 10:486–504

Raymond KM, Denz E (2004) Biochemical and physical properties of siderophores. In: Crosa et al (eds) Iron transport in bacteria. ASM Press, p 3–17

Raza W, Yousaf S, Rajer FU (2016) Plant growth promoting activity of volatile organic compounds produced by bio-control strains. Sci Lett 4:40–43

Reinhold-Hurek B, Hurek T (1998) Life in grasses: diazotrophic endophytes. Trends Microbiol 6:139–144. https://doi.org/10.1016/S0966-842X(98)01229-3

Reinhold-Hurek B, Hurek T (2011) Living inside plants: bacterial endophytes. Curr Opin Plant Biol 14:435–443

Reinhold-Hurek B, Bünger W, Burbano CS, Sabale M, Hurek T (2015) Roots shaping their microbiome: global hotspots for microbial activity. Annu Rev Phytopathol 53:403–424. https://doi.org/10.1146/annurev-phyto-082712-102342

Richardson AE, Barea J-M, McNeill AM, Prigent-Combaret C (2009) Acquisition of phosphorus and nitrogen in the rhizosphere and plant growth promotion by microorganisms. Plant Soil 321:305–339

Rillig MC, Wright SF, Nichols KA, Schmidt WF, Torn MS (2001) Large contribution of arbuscular mycorrhizal fungi to soil carbon pools in tropical forest soils. Plant Soil 233:167–177

Rillig MC, Wright SF, Eviner VT (2002) The role of arbuscular mycorrhizal fungi and glomalin in soil aggregation: comparing effects of five plant species. Plant Soil 238:325–333

Rillig MC, Ramsey PW, Morris S, Paul EA (2003) Glomalin, an arbuscular-mycorrhizal fungal soil protein, responds to land-use change. Plant Soil 253:293–299

Robin A, Vansuyt G, Hinsinger P, Meyer JM, Briat JF, Lemanceau P (2008) Iron dynamics in the rhizosphere: consequences for plant health and nutrition. Adv Agron 99:183–225

Rodríguez H, Fraga R (1999) Phosphate solubilizing bacteria and their role in plant growth promotion. Biotechnol Adv 17:319–339

Rodríguez H, Fraga R, Gonzalez T, Bashan T (2006) Genetics of phosphate solubilization and its potential applications for improving plant growth-promoting bacteria. Plant Soil 287:15–21

Rodriguez H, Vessely S, Shah S, Glick BR (2008) Effect of a nickel-tolerant ACC deaminase-producing *Pseudomonas* strain on growth of nontransformed and transgenic canola plants. Curr Microbiol 57:170–174

Römheld V (1987) Different strategies for iron acquisition in higher plants. Physiol Plant 70:231–234

Ryan PR, Dessaux Y, Thomashow LS, Weller DM (2009) Rhizosphere engineering and management for sustainable agriculture. Plant Soil 321:363–383

Sahu PK, Brahmaprakash GP (2016) Formulations of biofertilizers – approaches and advances. In: Singh DP, Singh HB, Prabha R (eds) Microbial inoculants in sustainable agricultural productivity. Vol 2 functional applications. Springer, New Delhi, pp 179–198

Sahu PK, Lavanya G, Brahmaprakash GP (2013) Fluid bed dried microbial inoculants formulation with improved survival and reduced contamination level. J Soil Biol Ecol 33:81–94

Salamone IEG, Hynes RK, Nelson LM (2005) Role of cytokinins in plant growth promotionby rhizosphere bacteria. In: Siddiqui ZA (ed) PGPR: biocontrol and biofertilization. Springer, Dordrecht, pp 173–195

Sandhya V, Ali SKZ, Grover M, Reddy G, Venkateswarlu B (2009) Alleviation of drought stress effects in sunflower seedlings by the exopolysaccharides producing *Pseudomonas putida* strain GAP-P45. Biol Fertil Soils 46:17–26

Sano K, Fukuhara H, Nakamura JJ (1999) Phytase of the yeast *Arxula adeninivorans*. Biotechnol Lett 21:33–38

Schalk IJ, Hannauer M, Braud A (2011) New roles for bacterial siderophores in metal trans-port and tolerance. Environ Microbiol 13:2844–2854

Scheffknecht S, Mammerler R, Steinkellner S, Vierheilig H (2006) Root exudates of mycorrhizal tomato plants exhibit a different effect on microconidia germination of *Fusarium oxysporum* f. sp.*lycopesici* than root exudates from nonmycorrhizal tomato plants. Mycorrhiza 16:365–370

Schreiter S, Ding GC, Heuer H, Neumann G, Sandmann M (2014) Effect of the soil type on the microbiome in the rhizosphere of field-grown lettuce. Front Microbiol 5:144

Scott JC, Greenhut IV, Leveau JH (2013) Functional characterization of the bacterial *iac* genes for degradation of the plant hormone indole-3-acetic acid. J Chem Ecol 39:942–951

Segueilha L, Lambrechts C, Boze H, Moulin G, Galzy P (1992) Purification and properties of the phytase from *Schwaniomyces castellii*. J Ferment Bioeng 74:7–11

Seguel A, Cumming J, Klugh-Stewart K, Cornejo P, Borie F (2013) The role of arbuscular mycorrhizas in decreasing aluminium phytotoxicity in acidic soils: a review. Mycorrhiza 23:167–183

Shaharoona B, Arshad M, Zahir ZA (2006) Effect of plant growth promoting rhizobacteria containing ACC-deaminase on maize (*Zea mays* L.) growth under axenic conditions and on nodulation in mung bean (*Vigna radiata* L.). Lett Appl Microbiol 42:155–159

Shanmugam V, Kanoujia N, Singh M et al (2011) Biocontrol of vascular wilt and corm rot of gladiolus caused by *Fusarium oxysporum* f. sp. *gladioli* using plant growth promoting rhizobacterial mixture. Crop Prot 30:807–813

Sharma A, Johri BN, Sharma AK, Glick BR (2013a) Plant growth-promoting bacterium *Pseudomonas* sp. strain GRP3 influences iron acquisition in mung bean (*Vignaradiata* L. Wilzeck). Soil Biol Biochem 35:887–894

Sharma SB, Sayyed RZ, Trivedi MH, Gobi TA (2013b) Phosphate solubilizing microbes: sustainable approach for managing phosphorus deficiency in agricultural soils. Springerplus 2:587–600. https://doi.org/10.1186/2193-1801-2-587

Shen J, Li C, Mi G, Li L, Yuan L, Jiang R, Zhang F (2013) Maximizing root/rhizosphere efficiency to improve crop productivity and nutrient use efficiency in intensive agriculture of China. J Exp Bot 64:1181–1192

Shidore T, Dinse T, Ohrlein J, Becker A, Reinhold-Hurek B (2012) Transcriptomic analysis of responses to exudates reveal genes required for rhizosphere competence of the endophyte *Azoarcus* sp. strain BH72. Environ Microbiol 14:2775–2787

Shilev S (2013) Soil Rhizobacteria regulating the uptake of nutrients and undesirable elements by plants. In: Arora NK (ed) Plant microbe symbiosis: fundamentals and advances, p 147–167

Shilev S, Naydenov M, Sancho Prieto M, Sancho ED, Vassilev N (2012a) PGPR as inoculants in management of lands contaminated with trace elements. In: Maheshwari DK (ed) Bacteria in agrobiology: stress management. Springer, Berlin/Heidelberg, pp 259–277

Shilev S, Sancho ED, Benlloch M (2012b) Rhizospheric bacteria alleviate salt-produced stress in sunflower. J Environ Manag 95(Issue SUPPL):S37–S41

Shimizu M (1992) Purification and characterization of phytase from *Bacillus subtillis* (nato) N-77. Biosci Biotechnol Biochem 56:1266–1269

Simpson CJ, Wise A (1990) Binding of zinc and calcium to inositol phosphates (phytate) in vitro. Br J Nutr 64:225–232

Singleton P, Keyser H, Sande E (2002) Development and evaluation of liquid inoculants. Inoculants and nitrogen fixation of legumes in Vietnam. Australian Centre for International Agricultural Research, Canberra, pp 52–66

Smith SE, Read DJ (2008) Mycorrhizal symbiosis, 3rd edn. Elsevier, City

Smith SE, Smith FA, Jakobsen I (2003) Mycorrhizal fungi can dominate phosphate supply to plants irrespective of growth responses. Plant Physiol 133:16–20

Somers E, Vanderleyden J, Srinivasan M (2004) Rhizosphere bacterial signalling: a love parade beneath our feet. Crit Rev Microbiol 30:205–240

Souza R, Beneduzi A, Ambrosini A, Costa PB, Meyer J, Vargas LK, Schoenfeld R, Passaglia LMP (2013) The effect of plant growth-promoting rhizobacteria on the growth of rice (*Oryza sativa* L.) cropped in southern Brazilian fields. Plant Soil 366:585–603

de Souza R, Ambrosini A, Passaglia LMP (2015) Plant growth-promoting bacteria as inoculants in agricultural soils. Gen Mol Biol 38:401–419

Srinivasan R, Yandigeri MS, Kashyap S, Alagawadi AR (2012) Effect of salt on survival and P-solubilization potential of phosphate solubilizing microorganisms from salt affected soils. Saudi J Biol Sci 19:427–434. https://doi.org/10.1016/j.sjbs.2012.05.004

Staddon PL, Bronk Ramsey C, Ostle N, Ineson P, Fitter AH (2003) Rapid turnover of hyphae of mycorrhizal fungi determined by AMS microanalysis of 14C. Science 300:1138–1140

Steinkellner S, Lendzemo V, Langer I, Schweiger P, Khaosaad T, Toussaint J-P, Vierheilig H (2007) Flavonoids and strigolactones in root exudates as signals in symbiotic and pathogenic plant–fungus interactions. Molecules 12:1290–1306. https://doi.org/10.3390/12071290

Stephens J, Rask H (2000) Inoculant production and formulation. Field Crop Res 65:249–258

Stoltzfus JR, So R, Malarvithi PP, Ladha JK, de Bruijn FJ (1997) Isolation of endophytic bacteria from rice and assessment of their potential for supplying rice with biologically fixed nitrogen. Plant Soil 194:25–36. https://doi.org/10.1023/a:1004298921641

Sturz AV, Christie BR, Nowak J (2000) Bacterial endophytes: potential role in developing sustainable systems of crop production. Crit Rev Plant Sci 19:1–30. https://doi.org/10.1080/07352680091139169

Sun Y-P, Unestam T, Lucas SD, Johanson KJ, Kenne L, Finlay R (1999) Exudation-reabsorption in a mycorrhizal fungus, the dynamic interface for interaction with soil and soil microorganisms. Mycorrhiza 9:137–144

Swain T (1974) Ethylene in plant biology. Academic, New York

Szilagyi-Zecchin VJ, Ikeda AC, Hungria M, Adamoski D, Kava-Cordeiro V, Glienke C, Galli-Terasawa LV (2014) Identification and characterization of endophytic bacteria from corn (*Zea mays* L.) roots with biotechnological potential in agriculture. AMB Express. https://doi.org/10.1186/s13568-014-0026

Tahat MM, Radziah O, Kamaruzaman S, Kadir J, Masdek NH (2008) Role of plant host in determining differential responses to *Ralstonia solanacearum* and *Glomus mosseae*. Plant Pathol J 7:140–147
Tahat MM, Sijam K, Othman R (2010) The role of tomato and corn root exudates on *Glomus mosseae* spores germination and *Ralstonia solanacearum* growth *in vitro*. Int J Plant Pathol 1:1–12
Tahat MM, Sijam K, Othman R (2011) Bio-compartmental in vitro system for *Glomus mosseae* and *Ralstonia solanacearum* interaction. Int J Bot 7:295–299
Tahir HAS, Gu Q, Wu H, Raza W, Safdar A et al (2017) Effect of volatile compounds produced by *Ralstonia solanacearum* on plant growth promoting and systemic resistance inducing potential of *Bacillus* volatiles. BMC Plant Biol., BMC series – open, inclusive and trusted 17:133
Tamasloukht MB, Séjalon-Delmas N, Kluever A (2003) Root factors induce mitochondrial-related gene expression and fungal respiration during the developmental switch from asymbiosis to presymbiosis in the arbuscular mycorrhizal fungus *Gigaspora rosea*. Plant Physiol 131:1–11
Tanimoto E (2005) Regulation of root growth by plant hormones-roles for auxin and gibberellin. Crit Rev Plant Sci 24:249–265
Tewari S, Arora NK (2014) Multifunctional exopolysaccharides from *Pseudomonas aeruginosa* PF23 involved in plant growth stimulation, biocontrol and stress amelioration in sunflower under saline conditions. Curr Microbiol 69:484–494
Thomas L, Larroche C, Pandey A (2013) Current developments in solid-state fermentation. Biochem Eng J 81:146–161
Tilak KVBR, Ranganayaki N, Pal KK, De R, Saxena AK, Nautiyal CS, Mittal S, Tripathi AK, Johri BN (2005) Diversity of plant growth and soil health supporting bacteria. Curr Sci India 89:136–150
Tkacz A, Cheema J, Chandra G, Grant A, Poole PS (2015) Stability and succession of the rhizosphere microbiota depends upon plant type and soil composition. ISME J 9:2349–2359. https://doi.org/10.1038/ismej.2015.41
Tortora ML, Díaz-Ricci JC, Pedraza RO (2011) *Azospirillum brasilense* siderophores with antifungal activity against *Colletotrichum acutatum*. Arch Microbiol 193:275–286
Trevathan-Tackett SM, Seymour JR, Nielsen DA, Macreadie PI, Jeffries TC, Sanderman J et al (2017) Sediment anoxia limits microbial-driven seagrass carbon remineralization under warming conditions. FEMS Microbiol Ecol 93:1–15. https://doi.org/10.1093/femsec/fix033
Ulloa-Ogaz AL, Munoz-Castellanos LN, Nevarez-Moorillon GV (2015) Biocontrol of phytopathogens: antibiotic production as mechanism of control, the battle against microbial pathogens. In: Mendez Vilas A (ed) Technological advance and educational programs 1. pp 305–309
Vacheron J, Desbrosses G, Bouffaud ML, Touraine B, Moenne-Loccoz Y, Muller D, Legendre L, Wisniewski-Dye F, Prigent-Combaret C (2013) Plant growth–promoting rhizobacteria and root system functioning. Front Plant Sci 4:1–19
Vance CP, Uhde-Stone C, Allan DL (2003) Phosphorus acquisition and use: critical adaptations by plants for securing a nonrenewable resource. New Phytol 157:423–447
Vassilev N, Mendes G (2018) Solid-state fermentation and plant-beneficial microorganisms. In: Pandey A, Larroche CH, Soccol C (eds) Current developments in biotechnology and bioengineering. Curr Adv Solid-State Ferment. Elsevier, p 435–450
Vassilev N, Vassileva M, Fenice M, Federici F (2001) Immobilized cell technology applied in solubilization of insoluble inorganic (rock) phosphates and P plant acquisition. Bioresour Technol 79:263–271
Vassilev N, Nikolaeva I, Vassileva M (2005) Polymer-based preparation of soil inoculants: applications to arbuscular mycorrhizal fungi. Rev Environ Sci Biotechnol 4:235–243
Vassilev N, Nikolaeva I, Vassileva M (2007) Indole-3-acetic acid production by gel-entrapped *Bacillus thuringiensis* in the presence of rock phosphate ore. Chem Eng Commun 194:441–445
Vassilev N, Martos E, Mendes G, Martos V, Vassileva M (2013) Biochar of animal origin: a sustainable solution of the high-grade rock phosphate scarcity. J Sci Food Agric 93:1799–1804
Vassilev N, Vassileva M, Lopez A et al (2015) Unexploited potential of some biotechnological techniques for biofertilizer production. Appl Microbiol Biotechnol 99:4983–4996

Vassilev N, Eichler Löbermann B, Flor Peregrin E, Martos V, Reyes A, Vassileva M (2017a) Production of a potential liquid plant bio-stimulant by immobilized *Piriformospora indica* in repeated-batch fermentation process. AMB Express 7(1):1–7

Vassilev N, Malusa E, Requena A et al (2017b) Potential application of glycerol in the production of plant beneficial microorganisms. J Ind Microbiol Biotechnol 44:735–743

Vassileva M, Azcon R, Barea JM, Vassilev N (1999) Effect of encapsulated cells of *Enterobacter* sp. on plant growth and phosphate uptake. Bioresour Technol 67:229–232

Vassileva M, Serrano M, Bravo V, Jurado E, Nikolaeva I, Martos V, Vassilev N (2010) Multifunctional properties of phosphate-solubilizing microorganisms grown on agro-industrial wastes in fermentation and soil conditions. Appl Microbiol Biotechnol 85:1287–1299

Vats P, Banerjee UC (2004) Production studies and catalytic properties of phytases (myoinositol-hexakisphosphate phosphohydrolase): an overview. Enzyme Micob Technol 35:3–4

Vejan PA, Khadiran RN, Salmah I, Amru NB (2016) Role of plant growth promoting Rhizobacteria in agricultural sustainability-a review. Molecules 21:573. https://doi.org/10.3390/molecules21050573

Vengavasi K, Pandey R (2018) Root exudation potential in contrasting soybean genotypes in response to low soil phosphorus availability is determined by photo-biochemical processes. Plant Physiol Biochem 124:1–9

Vengavasi K, Kumar A, Pandey R (2016) Transcript abundance, enzyme activity and metabolite concentration regulates differential carboxylate efflux in soybean under low phosphorus stress. Indian J Plant Physiol 21:179–188. https://doi.org/10.1007/s40502-016-0219-2

Vengavasi K, Pandey R, Abraham G, Yadav RK (2017) Comparative analysis of soybean root proteome reveals molecular basis of differential carboxylate efflux under low phosphorus stress. Genes 8:341. https://doi.org/10.3390/genes8120341

Vessey JK, Pawlowski K, Bergman B (2005) N_2-fixing symbiosis: legumes, actinorhizal plants, and cycads. Plant Soil 274:51–78

Vimal SR, Singh JS, Arora NK, Singh S (2017) Soil-plant-microbe interactions in stressed agriculture management: a review. Pedosphere 27:177–192

Vohra A, Satyanarayama T (2002) Purification characterization of a thermo stable and acid – stable phytase from Pichia anomala. World J Microbiol Biotechnol 18:687–691

Vohra A, Satyanarayana T (2003) Phytases: microbial sources, production, purification, and potential biotechnological applications. Crit Rev Biotechnol 23:29–60

Volfová O, Dvoráková J, Hanzlíková A, Jandera A (1994) Phytase from *Aspergillus niger*. Folia Microbiol 39:481–484

Wagner MR, Lundberg DS, del Rio TG, Tringe SG, Dangl JL, Mitchell-Olds T (2016) Host genotype and age shape the leaf and root microbiomes of a wild perennial plant. Nat Commun 7:12–51. https://doi.org/10.1038/ncomms12151

Walker TS, Bais HP, Grotewold E, Vivanco JM (2003) Root exudation and rhizosphere biology. Plant Physiol 132:44–51

Wang Q, Xiong D, Zhao P, Yu X, Tu B, Wang G (2011) Effect of applying an arsenic-resistant and plant growth-promoting rhizobacterium to enhance soil arsenic phytoremediation by *Populus deltoides* LH05-17. J Appl Microbiol 111:1065–1074

Wang X, Mavrodi DV, Ke L, Mavrodi OV, Yang M, Thomashow LS, Zheng N, Weller DM, Zhang J (2015a) Biocontrol and plant growth-promoting activity of rhizobacteria from Chinese fields with contaminated soils. Microb Biotechnol 8:404–418

Wang Y, Tang S, Jin H (2015b) Effect of glucose, root exudates and N forms in mycorrhizal symbiosis using *Rhizophagus intraradices*. J Soil Sci Plant Nutr 15:726–736. ISSN 0718-9516

Warren GP, Robinson JS, Someus E (2009) Dissolution of phosphorus from animal bone char in 12 soils. Nutr Cycl Agroecosyst 84:167–178

Wemheuer F, Kaiser K, Karlovsky P, Daniel R, Vidal S, Wemheuer B (2017) Bacterial endophyte communities of three agricultural important grass species differ in their response towards management regimes. Sci Rep 7:40914. https://doi.org/10.1038/srep40914

Whipps JM (2001) Microbial interactions and biocontrol in the rhizosphere. J Exp Bot 52:487–511

Widder S, Allen RJ, Pfeiffer T, Curtis TP, Wiuf C, Sloan WT et al (2016) Challenges in microbial ecology: building predictive understanding of community function and dynamics. ISME J 10:2557–2568. https://doi.org/10.1038/ismej.2016.45

Wu QS, Xia RX, Zou YN (2008) Improved soil structure and citrus growth after inoculation with three arbuscular mycorrhizal fungi under drought stress. Eur J Soil Biol 44:122–128

Yadav BK, Akhtar MS, Panwar J (2015) Rhizospheric plant-microbe interactions: key factors to soil fertility and plant nutrition. In: Arora N (ed) Plant microbes symbiosis: applied facets. Springer, India, pp 127–145

Yoon SJ, Choi YJ et al (1996) Isolation and identification of phytase producing bacterium, *Enterobacterium* sp. 4, and enzymatic properties of phytase enzyme. Enzym Microb Technol 18:449–454

Zhang FS, Shen JB, Zhang JL, Zuo YM, Li L, Chen XP (2010) Rhizosphere processes and management for improving nutrient use efficiency and crop productivity: implications for China. In: DL Sparks (ed) Adv agron 107:1–32

Zhang YX, Ruyter-Spira C, Bouwmeester HJ (2015) Engineering the plant rhizosphere. Curr Opin Biotechnol 32:136–142. https://doi.org/10.1016/j.copbio.2014.12.006. WOS:000353865700022. PMID: 255551

Zhao K, Penttinen P, Zhang X, Ao X, Liu M, Yu X et al (2014) Maize rhizosphere in Sichuan, China, hosts plant growth promoting *Burkholderia cepacia* with phosphate solubilizing and antifungal abilities. Microbiol Res 169:76–82. https://doi.org/10.1016/j.micres.2013.07.003

Zimmermann P (2003) Root-secreted phosphomonoesterases mobilizing phosphorus from the rhizosphere: a molecular physiological study in *Solanum tuberosum*. Ph.D. thesis, Swiss Federal Institute of Technology, Zurich

Microbe-Mediated Tolerance in Plants Against Biotic and Abiotic Stresses

7

Syed Sarfraz Hussain

7.1 Introduction: A Glimpse of Plant Productivity Under Environmental Stresses

A plethora of data suggested that significant climatic changes have welcomed the twenty-first century (Kumar and Verma 2018). Many research reports have pointed that environmental stresses constitute significant threat to future food security around the globe (Battisti and Naylor 2009) with the ever-increasing world population which would be at least nine billion by 2050 (Singh et al. 2011; Hussain et al. 2012, 2014). Current estimates have revealed that over 800 million people are experiencing food shortage and malnutrition worldwide. Agricultural sustainability is threatened by a multitude of factors including unpredictable climate variation, population and reduction in soil health (Cushman and Bohnert 2000). Global food production is limited by several reasons, primarily by extreme climatic stresses which cause 20–30% yield losses globally (Savary et al. 2012; Dikilitas et al. 2018). Similarly, diseases can significantly affect virtually all crop plants with the potential to reduce both yield and quality, and an estimated 20–40% global harvest is lost to diseases alone (Savary et al. 2012; Dikilitas et al. 2018). Indiscriminate and widespread use of pesticides and weedicides for disease eradication has negatively impacted the environment; therefore, development of resistant/tolerant crop plant is human friendly and an effective strategy to enhance productivity (Hussain et al. 2011), which causes major loss of beneficial microbial diversity from the soil (Kumar and Verma 2018). However, benefits of green revolution are now over

S. S. Hussain (✉)
Department of Biological Sciences, Forman Christian College (A Chartered University), Lahore, Pakistan

School of Agriculture, Food & Wine, Waite Campus, University of Adelaide, Adelaide, SA, Australia
e-mail: sarfrazhussain@fccollege.edu.pk

D. P. Singh et al. (eds.), *Microbial Interventions in Agriculture and Environment*,
https://doi.org/10.1007/978-981-13-8391-5_7

mainly due to uncontrolled world population, narrow range of germplasm resources, lengthy breeding process and extreme climatic stresses (Hussain et al. 2012).

Therefore, it is well conceived that conventional breeding alone cannot keep pace with future food needs of the world population. Selective breeding and genetic modifications have played a promising role for the improvement of all major crop plants in order to meet the human food requirements (Capell et al. 2004; Bartels and Hussain 2008; Hussain et al. 2012). Combining general breeding schemes and current molecular strategies have been wisely utilized to develop crop plant with enhanced stress tolerance (Capell et al. 2004; Hussain et al. 2012). Plants overexpressing several different genes have shown improved tolerance to different environmental stresses and promotion of plant health and yield (Roy et al. 2014; Hussain et al. 2016) under both laboratory and greenhouse conditions. Currently, plant engineering approaches have been designed to transfer important genes playing significant role (synthesis of osmolytes, antioxidants and stress-related proteins such as Lea, HSP) in biochemical pathways (Wang et al. 2003; Vinocur and Altman 2005; Valliyodan and Nyugen 2006; Sreenivasulu et al. 2007; Kathuria et al. 2007; Bartels and Hussain 2008; Hussain et al. 2012, 2014; Marasco et al. 2016; Thao and Tran 2016). However, identification and isolation of key genes and acceptance of transgenic products at community level pose the main bottleneck of this strategy. Similarly, several research reports have revealed that crop health, adaptation and tolerance to various stresses are not only linked to the genome of the plant but evidence suggest that these might also be intricately influenced by multiple environmental factors (Munns and Gilliham 2015; Tiwari et al. 2017). Potentially, plant-associated microbes represent possible strategies to decrease the negative effects of chemical fertilizers, pesticides, herbicides and abiotic stresses.

There is now overwhelming research evidence that plant microbiome including symbiotic associations through numerous mechanisms help (Vandenkoornhuyse et al. 2015) significantly to sustainable plant yield management strategies (Berendsen et al. 2012; Mendes et al. 2013; Wagg et al. 2014; Mueller and Sachs 2015; Gouda et al. 2018). Emerging plant-associated microbiome-based technologies have received attention which offer potential increase in plant growth and development, nutrient acquisition, health, enhanced biotic/abiotic stress tolerance and host immune regulation leading to enhanced crop yields (Mayak et al. 2004; Glick et al. 2007; Marulanda et al. 2009; Yang et al. 2009; Berendsen et al. 2012; Bakker et al. 2013; Mendes et al. 2013; Turner et al. 2013; Berg et al. 2014; Lakshmanan et al. 2014; Ngumbi and Kloepper 2016). Several researchers have reported the beneficial impact of integration and utilization of mycorrhizal fungi (Rodriguez and Redman 2008; Bonfante and Anca 2009; Singh et al. 2011; Aroca and Ruiz-Lozan 2012; Azcon et al. 2013), bacteria for atmospheric nitrogen fixation (Lugtenberg and Kamilova 2009) and PGPR (Kloepper et al. 2004; Mayak et al. 2004; Glick et al. 2007; Kim et al. 2009; Glick 2012; Pineda et al. 2013; Chauhan et al. 2015) on crop plants for enhanced tolerance to various biotic and abiotic stresses (Timmusk and Wagner 1999; Mayak et al. 2004; Dimpka et al. 2009; Sandhya et al. 2009; Grover et al. 2011; Kasim et al. 2013; Coleman-Derr and Tringe 2014; Nadeem et al. 2014; Hussain et al. 2018). However, this should be noted that this is a vast but still largely an untrapped area which calls for more systematic and intensive research efforts for completely realizing its potential in increasing yields in a changing climate (Hussain et al. 2018).

Similarly, very little is known of how plants strategically prioritize their requirements, such as investing energy resources into defence at the expense of other vital functions, to modify the internal system to enhance tolerance to different environmental stresses (Schenk et al. 2012a, b). With the availability of high-throughput molecular tools, several diverse and unexpected research discoveries have revealed the underlying responses of plant adaptation to stress tolerance using plant-related microbiome (Mendes et al. 2011; Bulgarelli et al. 2012; Lundberg et al. 2012; Berg et al. 2016; Timmusk et al. 2017; White et al. 2017; Hussain et al. 2018). Although several different PGPRs have helped plants in mitigating various stresses, the mechanisms involved remain mostly unexplored. Meanwhile, several plant-associated microbes have been characterized for improved growth, development and stress management which significantly contributed to our understanding to design strategies for the use of these PGPRs (Hayat et al. 2010; Lakshmanan et al. 2012; Mapelli et al. 2013; Vejan et al. 2016). Integration and exploitation of plant-associated microbes hold great promise which can play important roles in improving plant health, growth and development (Rolli et al. 2015; Wallenstein 2017), by managing plant tolerance to various environmental stresses (Mapelli et al. 2013; Vejan et al. 2016) and enhancing plant productivity for food security (Lugtenberg and Kamilova 2009; Celebi et al. 2010; Mengual et al. 2014; Rolli et al. 2015; Berg et al. 2016; Marasco et al. 2016). Overall, sustainable agriculture challenged by abiotic stresses needs nonconventional strategies like the use of plant-related microbiomes (Schaeppi and Bulgarelli 2015; Bulgarelli et al. 2015). Taken together, the identification, characterization and use of microbes which enhance plant abiotic stress tolerance by diverse mechanisms would help to sustain agriculture in the future (Jorquera et al. 2012; Bhardwaj et al. 2014; Nadeem et al. 2014).

Innumerable reviews have highlighted several plant traits which are used by microbes for developing stress tolerance (Rosenblueth and Martinez-Romero 2006; Hardoim et al. 2008; Lugtenberg and Kamilova 2009; Rodriguez et al. 2009; Yang et al. 2009; Grover et al. 2011; Friesen et al. 2011; Singh et al. 2011; de Zelicourt et al. 2013; Bulgarelli et al. 2013; Nadeem et al. 2014; Qiu et al. 2014; Wellenstein 2017). It is the need of the time to join hands for exploring microbial traits beneficial to both plants and the environment because this strategy has a huge potential for sustainable agriculture in the future (Lally et al. 2017). This chapter highlights the advantages of the plant-related microbial community approach, especially increasing plant tolerance to various environmental stresses which constitute a serious threat to food security around the globe.

7.2 Plants and Their Microbial Environment: Exploring Plant Microbiome Diversity

It is well established that virtually the whole plant is populated by an uncountable number of microorganisms (Quiza et al. 2015) and has been classified mainly on the basis of plant part they colonized such as endophyte (present inside the plant part), epiphyte (aerial plant part like leaves and twigs) and rhizosphere (on the roots under the soil) (Ali et al. 2012; Penuelas and Terradas 2014; Bai et al. 2015; Santoyo et al. 2016). Under natural conditions, plants establish multiple mutually beneficial interactions with these microbes (Schenk et al. 2012a, b) for improvement of plant characters such as seed germination and vigour, growth and development, plant health (environmental stress tolerance) and crop productivity (Mendes et al. 2013; Quiza et al. 2015).

Despite their potential utility for plant productivity and other traits, progress in identification, characterization and utilization of these extremely complex microbial communities has been hampered mainly due to technological limitations (Hussain et al. 2018). Historic documents that report the use of microbes in agriculture date back to 1800, and rhizobium bacteria were first recommended for use in legume crops to enhance growth, development and uptake of nutrients from the soil (Jones et al. 2014). Initial efforts to use microbes have focused on exploring the functional roles of few members of plant-associated microbial groups which met with limited success largely because of the fact that most microbes are not culturable (Amann et al. 1995; Andreote et al. 2009; Schenk et al. 2012a, b; Balbontin et al. 2014; Larimer et al. 2014; van der Heijden et al. 2015). However, several individual microbes helping in improving plant health, growth and development, such as atmospheric nitrogen-fixing microbes (Olivares et al. 2013; de Bruijin 2015) and mycorrhizal fungi (Smith and Read 2008; Chagnon et al. 2013; van der Heijden et al. 2015), have been successfully characterized. On the other hand, concerted efforts to study microbial system recognize the utility of saprophytic or symbiotic interactions with plants ranging from beneficial to pathogenic (Mendes et al. 2013; Quiza et al. 2015). It is further noticed that pathogenic microbes despite their detrimental effects may use plant-derived organic substances for growth, hence may indirectly play a functional role in nutrient cycling and modifying plant environment (Schenk et al. 2012a, b), while beneficial microbes promote plant growth by improving nutrient acquisition (Mishra et al. 2012; Santoyo et al. 2012; Bulgarelli et al. 2013; Santoyo et al. 2016; Calvo et al. 2017), synthesizing growth regulators (Glick 2012) and suppressing different stresses by biosynthesis of pathogen-inhibiting compounds (Glick 2012; Santoyo et al. 2012; Martinez-Absalon et al. 2014; Hernandez-Leon et al. 2015) and other mechanisms (Smith and Read 2008; Berg 2009; Schenk et al. 2012a, b; Chagnon et al., 2013; Olivares et al., 2013; de Bruijin 2015; van der Heijden et al. 2015; Orozco-Mosqueda et al. 2018).

Lederberg and McCray (2001) used for the first time plant microbiome representing microbes occupying plants with beneficial outcomes such as plant health and plant productivity. Technically speaking, the term microbiome has been broadly applied to microbial community composition and their interaction (Beneficial or pathogenic) with specific hosts or environment (Mendes et al. 2011; Lakshmanan et al. 2012; Boon et al. 2014; Ofek et al. 2014; Panke-Buisse et al. 2015; Lareen et al. 2016). The current focus of plant-microbe interaction research involves three aspects. These include microbes involved in nutrient acquisition by symbiosis between plants and arbuscular mycorrhizal fungi (AMF) (Smith and Smith 2011; Sessitsch and Mitter 2014) and atmospheric nitrogen-fixing rhizobia (Oldroyd et al. 2011; Lundberg et al. 2012), microbes improving plant tolerance to various stresses (Doornbos et al. 2011; Ferrara et al. 2012; Marasco et al. 2012; Kavamura et al. 2013; Zolla et al. 2013) and disease-causing microbes (Kachroo and Robin 2013; Mendes et al. 2011, 2013; Wirthmueller et al. 2013; Quecine et al. 2014). Previous research efforts considered the plant-microbe association initially in relation to plant diseases (Mendes et al. 2013). However, advanced research in this field demonstrated that a huge amount of microbes are involved in beneficial functions to plants (Mendes et al., 2013; Bhardwaj et al. 2014; Santoyo et al. 2016, 2017). However, apart from well-known mutualistic interactions among plant and microbes, other characterized or uncharacterized useful microbes often are not included in field-based plant production strategies.

7.3 Shaping Plant Microbiome: Technical Progress

Extensive research efforts have attributed several functions to plant-associated microbes. However, these microbial communities, comprising of several diverse microbial strains, represent an extremely complex and dynamic fraction of plant microbiome (Farrar et al. 2014; Mueller and Sachs 2015). Therefore, research studies have partitioned plant microbiome and targeted different fractions separately. A plant environment has been divided into three major components such as rhizosphere, endosphere and phyllosphere based on the microbial presence where these can live and develop (Hardoim et al. 2008; Hirsch and Mauchline 2012; Haney and Ausubel 2015; Haney et al. 2015; Nelson 2018; Orozco-Mosqueda et al. 2018). In fact, new developments and technical advances resulted in enhanced research in this unexplored field (Porras-Alfaro and Bayman 2011; Berendsen et al. 2012; Bakker et al. 2013; Bulgarelli et al. 2013; Philippot et al. 2013; Schlaeppi et al. 2013; Turner et al. 2013; Guttman et al. 2014; Berg et al. 2014; Knief 2014; Lebeis 2014; Schaeppi and Bulgarelli 2015; Santoyo et al. 2017). Keeping in view the plant nutrition, it is important to characterize microbes that are involved in nutrient recycling and uptake for plants under various extreme soil situations (Leveau et al. 2010; Mapelli et al. 2012; Tajini et al. 2012; Krey et al. 2013; Lally et al. 2017). The scientific literature provides several examples of well-characterized microbes like bacteria (PGPR) and fungi (PGPF) with both antagonistic and synergistic interactions which contribute to enrich plant growth (Verma et al. 2010; Murray 2011; Rout and Callaway 2012; Bhardwaj et al. 2014). Furthermore, these microbes produce different phytohormones like auxin and siderophores (Khalid et al. 2004; Cassan et al. 2009; Abbasi et al. 2011; Filippi et al. 2011; Yu et al. 2011) which play critical roles in host nutrition, growth and health and provide protection to plants from biotic and abiotic stresses (Berendsen et al., 2012; Bakker et al. 2013; Bulgarelli et al. 2013; Mendes et al. 2013; Rastogi et al. 2013; Berg et al. 2014; Lakshmanan et al. 2014; Prashar et al. 2014; Bell et al. 2014; Mueller and Sachs 2015; Wallenstein 2017; Hussain et al. 2018).

Well-explored systems for mutualistic interactions include *Rhizobia* spp. and arbuscular mycorrhizae (AM) that exchange plant carbohydrates and important amino acids (Moe 2013) for fixing atmospheric nitrogen and insoluble phosphate bioavailability (Spaink 2000, Luvizotto et al. 2010; Leite et al. 2014) for plants. Microbes inhabiting in rhizosphere also help plants by providing many trace elements such as iron (Zhang et al. 2009; Marschner et al. 2011; Shirley et al. 2011) and calcium (Lee et al. 2010). Likewise, plant microbiome also plays essential functions in degrading non-bioavailable organic compounds required not only for microbes own survival but also for plant's vital functions like growth and development in nutrient-poor and nutrient-contaminated soils (Leveau et al. 2010; Mapelli et al. 2012; Turner et al. 2013; Bhattacharyya et al. 2015). Taken together, shaping and strengthening plant microbiome will have a significant and positive effect on sustainable agriculture in the future (Mitter et al. 2017).

Many reports have designated microbiome as the second genome, while some other researchers treated microbiome as a holobiont to demonstrate the critical roles played by microbial communities associated with plants (Zilber-Rosenberg and Rosenberg 2008; Grice and Segre 2012; Agler et al. 2016; Clavel et al. 2016; Paredes and Lebeis 2016; Zmora et al. 2016). Currently, an effort to explore plant microbiome comprising of several different microbial communities is largely hindered due to several factors, mainly because of methodological constraints (Bulgarelli et al. 2013). Therefore, development and validation of protocols is essential for exploring the whole plant microbiome diversity (Calvo et al. 2017; Hussain et al. 2018). With the advent of next-generation sequencing, selection under artificial ecosystem and other molecular techniques like florescent tagging especially for studying unculturable species (endophytes) are now a gradually routine in research (Swenson et al. 2000; Bulgarelli et al. 2012; Hernández-Salmerón et al. 2016). A huge body of data are accumulated as a result of these technological advancements in the field (Martinez-Absalon et al. 2014; Hernandez-Leon et al. 2015; Hernández-Salmerón et al. 2016; Orozco-Mosqueda et al. 2018). On the other hand, integration of different computational models is also essential for dissection of this complex and dynamic hidden treasure (Farrar et al. 2014; Mendes and Raaijmakers 2015) with the aim of searching for new beneficial microbes and effectively manipulating plant microbiome for increasing plant productivity (Hussain et al. 2018).

Taken together, investment in research aimed at exploring microbial traits that are beneficial to plants and environment constitutes an ideal approach towards next-generation sustainable plant productivity (Schaeppi and Bulgarelli 2015; Goswami et al. 2016; Khan et al. 2016; Compant et al. 2016). Despite the above-mentioned facts, assessing and accessing the microbiome of important local plants and native habitats represents a yet unexplored field to exploit synergism between microbes and plant traits in modern agriculture. However, understanding microbe-microbe dynamics is critical to identify key factors that help to shape and establish microbial communities. Therefore, information extracted from the indigenous plant-associated microbiome constitutes an integral part for designing/engineering a microbiome to be used for sustaining agriculture in the future.

7.4 Reinstating a Functional Plant Microbiome: Smart Solution to a Complex Problem

Researchers are suffering from information gap which is negatively affecting the ability to manage and manipulate the rhizosphere microbiome (rhizobiome) while the strategy to use microbes for increasing plant productivity is not new. Current data reveal the potential of engineering rhizosphere microbiome which offers a unique opportunity to achieve maximum benefits in plant production despite different challenges (Bakker et al. 2012; Berendsen et al. 2012; Bainard et al. 2013; Qiu et al. 2014; Bulgarelli et al. 2015; Yadav et al. 2015). It is noteworthy that rhizosphere represents an extremely competitive environment for microbes, while these microbes play a critical role in plant growth and productivity (Berendsen et al. 2012; Ziegler et al. 2013; Chaparro et al. 2014). Therefore, plenty of progress has been achieved in engineering sustainable plant productivity through engaging microbial communities (Bakker et al. 2012; Lebeis et al. 2012; Bulgarelli et al. 2013; Su et al. 2015).

Huge research endeavours have resulted in isolation, identification and characterization of hundreds of bacterial/fungal strains with known beneficial effects and are currently being utilized in developing microbial consortia (Patel and Sinha 2011; Kim and Timmusk 2013; Dong and Zhang 2014). Several researches have demonstrated the importance of microbial consortia approach which has contributed significantly towards increased agricultural production with less chemical inputs, reduced emission of greenhouse gases and high tolerance to different stresses (Barka et al. 2006; Adesemoye et al. 2009; Yang et al. 2009; Singh et al. 2010; Bakker et al. 2012; Jha et al. 2012; Jorquera et al. 2012; Adesemoye and Egamberdieva 2013; Berg et al. 2013; Turner et al. 2013; Egamberdieva et al. 2017). This is crucial for keeping pace with the rapidly growing world population (Zolla et al. 2013; Nadeem et al. 2014). Another way to extract maximum benefits out of this approach is to exploit knowledge from microbes with publically available genome sequences and synthetically develop a microbiome that can help to improve plant traits as reported for a few important plants including wheat, rice, *Arabidopsis*, maize, *Brassica rapa*, potato, barley, sugarcane and rice (Rasche et al. 2006; Bulgarelli et al. 2013, 2015; Lundberg et al. 2012; Peiffer et al. 2013; Lebeis et al. 2015; Panke-Buisse et al. 2015; Raajimakers 2015; Yeoh et al. 2016). Furthermore, it is expected that microbes in their natural habitats have the potential to contribute significantly in the improvement of crop productivity under environmental challenges.

Consequently, research reports have demonstrated the potential of these microbes which have positively impacted many plant traits including growth, development and productivity under various environmental stresses (Bhattacharya and Jha 2012; Goh et al. 2013; Coleman-Derr and Tringe 2014; Schlaeppi et al. 2014; Tkacz et al. 2015;

Lebeis et al. 2015; Yeoh et al. 2016). It is extremely vital to understand thoroughly both way interactions (microbe-microbe and plant-microbe) for the successful engineering of beneficial soil microbiome (rhizosphere). Similarly, available data have revealed the genetic and molecular basis of these interactions (Bloemberg and Lugtenberg 2001; Wang et al. 2005; Lim and Kim 2013; Timmusk et al. 2014; Vargas et al. 2014; Kim et al. 2015; Busby et al. 2017; Lally et al. 2017; Iannucci et al. 2017), and this information can be used for genetically modifying either partner using genetic engineering protocols for enhanced plant productivity. However, it is worthy to note that microbiome interactions are dynamic and complex depending on several factors including soil biochemistry, plant genotypes and external environment which heavily influence the composition and colonization of several bacterial communities with plant roots. Additionally, these factors involved are in crucial functions such as triggering plant-genotype-specific physiological responses, resulting in different exudation patterns in roots (Hamel et al. 2005; Bais et al. 2006; Hartmann et al. 2009; Dumbrell et al. 2010; Oburger et al. 2013). As a result of increased interest in this research, these factors (different soil types, different native plant species and microbial communication) have been extensively reviewed on the rhizomicrobiome (Tarkka et al. 2008; Berg and Smalla 2009; De-la-Pena et al. 2012; Philippot et al. 2013; Bulgarelli et al. 2013, 2015; Lareen et al. 2016). A broader picture of these interactions revealed that these factors played significant roles in the selective enrichment of microbial communities in rhizosphere microbiome (Berendsen et al. 2012; Miller and Oldroyd 2012; Schenk et al. 2012a, b; Sugiyama and Yazaki 2012; Morel and Castro-Sowinski 2013; Oldroyd 2013), by coordinating the establishment and recruitment of diverse bacterial communities for engineering a specific rhizobiome with positive impact on plant productivity (Bulgarelli et al. 2013, 2015; Peiffer et al. 2013; Philippot et al. 2013; Schlaeppi et al. 2014; Su et al. 2015; Tkacz et al. 2015; Lebeis et al. 2015; Yadav et al. 2015; Yeoh et al. 2016).

Multiple studies have reported positive interaction between specific plants and belowground soil-dwelling microbial communities (Micallef et al. 2009; Inceoglu et al. 2013). This clearly highlighted the fact that plant root exudates play critical roles in identification and recruitment of specific microbes which result in changes in composition and diversity of microbes in the rhizosphere (Haichar et al. 2008; Badri et al. 2009, 2013; Moe 2013; Weston and Mathesius 2013). Based on the above discussion, some useful approaches have been devised to reinstate the root-associated microbiome and re-route microbial activity by enhancing root exudates through more systematic breeding efforts (Bakker et al. 2012; Huang et al. 2014a, b; Reyes-Darias et al. 2015; Yuan et al. 2015; Corral-Lugo et al. 2016; Webb et al. 2016). Root exudates not only serve as food for root-associated microbes but also act as a signal molecule for the initiation of diverse physical and chemical interactions around plant roots (Berendsen et al. 2012; Hawes et al. 2012; Baetz and Martinoia 2013; Chaparro et al. 2013; Vacheron et al. 2013; Huang et al. 2014a, b; Reyes-Darias et al. 2015; Yuan et al. 2015; Corral-Lugo et al. 2016; Webb et al. 2016). Significant growing evidence suggested that progress has been made towards the development of PGPR and/or PGPF consortia using knowledge derived from plant ecosystem for mimicking or partially reconstructing the plant microbiome/

rhizobiome (Adesemoye et al. 2009; Atieno et al. 2012; Masciarelli et al. 2014; Mengual et al. 2014). It is worthy to note that the success of a tailored design of a plant microbiome depends on several factors including identification of the genetic components of the microbiome control and smart integration of critical players in the system. Similarly, it is speculated that changes in root system architecture (RSA) through breeding techniques may help in the recruitment of beneficial plant-specific rhizobiome. However, more systematic and detailed investigations will be required to study these interactions.

7.5 Plant Microbiome and Biotic Stress

Pathogen-free plants present the important and most ignored trait of the plant-associated microbes. Different pathogens especially viruses, bacteria and fungi are responsible for biotic stresses, and crop productivity is significantly reduced (≥15%) by these stresses worldwide (Strange and Scott 2005; Haggag et al. 2015). Stress (both biotic and abiotic) is a major challenge to agricultural yield, and huge economic losses urgently require the development of resistant crop plants. Gusain et al. (2015) have revealed adverse impacts of biotic stress on plants in detail. Several microorganisms belonging to different genera (*Burkholderia*, *Pseudomonas*, *Bacillus*, *Azotobacter* and *Azospirillum*) are the major group of PGPR that are involved in eliciting induced systemic resistance (ISR) response in plants (Alstrom 1991; Van Peer et al. 1991; Wei et al. 1991; Riggs et al. 2001; Shaharoona et al. 2006; Lebeis 2015; Tiwari et al. 2017; Hussain et al. 2018). Similarly, some microbial species belonging to a symbiotic group of rhizobacteria are also involved directly or indirectly with different PGPRs and can evoke ISR in plants (Elbadry et al. 2006). Inoculation of plants or their parts with PGPR which exhibits resistance to different pathogens of biotic stress (Ngumbi and Kloepper 2016). Zamioudis et al. (2013) demonstrated that *P. fluorescens* WCS417 is able to promote important plant traits in *A. thaliana*. This report further revealed that the improvement of different traits occurs via an auxin-dependent and JA-independent mechanism resulting in ISR (Zamioudis et al., 2013). Thus, PGPR/PGPF interactions with their host plant revealed the power to unravel mechanisms which act as the prime barrier of plant defence (Badri et al. 2009; De-la-Pena et al. 2012; Dangl et al. 2013). In fact, PGPR and PGPF are also involved in induction of immune "priming", by secreting signalling compounds which do not result in direct immune activation, but just activate and govern the immune responses against different pathogens (Conrath 2006; Badri et al. 2009; De-la-Pena et al. 2012; Dangl et al. 2013), even in distal tissues.

The defensive capacity of plants represents a physiological condition which is evoked by different signalling molecules known as elicitors. Thus, elicitors are molecules that induce different plant immune responses. Several reports have described two mechanisms which constitute plant immune responses include induced systemic resistance (ISR) and systemic acquired resistance (SAR). Thus, rhizobacteria infection triggers induced systemic resistance (ISR; Ortiz-Castro et al. 2008), while arbuscular mycorrhizae (AM) can produce mycorrhizal-induced resistance (MIR; Pozo and Azcon-Aguilar, 2007; Zamioudis and Pieterse, 2012) suggesting that microbial exploitation is common which gives strength to plants to face pathogen attacks. PGPR-mediated ISR requires interaction between bacteria and plant root which renders plants resistance to some pathogenic microorganisms by the activation of plant natural defences (Raaijmakers et al. 1995; Lugtenberg and Kamilova 2009; Prathap and Ranjitha 2015). ISR is triggered by the interaction of usually non-pathogenic microorganisms in roots and further extending to shoots (Ramos-Solano et al. 2008). Activation of ISR primes the plants to respond faster and stronger against the attack of several pathogenic species including bacteria, protists, nematodes, virus, fungi, viroids and insects (Verhagen et al. 2004; Conrath 2006; Berendsen et al. 2012; Walters et al. 2013).

Therefore, it is known that ISR is a non-specific defence reaction, but it provides strength to the plants to fight different plant diseases (Kamal et al. 2014). Several reports have shown that root inoculation with several different PGPRs rendered the entire plant tolerant to lethal pathogens (Schuhegger et al. 2006; Choudhary et al. 2007; Tarkka et al. 2008). Hence, research has proved ISR as one of the PGPR-mediated mechanisms which reduce plant disease by bringing about critical changes in the host plants at physical and biochemical levels (Pieterse et al. 2002). Since then, the PGPR-elicited ISR is regarded as vital biocontrol mechanism and is under intensive research in plants such as maize, bean, *Arabidopsis*, wheat, tomato, rice, tobacco, radish, soybean, cucumber and carnation (Bevivino et al. 1998; van Loon et al. 1998; Ruy et al. 2004; Compant et al. 2005; Han et al. 2005; Landa et al. 2006; Rashedul et al. 2009; Senthilkumar et al. 2009; Filippi et al. 2011; Neeraj 2011; Pereira et al. 2011; Mavrodi et al. 2012; Martins et al. 2013). However, understanding the metabolic pathway participating in this method is not yet complete (Ramos Solano et al. 2008), which necessitates multidisciplinary intensive research efforts.

On the other hand, it is established that phytohormones such as ethylene and jasmonic acid behave as a signalling agent in ISR, and plant defence response is dependent on these molecules (van Loon 2007). In contrast to the above-mentioned two phytohormones, salicylic acid (SA) acts as a key determinant in SAR. However, a study has shown some overlap between ISR and SAR in some cases (Lopez-Baena et al. 2009). In fact, well-known biotic elicitors are cell wall polysaccharides, along with some others including different phytohormones and signalling molecules (Shuhegge et al. 2006; van Loon 2007; Ramos Solano et al. 2008; Berg 2009; Fouzia et al. 2015; Kanchiswamy et al. 2015; Ulloa-Ogaz et al. 2015; Wang et al. 2015; Meena et al. 2016; Goswami et al. 2016; Islam et al. 2016; Ramadan et al. 2016; Raza et al. 2016a, b; Santoro et al. 2016; Sharifi and Ryu 2016; Gouda et al. 2018).

7.6 Microbiome for Abiotic Stress Alleviation in Crop Plants

Virtually, stress is defined as any factor which negatively affects plant health, growth, and productivity (Foyer et al. 2016). Due to climate change, plants are frequently subjected to various environmental stresses (Hussain et al. 2018). Because expanding the agricultural land is near impossible, increasing demands for food place a serious threat to current crop production systems. Hence, a scientifically improved farming method is required for keeping pace with unprecedented demands and maintaining the soil fertility under intense farming. Currently, sustainable agriculture is based on several improved agricultural techniques (Kumar 2016; Mus et al. 2016; Passari et al. 2016; Perez et al. 2016; Shrestha 2016; Suhag 2016; Ubertino et al. 2016). On the other hand, heavy investment in stress-related research has increased our understanding of the molecular mechanisms implicated in environmental stress tolerance (Tripathi et al. 2015, 2016, 2017; Pontigo et al. 2017; Singh et al. 2017). Therefore, in the development of stress tolerance coupled with better nutritional value, crop plants significantly contributed towards sustainable agricultural development. Engaging beneficial microbes is one possible way to address stress tolerance in plants (Vejan et al. 2016). Following this, recent research has shown that a strain of *Bacillus amyloliquefaciens* living in rice rhizosphere is able to reduce various abiotic stresses via cross-talk with pathways regulating stresses and phytohormones (Tiwari et al. 2017). Similarly, it is known that several soil-inhibiting microbes such as *Paecilomyces formosus* can help reduce plant stress caused by different factors especially heavy metals such as nickel (Bilal et al. 2017). The advantages of using root-associated microbes include their capacity to alleviate negative effects of different abiotic stresses in a wide range of crop plants (Timmusk and Wagner 1999; Mayak et al., 2004; Sandhya et al. 2010; Kasim et al. 2013; Tkacz and Poole 2015) and also their capability to simultaneously tackle several biotic and/or abiotic stresses (Ramegowda et al. 2013; Sharma and Ghosh 2017). Consequently, these beneficial microorganisms are under intensive research as one of the most climate-friendly agents for safe crop management practices.

Currently, plant rhizobiome has attracted extreme attention for tackling plant stresses and enhancing plant yields by several mechanisms to fuel new innovations in sustainable crop production as part of the next green revolution (Marulanda et al. 2009; Yang et al. 2009; Mendes et al. 2011; Bulgarelli et al. 2012; Lau and Lennon 2012; Lundberg et al. 2012; Marasco et al. 2012, 2013; Bainard et al. 2013; Sugiyama et al. 2013; Berg et al. 2014; Bonilla et al. 2015; Panke-Buisse et al. 2015; Prosser 2015; Rolli et al. 2015; Jez et al. 2016; Premachandra et al. 2016; Fierer 2017; Goodrich et al. 2017; Hussain et al. 2018). In fact, isolation and characterization of microbes constitute an integral part to identify beneficial microbes. Extensively researched microbial communities include the symbiotic bacteria (Spaink 2000; Lugtenberg and Kamilova 2009; Luvizotto et al. 2010; Leite et al. 2014), mycorrhizal fungi (Khan et al. 2008; Ruiz-Lozano et al. 2011; Sheng et al. 2011; Singh et al. 2011; Aroca and Ruíz-Lozan 2012; Bashan et al. 2012; Azcon et al. 2013) and PGP rhizobacteria (Kloepper et al. 2004; Glick 2012; Rout and Callaway 2012; Bhardwaj et al. 2014; Gabriela et al. 2015). PGPR contains a huge range of well-studied rhizosphere bacteria (Gupta et al. 2015) which are able to produce several different enzymes and metabolites that play critical roles in host nutrition, growth and health and protect plants from environmental stresses (Dimpka et al. 2009; Kim et al. 2009; Yang et al. 2009; Grover et al. 2011; Timmusk and Nevo 2011; Berendsen et al. 2012; Bulgarelli et al. 2013; Mendes et al. 2013; Berg et al. 2014; Prashar et al. 2014; Rastogi et al. 2013; Ding et al. 2013; Kim et al. 2013; Pineda et al. 2013; Timmusk et al. 2014; Chauhan et al. 2015; Lidbury et al. 2016; Ofaim et al. 2017; Sanchez-Canizares et al. 2017; Syed Ab Rahman et al. 2018). Currently, efforts have been directed at exploring and utilizing naturally occurring, soil-inhibiting microbes for enhanced plant yield under changing climate (Yang et al. 2009; Nadeem et al. 2014; Bhattacharyya et al. 2016; Bashiardes et al. 2018; Jansson and Hofmockel 2018; Yuan et al. 2018). Convincing evidence has witnessed beneficial effects of plant-associated microbes, and this partnership has significantly contributed to establishing smart solutions under nutrient deficiency and mitigating other stresses using diverse mechanisms (Hayat et al. 2010; Mapelli et al. 2013; Vejan et al. 2016).

7.7 Drought Stress

Drought is one of the serious agricultural problems worldwide resulting in reduced growth, development and plant yield (Vinocur and Altman 2005; Hussain et al. 2012, 2014; Naveed et al. 2014a, b; Tiwari et al. 2016). It is also noteworthy that the frequency and intensity of water deficit are expected to increase in the future due to rapid environmental deterioration. Recent investigation revealed that different microbes have the power to support vital plant traits such as plant growth and development through interaction with plant root system under drought stress (Hussain et al. 2012; Huang et al. 2014a, b) to ensure tolerance to environmental stresses (Mendes et al. 2011; Ngumbi 2011; Lakshmanan et al. 2012; Marasco et al. 2012, 2013; Bainard et al. 2013; Sugiyama et al. 2013; Berg et al. 2014; Edwards et al. 2015; Rolli et al. 2015; Panke-Buisse et al. 2015; Hussain et al. 2018). Several approaches have been chalked out and applied to address the drought-associated negative impact on crop productivity. However, use of plant-associated microbes offers a sustainable solution to abiotic stresses by diverse mechanisms (Farooq et al. 2009; Budak et al. 2013; Cooper et al. 2014; Hussain et al. 2014; Porcel et al. 2014). Kang et al. (2014) reported that inoculated soybean with *Pseudomonas putida* H-2-3 mitigated drought impact by decreasing antioxidant activity, producing different osmolytes, enhancing chlorophyll contents, improving shoot length and productivity. Similarly, two maize cultivars inoculated with *Burkholderia phytofirmans* strain PsJN showed 70% and 58% increase in root biomass and with *Enterobacter* sp. strain FD, 47% and 40%, respectively, under water deficit (Naveed et al. 2014a, b). Similarly, several other researchers reported a positive impact of these microbes on roots in different plants like maize and wheat (Yasmin et al. 2013; Timmusk et al. 2013, 2014). Inoculated plants showed promising results compared to non-inoculated control plants under low water condition which led to the conclusion that an increase in root biomass resulted in enhanced water uptake by plants under water deficit stress. Timmusk et al. (2014) have also demonstrated the positive effects on shoot biomass in corn and wheat under drought when inoculated with PGPR.

Crop plants treated with PGPR demonstrated several adjustments at molecular, biochemical and physiological levels for improving several traits such as growth and development, nutrient and water use efficiency, high chlorophyll content for increased photosynthesis, biocontrol activity and ultimately enhanced crop yield by bringing about alterations in root and shoot, phytohormonal activity, high relative water content, EPS production, osmotic adjustment due to osmolyte accumulation, ACC deaminase activity and antioxidant defence (Bano et al. 2013; Kasim et al. 2013; Marasco et al. 2013; Huang et al. 2014a, b; Naveed et al. 2014a, b; Naseem and Bano 2014; Sarma and Saikia 2014; Timmusk et al. 2014; Cohen et al. 2015; Fasciglione et al. 2015; Ortiz et al. 2015; Rolli et al. 2015; Ma et al. 2016a, b; Tiwari et al. 2016; Yang et al. 2016a). PGPR treatment has improved the growth of important crops like rice, wheat, sorghum, maize, sunflower, soybean, pea, tomato, lettuce and pepper under water deficit (Alami et al. 2000; Creus et al. 2004; Mayak et al. 2004; Dodd et al. 2005; Cho et al. 2006; Marquez et al. 2007; Figueiredo et al. 2008; Arshad et al. 2008; Kohler et al. 2008; Sandhya et al. 2010; Castillo et al. 2013; Kasim et al. 2013; Kim et al. 2013; Lim and Kim 2013; Perez-Montano et al. 2014; Naseem and Bano 2014; Sarma and Saikia 2014; Timmusk et al. 2014, 2017; Marasco et al. 2016).

7.8 Salinity Stress

Salinity is a major environmental stress and globally challenging plant growth and productivity (Wicke et al. 2011; Hussain et al. 2014). Researchers have adopted several approaches for tackling salinity problem including agronomic practices, physiological adjustments and molecular (genetic) engineering. However, despite appreciated utility, these practices are not environmentally friendly and practically sustainable due to the incomplete understanding of stress tolerance mechanism and rapidly deteriorating climate. On the other hand, a growing evidence highlighted that different microbial communities improved plant health with enhanced productivity by altering the selectivity of Na^+, K^+ and Ca^{2+} and sustaining a higher K^+/Na^+ ratio in roots under high salinity stress (Barassi et al. 2006; Berendsen et al. 2012; Damodaran et al. 2013; Zuppinger-Dingley et al. 2014; Fasciglione et al. 2015; Sloan and Lebeis 2015; Bacilio et al. 2016; Bharti et al. 2016; Kasim et al. 2016; Mahmood et al. 2016; Sharma et al. 2016; Khan et al. 2017; Shahzad et al. 2017; Timmusk et al. 2017; de la Torre-Gonzalez et al. 2017). Consequently, engaging both PGPR and PGPF has demonstrated a promising success under salinity stress (Upadhyay et al. 2011; Shukla et al. 2012; Bharti et al. 2016; Yang et al. 2016b). Crop plants growing on salty soil which are inoculated with PGPR/PGPF performed better with optimal yield (Tiwari et al. 2011; Shabala et al. 2013; Paul and Lade 2014; Qin et al. 2014; Ruiz et al. 2015). Similarly, multiple reports have demonstrated practical utility of microbial communities where plants like rice, barley, wheat, canola, tomato, mung bean, maize, oat, lettuce and peanuts have developed significantly higher biomass in high salt condition (Mayak et al. 2004; Upadhyay et al. 2009; Ahmad et al. 2011; Jha et al. 2012; Shukla et al. 2012; Nautiyal et al. 2013; Ali et al. 2014; Chang et al. 2014; Jha and Subramanian 2014; Leite et al. 2014; Timmusk et al. 2014; Fasciglione et al. 2015; Suarez et al. 2015; Bharti et al. 2016; Mahmood et al. 2016; Sharma et al. 2016; Zhao et al. 2016).

It is well documented that microbes living in harsh environments modify their physiology accordingly and serve as potential candidates for enhancing plant growth and productivity under stress conditions (Rodriguez et al. 2008; Timmusk et al. 2014). Several researchers isolated bacterial strains from plant roots challenged with high salt stress. Researchers isolated 130 rhizobacterial strains from wheat roots facing salinity stress, and 24 out of 130 isolates showed good growth in culture at 8% of NaCl stress (Upadhyay et al. 2009; Siddikee et al. 2010; Upadhyay et al. 2011; Arora et al. 2014). Different PGPR strains mitigate stress using various mechanisms. For example, Korean halotolerant strain inoculation resulted in enhanced growth because bacterial ACC deaminase activity negatively affected ethylene production under stress (Siddikee et al. 2010). Wheat inoculated with EPS-producing PGPR demonstrated high biomass by binding with cations and zero negative effect on plants under salinity stress (Upadhyay et al. 2011; Vardharajula et al. 2011). A plethora of research has used several PGPR strains including *Hartmannibactor diazotrophicus* E19, *Pseudomonas alcaligenes* PsA15, *Bacillus polymyxa* BcP26, *Mycobacterium phlei* MbP18, *P. fluorescens*, *P. aeruginosa*, *P. stutzeri* and *B. amyloliquefaciens* that have been successfully utilized in different plant species for mitigating salinity stress (Egamberdiyeva 2007; Bano and Fatima 2009; Tank and Saraf 2010; Bal et al. 2013; Nautiyal et al. 2013; Suarez et al. 2015).

Plants inoculated with PGPF showed significant tolerance to high salinity condition (Giri and Mukerji 2004; Grover et al. 2011; Velazquez-Hernandez et al. 2011) due to diverse mechanisms like osmotic adjustment, root growth, increased phosphate and decreased Na^+ concentration in shoots, improved photosystem II efficiency and antioxidant systems and reduced ROS compared to un-inoculated controls (Shukla et al. 2012; Navarro et al. 2014; Ruiz-Lozano et al. 2016). Therefore, maize, rice, cucumber, mung bean, clover, citrus and tomato have shown improved salt tolerance after PGPF treatment which could serve as potential tool for alleviating salt stress especially in stress-sensitive crop plants (Jindal et al. 1993; Al-Karaki et al. 2001; Feng et al. 2002; Ben Khaled et al. 2003; Yang et al. 2009; Grover et al. 2011; Velazquez-Hernandez et al. 2011; Shukla et al. 2012; Navarro et al. 2014; Ruiz-Lozano et al. 2016).

7.9 Heavy Metal Stress

Researchers have shown that industrialization leads to heavy metal accumulation with a huge impact on plant and human health (Qin et al., 2015; Wu et al., 2015). However, the heavy metal problem has received research priority around the globe in recent years due to non-degradable nature of these contaminants (Duruibe et al. 2007; Kidd et al. 2009; Ma et al. 2011; Rajkumar et al. 2012; Ma et al. 2016a, b). Apart from heavy metals, some metalloids such as antimony (Sb) and arsenic (As) are also contributing a huge toxicity (Duruibe et al. 2007; Park 2010; Wuana and Okieimen 2011; Pandey 2012). Heavy metals also constitute significant threat to agricultural productivity and soil health. Many biophysiochemical approaches adopted to reclaim contaminated soils have failed because these were environmentally unsafe, deleterious to soil structure, and unacceptable to the community (Boopathy 2000; Vidali 2001; Doble and Kumar 2005; Glick 2010). Phytoremediation strategy uses different plants supported by microbial communities to clean up heavy metal contaminants in soil and is believed to be a sustainable and cost-effective technology with no negative impact on environment and accepted by the communities (Broos et al. 2004; Hadi and Bano 2010; Afzal et al. 2011; Beskoski et al. 2011; Chen et al. 2014; Fester et al. 2014; Arslan et al. 2017; Hussain et al. 2018). The only limitation of phytoremediation is that plants used for soil reclamation (heavy metals) suffer from negative effects on plant growth due to nutrient shortage and heavy metal-based oxidative stress (Gerhardt et al. 2009; Hu et al. 2016). However, microbe-assisted phytoremediation represents a novel and working alternative (Jamil et al. 2014), whereby microbial activities increase soil reclamation using several unique mechanisms such as efflux, volatilization, metal complexation and enzymatic detoxification (Rajkumar et al. 2010; Ma et al. 2011; Aafi et al. 2012; Yang et al. 2012; Fatnassi et al. 2015; Ghosh et al. 2015; Zhang et al. 2015; Kumar and Verma 2018). It is an established fact that microbes promote plant growth and development by restricting ethylene production and production of plant growth substances such as IAA, cytokinins and gibberellins, siderophores, EPS and ACC deaminase under different stresses including heavy metal stress (Ahmad et al. 2011; Babu and Reddy 2011; Luo et al. 2011, 2012; Wang et al. 2011; Verma et al. 2013; Bisht et al. 2014; Kukla et al. 2014; Waqas et al. 2015; Ijaz et al. 2016; Santoyo et al. 2016; Deng and Cao 2017).

Waqas et al. (2015) have mentioned a few PGPR genera among rhizosphere microbes which demand more intensive research because these can be actively involved in phytoremediation process. These microbes are able to enhance process efficiency by bringing changes in soil pH and other allied oxidation/reduction processes (Khan et al. 2009; Kidd et al. 2009; Uroz et al. 2009; Wenzel 2009; Rajkumar et al. 2010; Ma et al. 2011). Recently, it has been demonstrated that soybean inoculation with *Paecilomyces formosus* exhibited significantly improved growth in soils with Ni accumulation (Bilal et al. 2017).

Similarly, Jamil et al. (2014) reported the positive impact of *Bacillus licheniformis* strain NCCP-59 inoculation with rice, whereby rice seeds exhibited improved germination in Ni-accumulated soil compared to control plants, indicating the ability of *Bacillus licheniformis* strain NCCP-59 to confer protection against Ni toxicity. Recently, a huge data have demonstrated that common heavy metals that include mercury (Hg), manganese (Mn), chromium (Cr), arsenic (As), cadmium (Cd), lead (Pb), chromium (Cr), zinc (Zn), nickel (Ni), aluminium (Al) and copper (Cu) can be efficiently removed by microbes using a plethora of mechanisms in different crop plants such as rice, *Brassica*, maize, lettuce and others (Sheng et al. 2008; Hadi and Bano 2010; Mani et al. 2016; Jing et al. 2014; Adediran et al. 2015; Hristozkova et al. 2016; Mani et al. 2016; Stella et al. 2017; Hussain et al. 2018). A wide diversity of PGPRs including *Bacillus* sp., *Rhizobia*, *Serratia*, *Azospirillum*, *Enterobacter*, *Klebsiella*, *Burkholderia* sp. and *Agrobacterium* have efficiently improved the phytoremediation efficiency by enhancing biomass in heavily contaminated soils (Wani et al. 2008; Kumar et al. 2009; Mastretta et al. 2009; Luo et al. 2012; Nonnoi et al. 2012; Afzal et al. 2014; Glick 2014, 2015; Ghosh et al. 2015; Hardoim et al. 2015; Jha et al. 2015; Deng et al. 2016; Ijaz et al. 2016; Singh et al. 2016; Zheng et al. 2016; Feng et al. 2017).

7.10 Nutrient Deficiency Stress

Despite continuous depletion of soil fertility, soil microbes are playing a vital role in enhancing crop productivity in conventional agricultural production systems (Berendsen et al. 2012). Exploring and utilizing plant microbiome is one of the nonconventional solutions required for maintaining the sustainability of crop plants which are facing nutrient deficiency (Schaeppi and Bulgarelli 2015). The main challenge is the efficient monitoring of processes mediated by these microbes because global attention has been diverted towards their role in plant nutrition only recently (Lebeis et al. 2012; Bulgarelli et al. 2013; Turner et al. 2013; Wei et al. 2016). Plant symbiosis with microbes (rhizobia, bradyrhizobia and AMF) represents one of the widely researched plant-microbe interactions (Hawkins et al. 2000; Jefferies and Barea 2001; Richardson et al. 2009; Miransari 2011; Wu et al. 2016), where these microbes participate in crucial functions for maintaining adequate plant nutrient and high productivity by developing nitrogen-fixing nodules and mycorrhizal arbuscules, respectively (Adesemoye et al. 2009; Miao et al. 2011; Adesemoye and Egamberdieva 2013; Adhya et al. 2015).

Generally, rhizobial symbiosis only occurs in leguminous plants, while AMF-based symbiosis is widespread, and over 80% of land plants experience this symbiosis (Guimaraes et al. 2012; Oldroyd 2013; Hussain et al. 2018). It has been observed that apart from *Rhizobium* and *Bradyrhizobium*, several other bacterial endophytes establish symbiosis or symbiosis-like relationship with plants for bioavailable nitrogen fixation in unspecialized host tissues using nodule-less system (Zehr et al. 2003; Gaby and Buckley 2011; Guimaraes et al. 2012; Santi et al. 2013). *Cyanobacteria*, for example, establish symbiotic relationship with several plants and develop heterocysts instead of nodules which are suitable for BNF using nitrogenase (Berman-Frank et al. 2003; Santi et al. 2013). Leite et al. (2014) reported that sugarcane root-associated bacteria are helpful in fixing nitrogen and solubilizing phosphorus, respectively. Apart from the above reports, some algal genera such as *Anabaena*, *Aphanocapra* and *Phormidium* are also actively involved in fixing nitrogen in field-grown rice by some unknown mechanisms (Shridhar 2012; Hasan 2013).

Similarly, a recent work reported the benefits of mycorrhizal fungi-based symbiosis for making available nutrients and minerals such as phosphorous and essential minor elements (Hartmann et al. 2009; Gianinazzi et al. 2010; Tian et al. 2010; Adeleke et al. 2012; Jin et al. 2012; Carvalhais et al. 2013; Johnson and Graham 2013; Lareen et al. 2016; Salvioli et al. 2016) to many crop plants for meeting their nutritional requirements (Johnson et al. 2012; Philippot et al. 2013; Salvioli and Bonfante 2013; Schlaeppi et al. 2014). Furthermore, research reports also highlighted the significant role of AMF in improving soil structure and establishing beneficial microbes (Bulgarelli et al. 2012; Peiffer et al. 2013; Dell Fabbro and Prati 2014; Tkacz et al. 2015; Wu et al. 2016). Recently, Symanczik et al. (2017) demonstrated that naranjilla (*Solanum quitoense*) inoculated with AMF showed improved plant growth and enhanced nutrition and soil water retention due to the successful establishment of AMF symbiosis which led to the high acquisition of phosphorous (up to 104%) compared to control plants. Furthermore, this study proved that highly diverse belowground systems like AMF play a significant role in maintaining soil structure and aggregation by hyphae and exudates which is essential for sustainable soil productivity (Van der Heijden et al. 2008; De Vries et al. 2013; Wagg et al. 2014). On the other hand, there are published reports revealing many non-AMF involved in AMF like symbiotic benefits to plants (Cai et al. 2014; Ghanem et al. 2014; Pandey et al. 2016).

Keeping the importance of nutrients in plant life, it would be logical to identify bacterial and AMF strains that effectively increase macro- and micronutrient uptake in plants under nutrient deficiency stress (Leveau et al. 2010; Mapelli et al. 2012; Pankaj et al. 2016; Wang et al. 2017). As a matter of fact, rhizospheric microbes can also help in the uptake of many trace elements such as iron and calcium (Zhang et al. 2009; Lee et al. 2010; Marschner et al. 2011; Shirley et al. 2011) from the soil with improved plant root system (Cummings and Orr 2010; Qiang et al. 2016). Taken together, it is safe to conclude that microbes residing in the rhizosphere are especially playing a vital role in degrading insoluble organic compounds which are not only required for their own life but also needed for proper plant growth under nutrient deficiency stress (Leveau et al. 2010; Mapelli et al. 2012; Turner et al. 2013; Bhattacharyya et al. 2015; Pankaj et al. 2016; Wang et al. 2017).

7.11 Extreme Temperature Stress

Rapid climate changes have increased the frequency of global temperature fluctuations. As a result of these changes, extreme temperatures (hot and cold) have now been treated as significant abiotic stress (Hussain et al. 2018; Kumar and Verma 2018). Reports have predicted that global temperatures will increase by 1.8–3.6 °C by the end of this century due to extreme changes (International Panel on Climate Change (IPCC 2007)). High temperatures are not only considered a major obstacle in crop growth and productivity but also negatively impact microbial colonization (Carson et al. 2010). Both plants and microbes respond to high temperature by producing heat shock proteins (HSPs) which help to avoid major cellular damage such as protein degradation and aggregation (Rodell et al. 2009; Alam et al. 2017). Stress adaptation in microbes constitutes a complex regulatory mechanism that may comprise of many gene expressions (Srivastava et al. 2008), helping microbes in developing strategies to mitigate the stress (Kumar and Verma 2018; Yang et al. 2016a).

As have been mentioned, high soil temperature significantly affects the performance of plant-associated microbes. However, several microbes have been isolated from hot environments, and these microbes performed significantly under heat stress. And based on observation, these microbes may be suitable candidates to use with crop plants under high temperature. In a study, wheat cultivars Olivin and Sids1 were treated with *Bacillus amyloliquefaciens* UCMB5113 or *Azospirillum brasilense* NO40, and young seedlings were tested for effect of short-term heat stress (Abd El-Daim et al. 2014). Few stress-associated genes also showed raised transcripts in leaves of control plants. However, such genes exhibited much lower expression in plants inoculated with microbes compared to control plants (Abd El-Daim et al. 2014). Similarly, low reactive oxygen species production was observed with non-significant changes in metabolome in wheat seedling treated with bacteria under high temperature. Certain microbes mitigate heat stress by exopolysaccharide (EPS) synthesis. EPS has the ability to hold water and has cementing characteristics which lead to confer stress tolerance mainly by biofilm synthesis traits (Hussain et al. 2018). *Pseudomonas putida* strain NBR10987 was isolated from chickpea rhizosphere under drought. Inoculated chickpea plants exhibited thermotolerance. Detailed investigations showed that thermotolerance in chickpea was due to stress sigma factor (δs) overexpression as well as thick biofilm synthesis (Srivastava et al. 2008). Similarly, inoculation of sorghum seedlings with two *Pseudomonas* strains AKM-P6 and NBR10987 improved thermotolerance manifested by better physiological and metabolic performance through diverse mechanisms (Redman et al. 2002; Ali et al. 2009; Grover et al. 2011). McLellan et al. (2007) noticed induction of small heat shock HSP101 and HSP70 proteins and enhanced heat tolerance in *Arabidopsis* when inoculated with fungus *Paraphaeosphaeria quadriseptata*.

Plants primed with microbes adapted for low temperature show high growth and development under cold stress. Therefore, researchers used microbes to mitigate negative effects of low temperature stress. Various bacterial strains have been used to enhance cold stress tolerance in plants (Selvakumar et al. 2008a, b, 2009, 2010a, b). Several low temperature-adapted bacterial strains such as *Brevundimonas terrae*, *Pseudomonas cedrina*, and *Arthrobacter nicotianae* have demonstrated multifunctional plant growth-promoting attributes (Yadav et al. 2014). Similarly, *B. phytofirmans* PsJN conferred not only high stress tolerance to low non-freezing temperatures but also grapevine plants showed resistance to grey mold (Meena et al. 2015). Barka et al. (2006) used *Burkholderia phytofirmans* PsJN to inoculate grapevine roots and concluded that inoculated plants physiologically performed better as manifested by their fast root growth and high plant biomass at low temperature (4 °C). Theocharis et al. (2012) showed positive priming effect of endophyte on plant at low temperature mainly due to high accumulation of several stress proteins. It is known that soybean symbiotic activities are inhibited by low temperature but soybean seedlings inoculated with both *Bradyrhizobium japonicum* and *Serratia proteamaculans* responded to symbiosis at low temperature (15 °C) and showed higher growth (Zhang et al. 1995, 1996). Mishra et al. (2009) noted that wheat seedlings primed with *Pseudomonas* sp. strain PPERs23 exhibited higher root/shoot ratio with increased dry root/shoot biomass, and other physiological traits such as increased iron, anthocyanins, proline, protein and relative water contents and reduced Na^+/K^+ ratio and electrolyte leakage also contributed to enhanced cold tolerance (Mishra et al. 2009). The above-mentioned studies clearly highlighted the importance of cold-adapted microbes like *Burkholderia phytofirmans* PsJN inoculated in plant species such as grapevines, maize, soybean, sorghum, wheat and switch grass that seems to be promising agents for low-temperature stress tolerance (Kim et al. 2012).

7.12 Future Perspective

Feeding the growing population requires high and stable yields using smart crop production technologies. The current agriculture in developing countries apparently relied on the cultivation of high-yield, moderately stress-tolerant varieties further fuelled by agrochemicals. It is not surprising that abiotic stresses, especially high temperature, drought and salinity, are considered by researchers as the most significant threats to agriculture (Trabelsi and Mhamdi 2013; Busby et al. 2017). Given this, we have to either develop stress-tolerant crop plants or look for alternative and more realistic agricultural practices (Bulgarelli et al. 2012; Mengual et al. 2014). Developing more sustainable solutions to agricultural problems seems logical under rapidly changing global climate and uncontrolled human growth (Hussain et al. 2014). Opportunities for exploiting the plant-associated microbes for raising successful crops are uncountable and diverse which can play a promising role for effectively tackling stresses in sustainable next-generation agriculture (Vandenkoornhuyse et al. 2015; Hussain et al. 2018).

The development and integration of smart agricultural tools and practices will depend on the successful use of all players in the system. Moderate success has been achieved towards the development of model host-microbiome systems for poplar, rice, sorghum, maize, miscanthus, tomato and *Medicago truncatula* (Johnston-Monje and Raizada 2011; Sessitsch et al. 2011; Knief et al. 2012; Peiffer et al. 2013; Ramond et al. 2013; Spence et al. 2014; Edwards et al. 2015; Hacquard and Schadt 2015; Lakshmanan 2015; Tian et al. 2015; Li et al. 2016; Hussain et al. 2018). However, great variation and success depend on many factors including the individual plant species, genotype, native soil microbiota, microbiome and interplay between these players with their specific traits that interact with each other under given climatic conditions. Under such circumstances, it is recommended that established microbiomes are most likely suitable candidates for generating diverse but functionally variable associations to select on a trial basis (Mueller and Sachs 2015). Hence, novel methods to utilize the plant-associated microbiome in next-generation agriculture could be helpful in enhancing crop productivity under different stresses (Bakker et al. 2012; Marasco et al. 2012; Prudent et al. 2015; Celebi et al. 2010; Mengual et al. 2014; Nadeem et al. 2014; Rolli et al. 2015). Novel versions of the most recent and advanced technologies especially omics approaches, methods and techniques are also offering its open-ended services for generation and interpretation of data from the field level to assess the real impacts of the inoculants on crop plants (Baetz and Martinoia 2013; White et al. 2017).

7.13 Conclusion

Several reports have shown promising results of significant stress tolerance in crop plants primed with plant growth-promoting microbes (PGPM) under field conditions with some negative findings as well. Application of microbial consortium represents one promising strategy for beneficial outcome in field-based agriculture to collectively respond to specific environmental stresses with no apparent impact on plant growth and productivity. Development and application of multispecies consortia have the potential to address inconsistency in performance (Hernández-Salmerón et al. 2016). Therefore, the mechanisms by which microbes confer stress tolerance to their hosts need further exploration to develop ideal microbial consortia for use under different stresses. Recent strategies like the use of omics approaches in this field provide powerful insights to understand how different players interact with each other and establish the functional relationship among microbe-microbe and plant-microbe under stress.

References

Aafi NE, Brhada F, Dary M, Maltouf AF, Pajuelo E (2012) Rhizostabilization of metals in soils using *Lupinus luteus* inoculated with the metal resistant rhizobacterium *Serratia* sp. MSMC 541. Int J Phytoremediation 14:26174

Abbasi MK, Sharif S, Kazmi M, Sultan T, Aslam M (2011) Isolation of plant growth promoting rhizobacteria from wheat rhizosphere and their effect on improving growth, yield and nutrient uptake of plants. Plant Biosyst 145:159–168

Abd El-Daim IA, Bejai S, Meijer J (2014) Improved heat stress tolerance of wheat seedling by bacterial seed treatment. Plant Soil 379:337–350

Adediran GA, Ngwenya BT, Mosselmans JF, Heal KV, Harvie BA (2015) Mechanisms behind bacteria induced plant growth promotion and Zn accumulation in *Brassica juncea*. J Hazard Mater 283:490–499

Adeleke RA, Cloete TE, Bertrand A, Khasa DP (2012) Iron ore weathering potentials of ectomycorrhizal plants. Mycorrhiza 22:535–544

Adesemoye AO, Egamberdieva D (2013) Beneficial effects of plant growth-promoting rhizobacteria on improved crop production: prospects for developing economies. In: Maheshwari DK, Saraf M, Aeron A (eds) Bacteria in agrobiology: crop productivity. Springer-Berlin, Berlin, pp 45–63

Adesemoye AO, Torbert HA, Kloepper JW (2009) Plant growth promoting rhizobacteria allow reduced application rates of chemical fertilizers. Microb Ecol 58:921–929

Adhya TK, Kumar N, Reddy G, Podile AR, Bee H, Bindiya S (2015) Microbial mobilization of soil phosphorus and sustainable P management in agricultural soils. Curr Sci 108:1280–1287

Afzal M, Yousaf S, Reichenauer TG, Kuffner M, Sessitsch A (2011) Soil type affects plant colonization, activity and catabolic gene expression of inoculated bacterial strains during phytoremediation of diesel. J Hazard Mater 186:1568–1575

Afzal M, Khan QM, Sessitsch A (2014) Endophytic bacteria: prospects and applications for the phytoremediation of organic pollutants. Chemosphere 117:232–242

Agler MT, Ruhe J, Kroll S, Morhenn C, Kim ST, Weigel D, Kemen EM (2016) Microbial hub taxa link host and abiotic factors to plant microbiome variation. PLoS Biol 14:e1002352

Ahmad M, Zahir ZA, Asghar HN, Asghar M (2011) Inducing salt tolerance in mung bean through co-inoculation with rhizobia and plant-growth-promoting rhizobacteria containing 1-aminocyc lopropane-1-carboxylate deaminase. Can J Microbiol 57:578–589

Alam MA, Seetharam KM, Zaidi PH, Dinesh A, Vinayan MT, Nath UK (2017) Dissecting heat stress tolerance in tropical maize (*Zea mays* L.). Field Crop Res 204:110–119
Alami Y, Achouak W, Marol C, Heulin T (2000) Rhizosphere soil aggregation and plant growth promotion of sunflowers by exopolysaccharide producing *Rhizobium* sp. strain isolated from sunflower roots. Appl Environ Microbiol 66:3393–3398
Ali SKZ, Sandhya V, Grover M, Kishore N, Rao LV, Venkateswarlu B (2009) *Pseudomonas* sp. strain AKM-P6 enhances tolerance of sorghum seedlings to elevated temperatures. Biol Fert Soil 46:45–55
Ali S, Charles TC, Glick BR (2012) Delay of flower senescence by bacterial endophytes expressing 1-aminocyclopropane-1-carboxylate deaminase. J Appl Microbiol 113:1139–1144
Ali S, Duan J, Charles TC, Glick BR (2014) A bioinformatics approach to the determination of genes involved in endophytic behavior in *Burkholderia* sp. J Theor Biol 343:193–198
Al-Karaki GN, Ammad R, Rusan M (2001) Response of two tomato cultivars differing in salt tolerance to inoculation with mycorrhizal fungi under salt stress. Mycorrhiza 11:43–47
Alström S (1991) Induction of disease resistance in common bean susceptible to halo blight bacterial pathogen after seed bacterization with rhizosphere pseudomonads. J Gen Appl Microbiol 37:495–501
Amann RI, Ludwing W, Schleifer KH (1995) Phylogenetic identification and in situ detection of individual microbial cells without cultivation. Microbiol Rev 59:143–169
Andreote FD, Azevedo JL, Araújo WL (2009) Assessing the diversity of bacterial communities associated with plants. Braz J Microbiol 40:417–432
Aroca R, Ruíz-Lozan JM (2012) Regulation of root water uptake under drought stress conditions. In: Aroca R (ed) Plant responses to drought stress. Springer, Berlin-Germany, pp 113–128
Arora S, Patel PN, Vanza MJ, Rao GG (2014) Isolation and characterization of endophytic bacteria colonizing halophyte and other salt tolerant plant species from coastal Gujarat. Afr J Microbiol Res 8:1779–1788
Arshad M, Sharoona B, Mahmood T (2008) Inoculation with Pseudomonas spp. containing ACC deaminase partially eliminate the effects of drought stress on growth, yield and ripening of pea (*Pisum sativum* L.). Pedosphere 18:611–620
Arslan M, Imran A, Khan QM, Afzal M (2017) Plant–bacteria partnerships for the remediation of persistent organic pollutants. Environ Sci Pollut Res 24:4322–4336
Atieno M, Herrmann L, Okalebo R, Lesueur D (2012) Efficiency of different formulations of *Bradyrhizobium japonicum* and effect of co-inoculation of *Bacillus subtilis* with two different strains of *Bradyrhizobium japonicum*. World J Microbol Biotechnol 28:2541–2550
Azcon R, Medina A, Aroca R, Ruiz-Lozano JM (2013) Abiotic stress remediation by the arbuscular mycorrhizal symbiosis and rhizosphere bacteria/yeast interactions. In: de Bruijin FJ (ed) Molecular microbial ecology of the rhizosphere. Wiley Blackwell, Hoboken, pp 991–1002
Babu AG, Reddy S (2011) Dual inoculation of arbuscular mycorrhizal and phosphate solubilizing fungi contributes in sustainable maintenance of plant health in fly ash ponds. Water Air Soil Pollut 219:3–10
Bacilio M, Moreno M, Bashan Y (2016) Mitigation of negative effects of progressive soil salinity gradients by application of humic acids and inoculation with *Pseudomonas stutzeri* in a salt-tolerant and a salt-susceptible pepper. Appl Soil Ecol 107:394–404
Badri DV, Weir TL, Van Der Lelie D, Vivanco JM (2009) Rhizosphere chemical dialogues: plant-microbe interactions. Curr Opin Biotechnol 20:642–650
Badri DV, Chaparro JM, Zhang R, Shen Q, Vivanco JM (2013) Application of natural blends of phytochemicals derived from the root exudates of arabidopsis to the soil reveal that phenolic related compounds predominantly modulate the soil microbiome. J Biol Chem 288:4502–4512
Baetz U, Martinoia E (2013) Root exudates: the hidden part of plant defense. Trends Plant Sci 19:90–97
Bai Y, muller DB, Srinivas G, Garrido-Oter R, Potthoff E, Rott M, Huttel B (2015) Functional overlap of the Arabidopsis leaf and root microbiota. Nature 528:364

Bainard LD, Koch AM, Gordon AM, Klironomos JN (2013) Growth response of crops to soil microbial communities from conventional monocropping and tree-based intercropping systems. Plant Soil 363:345–356

Bais HP, Weir TL, Perry LG, Gilroy S, Vivanco JM (2006) The role of root exudates in rhizosphere interactions with plants and other organisms. Annu Rev Plant Biol 57:233–266

Bakker MG, Manter DK, Sheflin AM, Weir TL, Vivanco JM (2012) Harnessing the rhizosphere microbiome through plant breeding and agricultural management. Plant Soil 360:1–13

Bakker PA, Berendsen RL, Doornbos RF, Wintermans PC, Pieterse CM (2013) The rhizosphere is revisited: root microbiomics. Front Plant Sci 4:165

Bal HB, Nayak L, Das S, Adhya TK (2013) Isolation of ACC deaminase producing PGPR from rice rhizosphere and evaluating their plant growth promoting activity under salt stress. Plant Soil 366:93–105

Balbontin R, Vlamakis H, Kolter R (2014) Mutualistic interaction between *Salmonella enteric* and *Aspergillus niger* and its effects on *Zea mays* colonization. Microb Biotechnol 7:589–600

Bano A, Fatima M (2009) Salt tolerance in *Zea mays* (L.) following inoculation with rhizobium and Pseudomonas. Biol Fert Soil 45:405–413

Bano Q, Ilyas N, Bano A, Zafar N, Akram A, Hassan F (2013) Effect of Azospirillum inoculation on maize (*Zea mays* L.) under drought stress. Pak J Bot 45:13–20

Barassi CA, Ayrault G, Creus CM, Sueldo RJ, Sobrero MT (2006) Seed inoculation with Azospirillum mitigates NaCl effects on lettuce. Sci Hort 109:8–14

Barka EA, Nowak J, Clement C (2006) Enhancement of chilling resistance of inoculated grapevine plantlets with a plant growth-promoting rhizobacterium, *Burkholderia phytofirmans* strain PsJN. Appl Environ Microbiol 72:7246–7252

Bartels D, Hussain SS (2008) Current status and implications of engineering drought tolerance in plants using transgenic approaches. CAB Rev Persp Agric Vet Sci Nutri Nat Sci 3:020

Bashan Y, Salazar BG, Moreno M, Lopez BR, Lindermann RG (2012) Restoration of eroded soil in the sonorant desert with native leguminous trees using plant growth promoting microorganisms and limited amounts of compost and water. J Environ Manag 102:26–36

Bashiardes S, Godneva A, Elinav E, Segal E (2018) Towards utilization of the human genome and microbiome for personalized nutrition. Curr Opin Biotechnol 51:57–63

Battisti DS, Naylor RL (2009) Historical warnings of future food insecurity with unprecedented seasonal heat. Science 323:240–244

Bell TH, Joly S, Pitre FE, Yergeau E (2014) Increasing phytoremediation efficiency and reliability using novel omics approaches. Trends Biotechnol 32:271–180

Ben Khaled L, Gomez AM, Ourraqi EM, Oihabi A (2003) Physiological and biochemical responses to salt stress of mycorrhized and/or nodulated clover seedlings (*Trifolium alexandrinum* L.). Agronomie 23:571–580

Berendsen RL, Pieterse CMJ, Bakker P (2012) The rhizosphere microbiome and plant health. Trends Plant Sci 17:478–486

Berg G (2009) Plant-microbe interactions promoting plant growth and health: perspectives for controlled use of microorganisms in agriculture. Appl Microbial Biotechnol 84:11–18

Berg G, Smalla K (2009) Plant species and soil type cooperatively shape the structure and function of microbial communities in the rhizosphere. FEMS Microbiol Ecol 68:1–13

Berg G, Zachow C, Müller H, Philipps J, Tilcher R (2013) Next-generation bio-products sowing the seeds of success for sustainable agriculture. Agronomy 3:648–656

Berg G, Grube M, Schloter M, Small K (2014) Unraveling the plant microbe: looking back and future prospective. Front Microbiol 5:148

Berg G, Rybakova D, Grube M, Koberl M (2016) The plant microbioe explored: implications for experimental botany. J Exp Bot 67:995–1002

Berman-Frank I, Lundgren P, Falkowski P (2003) Nitrogen fixation and photosynthetic oxygen evolution in *Cyanobacteria*. Res Microbiol 154:157–164

Beskoski VP, Gojgic-Cvijovic G, Milic J, Ilic M, Miletic S, Solevic T, Vrvic MM (2011) Ex-situ bioremediation of a soil contaminated by mazut (heavy residual fuel oil), a field experiment. Chemosphere 83:34–40

Bevivino A, Sarrocco S, Dalmastri S, Tabacchioni S, Cantale C, Chiarini L (1998) Characterization of a free-living maize rhizosphere population of *Burkholderia cepacia*: effect of seed treatment on disease suppression and growth promotion of maize. FEMS Microbiol Ecol 27:225–237
Bhardwaj D, Ansari MW, Sshoo RK, Tuteja N (2014) Biofertilizers function as key player in sustainable agriculture by improving soil fertility, plant tolerance and crop productivity. Microbial Cell Fact 13:1–10
Bharti N, Pandey SS, Barnawal D, Patel VK, Kalra A (2016) Plant growth promoting rhizobacteria *Dietzia natronolimnaea* modulates the expression of stress responsive genes providing protection of wheat from salinity stress. Sci Rep 6:34768
Bhattacharya PN, Jha DK (2012) Plant growth-promoting rhizobacteria (PGPR): emergence in agriculture. World J Microbial Biotechnol 28:1327–1350
Bhattacharyya PN, Sarmah SR, Dutta P, Tanti AJ (2015) Emergence in mapping microbial diversity in tea (*Camellia sinensis* L.) soil of Assam, North-East India: a novel approach. Eur J Biotechnol Biosci 3:20–25
Bhattacharyya PN, Goswani MP, Bhattacharyya LH (2016) Perspective of beneficial microbes in agriculture under changing climatic scenario: a review. J Phytology 8:26–41
Bilal S, Khan AL, Shahzad R, Asaf S, Kang SM, Lee IJ (2017) Endophytic *Paecilomyces formosus* LHL10 augments *Glycine max* L. adaptation to Ni-contaimination through affecting endogenous phytohormones and oxidative stress. Front. Plant Sci 8:870
Bisht S, Pandey P, Kaur G, Aggarwal H, Sood A, Sharma S, Kumar V, Bisht NS (2014) Utilization of endophytic strain Bacillus sp. SBER3 for biodegradation of polyaromatic hydrocarbons (PAH) in soil model system. Eur J Soil Biol 60:67–76
Bloemberg GV, Lugtenberg BJJ (2001) Molecular basis of plant growth promotion and biocontrol by rhizobacteria. Curr Opin Plant Biol 4:343–350
Bonfante P, Anca IA (2009) Plant, mycorrhizal fungi and bacteria: a network of interactions. Annu Rev Microbiol 63:363–383
Bonilla N, Vida C, Martínez-Alonso M, Landa BB, Gaju N, Cazorla FM, de Vicente A (2015) Organic amendments to avocado crops induce suppressiveness and influence the composition and activity of soil microbial communities. Appl Environ Microbiol 81:3405–3418
Boon E, Meehan CJ, Whidden C, Wong DHJ, Langille MGI, Beiko RG (2014) Interactions in the microbiome: communities of organisms and communities of genes. FEMS Microbiol Rev 38:90–118
Boopathy R (2000) Factors limiting bioremediation technologies. Bioresour Technol 74:63–67
Broos K, Uyttebroek M, Mertens J, Smolders E (2004) A survey of symbiotic nitrogen fixation by white clover grown on metal contaminated soils. Soil Biol Biochem 36:633–640
Budak H, Kantar M, Yucebilgili Kurtoglu K (2013) Drought tolerance in modern and wild wheat. Sci World J 2013:548246
Bulgarelli D, Rott M, Schlaeppi K, Loren Ver van Themaat E, Ahmadinejad N, Assenza F, Rauf P, Huettel B, Reinhardt R, Schmelzer E, Peplies J, Gloeckner FO, Amann R, Eickhorst T, Schulze-Lefert P (2012) Revealing structure and assembly cues for *Arabidopsis* root-inhabiting bacterial microbiota. Nature 488:91–95
Bulgarelli D, Schlaeppi K, Spaepen S, Ver Loren van Themaat E, Schulze-Lefert P (2013) Structure and functions of the bacterial microbiota of plants. Annu Rev Plant Biol 64:807–838
Bulgarelli D, Garrido-Oter R, Munch PC, Weiman A, Droge J, Pan Y, McHardy AC, Schulze-Lefert P (2015) Structure and function of the bacterial root microbiota in wild and domesticated barley. Cell Host Microbe 17:392–403
Busby PE, Soman C, Wagner MR, Friesen ML, Kremer J, Bennett A, Morsy M, Eisen JA, Leach JE, Dangl JL (2017) Research priorities for harnessing plant microbiomes in sustainable agriculture. PLoS Biol 15:e2001793
Cai F, Chen W, Wei Z, Pang G, Li R, Ran W, Shen Q (2014) Colonization of *Trichoderma harzianum* strain SQR-T037 on tomato roots and its relationship to plant growth, nutrient availability and soil microflora. Plant Sci 388:337–350

Calvo P, Watts DB, Kloepper JW, Torbert HA (2017) Effect of microbial-based inoculants on nutrient concentrations and early root morphology of corn (*Zea mays*). J Plant Nutr Soil Sci 180:56–70

Capell T, Bassie L, Christou P (2004) Modulation of the polyamine biosynthetic pathway in transgenic rice confers tolerance to drought stress. Proc Natl Acad Sci U S A 101:9909–9914

Carson JK, Gonzalez-Quinones V, Murphy DV, Hinz MC, Shaw JA, Gleeson DB (2010) Low pore connectivity increases bacterial diversity in soil. Appl Environ Microbiol 76:3936–3942

Carvalhais LC, Dennis PG, Fan B, Fedoseyenko D, Kierul K, Becker A, von Wiren N, Borriss R (2013) Linking plant nutritional status to plant-microbe interactions. PLoS One 8:e68555

Cassán F, Perrig D, Sgroy V, Masciarelli O, Penna C, Luna V (2009) *Azospirillum brasilense* Az39 and *Bradyrhizobium japonicum* E109, inoculated singly or in combination, promote seed germination and early seedling growth in corn (*Zea mays* L.) and soybean (*Glycine max* L.). Eur J Soil Biol 45:28–35

Castillo P, Escalante M, Gallardo M, Alemano S, Abdala G (2013) Effects of bacterial single inoculation and co-inoculation on growth and phytohormone production of sunflower seedlings under water stress. Acta Physiol Plant 35:2299–2309

Celebi SZ, Demir S, Celebi R, Durak ED, Yilmaz IH (2010) The effect of arbuscular mycorrhizal fungi (AMF) applications on the silage maize (*Zea mays* L.) yield in different irrigation regimes. Eur J Soil Biol 46:302–305

Chagnon PL, Bradley RL, Maherali H, Klironomos JN (2013) A trait-based framework to understand life history of mycorrhizal fungi. Trends Plant Sci 18:484–491

Chang P, Gerhardt KE, Huang XD, Yu XM, Glick BR, Gerwing PD, Greenberg BM (2014) Plant growth-promoting bacteria facilitate the growth of barley and oats in salt-impacted soil: implications for phytoremediation of saline soils. Int J Phytoremedation 16:1133–1147

Chaparro JM, Badri DV, Vivanco JM (2013) Rhizosphere microbiome assemblage is affected by plant development. ISME J 8:790–803

Chaparro JM, Badri DV, Vivanco JM (2014) Rhizosphere microbiome assemblage is affected by plant development. ISMI J 8:790–803

Chauhan H, Bagyaraj DJ, Selvakumar G, Sundaram SP (2015) Novel plant growth promoting rhizobacteria prospects. Appl Soil Ecol 95:38–53

Chen L, Luo S, Li X, Wan Y, Chen J, Liu C (2014) Interaction of Cd-hyperaccumulator *Solanum nigrum* L. and functional endophyte Pseudomonas sp. Lk9 on soil heavy metals uptake. Soil Biol Biochem 68:300–308

Cho K, Toler H, Lee J, Ownley B, Stutz JC, Moore JL, Auge RM (2006) Mycorrhizal symbiosis and response of sorghum plants to combined drought and salinity stresses. J Plant Physiol 163:517–528

Choudhary DK, Prakash A, Johri BN (2007) Induced systemic resistance (ISR) in plants: mechanism of action. Indian J Microbiol 47:289–297

Clavel T, Lagkouvardos I, Blaut M, Stecher B (2016) The mouse gut microbiome revisited: from complex diversity to model ecosystems. Int J Med Microbiol 306:316–327

Cohen AC, Bottini R, Pontin M, Berli FJ, Moreno D, Boccanlandro H, Travaglia CN, Piccoli PN (2015) *Azospirillum brasilense* ameliorates the response of *Arabidopsis thaliana* to drought mainly via enhancement of ABA levels. Physiol Plant 153:79–90

Coleman-Derr D, Tringe SG (2014) Building the crops of tomorrow: advantages of symbiont-based approaches to improving abiotic stress tolerance. Front Microbiol 5:283

Compant S, Duffy B, Nowak J, Clément C, Barka EA (2005) Use of plant growth-promoting bacteria for biocontrol of plant diseases: principles, mechanisms of action, and future prospects. Appl Environ Microbiol 71:4951–4959

Compant S, Saikkonen K, Mitter B, Campisano A, Blanco JM (2016) Soil, plants and endophytes. Plant Soil 405:1–11

Conrath U (2006) Systemic acquired resistance. Plant Signal Behav 4:179–184

Cooper M, Gho C, Leafgren R, Tang T, Messina C (2014) Breeding drought-tolerant maize hybrids for the US corn-belt: discovery to product. J Exp Bot 65:6191–6204

Corral-Lugo A, De la Torre J, Matilla MA, Fernández M, Morel B, Espinosa-Urgel M, Krell T (2016) Assessment of the contribution of chemoreceptor-based signalling to biofilm formation. Environ Microbiol 18:3355–3372

Creus CM, Sueldo RJ, Barassi CA (2004) Water relations and yield in *Azospirillum*-inoculated wheat exposed to drought in the field. Can J Bot 82:273–281

Cummings SP, Orr C (2010) The role of plant growth promoting rhizobacteria in sustainable and low graminaceous crop production. In: Maheshwari DK (ed) Plant growth and health promoting bacteria. Springer, Berlin

Cushman JC, Bohnert HJ (2000) Genomic approaches to plant stress tolerance. Curr Opin Plant Biol 3:117–124

Damodaran T, Sah V, Rai RB, Sharma DK, Mishra VK, Jha SK, Kannan R (2013) Isolation of salt tolerant endophytic and rhizospheric bacteria by natural selection and screening for promising plant growth-promoting rhizobacteria (PGPR) and growth vigour in tomato under sodic environment. Afr J Microbiol Res 7:5082–5089

Dangl JL, Horvath DM, Staskawicz BJ (2013) Pivoting the plant immune system from dissection to deployment. Science 341:746–751

De Bruijin FJ (2015) Biological nitrogen fixation. In: Lugtenberg B (ed) Principles of plant-microbe interactions. Springer International Publishing Switzerland, Heidelberg, pp 215–224

de la Torre-González A, Navarro-Leon E, Albacete A, Blasco B, Ruiz JM (2017) Study of Phytohormone profile and oxidative metabolism as key process to identification of salinity response in tomato commercial genotypes. J Plant Physiol 216:164–173

De Vries FT, Thebault E, Liiri M, Birkhofer K, Tsiafouli MA, Bjornlund L, Jorgensen HB, Brady MV, Christensen S, de Ruiter PC, Hertefeldt T, Frouz J, Hedlund K, Hemerik L, Gera Hol WH, Hotes S, Mortimer SR, Setala H, Sgardelis SP, Uteseny K, der Putten WH, Wolters V, Bardgett RD (2013) Soil food web properties explain ecosystem sevices across European land use systems. Proc Natl Acad Sci U S A 110:14296–14301

de Zelicourt A, Al-Yousif M, Hirt H (2013) Rhizosphere microbes as essential partners for plant stress tolerance. Mol Plant 6:242–245

De-la-Peña C, Badri D, Loyola-Vargas V (2012) Plant root secretions and their interactions with neighbors. In: Vivanco JM, Baluška F (eds) Secretions and exudates in biological systems. Springer, Berlin, pp 1–26

Dell Fabbro C, Prati D (2014) Early responses of wild plant seedlings to arbuscular mycorrhizal fungi and pathogens. Basic Aggl Ecol 15:534–542

Deng Z, Cao L (2017) Fungal endophytes and their interactions with plants in phytoremediation: a review. Chemosphere 168:1100–1106

Deng B, Yang K, Zhang Y, Li Z (2016) Can heavy metal pollution defend seed germination against heat stress? Effect of heavy metals (Cu2C, Cd2C and Hg2C) on maize seed germination under high temperature. Environ Pollut 216:46–52

Dikilitas M, Karakas S, Ahmad P (2018) Predisposition of crop plants to stress is directly related to their DNA health. In: Egamberdieva D, Ahmad P (eds) Plant microbiome: stress response. Springer Nature, Singapore, pp 233–254

Dimpka C, Weinard T, Asch F (2009) Plant-rhizobacteria interactions alleviate abiotic stress conditions. Plant Cell Environ 32:1682–1694

Ding GC, Piceno YM, Heuer H, Weinert N, Dohrmann AB, Carrillo A, Andersen GL, Castellanos T, Tebbe CC, Small K (2013) Changes of soil bacterial diversity as a consequence of agricultural land use in a semi-arid ecosystem. PLoS One 8:e59497

Doble M, Kumar A (2005) Biotreatment of industrial effluents. Elsvier, Butterworth-Heinemann, Amsterdam/Boston, pp 1–5

Dodd IC, Belimov AA, Sobeih WY, Safronova VI, Grierson D, Davies WJ (2005) Will modifying plant ethylene status improve plant productivity in water-limited environments? 4th international crop science congress. http://www.cropscience.org.au/icsc2004/poster/1/3/4/510_doddicref.htm. Accessed 21 Aug 2017

Dong HN, Zhang DW (2014) Current development in genetic engineering strategies of *Bacillus* species. Microb Cell Factories 13:63

Doornbos RF, van Loon LC, Bakker PHM (2011) Impact of root exudates and plant defense signaling on bacterial communities in the rhizosphere: a review. Agron Sustain Dev 32:227–243

Dumbrell AJ, Nelson M, Helgason T, Dytham C, Fitter AH (2010) Relative roles of niche and neutral processes in structuring a soil microbial community. ISME J 4:337–345

Duruibe JO, Ogwuegbu MOC, Egwurugwu JN (2007) Heavy metal pollution and human biotoxic effects. Int J Phys Sci 2:112–118

Edwards J, Johnson C, Santos-Medellín C, Lurie E, Podishetty NK, Bhatnagar S, Eisen JA, Sundaresan V (2015) Structure, variation, and assembly of the root-associated microbiomes of rice. Proc Natl Acad Sci U S A 112:E911–EE20

Egamberdieva D, Wirth SJ, Alqarawi AA, Abd-Allah EF, Hashem A (2017) Phytohormones and beneficial microbes: essential components for plant to balance stress and fitness. Front Microbiol 8:2104

Egamberdiyeva D (2007) The effect of plant growth promoting bacteria on growth and nutrient uptake of maize in two different soils. Appl Soil Ecol 36:184–189

Elbadry M, Taha RM, Eldougdoug KA, Gamal-Eldin H (2006) Induction of systemic resis-tance in faba bean (*Vicia faba* L.) to bean yellow mosaic potyvirus (BYMV) via seed bacterization with plant growth promoting rhizobacteria. J Plant Dis Prot 113:247–251

Farooq M, Wahid A, Kobayashi N, Fujita D, Basra SMA (2009) Plant drought stress: effects, mechanisms, and management. Sustainable agriculture. Springer, Netherlands, pp 153–188

Farrar K, Bryant D, Cope-Selby N (2014) Understanding and engineering beneficial plant-microbe interactions: plant growth promotion in energy crops. Plant Biotechnol J 12:1193–1206

Fasciglione G, Casanovas EM, Quillehauquy V, Yommi AK, Goni MG, Roura SI, Barassi CA (2015) Azospirillum inoculation effects on growth, product quality and storage life of lettuce plants grown under salt stress. Sci Hortic 195:154–162

Fatnassi IC, Chiboub M, Saadani O, Jebara M, Jebara CA (2015) The impact of dual inoculation with Rhizobium and PGPR on growth and antioxidant status of *Vicia faba* L. under copper stress. Crit Rew Biol 338:241–254

Feng G, Zhang FS, Li XL, Tian CY, Tang C, Renegal Z (2002) Improved tolerance of maize plants to salt stress by *arbuscular mycorrhiza* is related to higher accumulation of leaf P-concentration of soluble sugars in roots. Mycorrhiza 12:185–190

Feng NX, Yu J, Zhao HM, Cheng YT, Mo CH, Cai QY, Li YW, Li H, Wng MH (2017) Efficient phytoremediation of organic contaminants in soils using plant-endophyte partnerships. Sci Total Environ 583:352–368

Ferrara FIS, Oliveira ZM, Gonzales HHS, Floh EIS, Barbosa HR (2012) Endophytic and rhizospheric enterobacteria isolated from sugar cane have different potentials for producing plant growth-promoting substances. Plant Soil 353:409–417

Fester T, Giebler J, Wick LY, Schlosser D, Kästner M (2014) Plant–microbe interactions as drivers of ecosystem functions relevant for the biodegradation of organic contaminants. Curr Opin Biotechnol 27:168–175

Fierer N (2017) Embracing the unknown: disentangling the complexities of the soil microbiome. Nat Rev Microbiol 15:579–590

Figueiredo MVB, Burity HA, Martinez CR, Chanway CP (2008) Alleviation of drought stress in common bean (*Phaseolus vulgaris* L.) by co-inoculation with *Paenibacillus polymyxa* and *Rhizobium tropici*. Appl Soil Ecol 40:182–188

Filippi MCC, da Silva GB, Silva-Lobo VL, Côrtes MVCB, Moraes AJG, Prabhu AS (2011) Leaf blast (*Magnaporthe oryzae*) suppression and growth promotion by rhizobacteria on aerobic rice in Brazil. Biol Control 58:160–166

Fouzia A, Allaoua S, Hafsa C, Mostefa G (2015) Plant growth promoting and antagonistic traits of indigenous fluorescent Pseudomonas spp. isolated from wheat rhizosphere and a thalamus endosphere. Eur Sci J 11:129–148

Foyer CH, Rasool B, Davey JW, Hancock RD (2016) Cross-tolerance to biotic and abiotic stresses in plants: a focus on resistance to aphid infestation. J Exp Bot 7:2025–2037

Friesen M, Porter S, Stark SC, von Wettberg EJ, Sachs JL, Martinez-Romero E (2011) Microbially mediated plant functional traits. Annu Rev Ecol Evol Syst 42:23–46

Gabriela F, Casanovas EM, Quillehauquy V, Yommi AK, Goni MG, Roura SI, Barassi CA (2015) Azospirillum inoculation effects on growth, product quality and storage life of lettuce plants grown under salt stress. Sci Hortic 195:154–162
Gaby JC, Buckley DH (2011) A global census of nitrogenase diversity. Environ Microbiol 13:1790–1799
Gerhardt KE, Huang XD, Glick BR, Greenberg BM (2009) Phytoremediation and rhizoremediation of organic soil contaminants: potential and challenges. Plant Sci 176:20–30
Ghanem G, Ewald A, Henning F (2014) Effect of root colonization with *Piriformospora indica* and phosphate availability on the growth and reproductive biology of a *Cyclamen persicum* cultivar. Sci Hortic 172:233–241
Ghosh P, Rathinasabapathi B, Ma LQ (2015) Phosphorus solubilization and plant growth enhancement by arsenic-resistant bacteria. Chemosphere 134:1–6
Gianinazzi S, Gollotte A, Binet MN, van Tuinen D, Redecker D, Wipf D (2010) Agroecology: the key role of *arbuscular mycorrhizae* in ecosystem services. Mycorrhiza 20:519–530
Giri B, Mukerji KG (2004) Mycorrhizal inoculant alleviate salt stress in *Sesbania aegyptiaca* and *Sesbania grandiflora* under field conditions: evidence for reduced sodium and improved magnesium uptake. Mycorrhiza 14:307–312
Glick BR (2010) Using soil bacteria to facilitate phytoremediation. Biotechnol Adv 28:367–374
Glick BR (2012) Plant growth-promoting bacteria: mechanisms and applications. Scientifica 2012:1–15
Glick BR (2014) Bacteria with ACC deaminase can promote plant growth and help to feed the world. Microbiol Res 169:30–39
Glick BR (2015) Phytoremediation: beneficial plant-bacterial interactions. Springer International Publishing, Cham, pp 191–221
Glick BR, Cheng Z, Czarny J, Duan J (2007) Promotion of plant growth by ACC deaminase-producing soil bacteria. Eur J Plant Pathol 119:329–339
Goh CH, Valiz Vallejos DF, Nicotra AB, Mathesius U (2013) The impact of beneficial plant-associated microbes on plant phenotypic plasticity. J Chem Ecol 39:826–839
Goodrich JK, Davenport ER, Clark AG, Ley RE (2017) The relationship between the human genome and microbiome comes into view. Annu Rev Genet 51:413–433
Goswami D, Thakker JN, Dhandhukia PC (2016) Portraying mechanics of plant growth-promoting rhizobacteria (PGPR): a review. Cogent Food Agric 2:1–19
Gouda S, Kerry RG, Das G, Paramithiotis S, Shin H-S, Patra JK (2018) Revitalization of plant growth promoting rhizobacteria for sustainable development in agriculture. Microbiol Res 206:131–140
Grice EA, Segre JA (2012) The human microbiome: our second genome. Annu Rev Genomics Hum Genet 13:151–170
Grover M, Ali Sk Z, Sandhya V, Rasul A, Venkateswarlu B (2011) Role of microorganisms in adaptation of agriculture crop to abiotic stresses. World J Bicrobiol Biotechnol 27:1231–1240
Guimaraes AA, Jaramillo PMD, Nobrega RSA, Florentino LA, Silva KB, de Souza Moreira FM (2012) Genetic and symbiotic diversity of nitrogen-fixing bacteria isolated from agricultural soils in the western Amazon by using cowpea as the trap plant. Appl Environ Microbiol 78:6726–6733
Gupta G, Parihar SS, Ahirwar NK, Snehi SK, Singh V (2015) Plant growth promoting rhizobacteria (PGPR): current and future prospects for development of sustainable agriculture. J Microbiol Biochem 7:96–102
Gusain YS, Singh US, Sharma AK (2015) Bacterial mediated amelioration of drought stress in drought tolerant and susceptible cultivars of rice (*Oryza sativa* L.). Afr J Biotechnol 14:764–773
Guttman D, McHardy AC, Schulze-Lefert P (2014) Microbial genome-enabled insights into plant-microorganism interactions. Nat Rev Genet 15:797–813
Hacquard S, Schadt CW (2015) Towards a holistic understanding of the beneficial interactions across the Populus microbiome. New Phytol 205:1424–1430
Hadi F, Bano A (2010) Effect of diazotrophs (*Rhizobium* and *Azotobacter*) on growth of maize (*Zea mays* L.) and accumulation of Lead (Pb) in different plant parts. Pak J Bot 42:4363–4370

Haggag WM, Abouziena HF, Abd-El-Kreem F, Habbasha S (2015) Agriculture biotechnology for management of multiple biotic and abiotic environmental stress in crops. J Chem Pharm 7:882–889

Haichar FZ, Marol C, Berge O, Rangel-Castro JI, Prosser JI, Balesdent J, Heutin T, Achouak W (2008) Plant host habitat and root exudates shape soil bacterial community structure. ISME J 2:1221–1230

Hamel C, Vujanovic V, Jeannotte R, Nakano-Hylander A, St-Arnaud M (2005) Negative feedback on perennial crop: fusarium crown and root rot of asparagus is related to changes in soil microbial community structure. Plant Soil 268:75–87

Han J, Sun L, Dong X, Cai Z, Sun X, Yang H, Wang Y, Song W (2005) Characterization of a novel plant growth-promoting bacteria strain *Delftia tsuruhatensis* HR4 both as a diazotroph and a potential biocontrol agent against various plant pathogens. Syst Appl Microbiol 28:66–76

Haney CH, Ausubel FM (2015) Plant microbiome blueprints. Science 349:788–789

Haney CH, Samuel BS, Bush J, Ausubel FM (2015) Associations with rhizosphere bacteria can confer an adaptive advantage to plants. Nat Plants 1:15051

Hardoim PR, van Overbeek L, van Elsas J (2008) Properties of bacterial endophytes and their proposed role in plant growth. Trends Microbiol 16:463–471

Hardoim PR, van Overbeek LS, Berg G, Pirtilla AM, Compant S, Campisano A, Doring M, Sessitsch A (2015) The hidden world within plants: ecological and evolutionary considerations for defining functioning of microbial endophytes. Microbiol Mol Biol Rev 79:293–320

Hartmann A, Schmid M, van Tuinen D, Berg G (2009) Plant-driven selection of microbes. Plant Sci 268:75–87

Hasan MA (2013) Investigation on the nitrogen fixing *Cyanobacteria* (BGA) in rice fields of North-West region of Bangladesh. J Environ Sci Nat Resour 6:253–259

Hawes MC, Curlango-Rivera G, Xiong Z, Kessler JO (2012) Roles of root border cells in plant defense and regulation of rhizosphere microbial population by extracellular DNA "trapping". Plant Soil 355:1–16

Hawkins HJ, Johansen A, George E (2000) Uptake and transport of organic and inorganic nitrogen by arbuscular mycorrhizal fungi. Plant Soil 226:275–285

Hayat R, Ali S, Amara U, Khalid R, Ahmed I (2010) Soil beneficial bacteria and their role in plant growth promotion: a review. Ann Microbiol 60:579–598

Hernandez-Leon R, Rojas-Solfs D, Contreras-Perez M, Orozco-Mosqueda M, Macias-Rodriguez L, Reyes-delaCruz H, Valencia-Cantero E, Santoyo G (2015) Characterization of antifungal and plant growth-promoting effects of diffusible and volatile organic compounds produced by *Pseudomonas fluorescens* strains. Biol Control 81:83–92

Hernández-Salmerón J, Hernández-León R, del Orozco-Mosqueda MAC, Moreno-Hagelsieb G, Valencia-Cantero E, Santoyo G (2016) Draft genome sequence of the biocontrol and plant growth-promoting rhizobacterium *Pseudomonas fluorescens* UM270. Stand Genom Sci 11:5

Hirsch PR, Mauchline TH (2012) Who's who in the plant root microbiome? Nat Biotechnol 30:961–962

Hristozkova M, Geneva M, Stanchova I, Boychinova M, Djonova E (2016) The contribution of arbuscular mycorrhizal fungi in attenuation of heavy metal impact on *Calendula officinalis* development. Appl Soil Ecol 101:57–63

Hu S, Gu H, Cui C, Ji R (2016) Toxicity of combined chromium (VI) and phenanthrene pollution on the seed germination, stem lengths, and fresh weights of higher plants. Environ Sci Pollut Res 23:15227–15235

Huang B, DaCosta M, Jiang Y (2014a) Research advances in mechanisms of turfgrass tolerance to abiotic stresses: from physiology to molecular biology. Crit Rev Plant Sci 33:141–189

Huang X-F, Chaparro JM, Reardon KF, Zhang R, Shen Q, Vivanco JM (2014b) Rhizosphere interactions: root exudates, microbes, and microbial communities. Botany 92:267–275

Hussain SS, Iqbal MT, Arif MA, Amjad M (2011) Beyond osmolytes and transcription factors: drought tolerance in plants via protective proteins and aquaporins. Biol Plant 55:401–413

Hussain SS, Raza H, Afzal I, Kayani MA (2012) Transgenic plants for abiotic stress tolerance: current status. Arch Agron Soil Sci 58:693–721

Hussain SS, Siddique KHM, Lopato S (2014) Towards integration of bacterial genomics in plants for enhanced abiotic stress tolerance: clues from transgenics. Adv Environ Res 33:65–122
Hussain SS, Asif MA, Sornaraj P, Ali M, Shi BJ (2016) Towards integration of system based approach for understanding drought stress in plants. In: Ahmad P, Rasool S (eds) Water stress and crop plants: a sustainable approach. Elsevier, Atlanta, pp 227–247
Hussain SS, Mehnaz S, Siddique KM (2018) Harnessing the plant microbiome for improved abiotic stress tolerance. In: Ahmad P, Egamberdieva D (eds) Microbiome: stress response and microbes for sustainable agriculture. Springer, Singapore, pp 21–43
Iannucci A, Fragasso M, Beleggic R, Nigo F, Papa R (2017) Evolution of crop rhizosphere: impact of domestication on root exudates in tetraploid wheat (*Triticum turgidum* L.). Front Plant Sci 8:2124
Ijaz A, Imran A, ul Haq MA, Khan QM, Afzal M (2016) Phytoremediation: recent advances in plant-endophytic synergistic interactions. Plant Soil 405:179–195
Inceoglu O, Overbeek LS, Salles JF, Elsas JD (2013) Normal operating range of bacterial communities in soil used for potato cropping. Appl Environ Microbiol 79:1160–1170
IPCC (2007) The physical science basis. Contribution of working group I to the fourth assessment report of the intergovernmental panel on climate change. Cambridge University Press, Cambridge
Islam S, Akanda AM, Prova A, Islam MT, Hossain M (2016) Isolation and identification of plant growth promoting rhizobacteria from cucumber rhizosphere and their effect on plant growth promotion and disease suppression. Front Microbiol 6:1–12
Jamil M, Zeb S, Anees M, Roohi A, Ahmed I, Rehman SU, Rha ES (2014) Role of *Bacillus licheniformis* in phytoremediation of nickel contaminated soil cultivated with rice. Int J Phytoremediation 16:554–571
Jansson JK, Hofmockel KS (2018) The soil microbiome: from metagenomics to metaphenomics. Curr Opin Microbiol 43:162–168
Jefferies P, Barea JM (2001) *Arbuscular mycorrhiza*: a key component of sustainable plant-soil ecosystems. In: Hock B (ed) The mycota (Vol. IX: fungal associations). Springer, Berlin
Jez JM, Lee SG, Sherp AM (2016) The next green movement: plant biology for the environment and sustainability. Science 353:1241–1244
Jha Y, Subramanian RB (2014) PGPR regulate caspase-like activity, programmed cell death, and antioxidant enzyme activity in paddy under salinity. Physiol Mol Biol Plant 20:201–207
Jha B, Gontia I, Hartmann A (2012) The roots of the halophyte *Salicornia brachiata* are a source of new halotolerant diazotrophic bacteria with plant growth-promoting potential. Plant Soil 356:265–277
Jha P, Panwar J, Jha PN (2015) Secondary plant metabolites and root exudates: guiding tools for polychlorinated biphenyl biodegradation. Int J Environ Sci Technol 12:789–802
Jin H, Liu J, Liu J, Huang X (2012) Forms of nitrogen uptake, translocation and transfer via arbuscular mycorrhizal fungi: a review. Sci China Life Sci 55:474–482
Jindal V, Atwal A, Sekhon BS, Rattan S, Singh R (1993) Effect of vesicular-arbuscular mycorrhiza on metabolism of moong plants under salinity. Plant Physiol Biochem 31:475–481
Jing YX, Yan YJ, He HD, Yang DJ, Xiao L, Zhong T, Yuan M, Cai XD, Li SB (2014) Characterization of bacteria in the rhizosphere soils of *Polygonum pubescens* and their potential in promoting growth and Cd, Pb, Zn uptake by *Brassica napus*. Int J Phytoremediation 16:321–333
Johnson NC, Graham JH (2013) The continuum concept remains a useful framework for studying mycorrhizal functioning. Plant Soil 363:411–419
Johnson D, Martin F, Cairney JWG, Anderson IC (2012) The importance of individuals: intraspecific diversity of mycorrhizal plants and fungi in ecosystems. New Phytol 194:614–628
Johnston-Monje D, Raizada MN (2011) Conservation and diversity of seed associated endophytes in zea across boundaries of evolution, ethnography and ecology. PLoS One 6:e20396
Jones MB, Finnan J, Hodkinson TR (2014) Morphological and physiological traits for higher biomass production in perennial rhizomatous grasses grown on marginal land. Glob Change Biol Bioenergy 7:375–385

Jorquera MA, Shaharoona B, Nadeem SM, de la Luz MM, Crowley DE (2012) Plant growth-promoting rhizobacteria associated with ancient clones of creosote bush (*Larrea tridentata*). Microb Ecol 64:1008–1017

Kachroo A, Robin GP (2013) Systemic signaling during plant defense. Curr Opin Plant Biol 16:527–533

Kamal R, Gusain YS, Kumar V (2014) Interaction and symbiosis of fungi, actinomycetes and plant growth promoting rhizobacteria with plants: strategies for the improvement of plants health and defense system. Int J Curr Microbial Appl Sci 3:564–585

Kanchiswamy CN, Malnoy M, Maffei ME (2015) Chemical diversity of microbial volatiles and their potential for plant growth and productivity. Front Plant Sci 6:151

Kang SM, Radhakrishnan R, Khan AL, Kim MJ, Park JM, Kim BR, Shim DH, Lee IJ (2014) Gibberellin secreting rhizobacterium, Pseudomonas putida H-2-3 modulates the hormonal and stress physiology of soybean to improve the plant growth under saline and drought conditions. Plant Physiol Biochem 84:115–124

Kasim W, Osman M, Omar M, Abd El-Daim I, Bejai S, Meijer J (2013) Control of drought stress in wheat using plant-growth promoting rhizobacteria. J Plant Growth Regul 32:122–130

Kasim WA, Gaafar RM, Abou-Ali RM, Omar MN, Hewait HM (2016) Effect of biofilm forming plant growth promoting rhizobacteria on salinity tolerance in barley. Ann Agric Sci 61:217–227

Kathuria H, Giri J, Tyagi H, Tyagi AK (2007) Advances in transgenic rice biotechnology. Crit Rev Plant Sci 26:65–103

Kavamura VN, Santos SN, Silva JL, Parma MM, Avila LA, Visconti A, Zucchi TD, Taketani RG, Andreote FD, Melo IS (2013) Screening of Brazilian cacti rhizobacteria for plant growth promotion under drought. Microbiol Res 168:183–191

Khalid A, Tahir S, Arshad M, Zahir ZA (2004) Relative efficiency of rhizobacteria for auxin biosynthesis in rhizosphere and non-rhizosphere soils. Aust J Soil Res 42:921–926

Khan IA, Ayub N, Mirza SN, Nizami SN, Azam M (2008) Yield and water use efficiency (WUE) of *Cenchrus ciliaris* as influenced by vesicular arbuscular mycorrhizae (VAM). Pak J Bot 40:931–937

Khan A, Jilani G, Akhtar MS, Naqvi SMS, Rasheed M (2009) Phosphorus solubilizing bacteria: occurrence, mechanisms and their role in crop production. J Agric Biol Sci 1:48–58

Khan Z, Rho H, Firrincieli A, Hung H, Luna V, Masciarelli O, Kim SH, Doty SL (2016) Growth enhancement and drought tolerance of hybrid poplar upon inoculation with endophyte consortia. Curr Plant Biol 6:38–47

Khan AL, Waqas M, Asaf S, Kamran M, Shahzad R, Bilal S, Khan MA, Kang SM, Kim YH, Yun BW, Al-Rwahi A, Al-Harassi A, Lee IJ (2017) Plant growth-promoting endophyte Sphingomonas sp.: LK11 alleviates salinity stress in *Solanum pimpinellifolium*. Environ Exp Bot 133:58–69

Kidd P, Barcelo J, Bernal MP, Navari-Izzo F, Poschenrieder C, Shilev S, Clemente R, Monterroso C (2009) Trace element behavior at the root-soil interface: implications in phytoremediation. Environ Exp Bot 67:243–259

Kim SB, Timmusk S (2013) A simplified method for gene knockout and direct screening of recombinant clones for application in *Paenibacillus polymyxa*. PLoS One 8:e68092

Kim YC, Glick BR, Bashan Y, Ryu CM (2009) Enhancement of plant drought tolerance by microbes. In: Aroca R (ed) Plant responses to drought stress. Springer, Berlin, pp 383–412

Kim S, Lowman S, Hou G, Nowak J, Flinn B, Mei C (2012) Growth promotion and colonization of switchgrass (*Panicum virgatum*) cv. Alamo by bacterial endophyte *Burkholderia phytofirmans* strain PsJN. Biotechnol Bioguels 5:37

Kim YC, Glick B, Bashan Y, Ryu CM (2013) Enhancement of plant drought tolerance by microbes. In: Aroca R (ed) Plant responses to drought stress. Springer, Berlin

Kim JS, Lee J, Seo SG, Lee C, Woo SY, Kim SH (2015) Gene expression profile affected by volatiles of new plant growth promoting rhizobacteria, *Bacillus subtilis* strain JS, in tobacco. Genes Genom 37:387–397

Kloepper JW, Ryu CM, Zhang S (2004) Induced systemic resistance and promotion of plant growth by Bacillus spp. Phytopathology 94:1259–1266

Knief C (2014) Analysis of plant microbe interactions in the era of next generation sequencing technologies. Front Plant Sci 5:216

Knief C, Delmotte N, Chaffron S, Stark M, Innerebner G, Wassmann R, von Mering C, Vorholt JA (2012) Metaproteogenomic analysis of microbial communities in the phyllosphere and rhizosphere of rice. ISME J 6:1378–1390

Kohler J, Hernandez JA, Caravaca F, Rolden A (2008) Plant growth promoting rhizobacteria and arbuscular mycorrhizal fungi modify alleviation biochemical mechanisms in water stressed plants. Funct Plant Biol 35:141–151

Krey T, Vassilev N, Baum C, Eichler-Löbermann B (2013) Effects of long-term phosphorus application and plant-growth promoting rhizobacteria on maize phosphorus nutrition under field conditions. Eur J Soil Biol 55:124–130

Kukla M, Płociniczak T, Piotrowska-Seget Z (2014) Diversity of endophytic bacteria in *Lolium perenne* and their potential to degrade petroleum hydrocarbons and promote plant growth. Chemosphere 117:40–46

Kumar A (2016) Phosphate solubilizing bacteria in agriculture biotechnology: diversity, mechanism and their role in plant growth and crop yield. Int J Adv Res 4:116–124

Kumar A, Verma JP (2018) Does plant-microbe interaction confer stress tolerance in plants: a review? Microbiol Res 207:41–52

Kumar KV, Srivastava S, Singh N, Behl HM (2009) Role of metal resistant plant growth promoting bacteria in ameliorating fly ash to the growth of *Brassica juncea*. J Hazard Mater 170:51–57

Lakshmanan V (2015) Root microbiome assemblage is modulated by plant host factors. In: Bais H, Sherrier J (eds) Plant microbe interactions. Advances in botanical research, 75. Academic, Amsterdam, pp 57–79

Lakshmanan V, Kitto SL, Caplan JL, Hsueh YH, Kearns DB, Wu YS, Bais HP (2012) Microbe-associated molecular patterns-triggered root responses mediate beneficial rhizobacterial recruitment in Arabidopsis. Plant Physiol 160:1642–1661

Lakshmanan V, Selvaraj G, Bais HP (2014) Functional soil microbiome: below ground solution to an above ground problem. Plant Physiol 166:689–700

Lally RD, Galbally P, Moreira AS, Spink J, Ryan D, Germaine KJ, Dowling DN (2017) Application of endophytic *Pseudomonas fluorescens* and a bacterial consortium to *Brassica napus* can increase plant height and biomass undergreenhouse and field conditions. Front Plant Sci 8:2193

Landa BB, Mavrodi OV, Schroeder KL, Allende-Molar R, Weller DM (2006) Enrichmentand genotypic diversity of phlD-containing fluorescent Pseudomonas spp. in two soils after a century of wheat and flax monoculture. FEMS Microbiol Ecol 55:351–368

Lareen A, Burton F, Schäfer P (2016) Plant root-microbe communication in shaping root microbiomes. Plant Mol Biol 90:575–587

Larimer AL, Clay K, Becer JD (2014) Synergism and context dependency of interactions between arbuscular mycorrhizal fungi and rhizobia with a prairie legume. Ecology 95:1045–1054

Lau JA, Lennon JT (2012) Rapid responses of soil microorganisms improve plant fitness in novel environments. Proc Natl Acad Sci U S A 109:14058–14062

Lebeis SL (2014) The potential for give and take in plant-microbiome relationships. Front Plant Sci 5:287

Lebeis SL (2015) Greater than the sum of their parts: characterizing plant microbiomes at the community level. Curr Opin Plant Biol 24:82–86

Lebeis SL, Rott M, Dangl JL, Schulze-Lefert P (2012) Culturing a plant microbiome community at the cross-Rhodes. New Phytol 196:341–344

Lebeis SL, Paredes SH, Lundberg DS, Breakfield N, Gehring J, McDonald M, Malfatti S, Glavina del Rio T, Jones CD, Tringe SG, Dangl JL (2015) Salicylic acid modulates colonization of the root microbiome by specific bacterial taxa. Science 349:860–864

Lederberg I, McCray AT (2001) Ome sweet omics: a genealogical treasury of words. Scientist 15:8

Lee SW, Ahn PI, Sy S, Lee SY, Seo MW, Kim S, Sy P, Lee YH, Kang S (2010) *Pseudomonas* sp. LSW25R antagonistic to plant pathogens promoted plant growth and reduced blossom red rot of tomato roots in a hydroponic system. Eur J Plant Pathol 126:1–11

Leite MCBS, de Farias ARB, Freire FJ, Andreote FD, Sobral JK, Freire MBGS (2014) Isolation, bioprospecting and diversity of salt-tolerant bacteria associated with sugarcane in soils of Pernambuco, Brazil. Rev Bras Eng Agrıc Amb 18:S73–S79

Leveau JHJ, Uroz S, de Boer W (2010) The bacterial genus Collimonas: mycophagy, weathering and other adaptive solutions to life in oligotrophic soil environments. Environ Microbiol 12:281–292

Li D, Voigt TB, Kent AD (2016) Plant and soil effects on bacterial communities associated with miscanthus X giganteus rhizosphere and rhizomes. GCB Bioenergy 8:183–193

Lidbury ID, Murphy ARJ, Scanlan DJ, Bending GD, Jones AME, Moore JD, Goodall A, Hammond JP, Wellington EM (2016) Comparative genomic, proteomic and exoproteomic analyses of three Pseudomonas strains reveals novel insights into the phosphorus scavenging capabilities of soil bacteria. Environ Microbiol 18:3535–3549

Lim JH, Kim SD (2013) Induction of drought stress resistance by multifunctional PGPR *Bacillus licheniformis* K11 in pepper. Plant Pathol J 29:201–208

López-Baena FJ, Monreal JA, Pérez-Montaño F, Guash-Vidal B, Bellogín RA, Vinardell JM, Ollero FJ (2009) The absence of Nops secretion in *Sinorhizobium fredii* HH103 increases *GmPR1* expression in William soybean. MPMI 22:1445–1454

Lugtenberg B, Kamilova F (2009) Plant-growth-promoting rhizobacteria. Annu Rev Microbiol 63:541–556

Lundberg DS, Lebeis SL, Paredes SH, Yourstone S, Gehring J, Malfatti S, Tremblay J, Engelbrektson A, Kunin V, del Rio TG, Edgar RC, Eickhorst T, Ley RE, Hugenholtz P, Tringe SG, Dangl JL (2012) Defining the core *Arabidopsis thaliana* root microbiome. Nature 488:86–90

Luo SL, Chen L, Chen JI, Xiao X, Xu TY, Wan Y, Rao C, Liu CB, Liu YT, Lai C, Zeng GM (2011) Analysis and characterization of cultivable heavy metal-resistant bacterial endophytes isolated from Cd-hyperaccumulator *Solanum nigrum* L. and their potential use for phytoremediation. Chemosphere 85:1130–1138

Luo S, Xu T, Chen L, Chen J, Rao C, Xiao X, Wan Y, Zeng G, Long F, Liu C, Liu Y (2012) Endophyte-assisted promotion of biomass production and metal-uptake of energy crop sweet sorghum by plant-growth-promoting endophyte *Bacillus* sp. SLS18. Appl Microbiol Biotechnol 93:1745–1753

Luvizotto DM, Marcon J, Andreote FD, Dini-Andreote F, Neves AAC, Araújo WL, Pizzirani-Kleiner AA (2010) Genetic diversity and plant-growth related features of *Burkholderia* spp. from sugarcane roots. World J Microbiol Biotechnol 26:1829–1836

Ma Y, Prasad MNV, Rajkumar M, Freitas H (2011) Plant growth promoting rhizobacteria and endophytes accelerate phytoremediation of metalliferous soils. Biotechnol Adv 29:248–258

Ma Y, Rajkumar M, Zhang C, Freitas H (2016a) Inoculation of *Brassica oxyrrhina* with plant growth promoting bacteria for the improvement of heavy metal phytoremediation under drought conditions. J Hazard Mater 320:36–44

Ma Y, Rajkumar M, Zhang C, Freitas H (2016b) The beneficial role of bacterial endophytes in heavy metal phytoremediation. J Environ Manag 174:14–25

Mahmood S, Daur I, Al-Solaimani SG, Ahmad S, Madkour MH, Yasir M, Hirt H, Ali S, Ali Z (2016) Plant growth promoting rhizobacteria and silicon synergistically enhance salinity tolerance of mung bean. Front Plant Sci 7:876

Mani D, Kumar C, Patel NK (2016) Integrated micro-biochemical approach for phytoremediation of cadmium and lead contaminated soils using *Gladiolus grandiflorus* L. cut flower. Ecotoxicol Environ Saf 124:435–446

Mapelli F, Marasco R, Rolli E, Cappitelli F, Daffonchio D, Borin S (2012) Mineral-microbe interactions: biotechnological potential of bio-weathering. J Biotechnol 157:473–481

Mapelli F, Marasco R, Rolli E, Barbato M, Cherif H, Guesmi A, Ouzari I, Daffonchio D, Borin S (2013) Potential for plant growth promotion of rhizobacteria associated with Salicornia growing in Tunisian hypersaline soils. Biomed Res 2013:248078

Marasco R, Rolli E, Attoumi B, Vigani G, Mapelli F, Borin S, Daffonchio D (2012) A drought resistance-promoting microbiome is selected by root system under desert farming. PLoS One 7:e48479

Marasco R, Rolli E, Vigani G, Borin S, Sorlini C, Ouzari H, Zocchi G, Daffonchio D (2013) Are drought-resistance promoting bacteria cross-compatible with different plant models? Plant Signal Behav 8:e26741

Marasco R, Mapelli F, Rolli E, Mosqueira MJ, Fusi M, Bariselli P, Reddy M, Cherif A, Tsiamis G, Borin S, Daffonchio D (2016) *Salicornia strobilacea* (synonum of *Halocnemum strobilaceum*) growth under different tidal regimes selects rhizosphere bacteria capable of promoting plant growth. Front Microbiol 7:1286

Marquez LM, Redman RS, Rodriguez RJ, Roosinck MJ (2007) A virus in a fungus in a plant: three-way symbiosis required for thermal tolerance. Science 315:513–515

Marschner P, Crowley D, Rengel Z (2011) Rhizosphere interactions between microorganisms and plants govern iron and phosphorus acquisition along the root axis–model and research methods. Soil Biol Biochem 43:883–894

Martinez-Absalon S, Rojas-Solis D, Hernandez-Leon R, Prieto-Barajas C, Orozco-Mosqueda MC, Pena-Cabriales JJ, Sakuda S, Valencia-Cantero E, Santoyo G (2014) Potential use and mode of action of the new strain *Bacillus thuringiensis* UM96 for the biological control of the grey-mould phytopathogen *Botrytis cinerea*. Biocontrol Sci Tech 24:1349–1362

Martins SJ, Vasconcelos de Medeiros FH, Magela de Souza R, Vilela de Resende ML, Martins Ribeiro Junior P (2013) Biological control of bacterial wilt of common bean by plant growth-promoting Rhizobacteria. Biol Control 66:65–71

Marulanda A, Barea JM, Azcón R (2009) Stimulation of plant growth and drought tolerance by native microorganisms (AM fungi and bacteria) from dry environments: mechanisms related to bacterial effectiveness. J Plant Growth Regul 28:115–124

Masciarelli O, Llanes A, Luna V (2014) A new PGPR co-inoculated with *Bradyrhizobium japonicum* enhances soybean nodulation. Microbiol Res 169:609–615

Mastretta C, Taghavi S, Van Der Lelie D, Mengoni A, Galardi F, Gonnelli C, Barac T, Boulet J, Weyens N, Vangronsveld J (2009) Endophytic bacteria from seeds of *Nicotiana tabacum* can reduce cadmium phytotoxicity. Int J Phytoremediation 11:251–267

Mavrodi OV, Walte N, Elateek S, Taylor CG, Okubara PA (2012) Suppression of Rhizocto-nia and Pythium root rot of wheat by new strains of Pseudomonas. Biol Control 62:93–102

Mayak S, Tirosh T, Glick BR (2004) Plant growth promoting bacteria that confer resistance to water stress in tomato and pepper. Plant Sci 166:525–530

McLellan CA, Turbyville TJ, Wijerante EMK, Kerschen EV, Queitsch C, Whitesell L, Gunatilaka AAL (2007) A rhizosphere fungus enhances Arabidopsis thermtolerance through production of an HSP90 inhibitor. Plant Physiol 145:174–182

Meena RK, Singh RK, Singh NP, Meena SK, Meena VS (2015) Isolation of low temperature surviving plant growth–promoting rhizobacteria (PGPR) from pea (*Pisum sativum* L.) and documentation of their plant growth promoting traits. Biocatal Agric Biotechnol 4:806–811

Meena MK, Gupta S, Datta S (2016) Antifungal potential of PGPR, their growth promoting activity on seed germination and seedling growth of winter wheat and genetic variability among bacterial isolates. Int J Curr Microbial Appl Sci 5:235–243

Mendes R, Raaijmakers JM (2015) Cross-kingdom similarities in microbiome functions. ISME J 9:1905–1907

Mendes R, Kruijt M, de Bruijn I, Dekkers E, van der Voort M, Schneider JH, Piceno YM, DeSantis TZ, Andersen GL, Bakker PA, Raaijmakers JM (2011) Deciphering the rhizosphere microbiome for disease-suppressive bacteria. Science 332:1097–1100

Mendes R, Garbeva P, Raajimakers JM (2013) The rhizosphere microbiome: significance of plant beneficial, plant pathogenic and human pathogenic microorganisms. FEMS Microbial Rev 37:634–663

Mengual C, Schoebitz M, Azcón R, Roldán A (2014) Microbial inoculants and organic amendment improves plant establishment and soil rehabilitation under semiarid conditions. J Environ Manag 134:1–7

Miao B, Stewart BA, Zhang F (2011) Long-term experiments for sustainable nutrient management in China: a review. Agron Sustain Dev 31:397–414

Micallef SA, Shiaris MP, Colón-Carmona A (2009) Influence of *Arabidopsis thaliana* accessions on rhizobacterial communities and natural variation in root exudates. J Exp Bot 60:1729–1742
Miller JB, Oldroyd GD (2012) The role of diffusible signals in the establishment of rhizobial and mycorrhizal symbioses. In: Perotto S, Baluška F (eds) Signaling and communication in plant symbiosis. Springer, Berlin, pp 1–30
Miransari M (2011) Hyperaccumulators, arbuscular mycorrhizal fungi and stress of heavy metals. Biotechnol Adv 29:645–653
Mishra PK, Mishra S, Selvakumar G, Kundub S, Gupta HS (2009) Enhanced soybean (*Glycine max* L) plant growth and nodulation by *Bradyrhizobium japonicum*-SB1 in presence of *Bacillus thuringiensis*-KR1. Acta Agric Scand B Soil Plant Sci 59:189–196
Mishra PK, Bisht S, Mishra S, Selvakumar G, Bisht J, Gupta HS (2012) Co-inoculation of rhizobium leguminosarum-PR1 with a cold tolerant Pseudomonas sp. improves iron acquisition, nutrient uptake and growth of fieldpea (*Pisum sativum* L.). J Plant Nutr 35:243–256
Mitter B, Pfaffenbichler N, Flavell R, Compant S, Antonielli L, Petric A, Berninger T, Naveed M, Sheibani-Tezerji R, von Maltzahn G, Sessitsch A (2017) A new approach to modify plant microbiomes and traits by introducing beneficial bacteria at flowering into progeny seeds. Front Microbiol 8:11
Moe LA (2013) Amino acids in the rhizosphere: from plants to microbes. Am J Bot 100:1692–1705
Morel M, Castro-Sowinski S (2013) The complex molecular signaling network in microbe-plant interaction. In: Arora NK (ed) Plant microbe symbiosis: fundamentals and advances. Springer, NewDelhi, pp 169–199
Mueller UG, Sachs JL (2015) Engineering microbiome to improve plant and animal health. Trends Microbiol 23:606–617
Munns R, Gilliham M (2015) Salinity tolerance of crops–what is the cost? New Phytol 208:668–673
Murray JD (2011) Invasion by invitation: rhizobial infection in legumes. Mol Plant-Microbe Interact 24:631–939
Mus F, Crook MB, Garcia K, Costas AG, Geddes BA, Kouri ED, Paramasivan P, Ryu MH, Oldroyd GED, Poole PS, Udvardi MK, Voigt CA, Ané JM, Peters JW (2016) Symbiotic nitrogen fixation and the challenges to its extension to nonlegumes. Appl Environ Microbiol 82:3698–3710
Nadeem SM, Ahmad M, Zahir ZA, Javaid A, Ashraf M (2014) The role of mycorrhizae and plant growth promoting rhizobacteria (PGPR) in improving crop productivity under stressful environments. Biotechnol Adv 32:429–448
Naseem H, Bano A (2014) Role of plant growth-promoting rhizobacteria and their exopolysaccharide in drought tolerance in maize. J Plant Interact 9:689–701
Nautiyal CS, Srivastava S, Chauhan PS, Seem K, Mishra R, Sopory SK (2013) Plant growth-promoting bacteria *Bacillus amyloliquefaciens* NBRISN13 modulates gene expression profile of leaf and rhizosphere community in rice during salt stress. Plant Physiol Biochem 66:1–9
Navarro JM, Pérez-Tornero O, Morte A (2014) Alleviation of salt stress in citrus seedlings inoculated with arbuscular mycorrhizal fungi depends on the rootstock salt tolerance. J Plant Physiol 171:76–85
Naveed M, Hussain B, Zahir A, Mitter B, Sessitsch A (2014a) Drought stress amelioration in wheat through inoculation with Burkholderia phytofirmans strain PsJN. Plant Growth Regul 73:121–131
Naveed M, Mitter B, Reichenauer TG, Wieczorek K, Sessitsch A (2014b) Increased drought stress resilience of maize through endophytic colonization by *Burkholderia phytofirmans* PsJN and *enterobacter* sp. FD17. Environ Exp Bot 97:30–39
Neeraj KS (2011) Organic amendments to soil inoculated arbuscular mycorrhizal fungi and *Pseudomonas fluorescens* treatments reduce the development of root-rot disease and enhance the yield of *Phaseolus vulgaris* L. Eur J Soil Biol 47:288–295
Nelson EB (2018) The seed microbiome: origins, interactions, and impacts. Plant Soil 422:7–34
Ngumbi EN (2011) Mechanisms of olfaction in parasitic wasps: analytical and behavioral studies of response of a specialist (Microplitis croceipes) and a generalist (Cotesia marginiventris) parasitoid to host-related odor. Ph.D. dissertation, Auburn University, Auburn

Ngumbi E, Kloepper J (2016) Bacterial-mediated drought tolerance: current and future prospects. Appl Soil Ecol 105:109–125

Nonnoi F, Chinnaswamy A, García de la Torre VS, Coba de la Peña T, Lucas MM, Pueyo JJ (2012) Metal tolerance of rhizobial strains isolated from nodules of herbaceous legumes *Medicago* sp. and *Trifolium* sp. growing in mercury-contaminated soils. Appl Soil Ecol 61:49–59

Oburger E, Dell'Mour M, Hann S, Wieshammer G, Puschenreiter M, Wenzel WW (2013) Evaluation of an ovel tool for sampling root exudates from soil-grown plants compared to conventional techniques. Environ Exp Bot 87:235–247

Ofaim S, Ofek-Lalzar M, Sela N, Jinag J, Kashi Y, Minz D, Freilich F (2017) Analysis of microbial functions in the rhizosphere using a metabolic-network based framework for metagenomics interpretation. Front Microbiol 8:1606

Ofek M, Voronov-Goldman M, Hadar Y, Minz D (2014) Host signature effect on plant root-associated microbiomes revealed through analyses of resident *vs* active communities. Environ Microbiol 16:2157–2167

Oldroyd GED (2013) Speak, friend, and enter: signaling systems that promote beneficial symbiotic associations in plants. Nat Rev Microbiol 11:252–263

Oldroyd GED, Murray JD, Poole PS, Downie JA (2011) The rules of engagement in the legume-rhizobial symbiosis. Annu Rev Genet 45:119–144

Olivares J, Bedmar EJ, Sanjuan J (2013) Biological nitrogen fixation in the context of global change. Mol Plant-Microbe Interact 26:486–494

Orozco-Mosqueda MDC, Rocha-Granados MDC, Glick BR, Santoyo G (2018) Microbiome engineering to improve biocontrol and plant growth-promoting mechanisms. Microbiol Res 208:25–31

Ortiz N, Armada E, Duque E, Roldan A, Azcon R (2015) The contribution of arbuscular mycorrhizal fungi and/or bacteria to enhancing plant drought tolerance under natural soil conditions: effectiveness of autochthonous or allochthonous strains. J Plant Physiol 174:87–96

Ortiz-Castro R, Martinez-Trujillo M, Lopez-Bucio J (2008) N-acyl-L- homoserinelactones:a class of bacterial quorum-sensing signals alter post-embryonic root development in *Arabidopsis thaliana*. Plant Cell Environ 31:1497–1509

Pandey VC (2012) Phytoremediation of heavy metals from fly ash pond by *Azolla caroliniana*. Ecotoxicol Environ Saf 82:8–12

Pandey VC, Ansari MW, Tula S, Yadav S, Sahoo RK, Shukla N, Bains G, Badal S, Chandra S, Gaur AK, Kumar A, Shukla A, Kumar J, Tuteja N (2016) Dose dependent response of *Trichoderma harzianum* in improving drought tolerance in rice genotypes. Planta 243:1251–1264

Pankaj U, Verma SK, Semwal M, Verma RK (2016) Assessment of natural mycorrhizal colonization and soil fertility status of lemongrass crop in subtropical India. J Appl Res Med Arom Plants 5:41–46

Panke-Buisse K, Poole A, Goodrich J, Ley R, Kao-Kniffin J (2015) Selection on soil microbiomes reveals reproducible impacts on plant function. ISME J 9:980–989

Paredes SH, Lebeis SL (2016) Giving back to the community: microbial mechanisms of plant-soil interactions. Funct Ecol 30:1–10

Park JD (2010) Heavy metal poisoning. Hanyang Med Rev 30:319–325

Passari AK, Chandra P, Mishra VK, Leo VV, Gupta VK, Kumar B, Singh BP (2016) Detection of biosynthetic gene and phytohormone production by endophytic actinobacteria associated with *Solanum lycopersicum* and their plant growth-promoting effect. Res Microbiol 167:692–705

Patel U, Sinha S (2011) Rhizobia species: a boon for "plant genetic engineering". Indian J Microbiol 51:521–527

Paul D, Lade H (2014) Plant-growth-promoting rhizobacteria to improve crop growth in saline soils: a review. Agron Sustain Dev 34:737–752

Peiffer JA, Spor A, Koren O, Jin Z, Tringe SG, Dangl JL, Buckler ES, Ley RE (2013) Diversity and heritability of the maize rhizosphere microbiome under field conditions. Proc Natl Acad Sci U S A 110:6548–6553

Penuelas J, Terradas J (2014) The foliar microbiome. Trends Plant Sci 19:278–280

Pereira P, Ibáñez SG, Agostini E, Miriam Etcheverry M (2011) Effects of maize inoculation with *Fusarium verticillioides* and with two bacterial biocontrol agents on seedlings growth and antioxidative enzymatic activities. Appl Soil Ecol 51:52–59

Perez YM, Charest C, Dalpe Y, Seguin S, Wang X, Khanizadeh S (2016) Effect of inoculation with arbuscular mycorrhizal fungi on selected spring wheat lines. Sustain Agric Res 5:24–29

Perez-Montano F, Alías-Villegas C, Bellogin RA, del Cerro P, Espuny MR, Jimenez-Guerrero I, Lopez-Baena FJ, Ollero FJ, Cubo T (2014) Plant growth promotion in cereal and leguminous agricultural important plants: from microorganism capacities to crop production. Microbiol Res 169:325–336

Philippot L, Raaijmakers JM, Lemanceau P, van der Putten WH (2013) Going back to the roots: the microbial ecology of the rhizosphere. Nat Rev Microbiol 11:789–799

Pieterse CMJ, Van Wees SCM, Ton J, Van Pelt JA, Van Loon LC (2002) Signalling in rhizobacteria-induced systemic resistance in *Arabidopsis thaliana*. Plant Biol 4:535–544

Pineda A, Dicke M, Pieterse CMJ, Pozo MJ (2013) Beneficial microbes in a changing environment: are they always helping plants deal with insects? Funct Ecol 27:574–586

Pontigo S, Godoy K, Jiménez H, Gutierrez-Moraga A, Mora MDLL, Cartes P (2017) Silicon-mediated alleviation of aluminum toxicity by modulation of Al/Si uptake and antioxidant performance in ryegrass plants. Front Plant Sci 8:642

Porcel R, Zamarreno AM, Garcia-Mina JM, Aroca R (2014) Involvement of plant endogenous ABA in *Bacillus megaterium* PGPR activity in tomato plants. BMC Plant Biol 14:36

Porras-Alfaro A, Bayman P (2011) Hidden fungi, emergent properties: endophytes and microbiomes. Annu Rev Phytopathol 49:291–315

Pozo MJ, Azcon-Aguilar C (2007) Unraveling mycorrhiza-induced resistance. Curr Opin Plant Biol 10:393–398

Prashar P, Kapoor N, Sachdeva S (2014) Rhizosphere: its structure, bacterial diversity and significance. Rev Environ Sci Biotechnol 13:63–77

Prathap M, Ranjitha KBD (2015) A critical review on plant growth-promoting rhizobacteria. J Plant Pathol Microbiol 6:1–4

Premachandra D, Hudek L, Brau L (2016) Bacterial modes of action for enhancing plant growth. J Biotechnol Biomater 6:3

Prosser JI (2015) Dispersing misconceptions and identifying opportunities for the use of “omics” in soil microbial ecology. Nat Rev Microbiol 13:439–446

Prudent M, Salon C, Souleimanov A, Emery RJN, Smith DI (2015) Soybean is less impacted by water stress using *Bradyrhizobium japonicum* and thuricin-17 from *Bacillus thuringiensis*. Agron Sustain Dev 35:749–757

Qiang SW, Ming QC, Ying NZ, Chu W, Xin HH (2016) Mycorrhizal colonization represents functional equilibrium on root morphology and carbon distribution of trifoliate orange grown in a split-root system. Sci Hortic 199:95–102

Qin S, Zhang YJ, Yuan B, Xu PY, Xing K, Wang J, Jiang JH (2014) Isolation of ACC deaminase-producing habitat-adapted symbiotic bacteria associated with halophyte *Limonium sinense* (Girard) Kuntze and evaluating their plant growth promoting activity under salt stress. Plant Soil 374:753–766

Qin YY, Leung CKM, Lin CK, Wong MH (2015) The associations between metals/metalloids concentrations in blood plasma of Hong Kong residents and their seafood diet, smoking habit, body mass index and age. Environ. Sci Pollut Res 22:13204–13211

Qiu M, Li S, Zhou X, Cui X, Vivanco J, Zhang N, Shen Q, Zhang R (2014) De-coupling of root-microbiome associations followed by antagonist inoculation improves rhizosphere soil suppressiveness. Biol Fertil Soils 50:217–224

Quecine MC, Araujo WL, Tsui S, Parra JRP, Azevedo JL, Pizzirani-Kleiner AA (2014) Control of *Diatraea saccharalis* by the endophytic *Pantoea agglomerans* 33.1 expressing *cry1Ac7*. Arch Microbiol 196:227–234

Quiza L, St-Arnaud M, Yergeau E (2015) Harnessing phytomicrobiome signaling for rhizosphere microbiome engineering. Front Plant Sci 6:507

Raaijmakers JM, Leeman M, MMP VO, Van der Sluis I, Schippers B, PAHM B (1995) Dose-response relationships in biological control of fusarium wilt of radish by *Pseudomonas* spp. Phytopathology 85:1075–1081

Raajimakers JM (2015) The minimal rhizosphere microbiome. In: Lugtenberg B (ed) Principles of plant-microbe interactions. Springer International Publishing Switzerland, Heidelberg, pp 411–417

Rajkumar M, Ae N, Prasad MNV, Freitas H (2010) Potential of siderophore-producing bacteria for improving heavy metal phytoextraction. Trends Biotechnol 28:142–149

Rajkumar M, Sandhya S, Prasad MNV, Freitas H (2012) Perspectives of plant-associated microbes in heavy metal phytoremediation. Biotechnol Adv 30:1562–1574

Ramadan EM, AbdelHafez AA, Hassan EA, Saber FM (2016) Plant growth-promoting rhizobacteria and their potential for biocontrol of phytopathogens. Afr J Microbiol Res 10:486–504

Ramegowda V, Senthil-Kumar M, Ishiga Y, Kaundal A, Udayakumay M, Mysore KS (2013) Drought stress acclimation imparts tolerance to *Sclerotinia sclerotiorum* and *Pseudomonas syringae* in *Nicotiana benthamiana*. Int J Mol Sci 14:9497–9513

Ramond JB, Tshabuse F, Bopda CW, Cowan DA, Tuffin MI (2013) Evidence of variability in the structure and recruitment of rhizospheric and endophytic bacterial communities associated with arable sweet sorghum (*Sorghum bicolor* (L.) Moench). Plant Soil 372:265–278

Ramos Solano B, Barriuso Maicas J, Pereyra de la Iglesia MT, Domenech J, GutiérrezManero FJ (2008) Systemic disease protection elicited by plant growth promoting rhizobacteria strains: relationship between metabolic responses, systemic disease protection, and biotic elicitors. Phytopathology 98:451–457

Rasche F, Velvis H, Zachow C, Berg G, van Elsas JD, Sessitsch A (2006) Impact of transgenic potatoes expressing antibacterial agents on bacterial endophytes is comparable with the effects of plant genotype, soil type and pathogen infection. J Appl Ecol 43:555–566

Rashedul IM, Madhaiyan M, Deka Boruah HP, Yim W, Lee G, Saravanan VS, Fu Q, Hu H, Sa T (2009) Characterization of plant growth-promoting traits of free-living diazotrophic bacteria and their inoculation effects on growth and nitrogen uptake of crop plants. Microbiol Biotechnol 19:1213–1222

Rastogi G, Coaker GL, Leaveu JHJ (2013) New insight into the structure and function of phyllosphere microbiota through high-throughput molecular approaches. FEMS Microbiol Lett 348:1–10

Raza W, Ling N, Yang L, Huang Q, Shen Q (2016a) Response of tomato wilt pathogen *Ralstonia solanacearum* to the volatile organic compounds produced by a biocontrol strain *Bacillus amyloliquefaciens* SQR-9. Sci Rep 6:24856

Raza W, Yousaf S, Rajer FU (2016b) Plant growth promoting activity of volatile organic compounds produced by bio-control strains. Sci Lett 4:40–43

Redman RS, Sheehan KB, Stout RG, Rodriguez RJ, Henson JM (2002) Thermotolerance generated by plant/fungal symbiosis. Science 298:1581

Reyes-Darias JA, García V, Rico-Jiménez M, Corral-Lugo A, Lesouhaitier O, Juárez-Hernández D, Yang Y, Bi S, Feuilloley M, Muñoz-Rojas J, Sourjik V, Krell T (2015) Specific gamma-aminobutyrate chemotaxis in pseudomonads with different lifestyle. Mol Microbiol 97:488–501

Richardson AE, Barea JM, McNeill AM, Prigent-Combaret C (2009) Acquisition of phosphorus and nitrogen in the rhizosphere and plant growth promotion by microorganisms. Plant Soil 321:305–339

Riggs PJ, Chelius MK, Iniguez AL, Kaeppler SM, Triplett EW (2001) Enhanced maize productivity by inoculation with diazotrophic bacteria. Aus J Plant Physiol 28:829–836

Rodell M, Velicogna I, Famiglietti JS (2009) Satellite-based estimates of groundwater depletion in India. Nature 460:999–1002

Rodriguez H, Redman RS (2008) Effect of a nickel-tolerant ACC deaminase-producing pseudomonas strain on growth of nontransformed and transgenic canola plants. Curr Microbiol 57:170–174

Rodriguez H, Vessely S, Shah S, Glick BR (2008) Effect of a nickel-tolerant ACC deaminase-producing pseudomonas strain on growth of nontransformed and transgenic canola plants. Curr Microbiol 57:170–174

Rodriguez RJ, White JF Jr, Arnold AE, Redman RS (2009) Fungal endophytes: diversity and functional roles. New Phytol 182:314–330

Rolli E, Marasco R, Vigani C, Ettoumi B, Mapelli F, Deangelis MI, Gandolfi C, Casati E, Previtali F, Gerbino R, Pierotti Cei F, Borin S, Sorlini C, Zocchi G, Daffonchio D (2015) Improved plant resistance to drought is promoted by the root-promoted microbiome as a water stress dependent trait. Environ Microbiol 17:316–331

Rosenblueth M, Martinez-Romero E (2006) Bacterial endophytes and their interactions with hosts. Soil Biol Biochem 21:373–378

Rout ME, Callaway RM (2012) Interactions between exotic invasive plants and soil microbes in the rhizosphere suggest that everything is not everywhere. Ann Bot 110:213–222

Roy SJ, Negrao S, Tester M (2014) Salt resistant crop plants. Curr Opin Biotechnol 26:115–124

Ruiz KB, Biondi S, Martínez EA, Orsini F, Antognoni F, Jacobsen SE (2015) Quinoa – a model crop for understanding salt-tolerance mechanisms in halophytes. Plant Biosyst 150:357–371

Ruiz-Lozano JM, Peralvarez MC, Aroca R, Azcon R (2011) The application of a treated sugar beet waste residue to soil modifies the responses of mycorrhizal and non mycorrhizal lettuce plants to drought stress. Plant Soil 346:153–166

Ruiz-Lozano JM, Aroca R, Zamarreno AM, Molina S, Andreo-Jimenez B, Porcel R, Garcia-Mina JM, Ruyter-Spira C, Lopez-Raez JA (2016) Arbuscular mycorrhizal symbiosis induces strigolactone biosynthesis under drought and improves drought tolerance in lettuce and tomato. Plant Cell Environ 39:441–452

Ruy CM, Murphy JF, Mysore KS, Kloepper JW (2004) Plant growth-promoting rhizobacteria systemically protect *Arabidopsis thaliana* against cucumber mosaic virus by a salicylic acid and NPR1-independent and jasmonic acid-dependent signaling pathway. Plant J 39:381–392

Salvioli A, Bonfante P (2013) Systems biology and "omics" tools: a cooperation for next-generation mycorrhizal studies. Plant Sci 203:107–114

Salvioli A, Ghignone S, Novero M, Navazio L, Venice F, Bagnaresi P, Bonfante P (2016) Symbiosis with an endobacterium increases the fitness of a mycorrhizal fungus, raising its bioenergetic potential. ISME J 10:130–144

Sánchez-Cañizares C, Jorrín B, Poole PS, Tkacz A (2017) Understanding the holobiont: the interdependence of plants and their microbiome. Curr Opin Microbiol 38:188–196

Sandhya V, Ali SZ, Grover M, Kishore N, Venkateswarlu B (2009) Pseudomonas sp. strain P45 protects sunflowers seedlings from drought stress through improved soil structure. J Oilseed Res 26:600–601

Sandhya V, Ali SZ, Grover M, Kishore N, Venkateswarlu B (2010) Pseudomonas sp. strain P45 protects sunflowers seedlings from drought stress through improved soil structure. J Oilseed Res 26:600–601

Santi C, Bogusz D, Franche C (2013) Biological nitrogen fixation in non-legume plants. Ann Bot 111:743–767

Santoro MV, Bogino PC, Nocelli N, Cappellari LR, Giordano WF, Banchio E (2016) Analysis of plant growth promoting effects of fluorescent pseudomonas strains isolated from *Mentha piperita* rhizosphere and effects of their volatile organic compounds on essential oil composition. Front Microbiol 7:1–17

Santoyo G, Orozco-Mosqueda MC, Govindappa M (2012) Mechanisms of bio-control and plant growth-promoting activity in soil bacterial species of Bacillus and Pseudomonas: a review. Biocontrol Sci Tech 22:855–872

Santoyo G, Moreno-Hagelsieb G, Orozco-Mosqueda MC, Glick BR (2016) Plant growth-promoting bacterial endophytes. Microbiol Res 183:92–99

Santoyo G, Hernandez-Pacheco C, Hernandez-Salmeron J, Hernandez-Leon R (2017) The role of abiotic factors modulating the plant-microbe-soil interactions: towards sustainable agriculture, a review. Span J Agric Res 15:e03R01

Sarma R, Saikia R (2014) Alleviation of drought stress in mung bean by strain *Pseudomonas aeruginosa* GGRJ21. Plant Soil 377:111–126

Savary S, Ficke A, Aubertot JN, Hollier C (2012) Crop losses due to disease and their implications for global food production losses and food security. Food Secur 4:519–537

Schaeppi K, Bulgarelli D (2015) The plant microbiome at work. Mol Plant-Microbe Interact 28:212–217

Schenk PM, Carvalhais LC, Kazan K (2012a) Unraveling plant-microbe interactions: can multispecies transcriptomics help. Trends Biotechnol 30:177–184

Schenk ST, Stein E, Kogel KH, Schikora A (2012b) Arabidopsis growth and defense are modulated by bacterial quorum sensing molecules. Plant Signal Behav 7:178–181

Schlaeppi K, van Themaat EVL, Bulgarelli D, Schulze-Lefert P (2013) *Arabidopsis thaliana* as model for studies on the bacterial root microbiota. In: de Bruijin FJ (ed) Molecular microbial ecology of the rhizosphere. Kluwer Academic, Dordrecht, pp 243–256

Schlaeppi K, Dombrowski N, Oter RG, Ver Loren van Themaat E, Schulze-Lefert P (2014) Quantitative divergence of the bacterial root microbiota in *Arabidopsis thaliana* relatives. Proc Natl Acad Sci U S A 111:585–592

Schuhegger R, Ihring A, Gantner S, Bahnweg G, Knappe C, Vogg G, Hutzler P, Schmid M, Van Breusegem F, Eberl L, Hartmann A, Langebartels C (2006) Induction of systemic resistance in tomato by N-acyl-L-homoserine lactone-producing rhizosphere bacteria. Plant Cell Environ 29:909–918

Selvakumar G, Kundu S, Joshi P, Nazim S, Gupta AD, Mishra PK, Gupta HS (2008a) Characterization of a cold-tolerant plant growth-promoting bacterium *Pantoea dispersa* 1A isolated from a sub-alpine soil in the North Western Indian Himalayas. World J Microbiol Biotechnol 24:955–960

Selvakumar G, Mohan M, Kundu S, Gupta AD, Joshi P, Nazim S, Gupta HS (2008b) Cold tolerance and plant growth promotion potential of *Serratia marcescens* strain SRM (MTCC 8708) isolated from flowers of summer squash (*Cucurbita pepo*). Lett Appl Microbiol 46:171–175

Selvakumar G, Joshi P, Nazim S, Mishra PK, Bisht JK, Gupta HS (2009) Phosphate solubilization and growth promotion by *Pseudomonas fragi* CS11RH1 (MTCC 8984) a psychrotolerant bacterium isolated from a high altitude Himalayan rhizosphere. Biologia 64:239245

Selvakumar G, Joshi P, Suyal P, Mishra PK, Joshi GK, Bisht JK, Bhatt JC, Gupta HS (2010a) *Pseudomonas lurida* M2RH3 (MTCC 9245), a psychrotolerant bacterium from the Uttarakhand Himalayas, solubilizes phosphate and promotes wheat seedling growth. World J Microbiol Biotechnol 5:1129–1135

Selvakumar G, Kundu S, Joshi P, Nazim S, Gupta AD, Gupta HS (2010b) Growth promotion of wheat seedlings by *Exiguobacterium acetylicum* 1P (MTCC 8707) a cold tolerant bacterial strain from the Uttarakhand Himalayas. Ind J Microbiol 50:50–56

Senthilkumar M, Swarnalakshmi K, Govindasamy V, Young KL, Annapurna K (2009) Bio-control potential of soybean bacterial endophytes against charcoal rot fungus, *Rhizoctonia bataticola*. Curr Microbiol 58:288–293

Sessitsch A, Mitter B (2014) 21st century agriculture: integration of plant microbiome for improved crop production and food safety. Microb Biotechnol 8:32–33

Sessitsch A, Hardoim P, Döring J, Weilharter A, Krause A, Woyke T, Mitter B, Hauberg-Lotte L, Friedrich F, Rahalkar M, Hurek T, Sarkar A, Bodrossy L, van Overbeek L, Brar D, van Elsas JD, Reinhold-Hurek B (2011) Functional characteristics of an endophyte community colonizing rice roots as revealed by metagenomic analysis. Mol Plant-Microbe Interact 25:28–36

Shabala S, Hariadi Y, Jacobsen SE (2013) Genotypic difference in salinity tolerance in quinoa is determined by differential control of xylem Na^+ loading and stomatal density. J Plant Physiol 170:906–914

Shaharoona B, Arshad M, Zahir ZA (2006) Effect of plant growth promoting rhizobacteriacontaining ACC-deaminase on maize (*Zea mays* L.) growth under axenic conditions and on nodulation in mung bean (*Vigna radiata* L.). Lett Appl Microbiol 42:155–159

Shahzad R, Khan AL, Bilal S, Waqas M, Kang SM, Lee IJ (2017) Inoculation of abscisic acid-producing endophytic bacteria enhances salinity stress tolerance in *Oryza sativa*. Environ Exp Bot 136:68–77
Sharifi R, Ryu CM (2016) Are bacterial volatile compounds poisonous odors to a fungal pathogen *Botrytis cinerea*, alarm signals to Arabidopsis seedlings for eliciting induced resistance, or both? Front Microbiol 7:1–10
Sharma M, Ghosh R (2017) Heat and soil moisture stress differently impact chickpea plant infection with fungal pathogens. In: Senthil-Kumar M (ed) Plant tolerance to individual and concurrent stresses. Springer Press, New Delhi, pp 47–57
Sharma S, Kulkarni J, Jha B (2016) Halotolerant rhizobacteria promote growth and enhance salinity tolerance in peanut. Front Microbiol 7:1600
Sheng X, He L, Wang Q, Ye H, Jiang C (2008) Effects of inoculation of biosurfactant-producing *Bacillus* sp. J119 on plant growth and cadmium uptake in a cadmium-amended soil. J Hazard Mater 155:17–22
Sheng M, Tang M, Zhang F, Huang Y (2011) Influence of arbuscular mycorrhiza on organic solutesin maize leaves under salt stress. Mycorrhiza 21:423–430
Shirley M, Avoscan L, Bernuad E, Vansuyt G, Lemanceau P (2011) Comparison of iron acquisition from Fe-pyoverdine by strategy I and strategy II plants. Botany 89:731–735
Shrestha J (2016) A review on sustainable agricultural intensification in Nepal. Int J Bus Soc Sci Res 4:152–156
Shridhar BS (2012) Review: nitrogen fixing microorganisms. Int J Microbiol Res 3:46–52
Shuhegge R, Ihring A, Gantner S, Bahnweg G, Knappe C, Vogg G, Hutzler P, Schmid M, Van Breusegem F, Eberl L, Hartmann A, Langebartels C (2006) Induction of systemic resistance in tomato by N-acyl-homoserine lactone-producing rhizosphere bacteria. Plant Cell Environ 29:909–918
Shukla PS, Agarwal PK, Jha B (2012) Improved salinity tolerance of *Arachis hypogaea* (L.) by the interaction of halotolerant plant growth-promoting rhizobacteria. J Plant Growth Regul 31:195–206
Siddikee MA, Chauhan PS, Anandham R, Han GH, Sa T (2010) Isolation, characterization and use for plant growth promotion under salt stress, of ACC deaminase producing halotolerant bacteria derived from coastal soil. J Microbiol Biotechnol 20:1577–1584
Singh BK, Bardgett RD, Smith P, Reay DS (2010) Microorganisms and climate change: terrestrial feedbacks and mitigation options. Nat Rev Microbiol 8:779–790
Singh LP, Gill SS, Tuteja N (2011) Unraveling the role of fungal symbionts in plant abiotic stress tolerance. Plant Signal Behav 6:175–191
Singh B, Kaur T, Kaur S, Manhas RK, Kaur A (2016) Insecticidal potential of an endophytic Cladosporiumvelox against *Spodoptera litura* mediated through inhibition of alpha glycosidases. Pestic Biochem Physiol 131:46–52
Singh S, Tripathi DK, Singh S, Sharma S, Dubey NK, Chauhan DK, Vaculík M (2017) Toxicity of aluminium on various levels of plant cells and organism: a review. Environ Exp Bot 137:177–193
Sloan SS, Lebeis S (2015) Exercising influence: distinct biotic interactions shape root microbiome. Curr Opin Plant Biol 26:32–36
Smith RE, Read DJ (2008) Mycorrhizal symbiosis, 3rd edn. Elsevier, Academic, New York
Smith SE, Smith FA (2011) Roles of arbuscular mycorrhizas in plant nutrition and growth: new paradigms from cellular to ecosystem scales. Annu Rev Plant Biol 62:227–250
Spaink HP (2000) Root nodulation and infection factors produced by rhizobial bacteria. Annu Rev Microbiol 54:257–288
Spence C, Alff E, Johnson C, Ramos C, Donofrio N, Sundaresan V, Bais H (2014) Natural rice rhizospheric microbes suppress rice blast infections. BMC Plant Biol 14:130
Sreenivasulu N, Sopory SK, Kavi Kishor PB (2007) Deciphering the regulatory mechanisms of abiotic stress tolerance in plants by genomic approaches. Gene 388:1–13

Srivastava S, Yadav A, Seem K, Mishra S, Chaudhary V, Srivastava CS (2008) Effect of high temperature on *Pseudomonas putida* NBRI0987 biofilm formation and expression of stress sigma factor RpoS. Curr Microbiol 56:453–457

Stella T, Covino S, Èvanèarová M, Filipová A, Petruccioli M, D'Annibale A, Cajthaml T (2017) Bioremediation of long-term PCB-contaminated soil by white-rot fungi. J Hazard Mater 324:701–710

Strange RN, Scott PR (2005) Plant disease: a threat to global food security. Annu Rev Phytopathol 43:83–116

Su J, Hu C, Yan X, Jin Y, Chen Z, Guan Q, Wang Y, Zhong D, Jansson C, Wang F, Schnurer A, Sun C (2015) Expression of barley SUSIBA2 transcription factor yields high-starch low-methane rice. Nature 523:602

Suarez C, Cardinale M, Ratering S, Jung S, Montoya AMZ, Geissler-Palaum R, Schnell S (2015) Plant growth-promoting effects of Hartmannibacter diazotrophic on summer barley (*Hordeum vulgare* L.) under salt stress. Appl Soil Ecol 95:23–30

Sugiyama A, Yazaki K (2012) Root exudates of legume plants and their involvement in interactions with soil microbes. In: Vivanco JM, Baluška F (eds) Secretions and exudates in biological systems. Springer, Berlin, pp 27–48

Sugiyama A, Bakker MG, Badri DV, Manter DK, Vivanco JM (2013) Relationships between *Arabidopsis* genotype-specific biomass accumulation and associated soil microbial communities. Botany-Botanique 91:123126

Suhag M (2016) Potential of biofertilizers to replace chemical fertilizers. Int Adv Res J Sci Eng Technol 3:163–167

Swenson W, Wilson DS, Elias R (2000) Artificial ecosystem selection. Proc Natl Acad Sci U S A 97:9110–9114

Syed Ab Rahman SF, Singh E, Pieterse CMJ, Schenk PM (2018) Emerging microbial biocontrol strategies for plant pathogens. Plant Sci 267:102–111

Symanczik S, Gisler M, Thonar C, Schlaeppi K, Van der Heijden MG, Kahmen A, Boller T, Maeder P (2017) Application of mycorrhiza and soil from a permaculture system improved phosphorus acquisition in Nanranjilla. Front Plant Sci 8:1263

Tajini F, Trabelsi M, Drevon JJ (2012) Combined inoculation with *Glomus intraradices* and *Rhizobium tropici* CIAT899 increases phosphorus use efficiency for symbiotic nitrogen fixation in common bean (*Phaseolus vulgaris* L.). Saudi J Biol Sci 19:157–163

Tank N, Saraf M (2010) Salinity-resistant plant growth promoting rhizobacteria ameliorates sodium chloride stress on tomato plants. J Plant Interact 5:51–58

Tarkka M, Schrey S, Hampp (2008) Plant associated soil microorganisms. In: Nautiyal C, Dion P (eds) Molecular mechanisms of plant and microbe coexistence. Springer, Berlin, pp 3–51

Thao NP, Tran LS (2016) Enhancement of plant productivity in the post genomic era. Curr Genomics 17:295–296

Theocharis A, Bordiec S, Fernandez O, Paquis S, Dhondt-Cordelier S, Baillieul F, Clement C, Barka EA (2012) *Burkholderia phytofirmans* PsJN primes *Vitis vinifera* L. and confers a better tolerance to low nonfreezing temperatures. Mol Plant-Microbe Interact 25:241–249

Tian C, Kasiborski B, Koul R, Lammers PJ, Bulcking H, Shachar-Hill Y (2010) Regulation of the nitrogen transfer pathway in the arbuscular mycorrhizal symbiosis: gene characterization and the coordination of expression with nitrogen flux. Plant Physiol 153:1175–1187

Tian BY, Cao Y, Zhang KQ (2015) Metagenomic insights into communities, functions of endophytes, and their associates with infection by root-knot nematode, *Meloidogyne Incognita*, in tomato roots. Sci Rep 5:17087

Timmusk S, Nevo E (2011) Plant root associated biofilms. In: Meshwari DK (ed) Bacteria in agrobiology. Plant nutrient management, vol 3. Springer, Berlin, pp 285–300

Timmusk S, Wagner EGH (1999) The plant growth-promoting rhizobacterium *Paenibacillus polymyxa* induces changes in *Arabidopsis thaliana* gene expression: a possible connection between biotic and abiotic stress responses. Mol Plant-Microbe Interact 12:951–959

Timmusk S, Timmusk K, Behers L (2013) Rhizobacterial plant drought stress tolerance enhancement: towards sustainable water resource management and food security. J Food Secur 1:6–9

Timmusk S, El-Daim IA, Cpolovici L, Tanilas T, Kannaste A, Behers L, Nevo E, Seisenbaeva G, Stenstrom E, Niinemets U (2014) Drought-tolerance of wheat improved y rhizosphere bacteria from harsh environments: enhanced biomass production and reduced emissions of stress volatiles. PLoS One 9:e96086

Timmusk S, Behers L, Muthoni J, Aronsson AC (2017) Perspectives and challenges of microbe application for crop improvement. Front Plant Sci 8:49

Tiwari S, Singh P, Tiwari R, Meena KK, Yandigeri M, Singh DP, Arora DK (2011) Salt-tolerant rhizobacteria-mediated induced tolerance in wheat (*Triticum aestivum* L.) and chemical diversity in rhizosphere enhance plant growth. Biol Fert Soils 47:907–916

Tiwari S, Lata C, Chauhan PS, Nautiyal CS (2016) *Pseudomonas putida* attunes morphophysiological: biochemical and molecular responses in *Cicer arietinum* L. during drought stress and recovery. Plant Physiol Biochem 99:108–117

Tiwari S, Prasad V, Chauhan PS, Lata C (2017) *Bacillus amyloliquefaciens* confers tolerance to various abiotic stresses and modulates plant response to phytohormones through osmoprotection and gene expression regulation in rice. Front Plant Sci 8:1510

Tkacz A, Poole P (2015) Role of root microbiota in plant productivity. J Exp Bot 66:2167–2175

Tkacz A, Cheema J, Chandra G, Grant A, Poole PS (2015) Stability and succession of the rhizosphere microbiota depends upon plant type and soil composition. ISME J 9:2349–2359

Trabelsi D, Mhamdi R (2013) Microbial inoculants and their impact on soil microbial communities: a review. Biomed Res 2013:e863240

Tripathi DK, Singh VP, Prasad SM, Chauhan DK, Dubey NK, Rai AK (2015) Silicon-mediated alleviation of Cr (VI) toxicity in wheat seedlings as evidenced by chlorophyll florescence, laser induced breakdown spectroscopy and anatomical changes. Ecotoxicol Environ Saf 113:133–144

Tripathi DK, Singh S, Singh VP, Prasad SM, Chauhan DK, Dubey NK (2016) Silicon nanoparticles more efficiently alleviate arsenate toxicity than silicon in maize cultivar and hybrid differing in arsenate tolerance. Front Environ Sci 4:46

Tripathi DK, Singh S, Singh VP, Prasad SM, Dubey NK, Chauhan DK (2017) Silicon nanoparticles more effectively alleviated UV-B stress than silicon in wheat (*Triticum aestivum*) seedlings. Plant Physiol Biochem 110:70–81

Turner TR, James EK, Poole PS (2013) The plant microbiome. Genome Biol 14:209

Ubertino S, Mundler P, Tamini LD (2016) The adoption of sustainable management practices by Mexican coffee producers. Sustain Agric Res 5:1–12

Ulloa-Ogaz AL, Munoz-Castellanos LN, Nevarez-Moorillon GV (2015) Biocontrol of phytopathogens: antibiotic production as mechanism of control, the battle against microbial pathogens. In: Mendez Vilas A (ed) Basic science, technological advance and educational programs, vol 1, pp 305–309

Upadhyay SK, Singh DP, Saikia R (2009) Genetic diversity of plant growth promoting rhizobacteria from rhizospheric soil of wheat under saline conditions. Curr Microbiol 59:489–496

Upadhyay SK, Singh JS, Singh DP (2011) Exopolysaccharide-producing plant growth-promoting rhizobacteria under salinity condition. Pedosphere 21:214–222

Uroz S, Dessaux Y, Oger P (2009) Quorum sensing and quorum quenching: the yin and yang of bacterial communication. Chembiochem 10:205–216

Vacheron J, Desbrosses G, Bouffaud ML, Touraine B, Moënne-Loccoz Y, Muller D, Legendre L, Wisniewski-Dyé F, Combaret CP (2013) Plant growth-promoting rhizobacteria and root system functioning. Front Plant Sci 4:1–19

Valliyodan B, Nguyen H (2006) Understanding regulatory networks and engineering for enhanced drought tolerance in plants. Curr Opin Plant Biol 9:1–7

Van der Heijden MG, Bardgett RD, Van Straalen NM (2008) The unseen majority: soil microbes as drivers of plant diversity and productivity in terrestrial ecosystems. Ecol Lett 11:296–310

Van der Heijden MGA, Martin FM, Selosse MA, Sanders IR (2015) Mycorrhizal ecology and evolution: the past, the present and the future. New Phytol 205:1406–1423

van Loon LC (2007) Plant responses to plant growth promoting bacteria. Eur J Plant Pathol 119:243–254 van Loon LC, Bakker PAHM, Pieterse CMJ (1998) Systemic resistance induced by rhizosphere bacteria. Annu Rev Phytopathol 36:453–483

van Peer R, Niemann GJ, Schippers B (1991) Induced resistance and phytoalexin accumulation in biological control of Fusarium wilt of carnation by Pseudomonas sp. strainWCS417r. Phytopathology 91:728–734

Vandenkoornhuyse P, Quaiser A, Duhamel M, LeVan A, Dufresne A (2015) The importance of the microbiome of the plant holobiont. New Phytol 206:1196–1206

Vardharajula S, Ali SZ, Grover M, Reddy G, Bandi V (2011) Drought-tolerant plant growth promoting Bacillus spp.: effect on growth osmolytes, and antioxidant status of maize under drought stress. J Plant Interact 6:1–14

Vargas L, Santa Brigida AB, Mota-Filho JP, de Carvalho TG, Rojas CA, Vaneechoutte D, Van Bel M, Farrinelli L, Ferreiro PCG, Vandepoele K, Hemerly A (2014) Drought tolerance conferred to sugarcane by association with *Gluconacetobacter diazotrophicus*: a transcriptomic view of hormone pathways. PLoS One 9:e114744

Vejan P, Abdullah R, Khadiran T, Ismail S, Nusrulhaq Boyce A (2016) Role of plant growth-promoting rhizobacteria in agricultural sustainability: a review. Molecules 21:573

Velazquez-Hernandez ML, Baizabal-Aguirre VM, Cruz-Vazquez F, Trejo-Contreras MJ, Fuentes-Ramırez LE, Bravo-Patino A, Valdez-Alarcon JJ (2011) Gluconacetobacter *diazotrophicus levansucrase* is involved in tolerance to NaCl, sucrose and desiccation, and in biofilm formation. Arch Microbiol 193:137–149

Verhagen BW, Glazebrook J, Zhu T, Chang HS, van Loon LC, Pieterse CM (2004) The transcriptome of rhizobacteria-induced systemic resistance in Arabidopsis. Mol Plant-Microbe Interact 17:895–908

Verma JP, Yadav J, Tiwari KN, Lavakush, Singh V (2010) Impact of plant growth promoting rhizobacteria on crop production. Int J Agric Res 5:954–983

Verma JP, Yadav LJ, Tiwari KN, Kumar A (2013) Effect of indigenous Mesorhizobium spp. and plant growth promoting rhizobacteria on yields and nutrients uptake of chickpea (*Cicer arietinum* L.) under sustainable agriculture. Ecol Eng 51:282–286

Vidali M (2001) Bioremediation: an overview. Pure Appl Chem 73:1163–1172

Vinocur B, Altman A (2005) Recent advances in engineering plant tolerance to abiotic stress: achievements and limitations. Curr Opin Biotechnol 16:123–132

Wagg C, Bender SF, Widmer F, Van der Heijden MG (2014) Soil biodiversity and soil community composition determine ecosystem multifunctionality. Proc Natl Acad Sci U S A 111:5266–5270

Wallenstein MD (2017) Managing and manipulating the rhizosphere microbiome for plant health: a system approach. Rhizosphere 3:230–232

Walters DR, Ratsep J, Havis ND (2013) Controlling crop diseases using induced resistance: challenges for the future. J Exp Bot 64:1263–1280

Wang WX, Vinocur B, Altman A (2003) Plant responses to drought, salinity and extreme temperatures: towards genetic engineering for stress tolerance. Planta 219:1–14

Wang Y, Ohara Y, Nakayashiki H, Tosa Y, Mayama S (2005) Microarray analysis of the gene expression profile induced by the endophytic plant growth-promoting rhizobacteria, Pseudomonas fluorescens FPT9601-T5 in Arabidopsis. Mol Plant-Microbe Interact 18:385–396

Wang HG, Zhnag XZ, Li H, He HB, Fang CX, Zhang AJ, Li QS, Chen RS, Guo XK, Lin HF, Wu LK, Lin S, Chen T, Lin RY, Peng XX, Lin WX (2011) Characterization of metaproteomics in crop rhizospheric soil. J Proteome Res 10:932–940

Wang X, Mavrodi DV, Ke L, Mavrodi OV, Yang M, Thomashow LS, Zheng N, Weller DM, Zhang J (2015) Biocontrol and plant growth-promoting activity of rhizobacteria from Chinese fields with contaminated soils. Microbial Biotechnol 8:404–418

Wang R, Dungait JAJ, Buss HL, Yang S, Zhang Y, Xu Z, Jiang Y (2017) Base cations and micronutrients in soil aggregates as affected by enhanced nitrogen and water inputs in semi-arid steppe grassland. Sci Total Environ 575:564–572

Wani PA, Khan MS, Zaidi A (2008) Effect of metal tolerant plant growth-promoting rhizobium on the performance of pea grown in metal-amended soil. Arch Environ Contam Toxicol 55:33–42

Waqas M, Khan AL, Hamayun M, Shahzad R, Kim YH, Choi KS, Lee IJ (2015) Endophytic infection alleviates biotic stress in sunflower through regulation of defense hormones, antioxidants and functional amino acids. Eur J Plant Pathol 141:803–824

Webb BA, Helm RF, Scharf BE (2016) Contribution of individual chemoreceptors to *Sinorhizobium meliloti* chemotaxis towards amino acids of host and non-host seed exudates. Mol Plant-Microbe Interact 29:231–239

Wei G, Kloepper JW, Tuzun S (1991) Induction of systemic resistance of cucumber to *Colletotrichum orbiculare* by select strains of plant-growth promoting rhizobacteria. Phytopathology 81:1508–1512

Wei Y, Su Q, Sun ZJ, Shen YQ, Li JN, Zhu XL, Hou H, Chen ZP, Wu FC (2016) The role of arbuscular mycorrhizal fungi in plant uptake, fractions and speciation of antimony. Appl Soil Ecol 107:244–250

Wenzel WW (2009) Rhizosphere processes and management in plant-assisted bioremediation (phytoremediation) of soils. Plant Soil 321:385–408

Weston LA, Mathesius U (2013) Flavonoids: their structure, biosynthesis and role in the rhizosphere including allelopathy. J Chem Ecol 39:283–297

White RA III, Borkum MI, Rivas-Ubach A, Bilbao A, Wendler JP, Colby SM, Koberl M, Jansson C (2017) From data to knowledge: the future of multi-omics data analysis for the rhizosphere. Rhizosphere 3:222–229

Wicke B, Smeets E, Dornburg V, Vashev B, Gaiser T, Turkenburg W, Faij A (2011) The global technical and economic potential of bioenergy from salt-affected soils. Energy Environ Sci 4:2669–2681

Wirthmueller L, Maqbool A, Banfield MJ (2013) On the front line: structural insights into plant–pathogen interactions. Nat Rev Microbiol 11:761–776

Wu Q, Cui Y, Li Q, Sun J (2015) Effective removal of heavy metals from industrial sludge with the aid of a biodegradable chelating ligand GLDA. J Hazard Mater 283:748–754

Wu QS, Wang S, Srivastava AK (2016) Mycorrhizal hyphal disruption induces changes in plant growth, glomalin-related soil protein and soil aggregation of trifoliate orange in a core system. Soil Tillage Res 160:82–91

Wuana RA, Okieimen FE (2011) Heavy metals in contaminated soil: a review of sources, chemistry, risks and best available strategies for bioremediation. ISRN Ecol 2011:402647

Yadav LJ, Verma JP, Jaiswal DK, Kumar A (2014) Evaluation of PGPR and different concentration of phosphorus level on plant growth, yield and nutrient content of rice (*Oryza sativa*). Ecol Eng 62:123–128

Yadav UP, Ayre BG, Bush DR (2015) Transgenic approaches to altering carbon and nitrogen partitioning in whole plants: assessing the potential to improve crop yields and nutritional quality. Front Plant Sci 6:275

Yang J, Kloepper JW, Ryu CM (2009) Rhizosphere bacteria help plants tolerate abiotic stress. Trends Plant Sci 14:1–4

Yang Q, Tu S, Wang G, Liao X, Yan X (2012) Effectiveness of applying arsenate reducing bacteria to enhance arsenic removal from polluted soils by *Pteris vittata* L. Int J Phytoremediation 14:89–99

Yang AZ, Akhtar SS, Amjad M, Iqbal S, Jacobsen SE (2016a) Growth and physiological responses of quinoa to drought and temperature stress. J Agron Crop Sci 202:445–453

Yang AZ, Akhtar SS, Iqbal S, Amjad M, Naveed M, Zahir ZA, Jacobsen SE (2016b) Enhancing salt tolerance in quinoa by halotolerant bacterial inoculum. Funct Plant Biol 43:632–642

Yasmin H, Bano A, Samiullah A (2013) Screening of PGPR isolates from semi-arid region and their implication to alleviate drought stress. Pak J Bot 45:51–58

Yeoh YK, Paungfoo-Lonhienne C, Dennis PG, Robinson N, Ragan MA, Schmidt S, Hugenholtz P (2016) The core root microbiome of sugarcanes cultivated under varying nitrogen fertilizer application. Environ Microbiol 18:1338–1135

Yu XM, Ai CX, Xin L, Zhou GF (2011) The siderophore-producing bacterium, *Bacillus subtilis* CAS15, has a biocontrol effect on Fusarium wilt and promotes the growth of pepper. Eur J Soil Biol 47:138–145

Yuan J, Zhang N, Huang Q, Raza W, Li R, Vivanco JM, Shen Q (2015) Organic acids from root exudates of banana help root colonization of PGPR strain *Bacillus amyloliquefaciens* NJN-6. Sci Rep 25:13438

Yuan Z, Jiang S, Sheng H, Liu X, Hua H, Liu X, Hua H, Liu X, Zhang Y (2018) Human perturbation of the global phosphorus cycle: changes and consequences. Environ Sci Technol 52:2438–2450

Zamioudis C, Pieterse CMJ (2012) Modulation of host immunity by beneficial microbes. Mol Plant-Microbe Interact 25:139–150

Zamioudis C, Mastranesti P, Donukshe P, Blilou I, Pieterse CM (2013) Unraveling root development programs initiated by beneficial *Pseudomonas* spp. Bacteria. Plant Physiol 162:304–318

Zehr JP, Jenkins BD, Short SM, Steward GF (2003) Nitrogenase gene diversity and microbial community structure: a cross-system comparison. Environ Microbiol 5:539–554

Zhang F, Lynch DH, Smith DL (1995) Impact of low root temperatures in soybean [*Glycine max* (L.) Merr.] on nodulation and nitrogen fixation. Environ Exp Bot 35:279–285

Zhang F, Dashti N, Hynes R, Smith DL (1996) Plant growth promoting rhizobacteria and soybean [*Glycine max* (L.) Merr.] nodulation and nitrogen fixation at suboptimal root zone temperatures. Ann Bot 77:453–460

Zhang H, Sun Y, Xie X, Kim MS, Dowd SE, Pare PW (2009) A soil bacterium regulates plant acquisition of iron via deficiency-inducible mechanisms. Plant J 58:568–577

Zhang J, Wang LH, Yang JC, Liu H, Dai JL (2015) Health risk to residents and stimulation to inherent bacteria of various heavy metals in soil. Sci Total Environ 508:29–36

Zhao S, Zhou N, Zhao ZY, Zhang K, Wu GH, Tian CY (2016) Isolation of endophytic plant growth-promoting bacteria associated with the halophyte Salicornia europea and evaluating their promoting activity under salt stress. Curr Microbiol 73:574–581

Zheng YK, Qiao XG, Miao CP, Liu K, Chen YW, Xu LH, Zhao LX (2016) Diversity, distribution and biotechnological potential of endophytic fungi. Ann Microbiol 66:529–542

Ziegler M, Engel M, Welzl G, Schloter M (2013) Development of a simple root model to study the effects of single exudates on the development of bacterial community structure. J Microbiol Methods 94:30–36

Zilber-Rosenberg I, Rosenberg E (2008) Role of microorganisms in the evolution of animals and plants: the hologenome theory of evolution. FEMS Microbiol Rev 32:723–735

Zmora N, Zeevi D, Korem T, Segal E, Elinav E (2016) Taking it personally: personalized utilization of the human microbiome in health and disease. Cell Host Microbe 19:12–20

Zolla G, Badri DV, Bakker MG, Manter DK, Vivanco JM (2013) Soil microbiomes vary in their ability to confer drought tolerance to *Arabidopsis*. Appl Soil Ecol 68:1–9

Zuppinger-Dingley D, Schmid B, Petermann JS, Yadav V, De Deyn GB, Flynn DF (2014) Selection for niche differentiation in plant communities increases biodiversity effects. Nature 515:108–111

Arbuscular Mycorrhizal Colonization and Activation of Plant Defense Responses Against Phytopathogens

8

Anupam Maharshi, Gagan Kumar, Arpan Mukherjee, Richa Raghuwanshi, Harikesh Bahadur Singh, and Birinchi Kumar Sarma

8.1 Introduction

Plant roots are colonized as well as surrounded by a diverse microbial population at root-soil interfaces known as rhizosphere. Rhizosphere microbes potentially regulate plant growth and health by interacting with plant roots as well as surrounding soil microorganisms. This interaction may be beneficial or harmful for the plants. Beneficial interactions of such microbes with the plants result in better health and growth of the plants. Several microbes, i.e., bacteria, fungus, blue green algae, etc., establish positive associations with the plants, and among all of them, mycorrhizal association is considered highly beneficial.

The term mycorrhiza comprises two Greek words: "myco" means fungi and "rhiza" means roots, and its complete meaning is association of fungus with roots of plants. The term "mycorrhiza" was coined by Bernhardt Frank by identifying unique structures in roots of tree in 1885. Mycorrhizas grow only on living hosts and establish an obligate symbiotic association with the host roots (Owen et al. 2015). More than 90% of the terrestrial plants are found to have mycorrhizal association. Mycorrhizas may appear in several forms of association such as ectomycorrhizas, endomycorrhizas with septate fungi, endomycorrhizas with aseptate fungi also known as arbuscular mycorrhizas, and ecto-endomycorrhizas. Among all types of symbiotic mycorrhizas, endotrophic arbuscular mycorrhizas (AM) are the most important occurring in 80% of terrestrial plant species (Cervantes-Gamez et al. 2015). Most of the arbuscular mycorrhizal fungus belongs to *Zygomycota* that

A. Maharshi · G. Kumar · H. B. Singh · B. K. Sarma (✉)
Department of Mycology and Plant Pathology, Institute of Agricultural Sciences, Banaras Hindu University, Varanasi, India

A. Mukherjee · R. Raghuwanshi
Department of Botany, Institute of Science, Banaras Hindu University, Varanasi, India

D. P. Singh et al. (eds.), *Microbial Interventions in Agriculture and Environment*, https://doi.org/10.1007/978-981-13-8391-5_8

account for more than 150 species in the order *Glomales* (Morton and Benny 1990). In the new phylogenetic classification, *Glomales* is placed under a new monophyletic phylum *Glomeromycota*. It has been found so far that AM colonization may occur in almost all environmental conditions such as tropical rainforests (Gaur and Adholeya 2002), deserts (Titus et al. 2002), aquatic environments (Khan 1993), sodic or gypsum soils (Landwehr et al. 2002), and strong saline conditions (Sengupta and Chaudhuri 2002). AM symbiosis is a beneficial interaction between host plants and fungi probably most widespread in nature (Smith and Read 2008). This interaction consists of nutrient exchange between host plants and the fungus. Plants provide carbon to the fungal symbiont, while arbuscular mycorrhizal fungus (AMF) enhances the water and nutrient absorption capabilities of the plants from the soil (Smith and Read 2008). AM potentially absorb and translocate plant nutrients beyond the rhizospheric depletion zones of plants and alter secondary metabolism in the associated plants (Rouphael et al. 2015). AM symbiosis thus plays a major role in increasing plant growth development and enhances plant resistance to abiotic stresses (Kapoor and Bhatnagar 2007) as well as to biotic stresses (Song et al. 2015). Hence, AMF is the major group among plant-microbe symbionts that has the potentiality to promote the plant ability to cope under various types of adverse environmental conditions.

8.2 Arbuscular Mycorrhiza (AM): Biotic Stress Management

Mycorrhizas can be considered to be the most important mutualistic symbiotic association of plants and microbes on earth. The arbuscular mycorrhizal fungi generally grow into root cortex and form intercellular hyphae from where other highly branched structures such as arbuscules originate within the cortex cells. Molecular modifications investigated in relation to mycorrhizal symbiosis were initially observed to be associated with plant defense responses. AM colonization affects the plant morphology and physiology leading to altered response towards the soil microorganisms. These altered responses are the key regulators of the plant defense mechanism as an outcome of biotic stress management. The mechanisms underlying behind biotic stress management involves improved nutrient status, compensation of damage of plants, strengthening of the plant, production of several secondary metabolites, altered soil microbial flora, induction of defense genes, and activation of systemic resistance in plants. Several reports have shown that AM also mediate biotic stress management in plants (Song et al. 2015). For example, in tomato, induction of defense response was much quicker and effective in AM-inoculated plants compared to the non-inoculated plants against the early blight-causing pathogen *Alternaria solani* (Song et al. 2015). The resistance mediated by AM colonization in plants is also known as mycorrhiza-induced resistance (MIR). In this chapter, we discussed the molecular aspects of AM-mediated biotic stress management (Fig. 8.1).

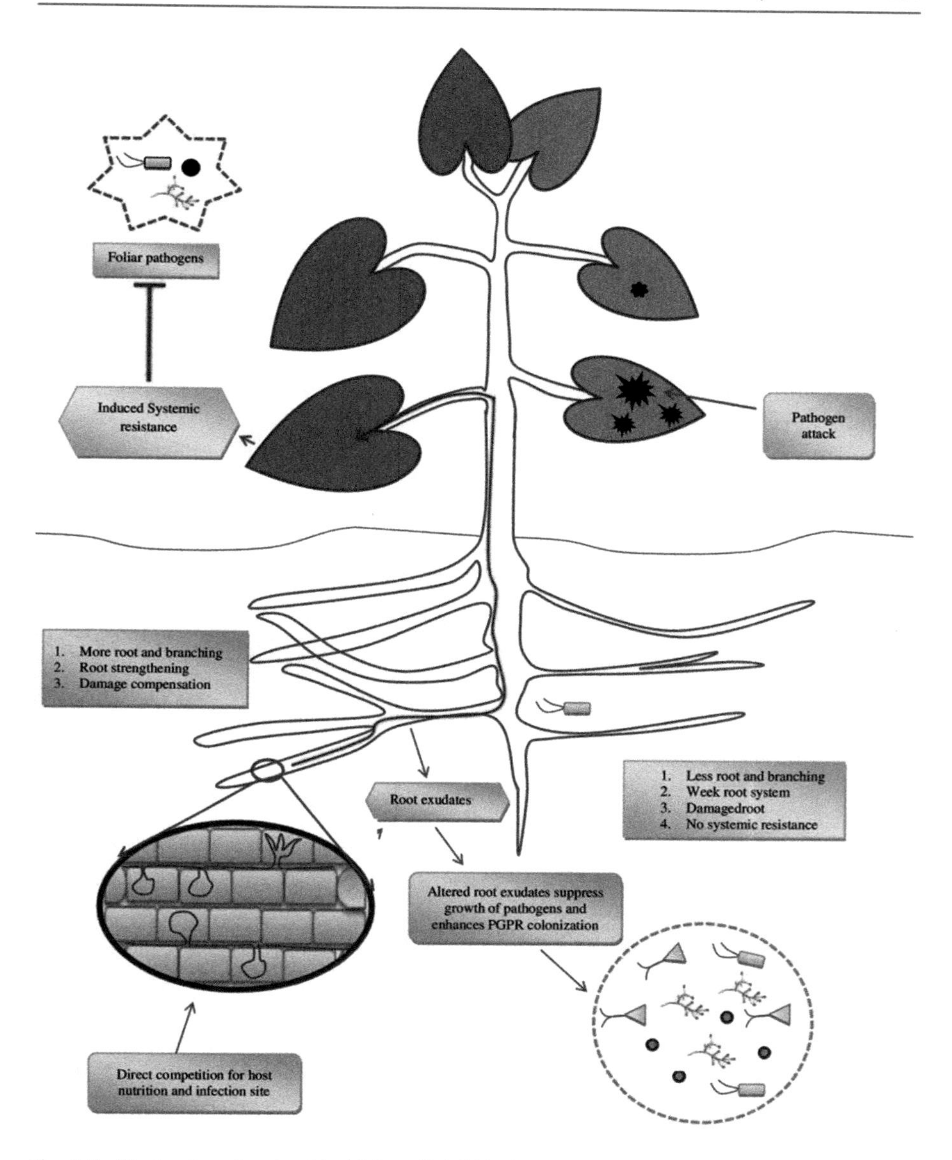

Fig. 8.1 Mechanisms involved in AM-mediated biotic stress management

8.2.1 Secondary Metabolites Increase During AM Colonization

AM colonization influences both primary and secondary metabolic activities in host plants (Schliemann et al. 2008). AM colonization triggers change in both enzyme activities (such as catalase, superoxide dismutase, etc.) (Marin et al. 2002) and physiological processes associated with secondary metabolite accumulation (such as polyphenols, carotenoids, etc.) (Toussaint et al. 2007). Earlier works showed that AM symbiosis with host plants induces plant primary metabolisms like photosynthesis and water uptake leading to tolerance for drought (Ruiz-Lozano 2003). After colonization with roots, secondary metabolism of host plants also changes such as synthesis of phytohormone, structural modifications, and activation of defense responses (Copetta et al. 2006). Detailed study of *Medicago truncatula* roots after colonization with AM showed activation of Krebs cycle and of plastidial metabolism and also helped in increasing endogenous levels of fatty acids and amino acids (tyrosine) that combine with phenylalanine, which is the key component of polyphenol synthesis in the phenylpropanoid metabolism (Lohse et al. 2005).

Phenylalanine ammonia lyase (PAL) activity increases in *Oryza sativa* plants after colonization of *Glomus mosseae* and in *Medicago truncatula* plants after colonization of *Glomus versiforme* (Blilou et al. 2000). Further research concludes that mycorrhizal colonization in artichoke plants showed a great effect on some important precursor enzymes of hydroxycinnamates biosynthesis, hydroxycinnamoyl-CoA:quinate hydroxycinnamoyl transferase (HQT), and hydroxycinnamoyl-CoA:shikimate/quinate hydroxycinnamoyl transferase (HCT) (Comino et al. 2009). AM colonization is also known to have regulatory effects on phenolic production in plants. Greater accumulation of antioxidant compounds like rosmarinic and caffeic acids was reported in *Ocimum basilicum* L. during colonization (sweet basil shoots) by different species of *Glomus* (Toussaint et al. 2007). Mycorrhizal colonization in plants showed a great change in many chemical compounds such as organic acids and sugars (Lioussanne et al. 2008), amino acids, some phenolics (McArthur and Knowles 1992), and flavonoids (Steinkellner et al. 2007) and in some plant hormones such as strigolactone (Lopez-Raez et al. 2011).

8.2.2 Signal Between AM and Host Plant

Signal transduction is a very important event during interaction of a plant with any microbial partner. Plant recognizes several molecular markers produced during plant-microbe interaction and transduces a downstream signaling, leading to activation of various metabolic as well as physiological pathways in plant system. It is the integral approach responsible for production of various defense compounds. Defense activation is the major focus here to know the role of AMF in induction of systemic resistance in plants. Recognition is the first step for any response during host-microbial interaction. Several volatile compounds are secreted by the plants roots including strigolactone that is sensitized by the AMF and induces its germination of spore and helps in forming more branches (Akiyama et al. 2005). During spore germination, an unidentified small molecule known as myc factor is produced. Host plant senses the presence of myc factor, leading to triggering of downstream responses including Ca^{2+} responses (Muller et al. 2000) that finally lead to form a symbiotic relationship with the host plant. This Ca-dependent signaling is mediated by a Ca-calmodulin-dependent protein kinase (ccaMK) essential for AMF symbiosis (Levy et al. 2004). This ccaMK has calmodulin-binding domain and Ca-binding EF hand motifs. It allows the protein to sense calcium, making it a prime candidate for the response to calcium signatures that are induced by AM fungi (Kosuta et al. 2008). This induces the elevated level of cytosolic calcium ($[Ca^{2+}]$cyt) (Navazio et al. 2007). This Ca elevation makes an increment in the ROS (reactive oxygen species) generation which is an initial step of active defense (Strange, 2003). Superoxide anion (O_2^-) and hydrogen peroxide (H_2O_2) are the main ROS produced with normal plant biochemical processes. At the site of pathogen infection, ROS are the major responsible factors for the membrane destruction by lipid peroxidation, protein inactivation, and DNA mutation (Torres et al. 2006) resulting in hypersensitive response (HR) or systemic resistance by inducing oxidative burst (Bolwell 2004). H_2O_2 generation exhibits direct antimicrobial activity by inhibiting fungal spore germination as well as participates in phenoxyl-radicals synthesis during phenol-polymerization within the plant cell wall (Lamb and Dixon 1997). Disintegration of cell membrane, tissue necrosis, and induced phytoalexin synthesis are outcomes of lipid peroxidation (El- Khallal 2007). AMF is found to have a potential role to induce several phytoalexins such as rishitin and solavetivone and PAL (Engstrom et al. 1999) as well as hydroxyproline-rich glycoproteins (Lambais, 2000). Genes of 9-LOX oxylipins pathway are found to be upregulated during interaction with plant pathogens in AM-colonized roots (Song et al. 2015) (Fig. 8.2).

LOX metabolites possess direct antimicrobial property as well as induce or alter defense gene expression in presence of any wound or pathogen (Künkel and Brooks 2002). This is mediated by an octadecanoid pathway (Hause et al. 2007) responsible for JA biosynthesis. Mycorrhizal-induced systemic resistance is mediated by endogenous signal of JA and its derivatives (El- Khallal 2007). At the same time ROS metabolism also leads to production of several antioxidants like POX, CAT, SOD, and APX that all are also antimicrobial in nature (El- Khallal 2007). AMF activity has a role in regulation of catalase activity as well as availability of phosphorus in roots of the bean (Lambais 2000).

SOD activity is found to be activated during pathogen attack to induce pathogen-related HR reaction, while microorganism-induced HRs reduce the catalase activity significantly (Delledonne et al. 2002). In the same way, *G. intraradices* regulates CAT and POX in bean and wheat (Blee and Anderson, 2000). Elevated level of cytosolic calcium ($[Ca^{2+}]$cyt) also induces MAPK and alterations in G-protein-mediated pathway after or parallel with ROS generation. These MAPKs and G-protein modifications regulate the phosphorylation or dephosphorylation of several antioxidant enzymes responsible for active defense mechanism (Strange 2003). Their role to transduce the external stimuli of the cell's machinery leads to bringing about a response against phytopathogens (Strange 2003). Taking consideration of the above facts, it could be stated that AMF has the potential and is involved in plant defense signaling through activation of certain precursors from a series of reactions which lead to their defensive end products, responsible for combating the plants with harmful invaders. But, there are many other aspects that are unknown like the exact reaction or pathway in the AMF induction, potential factors responsible for

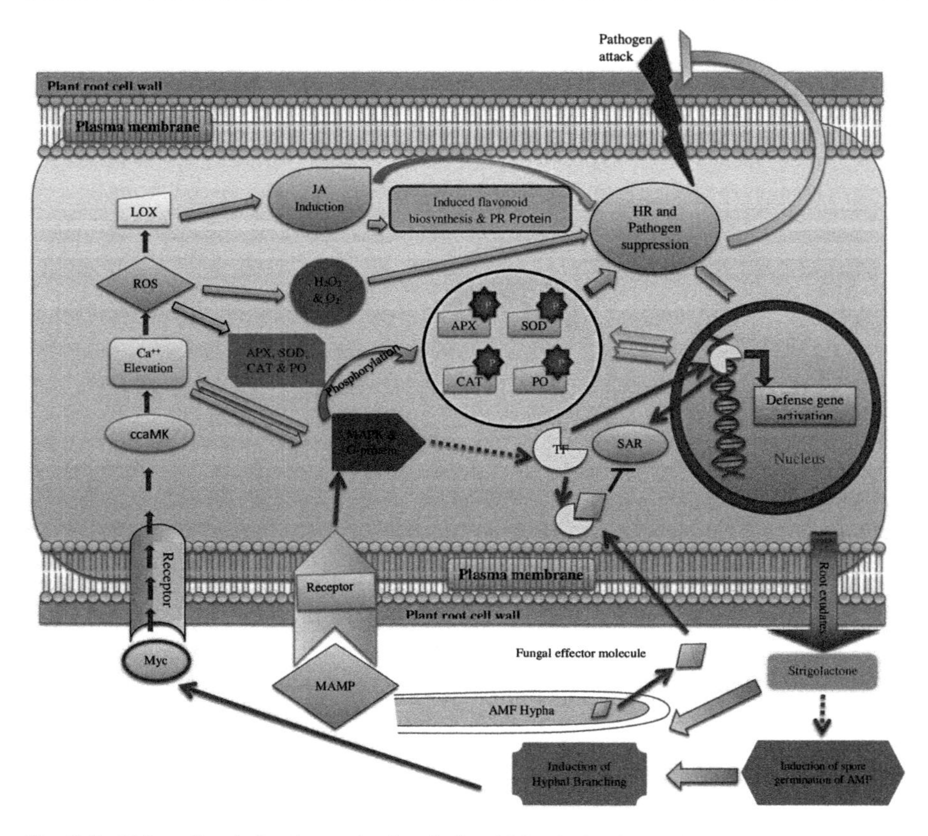

Fig. 8.2 AM-mediated signal transduction during AM colonization and subsequent plant pathogen suppression. In the initial step, the host root cell recognizes the microbe-associated molecular patterns (MAMPs) of AM hyphae leading to salicylic acid (SA)-mediated systemic acquired resistance (SAR) activation to suppress AM colonization via TF (transcription factor). At the same time, AM hyphae secrete some effector to suppress the SAR activation. In the later stage, the myc factor from the associated AMF triggers $[Ca^{2+}]$cyt which further induces ROS generation and mitogen-activated protein kinase (MAPK) and G-protein alterations. ROS also induces LOX, which mediates jasmonic acid (JA) biosynthesis. Several antioxidant enzymes such as superoxide dismutase (SOD), peroxidases (POX), catalases (CAT), and ascorbate peroxidase (APX) are synthesized, which play a significant role in ROS metabolism and get phosphorylated through MAPK and G-protein. MAPK and G-protein also trigger plant's defense genes. As pathogen enters into the host tissues, it triggers plant's defense genes. These defense-related genes encode proteins that attack the invading pathogens and try to neutralize them. However, the antioxidant enzymes and ROS act constitutively on the pathogen at the site of infection and initiate HR leading to the suppression of the pathogen

such activation of several genes, the changes in plant before and after the activation through these factors, and, finally, how plant reacts and up to which extent it comes up with these changes. This necessitates further exploration with the help of advanced phytochemical studies coupled with genomics and proteomics approaches.

8.2.3 Impact of AM Colonization on Jasmonic Acid-Mediated Defense Activation

Plant hormones have very important and essential roles in signaling of various morphological and physiological developmental processes as well as dealing with the adverse surroundings by activating plant defense mechanisms. Phytohormones such as ethylene (ET), cytokinins (CK), abscisic acid (ABA), gibberellins (GA), auxins, and jasmonic acid (JA) are the plant hormones believed to participate during interaction between AM fungus and plant (Ludwig-Muller 2000). JA, SA, and ET are the plant hormones that mainly participate in defense activation by activating systemic resistance. SA is a plant hormone that actively participates in SAR and provides systemic resistance to plants. But it is found that SA and ET negatively regulate during mycorrhizal colonization (Gutjahr and Paszkowski 2009). JA and its derivatives (jasmonates) play a central role in AM-mediated plant systemic defense activation. Jasmonates are well known for their ability to act as signal molecules involved in plant development processes as well as responses to both abiotic and biotic stresses (Wasternack and Hause 2002). The role of jasmonates is well established as part of a complex signal transduction pathway that is activated during wounding of leaves by insects (Schilmiller and Howe 2005) and when plants interact with microorganisms (Pozo et al. 2004). Endogenous increase in the level of jasmonates leads to the activation of defense-related genes in plants such as those coding for proteinase inhibitors, defensins, enzymes of phytoalexin synthesis, thionins, and vegetative storage proteins (Lorenzo and Solano 2005).

During AMF-plant interaction, a complex pathway leads to initial increase of SA to some extent that leads to systemic resistance followed by suppression of SA-mediated defense pathway after proper establishment of AMF colonization and then finally increases the level of JA that is responsible for systemic resistance in plants to various other plant pathogens. Regulation of SA and JA signaling is significantly dependent on AM colonization and explains the wide range of protection provided by this mutualistic relationship (Pozo and Azcon-Aguilar 2007). Plants recognize the AMF as foreign organism and activate some defense-associated responses that are subsequently suppressed during mycorrhization (Garcia-Garrido and Ocampo 2002). Rapid and transient increase in endogenous SA is observed in the roots along with concurrent accumulation of several defense compounds, such as ROS, specific isoforms of hydrolytic enzymes, and activation of the phenylpropanoid pathway (Jung et al. 2012). This initial activity is temporally and spatially limited compared to plant-pathogen interaction and responsible for establishment of colonization of AMF in roots (Garcia-Garrido and Ocampo 2002). AM is showed to secrete an effector protein, SP7, that suppress activation of defense-related genes mediated by ERF19 (a pathogenesis-related transcription factor in the plant nucleus) for successful colonization in roots (Kloppholz et al. 2011). So, successful colonization of AMF needs suppression of certain SA-regulated responses (Dumas-Gaudot et al. 2000).

Jasmonates are the lipid-derived molecules synthesized by the oxidative metabolism of fatty acid molecules. They are basically plant oxylipins. Several types of oxylipins generated by oxidative metabolism of fatty acid serve various functions in plants, among them jasmonates participate in plant's systemic defense activation. Oxylipins are synthesized by the coordinated activity of lipases, LOXs, and a group of cytochrome P450 (CYP74) by metabolizing hydroperoxy fatty acids in a specialized way (Howe and Schilmiller 2002). The initial step of oxylipin synthesis is addition of an oxygen molecule either at carbon 9 or 13 of the linoleic or linolenic acid (Siedow 1991). Depending upon the place of oxygenation, this pathway is of two types, i.e., either a 9-LOX or a 13-LOX pathway. The reaction is catalyzed by lipoxygenases (LOXs) which is a member of non-heme iron dioxygenases. The LOX-derived hydroperoxy acids are further metabolized by the enzymes divinyl ether synthase (DES), allene oxide synthase (AOS), hydroperoxide lyase (HPL), peroxygenase, alkyl hydroperoxide reductase, epoxy alcohol synthase, or LOX itself giving rise to different groups of oxylipins as the end products (Feussner et al. 2001). 13-LOX oxylipins are shown to play an important role in various physiological plant developmental processes, such as growth and fertility (Stintzi and Browse 2000). The oxylipins derived other than 13-LOX pathway are known to play a direct antimicrobial role against oomycetes, fungus, and bacterial pathogens (Prost et al. 2005) as well as capable of activating the expression of defense-related genes and regulate HR response (De Leon et al. 2002). However, several studies showed that the 9-LOX pathway has an essential role to play in plant defense activation against phytopathogens (Blee 2002) and activation of JA-mediated defense mechanisms (Leon Morcillo et al. 2012). AM colonization leads to accumulation of jasmonates within the roots of *Hordium vulgare* (Hause et al. 2002), *M. truncatula* (Stumpe et al. 2005), *Cucumis sativus* (Vierheilig and Piche, 2002), and *Glycin max* (Meixner et al. 2005). This increase may differ among various plant species. In *M. truncatula*, accumulation of jasmonates is relatively low of two to three times in mycorrhizal roots (Stumpe et al. 2005), whereas in *H. vulgare* and *C. sativus*, their accumulation may increase up to 5-fold and 14-fold, respectively (Hause et al. 2002). Jasmonates are the molecules that act as elicitor. Polyamines, alkaloids, phenylpropanoids, quinines, glucosinolates, and antioxidants are the several classes of secondary metabolites that are induced by JA (De Geyter et al. 2012). Induction of transcripts of enzyme encoding genes PAL and chalcone synthase (CHS) of the isoflavone biosynthetic pathway is found to induce significantly high specifically in *M. truncatula* cells containing arbuscules (Harrison and Dixon, 1994). The gene coding for β-tubulin (MtTubb1) was observed to be transcriptionally upregulated in mycorrhizal roots (Manthey et al. 2004) leading to strengthening of cytoskeleton. A plant shows more fitness by expressing increased resistance of mycorrhizal plants compared to pathogens and drought stress (Auge 2001; Cordier et al. 1998). This is possibly mediated through a JA-induced expression of defense-related genes and vegetative storage proteins (Wasternack and Hause 2002). There are several functions led through the accumulation of jasmonates in plant roots colonized with AM (Hause et al. 2007).

1. Induction of flavonoid biosynthesis
2. Reorganization of cytoskeleton
3. Alterations of sink status of roots
4. Increase in plant fitness

8.2.4 Genetic Basis of AM-Mediated Plant Defense Activation

AM colonization incites several molecular modifications at genetic level that leads to plant defense activation. In plants, the major defense mechanisms that are induced in plants during attack by a pathogen are cell wall modifications, enhancement of secondary metabolism, and accumulation of pathogenesis-related proteins (Dixon and Harrison 1990). Some events that are found in plant-pathogen interactions have also been found in the plants interacting with AMF. These events are signal perception, transduction, and activation of defense genes. Ectomycorrhizal and AM fungi secrete similar chitin elicitors, which could induce a defense response (Salzer and Boller 2000). Studies by different researchers showed that mycorrhizal colonization enhances plant's resistance against plant pathogenic fungi. It was showed that inoculation with the AM mycorrhizal fungi *Funneliformis mosseae* sufficiently alleviated the early blight disease of tomato caused by *Alternaria solani*. Pre-inoculation of AM fungi significantly enhances activities of chitinase, β-1,3-glucanase, phenylalanine ammonia lyase, and lipoxygenase in tomato leaves following pathogen inoculation. In AM fungi-inoculated plants, during pathogen attack, strong defense response through activation of pathogenesis-related (PR) proteins, PR1, PR2, and PR3, as well as many defense-related genes, like *AOC*, *LOX*, and *PAL*, in tomato leaves was stimulated. Induction of defense responses in AM fungi pre-inoculated plants was much higher and more rapid compared to uninoculated plants after pathogen challenge (Song et al. 2015). Activation of the phenylpropanoid pathway cannot be considered as a general response associated to AM development. PR proteins involved in the defense response of plants include the acidic or basic chitinases and β-1,3-glucanases, which are basically antimicrobial hydrolases (Collinge et al. 1994). Enhanced chitinase and β-1,3-glucanase activities are observed during early plant-AM fungal interactions and then strongly diminish as root colonization by AM fungi proceeds (Vierheilig et al. 1994). The site of activity and the role of mycorrhiza-related isozymes in the symbiosis are not yet clearly understood. However, in myc-mutants, the mycorrhiza-related chitinase isoform is weakly activated in incompatible interactions (Dumas-Gaudot et al. 1994). Modulation of expression of the genes encoding hydroxyproline-rich glycoproteins is also observed during AM interactions (Franken and Gnadinger 1994), and in tobacco the level of PR-1 gene transcript accumulation is considerably low during AM development compared to roots challenged with the fungal pathogen *Chalara elegans*. Interestingly, defense gene expression is highly restricted in fully mycorrhizal tissues. Further, transcripts of *PAL*, *CHS*, β-1,3-glucanase, chitinase, and pathogenesis-related protein *PR-1* accumulate in a discrete manner in arbuscule-containing cells (Blee and Anderson 1996), and transcripts of the *gstl* gene encoding glutathione-S-transferase are restricted to AM-colonized cortical cells only in transgenic potato roots. Such spatial pattern of expression of many defense genes is very different in plant-pathogen interactions, where PR proteins accumulate throughout fungal-infected root tissues and defense gene expression is also activated in cells other than those containing the pathogen (Strittmatter et al. 1996). The uncoordinated, transient, weak, and/or very localized expression of host defense responses to AM fungi differs in many respects in compatible or incompatible plant-pathogen interactions, where the expression pattern differs from each other only in timing and extent (Collinge et al. 1994). Colonization of plant roots by AM fungi enhances plant's resistance/tolerance to a variety of biotic stresses. During mycorrhiza establishment, plants recognize the AM fungi and modulate the host defense responses in a way to allow a functional symbiotic association. As a result of such modulation, a mild but effective activation of the plant immune system takes place not only at the interacting site but also systemically. Such activation transforms the host plants to a primed state so that an efficient activation of defense responses could take place during attack by potential enemies (Pozo et al. 2009) (Table 8.1).

Table 8.1 Genes induced after AMF colonization in host plant that are responsible for the plant's defense against phytopathogens

Gene	Product	Function	Host	References
TC104515	Cysteine-rich protein	Antifungal property	*M. truncatula*	Liu et al. (2007)
TC101060	Cysteine-rich protein	Antifungal property	*M. truncatula*	Liu et al. (2007)
TC98064	Cysteine-rich protein	Antifungal property	*M. truncatula*	Liu et al. (2007)
PR-1a	PR-1a protein	Pathogenesis-related (antimicrobial)	Tomato	Conrath et al. (2006)
BGL	β-1,3-Glucanase (PR protein)	Antifungal property	Tomato	Conrath et al. (2006)
VCH3	Chitinase protein (PR protein family)	Antifungal property against *Meloidogyne incognita*	*Vitis amurensis*	Li et al. (2006)
Pal	Phenylalanine ammonia lyase (PAL) enzyme	Leads to production of phytoalexins and phenolic substances	*Oryza sativa*	Blilou et al. (2000)
Ltp	Lipid transfer protein	Antimicrobial	*Oryza sativa*	Blilou et al. (2000).
PR10	PR 10	Pathogenesis-related protein (have RNases activity)	*Pisum sativum*; *Petroselinum crispum*	Ruiz-Lozano et al. (1999)
pI 49	pI 49	Member of multigene family PR 10 (have ribonuclease activity)	*Pisum sativum*	Ruiz-Lozano et al. (1999)
pI 176	pI 176	Member of multigene family PR 10	*Pisum sativum*	Ruiz-Lozano et al. (1999)

8.2.5 Mechanisms Involved in AM-Induced Biotic Stress Management

8.2.5.1 Competition for Colonization in Rhizosphere

AM colonization has been not reported as antibiotic production or mycoparasitism directly on other microorganisms in soil. It was found that microbes having similar physiological requirement in an ecological niche compete with each other for nutrients or for space at the site of infection (Vos et al. 2014). Dehne (1982) described the colonization patterns of AM fungi and root pathogens on the same host tissue, and reported that both usually colonize in different cortical cells indicating their competitiveness for space. It was reported that *Phytophthora* colonization reduced significantly in AMF-colonized and adjacent uncolonized regions (Cordier et al. 1998). This concludes that AMF creates a local competition against plant root pathogens even in the absence of systemic resistance.

8.2.5.2 Changes in Microbial Population in Mycorrhizosphere

Mycorrhizas have several types of interactions in soil leading to changes in microbial community in rhizosperic zones. This interaction has a positive role in managing the plant pathogenic agents. The interaction may be direct by suppressing plant pathogens in root zone through antagonism or may be indirect by positive interaction with other plant growth-promoting and pathogenic antagonist microbes leading to retarded growth of plant pathogens. Soil population of *Glomus fasciculatum* shows a positive interaction with actinomycetes population in soil (Barea et al. 2002). Phosphate-solubilizing bacteria (PSB) have a positive role in plant disease management and can survive for a longer period of time in soil when soil is inoculated with AMF (Barea et al. 2002). AM presence induces the changes in host physiology that is responsible for changes in root exudates (Jones et al. 2004) that consequently alters microbial communities in roots. AM colonization significantly changes in microbial population and promotes a specific group of microbes in the rhizospheric zone. This change in rhizospheric zone leads to reduced number of sporangia and zoospores formed by *Phytophthora cinnamomi* in rhizospheric soil (Meyer and Linderman 1986). Abundance of actinomycetes showing antagonist property against plant pathogens was seen in the rhizosphere of AM-colonized plants compared to nonmycorrhizal controls (Secilia and Bagyaraj 1987). The population count of *Trifolium subterraneum* in *Zea mays* rhizoplane was found to be significantly high in AM-colonized roots (Meyer and Linderman 1986).

A reduced population of *Fusarium oxysporum* has been reported in the soil surrounding tomato roots colonized with AM as compared with the soil of nonmycorrhizal controls (Johansson et al. 2004). AM fungi also have a positive effect on PGPR by establishing the mutual interaction in soil rhizosphere. This mutual interaction promotes enhancement in plant rooting, plant growth and nutrition, improved nodulation in the case of legumes, and biological control of root pathogens (Barea et al. 1996). All these studies elucidate how AM colonization regulates the soil microbial community to overcome the biotic stress of the plants.

8.2.5.3 Root Cell Morphology Change

Morphological studies have shown that rapid changes occur in AM fungi and plant roots during AM colonization in plant roots. Arbuscules form a trunk-like structure that helps in the formation of branching in hyphae, and the change in the branching patterns depends on the host plants. It is found that arbuscule formation completes within 2–4 days (Alexander et al. 1989) and they rapidly collapse and form clumps. Ultrastructural study of AM shows that major modifications also occur in the cell wall that becomes thinner during root colonization and in the organization of the cytoplasm (Bonfante and Scannerini 1992). Development of arbuscule in the cortical cell modifies the host cell interior structure like invagination of the cell plasmalemma, fragmentation of the vacuole, fading of amyloplasts, and increase in the number of Golgi bodies (Bonfante and Perotto 1992). The presence of the fungus causes increases in size of the plant nucleus due to unfolding of its chromatin and thereby impacts its morphology (Berta et al. 1990). AM colonization also affects the plant nucleus and generally a shift in its position is observed in root cells. Normally the root cell nucleus is present in the peripheral position of AM uninoculated cells, but shifts to the central position in AM-infected cells (Balestrini et al. 1992). Occurrence of such nuclear movement is possibly due to modifications in the organization of the plant cytoskeleton as it is observed in plant-pathogen interactions as well (Kobayashi et al. 1992).

8.2.5.4 Mycorrhizas on Heavy Metals

The AM fungal outer membrane serves as adsorption area of many heavy metals, and this process helps to prevent the entry of toxic materials into the host plants cell (Joner et al. 2000). Electrostatic interactions help in binding the toxic metals with the cell surface of AM. The negatively charged group on AM cell surface such as phosphoryl, carboxyl, hydroxyl, and phenolic works as ligand for the toxic heavy metals. *Glomus mosseae* binds Zn by the external hyphae 3% higher than that of hyphal dry weight which is the basic mechanism to detoxify Zn phytotoxicity in contaminated soils (Christie et al. 2004). The *G. mosseae* P2 strain isolated from contaminated area can accumulate four times of Cd (2000 μmol/g) in comparison to *G. mosseae* (450 μmol/g) in uncontaminated soils (Joner et al. 2000). AM binds with heavy metals on rhizospheric region and protects the roots; it also prevents the translocation of heavy metal into shoot tissues (Kaldorf et al. 1999). The number of AM spores and root colonization on plant roots are dependent on soil disturbances (Waaland and Allen 1987). However, some AM fungal species adapt to such situations by developing mechanisms that help them to grow in unfavorable environmental conditions (Gaur and Adholeya 2004). Thus, AM colonization has a positive role in detoxification of heavy metals leading to reduced damage to root tissue, and it would be a preventive mechanism for plants to check the entry of any plant pathogen.

8.2.5.5 Mycorrhiza as Biocontrol

Reduction of soilborne pathogen population after root colonization by AM fungi was studied, and it was observed that AM reduced host infection by other pathogenic fungi, bacteria, and nematodes (Whipps 2004). Many different hypotheses have been put forward to explain biocontrol activity of soilborne phytopathogens by AM fungi. Symptoms produced by pathogen infection are greatly reduced due to AM colonization and development of ISR (induced systemic resistance) activities (Pozo and Azcon-Aguilar, 2007). Enhanced synthesis of SOD, POX (Garmendia et al. 2006), and PR-1 proteins (Cordier et al. 1998) and higher accumulation of phenolics (Zhu and Yao 2004) were observed in plants following AM fungi colonization probably leading to biocontrol activities through ISR. Furthermore, the additional forms of defense-related enzymes such as chitinases, chitosanases, β-1,3-glucanases, POX, and SOD were detected in mycorrhiza-colonized plant roots possibly to reduce the load of pathogens (Pozo et al. 1999).

8.2.5.6 Improved Nutrient Status of Host

Arbuscular mycorrhizal fungi uptake water and mineral nutrients mainly phosphate and nitrogen but probably also micro-elements and supply to their host plants. In return, the host provides photosynthetic carbon to AM fungi. High uptake of phosphate has been correlated with the AMF-mediated biological control; however, increased supply of phosphate to nonmycorrhizal plants did not respond in the same way in reduction of pathogen infection (Bodker et al. 1998). Tomato plants when colonized by *Rhizophagus irregularis* also lowered disease symptoms caused by *A. solani* compared to the nonmycorrhizal plants and at the same time there was no increase in phosphate uptake (Fritz et al. 2006). Disease incidence was even higher when an additional amount of phosphate was supplied. Therefore, positive outcome of increased phosphate uptake cannot be expected in terms of disease management apart from its impact on plant growth promotion in mycorrhizal plants. Further, in some cases plant growth was suppressed as a result of AM fungi colonization, even though phosphate transport from the AM fungi to the host plant was taking place effectively (Smith and Smith 2011). Increased capability of nutrient uptake in the host plants due to AM symbiosis that results into promotion of plant growth possibly makes the plants more resistant or tolerant to pathogen attack. Although improved nutrition was demonstrated as a mechanism for disease control and enhanced P uptake could account for tolerance of mycorrhizal plants to pathogens (either fungus or nematode), there are contradictory reports as well (Linderman 1994). For example, P-tolerant AM fungi reduced nematode infestation in high-P conditions which indicates that non-P-mediated mechanisms are also involved possibly through physiological changes in the roots (Smith 1987).

8.2.5.7 Damage Compensation

Plant pathogens do always behave like a devil for plants because they parasitize on host plants and exhaust whole biomass from plant till death. But most AM fungi have the ability to increase host crop tolerance to pathogen attack mostly through compensating the loss of root biomass or function caused by pathogens (Linderman 1994), including soil microorganisms such as fungi (Cordier et al. 1996). Restoration of root system is an indirect benefit from biocontrol potential AM fungi where the fungal hyphae grow out into the soil and increase the absorbing surface of the roots as well as maintenance of root cell activities through arbuscule formation (Cordier et al. 1996). It is concluded that the AM fungi always play an important role in respect to tolerance towards pathogen attack on the host plants by developing resistance.

8.2.5.8 Competition for Host Photosynthates

Any living organism requires daily feed and energy for completing its own life cycle. Plant pathogens and AM fungi also utilize host photosynthates for growth and establishment and compete for carbon compounds available in the roots (Smith 1987; Linderman 1994). Generally, AM fungi have fast growing ability and therefore they utilize more photosynthates compared to pathogen and their higher carbon demand may cause inhibition of pathogen growth. So, they can suppress the growth of pathogens which can be termed as competitive inhibition of pathogen growth. But, there is no strong evidence which exists to conclude that competition for carbon compounds is a generalized mechanism for pathogen biocontrol activity through AM symbiosis.

8.2.6 Approaches to Enhance Arbuscular Mycorrhizal Colonization in Field

Effective crop management is essential to attain high yield and quality. Gosling et al. (2006) explained that crop management can include a variety of practices that can impact the AM fungi association either directly by damaging or killing AMF or indirectly by creating conditions that are unfavorable to AM fungi (Menendez et al. 2001).

8.2.6.1 Cover Cropping

Mycorrhizal fungi depend on host plants for their nutrition. Cover crops help them to maintain their population. Four cover crops were compared where it was observed that hairy vetch caused the highest AM fungi spore abundance. However, AM fungi species richness and diversity were observed to be the highest in tomato fields where a mixture of seven cover crops was used (Njeru et al. 2015). The symbiosis between plants and AM fungi can be beneficial to health, nutrition, and abiotic stress tolerance in host plants. Maintenance of potential AM fungal inoculums is essential in winter as it has a positive impact in the colonization process in the subsequent crop. Introduction of cover crops to replace the fallow period improved AM fungi development significantly in the subsequent crops such as sunflower and maize. It improved mycorrhizal colonization, extra-radical mycelium, and AM fungi spores as well as EE-GRSP and b-glucosaminidase activity, water-stable aggregates, etc. In general, it was observed that use of barley as cover crop has a positive impact on all AM fungi variables compared to fallow, whereas the vetch treatment was observed to be intermediate (Garcia-Gonzalez et al. 2016).

8.2.6.2 Reducing Tillage

Heavy tillage may have a negative impact on the symbiosis through reducing root colonization primarily by disrupting the hyphal network (Evans and Miller 1988). In fact, no tillage system is recommended for AM symbiosis and therefore practices such as no or minimum tillage would enhance the functions of mycorrhiza and aid to sustainability of the system. Tillage reduces inoculation potential of the soil and efficacy of mycorrhiza by disruption of the extra-radical hyphal network (Mozafar et al. 2000). Thus, breaking of the soil macrostructure leads to no-effective hyphal network. Disruption of hyphal network finally decreases the absorptive abilities of mycorrhizas due to reduction in the spanning of the hyphae surface area. Disruption of hyphal network thus lowers the quantity of phosphorus supplement to those plants which are connected via the hyphal network (McGonigle and Miller 1999). In contrast, in the reduced tillage system, heavy phosphorus fertilizer input is not required as it is in the case of heavy tillage systems. This is attributed to intact mycorrhizal network which provides access to greater surface area for the crop phosphorus uptake (Miller et al. 1995).

8.2.6.3 Judicial Fertilization Application

Existing literature on the impact of fertilizer on mycorrhizal colonization is contradictory. Soluble phosphate application tends to reduce spore germination of mycorrhiza and extent of AM fungi colonization (Miranda and Harris 1994). However, AM fungal species such as *Glomus intraradices* are not sensitive to fertilizer application. The use of organic fertilizer and slow release mineral fertilizers does not appear to suppress AM fungi; it rather stimulates those (Singh et al. 2011). However, some authors recommended careful selection of organic amendments and cautioned against their overuse. While selecting organic amendments, thorough consideration must be given to application of pesticide, humified organic matter, heavy metals, soluble phosphorus, salinity, and other inorganic nutrients. Phosphorus fertilizer in known to inhibit AM fungal colonization and growth. The benefits of AM fungi are highest in agricultural systems where test phosphorus in soil is low. With increase in plant available soil phosphorus, the plant tissue phosphorous and the plant carbon investment in mycorrhizas are no longer economically beneficial to the plant. Stimulus to mycorrhizal symbiosis can increase early phosphorus uptake and improve crop yield potential without starter phosphorus fertilizer applications (Grant et al. 2005).

8.2.6.4 Crop Breeding

In progressing agricultural systems, crop breeding is a tool to obtain new varieties. Mycorrhizal dependency not only varies among crops but also among plant species, and under natural environment, crop varieties that are highly responsive to mycorrhizal colonization may be exploited for low-input production system (Subramanian and Charest 1999). Breeding of crop plants is generally conducted at experimental stations where available mineral nutrients have never been limiting factors. It is well known that increasing soil fertility diminishes AM fungal development and thereby reduces the benefits of mycorrhizal fungi. Therefore, it was hypothesized that breeding in such stations could lead to selection of varieties with high phosphate requirements. In other words, breeders would be selecting crop varieties against mycorrhizal dependency (Plenchette et al. 1982).

8.2.6.5 Proper Crop Rotation

Crop rotation could have a strong impact on the population and activity of AM fungi. Low diversity of host plants in a geographic area appears to be associated with low diversity and benefits of AM fungi. Monoculture for a prolonged period could reduce soil quality in relation to microbial diversity and community structure (Jiao et al. 2011). However, in certain cases monoculture showed to have not any adverse effect on the number of fungi, as found for watermelon compared with watermelon inter-cropped with pepper (Sheng et al. 2012), such type of trend can be seen as exceptional. Length of rotations, however induce both the density and diversity of AM fungi (Vestberg et al. 2011). More diverse is the crop rotation more beneficial it can be for the AM fungi. Increasing crop diversity may include incorporating agricultural crops with other cover crops and weeds (Njeru et al. 2015). Within crop rotation, high mycorrhizal dependent crops seem to enhance the density and diversity of AM fungi (Bharadwaj et al. 2008).

8.3 Conclusion

AM has the potentiality to colonize in natural as well as manmade ecosystems and establish a mutual collaboration with plants. The principle behind successful AM colonization and functioning of plants is of great interest. This interaction plays an important role in protection of plants from various adverse environmental conditions. AM colonization significantly regulates the plant physiology and its metabolism leading to changes of host response to external harms. Experimental evidences confirm that this protection is based not only on improved nutrition or local changes within the roots and the rhizosphere but that plant defense mechanisms play a key role. Several genes of the hosts that are associated with production of secondary metabolites were reported to be triggered during AM colonization and actively participate in host defense responses. Successful colonization of AM causes suppression of initial induction of salicylic acid-mediated pathway and leads to induction of jasmonates-mediated pathway. Elevated JA levels enhance the defense level of plant tissues sensitive to infections by phytopathogens or to abiotic stresses. JA-induced pathway triggers induced systemic resistance (ISR) in hosts in a similar way induced by plant growth-promoting rhizobacteria. AM-mediated plant defense activation is mediated by succession of various events of signal transduction. Although AM colonization occurs in almost all environmental conditions, several strategies also proved to be helpful for maintaining the population of AMF in the soil. Finally, it can be concluded that scientists and industrialists should collaborate to develop understanding of all potential key missing points regarding this symbiosis and commercialize effective AMF as it is one of the most potential sustainable tools for improving yield and increasing quality of the crop produce.

References

Akiyama K, Matsuzaki K, Hayashi H (2005) Plant sesquiterpenes induce hyphal branching in arbuscular mycorrhizal fungi. Nature 435:824–827

Alexander T, Toth R, Meier R, Weber HC (1989) Dynamics of arbuscule development and degeneration in onion, bean, and tomato with reference to vesicular-arbuscular mycorrhizas in grasses. Can J Bot 67:2505–2513

Auge R (2001) Water relations, drought and vesicular-arbuscular mycorrhizal symbiosis. Mycorrhiza 11:3–42

Balestrini R, Berta G, Bonfante P (1992) The plant nucleus in mycorrhizal roots: positional and structural modifications. Biol Cell 75:235–243

Barea JM, Azcón-Aguilar C, Azcón R (1996) Interactions between mycorrhizal fungi and rhizosphere microorganisms within the context of sustainable soil-plant systems. In: Gange AC, Brown VK (eds) Multitrophic interactions in terrestrial systems. Blackwell, Oxford

Barea JM, Azcon R, Azcon-Aguilar C (2002) Mycorrhizosphere interactions to improve plant fitness and soil quality. Antonie Van Leeuwenhoek 81:343–351

Berta G, Sgorbati S, Soler V, Fusconi A, Trotta A, Citterio A, Scannerini S (1990) Variations in chromatin structure in host nuclei of a vesicular arbuscular mycorrhiza. New Phytol 114:199–205

Bharadwaj DP, Lundquist PO, Alstrom S (2008) Arbuscular mycorrhizal fungal spore-associated bacteria affect mycorrhizal colonization, plant growth and potato pathogens. Soil Biol Biochem 40:2494–2501

Blee E (2002) Impact of phyto-oxylipins in plant defense. Trends Plant Sci 7:315–322

Blee KA, Anderson AJ (1996) Defense-related transcript accumulation in *Phaseolus vulgaris* L. colonized by the arbuscular mycorrhizal fungus *Glomos intraradices* Schsnck & Smith. Plant Physiol 110:675–688

Blee KA, Anderson AJ (2000) Defence responses in plants to arbuscular mycorrhizal fungi. In: Podilla GK, Douds DD (eds) Current advances in mycorrhizas research. The American Phytopathological society, St. Paul, pp 45–59

Blilou I, Ocampo JA, García-Garrido JM (2000) Induction of Ltp (lipid transfer protein) and pal (phenylalanine ammonia-lyase) gene expression in rice roots colonized by the arbuscular mycorrhizal fungus *Glomus mosseae*. J Exp Bot 51:1969–1977

Bodker L, Kjoller R, Rosendahl S (1998) Effect of phosphate and the arbuscular mycorrhizal fungus Glomus intra radices on disease severity of root rot of peas (Pisum sativum) caused by Aphanomyces euteiches. Mycorrhiza 8:169–174

Bolwell GP (2004) Role of active oxygen species and NO in plant defence responses. Curr Opin Plant Biol 2:287–294

Bonfante-Fasolo P, Perotto S (1992) Plant and endomycorrhizal fungi: the cellular and molecular basis of their interaction. In: Verma DPS (ed) Molecular signal in plant microbe communication. CRC Press, Boca Raton, pp 445–470

Bonfante-Fasolo P, Scannerini S (1992) In: Allen MJ (ed) The cellular basis of plant-fungus interchanges in mycorrhizal associations. Chapman and Hall, New York, pp 65–101

Cervantes-Gamez RG, Bueno-Ibarra MA, Cruz-Mendivil A, Calderon- Vazquez CL, Ramirez-Douriet CM, Maldonado-Mendoza IE et al (2015) Arbuscular mycorrhizal symbiosis-induced expression changes in *Solanum lycopersicum* leaves revealed by RNA- seqanalysis. Plant Mol Biol Rep 23:1–14

Christie P, Li X, Chen B (2004) Arbuscular mycorrhiza can depress translocation of zinc to shoots of host plants in soils moderately polluted with zinc. Plant Soil 261(1–2):209–217

Collinge DB, Gregersen PL, Thordal-Christensen H (1994) The induction of gene expression in response to pathogenic microbes. In: Perspectives ASB (ed) Mechanisms of plant growth and Improved productivity: modern approaches and. Marcel Dekker, New York, pp 391–433

Comino C, Hehn A, Moglia A, Menin B, Bourgaud F, Lanteri S, Portis E (2009) The isolation and mapping of a novel hydroxycinnamoyltransferase in the globe artichoke chlorogenic acid pathway. BMC Plant Biol 9:30

Conrath U, Beckers GJ, Flors V, Garcia-Agustín P, Jakab G, Mauch F, Newman MA, Pieterse CM, Poinssot B, Pozo MJ, Pugin A (2006) Priming: getting ready for Battle. Mol Plant-Microbe Interact 19:1062–1071

Copetta A, Lingua G, Berta G (2006) Effects of three AM fungi on growth, distribution of glandular hairs, and essential oil production in *Ocimum basilicum* L. var. Genovese. Mycorrhiza 16:485–494

Cordier C, Gianinazzi S, Gianinazzi-Pearson V (1996) Colonisation patterns of roots tissues by *Phytophthora nicotianae* var. *parasitica* related to reduced disease in mycorrhizal tomato. Plant Soil 185:223–232

Cordier C, Pozo MJ, Barea JM, Gianinazzi S, Gianinazzi-Pearson V (1998) Cell defense responses associated with localized and systemic resistance to Phytophthora parasitica induced in tomato by an arbuscular mycorrhizal fungus. Mol Plant-Microbe Interact 11:1017–1028

De Geyter N, Gholami A, Goormachtig S, Goossens A (2012) Transcriptional machineries in jasmonate-elicited plant secondary metabolism. Trends Plant Sci 17:349–359

De Leon IP, Sanz A, Hamberg M, Castresana C (2002) Involvement of the *Arabidopsis* α-DOX1 fatty acid dioxygenase in protection against oxidative stress and cell death. Plant J 29:61–62

Dehne HW (1982) Interaction between vesicular-arbuscular mycorrhizal fungi and plant pathogens. Phytopathology 72:1115–1119

Delledonne M, Murgia I, Ederle D, Sbicego PF, Biondian A, Polverari A, Lamb C (2002) Reactive oxygen intermediates modulates nitric oxide signaling in the plant hypersensitive disease-resistance response. Plant Physiol Biochem 40:605–610

Dixon RA, Harrison MJ (1990) Activation, structure and organization of genes involved in microbial defense in plants. Adv Genet 28:165–234

Dumas-Gaudot E, Asselin A, Gianinaui-Pearson V, Gollotte A, Gianinaui S (1994) Chitinase isoforms in roots of various pea genotypes infected with arbuscular mycorrhizal fungi. Plant Sci 99:27–37

Dumas-Gaudot E, Gollotte A, Ordier C, Gianinazzi S, Gianinazzi-Pearson V (2000) Modulation of host defence systems. In Arbuscular mycorrhizas: physiology and function, Kluwer Academic Publishers, Dordrecht, pp 173–200

El- Khallal SM (2007) Induction and modulation of resistance in tomato plants against *Fusarium* wilt disease by bioagent fungi (arbuscular mycorrhiza) and/or hormonal elicitors (Jasmonic acid & Salicylic acid): 2-Changes in the antioxidant enzymes, phenolic compounds and pathogen related proteins. Aust J Basic Appl Sci 1:717–732

Engstrom K, Widmark AK, Brishammar S, Helmersoon S (1999) Antifungal activity to *Phytophthora infestans* of sesquiterpenoids from infected potato tubers. Potato Res 42:43–50

Evans DG, Miller MH (1988) Vesicular arbuscular mycorrhizas and the soil induced reduction of nutrient absorption in maize, casual relation. New Phytol 110:67–74

Feussner I, Kühn H, Wasternack C (2001) Lipoxygenase dependent degradation of storage lipids. Trends Plant Sci 6:268–273

Franken P, Gnadinger F (1994) Analysis of parsley arbuscular endomycorrhiza: Infection development and mRNA levels of defense related genes. Plant-Microbe Interact 7:612–620

Fritz M, Jakobsen I, Lyngkjaer MZ, Thordal-Christensen H, Pons Kuhnemann J (2006) Arbuscular mycorrhiza reduces susceptibility of tomato to *Alternaria solani*. Mycorrhiza 16:413–419

Garcia-Garrido JM, Ocampo JA (2002) Regulation of the plant defence response in arbuscular mycorrhizal symbiosis. J Exp Bot 53:1377–1386

Garcia-Gonzalez I, Quemada M, Gabriel JL, Hontoria C (2016) Arbuscular mycorrhizal fungal activity responses to winter cover crops in a sunflower and maize cropping system. Appl Soil Ecol 102:10–18

Garmendia I, Aguirreolea J, Goicoechea N (2006) Defence-related enzymes in pepper roots during interactions with arbuscular mycorrhizal fungi and/or *Verticillium dahliae*. BioControl 51:293–310
Gaur A, Adholeya A (2004) Prospects of arbuscular mycorrhizal fungi in phytoremediation of heavy metal contaminated soils. Curr Sci 86:528–534
Gaur A, Adholeya A (2002) Arbuscular–mycorrhizal inoculation of five tropical fodder crops and inoculum production in marginal soil amended with organic matter. Biol Fertil Soils 35:214–218
Gosling P, Hodge A, Goodlass G, Bending GD (2006) Arbuscular mycorrhizal fungi and organic farming. Agric Ecosyst Environ 113:17–35
Grant C, Bittman S, Montreal M, Plenchette C, Morel C (2005) Soil and fertilizer phosphorus: effects on plant P supply and mycorrhizal development. Can J Plant Sci 85:3–14
Gutjahr C, Paszkowski U (2009) Weights in the balance: jasmonic acid and salicylic acid signaling in root-biotroph interactions. Mol Plant-Microbe Interact 22:763–772
Hause B, Maier W, Miersch O, Kramell R, Strack D (2002) Induction of jasmonate biosynthesis in arbuscular mycorrhizal barley roots. Plant Physiol 130:1213–1220
Hause B, Mrosk C, Isayenkov S, Strack D (2007) Jasmonates in arbuscular mycorrhizal interactions. Phytochemistry 68:101–110
Howe GA, Schilmiller AL (2002) Oxylipin metabolism in response to stress. Curr Opin Plant Biol 5:230–236
Jiao H, Yinglong C, Lin X, Liu R (2011) Diversity of arbuscular mycorrhizal fungi in greenhouse soils continuously planted to watermelon in North. China. Mycorrhiza 21:681–688
Johansson JF, Paul LR, Finlay RD (2004) Microbial interactions in the mycorrhizosphere and their significance for sustainable agriculture. FEMS Microbiol Ecol 48:1–13
Joner EJ, Briones R, Leyval C (2000) Metal-binding capacity of arbuscular mycorrhizal mycelium. Plant Soil 226(2):227–234
Jones DL, Hodge A, Kuzyakov Y (2004) Plant and mycorrhizal regulation of rhizo deposition. New Phytol 163:459–480
Jung SC, Martinez-Medina A, Lopez-Raez JA, Pozo MJ (2012) Mycorrhiza-induced resistance and priming of plant defenses. J Chem Ecol 38:651–664
Kaldorf M, Kuhn AJ, Schroder WH, Hildebrandt U, Bothe H (1999) Selective element deposits in maize colonized by a heavy metal tolerance conferring arbuscular mycorrhizal fungus. J Plant Physiol 154:718–728
Kapoor R, Bhatnagar AK (2007) Attenuation of cadmium toxicity in mycorrhizal celery (*Apium graveolens* L.). World J Microbiol Biotechnol 23:1083–1089
Khan AG (1993) Occurrence and importance of mycorrhizas in aquatic trees of New South Wales Australia. Mycorrhiza 3:31–38
Kloppholz S, Kuhn H, Requena N (2011) A secreted fungal effector of *Glomus intraradices* promotes symbiotic biotrophy. Curr Biol 21:1204–1209
Kobayashi I, Kobayashi Y, Yamaoka N, Kunoh H (1992) Recognition of a pathogen and a nonpathogen by barley coleoptile cells. III. Responses of microtubules and actin filaments in barley coleoptile cells to penetration attempts. Can J Bot 70(9):1815–1823
Kosuta S, Hazledine S, Sun J, Miwa H, Morris RJ, Downie JA, Oldroyd GED (2008) Differential and chaotic calcium signatures in the symbiosis signaling pathway of legumes. Proc Natl Acad Sci U S A 105:9823–9828
Künkel BN, Brooks DM (2002) Cross talk between signaling pathway in pathogen defense. Curr Opin Plant Biol 5:325–331
Lamb C, Dixon RA (1997) The oxidative burst in plant disease resistance. Annu Rev Plant Physiol Plant Mol Biol 48:251–257
Lambais MR (2000) Regulation of plant defence-related genes in arbuscular mycorrhizas. In: Podilla GK, Douds DD (eds) Current advances in mycorrhizas research. The American Phytopathological Society, St. Paul, pp 45–59

Landwehr M, Hildebrandt U, Wilde P, Nawrath K, Tóth T, Biró B, Bothe H (2002) The arbuscular mycorrhizal fungus *Glomus geosporum* in European saline, sodic and gypsum soils. Mycorrhiza 12:199–211

Leon Morcillo RJ, Ocampo JA, Garcia Garrido JM (2012) Plant 9-lox oxylipin metabolism in response to arbuscular mycorrhiza. Plant Signal Behavior 7:1584–1588

Levy J, Bres C, Geurts R, Chalhoub B, Kulikova O, Duc G, Journet EP, Ane JM, Lauber E, Bisseling T, Denarie J, Rosenberg C, Debelle F (2004) A putative Ca^{2+} and calmodulin dependent protein kinase required for bacterial and fungal symbioses. Science 303:1361–1364

Li HY, Yang GD, Shu HR, Yang YT, Ye BX, Nishida I, Zheng CC (2006) Colonization by the arbuscular mycorrhizal fungus *Glomus versiforme* induces a defense response against the root-knot nematode *Meloidogyne incognita* in the grapevine (*Vitis amurensis* Rupr.), which includes transcriptional activation of the class III chitinase gene VCH3. Plant Cell Physiol 47:154–163

Linderman RG (1994) Role of VAM fungi in biocontrol. In: Pfleger FL, Linderman RG (eds) Mycorrhizae and plant health. APS, St Paul, pp 1–26

Lioussanne L, Jolicoeur M, St-Arnaud M (2008) Mycorrhizal colonization with *Glomus intraradices* and development stage of transformed tomato roots significantly modify the chemotactic response of zoospores of the pathogen *Phytophthora nicotianae*. Soil Biol Biochem 40:2217–2224

Liu J, Maldonado-Mendoza I, Lopez-Meyer M, Cheung F, Town CD, Harrison MJ (2007) Arbuscular mycorrhizal symbiosis is accompanied by local and systemic alterations in gene expression and an increase in disease resistance in the shoots. Plant J 50:529–544

Lohse S, Schliemann W, Ammer C, Kopka J, Strack D, Fester T (2005) Organization and metabolism of plastids and mitochondria in arbuscular mycorrhizal roots of *Medicago truncatula*. Plant Physiol 139:329–340

Lopez-Raez JA, Charnikhova T, Fernández I, Bouwmeester H, Pozo MJ (2011) Arbuscular mycorrhizal symbiosis decreases strigolactone production in tomato. J Plant Physiol 168:294–297

Lorenzo O, Solano R (2005) Molecular players regulating the jasmonate signalling network. Curr Opin Plant Biol 8:532–540

Ludwig-Muller J (2000) Hormonal balance in plants during colonization be mycorrhizal fungi. In: Kapulnik Y, Douds D (eds) Arbuscular mycorrhizas: physiology and function. Kluwer Academic Publishers, Amsterdam, pp 263–283

Manthey K, Krajinski F, Hohnjec N, Firnhaber C, Puhler A, Perlick A, Kuster H (2004) Transcriptome profiling in root nodules and arbuscular mycorrhiza identifies a collection of novel genes induced during *Medicago truncatula* root endosymbioses. Mol Plant-Microbe Interact 17:1063–1077

Marin M, Ybarra M, Fe A, Garcia-Ferriz L (2002) Effect of arbuscular mycorrhizal fungi and pesticides on *Cynara cardunculus* growth. Agric Food Sci Finland 11:245–251

McArthur DA, Knowles NR (1992) Resistance responses of potato to vesicular-arbuscular mycorrhizal fungi under varying abiotic phosphorus levels. Plant Physiol 100:341–351

McGonigle TP, Miller MH (1999) Winter survival of extra radical hyphae and spores of arbuscular mycorrhizal fungi in the field. Appl Soil Ecol 12:41–50

Meixner C, Ludwig-Muller J, Miersch O, Gresshoff P, Staehelin C, Vierheilig H (2005) Lack of mycorrhizal autoregulation and phytohormonal changes in the supernodulating soybean mutant nts1007. Planta 222:709–715

Menendez AB, Scervino JM, Godeas AM (2001) Arbuscular mycorrhizal populations associated with natural and cultivated vegetation on a site of Buenos Aires province. Argent. Biol. Fertil Soil 33:373–381

Meyer JR, Linderman RG (1986) Response of subterranean clover to dual inoculation with vesicular-arbuscular mycorrhizal fungi and a plant growth-promoting bacterium, *Pseudomonas putida*. Soil Biol Biochem 18:185–190

Miller MH, McGonigle TP, Addy HD (1995) Functional ecology if vesicular arbuscular mycorrhizas as influenced by phosphate fertilization and tillage in an agricultural ecosystem. Crit Rev Biotechnol 15:241–255

Miranda JCC, Harris PJ (1994) Effects of soil P on spore germination and hyphal growth of fungi. New Phytol 128:103–108
Morton JB, Benny GL (1990) Revised classification of arbuscular mycorrhizal fungi (Zygomycetes): a new order, Glomales, two new suborders, Glomineae and Gigasporineae, and two new families, Acaulosporaceae and Gigasporaceae with an emendation of Glomaceae. Mycotaxon 37:471–492
Mozafar A, Anken T, Ruh R, Frossard E (2000) Tillage intensity, mycorrhizal and non mycorrhizal fungi and nutrient concentrations in maize, wheat and canola. Agron J 92:1117–1124
Muller J, Staehelin C, Xie JP, Neuhaus-Url G, Boller T (2000) Nod factors and chitooligomers elicit an increase in cytosolic calcium in aequorin expressing soybean cells. Plant Physiol 124:733–740
Navazio L, Baldan B, Moscatiello R, Zuppini A, Woo SL, Mariani P, Lorito M (2007) Calcium-mediated perception and defense responses activated in plant cells by metabolite mixtures secreted by the biocontrol fungus *Trichoderma atroviride*. BMC Plant Biol 7:41–49
Njeru EM, Avio L, Bocci G, Sbrana C, Turrini A, Barberi P, Giovannetti M, Oehl F (2015) Contrasting effects of cover crops on 'hot spot' arbuscular mycorrhizal fungal communities in organic tomato. Biol Fertil Soils 51:151–166
Owen D, Williams AP, Griffith GW, Withers PJA (2015) Use of commercial bio-inoculants to increase agricultural production through improved phosphorus acquisition. Appl Soil Ecol 86:41–54
Plenchette C, Furlan V, Fortin JA (1982) Comparative effects of different endomycorrhizal fungi on five host plants grown on calcined montmorillonite clay. J Am Soc Hortic Sci 107:535–538
Pozo MJ, Azcon-Aguilar C (2007) Unraveling mycorrhiza-induced resistance. Curr Opin Plant Biol 10:393–398
Pozo MJ, Azcón-Aguilar C, Dumas-Gaudot E, Barea JM (1999) β-1, 3-glucanase activities in tomato roots inoculated with arbuscular mycorrhizal fungi and/or *Phytophthora parasitica* and their possible involvement in bioprotection. Plant Sci 141:149–157
Pozo MJ, Loon LCV, Pieterse CMJ (2004) Jasmonates – signals in plant–microbe interactions. J Plant Growth Regul 23:211–222
Pozo MJ, Verhage A, García-Andrade J, García JM, Azcón-Aguilar C (2009) Priming plant defence against pathogens by arbuscular mycorrhizal fungi. In: Mycorrhizas-functional processes and ecological impact. Springer, Berlin/Heidelberg, pp 123–135
Prost I, Dhondt S, Rothe G, Vicente J, Rodriguez MJ, Kift N (2005) Evaluation of the antimicrobial activities of plant oxylipins supports their involvement in defense against pathogens. Plant Physiol 139:1902–1913
Rouphael Y, Cardarelli M, Colla G (2015) Role of arbuscular mycorrhizal fungi in alleviating the adverse effects of acidity and aluminium toxicity in zucchini squash. Sci Horticult 188:97–105
Ruiz-Lozano JM (2003) Arbuscular mycorrhizal symbiosis and alleviation of osmotic stress. New perspectives for molecular studies. Mycorrhiza 13:309–317
Ruiz-Lozano JM, Roussel H, Gianinazzi S, Gianinazzi-Pearson V (1999) Defense genes are differentially induced by a mycorrhizal fungus and *Rhizobium* sp. in wild-type and symbiosis-defective pea genotypes. Mol Plant-Microbe Interact 12:976–984
Salzer P, Boller T (2000) Elicitor-induced reactions in mycorrhizas and their suppression. In: Podila GK, Douds DD (eds) Current advances in mycorrhizas research. The American Phytopathological Society, St. Paul, pp 1–10
Schilmiller AL, Howe GA (2005) Systemic signaling in the wound response. Curr Opin Plant Biol 8:369–377
Schliemann W, Ammer C, Strack D (2008) Metabolite profiling of mycorrhizal roots of *Medicago truncatula*. Phytochemistry 69:112–146
Secilia J, Bagyaraj DJ (1987) Bacteria and actinomycetes associated with pot cultures of vesicular-arbuscular mycorrhizas. Can J Microbiol 33:1069–1073
Sengupta A, Chaudhuri S (2002) Arbuscular mycorrhizal relations of mangrove plant community at the Ganges river estuary in India. Mycorrhiza 12:169–174

Sheng PP, Liu RJ, Li M (2012) Inoculation with an arbuscular mycorrhizal fungus and intercropping with pepper can improve soil quality and watermelon crop performance in a system previously managed by monoculture. Am Eurasian J Agri Environ Sci 12:1462–1468

Siedow JN (1991) Plant lipoxygenase: structure and function. Annu Rev Plant Physiol Plant Mol Biol 42:145–188

Singh RK, Dai O, Nimasow G (2011) Effect of arbuscular mycorrhizal (AM) inoculation on growth of chili plant in organic manure amended soil. Afr J Micorbiol Res 28:5004–5012

Smith GS (1987) Interactions of nematodes with mycorrhizal fungi. In: Veech JA, Dickon DW (eds) Vistas on nematology. Society of Nematology, Hyattsville, pp 292–300

Smith SE, Read DJ (2008) Mycorrhizal symbiosis, 3rd edn. Academic, London

Smith FA, Smith SE (2011) What is the significance of the arbuscular mycorrhizal colonisation of many economically important crop plants. Plant Soil 348:63–79

Song Y, Chen D, Lu K, Sun Z, Zeng R (2015) Enhanced tomato disease resistance primed by arbuscular mycorrhizal fungus. Front Plant Sci 6:786

Steinkellner S, Lendzemo V, Langer I, Schweiger P, Khaosaad T, Toussaint JP, Vierheilig H (2007) Flavonoids and strigolactones in root exudates as signals in symbiotic and pathogenic plant-fungus interactions. Molecules 12:1290–1306

Stintzi A, Browse J (2000) The Arabidopsis male-sterile mutant, opr3, lacks the 12-oxo-phytodienoic acid reductase required for jasmonate synthesis. Proc Natl Acad Sci U S A 97:10625–10630

Strange RN (2003) Introduction to plant pathology. Wiley, England

Strittmatter G, Gheysen G, Gianinaui-Pearson V, Hahn K, Niebel A, Rohde W, Tacke E (1996) Infections with various types of organisms stimulate transcription from a short promoter fragment of the potato gstl gene. Mol Plant-Microbe Interact 9:68–73

Stumpe M, Carsjens JG, Stenzel I, Gobel C, Lang I, Pawlowski K, Hause B, Feussner I (2005) Lipid metabolism in arbuscular mycorrhizal roots of *Medicago truncatula*. Phytochemistry 66:781–791

Subramanian KS, Charest C (1999) Acquisition of N by external hyphae of an arbuscular mycorrhizal fungus and its impact on physiological response in maize under drought-stressed and well-watered conditions. Mycorrhiza 9:69–75

Titus JH, Titus PJ, Nowak RS, Smith SD (2002) Arbuscular mycorrhizas of Mojave Desert plants. Western N Amer Naturalist 62:327–334

Torres MA, Jonathan DG, Dangl JL (2006) Reactive oxygen species signalling in response to pathogen. Plant Physiol 141:373–378

Toussaint J, Smith FA, Smith SE (2007) Arbuscular mycorrhizal fungi can induce the production of phytochemicals in sweet basil irrespective of phosphorus nutrition. Mycorrhiza 17:291–297

Vestberg M, Kahiluoto H, Wallius E (2011) Arbuscular mycorrhizal fungal diversity and species dominance in a temperate soil with long-term conventional and low-input cropping systems. Mycorrhiza 21:351–361

Vierheilig H, Piche Y (2002) Signalling in arbuscular mycorrhiza: facts and hypotheses. In: Buslig B, Manthey J (eds) Flavonoids in cell functions. Kluwer Academic/Plenum Publishers, New York, pp 23–39

Vierheilig H, Alt M, Mohr U, Boller T, Wiemken A (1994) Ethylene biosynthesis and activities of chitinase and β-1,3-glucanase in the roots of host and non-host plants of vesicular-arbuscular mycorrhizal fungi after inoculation with *Glomus mosseae*. J Plant Physiol 143:337–343

Vos CM, Yang Y, DeConinck B, Cammue BPA (2014) Fungal (−like) biocontrol organisms in tomato disease control. Biol Control 74:65–81

Waaland ME, Allen EB (1987) Relationships between VA mycorrhizal fungi and plant cover following surface mining in Wyoming. J Range Manag 40:271–276

Wasternack C, Hause B (2002) Jasmonates and octadecanoids: signals in plant stress responses and development. Prog Nucleic Acid Res Mol Biol 72:165–221

Whipps JM (2004) Prospects and limitations for mycorrhizas in biocontrol of root pathogens. Canad J Bot 82:1198–1227

Zhu HH, Yao Q (2004) Localized and systemic increase of phenols in tomato roots induced by *Glomus versiforme* inhibit *Ralstonia Solanacearum*. J Phytopathol 152:537–546

Microbes as Resource of Biomass, Bioenergy, and Biofuel

9

Vincent Vineeth Leo, Lallawmsangi, Lalrokimi, and Bhim Pratap Singh

9.1 Introduction

Decades of dependence on fossil fuels have helped in the realization that an impending diminution of the existing nonrenewable resources is inevitable, even though bulk of the countries globally still depends on crude oil as its major energy source (Makishah 2017). Given its frequent usage and growing environmental concerns over greenhouse gas emissions and global warming research and study on alternative energy or fuel sources has been subjected to immense importance resulting in the improvements and innovation of numerous approaches (Leo et al. 2016, 2018). Alternative energy especially biodiesel and bioethanol has been known to be utilized partially in some parts of world. Other bio-based products like biogas, biohydrogen, biobutanol, syngas, bio-propanol, etc. have also known to be pursued lately (Farrell et al. 2006; Lynd et al. 2008 from Jang et al. 2012). The sources for such alternative energy feedstocks capable of replacing the nonrenewable sources have ranged from plant and animal resources to the smaller microbes.

Biofuel industry as such has been focused primarily on fuel derived from lignocellulosic or fatty acids in plants or animal oils or waste. The advancements in microbial biotechnology and allied fields have enhanced the prospective of microorganisms in utilizing various crude materials for biofuel production (Makishah 2017). Recently, the work on utilizing microbes itself as a prospective biomass for biofuel or bioenergy production has gained immense momentum. Microbial fuel cells (MFC) capable of growing in bulk, with faster growth rate, higher metabolisms, and rich in its biomass (cellulosic, hemi-cellulosic, disaccharides,

V. V. Leo · Lallawmsangi · Lalrokimi · B. P. Singh (✉)
Molecular Microbiology and Systematic Laboratory, Department of Biotechnology, Mizoram University, Aizawl, Mizoram, India

D. P. Singh et al. (eds.), *Microbial Interventions in Agriculture and Environment*,
https://doi.org/10.1007/978-981-13-8391-5_9

monosaccharides, fatty acids, etc.) are ideal sources for such energy production usage. Among the microorganisms explored, though fungal, bacterial, actinobacterial cultures of mesophilic or extremophilic nature were reported as microbial fuel cells, a substantial research was also extended to the highly efficient algae (microalgae to macroalgae).

In this work, we intend to look up onto the usage of microorganisms in the field of alternative energy, its role as a biomass and its effectiveness in biofuel or bioenergy production from the past, current and potential future scenarios. Few well-explored microorganisms used as a resource for biofuel and bioenergy productions are represented in Table 9.1.

9.2 Microbes as Resource for Biomass

Biomass is defined as living or lately deceased parts of plants, animals, or microorganisms and any subsequent materials produced or excreted by these organisms. They mostly encompass bioenergy crops, crop residues dominant in lignocellulose, algal growth mater, animals waste, etc. which are sourced for biofuel or bioenergy productivity. Living biomass uses carbon during growth and discharges this carbon for energy supplies and metabolisms, which in turn helps in ensuring a carbon-neutral cycle thereby regulating the greenhouse gases concentrations (Naithani et al. 2011). Though differential approaches were in development for the past years for utilization of biomass for bioenergy, a sustainable, cost effective and efficient system is still facing its challenges. In this scenario the outlook towards microorganisms as a biomass resource finds its profound relevance.

Among the microbial biomass, algae represents the most explored and competent bio-resource for biofuel or bioenergy industry. Given its versatility to acclimate to diverse range of aquatic ecosystems ranging from extreme saline, marine to freshwater conditions, and its capability to utilize CO_2 and to do carbon fixation makes them microbes of huge potential. These microorganisms have capability to enhance its biomass quantity very rapidly (Hannon et al. 2010). Numerous algal varieties have been reported to show oil-producing capability within their total dry biomass making them ideal (Rodolfi et al. 2009). One of its major advantages is its nominal land usage and its potential used as a bioremediation tool while culturing them in wastewater streams (Douskova et al. 2009). *Botryococcus braunii*, *Chlorella* sp., *Nannochloropsis* sp., *Schizochytrium* sp. are some of the renowned oleaginous microalgae used as a microbial biomass (Chisti 2007).

Recent study on biomass enhancement for bioenergy by Arias et al. (2018) was a mix of microalgae digest was grown on a tertiary mode for wastewater treatment plant along with secondary effluents. This biomass generated was then co-digested that yielded more methane quantity. In the above mentioned study, the addition of NaOH as biocatalyst with steam-to-biomass (S/B) ratio of 0.8 (wt/wt): 0.5 (mm of biomass size) resulted in enhanced hydrogen and subsequent syngas productions.

Table 9.1 Overview of some of the most explored microorganisms used as biomass for biofuel and bioenergy productions

Microbes	Uses	References
Arthrospira maxima	Used in the production of biomass using photo-bioreactors	Ullah et al. (2014)
		Raja et al. (2014)
Chlorella vulgaris	Used as biomass for generation of biofuels	Khandelwal et al. (2018)
		De Farias Silva and Bertucco (2017)
		El-Dalatony et al. (2017)
		Ullah et al. (2014)
		Raja et al. (2014)
Chlorella sarokiniana	Used for biomass production to generate bioethanol	De Farias Silva and Bertucco (2017)
Cladophora glomerata	Used as biomass for production of biofuel	Ebadi et al. (2018)
Dunaliella tertiolecta	Used as biomass through mass oil production	Ullah et al. (2014)
Botryococcus braunii	Used as biomass due to its high lipid content	Ullah et al. (2014)
		Meng et al. (2009)
Scenedesmus quadricauda	Used as a biomass to produce biofuels	El- Dalatony et al. (2017) and Ullah et al. (2014)
Chaetoceros muelleri	Used as biomass due to high lipid content for production of biofuels	Ullah et al. (2014)
		Raja et al. (2014)
Bacillus cereus	Used to produce biomass under defined conditions	Seel et al. (2016)
Bacillus alcalophilus	Used as biomass due to high oil/lipid content	Meng et al. (2009)
Listeria monocytogenes	Used to produce biomass at low growth temperature	Seel et al. (2016)
Staphylococcus xylosus	Mesophilic bacteria used in production of biomass	Seel et al. (2016)
Arachniotus spp.	Used in biomass protein production	Ahmed et al. (2010)
Candida utilis	Microbial biomass protein production	Ahmed et al. (2010)

(continued)

Table 9.1 (continued)

Microbes	Uses	References
Mucor circinelloides	Production of lipid and protein-rich fungal biomass	Carvalho et al. (2015)
		Mitra et al. (2012)
Aspergillus niger	Production of biomass by de novo lipid synthesis	Chuppa-Tostain et al. (2018)
		Donot et al. (2014)
Aspergillus oryzae	Production of fungal biomass	Mahboubi et al. (2017)
		Donot et al. (2014)
Neurospora intermedia	Production of biomass for feed	Mahboubi et al. (2017)
Chlamydomonas spp.	Production of biomass for biofuel production	El- Dalatony et al. (2017)
Escherichia coli	Used in the production of bioethanol, butanol and isobutanol	Koppolu and Vasilaga (2016)
		Dunlop (2011)
		Antoni et al. (2007)
Saccharomyces cerevisae	Used in bioethanol production	Selim et al. (2018)
		De Farias Silva and Bertucco (2017)
		Koppolu and Vasilaga (2016)
		Antoni et al. (2007)
Cornebacteriumglucamitam	Used in bioethanol production	Koppolu and Vasiliga (2016)
Zymomonas mobilis	Used as front runner in the production of ethanol	Soleymani and Rosentrater (2017)
		Koppolu and Vasilaga (2016)
		Antoni et al. (2007)
Clostridium acetobutylicum	Used in production of acetone and butanol	Antoni et al. (2007)
Clostridium butyricum	Used in the production of biohydrogen	Elshahed (2010)
Enterobacter androgens	Used in the production of biohydrogen	Elshahed (2010)
Schizochytrium mangrovei	Used in the production of biodiesel	Hong et al. (2013)
Chlorococcum parinum	Used in the production of biodiesel	Feng et al. (2014)
Yarrowia lipolytica	Used in the production of biodiesel	Marella et al. (2018)
Gelidium amansii	Microalgae used in the production of biohydrogen	Park et al. (2011)

Clostridium butyricum	Used in fermentive hydrogen production	Sharma and Arya (2017)
		Cabrol et al. (2017)
Cladophora glomerata	Used in the production of syngas and biohydrogen	Ebadi et al. (2018)
Clostridium autoethanogenum	Used in synthetic gas fermentation that can produce 2,3- butanediol and hexanol	Phillips et al. (2017)
		Bengelsdorf et al. (2013)
Eubacterium limosum	Acetogenic bacteria used mostly in syngas fermentation	Bengeldorf et al. (2013)
Clostridium ljungdahlii	Used in syngas fermentation for production of biobutanol	Phillips et al. (2017)
		Bengelsdorf et al. (2013)
Butyribacterium methylotrophicum	Used in syngas fermentation and can produce biobutanol and bioethanol	Phillips et al. (2017)
		Bengelsdorf et al. (2013)
Clostridium carboxidivorans	Can synthesize butanol and hexanol through syngas fermentation	Phillips et al. (2017)
Pichia stipitis	Used in fermentation process for generation of bioethanol	Selim et al. (2018)
		Obata et al. (2016)
Scenedesmus acutus	Used in photosynthetic microbial fuel cells (PMFC)	Angioni et al. (2018)
Trichococcus pasteurii	Used in microbial fuel cells (MFC) by reduction of hexavalent chromium Cr(VI)	Tandukar et al. (2009)
Pseudomonas aeruginosa	Production of electricity in MFC	Chaturvedi and Verma (2016)
Clostridium cellulolyticum	Cellulose-degrading bacteria used in MFC for electricity generation	Chaturvedi and Verma (2016)
		Ren et al. (2008)
Geobacter sulfurreducens	Electrochemically active bacteria used in MFC	Ren et al. (2008)
Enterobacter cloacae	Cellulose-degrading bacteria that uses cellulose as a substrate for electricity generation using dual chambered MFC	Rezaei et al. (2009)
Klebsiella pneumoniae	Production of electricity by MFC using azo dyes as the cathode oxidants	Liu et al. (2009)
Chorella vulgaris	Used in MFC for energy generation by the lipid biomass produced by the algae itself	Khandelwal et al. (2018)
Synechococcus leopoliensis	Primary strain used in artificial cascade of MFC systems	Walter et al. (2015)
Gammaproteoba-cteria	Dominant strain found on the biocathode in a MFC	Milner et al. (2016)

Among the bacterial cultures used for biohydrogen production, the most studied and used biomass inoculums are *Clostridium* and *Enterobacter* (Ginkel et al. 2005). Ebadi et al. (2018) worked on a macroalgal variety *Cladophora glomerata* L. as the major biomass feedstock by using steam gasification along with a catalyst for production of syngas and biohydrogen. Here, the effect of steam-to-biomass (S/B) ratio and the particle size of biomass used for biohydrogen productivity were verified. The outcome showed that in presence of NaOH as catalyst, the best S/B ratio was 0.8 (wt/wt), and 0.5 mm of biomass size was deemed appropriate for the subsequent syngas and biohydrogen productions.

Recent studies on bioethanol production from algae as major resource was reported by De-Farias-Silva and Bertucco (2017). *Chlorella vulgaris* was subjected to hydrolysis by dilute acid for biomass solubilization and subsequent conversion to bioethanol using *Saccharomyces cerevisiae*. The biomass load initially used was within higher proportion of 10% to 100 g L^{-1}, which was comparable to that of normal lignocellulosic load used for such purposes. Interestingly with almost 90% of biomass conversion by hydrolysis and fermentation into bioethanol, using the process envisaged within the above said study, this work could easily be recommended for a scale-up.

Single cell oil termed as SCO like triglycerides are reportedly synthesized and produced by quite a few number of microbes of oleaginous nature like the molds, yeasts, etc. (Donot et al. 2014). It has been understood that the lipid accretion usually occurs under limited nitrogen resources and within enhanced carbon sources (Economou et al. 2010). Hena et al. (2015) utilized *Chlorella sorokiniana* for biomass enrichment and lipid augmentation by treating the algae on dairy farm effluents (DFE). It was at the heterotrophic condition (7th day), *C. Sorokiniana* DS6 showed its maximum biomass of 3.93 gL^{-1} with biomass productivity of 280.72 mg L^{-1} d^{-1} which enhanced also its lipid content up to 1.23 gL^{-1}. Ji et al. (2015) employed food wastewater as a biomass multiplier (0.41 gL^{-1}) source for *Scenedesmus obliquus* (green microalgae) growth, which resulted in enhanced metabolites production (6th day) that had potential for exploitation in bioenergy and biofuel production based studies. Huy et al. (2018) used a consortium of microalgae which was grown using different organic wastewaters like textile effluents for biomass enhancement and bio-resource for biofuel production. The biomass productivity peaked to 0.4 g/L of volatile solids with complete reduction of phosphate and partial removal of nitrogen. Tan et al. (2018) used alcoholic wastewater as the major carbon resource for *Chlorella pyrenoidosa* biomass enrichment and for its subsequent lipid enhancement. Here, they grew this microalga in wastewater that contained starch also that was previously anaerobically digested. The addition of alcoholic wastewater on addition to starch wastewater in the ratio of 1:1.5 enhanced the biomass content by 35.29%, with the lipid production reaching up to 102.68% in contrast to the study done in wastewater containing starch alone.

Quite a few of aerobic bacterial cultures as biomass feedstock *Alphaproteobacteria*, *Betaproteobacteria*, *Gammaproteobacteria*, *Bacteroidetes*, etc. are known to be involved in bioelectricity production as biocathode (Reimers et al. 2006; De-Schamphelaire et al. 2010; Milner et al. 2016). Microbial fuel cells (MFC) are another field where microbes are used directly as a biomass for energy production. Gajda et al. (2015) worked on a proficient method for electricity generation that was self-sustainable using raw algal biomass dominant wastewater in MFC type system. It was further known that the microbial biomass feedstock contained majorly green algae and other microorganisms like cyanobacteria, heterotrophic bacteria and protozoans. Interestingly, here the anode was the source for electricity productivity, which was seen to be reliant on the biomass regeneration within the cathode, which in turn was dependent on the various nutrients from anodes feedstock. Thus this closed loop system ensured that electric power resulted in enrichment of biomass within cathode by ensuring nutrient recovery within the media. A similar study was conducted by Walter et al. (2015) in which algal biomass was employed as the organic carbon fuel for electricity development. Here also there was no pre-treatment like acid or thermal pre-treatment of the fresh algal biomass, instead a flow through system in which a continuous algal biomass feedstock is maintained. Huarachi-Olivera et al. (2018) recently worked on a microalgal *Chlorella vulgaris* biomass depended MFC. Here the microalgae formed the substantial power house in cathode for electricity development, while a consortium of bacterial and archaea formed the anaerobic anode source for the combined bioelectricity development. While Khandelwal et al. (2018) used the cathode chamber for *Chlorella vulgaris* growth initially for the lipid harvesting soon after the lipid harvest, the remaining biomass was powered up as an electron donor for electricity generation.

9.3 Microbial Biofuel

Biofuels are majorly hydrocarbon transportation fuels or additives to these liquid fuels of biological origins. Given the greenhouse emission alarms set by the nonrenewable usage, such alternative fuel sources have become useful entity. Currently, it is the first-generation fuel sources (corn, sugarcane, soy, palm oil, etc.) that derived biodiesel and bioethanol and dominate the biofuel industry. However, work related to second-generation fuel sources (lignocellulosic matters, sewage matters, algae, etc.) has picked up more prominence recently. Such fuel sources are mostly non-edible biomass feedstocks and are available abundantly (Gomiero 2015; Dragone et al. 2010). It is in this context microbial sourced biofuels have started to make a niche for itself a reliable and efficient fuel source. Some of the biofuels available are biodiesel, bioethanol, biogas, biohydrogen, syngas, and bioforms of butanol, propanol, etc. (Behera et al. 2015). The usefulness of various forms of energy particularly as a transportation fuel will be relevant until an alternative bioenergy sourced vehicles like electric vehicles are proficiently manufactured and marketed. However, the hybrid variants that are capable of using both electric and nonrenewable fuels are currently gaining momentum in the transportation fields.

9.3.1 Microbial Biodiesel

Diesel fuels are hydrocarbons containing C8–C25 carbon containing liquid fuel that is a subsidiary of petroleum. But the discovery of biodiesel has revolutionized this concept with this nonrenewable fuel made available from more lipid-containing sources like animal fats, algal and other microbial lipids, vegetable oils, etc., by the process of transesterification accompanied with alkaline and alcoholic catalyst (Guo et al. 2015).

Guo et al. (2015) in their comprehensive review reported that high lipid or oil-yielding algal species like *Skeletonema costatum*, *Calluna vulgaris*, *Neochloris oleoabundans*, *Scenedesmus obliquus*, etc. with more that 20% lipid content and biomass feedstock of more than (20 dry t ha^{-1} $year^{-1}$) were considered for biodiesel production. Microalgae are known as efficient source for lipid or oil containing microbes especially for biodiesel production with both heterotrophic and autotrophic varieties pursued for this purpose successfully (Elshahed 2010). Heterotrophic conditions for the growth of *Auxenochlorella protothecoides* using plant material birch hydrolysates as carbon source yielded 5.42 ± 0.32 g/L of lipids with 64.52 ± 0.53% lipid content within 120 h of incubation (Patel et al. 2018). Lipid extraction that is a predominant feature for robust biodiesel productivity has been practiced by Shin et al. (2018) in which a combination of hexane and methanol (1:1) on a *Tetraselmis* sp. dry biomass of 10 mL/g resulted in maximal fatty acid methyl ester (FAME). Salgueiro et al. (2015) had recovered biodiesel with 52% conversion efficiency from *Phaeodactylum tricornutum* (a marine diatom). Here, this microalga was harvested using $CuSO_4$, with a biomass recovery of 83% from a 200 mg/L culture dosage; from which the lipid components was extracted using ultrasound and ethanol followed by microwave-assisted transesterification (2.45 GHz, 800 W, 1 bar, 4 min) along with 60 mL methoxide and 2% NaOH catalyst in methanol.

Siddiqua et al. (2015) optimized conditions for production of biodiesel from coastal macroalgae *Chara vulgaris* of 12 g dry biomass feedstock a biodiesel yield of 3.6 ml with 9255.106 kcal/kg was obtained. They used 198 ml of chloroform, 0.75% NaCl at 65 °C temperature for the biodiesel recovery. Here using Box-Behnken design, by response surface methods (RSM) an optimized condition for biodiesel yield from macroalgae *Chara vulgaris* was predicted. This RSM-based study predicted biodiesel of 9255.106 kcal/kg and yield of 3.6 ml from a 12 g of dry algal biomass using 198 ml of chloroform and 0.75% NaCl at 65°C.

Kuan et al. (2018) worked on oleaginous yeast, *Rhodotorula glutinis* was sourced as the feedstock for biodiesel production. Here they never used the lipid extraction; instead direct transesterification was carried on 1 g of *R. glutinis* biomass using catalytic agent of 0.6 M sulfuric acid (70 °C, 20 h) that yielded 111% FAME in contrast to conventional transesterification process. This process was effective even when the biomass moisture content was as high as 70% were a 43% FAME yield was reported from his acid based catalytical method.

Vicente et al. (2009, 2010) introduced oleaginous fungi *Mucor circinelloides* whose lipid content was shown as the driving force for bio-diesel production, without any prior lipid extraction process requirement. Here direct transformation by

SmF (submerged fermentation) of *M. circinelloides* catalyzed by acid was successful in transforming the lipids into FAME (fatty acid methyl esters). This was certainly a desirable character for biodiesel as FAME-containing biodiesel could bring high energy density, better lubricity, etc. Du et al. (2018) in a recent report worked on a combination of marine algal varieties *Nannochloropsis oceanica* and oleaginous fungi *Mortierella elongata*. Here the cultures were grown separately under individual optimized conditions by bio-flocculation for accretion of triacylglycerol (TAG) ~ 15%, total fatty acids 22% of total dry weight (DW), and polyunsaturated fatty acids (PUFAs). This amount of recovery of lipids that has potential usage in biodiesel was substantial given that the recovery was from nutrient replete conditions (~ 10% of DW).

9.3.2 Microbial Bioethanol

Bioethanol is systematic microbial saccharification and fermentation of carbohydrate rich biomass to ethyl alcoholic form. Among the microbial sources rich in carbohydrate sources macroalgal cultures are the most popular feedstock for bioethanol production, with large quantities and simple growth requirements (Kang and Lee 2015). Korzen et al. (2015) work on the popular seaweed *Ulvarigida* in utilizing it as a biomass feedstock (196 ± 2.5 mg glucose) for bioethanol production yielded 333.3 ± 4.7 mg bioethanol per gram of glucose utilized within 3 h. Here simultaneous saccharification and fermentation (SSF) of *Saccharomyces cerevisiae* is employed with sonication. Obata et al. (2016) used brown algae *Laminaria digitata* which are rapid in their growth was pre-treated with acid and subjected to action with commercially available enzymes to obtain simple sugars. This was further subjection to action using non-conventional yeast like *Kluyveromyces marxianus* resulting in an ethanol yield of 6 g/L. Sunwoo et al. (2017) study on waste seaweed ranging from red, brown, and green (26%, 46% and 28%) in the Gwangalli beach, Korea, was used as major biomass feedstock for bioethanol productivity. Here they employed enzymatic saccharification and acid pre-treatments to obtain monosaccharide of 30.2 g/L which was converted to ethanol.

Among the microalgal sourced bioethanol production reported Ashokkumar et al. (2015), work on *Scenedesmus bijugatus* showed that from a biomass feedstock of 20 g L^{-1} (130 °C, 2% of acidic hydrolysis) saccharification of 85% was achieved with 0.158 g of bioethanol yield per gram of lipid extracted biomass residue. De-Farias-Silva and Bertucco (2017), in a comprehensive analysis on *Chlorella vulgaris* biomass conversion parameters of acid hydrolysis and varying conditions (100–130 °C, 0–60 min) with biomass loads of 10% to 100 g L^{-1} exhibited 90% sugar recovery. This was further subjected to *Saccharomyces cerevisiae* fermentation for a bioethanol yield of 60% was achieved, with a final ethanol concentration of 4.97 ± 0.09 g/L.

Aikawa et al. (2018) work on cyanobacterial culture *Arthrospira platensis* for an enhanced conversion of its carbohydrate biomass to bioethanol with an ethanol titer 48 g L^{-1}, bioethanol yield of 93% and productivity of 1.0 g L^{-1} h^{-1} was successful while employing lysozyme and $CaCl_2$. Here, the glycogen produced by *A. platensis* was extracted and subjected to fermentation using recombinant *Saccharomyces cerevisiae* capable of amylase activity a direct transformation of glycogen to ethanol that was made plausible, which has made this technology of huge significance in reducing the time intervals and price ranges.

9.3.3 Microbial Biogas

Biogas quintessentially a combination of energy efficient gases of methane (CH_4) and carbon dioxide (CO_2), along with a mix of other small gases; capable of being supplement for natural gas from biological origins by microbial breakdown of organic material under anaerobic conditions. Hence they are deemed Biogas or if they are dominated by the methane gas as such then they are termed "Biomethane" (Bio-CH_4) based on its composition (Kovacs et al. 2014).

Van-der-Ha et al. (2012) performed one of the landmark research works for biogas production, implementing the usage of two microorganisms concomitantly for this purpose. According to this report, they used synthetic CO_2 which was photosynthetically fixed by microalgae *Scenedesmus* sp. whose oxygen (O_2) generated was found to be responsible for the utilization of CH_4 by methanotrophic bacteria *Methylocystis parvus* for usable energy producing by-products (lipids and polyhydroxybutyrate). Nolla-Ardèvol et al. (2015) tried biogas rich in methane production up to 96% by optimizing the conditions using microalgal feedstock of *Spirulina* and its fermentation using mixed alkaline sludge (pH 10, 2.0 M Na^+). Here at the hydraulic retention time (HRT) of 15 days, CH_4 of 83 + 9% and CO_2 of only 14 + 6% was obtained. Further metagenomics and metatransciptomic analysis revealed that the sludge was majorly populated by a methanogenic bacterial community of *Methanocalculus*. Bassani et al. (2015) assembled a two-step reactor, where biogas formation will be concomitantly coupled with CO_2 reduction in the first reactor with additional hydrogen feed added in the second reactor. The methane then enhanced to 89% in the initial mesophilic reactor, which was almost maintained in the thermophilic second reactor.

Natural methane producers are also of huge interest like that *Methanomassiliicoccus* and *Methanococcus* which are archaeal species are one of the most explored methanogen. Goyal et al. (2016) stated that *Methanococcus maripaludis* converted the CO_2 to methane by a metabolic pathway termed "Wolfe cycle". Kröninger et al. (2017) tried biomethane production from *Methanomassiliicoccus luminyensis* which showed capability in utilizing of methylamines along with methanol. It was postulated that within anoxic settings are most favourable for microbial methane productions. But Angle et al. (2017) contradicted this understanding by conducting a systematic study over oxygenated soil samples of freshwater wetlands. The future works linked to the methanogenic pathways regarding this microbe will open up interesting avenues for methane gas harvesting. Ding et al. (2016) in attempting to enhance and maintain the ratio of carbon to nitrogen in the biomass feed for methane gas productions, carbon rich *Laminaria digitata* and nitrogen rich *Chlorella pyrenoidosa* and *Nannochloropsis oceanica* were mixed and subjected to fermentation. Here after the biohydrogen production the hydrogenic effluents were used to biomethane productions, with the energy conversion of 57–70% of H/CH_4 observed.

Azadi et al. (2014) in the formulation for a new dual fluidized bed (DFB) gasifier. The authors hypothesized that this gasifier was ideal for both biohydrogen as well for syngas production by using algal feedstock. They calculated a syngas yield of 17–24 MJ/kg dry feed. Ebadi et al. (2018) further reported syngas production by using NaOH as a catalyst for gasification of algal biomass *Cladophora glomerata* L.

9.3.4 Microbial Biohydrogen

Nicknamed as the cleanest fuel, "hydrogen" (H_2) is deemed as the fuel for the future, given that it is simply oxidized to H_2O and has minimal CO_2 emission. Biologically sourced hydrogen has been popularly explored using photosynthetic microorganisms dominated by micro- and macroalgal communities (Elshahed et al. 2010). From the micro- and macroalgal perspective, there are two ways of biohydrogen productions, which are biophotolysis (light energy is utilized for H_2 production) and dark fermentation (various genera of bacteria possessing capability in utilizing the algal biomass to produce H_2) techniques (Buitrón et al. 2017).

In early studies Azadi et al. (2014) did a simulation work on dual fluidized bed for biohydrogen productivity from an algal origin. The work gave an insight into the parameters that could influence the final product, like the algae oil content, cold gas efficiency, and H_2:CO ratio within the feed water as the major regulatory criteria. Batista et al. (2014) explored the possibilities of using known freshwater microalgae *Scenedesmus obliquus* as biomass source for biohydrogen production using fermentative bacterial cultures *Enterobacter aerogenes* and *Clostridium butyricum*. The work helped to reveal that dry biomass could be avoided with wet version of the algal biomass (69%) helping produce more biohydrogen. Among the fermentative bacteria used the *C. butyricum* was most efficient with bioH_2 of the rate 113.1 mL H_2/g VS_{alga} from 50.0g_{alga}/L. Xia et al. (2016) concentrated in another component that could influence the bioH_2 which was the galactose content, which dominated various biomasses like marine red algae. Here also the action of fermentative bacteria was utilized for final biomass conversion from a substrate concentration of 5 g volatile solid/L a H_2 yield and production of 278.1 mL/g galactose and 33.6 mL/g galactose/h, correspondingly was observed under a yeast extract/galactose ratio of 0.56.

With respect to the usage of pre-treatment techniques for enhancing biohydrogen production, quite a few works have been pursued in recent times. *Cladophora glomerata* L. whose microalgal biomass was steamed for gasification with the highest hydrogen production reported with alkali and alkaline-earth metal (AAEM) compounds catalysts, NaOH. The catalyst at 900 °C ensuring the syngas yield enhancement, with char conversion and tar diminishment (Ebadi et al. 2018). Kumar et al. (2016a) had explored the potential of microalgal consortia dominated by *Scenedesmus* and *Chlorella* species for biohydrogen production by initially verifying the best pre-treatment conditions for this purpose. The electrolysis pre-treatment resulted in hydrogen yield of 236 ± 14 mL/L/d and 37.7 ± 0.4 mL/g (volatile solids) VS_{added}, while in the later study of Kumar et al. (2016b), a hydrogen yield of 210 mL/L/d and 29.5 mL/g VS_{added}, respectively, were obtained under controlled conditions of pH 5.5 and in presence of methanogenic inhibitor (BESA). The findings correlated to the efficacy of this process by four times higher hydrogen productivity in comparison to untreated biomass subjected to hydrogen production.

Kidanu et al. (2017) worked on prospective capabilities of three macroalgal types *Saccharina japonica* (Brown algae), *Cladophora glomerata* (Red algae), and *Enteromorpha crinite* (Green algae) to be used as biomass feedstock for biohydrogen production. They found that the marine macroalgae *S. japonica* with a final biomass of 35 g/L produced the maximum biohydrogen and volatile fatty acids (VFA) as by-product at the rate of 179 mL/g of volatile solids and 9.8 g/L, respectively, by the additional action of anaerobic fermentation carried out on the municipal wastewater sludge. This work had some interesting indicators for hydrogen and VFA productivity and yield, with productivity increasing with the use of methanogen inhibitor, inoculum heat treatments and carbon sources.

9.3.5 Microbial Biobutanol

Butanol given its low heating value (LHV) has gained immense interest recently, mainly because they could easily be blended with gasoline, and used for running the combustion engines. Biobutanol which shares similar compositions to that of chemically synthesized butanol, can also be generated by fermentation of microbial biomasses (Swana et al. 2011).

Gao et al. (2016) verified the usage of ionic liquid extracted algae termed as ILEA and that of hexane extracted algae capability in converting *Chlorella vulgaris* strain UTEX 2714 starch components to direct butanol production. They recovered butanol in the amount of 4.99 and 6.63 g/L, from ILEA and HEA, respectively, without any detoxification process. While Wang et al. (2016) modified well-known pre-treatment technique of alkali (NaOH, 1%) and acid (H_2SO_4, 3%) treatment of the microalgae *Chlorella vulgaris* JSC-6 biomass for butanol production of 0.66/L/h, with a yield of 0.58 mol/mol sugar and a concentration of 13.1 g/L. It was highlighted that this process ensured the lack of inhibitors and the enhancement of nitrogenic by products had adverse effect on butanol yield.

Hou et al. (2017) explored enzyme-hydrolyzed sugar-rich brown seaweed (macroalgae) *Laminaria digitata* garnered out of the Danish North Sea, for its biobutanol fermentation using *Clostridium beijerinckii* DSM-6422. A biobutanol yield of 0.42 g/g of the biomass, with a concentration of butanol at 7.16 g/L in batch fermentation. In a recent study Al-Shorgani et al. (2018) batch fermentation of *Clostridium acetobutylicum* YM1 was studied for its biobutanol production and the influence of pH and butyric acid's effect on this process. Under unregulated pH conditions along with butyric acid in batch fermentation enhanced the activity butanol dehydrogenase that was of NADH-dependent; which resulted in butanol yield of 0.345 g/g, with productivity of 0.163 g/L h and concentrations of 16.50 + 0.8 g/L.

9.4 Microbial Bioenergy

Biomass of an organism like that of plant or microorganisms has a repertoire of energy in its biomass, which could be used for generation of renewable power like electricity or heat (Hannon et al. 2010). Microorganisms capable of producing electricity are of great interest and such microbes form the major component of MFC. Logan and Regan (2006) defined the process of MFC action were predominantly the microorganism used in cathode is separated at its terminal end from an electron acceptor, so that for microbe to respire it needs to transfer its electrons. Electro-neutrality between the cathode and anode ends are maintained by this electron transfer which is equaled by the protons that goes in the opposite direction. Numerous algal as well as bacterial communities have been reported as microbial fuel cell components.

According to Logan and Regan (2006), some of the known bacterial cultures capable of being involved in MFC are *Alcaligenes faecalis*, *Enterococcus gallinarum*, *Pseudomonas aeruginosa*, *Brevibacillus agri*, *Shewanella affinis*, and *Pseudoalteromonas* spp. Their role has been majorly to act as fermentative agents usually seen in mixed cultures for converting other microbial biomasses. Chaudhuri and Lovley (2003) made a MFC using psychrotolerant bacteria *Rhodoferax ferrireducens* which utilized glucose by oxidizing it and used this to power through electrons to the electrodes made of graphite. Here, without electron proton shuttler agent, durable power generation was made plausible. Graphite electrodes of porous foam nature produced current as much as 74 mA/m^2 and 445 mV using this MFC.

Gajda et al. (2015) successfully envisaged a MFC having anaerobic bacterial anode being powered by the photosynthetic mixed culture of photosynthetic algae acting as cathode; were maximum power generation was attained from the cathode of 128 μW. Xafenias et al. (2015), made a anaerobic sludge fuelled MFC, with alkaline cathodes, were Cr(VI) containing water at a concentration of 100 mg/L was completely bioremediated (48 h) and the power density recorded was in higher amounts of 767.01 mW/m^2 and 2.08 mA/m^2. Walter et al. (2015) developed MFC with a flow through system where the cells were continuously fed with algal culture of *Synechococcus leopoliensis* which could potentially generate electricity of 42 W per cubic meter of culture feed in the range of 6×10^5 cells mL^{-1}.

Angioni et al. (2018) explored *Scenedesmus acutus* PVUW12, an alga with photosynthetic potential concurrently used along with bacterial cultures for wastewater removal and to act as MFC with the assistance of polybenzimidazole membrane. Platinum (Pt) was used as the carbon capture device from microbial fuel cell compartment of *S. acutus* (cathode) and domestic wastewater containing bacteria (anode) under continuous illumination. This Pt cell that was thus electrocatalyzed produced power density of ~400 mW m^{-3} and energy recovery of NER > 0.19 kWh kg_{COD}^{-1} after 100 days of treatment. Moreover, the CO_2 produced by the bacteria in anode were successfully fixed by PMFC-grown *S. acutus* helped in production of various fatty acid and pigment by products.

Huarachi-Olivera et al. (2018) used microalga *Chlorella vulgaris* and bacterial cultures to power up MFC and subsequent bioremediation of blue dye containing effluents. The electricity generated accounted to be 327.67 mW/m^2 of bielectrogenic activity with a potential charge of 954 mV within 32 days of formulating the MFC of algal cathode and bacterial anode. Here 95% of fats and oils separation within *C. vulgaris* cathode, followed by chemical oxygen demand of 71% in anode, was observed. Interestingly at the anode the bacterial community dominated by *Deltaproteobacteria* and *Actinobacteria* showed biofilm formations which showed the power struggle among communities for the substrates (blue dye brl) and for electron transfer from cathode in energy productivity. Khandelwal et al. (2018) utilized the lipid free *C. vulgaris* in cathode chamber as the electron donor substrate for the bacterial anode to act on for the production of power 2.7 W m^{-3}. This MFC was capable of producing electricity accounted by 0.0136 kWh Kg^{-1} COD day^{-1} and 0.0782 kWh m^{-3} per day of algal oil energy.

Macroalgae *Saccharina japonica* as a major fuel source for MFC for electricity production was explored by Gebresemati et al. (2017) where nickel-based nanoparticles were used as catalytical agent in cathode. The best energy productivity was obtained when anaerobically fermented (AF) (anaerobically digested sludge from municipal wastewater) macroalgae was used along with Ni-based cathode with power density of 540 mW/m^2. Though this was marginally lesser than that reported (560 mW/m^2) using acetate instead of AF, AF was considered efficient given its potential in using natural sources for this energy production. Here, the AF sludge was dominated by *Firmicutes* and *Proteobacteria* that dominated both in inoculum and in biofilm production.

In a new application study Bateson et al. (2018) devised a MFC termed as bio-bottle-voltaic (BBV) from recycled plastic bottle that utilized recycled aluminum as its anode while the green algal variety of *Chlorella sorokiniana* was the cathode source. At its maximal run it generated electricity up to~2000 mC·$bottle^{-1}$·day^{-1}. This work is planning to apply this technology to power small energy requiring electrical devices.

9.5 Conclusion

The recent advancements listed above will need to be perfected for large scale application with socio-economic feasibilities. With the advent of metagenomics, proteomics and metatransciptomic developments coupled with more durable biomass retention technologies these innovations could enhance the understandings into the mechanism involved in specific biofuel or bioenergy productions. This in turn could help in more robust utilization of these microbial substrates and recycling of the by-products. Recently, work has been focused on utilizing the entire biomass with emphasis given on utilization or recycling of even the by-products. With developed countries emphasizing on implementation of usage of alternative fuels with prime focus on reduction in CO_2 emissions, the focus has now shifted to non-edible biomasses like microbial biomass. Hence, exemplifying the existing technologies for harvesting and extracting fuel will be the exertion for the future.

Acknowledgment This work was supported under different research projects funded to Bhim Pratap Singh by the Department of Biotechnology (DBT), New Delhi.

References

Ahmed S, Ahmad F, Hashmi AS (2010) Production of microbial biomass protein by sequential culture fermentation of *Arachniotusap* and *Candida utilis*. Pak J Bot 42:1225–1234

Aikawa S, Inokuma K, Wakai S, Sasaki K, Ogino C, Chang J-S et al (2018) Direct and highly productive conversion of cyanobacteria *Arthrospira platensis* to ethanol with $CaCl_2$ addition. Biotechnol Biofuels 11:50. https://doi.org/10.1186/s13068-018-1050-y

Al-Shorgani NKN, Kalil MS, Yusoff WMW, Hamid AA (2018) Impact of pH and butyric acid on butanol production during batch fermentation using a new local isolate of *Clostridium acetobutylicum* YM1. Saudi J Biol Sci 25(2):339–348. https://doi.org/10.1016/j.sjbs.2017.03.020

Angioni S, Millia L, Mustarelli P, Doria E, Temporiti ME, Mannucci B et al (2018) Photosynthetic microbial fuel cell with polybenzimidazole membrane: synergy between bacteria and algae for wastewater removal and biorefinery. Heliyon 4:e00560. https://doi.org/10.1016/j.heliyon.2018.e00560

Angle JC, Morin TH, Solden LM, Narrowe AB, Smith GJ, Borton MA et al (2017) Methanogenesis in oxygenated soils is a substantial fraction of wetland methane emissions. Nat Commun 8(1):1567. https://doi.org/10.1038/s41467-017-01753-4

Antoni D, Zverlov VV, Schwarz WH (2007) Biofuels from microbes. Appl Microbiol Biotechnol 77:23–35. https://doi.org/10.1007/s00253-007-1163-x

Arias DM, Solé-Bundó M, Garfí M, Ferrer I, García J, Uggetti E (2018) Integrating microalgae tertiary treatment into activated sludge systems for energy and nutrients recovery from wastewater. Bioresour Technol 247:513–519. https://doi.org/10.1016/j.biortech.2017.09.123

Ashokkumar U, Salom Z, Tiwari ON, Chinnasamy S, Mohamed S, Ani FN (2015) An integrated approach for biodiesel and bioethanol production from *Scenedesmus bijugatus* cultivated in a vertical tubular photobioreactor. Energy Convers Manag 101:778–786. https://doi.org/10.1016/j.enconman.2015.06.006

Azadi P, Brownbridge GPE, Mosbach S, Inderwildi OR, Kraft M (2014) Production of biorenewable hydrogen and syngas via algae gasification: a sensitivity analysis. Energy Procedia 61:2767–2770. https://doi.org/10.1016/j.egypro.2014.12.302

Bassani I, Kougias PG, Treu L, Angelidaki I (2015) Biogas upgrading via hydrogenotrophic methanogenesis in two-stage continuous stirred tank reactors at mesophilic and thermophilic conditions. Environ Sci Technol 49:12585–12593. https://doi.org/10.1021/acs.est.5b03451

Bateson P, Fleet J, Riseley A, Janeva E, Marcella A, Farinea C et al (2018) Electrochemical characterization of Bio-Bottle-Voltaic (BBV) systems operated with algae and built with recycled materials. Biology 7:26. https://doi.org/10.3390/biology7020026

Batista AP, Moura P, Marques PASS, Ortigueira J, Alves L, Gouveia L (2014) *Scenedesmus obliquus* as feedstock for biohydrogen production by *Enterobacteraerogenes* and *Clostridium butyricum*. Fuel 117:537–543. https://doi.org/10.1016/j.fuel.2013.09.077

Behera S, Singh R, Arora R, Sharma NK, Shukla M, Kumar S (2015) Scope of algae as third generation biofuels. Front Bioeng Biotechnol 2:90. https://doi.org/10.3389/fbioe.2014.00090

Bengelsdorf FR, Straub M, Dürre P (2013) Bacterial synthesis gas (syngas) fermentation. Environ Technol 34:1639–1651
Buitrón G, Carrillo-Reyes J, Morales M, Faraloni C, Torzillo G (2017) Biohydrogen production from microalgae. In: Muñoz R, González C (eds) Microalgae-based biofuels and bioproducts. Elsevier, Amsterdam, pp 210–234. ISBN 9780081010235
Cabrol L, Marone A, Venegas ET, Steyer JP, Filippi GR, Trably E (2017) Microbial ecology of fermentative hydrogen producing bioprocesses: useful insights for driving the ecosystem function. FEMS Microbiol Rev 41:158–181
Carvalho AKF, Rivaldi JD, Barbosa JC, de Castro HF (2015) Biosynthesis, characterization and enzymatic transesterification of single cell oil of *Mucor circinelloides* – a sustainable pathway for biofuel production. Bioresour Technol 181:47–53. https://doi.org/10.1016/j.biortech.2014.12.110
Chaturvedi V, Verma P (2016) Microbial fuel cell: a green approach for the utilization of waste for the generation of bioelectricity. Bioresour Bioprecess 3:38
Chaudhuri SK, Lovley DR (2003) Electricity generation by direct oxidation of glucose in mediatorless microbial fuel cells. Nat Biotechnol 21:1229–1232. https://doi.org/10.1038/nbt867
Chisti Y (2007) Biodiesel from microalgae. Biotechnol Adv 25:294–306
Chuppa-Tostain G, Hoarau J, Watson M, Adelard L, Shum Cheong Sing A, Caro Y et al (2018) Production of *Aspergillus niger* biomass on sugarcane distillery wastewater: physiological aspects and potential for biodiesel production. Fungal Biol Biotechnol 5:1. https://doi.org/10.1186/s40694-018-0045-6
De Farias Silva CE, Bertucco A (2017) Dilute acid hydrolysis of microalgal biomass for bioethanol production: an accurate kinetic model of biomass solubilization, sugars hydrolysis and nitrogen/ash balance. React Kinet Mechan Catal 122:1095–1114. https://doi.org/10.1007/s11144-017-1271-2
De Schamphelaire L, Boeckx P, Verstraete W (2010) Evaluation of biocathodes in freshwater and brackish sediment microbial fuel cells. Appl Microbiol Biotechnol 87:1675–1687
Ding L, Cheng J, Xia A, Jacob A, Voelklein M, Murphy JD (2016) Co-generation of biohydrogen and biomethane through two-stage batch co-fermentation of macro- and micro-algal biomass. Bioresour Technol 218:224–231. https://doi.org/10.1016/j.biortech.2016.06.092
Donot F, Fontana A, Baccou JC, Strub C, Galindo SS (2014) Single cell oils (SCOs) from oleaginous yeasts and moulds: production and genetics. Biomass Bioenergy 68:135–150
Douskova I, Doucha J, Livansky K, Machat J, Novak P, Umaysova D et al (2009) Simultaneous flue gas bioremediation and reduction of microalgal biomass production costs. Appl Microbiol Biotechnol 82:179–185
Dragone G, Fernandes B, Vicente AA, Teixeira JA (2010) Third generation biofuels from microalgae in current research. In: Mendez-Vilas A (ed) Technology and education topics in applied microbiology and microbial biotechnology. Formatex, Badajoz, pp 1355–1366
Dunlop MJ (2011) Engineering microbes for tolerance to next-generation biofuels. Biotechnol Biofuels 4:32. https://doi.org/10.1186/1754-6834-4-32
Du ZY, Alvaro J, Hyden B, Zienkiewicz K, Benning N, Zienkiewicz A et al (2018) Enhancing oil production and harvest by combining the marine alga *Nannochloropsis oceanica* and the oleaginous fungus *Mortierella elongata*. Biotechnol Biofuels 11(1):174. https://doi.org/10.1186/s13068-018-1172-2
Ebadi AG, Hisoriev H, Zarnegar M, Ahmadi H (2018) Hydrogen and syngas production by catalytic gasification of algal biomass (*Cladophora glomerata* L.) using alkali and alkaline-earth metals compounds. Environ Technol 40(9):1178–1184. https://doi.org/10.1080/09593330.2017.1417495
Economou CN, Makri A, Aggelis G, Pavlou S, Vayenas DV (2010) Semi-solid state fermentation of sweet sorghum for the biotechnological production of single cell oil. Bioresour Technol 101:1385–1388
El-Dalatony MM, Salama ES, Kurade MB, Hassan SHA, Oh SE, Kim S et al (2017) Utilization of microalgal biofractions for bioethanol, higher alcohols, and biodiesel production: a review. Energies 10:2110. https://doi.org/10.3390/en10122110

Elshahed MS (2010) Microbiological aspects of biofuel production: current status and future directions. J Adv Res 1:103–111. https://doi.org/10.1016/j.jare.2010.03.001

Farrell AE, Plevin RJ, Turner BT, Jones AD, O'Hare M, Kammen DM (2006) Ethanol can contribute to energy and environmental goals. Science 311:506–508

Feng P, Deng Z, Huc Z, Wangb Z, Fan L (2014) Characterization of *Chlorococcum pamirum* as a potential biodiesel feedstock. Bioresour Technol 162:115–122. https://doi.org/10.1016/j.biortech.2014.03.076

Gajda I, Greenman J, Melhuish C, Ieropoulos I (2015) Self-sustainable electricity production from algae grown in a microbial fuel cell system. Biomass Bioenergy 82:87–93. https://doi.org/10.1016/j.biombioe.2015.05.017

Gao K, Orr V, Rehmann L (2016) Butanol fermentation from microalgae-derived carbohydrates after ionic liquid extraction. Bioresour Technol 206:77–85. https://doi.org/10.1016/j.biortech.2016.01.036

Gebresemati M, Das G, Park BJ, Yoon HH (2017) Electricity production from macroalgae by a microbial fuel cell using nickel nanoparticles as cathode catalysts. Int J Hydrog Energy 42:29874–29880. https://doi.org/10.1016/j.ijhydene.2017.10.127

Ginkel SWV, Oh SE, Logan BE (2005) Biohydrogen gas production from food processing and domestic wastewaters. Int J Hydrog Energy 30:1535–1542. https://doi.org/10.1016/j.ijhydene.2004.09.017

Gomiero T (2015) Are biofuels an effective and viable energy strategy for industrialized societies? A reasoned overview of potentials and limits. Sustainability 7:8491–8521

Goyal N, Zhou Z, Karimi IA (2016) Metabolic processes of *Methanococcus maripaludis* and potential applications. Microb Cell Factories 15:107. https://doi.org/10.1186/s12934-016-0500-0

Guo M, Song W, Buhain J (2015) Bioenergy and biofuels: history, status, and perspective. Renew Sust Energ Rev 42:712–725. https://doi.org/10.1016/j.rser.2014.10.013

Hannon M, Gimpel J, Tran M, Rasala B, Mayfield S (2010) Biofuels from algae: challenges and potential. Biofuels 1:763–784

Hena S, Fatihah N, Tabassum S, Ismail N (2015) Three stage cultivation process of facultative strain of *Chlorella sorokiniana* for treating dairy farm effluent and lipid enhancement. Water Res 80:346–356. https://doi.org/10.1016/j.watres.2015.05.001

Hong DD, Mai DTN, Thom LT, Ha NC, Lam BD, Tam LT et al (2013) Biodiesel production from Vietnam heterotrophic marine microalga *Schizochytrium mangrovei* PQ6. J Biosci Bioeng 116(2):180–185. https://doi.org/10.1016/j.jbiosc.2013.02.002

Hou X, From N, Angelidaki I, Huijgen WJJ, Bjerre A-B (2017) Butanol fermentation of the brown seaweed *Laminaria digitata* by *Clostridium beijerinckii* DSM-6422. Bioresour Technol 238:16–21. https://doi.org/10.1016/j.biortech.2017.04.035

Huarachi-Olivera R, Dueñas-Gonza A, Yapo-Pari U, Vega P, Romero-Ugarte M, Tapia J et al (2018) Bioelectrogenesis with microbial fuel cells (MFCs) using the microalga *Chlorella vulgaris* and bacterial communities. Electron J Biotechnol 31:34–43. https://doi.org/10.1016/j.ejbt.2017.10.013

Huy M, Kumar G, Kim H-W, Kim S-H (2018) Photoautotrophic cultivation of mixed microalgae consortia using various organic waste streams towards remediation and resource recovery. Bioresour Technol 247:576–581. https://doi.org/10.1016/j.biortech.2017.09.108

Jang YS, Park JM, Choi S, Choi YJ, Seung DY, Cho JH et al (2012) Engineering of microorganisms for the production of biofuels and perspectives based on systems metabolic engineering approaches. Biotechnol Adv 30:989–1000

Ji M-K, Yun H-S, Park S, Lee H, Park Y-T, Bae S et al (2015) Effect of food wastewater on biomass production by a green microalga *Scenedesmus obliquus* for bioenergy generation. Bioresour Technol 179:624–628. https://doi.org/10.1016/j.biortech.2014.12.053

Kang A, Lee TS (2015) Converting sugars to biofuels: ethanol and beyond. Bioengineering 2:184–203. https://doi.org/10.3390/bioengineering2040184

Khandelwal A, Vijay A, Dixit A, Chhabra M (2018) Microbial fuel cell powered by lipid extracted algae: a promising system for algal lipids and power generation. Bioresour Technol 247:520–527. https://doi.org/10.1016/j.biortech.2017.09.119

Kidanu WG, Trang PT, Yoon HH (2017) Hydrogen and volatile fatty acids production from marine macroalgae by anaerobic fermentation. Biotechnol Bioprocess Eng 22:612–619. https://doi.org/10.1007/s12257-017-0258-1

Koppolu V, Vasigala VK (2016) Role of *Escherichia coli* in biofuel production. Microbiol Insights 9:29–35. https://doi.org/10.4137/Mbi.s10878

Korzen L, Pulidindi IN, Israel A, Abelson A, Gedanken A (2015) Single step production of bioethanol from the seaweed *Ulvarigida* using sonication. RSC Adv 5:16223–16229. https://doi.org/10.1039/c4ra14880k

Kovács KL, Ács N, Böjti T, Kovács E, Strang O, Wirth R et al (2014) Biogas producing microbes and biomolecules. In: Lu X (ed) Biofuels: from microbes to molecules. Caister Academic Press, Norfolk

Kröninger L, Gottschling J, Deppenmeier U (2017) Growth characteristics of *Methanomassiliicoccus luminyensis* and expression of methyltransferase encoding genes. Archaea 2017:1–12. https://doi.org/10.1155/2017/2756573

Kuan I-C, Kao W-C, Chen C-L, Yu C-Y (2018) Microbial biodiesel production by direct transesterification of *Rhodotorula glutinis* biomass. Energies 11:1036. https://doi.org/10.3390/en11051036

Kumar G, Sivagurunathan P, Thi NBD, Zhen G, Kobayashi T, Kim S-H et al (2016a) Evaluation of different pretreatments on organic matter solubilization and hydrogen fermentation of mixed microalgae consortia. Int J Hydrog Energy 41(46):21628–21640. https://doi.org/10.1016/j.ijhydene.2016.05.195

Kumar G, Zhen G, Sivagurunathan P, Bakonyi P, Nemestóthy N, Bélafi-Bakó K et al (2016b) Biogenic H_2 production from mixed microalgae biomass: impact of pH control and methanogenic inhibitor (BESA) addition. Biofuel Res J 3:470–474. https://doi.org/10.18331/BRJ2016.3.3.6

Leo VV, Passari AK, Joshi B, Mishra VK, Uthandi S, Ramesh N et al (2016) A novel triculture system (CC3) for simultaneous enzyme production and hydrolysis of common grasses through submerged fermentation. Front Microbiol 7:1–13. https://doi.org/10.3389/fmicb.2016.00447

Leo VV, Asem D, Zothanpuia, Singh BP (2018) Actinobacteria: a highly potent source for holocellulose degrading enzymes. In: Singh BP, Gupta VK, Passari AK (eds) New and future developments in microbial biotechnology and bioengineering actinobacteria: diversity and biotechnological applications. Elsevier, The Boulevard, pp 191–205. https://doi.org/10.1016/B978-0-444-63994-3.00013-8

Liu L, Li F, Feng C, Li X (2009) Microbial fuel cell with an azo-dye-feeding cathode. Appl Microbiol Biotechnol 85:175–183. https://doi.org/10.1007/s00253-009-2147-9

Logan BE, Regan JM (2006) Electricity-producing bacterial communities in microbial fuel cells. Trends Microbiol 14:512–518. https://doi.org/10.1016/j.tim.2006.10.003

Lynd LR, Laser MS, Bransby D, Dale BE, Davison B, Hamilton R et al (2008) How biotech can transform biofuels. Nat Biotechnol 26:169–172

Mahboubi A, Ferreira JA, Taherzadehand MJ, Lennartsson PR (2017) Production of fungal biomass for feed, fatty acids, and glycerol by *Aspergillusoryzae* from fat-rich dairy substrates. Fermentation 3:48. https://doi.org/10.3390/fermentation3040048

Makishah NHA (2017) Bioenergy: microbial biofuel production advancement. Int J Pharm Res 6:93–106

Marella ER, Holkenbrink C, Siewers V, Borodina I (2018) Engineering microbial fatty acid metabolism for biofuels and biochemical. Curr Opin Biotechnol 50:39–46

Meng X, Yang J, Xu X, Zhang L, Nie Q, Xian M (2009) Biodiesel production from oleaginous microorganisms. Renew Energ 34(1):1–5. https://doi.org/10.1016/j.renene.2008.04.014

Milner EM, Popescu D, Curtis T, Head IM, Scott K, Yu EH (2016) Microbial fuel cells with highly active aerobic biocathodes. J Power Sources 324:8–16

Mitra D, Rasmussen ML, Chand P, Chintareddy VR, Yao L, Grewell D et al (2012) Value-added oil and animal feed production from corn-ethanol stillage using the oleaginous fungus *Mucor circinelloides*. Bioresour Technol 107:368–375. https://doi.org/10.1016/j.biortech.2011.12.031

Naithani S, Dubey AK, Joshi G, Gupta PK (2011) Potential and prospects of biofuels from lignocellulosic forestry residues. In: Pandey AK, Mandal AK (eds) Biofuels: potential and challenges. Scientific Publishers, New Delhi. ISBN: 978-81-7233-696-7, pp 77–82

Nolla-Ardèvol V, Strous M, Tegetmeyer HE (2015) Anaerobic digestion of the microalga *Spirulina* at extreme alkaline conditions: biogas production, metagenome, and metatranscriptome. Front Microbiol 6:597. https://doi.org/10.3389/fmicb.2015.00597

Obata O, Akunna J, Bockhorn H, Walker G (2016) Ethanol production from brown seaweed using non- conventional yeasts. Bioethanol 2:134–245

Park J-H, Yoon J-J, Park H-D, Kim YJ, Lim DJ, Kim S-H (2011) Feasibility of biohydrogen production from *Gelidium amansii*. Int J Hydrog Energy 36:13997–14003. https://doi.org/10.1016/j.ijhydene.2011.04.003

Patel A, Matsakas L, Rova U, Christakopoulos P (2018) Heterotrophic cultivation of *Auxenochlorella protothecoides* using forest biomass as a feedstock for sustainable biodiesel production. Biotechnol Biofuels 11:169. https://doi.org/10.1186/s13068-018-1173-1

Phillips JR, Huhnke RL, Atiyeh HK (2017) Syngas fermentation: a microbial conversion process of gaseous substrates to various products. Fermentation 3:28. https://doi.org/10.3390/fermentation3020028

Raja R, Shanmugam H, Ganesan V, Carvalho IS (2014) Biomass from microalgae: an overview. Oceanography 2(1):1–7. https://doi.org/10.4172/2332-2632.1000118

Reimers C, Girguis P, Stecher H, Tender L, Ryckelynck N, Whaling P (2006) Microbial fuel cell energy from an ocean cold seep. Geobiology 4:123–136

Ren Z, Steinberg LM, Regan JM (2008) Electricity production and microbial biofilm characterization in cellulose-fed microbial fuel cells. W Sci Technol 58:617–622. https://doi.org/10.2166/wst.2008.431

Rezaei F, Xing D, Wagner R, Regan JM, Richard TL, Logan BE (2009) Simultaneous cellulose degradation and electricity production by Enterobacter cloacae in a microbial fuel cell. Appl Environ Microbiol 75(11):3673–3678. https://doi.org/10.1128/AEM.02600-08

Rodolfi L, Chini-Zittelli G, Bassi N, Padovani G, Blondi N, Bonini G et al (2009) Microalgae for oil: strain selection, induction of lipid synthesis and outdoor mass cultivation in a low-cost photobioreactor. Biotechnol Bioeng 102:100–112

Salgueiro JL, Cancela Á, Sánchez Á, Maceiras R, Pérez L (2015) Analysis of extraction and transesterification conditions for *Phaeodactylum tricornutum* microalgae. Eur J Sustain Dev 4:89–96. https://doi.org/10.14207/ejsd.2015.v4n2p89

Seel W, Derichs J, Lipski A (2016) Increased biomass production by mesophilic food-associated bacteria through lowering the growth temperature from 30°C to 10°C. Appl Environ Microbiol 82:3754–3764. https://doi.org/10.1128/aem.00211-16

Selim KA, Ghwas DEE, Easa SM, Hassan MIA (2018) Bioethanol a microbial biofuel metabolite; new insights of yeasts metabolic engineering. Fermentation 4:16. https://doi.org/10.3390/fermentation4010016

Sharma A, Arya SK (2017) Hydrogen from algal biomass: a review of production process. Biotechnol Rep 15:63–69. https://doi.org/10.1016/j.btre.2017.06.001

Shin H-Y, Shim S-H, Ryu Y-J, Yang J-H, Lim S-M, Lee C-G (2018) Lipid extraction from *Tetraselmis* sp. microalgae for biodiesel production using hexane-based solvent mixtures. Biotechnol Bioprocess Eng 23:16–22. https://doi.org/10.1007/s12257-017-0392-9

Siddiqua S, Mamun AA, Enayetul Babar SM (2015) Production of biodiesel from coastal macroalgae (*Chara vulgaris*) and optimization of process parameters using box-Behnken design. Springer Plus 4:720. https://doi.org/10.1186/s40064-015-1518-1

Soleymani M, Rosentrater KA (2017) Techno-economic analysis of biofuel production from macroalgae (seaweed). Bioengineering 4:92. https://doi.org/10.3390/bioengineering4040092

Sunwoo IY, Kwon JE, Nguyen TH, Ra CH, Jeong G-T, Kim S-K (2017) Bioethanol production using waste seaweed obtained from Gwangalli beach, Busan, Korea by co-culture of yeasts with adaptive evolution. Appl Biochem Biotechnol 183:966–979. https://doi.org/10.1007/s12010-017-2476-6

Swana J, Yang Y, Behnam M, Thompson R (2011) An analysis of net energy production and feedstock availability for biobutanol and bioethanol. Bioresour Technol 102:2112–2117
Tandukar M, Huber SJ, Onodera T, Pavlostathis SG (2009) Biological chromium(VI) reduction in the cathode of a microbial fuel cell. Environ Sci Technol 43:8159–8165. https://doi.org/10.1021/es9014184
Tan X-B, Zhao X-C, Zhang Y-L, Zhou Y-Y, Yang L-B, Zhang W-W (2018) Enhanced lipid and biomass production using alcohol wastewater as carbon source for *Chlorella pyrenoidosa* cultivation in anaerobically digested starch wastewater in outdoors. Bioresour Technol 247:784–793. https://doi.org/10.1016/j.biortech.2017.09.152
Ullah K, Ahmad M, Sofia Sharma VK, Lu P, Harvey A et al (2014) Algal biomass as a global source of transportation fuels: overview and development perspectives. Prog Nat Sci Mater 24:329–339
Van-der-Ha D, Nachtergaele L, Kerckhof FM, Rameiyanti D, Bossier P, Verstraete W et al (2012) Conversion of biogas to bioproducts by algae and methane oxidizing bacteria. Environ Sci Technol 46(24):13425–13431. https://doi.org/10.1021/es303929s
Vicente G, Bautista LF, Rodríguez R, Gutiérrez FJ, Sádaba I, Ruiz-Vázquez RM et al (2009) Biodiesel production from biomass of an oleaginous fungus. Biochem Eng J 48(1):22–27. https://doi.org/10.1016/j.bej.2009.07.014
Vicente G, Bautista LF, Gutiérrez FJ, Rodríguez R, Martínez V, Rodríguez-Frómeta RA et al (2010) Direct transformation of fungal biomass from submerged cultures into biodiesel. Energ Fuel 24(5):3173–3178. https://doi.org/10.1021/ef9015872
Walter XA, Greenman J, Taylor B, Ieropoulus IA (2015) Microbial fuel cells continuously fuelled by untreated fresh algal biomass. Algal Res 11:103–107
Wang Y, Guo W, Cheng C-L, Ho S-H, Chang J-S, Ren N (2016) Enhancing bio-butanol production from biomass of *Chlorella vulgaris* JSC-6 with sequential alkali pretreatment and acid hydrolysis. Bioresour Technol 200:557–564. https://doi.org/10.1016/j.biortech.2015.10.056
Xafenias N, Zhang Y, Banks CJ (2015) Evaluating hexavalent chromium reduction and electricity production in microbial fuel cells with alkaline cathodes. Int J Env Sci Technol 12:2435–2446
Xia A, Jacob A, Herrmann C, Murphy JD (2016) Fermentative bio-hydrogen production from galactose. Energy 96:346–354. https://doi.org/10.1016/j.energy.2015.12.087

Microbe-Mediated Reclamation of Contaminated Soils: Current Status and Future Perspectives

10

Muhammad Shahid, Temoor Ahmed, Muhammad Noman, Natasha Manzoor, Sabir Hussain, Faisal Mahmood, and Sher Muhammad

10.1 Introduction

Soil is the major life-supporting natural resource for plant growth and crop productivity. It is assumed that approximately 30% of land is degraded or polluted by several anthropogenic activities and this proportion is continuously increasing globally (Abhilash et al. 2012). Today, different pollutants, e.g., heavy metals, salts, pesticides, and organic pollutants, pose serious threats to arable lands (Dixit et al. 2015). These toxic materials are harmful for agricultural system; moreover, they cause serious toxicity to life when they are added to the environment through different agricultural practices (Sarwar et al. 2017). Environmental contamination of heavy metals is caused by many sources including medical waste; combustion of leaded batteries; fertilizers, coal, and petrol combustion; smelting; industrializations; and mining (Liu et al. 2018). Heavy metals adversely affect plant growth by hindering the photosynthetic process. Toxic metals also decrease the leaf water content, transpiration rate, and stomatal conductance by reducing the number and size of xylem vessels in plants (Nagajyoti et al. 2010; Ruperao et al. 2014). These heavy metals also contaminate the food chain through their accumulation in the edible portion of the plants. Hence, it is necessary to eliminate these metals from agro-ecosystem to attain a sustainable yield and ecological safety. In agro-ecosystems, beneficial microorganisms are known to play an important function in the

M. Shahid (✉) · T. Ahmed · M. Noman · S. Muhammad
Department of Bioinformatics & Biotechnology, Government College University, Faisalabad, Pakistan
e-mail: mshahid@gcuf.edu.pk

N. Manzoor
Department of Soil and Water Sciences, China Agricultural University, Beijing, China

S. Hussain · F. Mahmood
Department of Environmental Sciences & Engineering, Government College University, Faisalabad, Pakistan

D. P. Singh et al. (eds.), *Microbial Interventions in Agriculture and Environment*,
https://doi.org/10.1007/978-981-13-8391-5_10

reclamation of heavy metal-contaminated soil by triggering the stress-tolerant mechanisms in plants (Akram et al. 2016).

Under stress conditions, plants utilize various physiological and morphological mechanisms to neutralize the heavy metal stress. The sensitive and resistant crops have different stress tolerance levels (Chaves et al. 2009; Munns 2002). A variety of microorganisms are found to develop an interaction with crop plants in soil. However, their diversity is different for rhizosphere and bulk soil because of the presence of different mechanisms of adaptation in crop plants. Crop plants generally develop a healthy plant microbe interaction from the available soil microbiome (Miransari 2017; Sobariu et al. 2017). A broad range of beneficial microbes including plant growth-promoting rhizobacteria (PGPRs) and arbuscular mycorrhizal fungi (AMF) have ability to develop a symbiotic or non-symbiotic association with host plant to improve the plant growth (Shahid et al. 2017). The coordination mechanisms are important in order to develop symbiotic association between the microbes and host plant (Shahid et al. 2017).

Presence of microorganisms in the soil is important for alleviation of stress effect in contaminated soils. For example, soil biochemical properties are positively affected by different microbial metabolites such as polysaccharides, organic acids, enzymes, osmolytes, etc. (Miransari 2017). Plenty of interactions occur in the interactive microenvironment of the soil and plant between the soil-inhabiting microbes and the plants affecting the physicochemical and biological characteristics of the soil. Various plant beneficial microbes have the ability to promote plant growth and development by various mechanisms of direct and indirect nature, including increased nutrients and water uptake, synthesis of various phytohormones, alleviation of stress, and siderophore production (Jha and Subramanian 2018; Mittal et al. 2017; Novo et al. 2018). Several physical, biological, and chemical practices can also be utilized for the remediation of the metal- and salt-contaminated soils. But microbe-mediated methods have emerged as cost-efficient and environmental friendly methods. This approach is one of the significant approaches for decontamination of soils, specifically in case of metal and salt contamination (Sarwar et al. 2017).

This chapter discusses about the impact of heavy metals and salts in the agro-ecosystems and effective microbial processes used for the reclamation of these contaminated soils.

10.2 Contaminated Soils

Worldwide, the soil is mostly contaminated with metal ions, salts, and organic residues due to urbanization and weathering processes (Cristaldi et al. 2017). Due to industrial revolution, the amount of metals in soil has increased exponentially (Alloway 2013). The heavy metals are naturally found throughout the earth's crust, but many heavy metals like metalloid mercury (Hg), cadmium (Cd), zinc (Zn), arsenic (As), chromium (Cr), nickel (Ni), lead (Pb), and copper (Cu) are mostly utilized in different industrial processes and subsequently discharged into the environment as a waste (Alloway 2013; Cristaldi et al. 2017). Wastewater from the tanneries carrying many malodorous chemical materials such as ammonia, dyes, and hydrogen sulfide is discharged into the agricultural soils, especially in the areas adjacent to cities (Karabay 2008). In addition, petrochemical industries and other vehicles release numerous metals into the environment (Manno et al. 2006). Similarly, soils have also been contaminated with sodium salts due to the water logging and excessive salt runoff from nearby areas.

10.3 Types of Contaminated Soils

Generally, contamination can be categorized based on the source of the contaminant. Mostly, the point of the contamination is related to the improper disposal of the waste, industrial effluents, and accidental spillage of toxic materials during transportation (Valentín et al. 2013). Few domestic and industrial sources of contamination are usage of septic tanks in an inconsiderate way, leakage of the underground oil tanks, and industrial effluents. Diffuse contamination is related to the agricultural practices, improper waste disposal, spillage during transportation, forestry, and management of wastewater (Loehr 2012; Valentín et al. 2013). Contamination of the soil is mostly connected with the contamination of the groundwater. Water in the soil pores moves in a vertical way at the rate mainly driven by soil texture (Mulligan et al. 2001). In contrast, the flow of horizontal groundwater is determined by the lakes and rivers. Transportation of several water-soluble contaminants is carried out by the horizontal and the vertical flow of the groundwater. Contamination of groundwater is much problematic and expensively managed. So, it is important to avoid the seepage of the contaminants into the underground water resources (Meffe and de Bustamante 2014). Furthermore, the chemicals that are insoluble in water also influence the higher organisms as a component of biodegradable lipophilic compounds that have tendency to gather in food chain (Duruibe et al. 2007).

10.4 Agriculture in Contaminated Soils

Agricultural productivity is adversely affected by the discharge of the industrial effluents as well as by the heavy metals and salts. Stresses due to the salinity and heavy metals suppress the plant growth and result in a decrease in yield (Nicholson et al. 2003). For plant survival, these abiotic stresses must be suppressed or plants must be able to tolerate these stresses (Vimal et al. 2017). The stresses affect the plant growth by producing nutritional and the hormonal imbalance together with some physiological disorders like abscission, senescence, epinasty, and vulnerability to diseases (Nicholson et al. 2003). In stressed conditions, plant produces high levels of ethylene (C_2H_4) which negatively affects the growth and health of the plant (Vimal et al. 2017). Furthermore, the factors that limit the germination, seedling vigor, and agricultural productivity are frequently found in arid and semiarid regions of the world (Amato et al. 2014). High levels of salts, organic residues, and heavy metals can cause a disparity of the cellular ions, resulting in osmotic stress, ion toxicity, and synthesis of reactive oxygen species (ROS) (Rodríguez-Serrano et al. 2009; Tuteja 2007). The abiotic stresses directly influence the biochemical and physiological properties and decrease plant productivity.

10.5 Significance of Microbes in Contaminated Soils

Plant-associated microbes have a proficient role in nutrient recycling and alleviation of stress consequences (Abriouel et al. 2011; Burd et al. 2000). Recent studies depicted the importance of plant-associated microbes in conferring stress tolerance in plants. The successful colonization of PGPR and AMF with host plants results in better nutrient acquisition and improved plant biomass under stress conditions. It is indicated that heavy metals hinder the uptake of other metals and nutrients essential for plant growth like Fe, Ca, P, and Zn and thus retard the plant growth (Glick 2010). Under such circumstances, plant-associated microbes are known to enhance the plant nutrition by mobilization of fixed nutrients, thus making these accessible to the plant roots. The N_2 fixation by rhizobacterial genera such as *Mesorhizobium*, *Rhizobium*, and *Bradyrhizobium* may contribute to enhance the growth of the legume crops in metal-contaminated and salt-affected soils by providing N to plants (Adams et al. 2004; Geddes et al. 2015). The improved P uptake is also reported after treatment of crops with phosphate-solubilizing microbes (Rajkumar et al. 2012). Moreover, siderophore-producing microbes have the ability to provide plants with iron under iron scarcity. Previous studies showed that plants inoculated with ACC deaminase-producing microbes have better germination and growth of seedlings due to the reduced ethylene concentration under abiotic stress conditions (Rodriguez et al. 2008). Madhaiyan et al. (2007) described that tomato seeds treated with ACC deaminase-containing *Methylobacterium oryzae* showed improved growth when grown in soils contaminated with Ni and Cd than non-inoculated plants. The microbes decreased the synthesis of ethylene under heavy metal-induced stress via ACC deaminase activity (Ali et al. 2014; Madhaiyan et al. 2007). Some studies also reported the presence of the cumulative influence of microbes on the plant growth under stressed environmental conditions (Li et al. 2014; Meffe and de Bustamante 2014).

10.5.1 Role of Microbes in Bioremediation

The plant growth and productivity, in salt- and metal-contaminated lands, is influenced by a diverse group of microorganisms having the ability to endure high metal concentration and confer stress tolerance on plants (Zhuang et al. 2007). Many PGPRs and AMF have been found to remove salts, organic residues, and heavy metals from contaminated soils. They improve the plant growth and soil health by a variety of mechanisms such as metal bioavailability, bioleaching, biotransformation, biovolatilization, bioaccumulation, release of chelators such as siderophores, ACC deaminase activity, solubilization of inorganic phosphate, exopolysaccharide synthesis, nitrogen fixation, phytohormones production, etc. (Fig. 10.1) (Barriuso et al. 2008; Bhattacharyya and Jha 2012; Gupta and Verma 2015; Novo et al. 2018). Table 10.1 describes the microbial species, their metabolites, and mechanisms of heavy metal bioremediation.

10.5.2 Role of Microbes in Plant Growth Promotion

10.5.2.1 Nitrogen Fixation

The essential plant nutrients have important role in ameliorating the heavy metal-induced toxicity in plants. Nitrogen (N) is the main constituent of many biomolecules like vitamins, proteins, hormones, and nucleic acids and thus essential for plant growth (Defez et al. 2017). Nitrogen supply increases tolerance against heavy metals in plants by improving the activity of ribulose 1,5-bisphosphate carboxylase as well as photosynthetic capacity (Ahmad et al. 2016; Rajkumar et al. 2012). Sufficient level of N is required by plants to tolerate the heavy metals in the form of N metabolites (Schutzendubel and Polle 2002). Several PGPRs and AMF have potential to fix atmospheric nitrogen and improve nitrogen availability to plants. PGPRs and AMF strains *Klebsiella mobilis* and *Glomus mosseae* have been reported to tolerate heavy metals and improve the grain yield (Meng et al. 2015; Pishchik et al. 2002; Rajkumar

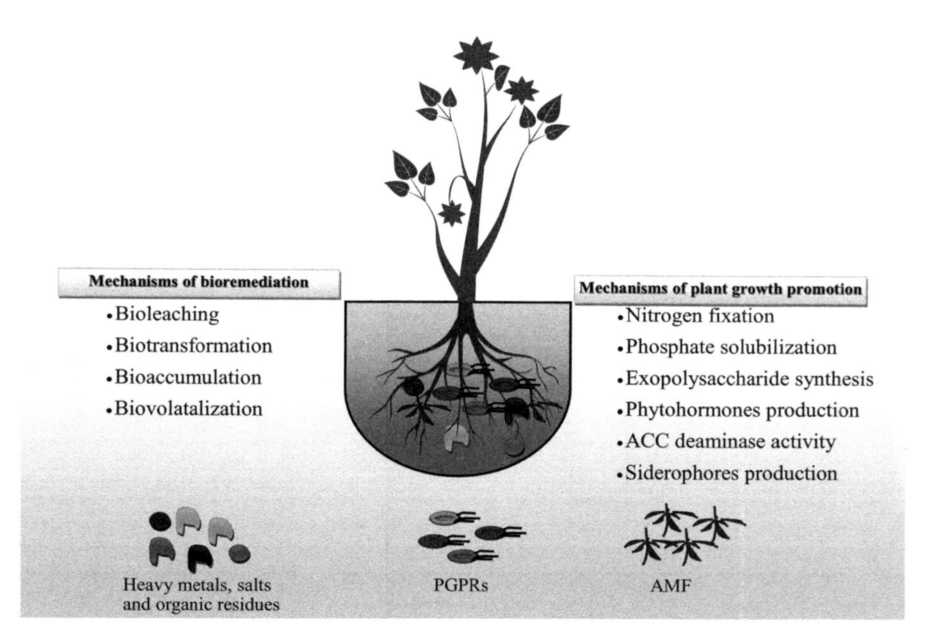

Fig. 10.1 Schematic representation of mechanisms involved in bioremediation and plant growth promotion

Table 10.1 Microorganisms and their reported mechanisms involved stress alleviation and plant growth promotion in contaminated soils

Microorganisms	Plant	Heavy metals	Mechanism	References
Azotobacter chroococcum HKN-5 *Bacillus megaterium* HKP-1	*Brassica juncea*	Lead and zinc	Nitrogen fixation, phosphate solubilization, and potassium solubilization	Niu et al. (2006)
Pseudomonas sp.	*Brassica napus*	Cadmium	IAA production	Sheng and Xia (2006)
Methylobacterium oryzae	*Lycopersicon esculentum*	Cadmium	ACC deaminase activity, bioaccumulation, and biotransformation	Madhaiyan et al. (2007)
B. edaphicus	*Brassica juncea*	Lead	Siderophore production	Sheng et al. (2008)
Streptomyces tendae F4	*Helianthus*	Cadmium	Siderophore production	Dimkpa et al. (2009)
S. acidiscabies E13	*Vigna unguiculata*	Copper, iron, and magnesium	Siderophore production	Dimkpa et al. (2009)
B. cereus	*Trifolium repens*	Iron, magnesium, zinc, and cadmium	IAA production, siderophore production, and bioleaching	Azcón et al. (2010)
Bacillus sp. SLS18	*Sweet sorghum*	Cadmium	IAA production, ACC deaminase, and siderophore production	Couillerot et al. (2011)
Agrobacterium radiobacter	*Populus deltoides*		IAA production and siderophore production	Cui et al. (2011)
Pseudomonas sp.	*Alyssum serpyllifolium*	Nickel	Phosphate solubilization, IAA production, ACC deaminase, and siderophore production	Abriouel et al. (2011)
P. aeruginosa OSG41	*Cicer arietinum*	Chromium	Phosphate solubilization, IAA production, ACC deaminase, and siderophore production	Oves et al. (2013)
Bradyrhizobium sp.	*Glycine max*	Cadmium	Siderophore production	Guo and Chi (2014)
Bacillus sp. SC2b	*Brassica napus*	Lead, zinc, and cadmium	Phosphate solubilization, exopolysaccharides production, and IAA production	Ahemad (2015)
Paecilomyces lilacinus NH 1	*Solanum nigrum* L	Cadmium	Antioxidative defense system	Gao et al. (2010)
Glomus etunicatum	*Glycine max*	Manganese	Siderophore production and biovolatilization	Nogueira et al. (2007)
G. mosseae	*Cajanus cajan*	Lead, and cadmium	Biosorption	Garg and Aggarwal (2011)
G. clarum and *Gigaspora margarita*	*Coffea arabica*	Copper and zinc	Nutrient acquisition	Andrade et al. (2010)

et al. 2012). Similarly, nitrogen fixation by soil microbes provided Cd tolerance to *Glycine max* through enhanced photosynthesis and decreased levels of heavy metals when grown in contaminated soil (Guo and Chi 2014).

10.5.2.2 Phosphate Solubilization

One of the important characteristics of phosphate-solubilizing microbes is to convert the unavailable inorganic and organic phosphate into the forms which are accessible to the plant (Chen et al. 2006; Qiu et al. 2011). Phosphate-solubilizing microbes are indigenous to all types of environment such as bulk and rhizosphere soils, soil rock phosphate dumping site, phyllosphere, rhizoplane, and stressed soils (Ahemad 2015). Phosphate-solubilizing microbes help in mobilization of unavailable soil phosphates chelated to metal ions (e.g., Fe-P, Ca-P, and Al-P) and increase the phosphate availability to plants (Etesami 2018). The known mechanism of solubilization of soil phosphates is the synthesis of organic acids, lowering the pH of surroundings and, subsequently, detaching the bound phosphates (Sharma et al. 2013; Kumar and Shastri 2017; Qiu et al. 2011). The available phosphate rapidly stops the mobilization of salt and heavy metals from soil to plant and increase the plant's ability to resist metals through the production of non-soluble complexes with heavy metals (Etesami 2018). It is concluded that both PGPR and AMF help plants to alleviate heavy metal and salt stress through the process of phosphate solubilization. Soil microbes enhance the plant growth by phosphate uptake after inoculation. Species such as *Rhizophagus irregularis*, *Pseudomonas* sp. (wheat), *Pantoea* J49 (peanut), and *Psychrobacter* sp. (sunflower) are reported to improve plant growth through phosphate solubilization under stressed environments (Rajkumar et al. 2012; Taktek et al. 2015).

10.5.2.3 Exopolysaccharides Synthesis

Exopolysaccharides (EPS) are carbohydrate polymers that are produced by rhizobacteria (Dhole et al. 2015). They form a capsule layer on the cell wall or act as a slime layer when released from the cell. In bacteria, EPS perform variety of functions including maintenance of cellular functions, production of biofilms, antibacterial activity, protection of bacteria from dry environment, gelling, and bioremediation activity (Bogino et al. 2013; Costerton et al. 2003). The production of EPS is directly associated with stress alleviation. The EPS-producing bacteria improve plant growth and development under stressed environment. In a study, wheat plants showed improved growth when inoculated with exopolysaccharide-producing bacteria belonging to *Bacillus* and *Enterobacter* genera in contrast to non-inoculated plants. However, the mechanism of EPS synthesis and their stimulatory effect on plant growth under saline environment is not well understood (Chen et al. 2016). The strains like *Bacillus subtilis*, *Pseudomonas aeruginosa,* and *Streptococcus mutans* have been best known for their exopolysaccharide production potential (Chen et al. 2016; Vimala and Lalithakumari 2003).

10.5.2.4 Phytohormone Production

Phytohormones (auxins, gibberellins, and cytokinins) produced by soil microbes are the direct plant growth-promoting agents under salinity and heavy metal stress conditions (Nagel et al. 2017). Under heavy metal and salt stress, plant cells start producing ROS which further induce the MAP kinase (MAPK) signalling pathway (Abdel-Lateif et al. 2012; Passari et al. 2016). The heavy metal-induced stress interrupts auxin physiology in *Arabidopsis* and poplar through decrease in auxin levels (Elobeid et al. 2011). Soil microbes that produce indole-3-acetic acid (IAA) can prevent the negative effects of heavy metals on plant development. Under Cd stress, rhizobacteria such as *Agrobacterium radiobacter*, *Azospirillum lipoferum*, *Flavobacterium* sp., and *Arthrobacter mysorens* were able to produce IAA and improve the development and growth of barley (Azcón et al. 2010; Gontia-Mishra et al. 2016). IAA reduces the toxicity caused by heavy metals in plants by decreasing heavy metal translocation and sorption or by enhancing the antioxidative enzymatic machinery. The root growth and development induced by auxins help in nutrient acquisition and root proliferation (Passari et al. 2016).

10.5.2.5 ACC Deaminase Activity

At a high concentration, ethylene has negative effects on plants in terms of reduced root and shoot proliferation (Ali et al. 2014; Etesami 2018). The stressed environment stimulates the production of 1-aminocyclopropane-1-carboxylate (ACC) which is the ethylene precursor. A bacterial enzyme ACC deaminase has a substantial potential to promote plant growth under heavy metal and salt stress due to its ability to cleave ACC into ammonia and α-ketobutyrate (Ali et al. 2014; Glick et al. 2007). The production of ACC deaminase by PGPRs and AMF is the fundamental mechanism to cope with the environmental stresses, especially the heavy metals and salts stress conditions (Glick 2005). The ACC deaminase producing PGPR are native to various soil environments and can act as ACC reservoir for maintaining the optimum level of ethylene required for the normal plant growth. The microbes are responsible for proliferation of root system in plants in order to acquire more nutrients to ameliorate abiotic stress (Frazier et al. 2011). In general, various studies have reported the proficient role of ACC deaminase-producing microbes (PGPRs and AMF) in improving plant growth in heavy metal- and salt-contaminated soils (Etesami 2018). The widely studied species of ACC deaminase-producing microbes are *Pseudomonas* sp., *M. oryzae, P. brassicacearum, P. fluorescens, P. koreensis, E. aerogenes, Achromobacter*, *B. megaterium, Burkholderia* sp., *Actinobacteria* sp., and *G. mosseae*, etc. (Etesami 2018; Rajkumar et al. 2012). To screen the best microbe for ameliorating stress in plants grown in heavy metals and salt contaminated soils, it's recommended that they are initially investigated for their ACC deaminase producing potential (Glick 2010).

10.5.2.6 Siderophore Production

Iron (Fe) is one of the most important elements for the normal plant growth and to carry out the healthy plant and microbial cellular activities specifically during heavy metal and salt stress (Parida et al. 2003). Berg et al. (2002) indicated that high levels of heavy metals and salt within plant body negatively affect the biosynthesis of chlorophyll and decrease the Fe uptake by the plant. Plants contain mechanisms to survive in Fe-deficient environment and enhance the Fe uptake either through acidification of rhizosphere or through the synthesis of phytosiderophores. Some rhizospheric microbes including PGPRs and AMF are known to provide plant with Fe by producing siderophores (low molecular weight Fe-chelating secondary metabolites) (Glick 2010; Rajkumar et al. 2010). It has been documented that in contrast to phytosiderophores, the microbe-oriented siderophores have high metal chelating ability. Therefore, the PGPRs and AMF may prove to be the solution of metal solubilization to improve the efficacy of plants (Rajkumar et al. 2010). Studies suggested that siderophore-producing microbes should improve the chlorophyll content and growth of the plant under metal-contaminated conditions by triggering the mobilization of Fe from heavy metal cations complex. Another study reveals that siderophore-producing microbes (SPM) have the ability for chlorophyll production in plant by provision of additional Fe and N (Rasouli-Sadaghiani et al. 2010). It is recognized that SPM have the capability to save plants against heavy metal toxicity. The reported PGPR strains involved in the production of siderophores are *P. fluorescens* and *P. putida*, whereas siderophore-producing AMF mostly belong to genus *Aspergillus*. These SPM can improve the nutrient status of the plant under heavy metal-induced stress conditions (Machuca and Milagres 2003; Rajkumar et al. 2012).

10.5.3 Role of Microbes in Stress Alleviation

The ameliorative role of soil microbes against stressed environments has been well documented. The soilborne microbial community improves the soil structure by producing exopolysaccharides (EPS), osmolytes, stress-related proteins, etc. The diverse interactions in plant rhizosphere positively affect the biological and physicochemical characteristics of the soil (Flemming and Wingender 2010; Singh and Satyanarayana 2011). The EPS-producing microbes have the potential to improve the soil structure by increasing the macropore volume within soil and aggregation of rhizospheric soil. This helps the plants to uptake more water and nutrients from soil under stress conditions. The mycorrhizal hyphae anchor deeply into the soil micropores and improve the nutrient and water availability to host plants under stressed environments (Sandhya et al. 2009a, b). The PGPRs also harbor the substantial potential to mitigate stress and plant growth promotion by producing ROS scavenging enzymes (Berg 2009; Grover et al. 2011). Microbes generally use different mechanisms to alleviate soil stress. Various mechanisms of heavy metal stress alleviation are known such as bioleaching, biotransformation, bioaccumulation, biovolatilization, etc. (Fig. 10.1). Microorganisms utilize heavy metals present in soil as chemicals for their own growth and development (Gupta and Verma 2015; Novo et al. 2018). The role of AMF and PGPRs in stress mitigation is described below.

10.5.3.1 Role of AMF in Stress Alleviation

The AMF are capable of improving the host plant growth under abiotic stress conditions. Various mechanisms that are employed by mycorrhizal fungi to enhance the growth of the plants under stress conditions are diverse hyphal framework, phytohormone synthesis, interactions with the neighboring microbes, etc. (Miransari 2017). The alleviating role of AMF under salinity in different crops has been widely investigated. It is reported that increased proline production, phosphorous uptake, and high sugar concentration provided plants with enhanced tolerance against induced salt stress (Daei et al. 2009; Garg and Singla 2016; Talaat and Shawky 2014). Previous studies showed the beneficial effects of mycorrhizal fungi, especially the species belonging to the genus *Glomus,* on plants under numerous stresses such as water logging, heavy metals, low temperature, salinity, drought, compaction, and acidity. The strains of *G. intraradices, Trichoderma koningii*, and *G. deserticola* were found to enhance the heavy metal tolerance in various crop plants like corn, tomato, eucalyptus, and *Medicago tranculata.* Nogueira et al. (2007) reported that inoculated *Glycine max* plants with *G. etunicatum* and *G. macrocarpum* exhibit better growth and P content than the control plants when grown in heavy metal-contaminated soil. The fungal hyphae can penetrate into fine micropores where plant roots are unable to grow and provide more water and nutrients to the host plant. Under compaction of soil, this feature promotes the plant growth and development (Garg and Chandel 2010; Miransari 2017). The plant growth is improved by AMF under drought and osmotic stress due to different mechanisms such as induction of antioxidative enzymatic machinery (i.e., catalases, superoxide dismutases, and peroxidases), limiting the malondialdehyde (MDA) synthesis, increase in the non-structural carbohydrate content, and high uptake of Ca^{+2}, Mg^{+2}, and K^+ (Wu and Xia 2006).

10.5.3.2 Role of PGPR in Stress Alleviation

It has been reported that PGPRs have growth-promoting effects on plants under stress ambience. To mitigate stress and promotion of plant growth, PGPRs use various mechanisms such as modification in plant structure and function, synthesis of phytohormones, high proline content, improved nutrient uptake, maintenance of low ethylene level by ACC deaminase activity, and interaction with soil microbiota (Miransari 2014). The ameliorative effects of *Azospirillum* spp. and *Rhizobium* spp. against drought and salinity have been extensively investigated (Arzanesh et al. 2011; Hamaoui et al. 2001). Ali et al. (2009) reported that *Pseudomonas* spp. strain AMK-P6-treated sorghum plants showed improved thermotolerance than non-inoculated plants under induced heat stress. The PGPRs are also reported to boost antioxidative defense mechanisms in plants due to the upregulation of stress-responsive genes to cope with abiotic stress conditions (Chakraborty et al. 2015; Akram et al. 2016; Shahid et al. 2019). The PGPRs resist various stresses by enhancing the K^+ ion uptake along with the accumulation of other solutes like polyols, amino acids, saccharides, and betaines. These solutes are either produced by bacteria or uptaken from its surroundings (Miransari 2017). It has been reported that *P. fluorescens* MSP-393 showed enhanced tolerance against salt stress by accumulating various solutes like aspartic acid, alanine, serine, glycine, glutamic acid, and threonine. Under stress, these solutes prevent proteins from denaturation (Paul and Nair 2008; Street et al. 2006). Moreover, PGPRs are capable of producing siderophores in the rhizosphere to enhance iron availability to plants under iron-deficient conditions. The PGPR-originated siderophores can reduce the heavy metal mobilization from contaminated soils and therefore can be utilized for soil reclamation purposes (Miransari 2017).

10.6 Current Perspectives of Microbes as Soil Reclamants

The exhaustive cropping systems have depleted the soil fertility and quality of arable lands. It is estimated that the intensive agricultural practices might convert the 30% of the world's total cultivated land to arid land by 2020 (Patel et al. 2015; Rashid et al. 2016). The decrease in soil fertility is one of the major global issues today. It can pose serious threats to crop cultivation and food security for the future generation. The agriculture system is mainly affected by the drastic changes in the environmental abiotic factors and reduction in the diversity and activities of soil microorganisms. The potential microbial populations can play pivotal roles to stabilize the degraded environment and agricultural soils (Patel et al. 2015; Singh 2015). The bacteria-releasing EPS have a significant role in the aggregation of rhizospheric soil under stressed conditions and is well documented. The microbially synthesized EPS in synergism with fungal hyphae stabilize the soil by forming macro- and micro-aggregates of soil particles (Grover et al. 2011; Nunkaew et al. 2015). The EPS-producing PGPR strains can also chelate cations such as Na^+, which enhance the ability of plant to survive under saline environmental conditions (Alami et al. 2000). Phytohormones of rhizobacterial origin induce certain physiological responses in the associated plants. The plant growth is enhanced by PGPR as a result of altering root morphology under abiotic stresses by producing different plant hormones like IAA, cytokinins, and gibberellic acid (Paul and Lade 2014). It has been revealed that wheat plants inoculated with IAA-producing *Streptomyces* showed enhanced growth than the control plants under induced salinity (Sadeghi et al. 2012).

The proliferating hyphal network of AMF helps plants to uptake water and mineral nutrients from the soil. The AMF-oriented glomalin (i.e., insoluble glycoprotein) stabilizes the soil structure by binding with macro- and micro-aggregates of soil particles (Li et al. 2015; Ortiz et al. 2015). The hyphal network of AMF not only facilitates plants in water and nutrient uptake but also restricts the heavy metal bioavailability to plants through biofiltration (Vimal et al. 2017). The restoration of agricultural land is associated with the efficient use of AMF and PGPR. The mixed culture of mycorrhizal fungi and bacteria was found to be more efficient in restoring soil fertility and organic matter profile in spite of their individual application (Rashid et al. 2016). However, extensive research insights are required in order to exploit microbial interactions for reconstructing the degraded agricultural lands.

10.7 Advantage of Microbes as Soil Reclamants

The economy of many developing countries depends on their agricultural production due to the significance of agriculture sector in those countries which provides food security, income, and employment to the people. However, poor soil fertility leads to soil erosion, lower crop yield, etc. Therefore, improvement of soil fertility is on the top of the development policy agenda list and should be dealt on a priority basis. Soil microbiota through various mechanisms like recycling of nutrients, regulation of soil organic matter, sequestration of carbon in soil, soil structure improvement, and enhancing the nutrient uptake efficiency and growth of plants contribute not only to improve the fertility status of soils but also sustainability of various ecosystems. Such services by soilborne microbes help in sustainability of the soil system along with the regulation of normal functioning of natural ecosystems (Singh et al. 2011). Microbiological reclamation of soil reduces the capital investment by improving resource utilization potential specifically cycling of nutrients, P bioavailability, N fixation, decomposition, and water uptake. The management of soil by microbes is an eco-friendly approach as it prevents land degradation and pollution by reducing the use of different chemical fertilizers. Hence, this microbe-based strategy improves crop productivity and quality by pest and disease controlling mechanisms, thus ensuring healthy food and food security for the continuously increasing population. Moreover, such techniques are helpful in the reclamation and restoration of non-fertile wastelands into fertile arable lands.

10.8 Problems in Microbe-Based Formulations

The potential application of microorganisms for various purposes (i.e., phytoremediation, bioremediation, bio-control, bio-fertilization, etc.) is directly associated with their capability to grow along with the native soil microbes and different environmental stresses in the field. Numerous reports are available indicating the efficient roles of microbes (i.e., PGPRs and AMF) in improving plant and soil health under stresses induced by various abiotic and biotic factors in laboratory and greenhouse experiments but are unable to display the same potential in the field conditions. For example, phosphate-solubilizing microbes were not found efficient under field experiments for phosphate solubilization but were found to be excellent P solubilizers under laboratory conditions (Compant et al. 2010; Gyaneshwar et al. 2002). Moreover, the beneficial activities of microorganisms have affected characteristics of the land, natural selection, and conventional agricultural practices such as crop rotation, application of agro-chemicals, and pesticides (Pishchik et al. 2002). The successful symbiotic association of microbial strains is based on the first-come-first-served principle. So, beneficial microbes must bypass others for their successful attachment with plant roots in order to help plants to grow under stressed environment. The bacteria are also present in the embryo or seeds of some weeds. These bacteria have to fight with other phyto-beneficial bacteria in soil. The successful establishment of plant beneficial bacteria directly depends on the uninterrupted supply of energy and carbon resources (Compant et al. 2010; Etesami and Maheshwari 2018). In field, the microbes that were found to be efficient under laboratory conditions have to compete with the natural soil microorganisms for nutrient uptake which limit their efficacy under natural environmental conditions. Furthermore, the microbiological soil management is also a time-consuming strategy because pre-selected microbes require time to adjust themselves according to the soil conditions. All the aforementioned limitations are required to be improved in the near future.

10.9 Future Prospects

Microorganisms help in the revitalization of nutrient-deficient soil and increase the growth and resistance of crop plants under various biotic and abiotic stress conditions. The complex and dynamic plant-microbe interactions affect plants as well as structural and physicochemical characteristics of soil under different stresses. The future of microbe-assisted reclamation of contaminated soil demands strong collaboration of bioremediation with nanobiotechnology (for improved remediation of affected soils), conventional biotechnology, and remediation of environmental stresses (such as nutrient scarcity and toxicity of contaminated site) through microbe-assisted phytoremediation. The remediation of contaminated soils using microbes can provide economical, agricultural, and environmental benefits to the world. However, it is difficult to achieve this goal.

Various challenges are also present that are not easy to handle, e.g., many contaminated lands require specific approaches and design for reclamation because of the uniqueness and complexity of the local conditions. However, we tried to summarize the roles and mechanisms of microorganisms that are used for remediation and reclamation purposes. Furthermore, nanotechnological and transgenic approaches due to their drastic environmental effects are not accepted publically. Fruitful research efforts are required to win the public interest and regulatory permission in order to implement these technologies on large scale. Till then, the conventional biotechnological techniques exploiting the interactions between soil microbiota and plant roots may serve for the remediation and reclamation of contaminated soils.

10.10 Conclusion

Microbe-based soil reclamation techniques present tremendous potential of implementing as alternatives or supplements to chemical solutions. More research investigations are needed to develop promising single strain or consortial bioinoculants for stressed soils in order to produce sustainable yield and improve soil health in a sustainable and cost-effective manner. Moreover, significant improvement in microbial strains, in terms of genetic and metabolic manipulations, is required to make the microbial solutions of polluted soils at par with the use of chemicals.

References

Abdel-Lateif K, Bogusz D, Hocher V (2012) The role of flavonoids in the establishment of plant roots endosymbioses with arbuscular mycorrhiza fungi, rhizobia and *Frankia* bacteria. Plant Signal Behav 7:636–641

Abhilash P, Powell JR, Singh HB et al (2012) Plant–microbe interactions: novel applications for exploitation in multipurpose remediation technologies. Trends Biotechnol 30:416–420

Abriouel H, Franz CM, Omar NB et al (2011) Diversity and applications of *Bacillus* bacteriocins. FEMS Microbiol Rev 35:201–232

Adams M, Zhao F, McGrath S et al (2004) Predicting cadmium concentrations in wheat and barley grain using soil properties. J Environ Qual 33:532–541

Ahemad M (2015) Phosphate-solubilizing bacteria-assisted phytoremediation of metalliferous soils: a review. 3 Biotech 5:111–121

Ahmad M, Nadeem SM, Naveed M et al. (2016) Potassium-solubilizing bacteria and their application in agriculture. In: Potassium solubilizing microorganisms for sustainable agriculture. Springer, New Dehli, pp 293–313

Akram MS, Shahid M, Tariq M, Azeem M, Javed T, Saleem S, Riaz S (2016) Deciphering *Staphylococcus sciuri* SAT-17 mediated anti-oxidative defense mechanisms and growth modulations in salt stressed maize (*Zea mays* L.). Front Microbiol 7

Alami Y, Achouak W, Marol C et al (2000) Rhizosphere soil aggregation and plant growth promotion of sunflowers by an exopolysaccharide-producing *Rhizobium* sp. strain isolated from sunflower roots. Appl Environ Microbiol 66:3393–3398

Ali SZ, Sandhya V, Grover M et al (2009) *Pseudomonas* sp. strain AKM-P6 enhances tolerance of sorghum seedlings to elevated temperatures. Biol Fertil Soils 46:45–55

Ali S, Charles TC, Glick BR (2014) Amelioration of high salinity stress damage by plant growth-promoting bacterial endophytes that contain ACC deaminase. Plant Physiol Biochem 80:160–167

Alloway BJ (2013) Sources of heavy metals and metalloids in soils. In: Heavy metals in soils. Springer, New York, pp 11–50

Amato P, Tachibana M, Sparman M et al (2014) Three-parent in vitro fertilization: gene replacement for the prevention of inherited mitochondrial diseases. Fertil Steril 101:31–35

Andrade S, Silveira A, Mazzafera P (2010) Arbuscular mycorrhiza alters metal uptake and the physiological response of *Coffea arabica* seedlings to increasing Zn and Cu concentrations in soil. Sci Total Environ 408:5381–5391

Arzanesh MH, Alikhani H, Khavazi K et al (2011) Wheat (*Triticum aestivum* L.) growth enhancement by *Azospirillum* sp. under drought stress. World J Microbiol Biotechnol 27:197–205

Azcón R, del Carmen Perálvarez M, Roldán A et al (2010) Arbuscular mycorrhizal fungi, *Bacillus cereus*, and *Candida parapsilosis* from a multicontaminated soil alleviate metal toxicity in plants. Microb Ecol 59:668–677

Barriuso J, Solano BR, Gutierrez Manero F (2008) Protection against pathogen and salt stress by four plant growth-promoting rhizobacteria isolated from *Pinus* sp. on *Arabidopsis thaliana*. Phytopathology 98:666–672

Berg G (2009) Plant–microbe interactions promoting plant growth and health: perspectives for controlled use of microorganisms in agriculture. Appl Microbiol Biotechnol 84:11–18

Berg G, Roskot N, Steidle A et al (2002) Plant-dependent genotypic and phenotypic diversity of antagonistic rhizobacteria isolated from different *Verticillium* host plants. Appl Environ Microbiol 68:3328–3338

Bhattacharyya P, Jha D (2012) Plant growth-promoting rhizobacteria (PGPR): emergence in agriculture. World J Microbiol Biotechnol 28:1327–1350

Bogino PC, Oliva MM, Sorroche FG et al (2013) The role of bacterial biofilms and surface components in plant-bacterial associations. Int J Mol Sci 14:15838–15859

Burd GI, Dixon DG, Glick BR (2000) Plant growth-promoting bacteria that decrease heavy metal toxicity in plants. Can J Microbiol 46:237–245

Chakraborty U, Chakraborty B, Dey P et al (2015) Role of microorganisms in alleviation of abiotic stresses for sustainable agriculture. In: Abiotic stresses in crop plants. CABI, Wallingford, pp 232–253
Chaves MM, Flexas J, Pinheiro C (2009) Photosynthesis under drought and salt stress: regulation mechanisms from whole plant to cell. Ann Bot 103:551–560
Chen YP, Rekha PD, Arun AB, Shen FT, Lai WA, Young CC (2006) Phosphate solubilizing bacteria from subtropical soil and their tricalcium phosphate solubilizing abilities. Appl Soil Ecol 34(1):33–41
Chen L, Cheng X, Cai J et al (2016) Multiple virus resistance using artificial trans-acting siRNAs. J Virol Methods 228:16–20
Compant S, Clément C, Sessitsch A (2010) Plant growth-promoting bacteria in the rhizo-and endosphere of plants: their role, colonization, mechanisms involved and prospects for utilization. Soil Biol Biochem 42:669–678
Costerton W, Veeh R, Shirtliff M et al (2003) The application of biofilm science to the study and control of chronic bacterial infections. J Clin Invest 112:1466–1477
Couillerot O et al (2011) The role of the antimicrobial compound 2, 4-diacetylphloroglucinol in the impact of biocontrol *Pseudomonas fluorescens* F113 on *Azospirillum brasilense* phytostimulators. Microbiology 157:1694–1705
Cristaldi A, Conti GO, Jho EH et al (2017) Phytoremediation of contaminated soils by heavy metals and PAHs: a brief review. Environ Technol Innov 8:309–326
Cui D, Kong F, Liang B et al (2011) Decolorization of azo dyes in dual-chamber biocatalyzed electrolysis systems seeding with enriched inoculum. J Environ Anal Toxicol S 3:001
Daei G, Ardekani M, Rejali F et al (2009) Alleviation of salinity stress on wheat yield, yield components, and nutrient uptake using arbuscular mycorrhizal fungi under field conditions. J Plant Physiol 166:617–625
Defez R, Andreozzi A, Bianco C (2017) The overproduction of indole-3-acetic acid (IAA) in endophytes upregulates nitrogen fixation in both bacterial cultures and inoculated rice plants. Microb Ecol 74:441–452
Dhole A, Shelat H, Panpatte D et al (2015) Biofertilizer formulation with absorbent polymers to surmount the drought stress. Pop Kheti 3(3):89–93
Dimkpa C, Merten D, Svatoš A et al (2009) Siderophores mediate reduced and increased uptake of cadmium by *Streptomyces tendae* F4 and sunflower (*Helianthus annuus*), respectively. J Appl Microbiol 107:1687–1696
Dixit R et al (2015) Bioremediation of heavy metals from soil and aquatic environment: an overview of principles and criteria of fundamental processes. Sustainability 7:2189–2212
Duruibe JO, Ogwuegbu M, Egwurugwu J (2007) Heavy metal pollution and human biotoxic effects. Int J Phys Sci 2:112–118
Elobeid M, Göbel C, Feussner I et al (2011) Cadmium interferes with auxin physiology and lignification in poplar. J Exp Bot 63:1413–1421
Etesami H (2018) Bacterial mediated alleviation of heavy metal stress and decreased accumulation of metals in plant tissues: mechanisms and future prospects. Ecotoxicol Environ Saf 147:175–191
Etesami H, Maheshwari DK (2018) Use of plant growth promoting rhizobacteria (PGPRs) with multiple plant growth promoting traits in stress agriculture: action mechanisms and future prospects. Ecotoxicol Environ Saf 156:225–246
Flemming H-C, Wingender J (2010) The biofilm matrix. Nat Rev Microbiol 8:623
Frazier TP, Sun G, Burklew CE et al (2011) Salt and drought stresses induce the aberrant expression of microRNA genes in tobacco. Mol Biotechnol 49:159–165
Gao M, Liang F, Yu A et al (2010) Evaluation of stability and maturity during forced-aeration composting of chicken manure and sawdust at different C/N ratios. Chemosphere 78:614–619
Garg N, Aggarwal N (2011) Effects of interactions between cadmium and lead on growth, nitrogen fixation, phytochelatin, and glutathione production in mycorrhizal *Cajanus cajan* (L.) Mill sp. J Plant Growth Regul 30:286–300

Garg N, Chandel S (2010) Arbuscular mycorrhizal networks: process and functions. A review. Agron Sustain Dev 30:581–599
Garg N, Singla P (2016) Stimulation of nitrogen fixation and trehalose biosynthesis by naringenin (Nar) and arbuscular mycorrhiza (AM) in chickpea under salinity stress. Plant Growth Regul 80:5–22
Geddes BA, Ryu M-H, Mus F et al (2015) Use of plant colonizing bacteria as chassis for transfer of N 2-fixation to cereals. Curr Opin Biotechnol 32:216–222
Glick BR (2005) Modulation of plant ethylene levels by the bacterial enzyme ACC deaminase. FEMS Microbiol Lett 251:1–7
Glick BR (2010) Using soil bacteria to facilitate phytoremediation. Biotechnol Adv 28:367–374
Glick BR, Cheng Z, Czarny J et al (2007) Promotion of plant growth by ACC deaminase-producing soil bacteria. Eur J Plant Pathol 119:329–339
Gontia-Mishra I, Sapre S, Sharma A et al (2016) Alleviation of mercury toxicity in wheat by the interaction of mercury-tolerant plant growth-promoting rhizobacteria. J Plant Growth Regul 35:1000–1012
Grover M, Ali SZ, Sandhya V et al (2011) Role of microorganisms in adaptation of agriculture crops to abiotic stresses. World J Microbiol Biotechnol 27:1231–1240
Guo J, Chi J (2014) Effect of Cd-tolerant plant growth-promoting rhizobium on plant growth and Cd uptake by *Lolium multiflorum* Lam. and *Glycine max* (L.) Merr. in Cd-contaminated soil. Plant Soil 375:205–214
Gupta A, Verma JP (2015) Sustainable bio-ethanol production from agro-residues: a review. Renew Sust Energ Rev 41:550–567
Gyaneshwar P, Kumar GN, Parekh L et al (2002) Role of soil microorganisms in improving P nutrition of plants. Plant Soil 245:83–93
Hamaoui B, Abbadi J, Burdman S et al (2001) Effects of inoculation with *Azospirillum brasilense* on chickpeas (*Cicer arietinum*) and faba beans (*Vicia faba*) under different growth conditions. Agronomie 21:553–560
Jha Y, Subramanian R (2018) From interaction to gene induction: an eco-friendly mechanism of pgpr-mediated stress management in the plant. In: Plant microbiome: stress response. Springer, Singapore, pp 217–232
Karabay S (2008) Waste management in leather industry. DEÜ Fen Bilimleri Enstitüsü, Buca, Turkey
Kumar R, Shastri B (2017) Role of phosphate-solubilising microorganisms in: *Sustainable Agricultural Development*. In: Agro-environmental sustainability. Springer, Cham, pp 271–303
Li W-W, Yu H-Q, He Z (2014) Towards sustainable wastewater treatment by using microbial fuel cells-centered technologies. Energy Environ Sci 7:911–924
Li X, Zhang J, Gai J et al (2015) Contribution of arbuscular mycorrhizal fungi of sedges to soil aggregation along an altitudinal alpine grassland gradient on the T ibetan P lateau. Environ Microbiol 17:2841–2857
Liu L, Li W, Song W et al (2018) Remediation techniques for heavy metal-contaminated soils: principles and applicability. Sci Total Environ 633:206–219
Loehr R (2012) Agricultural waste management: problems, processes, and approaches. Elsevier, Amsterdam
Machuca A, Milagres A (2003) Use of CAS-agar plate modified to study the effect of different variables on the siderophore production by *Aspergillus*. Lett Appl Microbiol 36:177–181
Madhaiyan M, Poonguzhali S, Sa T (2007) Metal tolerating methylotrophic bacteria reduces nickel and cadmium toxicity and promotes plant growth of tomato (*Lycopersicon esculentum* L.). Chemosphere 69:220–228
Manno E, Varrica D, Dongarra G (2006) Metal distribution in road dust samples collected in an urban area close to a petrochemical plant at Gela, Sicily. Atmos Environ 40:5929–5941
Meffe R, de Bustamante I (2014) Emerging organic contaminants in surface water and groundwater: a first overview of the situation in Italy. Sci Total Environ 481:280–295
Meng L, Zhang A, Wang F et al (2015) Arbuscular mycorrhizal fungi and rhizobium facilitate nitrogen uptake and transfer in soybean/maize intercropping system. Front Plant Sci 6:339

Miransari M (2014) Plant growth promoting rhizobacteria. J Plant Nutr 37:2227–2235
Miransari M (2017) The interactions of soil microbes affecting stress alleviation in agroecosystems. In: Probiotics in agroecosystem. Springer, Singapore, pp 31–50
Mittal P, Kamle M, Sharma S et al. (2017) 22 Plant growth-promoting rhizobacteria (PGPR): mechanism, role in crop improvement and sustainable agriculture. Adv PGPR Res 386–397
Mulligan C, Yong R, Gibbs B (2001) Remediation technologies for metal-contaminated soils and groundwater: an evaluation. Eng Geol 60:193–207
Munns R (2002) Comparative physiology of salt and water stress. Plant Cell Environ 25:239–250
Nagajyoti P, Lee K, Sreekanth T (2010) Heavy metals, occurrence and toxicity for plants: a review. Environ Chem Lett 8:199–216
Nagel R, Turrini PC, Nett RS et al (2017) An operon for production of bioactive gibberellin A4 phytohormone with wide distribution in the bacterial rice leaf streak pathogen *Xanthomonas oryzae* pv. *oryzicola*. New Phytol 214:1260–1266
Nicholson F, Smith S, Alloway B et al (2003) An inventory of heavy metals inputs to agricultural soils in England and Wales. Sci Total Environ 311:205–219
Niu Q-W, Shih-Shun L, Reyes JL et al (2006) Expression of artificial microRNAs in transgenic *Arabidopsis thaliana* confers virus resistance. Nat Biotechnol 24:1420
Nogueira M, Nehls U, Hampp R et al (2007) Mycorrhiza and soil bacteria influence extractable iron and manganese in soil and uptake by soybean. Plant Soil 298:273–284
Novo LA, Castro PM, Alvarenga P et al (2018) Plant growth–promoting rhizobacteria-assisted phytoremediation of mine soils. In: Bio-geotechnologies for mine site rehabilitation. Elsevier, Amsterdam, pp 281–295
Nunkaew T, Kantachote D, Nitoda T et al (2015) Characterization of exopolymeric substances from selected *Rhodopseudomonas palustris* strains and their ability to adsorb sodium ions. Carbohydr Polym 115:334–341
Ortiz N, Armada E, Duque E et al (2015) Contribution of arbuscular mycorrhizal fungi and/or bacteria to enhancing plant drought tolerance under natural soil conditions: effectiveness of autochthonous or allochthonous strains. J Plant Physiol 174:87–96
Oves M, Khan MS, Zaidi A (2013) Chromium reducing and plant growth promoting novel strain *Pseudomonas aeruginosa* OSG41 enhance chickpea growth in chromium amended soils. Eur J Soil Biol 56:72–83
Parida B, Chhibba I, Nayyar V (2003) Influence of nickel-contaminated soils on fenugreek (*Trigonella corniculata* L.) growth and mineral composition. Sci Hortic 98:113–119
Passari AK, Mishra VK, Leo VV et al (2016) Phytohormone production endowed with antagonistic potential and plant growth promoting abilities of culturable endophytic bacteria isolated from *Clerodendrum colebrookianum* Walp. Microbiol Res 193:57–73
Patel JS, Singh A, Singh HB et al (2015) Plant genotype, microbial recruitment and nutritional security. Front Plant Sci 6:608
Paul D, Lade H (2014) Plant-growth-promoting rhizobacteria to improve crop growth in saline soils: a review. Agron Sustain Dev 34:737–752
Paul D, Nair S (2008) Stress adaptations in a plant growth promoting rhizobacterium (PGPR) with increasing salinity in the coastal agricultural soils. J Basic Microbiol 48:378–384
Pishchik V, Vorobyev N, Chernyaeva I et al (2002) Experimental and mathematical simulation of plant growth promoting rhizobacteria and plant interaction under cadmium stress. Plant Soil 243:173–186
Qiu Q, Wang Y, Yang Z, Yuan J (2011) Effects of phosphorus supplied in soil on subcellular distribution and chemical forms of cadmium in two Chinese flowering cabbage (*Brassica parachinensis* L.) cultivars differing in cadmium accumulation. Food Chem Toxicol 49(9):2260–2267
Rajkumar M, Ae N, Prasad MNV et al (2010) Potential of siderophore-producing bacteria for improving heavy metal phytoextraction. Trends Biotechnol 28:142–149
Rajkumar M, Sandhya S, Prasad M et al (2012) Perspectives of plant-associated microbes in heavy metal phytoremediation. Biotechnol Adv 30:1562–1574
Rashid MI, Mujawar LH, Shahzad T et al (2016) Bacteria and fungi can contribute to nutrients bioavailability and aggregate formation in degraded soils. Microbiol Res 183:26–41

Rasouli-Sadaghiani M, Hassani A, Barin M et al (2010) Effects of AM fungi on growth, essential oil production and nutrients uptake in basil. J Med Plant Res 4:2222–2228

Rodriguez H, Vessely S, Shah S et al (2008) Effect of a nickel-tolerant ACC deaminase-producing *Pseudomonas* strain on growth of nontransformed and transgenic canola plants. Curr Microbiol 57:170–174

Rodríguez-Serrano M, Romero-Puertas MC, Pazmino DM et al (2009) Cellular response of pea plants to cadmium toxicity: cross talk between reactive oxygen species, nitric oxide, and calcium. Plant Physiol 150:229–243

Ruperao P et al (2014) A chromosomal genomics approach to assess and validate the desi and kabuli draft chickpea genome assemblies. Plant Biotechnol J 12:778–786

Sadeghi A, Karimi E, Dahaji PA et al (2012) Plant growth promoting activity of an auxin and siderophore producing isolate of *Streptomyces* under saline soil conditions. World J Microbiol Biotechnol 28:1503–1509

Sandhya V, Ali S, Grover M et al (2009a) *Pseudomonas* sp. strain P45 protects sunflowers seedlings from drought stress through improved soil structure. J Oilseed Res 26:600–601

Sandhya V, Grover M, Reddy G et al (2009b) Alleviation of drought stress effects in sunflower seedlings by the exopolysaccharides producing *Pseudomonas putida* strain GAP-P45. Biol Fertil Soils 46:17–26

Sarwar N et al (2017) Phytoremediation strategies for soils contaminated with heavy metals: modifications and future perspectives. Chemosphere 171:710–721

Schutzendubel A, Polle A (2002) Plant responses to abiotic stresses: heavy metal-induced oxidative stress and protection by mycorrhization. J Exp Bot 53:1351–1365

Shahid M, Hussain B, Riaz D et al (2017) Identification and partial characterization of potential probiotic lactic acid bacteria in freshwater *Labeo rohita* and *Cirrhinus mrigala*. Aquac Res 48:1688–1698

Shahid M, Javed MT, Mushtaq A, Akram MS, Mahmood F, Ahmed T, Noman M, Azeem M (2019) Microbe-mediated mitigation of cadmium toxicity in plants. In: Cadmium toxicity and tolerance in plants. Academic Press, pp 427–449

Sharma SB, Sayyed RZ, Trivedi MH, Gobi TA (2013) Phosphate solubilizing microbes: sustainable approach for managing phosphorus deficiency in agricultural soils. SpringerPlus 2(1)

Sheng X-F, Xia J-J (2006) Improvement of rape (*Brassica napus*) plant growth and cadmium uptake by cadmium-resistant bacteria. Chemosphere 64:1036–1042

Sheng XF, Xia JJ, Jiang CY, He CY, Qian M (2008) Characterization of heavy metal-resistant endophytic bacteria from rape (*Brassica napus*) roots and their potential in promoting the growth and lead accumulation of rape. Environ Pollut 156(3):1164–1170

Singh JS (2015) Microbes: the chief ecological engineers in reinstating equilibrium in degraded ecosystems. Agric Ecosyst Environ 203:80–82

Singh B, Satyanarayana T (2011) Microbial phytases in phosphorus acquisition and plant growth promotion. Physiol Mol Biol Plants 17:93–103

Singh JS, Pandey VC, Singh D (2011) Efficient soil microorganisms: a new dimension for sustainable agriculture and environmental development. Agric Ecosyst Environ 140:339–353

Sobariu DL et al (2017) Rhizobacteria and plant symbiosis in heavy metal uptake and its implications for soil bioremediation. New Biotechnol 39:125–134

Street TO, Bolen DW, Rose GD (2006) A molecular mechanism for osmolyte-induced protein stability. Proc Natl Acad Sci 103:13997–14002

Taktek S, Trépanier M, Servin PM et al (2015) Trapping of phosphate solubilizing bacteria on hyphae of the arbuscular mycorrhizal fungus *Rhizophagus irregularis* DAOM 197198. Soil Biol Biochem 90:1–9

Talaat NB, Shawky BT (2014) Protective effects of arbuscular mycorrhizal fungi on wheat (*Triticum aestivum* L.) plants exposed to salinity. Environ Exp Bot 98:20–31

Tuteja N (2007) Mechanisms of high salinity tolerance in plants. In: Methods in enzymology, vol 428. Elsevier, Amsterdam, pp 419–438

Valentín L, Nousiainen A, Mikkonen A (2013) Introduction to organic contaminants in soil: concepts and risks. In: Emerging organic contaminants in sludges. Springer, Berlin, pp 1–29

Vimal SR, Singh JS, Arora NK, Singh S (2017) Soil-plant-microbe interactions in stressed agriculture management: a review. Pedosphere 27:177–192

Vimala P, Lalithakumari D (2003) Characterization of exopolysaccharide (EPS) produced by *Leuconostoc* sp. V 41. Asian J Microbiol Biotechnol Environ Sci 5:161–165

Wu Q-S, Xia R-X (2006) Arbuscular mycorrhizal fungi influence growth, osmotic adjustment and photosynthesis of citrus under well-watered and water stress conditions. J Plant Physiol 163:417–425

Zhuang X, Chen J, Shim H et al (2007) New advances in plant growth-promoting rhizobacteria for bioremediation. Environ Int 33:406–413

Plant Growth-Promoting Rhizobacteria (PGPR) and Fungi (PGPF): Potential Biological Control Agents of Diseases and Pests

11

Pankaj Prakash Verma, Rahul Mahadev Shelake, Suvendu Das, Parul Sharma, and Jae-Yean Kim

11.1 Introduction

Microorganisms distressing plant health, i.e., plant pathogens are one of the key threats for sustainable global food production and ecosystem sustainability. These pathogenic microbes cause approximately 25% reduction in the global crop yield every year (Lugtenberg 2015). To increase the food production, fiber and biomaterial, strategies of plant pests and diseases (DP) management are crucial. Recently, the concern about global food security is growing and the total world production of food has to be increased by 70% until 2050 (Ingram 2011; Keinan and Clark 2012). The total food requirement in the world will keep on rising for upcoming 40 years with increasing human populations (Rahman et al. 2018). Globally, the food production system is accountable for loss of terrestrial biodiversity about 60% and increasing greenhouse gas emissions by 25% (Westhoek et al. 2016). There is a need to develop relatively reliable and more sustainable agricultural methods that can reduce the dependence on chemical pesticides.

P. P. Verma (✉)
Institute of Agriculture and Life Science, Gyeongsang National University, Jinju, Republic of Korea

Department of Basic Sciences, Dr. Y.S. Parmar University of Horticulture and Forestry, Solan, Himachal Pradesh, India

R. M. Shelake · J.-Y. Kim
Division of Applied Life Science (BK21 Plus Program), Plant Molecular Biology and Biotechnology Research Center, Gyeongsang National University, Jinju, South Korea

S. Das
Institute of Agriculture and Life Science, Gyeongsang National University, Jinju, Republic of Korea

P. Sharma
Department of Basic Sciences, Dr. Y.S. Parmar University of Horticulture and Forestry, Solan, Himachal Pradesh, India

D. P. Singh et al. (eds.), *Microbial Interventions in Agriculture and Environment*, https://doi.org/10.1007/978-981-13-8391-5_11

The microbes demonstrate different modes of antagonistic properties (Table 11.1) by producing antimicrobial compounds or by competing with phytopathogens commonly known as biocontrol agents or biological control agents (BCAs). There is a growing interest in BCAs as viable alternatives for DP management because of the harmful effects of chemical pesticides (Waghunde et al. 2016). The recent findings provide evidence of some bacterial and fungal endophytes which act as a nutrient distributor, tolerance enhancer under drought and abiotic stress, and promoter of growth and yield in plants (Jaber and Araj 2017; Waghunde et al. 2017; Bamisile et al. 2018). The application of entomopathogenic fungi as BCAs has been effective in DP management that also supports plant growth-promoting (PGP) activity (summarized in Tables 11.2 and 11.3). Therefore, more attention has been given to plant growth-promoting rhizobacteria and fungi (PGPR and PGPF, respectively) to replace or supplement agrochemicals in recent times. Their interactions with plants and phytopathogens lead to the activation of plant defense mechanisms such as induced systemic or systemic acquired resistance (ISR or SAR) pathways. The PGPR and PGPF help plants by many other ways such as decomposition of organic matter, increasing availability of nutrients, mineral solubilization, producing numerous phytohormones, and biocontrol of phytopathogens (Sivasakhti et al. 2014). The application of PGPR/PGPF is progressively increasing in agriculture and also offers a smart and economical way to substitute chemically synthesized pesticides and fertilizers (Borah et al. 2018).

This chapter is presented as the advanced survey of the literature currently available on the BCAs for DP management. The application of beneficial PGPR/PGPF reported in different host plants for the plant health management (PHM) are summarized. This work reviews the effects of PGPR and PGPF on host plants and their active role in plant DP management. It also addresses the possible mechanisms of protection and recent advancement conferred by these beneficial microbes as BCAs. Moreover, this chapter addresses the current trends in application and overall adoption of bacterial, fungal, and other microbials for DP management.

Table 11.1 Antagonisms exhibited by biological control agents

Type	Mechanism
Direct antagonism	Parasitism—symbiotic interaction between two phylogenetically unrelated organisms
	Hyperparasitism—parasites using other parasites as their host
	Commensalism—one partner benefits while other is neither benefited nor harmed
Indirect antagonism	Competition—interaction harmful to both the partners
	SAR—systemic acquired resistance
	ISR—induced systemic resistance
Mixed path antagonism	Antibiosis, lytic enzyme production, siderophore production, organic, and inorganic volatile substances

Table 11.2 Recent studies reporting biocontrol activities of PGPR, PGPF and other microbes against different phytopathogens are summarized

Biological control agents	Phytopathogens	Mechanism	Plant species	References
Bacteria				
Pseudomonas stutzeri (E25), *S. maltophilia* (CR71)	*Botrytis cinerea* (gray mold)	VOCs	Tomato	Rojas-Solís et al. (2018)
Azotobacter salinestris	*Fusarium* sp.	Antifungal substances, HCN, siderophores	—	Chennappa et al. (2018)
Bacillus amyloliquefaciens and *B. pumilus*	*P. syringae* pv. aptata (leaf spot)	Surfactin, fengycin A, iturin A	Sugar beet	Nikolić et al. (2018)
Azotobacter salinestris	*Fusarium species*	Antifungal substances, HCN, siderophores	—	Chennappa et al. (2018)
B. amyloliquefaciens ALB 629	*Rhizoctonia solani* (damping-off and web blight)	—	Common bean	Martins et al. (2018)
Bacillus mojavensis RRC101	*Fusarium verticillioides*	VOCs	Maize	Rath et al. (2018)
P. aeruginosa, *Bacillus* sp.	*Rhizoctonia solani* (root rot)	Antibiotic production	Mung	Kumari et al. (2018)
B. amyloliquefaciens subsp. *plantarum* 32a	*Agrobacterium tumefaciens* (crown gall)	Surfactin, iturin, difficidin polyketide, bacilysin dipeptide	Tomato	Abdallah et al. (2018)
Pseudomonas sp. CWD B, D, N; *Serratia* sp. CWD C	*Botrytis cinerea* and *Aspergillus niger*	Hydrolytic enzymes, siderophores, HCN	Pea and tomato	Tabli et al. (2018)
Pseudomonas putida, *P. fluorescens*, *P. aeruginosa*	*Dematophora necatrix*, *Fusarium oxysporum*, *Phytophthora cactorum* and *Pythium ultimum*	Phenazine, Pyrrolnitrin, DAPG	Apple	Sharma et al. (2017a, b, c)
Bacillus velezensis CC09	*Blumeria graminis* (powdery mildew)	—	Wheat	Cai et al. (2017)
Bacillus velezensis S3-1	*Botrytis cinerea*	Surfactin, iturin and fengycin	Tomato	Jin et al. (2017a, b)
Bacillus amyloliquefaciens CPA-8	*Monilinia laxa*, *M. fructicola*, *Botrytis cinerea*	VOCs	Cherry	Gotor-Vila et al. (2017)

(continued)

Table 11.2 (continued)

Biological control agents	Phytopathogens	Mechanism	Plant species	References
Pseudomonas sp. MCC 3145	*C. circinans, C. dematium, Fusarium oxysporum, R. solani, S. sclerotiorum*	PCA	Rice	Patil et al. (2017)
Bacillus amyloliquefaciens PGPBacCA1	*Sclerotium rolfsii, Sclerotinia sclerotiorum, Rhizoctonia solani, Fusarium solani, Penicillium* sp.	Surfactin, iturin and fengycin	Common bean	Torres et al. (2017)
Rhizobium sp.	*Fusarium, Rhizoctonia, Sclerotium, Macrophomina*	Antibiotics, HCN, mycolytic enzymes, siderophore	Legume plants	Das et al. (2017)
Bacillus amyloliquefaciens	*Sclerotium rolfsii, Sclerotinia sclerotiurum, Rhizoctonia solani, Fusarium solani, Macrophomina phaseolina,*	Surfactin, iturin, fengycin, kurstatin, polymyxin,	Common bean	Sabaté et al. (2017)
B. subtilis LHS11 and FX2	*Sclerotinia sclerotiorum* (stem rot)	Antibiosis	Rapeseed	Sun, et al. (2017)
Bacillus subtilis 9407	*Botryosphaeria dothidea* (apple ring rot)	Fengycin	Apple	Fan et al. (2017)
Paraburkholderia phytofirmans PsJN	*Pseudomonas syringae* pv. tomato DC3000	ISR	*Arabidopsis*	Timmermann et al. (2017)
Bacillus velezensis RC 218	*Fusarium graminearum* (head blight)	Fengycin, iturin, ericin	Wheat	Palazzini et al. 2016
Pseudomonas protegens S4LiBe and S5LiBe	*Botrytis cinerea, Verticillium dahliae, F. graminearum, Aspergillus niger, A. flavus*	Siderophores, chitinase, polymer degrading enzymes	—	Bensidhoum et al. (2016)
Pseudomonas chlororaphis MCC2693	*Alternaria alternata, Phytophthora* sp., *Fusarium solani, F. oxysporum*	PCA,, HCN, ammonia, siderophores, lytic enzymes	Wheat	Jain and Pandey (2016)
Bacillus amyloliquefaciens ZM9	*Ralstonia solanacearum* (bacterial wilt)	Surfactin	Tobacco	Wu et al. (2016)
Bacillus amyloliquefaciens SB14	*Rhizoctonia solani* AG-4 and AG2-2	Antibiotics, lytic enzymes, VOCs	Sugar beet	Karimi et al. (2016)
Bacillus sp.	*Fusarium oxysporum, R. solani, Botrytis cinerea* R16, *Galactomyces geotrichum* MUCL 28959, *Verticillium longisporum* O1	Fengycins, surfactins, mycosubtilin, bacillomycin, kurstakins	Date palm	El Arbi et al. (2016)

Bacillus subtilis B1	*Lasiodiplodia theobromae* (bluish black discoloration)	Surfactin, fengycin	Rubber wood	Sajitha and Dev (2016)
B. cepacia, *B. amyloliquefaciens*, *S. marcescens*, *S. marcescens*, *P. aeruginosa*	*Pythium myriotylum* (soft rot)	—	Ginger	Dinesh et al. (2015)
Burkholderia pyrrocinia 2327	*Rhizoctonia solani*, *Trichophyton*	Antibiotic production (pyrrolnitrin)	—	Kwak and Shin (2015)
Pseudomonas fluorescens	*Botrytis cinerea* (gray mold)	Phenazines, cyanogens, siderophores, proteases	*Medicago truncatula*	Hernández-león et al. (2015)
Arthrobacter, *Curtobacterium*, *Enterobacter*, *Microbacterium*, *Pseudomonas*	*Xanthomonas axonopodis* pv. *passiflorae*	Siderophore	Passion fruit	Halfeld-vieira et al. (2015)
B. amyloliquefaciens GB1	*Valsa mali* (apple valsa canker)	—	Apple	Zhang et al. (2015)
Pseudomonas sp., *Paenibacillus* sp.Pb28, *Enterobacter* sp. En38, *Serratia* sp. Se40	*Ralstonia solanacearum* (wilt)	Siderophore, HCN, protease production	Potato	Kheirandish and Harighi (2015)
Bacillus thuringiensis UM96	*Botrytis cinerea* (gray mold)	Chitinase	*Medicago truncatula*	Martínez-Absaló et al. 2014
Bacillus amyloliquefaciens S20	*F. oxysporum*, *R.solanacearum* (Wilt)	Iturins A	Eggplant	Chen et al. (2014)
Pseudomonas fluorescens	*Athelia rolfsii* (southern blight)	VOCs, dimethyl disulfide	*Atractylodes*	Zhou et al. (2014)
Pseudomonas fluorescens MGR12	*Fusarium proliferatum* (head blight)	VOCs	Cereals	Cordero et al. (2014)
Bacillus subtilis NCD-2	*Rhizoctonia solani* (damping-off)	Fengycin	Cotton	Guo et al. (2014)
Pseudomonas sp. LBUM223	*Streptomyces scabies* (potato scab)	PCA	Potato	Arseneault et al. (2014)
Bacillus licheniformis, *Pseudomonas fluorescens*	*Botrytis cinerea*	—	Grape	Salomon et al. (2014)

(continued)

Table 11.2 (continued)

Biological control agents	Phytopathogens	Mechanism	Plant species	References
Bacillus amyloliquefacien CM-2 and T-5	*Ralstonia solanacearum* (vascular wilt)	—	Tomato	Tan et al. (2013)
Pseudomonas putida PP3WT, *Bacillus cereus* SC1AW	*Ralstonia solanacearum* (wilt)	—	Tomato	Kurabachew and Wydra (2013)
Pseudomonas chlororaphis subsp. *aurantiaca* StFRB508	*Fusarium oxysporum* f. sp. *conglutinans*	PCA	Potato	Morohoshi et al. (2013)
Bacillus amylolequifaciens	*Polymyxa betae* (Rhizomania disease)	ISR	Sugar beet	Desoignies et al. (2013)
Acinetobacter lwoffii, *B. subtilis*, *Pantoea agglomeran*, *P. fluorescens*	*Botrytis cinerea*	ISR	Grapevine	Magnin-Robert et al. (2013)
Pseudomonas fluorescens	*P. syringae* pv. *tomato* (bacterial speck)	DAPG, ISR	*Arabidopsis*	Weller et al. (2012)
Pantoea agglomerans	*Erwinia amylovora* (fire blight)	Pantocin A, herbicolins, microcins, phenazines	Pome fruits	Braun-Kiewnick et al. (2012)
Pseudomonas brassicacearum J12	*Ralstonia solanacearum* (wilt)	DAPG, HCN, siderophore, protease	Tomato	Zhou et al. (2012)
Chryseobacterium wanjuense KJ9C8	*Phytophthora capsici* (blight)	Proteinase and HCN production	Pepper	Kim et al. 2012
Pseudomonas fluorescens	*P. syringae* pv. *tomato* (speck)	DAPG, ISR	*Arabidopsis*	Weller et al. (2012)
Pseudomonas brassicacearum J12	*Ralstonia solanacearum* (wilt)	DAPG, HCN, siderophore, protease	Tomato	Zhou et al. (2012)
Bacillus sp., *Stenotrophomonas maltophilia* 2JW6	*Ralstonia solanacearum*	—	Ginger	Yang et al. (2012)
Pseudomonas sp., *Bacillus* sp.	*Ralstonia solanacearum* (bacterial wilt)	—	Eggplant	Ramesh and Phadke (2012)
Pseudomonas fluorescens Psd	*Fusarium oxysporum*	PCA, pyrrolnitrin	Tomato	Upadhya, Srivastava 2011

Pseudomonas protegens	—	DAPG, pyoluteorin	—	Ramette et al. (2011)
Bacillus subtilis CAS15	*Fusarium oxysporum* (wilt)	2,3-dihydroxybenzoate and 2,3 dihydroxybenzoyl glycine	Pepper	Yu et al. (2011)
Bacillus cereus AR156	*Pseudomonas syringae* pv. *tomato* DC3000	ISR–SA, jasmonic acid and ethylene	*Arabidopsis*	Niu et al. (2011)
Fungus				
Aspergillus terreus JF27	*Pseudomonas syringae* pathovar (pv.) *tomato* DC3000 (speck)	ISR	Tomato	Yoo et al. (2018)
T. asperellum CWD CHF 78	*Fusarium oxysporum* f. sp. *lycopersici* (wilt)	Chitinases, proteases, siderophores	Tomato	Li et al. (2018)
Clonostachys rosea	*Rhizoctonia solani* AG-3 (black scurf)	—	Potato	Salamone et al. (2018)
T. harzianum Ths97	*Fusarium solani* (root rot)	Mycoparasitism	Olive trees	Amira et al. (2017)
T. harzianum ThHP-3	*F. oxysporum*, *C. capsici*, *C. truncatum*, *Gloesercospora sorghi*	Mycoparasitism	—	Sharma et al. (2017a, b, c)
T. harzianum T1A	*Guignardia citricarpa* (citrus black spot)	Mycoparasitism	—	de Lima et al. (2017)
T. harzianum	*Fusarium oxysporum*	ISR	Soybean	Zhang et al. (2017)
P. simplicissimum, *Leptosphaeria* sp., *Talaromyces flavus*, *Acremonium* sp.	*V. dahliae* Kleb. (Verticillium wilt)	ISR	Cotton	Yuan et al. (2017)
Trichoderma M10	*Uncinula necator* (powdery mildew)	Secondary metabolites, SIR	*Vitis vinifera*	Pascale et al. (2017)
Trichoderma atroviridae	*Fusarium solani* (root rot, damping off)	Mycoparasitism, secretion of toxic secondary metabolites, competition	Common bean	Toghueo et al. (2016)

(continued)

Table 11.2 (continued)

Biological control agents	Phytopathogens	Mechanism	Plant species	References
Trichoderma harzianum T-aloe	*Sclerotinia sclerotiorum* (stem rot)	Hyphal parasitism, 1,3-β-glucanase, chitinase	Soybean	Zhang et al. (2016)
T. asperellum CCTCC-RW0014	*Fusarium oxysporum* f. sp. *cucumerinum*	CWD enzymes (chitinase, protease, glucanase)	Cucumber	Saravanakumar et al. (2016)
T. polysporum	*Fusarium oxysporum* f. sp. *melonis* (melon wilt)	Competition, antibiosis, mycoparasitism	Water melon	Gava and Pinto (2016)
Trichoderma sp.	*Sclerotinia sclerotiorum* (white mold)	CWD enzymes, parasitism	Beans	Geraldine et al. (2013)
Chaetomium globosum	*Fusarium sulphureum, A. alternate, C. sorghi, F. oxysporum* f. sp. *vasinfectum, Botrytis cinerea, F. graminearum*	Antibiotic production (gliotoxin)	Ginkgo biloba	Li et al. (2011)
T. viride	*Fusarium oxysporum* f. sp. *adzuki* and *Pythium arrhenomanes,*	Mycoparasitism	Soybean	John et al. (2010)
Actinomycetes				
Streptomyces sp. CB-75	*Colletotrichum musae, C. gloeosporioides*	Type I polyketide synthase, nonribosomal peptide synthetase	Banana	Chen et al. (2018)
Streptomyces sp.	*Sclerotium rolfsii* (collar rot)	Host defense enzymes/genes and accumulation of phenolic compounds	Chickpea	Singh and Gaur (2017)
Streptomyces sp.	*Magnaporthe oryzae* (blast)	—	Rice	Law et al. (2017)
Streptomyces PM5	*Pectobacterium carotovorum subsp. brasiliensis* (soft rot)	Lipase and VOCs	Tomato	Dias et al. (2017)
Streptomyces UPMRS4	*Pyricularia oryzae* (blast)	Chitinase, glucanase and PR1	Rice	Awla et al. (2017)
Streptomyces sp.	*Phytophthora capsici, Sclerotium rolfsii*	Hydrolytic enzymes (amylases, proteases, lipases, cellulases)	Black pepper	Thampi and Bhai (2017)
S. corchorusii UCR3-16	Several pathogens	Antifungal metabolites, VOCs, siderophores, CWD enzymes	Rice	Tamreihao et al. (2016)

S. plicatus isolate B4-7	*Phytophthora capsici* (damping off, root rot, leaf blight)	Antibiotic (borrelidin)	Bell pepper	Chen et al. (2016)
Streptomyces sp.	*Sclerotium rolfsii* (stem rot)	—	Groundnut	Adhilakshmi et al. (2014)
Streptomyces sp. NSP (1–6)	*Fusarium oxysporum* f.sp. *capsici* (wilt)	Chitinase	Chili plants	Saengnak et al. (2013)
Streptomyces sp.	*Fusarium oxysporum* f.sp. *zingiberi* (rhizome rot)	—	Ginger	Manasa et al. (2013)
Streptomyces sp.	*Xanthomonas oryzae* pv. *oryzae (Xoo)*, (leaf blight)	Chitinase, phosphatase, and siderophore	Rice	Hastuti et al. (2012)
Streptomyces sp.	*Rhizoctonia solani* AG-2, *Fusarium solani*, *Phytophthora drechsleri* (root rot)	Protease, chitinase, α-amylase activity	Sugar beet	Karimi et al. (2012)
S. toxytricini vh6, *S. flavotricini* vh8, *S. toxytricini* vh22, *S. avidinii* vh32, *S. tricolor* vh85	*Rhizoctonia solani* (root rot)	ISR (antimicrobial phenolics and SA)	Tomato	Patil et al. (2011)

CWD enzymes cell wall-degrading enzymes, *DAPG* 2,4-diacetylphloroglucinol, *HCN* hydrocyanic acid, *ISR* induced systemic resistance, *PCA* phenazine-1-carboxylic acid, *SA* salicylic acid, *VOCs* volatile organic compounds

Table 11.3 Recent studies reporting biocontrol activities of PGPR, PGPF, and other microbes against different pests are summarized

Biological control agents	Pest	PGP traits	Plant species	References
Bacteria				
Pseudomonas, *Bacillus* sp.	*Meloidogyne javanica*, *Ditylenchus* sp	Production of phytohormones, antibiotic production	Garlic, soybean	Turatto et al. (2018)
Bacillus cereus, *B. licheniformis*, *Lysinibacillus sphaericus*, *P. fluorescens*, *P. brassicacearum*	*Meloidogyne incognita*	–	Tomato	Colagiero et al. (2018)
Bacillus sp., *Pseudomonas* sp.	Aphid	Yield enhancement	Wheat	Naeem et al. (2018)
Pseudomonas fluorescens, *Bacillus subtilis*	*Plutella xylostella*	–	Chinese kale	Rahardjo and Tarno (2018)
Serratia proteamaculans	*Meloidogyne incognita*	Increase in root and shoot growth	Tomato	Zhao et al. (2018)
Pseudomonas putida strain, BG2 and *Bacillus cereus* BC1	*Meloidogyne incognita*	Increase in plant growth and essential oil	Patchouli	Borah et al. (2018)
Kosakonia radicincitans	*Brevicoryne brassicae* and *Myzus persicae*	–	*Arabidopsis*	Brock et al. (2018)
Bacillus velezensis, *B. mojavensis*, *B. safensis*	*Heterodera glycine* (cyst nematode)	Increased in plant height, plant biomass and yield	Soybean	Xiang et al. (2017)
Bacillus sp. BC27 and BC29	*Meloidogyne javanica*	Increase in shoot weight	Soybean	Chinheya et al. (2017)
Pseudomonas putida and *Rothia sp.*	*Spodoptera litura*	Increase in plant biomass and yield	Tomato	Bano and Muqarab (2017)
Bacillus methylotrophicus strain R2-2	*Meloidogyne incognita*	Yield enhancement	Tomato	Zhou et al. (2016)
Lysobacter antibioticus strain 13-6				
Bacillus subtilis isolates Sb4–23, Mc5-Re2, and Mc2-Re2,	*Meloidogyne incognita*	–	Tomato	Adam et al. (2014)

(continued)

Table 11.3 (continued)

Biological control agents	Pest	PGP traits	Plant species	References
Fungus				
Beauveria bassiana	*Spodoptera littoralisn*	Boosted spike production	Wheat	Sánchez-Rodríguez et al. (2018)
Purpureocillium lilacinum	*Meloidogyne javanica*, *Meloidogyne incognita*	Increase in yield	Tomato	Kepenekci et al. (2018)
Beauveria bassiana GHA	–	Enhance the root sett	Sugarcane	Donga et al. (2018)
Metarhizium brunneum CB15	–	Biomass, leaf area, nitrogen and phosphorus contents were enhanced	Potato	Krell et al. (2018)
Beauveria bassiana, *Isaria fumosorosea*, and *Metarhizium brunneum*	–	Positive effect on survival, growth, health, length, and dry weight of cabbage	Cabbage	Dara et al. (2017)
Syncephalastrum racemosum, *Paecilomyces lilacinus*	*Meloidogyne incognita*	Stimulated root length, shoot length and increased the cucumber yield	Cucumber	Huang et al. (2016)
Beauveria bassiana and *Metarhizium brunneum*	–	Plant growth enhancement	*Vicia faba*	Jaber and Enkerli (2016)
Beauveria bassiana and *Purpureocillium lilacinum*	*Helicoverpa zea* (cotton bollworm)	Plant growth enhancement	Cotton	Lopez and Sword (2015)
Metarhizium robertsii	Several insects	Induced root hair proliferation and plant root growth	Switchgrass, haricot beans	Sasan and Bidochka (2012)
Metarhizium anisopliae LHL07	–	Higher shoot length, shoot fresh and dry biomass, chlorophyll contents, transpiration rate, photosynthetic rate and leaf area	Soybean	Khan et al. (2012)
Metarhizium anisopliae	–	Increased plant height, root length, shoot and root dry weigh	Tomato	Elena et al. (2011)
Actinomycetes				
S. rubrogriseus HDZ-9-47	*Meloidogyne incognita*	Increase in yield	Tomato	Jin et al. (2017a, b)
S. galilaeus strain KPS-C004	*Meloidogyne incognita*	Increase in plant biomass, shoot-root length	Chili	Nimnoi et al. (2017)

11.2 Plant Growth-Promoting Rhizobacteria (PGPR)

The term "PGPR" was first used for soil-borne bacteria supporting PGP activity by root colonization in plants (Kloepper and Schroth 1978). The PGPR comprises the heterogeneous group of nonpathogenic, root-colonizing bacteria that ameliorate plant growth. This group of rhizobacteria found in the narrow region of soil around plant root, known as the rhizosphere, primarily influenced by the plant root system. Lorenz Hiltner was the first to use term "rhizosphere," a word primarily originating from the Greek word "rhiza" (Hiltner 1904). The rhizosphere is a highly competitive microenvironment for diverse groups of microbes to obtain nutrients and proliferative growth that helps plants in development and PGP activity.

The growth promotion by PGPR occurs by the modification of the rhizospheric microbial community. Generally, PGPR affect plant growth by exhibiting a variety of direct and indirect mechanisms. The direct PGP activity entails either facilitating the resource acquisition (essential minerals and nutrients) from the surrounding environment or by providing synthesized compounds. The indirect mechanisms are related to reduce the harmful effects of phytopathogens by synthesis of antibiotics, lytic enzymes (chitinases, cellulases, 1,3-glucanases, proteases, and lipases), and chelation of available iron in the plant-root interface.

11.2.1 Categories of PGPRs

The PGPR are categorized into extracellular (ePGPR-symbiotics) and intracellular (iPGPR-free-living) PGPR depending on their habitat in plant compartment (Gray and Smith 2005). The ePGPR exists among the spaces in the root cortex cells, rhizosphere and rhizoplane, whereas iPGPRs reside in the nodular structures of root cells (Figueiredo et al. 2010). The ePGPR include different bacterial genera such as *Erwinia*, *Flavobacterium*, *Arthrobacter*, *Agrobacterium*, *Azotobacter*, *Azospirillum*, *Burkholderia*, *Bacillus*, *Caulobacter*, *Chromobacterium*, *Micrococcous*, *Pseudomonas*, and *Serratia* (Ahemad and Kibret 2014). The iPGPR includes the members of Rhizobiaceae family (such as *Rhizobium*, *Bradyrhizobium*, *Allorhizobium*, *Mesorhizobium*), *Frankia* species, and endophytes (Bhattacharyya and Jha 2012).

The PGPR can be also classified on the bases of their functional activities. This classification includes biofertilizer (enhances the availability of primary nutrients and growth of host plant), biopesticide (suppress or control diseases, mainly by antifungal metabolites and antibiotic production), phytostimulators (the ability to produce phytohormones like IAA, GAs, etc.), and rhizoremediators (degrading organic pollutants) (Bhardwaj et al. 2014). The PGPRs employ number of mechanisms to interact with their host plants either simultaneously or separately under different time and conditions.

11.3 Plant Growth-Promoting Fungi (PGPF)

Most of the previous studies have focused on PGPR and their association with phytopathogens whereas little is known about the PGPF. The PGPF are nonpathogenic saprophytes that exert advantageous effects on plants. They are known to enhance plant growth, suppress plant diseases, and induce ISR. Some PGPFs species reported to suppress the bacterial and fungal diseases of some crop plants. The well-known nonpathogenic fungal genera include *Aspergillus*, *Piriformospora*, *Fusarium*, *Penicillium*, *Phoma*, *Rhizoctonia*, and *Trichoderma* and stimulate different plant traits helpful for higher yields (Jaber and Enkerli 2017; Lopez and Sword 2015).

Some examples of PGPF with BCA activity include endophytes, ectomycorrhizas (EcM), arbuscular mycorrhizae (AMF), yeasts, *Trichoderma* sp., and certain avirulent strains of phytopathogens like *Fusarium oxysporum*, *Cryphonectria parasitica*, and *Muscodor albus* (Waghunde et al. 2017). These beneficial fungi have been produced in large quantities and widely applied for management of plant diseases (Ghorbanpour et al. 2017). The PGPF and plant root association has shown to modulate plant growth, mineral nutrient uptake, increased biomass, and yield of crop plants (Deshmukh et al. 2006). Plant beneficial microorganisms are of great interest for applications in agriculture as biofertilizers and biopesticides and for phytoremediation (Berg 2009; Weyens et al. 2009; Shelake et al. 2018).

11.4 Biological Control by PGPR and PGPF

The term "biological control" was first coined to describe the use of natural enemies (introduced or manipulated) to control insect pests by Harry Scott Smith (1919). Later, Paul H. DeBach and Hagen (1964), an entomologist, redefined "natural control" from "biological control." The natural control includes biotic (such as food availability and competition) and abiotic (like weather and soil) factors, and also the natural enemies (like predators, parasites, and pathogens) mediated effects. The natural enemies are affecting or regulating the pest populations. The biological control or biocontrol is a part of the natural control and described as the use of natural or living organisms to inhibit pathogen and suppress plant diseases. The chief mode of action of biocontrol in PGPR/PGPF implicates competition for nutrients, SAR/ISR induction, niche exclusion, and production of antifungal/antibacterial metabolites like antibiotics, bacteriocins, and lytic enzymes (Salomon et al. 2017). The biological control is generally separated into three types: classical biological control (CBC), conservation, and augmentation. Each of these approaches can be used separately or in combination with each other in the biological control program.

11.4.1 Classical Biological Control

The importation of natural enemies to control an introduced or "exotic" pest is known as CBC. The initial step in CBC involves the determination of the pest origin, and then an exploration for its natural enemies in its habitat. The potential BCAs then introduced to the new pest location and released for its establishment. For example, in the late 1800s, the cottony cushion scale, a pest which is native to Australia devastated California citrus industry. The Vedalia beetle (predatory insect) was then introduced from Australia, and the pest control achieved in short time. Three exotic encyrtid parasitoids (*Anagyrus loecki*, *Acerophagus papayae*, and *P. Mexicana*) were introduced in Southern state of India (Tamil Nadu) against a papaya mealybug *Paracoccus marginatus*, causing damage to mulberry fields (Sakthivel 2010).

11.4.2 Conservation

Conservation involves the practices that protect, maintain, and enhance the existing natural enemies. Conservation practices include either reducing or eliminating the factors which interfere with or destroy the natural enemies, for example, use of selective chemical pesticides or providing resources that natural enemies need in their environment.

11.4.3 Augmentation

Augmentation involves the mass culture and release of natural enemies. It consists of two types: inoculative and inundative. The inoculative involves the release of few natural enemies seasonally and suppresses pest outbreaks whereas inundative involves the release of enormous numbers of natural enemies to outcompete the pest population completely. In inundative release, immediate control of pest population is achieved by massive release of their natural enemies.

11.5 PGPR and PGPF as Biological Control Agents (BCAs)

The term BCA generally used in broader sense that includes naturally occurring materials (biochemical pesticides), microbes (microbial pesticides), and plants-produced materials consisting genetic material or plant-incorporated protectants (US EPA 2012). The biochemical pesticides include organic acids, plant and insect growth regulators, plant extracts, pheromones, minerals, and other substances. The Association of Southeast Asian Nations (ASEAN) Sustainable Agrifood Systems (Biocontrol) Project (ABC) classified BCA into four product categories to accommodate living and nonliving active agents: microbial control agents (microbial), macroorganisms (macrobials), semiochemicals, and natural products. Microbial control agents often called as "biopesticides" include a variety of microbes, viz., bacteria, fungi, protozoa, nematodes, and viruses. Among these, bacteria and fungi dominate the commercial BCA formulations including PGPR/PGPF.

The macrobials agents include the mites and insects. Their mode of deployment includes the conservation and CBC. A more recent example includes the release of the *Anagyrus lopezi* (wasp) from Benin to control *Phenacossus manihoti* (mealybug) of pink cassava in Thailand (Winotai et al. 2012). The semiochemcials refers to the biochemical molecules or mixtures that carry specific messages between individuals of the same or different species. These semiochemicals often used as insect attractants (pheromones) and repellents in extremely low dosage. The last one includes the natural plant extracts or "botanicals" which cover diverse natural substances like azadirachtin, pyrethrum, ginseng extract, etc. with different biological activity (Regnault-Roger et al. 2005). In this work, microbials that include PGPR/PGPF are discussed in detail and other BCA categories.

11.6 Mechanisms of Biological Control by PGPR and PGPF

Prediction of disease epidemiology in plants is determined by the associations among the constituents of disease triangle, i.e., pathogen, susceptible host and environment. The interactions among these three components show the severity and occurrence of the disease. The BCAs interact with all the three components of the disease triangle. The BCAs-pathogens interactions studies have revealed the multiple mechanisms of biological control (Table 11.1). The BCAs act on phytopathogens through one or more multifarious mechanisms resulting in plant growth inhibition and spread of phytopathogens (summarized in Tables 11.2 and 11.3). The various mechanisms employed in controlling the plant diseases can broadly classified into direct, indirect, and mixed path antagonism.

11.6.1 Direct Antagonism

11.6.1.1 Parasitism and Hyperparasitism

Parasitism is a type of interaction between two phylogenetically unrelated organisms in which one organism, the parasite, is usually benefitted and the other called the "host" is harmed. For example, *Trichoderma* spp. have a parasitic activity toward a wide variety of phytopathogens such as *Botrytis cinerea*, *Rhizoctonia solani*, *Pythium* spp., *Sclerotium rolfsii*, *Sclerotinia sclerotiorum*, and *Fusarium* spp. (summarized in Waghunde et al. 2016). The *Rhizoctonia solani* cause several plant diseases like rice blight and black scurf of potato and *Trichoderma* spp. is being used as a potential BCA for all these diseases (Jia et al. 2013; Rahman et al. 2014).

The terms mycoparasitism and hyperparasitism have been used for fungal species parasitic on another fungus. The involved pathogen is known as hyperparasite or mycoparasite, or parasite. The mycoparasitism involves the chemotropic growth of the BCA toward the pathogen, recognition through the host lectins and carbohydrate receptors present on the biocontrol fungus. The next step involves the coiling and making of cell wall-degrading (CWD) enzymes and penetration. Some examples include the powdery mildew pathogen parasitized by multiple hyperparasites like *Ampelomyces quisqualis*, *Acrodontium crateriforme*, *A. alternatum*, *Cladosporium oxysporum*, and *Gliocladium virens* (Kiss 2003; Heydari and Pessarakli 2010). An additional case is the virus causing hypovirulence on *Cryphonectria parasitica*, an ascomycete causing chestnut blight (Tjamos et al. 2010).

11.6.1.2 Commensalism

Commensalism is a type of symbiotic interaction benefiting one partner while the other is neither harmed nor benefited. The benefited organism is known as commensal and obtains its nutrients and shelter from its host species. A good example of commensals comprises rhizobacteria. The rhizobacteria such as PGPR control soil-borne phytopathogens through antibiotic production, nutrient competition thereby helping plants to survive from phytopathogens.

11.6.2 Indirect Antagonism

11.6.2.1 Competition

Competition is an indirect mechanism and plays a significant role in the biocontrol of pathogens. Biocontrol by competition occurs when nonpathogenic microbes compete for organic nutrients with pathogens to proliferate and survive in host plant. Predominantly, the BCAs have more competent nutrient uptake system than phytopathogens. One of the examples includes control of *Fusarium* wilt due to carbon competition between pathogenic and nonpathogenic strains of *F. oxysporum* (Alabouvette et al. 2009). Fire blight, a contagious disease caused by *Erwinia amylovora* is suppressed by its closely related saprophytic species *E. herbicola* due to nutrient competition on the leaf surface.

11.6.2.2 Systemic Acquired Resistance (SAR)

During the biotic or abiotic stress, the plant produces chemical signals like glutamate thereby activating the plant defense pathways (Toyota et al. 2018). In order to tackle abiotic and biotic stresses plants express a variety of active defense system. The PGPR and PGPF produce chemical stimuli which can induce a persistent variation in plants increasing its capacity to tolerate pathogenic infection and induce systemic host defense against wide-ranging pathogens, known as induced resistance. The induced resistance is of two different forms: the SAR and ISR represent the plant defense response active against phytopathogens. The SAR is the inherent resistance capacity of a plant which activates after being exposed to chemical elicitors from nonpathogenic, virulent, or avirulent microbes or artificial chemical stimuli (Gozzo and Faoro 2013). It remains active against broad-spectrum pathogens for a prolonged time. The SAR induction is mediated by the buildup of accumulated chemical stimuli like salicylic acid (SA) generally secreted after pathogen attack. The SA is the first chemical signal inducing the production of pathogenesis-related (PR) proteins, for example, chitinase, β-1, 3 glucanse. The PR genes code for chitinases and β-1, 3-glucanases which play a significant role in reducing or preventing the pathogen colonization (Sudisha et al. 2012). The SAR has been showed against some pathogens and pests, including *Uromyces viciae-fabae*, *Ascochyta fabae*, *M. incognita*, and *R. solanacearum* (Pradhanang et al. 2005; Molinari and Baser 2010; Sillero et al. 2012).

11.6.2.3 Induced Systemic Resistance (ISR)

The ISR naturally exists in plants and is generally associated to stimulation by nonpathogenic plant-associated rhizobacteria (Pieterse and Van Wees 2015). The ISR is independent of the SA-mediated pathway, and PR proteins are not involved. It is plant specific and depends upon the plant genotype. The applications of nonpathogenic PGPR/PGPF induce ISR facilitated by phytohormones production (viz., jasmonic acid and ethylene). The PGPRs induces ISR in several plants against numerous environmental stressors. The plant defense system produces an enormous number of enzymes involved in plant defense, like polyphenol oxidase, β-1, 3-glucanase, chitinase, phenylalanine ammonia lyase, peroxidase, etc. Even though ISR is not precisely against a specific pathogen, it plays a major role in control of a range of diseases in plant (Kamal et al. 2014). For example, the ISR activity induced by application of *Trichoderma* strains in the leaves was found effective against several diseases in tomato plants (Saksirirat et al. 2009). Rice plant treated with *Bacillus* sp. showed resistance against bacterial leaf blight (Udayashankar et al. 2011).

11.6.3 Mixed Path Antagonism

11.6.3.1 Antibiosis

Antibiosis is defined as the interactions involving a low-molecular-weight compound or an antibiotic that is detrimental to another microorganism. Antibiosis plays a significant role in the suppression of plant diseases and pathogens (Nikolić et al. 2018; Kumari et al. 2018). The PGPR like *Bacillus* sp. and *Pseudomonas* sp., produces a diverse range of antibiotics against different phytopathogens and is significantly more efficient biocontrol mechanism over the past decade (Ulloa-Ogaz et al. 2015). The antibiotics such as phenazine-1-carboxylic acid (PCA), phenazine-1-carboxamide, *N*-butylbenzene sulfonamide, pyrrolnitrin, pyoluteorin, rhamnolipids, oomycin A, cepaciamide A, 2,4-diacetylphloroglucinol, ecomycins, viscosinamide, butyrolactones, pyocyanin (antifungal), azomycin, pseudomonic acid, cepafungins, and Karalicine are produced by *Pseudomonas* sp. (Ramadan et al. 2016). *Bacillus* sp. also produces subtilintas A, subtilosin A, bacillaene, sublancin, difficidin, mycobacillin, chlorotetain bacilysin, rhizocticins, iturins, surfactin, and bacillomycin (Wang et al. 2015). The antibiotic 2,4-diacetyl phloroglucinol produced by *Pseudomonas* sp. is reported to inhibit *Pythium* sp. Similarly, iturin is reported to suppress *B. cinerea* and *R. solani* (Padaria et al. 2016).

11.6.3.2 Siderophores

In addition to water, carbon dioxide, and oxygen, all living plants need total 14 essential elements that include iron (Shelake et al. 2018). The PGPR produces low-molecular-weight (500–1500 Da) organic compounds called siderophores to competitively capture ferric ion under iron-lacking conditions. Siderophore-producing PGPRs gain more attention because of their distinctive property to extract iron from their surrounding (Saha et al. 2016). They sequester iron from their micro-environment, forming a ferric-siderophore complex that progress through diffusion and reverted to the cell surface (Andrews et al. 2003). The bacterial siderophores are of four classes depending on their iron coordinating functional groups: hydroxamates, carboxylate, pyoverdines and phenol catecholates (Crowley 2006).

The PGPRs exert their antagonism to several phytopathogens using secreted siderophores (Tables 11.2 and 11.3). They function by sequestering iron in the root zone, making it unavailable to phytopathogens and inhibiting their growth. Also, PGPR-secreted siderophores augment plant uptake of iron that can distinguish the bacterial ferric-siderophore complex (Katiyar and Goel 2004; Dimkpa et al. 2009). Siderophores produced by *Pesudomonas* group suppress several fungal pathogens and also enhanced growth of numerous crops (Bensidhoum et al. 2016; Sharma et al. 2017a, b, c; Tabli et al. 2018).

11.6.3.3 Volatile Substances

Soil microbes including PGPR produce and release various organic and inorganic volatile compounds (Audrain et al. 2015). The volatile compounds synthesized by PGPR suppressed diverse kind of phytopathogens, indicating their role in biocontrol of soil-borne pathogens (Karimi et al. 2016; Gotor-Vila et al. 2017; Rath et al. 2018). The volatile compounds from PGPR, for instance, *Pseudomonas*, *Bacillus*, and *Arthrobacter*, directly or indirectly facilitate enhanced resistance against diseases, tolerance against abiotic stress, and higher biomass production. The *Bacillus* sp. produces acetoin and 2, 3-butanediol, effective against fungal pathogens (Santoro et al. 2016). *Bacillus megaterium* was found to produce ammonia which inhibits *Fusarium oxysporum* (Shobha and Kumudini 2012). Several other studies on *Pseudomonas* sp. reported the production of ammonia and hydrocyanic acid serving PGP and biocontrol activity (Verma et al. 2016; Sharma et al. 2017a, b, c).

11.6.3.4 Lytic Enzyme Production

The PGPR/PGPF can suppress the growth and activities of phytopathogens by secreting lytic enzymes. The PGPR produces a diverse number of enzymes like ACC-deaminase, cellulases, chitinase, lipases, proteases, β-1,3-glucanase which are involved in the lysis of fungal cell wall (Goswami et al. 2016). The fungal cell wall primarily consists of chitin, glucans, and polysaccharides; hence β-1,3-glucanase- and chitinase-producing bacteria are effective to suppress their growth. The expression of lytic enzymes by PGPR can enhance the suppression of phytopathogens. For instance, chitinase produced by *S. plymuthica* strain C48 inhibits germ-tube elongation and spore germination in *Botrytis cinerea* (Frankowski et al. 2001). Chitinase secreted by *Paenibacillus* sp., *Streptomyces* sp., and *Serratia marcescens* was found to constrain the growth of *Botrytis cinerea*, *Sclerotium rolfsii*, and *Fusarium oxysporum* f. sp. *cucumerinum*. *Lysobacter* produces enzyme glucanase which inhibits *Bipolaris* and *Pythium* sp. (Palumbo et al. 2005). *Micromonospora chalcea* and

Actinoplanes philippinensis inhibit *Pythium aphanidermatum* in cucumber through the secretion of β-1, 3-glucanase (El-Tarabily 2006).

11.7 Advantages of PGPR and PGPF as BCAs

The agrochemicals and genetic approaches used as tools to control plant diseases, but they are not always effective. Moreover, several agrochemicals are nonbiodegradable and exert a harmful effect on the environment. The excessive usage of pesticides for plant disease management has increased pathogen-resistant strains (Burketova et al. 2015). In this regard, PGPR have been seen as an attractive strategy and a sustainable means of controlling soil-borne pathogens and diseases. The application of PGPR and PGPF in sustainable agriculture has been increased in several regions. The PGPR with biocontrol efficacy often provides long-term protection against soil-borne phytopathogens because of their rhizosphere competency, i.e., capacity to rapidly colonize the rhizosphere.

The PGPR/PGPF utilizes the plant's rhizodeposits as a chief carbon source for their development (Denef et al. 2007). The PGPF protect plants from harmful microbes by producing antibiotics while some act as a parasite and some compete for space and food with pathogens (described in earlier sections). They also protect plants by ISR against pathogenic bacteria (Yoshioka et al. 2012; Hossain and Sultana 2015), fungi (Murali et al. 2013; Tohid and Taheri 2015, Nassimi and Taheri 2017), viruses (Elsharkawy et al. 2013), and nematodes (Vu et al. 2006). The arbuscular mycorrhiza fungi (AMF) also help plants in resource acquisition, suppression of diseases, and tolerance to soil pollution and development (Wani et al. 2017). Many studies suggested AMF as an efficient BCAs against phytopathogens and nematodes (Veresoglou and Rillig 2012; Vos et al. 2012, 2013; Akhtar and Panwar 2013). The use of PGPR/PGPF as BCAs reduces the burden of agrochemicals (fertilizers and pesticides) in agricultural ecosystem thus preventing environmental pollution. The BCAs have several other advantages as compared to pesticides mentioned as follows:

1. The PGPR enhances growth and protects plants against phytopathogens.
2. The PGPR can act as a biofertilizer, biopesticide, phytostimulators, and rhizoremediators.
3. The PGPR multiply in soil, leaving no residual problem.
4. A single PGPR can protect against multiple plant pathogens.
5. The PGPR possess multifarious mechanisms including antibiosis, CWD enzymes and siderophore production and also induce SAR/ISR in plants.
6. They are nontoxic to plants and humans.
7. They are ecofriendly and easy to manufacture.
8. BCAs are cheaper as compared to the agrochemicals.
9. The PGPR can be handled easily and applied in the field.
10. The use of PGPR is sustainable in long-term.

11.8 Global Status of Biopesticides

The biopesticides have attracted more interest of global research community due to the harmful effects of chemical pesticides on human health through produced food and environmental safety. Consequently, the global crop protection chemical and conventional pesticide market have experienced major variations over the recent years (Pelaez and Mizukawa 2017). At present, biopesticides comprise only 5% of the total global crop protection market, with 3 billion dollars in revenue worldwide (Damalas and Koutroubas 2018). In the market of the United States, there are more than 200 products registered for use in comparison with 60 similar products in the market of European Union (EU). The global consumption of biopesticides is rising at a rate of 10% every year and is projected to increase further in the future (Kumar and Singh 2015).

The biopesticide development has prompted to replace the chemical pesticide for crop protection. The PGPR/PGPF seems effective in small amounts and much more specific to their target as compared to the conventional pesticides. A large number of biopesticides have already been registered and released in the market. Recently, novel substances have been formulated and reported for use as a biopesticides, like the products derived from plants (*Clitoria ternatea*), fungus (*Talaromyces flavus* SAY-Y-94-01, *Trichoderma harzianum*), bacteria (*Bacillus thuringiensis* var. *tenebrionis* strain Xd3, *Lactobacillus casei* LPT-111), oxymatrine (an alkaloid) (Damalas and Koutroubas 2018). It is anticipated that between the middle of 2040s and 2050s, biopesticide market will equalize with synthetic pesticides and major uncertainties will be due to its uptake in African and Southeast Asian countries (Olson 2015).

The biopesticide market development have improved the management practices and reduced the use of chemical pesticides. Various products have been certificated and commercialized for use in crop protection in different countries. However, in EU, there are very fewer biopesticides being registered as compared to Brazil, China, India, and the United States because of the complex and time-consuming registration processes. The main problem of the biopesticide industry is the lengthy submission process at the EU and other member state levels. The quicker implementation of registration procedures and time limits are essential if more new products have to be commercialized.

Furthermore, the high cost of registering a new BA or product is another limiting factor in its commercialization (Pavela 2014). Therefore, the regulatory authorities must try to ensure smooth and fast biopesticide registration processes and help to promote the safe technologies for product development. The small- and medium-sized firms should be developed to provide farmers with the reliable tools and products for pest management (Damalas and Koutroubas 2018).

11.9 Status of Biopesticide in India

In India, the organic pesticides market has generated total revenue of $102 million in 2016 and is projected to contribute $778 million by 2025. According to the market research report published by Inkwood research (2017) the market for biopesticides in India is anticipated to rise at a growth rate of 25.4% compounded annually during the 2017–2025 forecast period. The biopesticide industry in India represents only 4.2% of the entire pesticide market and is immensely driven by the sale of *Trichoderma viride*, *Pseudomonas fluorescens*, *Bacillus thuringinsis*, *Beauveria bassiana*, *Metarhizium anisopliae*, *Verticillium lecanii*, *Paecilomyces lilacinus*. The Indian biopesticides market, to a high degree, is dominated by numerous unorganized and organized companies like Pest Control India (PCI) and International Panaacea Ltd. (2015). There are around 150 companies involved in biopesticide manufacturing and 12 different types of bioinsecticides registered under the Insecticides Act, 1968 (Gautam et al. 2018).

11.10 Conclusion and Future Perspectives

There has been a considerable rise in the crop yields over the last century, which is mainly attributed to the utilization of chemical pesticides and agrochemicals. Globally, these agrochemicals have become a significant component of agriculture systems. Because of public concern about the damage caused by the intensive use of agrochemicals, an alternative path to their usage in agriculture production system has to be developed. Over the past decade, the use of BCAs has significantly increased in agriculture and is being recommended as an alternative.

Understanding the stimulation of plant responses by PGPR, PGPF, and other microbials is crucial for developing novel methodologies to regulate plant diseases and growth. The exploitation of these microbials relates to their use in PGP activity and mode of action against a variety of pathogens. Future research needs to focus on attaining integrated management of microbial communities in the rhizospheric soil. The advances in biotechnological and molecular approaches will provide more understanding of the cellular processes and signaling pathways linked to growth and DP resistance, resulting from plant-microbe interactions. Recently, genome editing, a modern genetic tool was used to study different aspects of plant-microbe interactions in two species, *Bacillus subtilis* HS3 and *B. mycoides* EC18 (Yi et al. 2018). Such studies will help to understand molecular mechanisms that support plant growth and to identify the superior PGPR/PGPF species in the future. The new alternatives should be discovered to be used as bioinoculants for different crops such as fruits, vegetables, pulses, and flowers. The application of compatible PFPR and PGPF consortium over single strain could be an effective method for reducing plant diseases. Also, compatible combinations of PGP microbes with the agrochemicals or organic amendments needed in the near future.

Many agricultural companies are working in crop protection especially in BCA products. The PGPR, PGPF, and other microbials are already being used in different countries under a different name and are expecting to grow at enormous speed. Eventually, for effective use of these microbes as BCAs, practical techniques for its mass culturing, formulation development, and storage need to be addressed and established. Additionally, an effort is needed to educate the farmers about the BCAs. We advocate the application of multifarious PGP microbial singly or in consortia for development of ecofriendly sustainable agriculture.

Acknowledgments Authors gratefully acknowledge financial support from the National Research Foundation of Korea, Republic of Korea (Grant #2017R1A4A1015515).

References

Abdallah DB, Frikha-Gargouri O, Tounsi S (2018) Rizhospheric competence, plant growth promotion and biocontrol efficacy of *Bacillus amyloliquefaciens* subsp. plantarum strain 32a. Biol Control. https://doi.org/10.1016/j.biocontrol.2018.01.013

Adam M, Heuer H, Hallmann J (2014) Bacterial antagonists of fungal pathogens also control root-knot nematodes by induced systemic resistance of tomato plants. PLoS One 9(2):e90402

Adhilakshmi M, Latha P, Paranidharan V et al (2014) Biological control of stem rot of groundnut (*Arachis hypogaea* L.) caused by *Sclerotium rolfsii* Sacc. with actinomycetes. Arch Phytopathol Plant Protect 47(3):298–311

Ahemad M, Kibret M (2014) Mechanisms and applications of plant growth promoting rhizobacteria: current perspective. J King Saud Univ-Sci 26(1):1–20

Akhtar MS, Panwar J (2013) Efficacy of root-associated fungi and PGPR on the growth of *Pisum sativum* (cv. Arkil) and reproduction of the root-knot nematode *Meloidogyne incognita*. J Basic Microbiol 53(4):318–326

Alabouvette C, Olivain C, Migheli Q et al (2009) Microbiological control of soil-borne phytopathogenic fungi with special emphasis on wilt-inducing *Fusarium oxysporum*. New Phytol 184(3):529–544

Amira MB, Lopez D, Mohamed AT et al (2017) Beneficial effect of *Trichoderma harzianum* strain Ths97 in biocontrolling *Fusarium solani* causal agent of root rot disease in olive trees. Biol Control 110:70–78

Andrews SC, Robinson AK, Rodríguez-Quiñones F (2003) Bacterial iron homeostasis. FEMS Microbiol Rev 27(2–3):215–237

Arseneault T, Pieterse CM, Gérin-Ouellet M et al (2014) Long-term induction of defense gene expression in potato by *Pseudomonas* sp. LBUM223 and *Streptomyces scabies*. Phytopathology 104(9):926–932

Audrain B, Farag MA, Ryu CM et al (2015) Role of bacterial volatile compounds in bacterial biology. FEMS Microbiol Rev 39(2):222–233

Awla HK, Kadir J, Othman R et al (2017) Plant growth-promoting abilities and biocontrol efficacy of *Streptomyces* sp. UPMRS4 against *Pyricularia oryzae*. Biol Control 112:55–63

Bamisile B, Dash CK, Akutse KS et al (2018) Fungal endophytes: beyond herbivore management. Front Microbiol 9:544

Bano A, Muqarab R (2017) Plant defence induced by PGPR against *Spodoptera litura* in tomato (*Solanum lycopersicum* L.). Plant Biol 19(3):406–412

Bensidhoum L, Nabti E, Tabli N et al (2016) Heavy metal tolerant *Pseudomonas protegens* isolates from agricultural well water in northeastern Algeria with plant growth promoting, insecticidal and antifungal activities. Eur J Soil Biol 75:38–46

Berg G (2009) Plant-microbe interactions promoting plant growth and health: perspectives for controlled use of microorganisms in agriculture. Appl Microbiol Biotechnol 84(1):11–18

Bhardwaj D, Ansari MW, Sahoo RK et al (2014) Biofertilizers function as key player in sustainable agriculture by improving soil fertility, plant tolerance and crop productivity. Microb Cell Factories 13(66):1–10

Bhattacharyya PN, Jha DK (2012) Plant growth-promoting rhizobacteria (PGPR): emergence in agriculture. World J Microbiol Biotechnol 28(4):1327–1350

Borah B, Ahmed R, Hussain M et al (2018) Suppression of root-knot disease in *Pogostemon cablin* caused by *Meloidogyne incognita* in a rhizobacteria mediated activation of phenylpropanoid pathway. Biol Control 119:43–50

Braun-kiewnick A, Lehmann A, Rezzonico F (2012) Development of species-, strain- and antibiotic biosynthesis-specific quantitative PCR assays for *Pantoea agglomerans* as tools for biocontrol monitoring. J Microbiol Methods 90(3):315–320

Brock AK, Berger B, Schreiner M et al (2018) Plant growth-promoting bacteria *Kosakonia radicincitans* mediate anti-herbivore defense in *Arabidopsis thaliana*. Planta:1–10

Burketova L, Trda L, Ott PG et al (2015) Bio-based resistance inducers for sustainable plant protection against pathogens. Biotechnol Adv 33(6):994–1004

Cai XC, Liu CH, Wang BT et al (2017) Genomic and metabolic traits endow *Bacillus velezensis* CC09 with a potential biocontrol agent in control of wheat powdery mildew disease. Microbiol Res 196:89–94

Chen D, Liu X, Li C et al (2014) Isolation of *Bacillus amyloliquefaciens* S20 and its application in control of eggplant bacterial wilt. J Environ Manag 137:120–127

Chen YY, Chen PC, Tsay TT (2016) The biocontrol efficacy and antibiotic activity of *Streptomyces plicatus* on the oomycete *Phytophthora capsici*. Biol Control 98:34–42

Chen Y, Zhou D, Qi D et al (2018) Growth promotion and disease suppression ability of a *Streptomyces* sp. CB-75 from banana rhizosphere soil. Front Microbiol 8:2704

Chennappa G, Sreenivasa MY, Nagaraja H (2018) *Azotobacter salinestris*: a novel pesticide-degrading and prominent biocontrol PGPR bacteria. In: Panpatte D, Jhala Y, Shelat H et al (eds) Microorganisms for green revolution. Microorganisms for sustainability, vol 7. Springer, Singapore, pp 23–43

Chinheya CC, Yobo KS, Laing MD (2017) Biological control of the rootknot nematode, *Meloidogyne javanica* (Chitwood) using *Bacillus* isolates, on soybean. Biol Control 109:37–41

Colagiero M, Rosso LC, Ciancio A (2018) Diversity and biocontrol potential of bacterial consortia associated to root-knot nematodes. Biol Control 120:11–16

Cordero P, Príncipe A, Jofré E et al (2014) Inhibition of the phytopathogenic fungus *Fusarium proliferatum* by volatile compounds produced by *Pseudomonas*. Arch Microbiol 196(11):803–809

Crowley DE (2006) Microbial siderophores in the plant rhizosphere. In: Barton LL, Abadía J (eds) Iron nutrition in plants and rhizospheric microorganisms. Springer, Dordrecht, pp 169–198

Damalas CA, Koutroubas SD (2018) Current status and recent developments in biopesticide use. Agriculture 8:1–6

Dara SK, Dara SS, Dara SS (2017) Impact of entomopathogenic fungi on the growth, development, and health of cabbage growing under water stress. Am J Plant Sci 8(06):1224

Das K, Prasanna R, Saxena AK (2017) Rhizobia: a potential biocontrol agent for soilborne fungal pathogens. Folia Microbiol 62(5):425–435

de Lima FB, Félix C, Osório N et al (2017) *Trichoderma harzianum* T1A constitutively secretes proteins involved in the biological control of *Guignardia citricarpa*. Biol Control 106:99–109

DeBach P, Hagen KS (1964) Manipulation of entomophagous species. In: DeBach P (ed) Biological control of insect pests and weeds. Chapman and Hall, London, pp 429–458

Denef K, Bubenheim H, Lenhart K et al (2007) Community shifts and carbon translocation within metabolically-active rhizosphere microorganisms in grasslands under elevated CO_2. Biogeosciences 4(5):769–779

Deshmukh S, Hückelhoven R, Schäfer P et al (2006) The root endophytic fungus *Piriformospora indica* requires host cell death for proliferation during mutualistic symbiosis with barley. PNAS 103(49):18450–18457

Desoignies N, Schramme F, Ongena M et al (2013) Systemic resistance induced by *Bacillus* lipopeptides in *Beta vulgaris* reduces infection by the rhizomania disease vector *Polymyxa betae*. Mol Plant Pathol 14(4):416–421
Dias MP, Bastos MS, Xavier VB et al (2017) Plant growth and resistance promoted by *Streptomyces* sp. in tomato. Plant Physiol Biochem 118:479–493
Dimkpa CO, Merten D, Svatos A et al (2009) Siderophores mediate reduced and increased uptake of cadmium by *Streptomyces tendae* F4 and sunflower (*Helianthus annuus*), respectively. J Appl Microbiol 107:1687–1696
Dinesh R, Anandaraj M, Kumar A et al (2015) Isolation, characterization, and evaluation of multi-trait plant growth promoting rhizobacteria for their growth promoting and disease suppressing effects on ginger. Microbiol Res 173:34–43
Donga TK, Vega FE, Klingen I (2018) Establishment of the fungal entomopathogen *Beauveria bassiana* as an endophyte in sugarcane, *Saccharum officinarum*. Fungal Ecol 35:70–77
El Arbi A, Rochex A, Chataigné G et al (2016) The Tunisian oasis ecosystem is a source of antagonistic *Bacillus* sp. producing diverse antifungal lipopeptides. Res Microbiol 167(1):46–57
Elena GJ, Beatriz PJ, Alejandro P et al (2011) *Metarhizium anisopliae* (Metschnikoff) Sorokin promotes growth and has endophytic activity in tomato plants. Adv Biol Res 5(1):22–27
Elsharkawy MM, Shimizu M, Takahashi H et al (2013) Induction of systemic resistance against cucumber mosaic virus in *Arabidopsis thaliana* by *Trichoderma asperellum* SKT-1. Plant Pathol J 29(2):193–200
El-Tarabily KA (2006) Rhizosphere-competent isolates of streptomycete and non-streptomycete actinomycetes capable of producing cell-wall degrading enzymes to control *Pythium aphanidermatum* damping-off disease of cucumber. Can J Bot 84:211–222
Environmental Protection Agency of the USA (2012) Regulating biopesticides. http://www.epa.gov/opp00001/biopesticides
Fan H, Ru J, Zhang Y (2017) Fengycin produced by *Bacillus subtilis* 9407 plays a major role in the biocontrol of apple ring rot disease. Microbiol Res 199:89–97
Figueiredo MDVB, Seldin L, de Araujo FF et al (2010) Plant growth promoting rhizobacteria: fundamentals and applications. In: Maheshwari DK (ed) Plant growth and health promoting bacteria. Springer, Berlin/Heidelberg, pp 21–43
Frankowski J, Lorito M, Scala F (2001) Purification and properties of two chitinolytic enzymes of *Serratia plymuthica* HRO-C48. Arch Microbiol 176:421–426
Gautam NK, Kumar A, Singh VK (2018) Bio-pesticide: a clean approach to healthy agriculture. Int J Curr Microbiol App Sci 7(3):194–197
Gava CAT, Pinto JM (2016) Biocontrol of melon wilt caused by *Fusarium oxysporum* Schlect f. sp. melonis using seed treatment with *Trichoderma* sp. and liquid compost. Biol Control 97:13–20
Geraldine AM, Lopes FAC, Carvalho DDC et al (2013) Cell wall-degrading enzymes and parasitism of sclerotia are key factors on field biocontrol of white mold by *Trichoderma* sp. Biol Control 67(3):308–316
Ghorbanpour M, Omidvari M, Abbaszadeh-Dahaji P et al (2017) Mechanisms underlying the protective effects of beneficial fungi against plant diseases. Biol Control. https://doi.org/10.1016/j.biocontrol.2017.11.006
Goswami D, Thakker JN, Dhandhukia PC (2016) Portraying mechanics of plant growth promoting rhizobacteria (PGPR): a review. Cogent Food Agric 2(1):1–19
Gotor-Vila, Teixidó N, Di Francesco A et al (2017) Antifungal effect of volatile organic compounds produced by *Bacillus amyloliquefaciens* CPA-8 against fruit pathogen decays of cherry. Food Microbiol 64:219–225
Gozzo F, Faoro F (2013) Systemic acquired resistance (50 years after discovery): moving from the lab to the field. J Agric Food Chem 61(51):12473–12491
Gray EJ, Smith DL (2005) Intracellular and extracellular PGPR: commonalities and distinctions in the plant-bacterium signaling processes. Soil Biol Biochem 37(3):395–412
Guo Q, Dong W, Li S et al (2014) Fengycin produced by *Bacillus subtilis* NCD-2 plays a major role in biocontrol of cotton seedling damping-off disease. Microbiol Res 169(7–8):533–540

Halfeld-vieira BDA, Luis W, Augusto D et al (2015) Understanding the mechanism of biological control of passion fruit bacterial blight promoted by autochthonous phylloplane bacteria. Biol Control 80:40–49

Hastuti RD, Lestari Y, Suwanto A et al (2012) Endophytic *Streptomyces* sp. as biocontrol agents of rice bacterial leaf blight pathogen (*Xanthomonas oryzae* pv. *oryzae*). HAYATI J Biosci 19(4):155–162

Hernández-león R, Rojas-solís D, Contreras-pérez M et al (2015) Characterization of the antifungal and plant growth-promoting effects of diffusible and volatile organic compounds produced by *Pseudomonas fluorescens* strains. Biol Control 81:83–92

Heydari A, Pessarakli M (2010) A review on biological control of fungal plant pathogens using microbial antagonists. J Biol Sci 10:273–290

Hiltner LT (1904) Uber nevere Erfahrungen und Probleme auf dem Gebiet der Boden Bakteriologie und unter besonderer Beurchsichtigung der Grundungung und Broche. Arbeit Deut Landw Ges Berlin 98:59–78

Hossain MM, Sultana F (2015) Genetic variation for induced and basal resistance against leaf pathogen *Pseudomonas syringae* pv. tomato DC3000 among *Arabidopsis thaliana* accessions. Springer Plus 4(1):296

Huang WK, Cui JK, Liu SM et al (2016) Testing various biocontrol agents against the root-knot nematode (*Meloidogyne incognita*) in cucumber plants identifies a combination of *Syncephalastrum racemosum* and *Paecilomyces lilacinus* as being most effective. Biol Control 92:31–37

India Biopesticide Market Forecast 2017–2025 (2017) Pages 1–48, Report ID: 4773059

India biopesticides market outlook to 2020- Trichoderma and Bacillus thuringiensis (bt) biopesticides to lead the future growth (2015) Products IdD- KR332

Ingram J (2011) A food systems approach to researching food security and its interactions with global environmental change. Food Secur 3:417–431

Jaber LR, Araj SE (2017) Interactions among endophytic fungal entomopathogens (Ascomycota: Hypocreales), the green peach aphid *Myzus persicae* Sulzer (Homoptera: Aphididae), and the aphid endoparasitoid *Aphidius colemani* Viereck (Hymenoptera: Braconidae). Biol Control 116:53–61

Jaber LR, Enkerli J (2016) Effect of seed treatment duration on growth and colonization of *Vicia faba* by endophytic *Beauveria bassiana* and *Metarhizium brunneum*. Biol Control 103:187–195

Jaber LR, Enkerli J (2017) Fungal entomopathogens as endophytes: can they promote plant growth? Biocontrol Sci Tech 27:28–41

Jain R, Pandey A (2016) A phenazine-1-carboxylic acid producing polyextremophilic *Pseudomonas chlororaphis* (MCC2693) strain, isolated from mountain ecosystem, possesses biocontrol and plant growth promotion abilities. Microbiol Res 190:63–71

Jia Y, Liu G, Park D, Yang Y (2013) Inoculation and scoring methods for rice sheath blight disease. In: Yang Y (ed) Rice protocols. Methods in molecular biology (methods and protocols), vol 956. Humana Press, Totowa, pp 257–268

Jin N, Hui X, Wen-jing LI et al (2017a) Field evaluation of *Streptomyces rubrogriseus* HDZ-9-47 for biocontrol of *Meloidogyne incognita* on tomato. J Integr Agric 16(4):1347–1357

Jin Q, Jiang Q, Zhao L et al (2017b) Complete genome sequence of *Bacillus velezensis* S3-1, a potential biological pesticide with plant pathogen inhibiting and plant promoting capabilities. J Biotechnol 259:199–203

John RP, Tyagi RD, Prévost D et al (2010) Mycoparasitic *Trichoderma viride* as a biocontrol agent against *Fusarium oxysporum* f. sp. adzuki and *Pythium arrhenomanes* and as a growth promoter of soybean. Crop Prot 29:1452–1459

Kamal R, Gusain YS, Kumar V (2014) Interaction and symbiosis of AM fungi, actinomycetes and plant growth promoting rhizobacteria with plants: strategies for the improvement of plants health and defense system. Int J Curr Microbiol Appl Sci 3:564–585

Karimi E, Sadeghi A, Abbaszade Dehaji P et al (2012) Biocontrol activity of salt tolerant *Streptomyces* isolates against phytopathogens causing root rot of sugar beet. Biocontrol Sci Tech 22:333–349

Karimi E, Safaie N, Shams-Baksh M et al (2016) *Bacillus amyloliquefaciens* SB14 from rhizosphere alleviates *Rhizoctonia* damping-off disease on sugar beet. Microbiol Res 192:221–230
Katiyar V, Goel R (2004) Siderophore-mediated plant growth promotion at low temperature by mutant of fluorescent pseudomonad. Plant Growth Regul 42:239–244
Keinan A, Clark AG (2012) Recent explosive human population growth has resulted in an excess of rare genetic variants. Science 336:740–743
Kepenekci I, Hazir S, Oksal E et al (2018) Application methods of *Steinernema feltiae*, *Xenorhabdus bovienii* and *Purpureocillium lilacinum* to control root-knot nematodes in greenhouse tomato systems. Crop Prot 108:31–38
Khan AL, Hamayun M, Khan SA et al (2012) Pure culture of *Metarhizium anisopliae* LHL07 reprograms soybean to higher growth and mitigates salt stress. World J Microbiol Biotechnol 28:1483–1494
Kheirandish Z, Harighi B (2015) Evaluation of bacterial antagonists of *Ralstonia solanacearum*, causal agent of bacterial wilt of potato. Biol Control 86:14–19
Kim HS, Sang MK, Jung HW et al (2012) Identification and characterization of *Chryseobacterium wanjuense* strain KJ9C8 as a biocontrol agent of *Phytophthora* blight of pepper. Crop Prot 32:129–137
Kiss L (2003) A review of fungal antagonists of powdery mildews and their potential as biocontrol agents. Pest Manag Sci 59:475–483
Kloepper JW, Schroth MN (1978) Plant growth-promoting rhizobacteria on radishes. In: Station de Pathologie, Proceedings of the 4th international conference on plant pathogenic bacteria, Tours, France, Végétale et Phyto-Bactériologie, Ed, pp 879–882
Krell V, Unger S, Jakobs-Schoenwandt D et al (2018) Endophytic *Metarhizium brunneum* mitigates nutrient deficits in potato and improves plant productivity and vitality. Fungal Ecol 34:43–49
Kumar S, Singh A (2015) Biopesticides: present status and the future prospects. J Fertil Pestic 6(2):2
Kumari P, Meena M, Upadhyay RS (2018) Characterization of plant growth promoting rhizobacteria (PGPR) isolated from the rhizosphere of *Vigna radiata* (mung bean). Biocatal Agric Biotechnol 16:155–162
Kurabachew H, Wydra K (2013) Characterization of plant growth promoting rhizobacteria and their potential as bioprotectant against tomato bacterial wilt caused by *Ralstonia solanacearum*. Biol Control 67(1):75–83
Kwak Y, Shin J (2015) Complete genome sequence of *Burkholderia pyrrocinia* 2327 T, the first industrial bacterium which produced antifungal antibiotic pyrrolnitrin. J Biotechnol 211:3–4
Law JWF, Ser HL, Khan TM et al (2017) The potential of *Streptomyces* as biocontrol agents against the rice blast fungus, *Magnaporthe oryzae* (*Pyricularia oryzae*). Front Microbiol 8:3
Li H, Li X, Wang Y (2011) Antifungal metabolites from *Chaetomium globosum*, an endophytic fungus in *Ginkgo biloba*. Biochem Syst Ecol 4(39):876–879
Li YT, Hwang SG, Huang YM (2018) Effects of *Trichoderma asperellum* on nutrient uptake and *Fusarium* wilt of tomato. Crop Prot 110:275–282
Lopez DC, Sword GA (2015) The endophytic fungal entomopathogens *Beauveria bassiana* and *Purpureocillium lilacinum* enhance the growth of cultivated cotton (*Gossypium hirsutum*) and negatively affect survival of the cotton bollworm (*Helicoverpa zea*). Biol Control 89:53–60
Lugtenberg BJJ (2015) Introduction to plant-microbe-interactions. In: Lugtenberg B (ed) Principles of plant-microbe interactions microbes for sustainable agriculture. Springer, Berlin, pp 1–2
Magnin-Robert M, Quantinet D, Couderchet M et al (2013) Differential induction of grapevine resistance and defense reactions against *Botrytis cinerea* by bacterial mixtures in vineyards. BioControl 58:117–131
Manasa M, Yashoda K, Pallavi S et al (2013) Biocontrol potential of *Streptomyces* species against *Fusarium oxysporum* f. sp. zingiberi (causal agent of rhizome rot of ginger). J Adv Sci Res 4:1–3
Martínez-Absalón S, Rojas-Solís D, Hernández-León R et al (2014) Potential use and mode of action of the new strain *Bacillus thuringiensis* UM96 for the biological control of the grey mould phytopathogen *Botrytis cinerea*. Biocontrol Sci Tech 24:1349–1362

Martins SA, Schurt DA, Seabra SS et al (2018) Common bean (*Phaseolus vulgaris* L.) growth promotion and biocontrol by rhizobacteria under *Rhizoctonia solani* suppressive and conducive soils. Appl Soil Ecol 127:129–135

Molinari S, Baser N (2010) Induction of resistance to root-knot nematodes by SAR elicitors in tomato. Crop Prot 29:1354–1362

Morohoshi T, Wang W, Suto T et al (2013) Phenazine antibiotic production and antifungal activity are regulated by multiple quorum-sensing systems in *Pseudomonas chlororaphis* subsp. *aurantiaca* StFRB508. J Biosci Bioeng 116:580–584

Murali M, Sudisha J, Amruthesh KN et al (2013) Rhizosphere fungus *Penicillium chrysogenum* promotes growth and induces defence-related genes and downy mildew disease resistance in pearl millet. Plant Biol 15:111–118

Naeem M, Aslam Z, Khaliq A et al (2018) Plant growth promoting rhizobacteria reduce aphid population and enhance the productivity of bread wheat. Braz J Microbiol 49:9–14

Nassimi Z, Taheri P (2017) Endophytic fungus *Piriformospora indica* induced systemic resistance against rice sheath blight via affecting hydrogen peroxide and antioxidants. Biocontrol Sci Tech 27:252–267

Nikolić I, Berić T, Dimkic I et al (2018) Biological control of *Pseudomonas syringae* pv. aptata on sugar beet with *Bacillus pumilus* SS10.7 and *Bacillus amyloliquefaciens* (SS12.6 and SS38.4) strains. J Appl Microbiol 126:165–176

Nimnoi P, Pongsilp N, Ruanpanun P (2017) Monitoring the efficiency of *Streptomyces galilaeus* strain KPS-C004 against root knot disease and the promotion of plant growth in the plant-parasitic nematode infested soils. Biol Control 114:158–166

Niu DD, Liu HX, Jiang CH et al (2011) The plant growth-promoting rhizobacterium *Bacillus cereus* AR156 induces systemic resistance in *Arabidopsis thaliana* by simultaneously activating salicylate- and jasmonate/ethylene-dependent signaling pathways. Mol Plant-Microbe Interact 24:533–542

Olson S (2015) An analysis of the biopesticide market now and where is going. Outlooks Pest Manag 26:203–206

Padaria JC, Tarafdar A, Raipuria R et al (2016) Identification of phenazine-1-carboxylic acid gene (phc CD) from *Bacillus pumilus* MTCC7615 and its role in antagonism against *Rhizoctonia solani*. J Basic Microbiol 56:999–1008

Palazzini JM, Dunlap CA, Bowman MJ et al (2016) *Bacillus velezensis* RC 218 as a biocontrol agent to reduce *Fusarium* head blight and deoxynivalenol accumulation: genome sequencing and secondary metabolite cluster profiles. Microbiol Res 192:30–36

Palumbo JD, Yuen GY, Jochum CC (2005) Mutagenesis of Beta-1,3-glucanase genes in *Lysobacter enzymogenes* strain C3 results in reduced biological control activity toward bipolaris leaf spot of tall fescue and *Pythium* damping-off of sugar beet. Phytopathology 95:701–707

Pascale A, Vinale F, Manganiello G et al (2017) *Trichoderma* and its secondary metabolites improve yield and quality of grapes. Crop Prot 92:176–181

Patil HJ, Srivastava AK, Singh DP et al (2011) Actinomycetes mediated biochemical responses in tomato (*Solanum lycopersicum*) enhances bioprotection against *Rhizoctonia solani*. Crop Prot 30:1269–1273

Patil S, Nikama M, Anokhinab T et al (2017) Multi-stress tolerant plant growth promoting *Pseudomonas* sp. MCC 3145 producing cytostatic and fungicidal pigment. Biocatal Agric Biotechnol 10:53–63

Pavela R (2014) Limitation of plant biopesticides. In: Singh D (ed) Advances in plant biopesticides. Springer, New Delhi, pp 347–359

Pelaez V, Mizukawa G (2017) Diversification strategies in the pesticide industry: from seeds to biopesticides. Ciênc Rural 47:e20160007

Pieterse CMJ, Van Wees SCM (2015) Induced disease resistance. In: Lugtenberg B (ed) Principles of plant-microbe interactions: microbes for sustainable agriculture. Springer, Cham, pp 123–134

Pradhanang PM, Ji P, Momol MT et al (2005) Application of acibenzolar-S-methyl enhances host resistance in tomato against *Ralstonia solanacearum*. Plant Dis 89(9):989–993

Rahardjo BT, Tarno H (2018) Diamondback Moth, *Plutella xylostella* (Linnaeus) Responses on Chinese Kale (*Brassica oleracea* var. *alboglabra*) treated by plant growth promoting rhizobacteria. Asian J Crop Sci 10(2):73–79

Rahman M, Ali AM, Dey TK et al (2014) Evolution of disease and potential biocontrol activity of *Trichoderma* sp. against *Rhiozctonia solani* on potato. Biosci J 30:1108–1117

Rahman SFSA, Singh E, Pieterse CMJ, Schenk PM (2018) Emerging microbial biocontrol strategies for plant pathogens. Plant Sci. https://doi.org/10.1016/j.plantsci.2017.11.012

Ramadan EM, AbdelHafez AA, Hassan EA et al (2016) Plant growth promoting rhizobacteria and their potential for biocontrol of phytopathogens. Afr J Microbiol Res 10:486–504

Ramesh R, Phadke GS (2012) Rhizosphere and endophytic bacteria for the suppression of eggplant wilt caused by *Ralstonia solanacearum*. Crop Prot 37:35–41

Ramette A, Frapolli M, Saux MF (2011) *Pseudomonas protegens* sp *nov*. widespread plant-protecting bacteria producing the biocontrol compounds 2,4-diacetylphloroglucinol and pyoluteorin. Syst Appl Microbiol 34:180–188

Rath M, Mitchell TR, Gold SE (2018) Volatiles produced by *Bacillus mojavensis* RRC101 act as plant growth modulators and are strongly culture-dependent. Microbiol Res 208:76–84

Regnault-Roger C, Philogène BJR, Vincent C (2005) Biopesticides of plant origin. Intercept, Paris, p 313

Rojas-Solís D, Zetter-Salmon E, Contreras-Perez M et al (2018) *Pseudomonas stutzeri* E25 and *Stenotrophomonas maltophilia* CR71 endophytes produce antifungal volatile organic compounds and exhibit additive plant growth-promoting effects. Biocatal Agric Biotechnol 13:46–52

Sabaté DC, Brandan CP, Petroselli G et al (2017) Decrease in the incidence of charcoal root rot in common bean (*Phaseolus vulgaris* L.) by *Bacillus amyloliquefaciens* B14, a strain with PGPR properties. Biol Control 113:1–8

Saengnak V, Chaisiri C, Nalumpang S (2013) Antagonistic *Streptomyces* species can protect chili plants against wilt disease caused by *Fusarium*. J Agric Technol 9:1895–1908

Saha M, Sarkar S, Sarkar B et al (2016) Microbial siderophores and their potential applications: a review. Environ Sci Pollut Res 23:3984–3999

Sajitha KL, Dev SA (2016) Quantification of antifungal lipopeptide gene expression levels in *Bacillus subtilis* B1 during antagonism against sapstain fungus on rubberwood. Biol Control 96:78–85

Saksirirat W, Chareerak P, Bunyatrachata W (2009) Induced systemic resistance of bio control fungus, *Trichoderma* sp. against bacterial and gray leaf spot in tomatoes. Asian J Food Agro-Ind (Special issue) 2:99–104

Sakthivel N (2010) Effectiveness of three introduced encyrtid parasitic wasps (*Acerophagus papayae*, *Anagyrus loecki* and *Pseudleptomastix mexicana*) against papaya mealybug, *Paracoccus marginatus*, infesting mulberry in Tamil Nadu. J Biopest 6:71–76

Salamone AL, Gundersen B, Inglis DA (2018) *Clonostachys rosea*, a potential biological control agent for *Rhizoctonia solani* AG-3 causing black scurf on potato. Biocontrol Sci Tech 28:895–900

Salomon MV, Bottini R, de Souza Filho GA et al (2014) Bacteria isolated from roots and rhizosphere of *Vitis vinifera* retard water losses, induce abscisic acid accumulation and synthesis of defense-related terpenes in in vitro cultured grapevine. Physiol Plant 151:359–374

Salomon MV, Pinter IF, Piccoli P et al (2017) Use of plant growth-promoting rhizobacteria as biocontrol agents: induced systemic resistance against biotic stress in plants. In: Kalia V (ed) Microbial applications, vol 2. Springer, New Delhi, pp 133–152

Sánchez-Rodríguez AR, Raya-Díaz S, Zamarreño ÁM et al (2018) An endophytic *Beauveria bassiana* strain increases spike production in bread and durum wheat plants and effectively controls cotton leafworm (*Spodoptera littoralis*) larvae. Biol Control 116:90–102

Santoro MV, Bogino PC, Nocelli N et al (2016) Analysis of plant growth promoting effects of fluorescent *Pseudomonas* strains isolated from *Mentha piperita* rhizosphere and effects of their volatile organic compounds on essential oil composition. Front Microbiol 7(1085):1–17

Saravanakumar K, Yu C, Dou K et al (2016) Synergistic effect of *Trichoderma*-derived antifungal metabolites and cell wall degrading enzymes on enhanced biocontrol of *Fusarium oxysporum* f. sp. *cucumerinum*. Biol Control 94:37–46

Sasan RK, Bidochka MJ (2012) The insect-pathogenic fungus *Metarhizium robertsii* (Clavicipitaceae) is also an endophyte that stimulates plant root development. Am J Bot 99:1483–1494

Sharma P, Verma PP, Kaur M (2017a) Identification of secondary metabolites produced by fluorescent pseudomonads applied for controlling fungal pathogens of apple. Indian Phytopathol 70:452–456

Sharma P, Verma PP, Kaur M (2017b) Phytohormones production and phosphate solubilization capacities of fluorescent *Pseudomonas* sp. isolated from Shimla Dist. of Himachal Pradesh. IJCMAS 6:2447–2454

Sharma V, Salwan R, Sharma PN et al (2017c) Elucidation of biocontrol mechanisms of *Trichoderma harzianum* against different plant fungal pathogens: universal yet host specific response. Int J Biol Macromol 95:72–79

Shelake RM, Waghunde RR, Morita EH et al (2018) Plant-microbe-metal interactions: basics, recent advances, and future trends. In: Egamberdieva D, Ahmad P (eds) Plant microbiome: stress response. Microorganisms for sustainability, vol 5. Springer, Singapore, pp 1–5

Shobha G, Kumudini BS (2012) Antagonistic effect of the newly isolated PGPR *Bacillus* sp. on *Fusarium oxysporum*. Int J Appl Sci Eng Res 1:463–474

Sillero JC, Rojas-Molina MM, Ávila CM et al (2012) Induction of systemic acquired resistance against rust, ascochyta blight and broomrape in faba bean by exogenous application of salicylic acid and benzothiadiazole. Crop Prot 34:65–69

Singh SP, Gaur R (2017) Endophytic *Streptomyces* sp. underscore induction of defense regulatory genes and confers resistance against *Sclerotium rolfsii* in chickpea. Biol Control 104:44–56

Sivasakhti S, Usharani G, Saranraj P (2014) Biocontrol potentiality of plant growth promoting bacteria (PGPR)- *Pseudomonas fluorescence* and *Bacillus subtilis*: a review. Afr J Agric Res 9:1265–1277

Smith HS (1919) On some phases of insect control by the biological method1. J Econ Entomol 12(4):288–292

Sudisha J, Sharathchandra RG, Amruthesh KN et al (2012) Pathogenesis related protiens in plant defence response. In: Plant defence: biological control. Springer, Dordrecht, pp 379–403

Sun G, Yao T, Feng C et al (2017) Identification and biocontrol potential of antagonistic bacteria strains against *Sclerotinia sclerotiorum* and their growth-promoting effects on *Brassica napus*. Biol Control 104:35–43

Tabli N, Rai A, Bensidhoum L (2018) Plant growth promoting and inducible antifungal activities of irrigation well water-bacteria. Biol Control 117:78–86

Tamreihao K, Ningthoujam DS, Nimaichand S et al (2016) Biocontrol and plant growth promoting activities of a *Streptomyces corchorusii* strain UCR3-16 and preparation of powder formulation for application as biofertilizer agents for rice plant. Microbiol Res 192:260–270

Tan S, Jiang Y, Song S et al (2013) Two *Bacillus amyloliquefaciens* strains isolated using the competitive tomato root enrichment method and their effects on suppressing *Ralstonia solanacearum* and promoting tomato plant growth. Crop Prot 43:134–140

Thampi A, Bhai RS (2017) Rhizosphere actinobacteria for combating *Phytophthora capsici* and *Sclerotium rolfsii*, the major soil borne pathogens of black pepper (*Piper nigrum* L.). Biol Control 109:1–13

Timmermann T, Armijo G, Donoso R et al (2017) *Paraburkholderia phytofirmans* PsJN protects *Arabidopsis thaliana* against a virulent strain of *Pseudomonas syringae* through the activation of induced resistance. Mol Plant-Microbe Interact 30:215–230

Tjamos EC, Tjamos SE, Antoniou PP (2010) Biological management of plant diseases: highlights on research and application. J Plant Pathol 92:17–21

Toghueo RMK, Eke P, Zabalgogeazcoa Í et al (2016) Biocontrol and growth enhancement potential of two endophytic *Trichoderma* sp. from *Terminalia catappa* against the causative agent of common bean root rot (*Fusarium solani*). Biol Control 96:8–20

Tohid VK, Taheri P (2015) Investigating binucleate *Rhizoctonia* induced defence responses in kidney bean against *Rhizoctonia solani*. Biocontrol Sci Tech 25(4):444–459

Torres MJ, Brandan CP, Sabaté DC et al (2017) Biological activity of the lipopeptide-producing *Bacillus amyloliquefaciens* PGPBacCA1 on common bean *Phaseolus vulgaris* L. pathogens. Biol Control 105:93–99

Toyota M, Spencer D, Sawai-Toyota S et al (2018) Glutamate triggers long-distance, calcium-based plant defense signaling. Science 361:1112–1115

Turatto MF, Dourado FDS, Zilli JE et al (2018) Control potential of *Meloidogyne javanica* and *Ditylenchus* sp. using fluorescent *Pseudomonas* and *Bacillus* sp. Braz J Microbiol 49:54–58

Udayashankar AC, Nayaka SC, Reddy MS et al (2011) Plant growth-promoting rhizobacteria mediate induced systemic resistance in rice against bacterial leaf blight caused by *Xanthomonas oryzae* pv. *oryzae*. Biol Control 59:114–122

Ulloa-Ogaz AL, Munoz-Castellanos LN, Nevarez-Moorillon GV (2015) Biocontrol of phyto-pathogens: antibiotic production as mechanism of control. In: Mendez Vilas A (ed) The battle against microbial pathogens, basic science, technological advance and educational programs. Formatex Research Center, Spain, pp 305–309

Upadhyay A, Srivastava S (2011) Phenazine-1-carboxylic acid is a more important contributor to biocontrol *Fusarium oxysporum* than pyrrolnitrin in *Pseudomonas fluorescens* strain Psd. Microbiol Res 166:323–335

Veresoglou SD, Rillig MC (2012) Suppression of fungal and nematode plant pathogens through arbuscular mycorrhizal fungi. Biol Lett 8:214–217

Verma PP, Thakur S, Kaur M (2016) Antagonism of *Pseudomonas putida* against *Dematophora nectarix* a major apple plant pathogen and its potential use as a biostimulent. J Pure Appl Microbiol 10:2717–2726

Vos C, Geerinckx K, Mkandawire R et al (2012) Arbuscular mycorrhizal fungi affect both penetration and further life stage development of root-knot nematodes in tomato. Mycorrhiza 22:157–163

Vos C, Schouteden N, Van Tuinen D et al (2013) Mycorrhiza-induced resistance against the root-knot nematode *Meloidogyne incognita* involves priming of defense gene responses in tomato. Soil Biol Biochem 60:45–54

Vu TT, Hauschild R, Sikora RA (2006) *Fusarium oxysporum* endophytes induced systemic resistance against *Radopholus similis* on banana. Nematology 8:847–852

Waghunde RR, Shelake RM, Sabalpara AN (2016) *Trichoderma*: a significant fungus for agriculture and environment. Afr J Agric Res 11:1952–1965

Waghunde RR, Shelake RM, Shinde MS et al (2017) Endophyte microbes: a weapon for plant health management. In: Microorganisms for green revolution. Springer, Singapore, pp 303–3025

Wang X, Mavrodi DV, Ke L et al (2015) Biocontrol and plant growth-promoting activity of rhizobacteria from Chinese fields with contaminated soils. Microb Biotechnol 8:404–418

Wani KA, Manzoor J, Shuab R (2017) Arbuscular mycorrhizal fungi as biocontrol agents for parasitic nematodes in plants. In: Varma A, Prasad R, Tuteja N (eds) Mycorrhiza – nutrient uptake, biocontrol, ecorestoration. Springer, Cham, pp 195–210

Weller DM, Mavrodi DV, van Pelt JA et al (2012) Induced systemic resistance in *Arabidopsis thaliana* against *Pseudomonas syringae* pv. tomato by 2,4-diacetylphloroglucinol-producing *Pseudomonas fluorescens*. Biol Control 102:403–412

Westhoek H, Ingram J, van Berkum S et al (2016) Food systems and natural resources, United Nations environment rogramme: United Nations Environment Programme

Weyens N, van der Lelie D, Taghavi S et al (2009) Phytoremediation: plant-endophyte partnerships take the challenge. Curr Opin Biotechnol 20(2):248–254

Winotai, A et al. (2012) Introduction of *Anagyrus lopezi* for biological control of the pink cassava mealybug, *Phenacoccus manihoti*, in Thailand. A paper presented in the XXIV international congress of entomology, Daegu, Korea, August 19–25

Wu B, Wang X, Yang L et al (2016) Effects of *Bacillus amyloliquefaciens* ZM9 on bacterial wilt and rhizosphere microbial communities of tobacco. Appl Soil Ecol 103:1–12

Xiang N, Lawrence KS, Kloepper JW et al (2017) Biological control of *Heterodera glycines* by spore-forming plant growth-promoting rhizobacteria (PGPR) on soybean. PLoS One 12(7):e0181201

Yang W, Xu Q, Liu HX et al (2012) Evaluation of biological control agents against *Ralstonia* wilt on ginger. Biol Control 62:144–151

Yi Y, Li Z, Song C, Kuipers OP (2018) Exploring plant-microbe interactions of the rhizobacteria *Bacillus subtilis* and *Bacillus mycoides* by use of the CRISPR-Cas9 system. Environ Microbiol. https://doi.org/10.1111/1462-2920.14305

Yoo SJ, Shin DJ, Won HY et al (2018) *Aspergillus terreus* JF27 promotes the growth of tomato plants and induces resistance against *Pseudomonas syringae* pv. tomato. Mycobiology:1–7

Yoshioka Y, Ichikawa H, Naznin HA et al (2012) Systemic resistance induced in *Arabidopsis thaliana* by *Trichoderma asperellum* SKT-1, a microbial pesticide of seed borne diseases of rice. Pest Manag Sci 68:60–66

Yu X, Ai C, Xin L et al (2011) The siderophore-producing bacterium, *Bacillus subtilis* CAS15, has a biocontrol effect on *Fusarium* wilt and promotes the growth of pepper. Eur J Soil Biol 47:138–145

Yuan Y, Feng H, Wang L et al (2017) Potential of endophytic fungi isolated from cotton roots for biological control against *verticillium* wilt disease. PLoS One 12(1):e0170557

Zhang JX, Gu YB, Chi FM et al (2015) *Bacillus amyloliquefaciens* GB1 can effectively control apple valsa canker. Biol Control 88:1–7

Zhang F, Ge H, Zhang F (2016) Biocontrol potential of *Trichoderma harzianum* isolate T-aloe against *Sclerotinia sclerotiorum* in soybean. Plant Physiol Biochem 100:64–74

Zhang F, Chen C, Zhang F et al (2017) *Trichoderma harzianum* containing 1-aminocyclopropane-1-carboxylate deaminase and chitinase improved growth and diminished adverse effect caused by *Fusarium oxysporum* in soybean. J Plant Physiol 210:84–94

Zhao D, Zhao H, Zhao D et al (2018) Isolation and identification of bacteria from rhizosphere soil and their effect on plant growth promotion and root-knot nematode disease. Biol Control 119:12–19

Zhou T, Chen D, Li C et al (2012) Isolation and characterization of *Pseudomonas brassicacearum* J12 as an antagonist against *Ralstonia solanacearum* and identification of its antimicrobial components. Microbiol Res 167:388–394

Zhou JY, Zhao XY, Dai CC (2014) Antagonistic mechanisms of endophytic *Pseudomonas fluorescens* against *Athelia rolfsii*. J Appl Microbiol 117:1144–1158

Zhou L, Yuen G, Wang Y et al (2016) Evaluation of bacterial biological control agents for control of root-knot nematode disease on tomato. Crop Prot 84:8–13

Biofortification: A Promising Approach Toward Eradication of Hidden Hunger

12

Amita Sharma and Rajnish Kumar Verma

12.1 Introduction

Globally, hunger affects about 11% of the population. The count of undernourished people raised from 777 to 815 million (2015–2016). More than two million people are suffering from malnutrition (FAO, IFAD, UNIICEF, WHO report 2017). Eradication of every type of hunger including chronic hunger (calorie deficiencies) as well as hidden hunger (micronutrient deficiencies) could be accomplished by attainment of a ground-level insight into the consequences and determinants of the problem. In the few preceding decades, the research community has shifted their focus progressing toward development of a sustainable agriculture with elevated quantities of cereal production. Moreover the concern is not merely about production of calorific food but also about production of nutrient-rich food. The latest innovations in food research lead to fortification of food with indispensable minerals, vitamins, fatty acids, fibers, phytonutrients, etc. Biofortification is a practice involving breeding of nutrients into crops, thus providing an approach to deliver nutrients with sustainability, long-term effect, and cost efficiency (Sharma et al. 2017). One should not expect that biofortification is capable of completely eliminating malnutrition, but it is definitely a groundbreaking methodology to meet the daily requirements of nutrient uptake among people (Saltzman et al. 2013; Blancquaert et al. 2017; Cakmak and Kutman 2017).

Both the authors are contributed equally to this chapter.

A. Sharma
Department of Agriculture, Shaheed Udham Singh College of Research and Technology, Tangori, Punjab, India

R. K. Verma (✉)
Department of Botany, Dolphin PG College of Science and Agriculture, Chunni Kalan, Punjab, India

D. P. Singh et al. (eds.), *Microbial Interventions in Agriculture and Environment*, https://doi.org/10.1007/978-981-13-8391-5_12

Biofortification is an advanced approach to complement existing approaches providing nutrients sustainable to people in need at a low cost. It's a feasible methodology to nourish nutrient-deprived people with limited access to supplements, varied foods, and expensive commercially modified food items. Considerable advancement has been witnessed regarding methodology, analysis, and effectiveness of biofortification during recent times (Bouis and Saltzman 2017). But still the process is in the early stages, if scaling and influence of biofortification are taken into consideration. The effect of biofortification on human's health and related disease burden can be estimated with the help of ex ante simulation models (Lividini et al. 2018). In general, these models involve the study of several factors typically estimating the decrease in the incidences of insufficient micronutrient consumption and the number of disability-adjusted life years (DALYs) saved. DALYs represent a method to measure the load of health-linked complications expressed as total healthful life years lost. DALYs actually provide an interpretation of the period and extent of health disorders thereby estimating the burden by counting the total life years (healthy) lost in concerned population because of disabilities and untimely demise (Murray et al. 2012). They can be counted by taking different risk factors into consideration like iron deficiency, protein malnutrition, vitamin A deficiency, childhood underweight, etc. (IHME report 2018). Recently, DALYs metric is employed to perform comparable studies of estimating load of hidden hunger, chronic hunger, and health-related problems (Gödecke et al. 2018). Though economic progress in current times has resulted in a decline of chronic hunger cases, malnutrition is still a challenging situation globally.

12.2 Technologies for Biofortification

To lessen the instances of hidden hunger, interventions include direct as well as indirect methods (Ruel and Alderman 2013). The direct methods (nutrition specific) include supplementation of micronutrients, food modification, dietary diversification, etc. The indirect interventions (nutrient sensitive) aim at fundamental reasons of undernourishment and involve biofortification. Different studies have revealed the point that fortification of foods is a safer technique and have potential impact to confront the challenge of malnutrition among human population. Biofortification could be realized through different technologies like agronomic method, crop breeding, and genetic engineering (Cakmak and Kutman 2017; Blancquaert et al. 2017) (Fig. 12.1).

12.2.1 Agronomic Biofortification

Agronomic biofortification is a practice wherein fertilizer enriched with micronutrients is added to soil and leaves (also known as foliar application). Generally, biofortification of the fundamental foods (sorghum, millet, sweet potato, legumes, wheat, rice, etc.) is the main focus of researchers worldwide, because these crops are

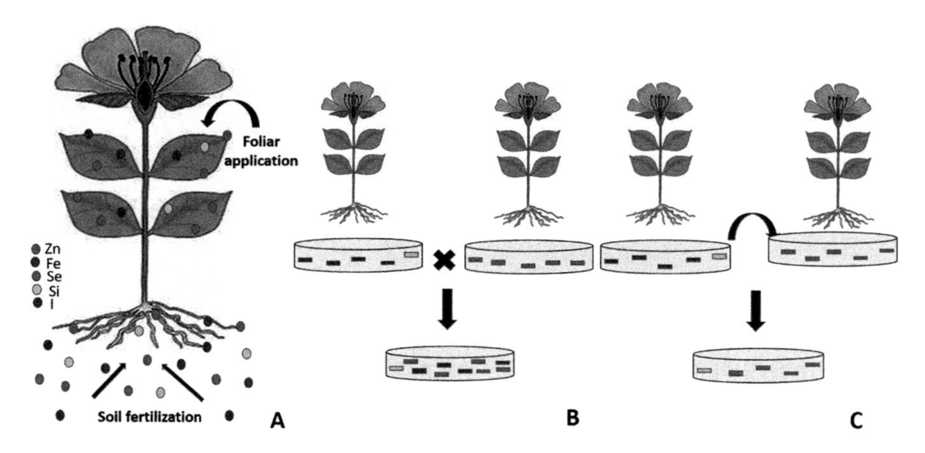

Fig. 12.1 Types of biofortification. (**a**) Agronomic. (**b**) Genetic biofortification (crop breeding). (**c**) Genetic biofortification (gene engineering)

dominant elements of diets especially among the populations at risk or vulnerable populations. Biofortification represents a practical method to target malnourished people with restricted approach to nutrient-rich diverse supplies or food supplements, etc. The best results of agronomic biofortification have been obtained with selenium and zinc (Cakmak 2014). The application of fertilizers enriched with Fe to soil is a challenging methodology as compared to application of Se and Zn, because iron shows precipitation in insoluble forms. The insoluble form is not absorbed by plants, and therefore a successful scheme for Fe enrichment is via foliar treatment or litter fertilization. The victory of agronomic biofortification is determined by many critical aspects which depend upon the availability of nutrients at different stages: accessibility in soil for plant uptake, allocation of nutrients in soil, re-translocation of nutrients into edible plant products, availability of nutrition in foodstuff prepared for man, and physiological stage of individual (Valença et al. 2017).

The intensity of crop enrichment with micronutrients by agronomic biofortification is dependent on the efficacy of fertilizer treatment and type of fertilizer. Fertilizer preparation greatly regulates nutrients' form and their accessibility to crops. Foliar application of micronutrient fertilizers facilitates more uptake of nutrients along with their allocation in crop parts as compared to the fertilizer treatment of soil, especially in case of leafy vegetables and cereals (Lawson et al. 2015). Efficiency of the nutrient uptake is maximum when a combination of foliar and soil micronutrient application is carried out. Coating of seeds with fertilizers is also another successful scheme for micronutrient application. Good soil condition is an additional significant factor to increase the availability of minerals in soil for uptake by the plants (Duffner et al. 2014; Valença et al. 2017).

Application of fertilizers enriched with micronutrients is a considerable approach with least undesirable environmental consequences. Majority of micronutrients don't show susceptibility to leaching as they show strong binding in the soil.

A drawback of this methodology is that these micronutrients got accumulated with time and may be toxic if higher concentrations accumulate repeatedly. Agronomic biofortification is a considerable strategy but represents a temporary way out to enhance micronutrient availability and to match genetic method of biofortification, which is recognized as a better ecological methodology. Agronomic biofortification ensures enhancement in plant yields with improvement in nutritive value when particular micronutrient-crop combinations were utilized (Valença et al. 2017).

12.2.2 Genetic Biofortification

It comprises both the traditional method of breeding and modern methods of engineering at gene level to improvise the nutritional status of the staple food. This strategy represents a single-time venture to raise plants with improved content of indispensable nutrients that can reach the poor and at-risk populations. The benefits of genetic fortification include low costs, one-time investment approach, and distribution of their germplasm at international levels (Melash et al. 2016). The gene-level alteration of crops is considered a justifiable answer to the question of micronutrient insufficiency, but development of novel nutrient-enriched plants is a long process. The success of the methodology is subjected to many factors and their commercialization is greatly affected by degree of public acceptance.

12.3 Advances in Food Biofortification

Biofortification of edible crops with essential nutrients represents an appreciable methodology with immense potential to solve the puzzle of global undernourishment (Fig. 12.2). A myriad of investigations are available in literature utilizing

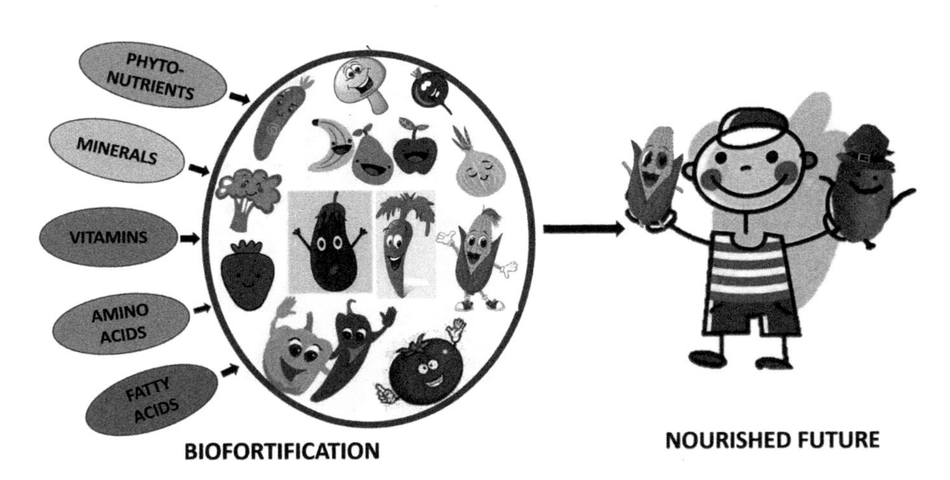

Fig. 12.2 Diagrammatic representation of biofortification approach

different technologies to enrich our cereal crops and other foods with vital nutrients. Till date, numerous cereal crops, vegetables, and fruits are fortified with essential fatty acids, amino acids, minerals, vitamins, phytonutrients, etc., and many investigations are aimed at aspects like bioavailability of nutrients to consumers, public acceptance, etc.

12.3.1 Biofortification with Micronutrients

Modernized agricultural approaches brought about advancement in diversity of crops with high yields. But the present-day scenario demands nutrient-rich crops along with the high-yielding crops. Majority of populations rely on few crops for survival, mainly maize, rice, and wheat. Exploitation of the crops' wild relatives will provide a rich gene supply to modify the crops at genetic level enhancing nutrient status. Moreover, the selective treatment of soil with varied fertilizers also changes the concentration of some micronutrients and their bioavailability to vegetation. Once the plant is ingested, the assimilation of the micronutrients depends on the dietary phytate. The phytate usually shows interactions with nutrients and results in production of insoluble complexes that could be digested or absorbed. So these anti-nutritional compounds should either be absent or be there in least concentrations in the diet. Besides these factors, some cultural practices employed for food process also results in the loss of crucial micronutrients, i.e., vitamins and minerals at different stages. The cultural practices include milling, dehulling, fermentation, cooking, etc. (Melash et al. 2016). Therefore to enrich the staple foods with essential nutrients that can reach our poor and vulnerable populations, biofortification have come up as a substantial approach.

Micronutrients are the compounds that must be supplemented to the human body in minute quantities. It includes both vitamins and minerals. Thus improving the micronutrient content in edible crops is a substantial approach to nourish the poor people with malnourishment. Deficiency of vitamin D leads to various bone diseases and nonskeletal metabolic disorders in various phases of life. Therefore approaches employed for preventing these ailments are of chief importance. Biofortification of edible feedstock with vitamin D have wider impact on the population as compared to supplements (Cashman 2015). Recently, it was found that irradiation of UV leads to biosynthesis of considerable quantities of vitamin D_2. Some cultivated species of mushroom like *Lentinula edodes*, *Pleurotus ostreatus*, and *Agaricus bisporus* have been proven beneficial to meet the demands of vitamin D2 (Taofiq et al. 2017).

Vitamin A deficiency represents a major micronutrient deficiency globally, influencing poor populations excessively in the developing countries. Biofortification by breeding methods and modern engineering methods has proven its capability to improve bioaccessibility of nutrients in crops (De Moura et al. 2015; Beswa et al. 2016a, b; Kamotho et al. 2017; Amah et al. 2018). The Biofortification of maize, sweet potato, and cassava greatly enhanced the retention levels of provitamin A carotenoid (pVAC) in these plants afterwards cooking and storage also (De Moura et al. 2015). Thus, these biofortified crops provided a higher vitamin A content to the consumers. Beswa et al. (2016a, b) reported that nutrient (provitamin A, amino acids, and iron) content of the biofortified maize has been improved with addition of vegetable amaranth in the powdered form. This vegetable powder not only enhanced the provitamin level in fortified maize snacks but also improved the level of phenolic compounds and thus the antioxidant activity. Chaudhary et al. (2016) reported raising biofortified sweet potato varieties with orange or yellow flesh. The color of the flesh is due to the higher content of vitamin A precursor, β-carotene. These sweet potato varieties also provide required dietary fiber and potassium. These varieties have been grown and consumed in Uttar Pradesh, India, reaching the target population to take care of vitamin A deficiency. In a recent study, maize flour biofortification with amaranth enhanced the nutrient value appreciably ($p \leq 0.05$). The levels of protein, calcium, zinc, and iron have been increased in particular. This method of "food to food" biofortification signifies a considerable technology toward fulfilling the requirements of mankind (Kamotho et al. 2017). Globally, banana (*Musa* spp.) is cultivated as an economically principal fruit crop including regions with prevalent VAD conditions. Therefore, biofortification of bananas signifies a noteworthy methodology to alleviate shortage of vitamin A in the vulnerable population (Amah et al. 2018).

Vitamin B_6 includes a group of interrelated complexes which can only be produced de novo by plants and microbes. Insufficiency of vitamin B6 in the diet leads to genetic defects and inflammatory and neurological disorders in vulnerable populations. Thus, vitamin B6 biofortification of food represents a great prospect to lessen load of diseases in poor and vulnerable people, alongside improving stress tolerance (Fudge et al. 2017).

Vitamin B1 (thiamine) deficiency is very common among people dependent on processed rice as the main carbohydrate source. Thiamine is essential to biosynthesize TPP (thiamine pyrophosphate), which acts as an important cofactor of many crucial enzymes. Thiamine buildup in rice crop could be enhanced by overexpression of some genes like *thi4* and *thiC* (Pourcel et al. 2013; Dong et al. 2015). Moreover, genetic engineering of the thiamine production route provides a platform to improvise thiamine content in rice (Minhas et al. 2018). In this study, the editing of the regulatory elements (*cis*-acting) present in gene promoter shifted the biosynthesis of thiamine-binding proteins and transporters to the endosperm region. Thus, the ability to obtain vitamin B1 in rice grains was enhanced by this method of genetic biofortification making the diet wholesome.

Globally, iron deficiency represents a key threat to whole community health. Iron fortification of common bean resulted in enhanced iron buildup in eatable portions with increase in yield, biomass, and expression of antioxidant enzymes (Sida-Arreola et al. 2015). Biofortification of foods with iron enhances its bioavailability to confront the challenge of iron deficiency, in resource-limited populations (Petry et al. 2016). Recently, the efficiency of Fe-biofortified edible crops like beans, rice, pearl millet, etc. was evaluated for improvising iron content in high-risk populations (Finkelstein et al. 2017). In this study random trials suggested that plants biofortified using iron are an effective intervention to develop a good iron status. Outcomes of experiments suggested influence of fortified crops was maximum among people who showed deficiency of iron at the baseline. Recently, iron biofortification of brown rice not only enhanced the iron status but also improved the concentrations of flavonoids and phenolic compounds. The biofortified brown rice showed effective antioxidant properties and enhanced the germination rate of rice (Li et al. 2018).

Zinc is considered as a crucial micronutrient for human health. Witkowska et al. (2015) reported the biofortification of cheese and milk with some micronutrients like Fe(II), Cu(II), Mn(II), and Zn(II), which are considered indispensable for human health. The goats were provided food augmented by soya-based formulations that carried the micronutrients. The fortified milk showed boosted quantities of microelements, i.e., Zn(II) 14.6%, Mn(II) 29.2%, and Cu(II) 8.2%, as compared to control. This designer milk represents the latest version of functional foods to deal with the micronutrient deficiencies prevalent in vulnerable population. In China, wheat was biofortified with zinc by agronomical method, and success of the technology was measured in terms of "disability-adjusted life years" determining health burden. In major wheat-growing regions, appreciable results were achieved with the biofortified wheat where diseases due to deficiency of Zn showed reduction up to 56.6% (Wang et al. 2016). In a similar study, Zn biofortification programs were designed with a leafy vegetable, like *Brassica oleracea* cv. Bronco (Barrameda-Medina et al. 2017). Zinc supplementation (80–100 μM) was observed to be optimal for sustaining normal plant development with promotion of Zn concentration in *B. oleracea* edible parts. Further enhancing the Zn concentration leads to induction of amino acid buildup with increase in biosynthesis of phenolics and glucosinolates in leaves. The agronomical fortification of edible plants, viz., wheat and rice, with Zn is a viable strategy for a healthy and nourished future. It not only provides food and health security but also reduces the load of diseases from the human population (Kadam et al. 2018).

Iodine represents a crucial micronutrient indispensable for human health. Biofortification of lettuce by iodine fertilization of soil was observed to be a successful approach to improvise the iodine levels in individuals of target populations (Kopec et al. 2015). In this study besides biofortification of crop with iodine, its effect on Wistar rats was also studied to gain insights into different aspects of accumulation of iodine in animal tissues and health-related benefits. The rat serum was examined for different biochemical parameters and majority of tissues showed iodine concentrations higher than the control rats. This highlights victory of the biofortified crops in dealing with micronutrient deficiency. In a similar study, soil and foliar fertilization of vegetables resulted in an increase in iodine accumulation in various parts of plants (Lawson et al. 2015). The foliar application of iodine provided superior outcomes than the soil fertilization in this experiment.

Selenium has been established as a crucial constituent of a balanced human diet owing to its antioxidative and anti-oncogenic properties. Selenium-enriched *Brassica* crops were developed by agronomic biofortification and biofortified cabbage seedlings showed boosted quantities of antitumor activity (Oancea et al. 2015). Comparable findings were observed with another *Brassica* crop, i.e., broccoli (*Brassica oleracea* Italica), which was biofortified with Se. In this investigation, the mature stages of biofortified broccoli possessed considerably higher contents of phenolic complexes and showed better antiproliferative and antioxidant activities (Bachiega et al. 2016). Mushrooms represent another fundamental crop that can be targeted for selenium biofortification. An edible medicinal mushroom *Cordyceps militaris* was utilized for biofortification, resulting in enhanced selenium content in fruiting bodies. These reproductive bodies exhibited enhanced biological efficiency and antioxidative properties (Hu et al. 2018). In a different study, selenium content in biofortification of edible mushrooms (*Pleurotus* sp.) was performed to analyze its efficacy (Kaur et al. 2018). Wheat straw with high selenium levels was used for successful farming of *Pleurotus* sp., i.e., *P. florida*, *P. sajorcaju*, and *P. ostreatus*. The mushroom extracts showed considerable upgrading in the antioxidant profiles and total protein content. The utilization of wheat straw (Se-rich) not only enhanced the selenium contents in the cultivated mushrooms but also presented a way to make use of this underutilized substrate. Moreover, utilization of wheat straw will reduce air pollution as this waste is mostly burnt by the farmers if not utilized. Mushrooms represent a healthy food for mankind and selenium fortification will enhance their nutritional potential to target vulnerable and at-risk populations.

Biofortification of foods with silicon represents an innovative implement to develop nutritionally valuable diets with good consequences on bone strength. D'Imperio et al. (2017) studied the success of fortification of leafy vegetables (Swiss chard, mizuna, tatsoi, chicory, and purslane) with silicon as a health-promoting strategy. In vitro analysis was performed to analyze the probable health-supporting results of biofortified vegetables on mineralization of bones as compared to market-available silicon supplement.

12.3.2 Fortification with Amino Acids

Man and animals don't possess the ability to biosynthesize many amino acids de novo which are crucial for their health. Therefore indispensable amino acids should be taken by man and animals in their diets. Nine crucial amino acids are phenylalanine, methionine, valine, lysine, tryptophan, histidine, threonine, isoleucine, and leucine. As these compounds have high nutritional value, amino acid biofortification of edible foods is considered a foreseeable methodology owing to their very less concentrations in major staple crops (Galili and Amir 2012; Galili et al. 2016; Yang et al. 2016).

Among the principal crops like cereals and legumes worldwide, the amino acids methionine and lysine are present in less concentrations. So biofortification with these amino acids represents a successful strategy to make the crops nutritionally favorable (Galili and Amir 2012). In a similar study, a major staple crop, rice, was modified to boost up concentration of an amino acid, lysine. The rice crop was genetically engineered to overexpress enzymes dihydrodipicolinate synthase and aspartate kinase to get enhanced accumulation of lysine. Higher levels of lysine were obtained by impeding enzymatic action of lysine ketoglutarate reductase in rice (Yang et al. 2016). Recently, combination of varied scientific strategies like biochemical approaches, reverse genetics, and transgenic methods was exercised to study genes expressing enzymes crucial for biosynthesis, degradation, and regulation of crucial amino acids (Galili et al. 2016). Despite the employment of varied approaches, very little success has been achieved in biofortification of edible foods with amino acids as limited genetic resources for breeding methods are available and also high levels of indispensable amino acids generally restrict the plant growth. To design better transgenic plants, an improved insight into amino acid biosynthetic and regulatory pathways is need of the hour, to boost up indispensable amino acid content of cereal crops and horticultural plants (Wang et al. 2017).

12.3.3 Fortification with Essential Fatty Acids

Omega-3 acid (α-linolenic acid [ALA]) and omega-6 acid (linoleic acid [LA]) are indispensable fatty acids as animals or human beings are incapable to biosynthesize them. These fatty acids produce crucial fatty acids like docosahexaenoic acid (DHA), arachidonic acid (ARA), and eicosapentaenoic acid (EPA) owing to their important contribution in management of homeostasis (Saini and Keum 2018). The scarcity of crucial fatty acids is the principal reason of prevalence of autoimmune/inflammatory and cardiovascular diseases. The evidences throw light on the crucial importance of polyunsaturated fatty acids (PUFAs). The long-chain PUFAs (omega-3) are significantly important for controlling brain growth and functioning along with the maintenance of cardiovascular health (Hefferon 2015; Saini et al. 2018). Although fish is a rich source of PUFAs, consumption of fish is limited due to many factors. Thus alternative strategies must be employed to find other PUFA-rich sources for a sustainable supply to target population.

Among efficient approaches, one is genetic biofortification of plants to synthesize PUFAs. Genetic biofortification of oilseed crops and *Arabidopsis thaliana* was performed for expressing omega-3 PUFA at concentrations equivalent to those present in marine systems (Ruiz-Lopez et al. 2012; 2014). Seed oil plants were biofortified with other crucial fatty acids like arachidonic acid, γ-linolenic acid, and stearidonic acid (Haslam et al. 2013). The engineering of the metabolic pathway involved in biosynthesis of omega-3 fatty acid was reconstructed in plants like false flax, and its increased quantities have been obtained (Adarme-Vega et al. 2014). In a recent study, genetically biofortified safflower was developed that produced enhanced concentrations of alpha-linolenic acid (ALA). The seeds of the crop accumulated ~78% of the linoleic acid which act as direct precursor of ALA (Rani et al. 2018).

12.3.4 Fortification with Phytonutrients

Phytonutrients are the bioactive compounds extracted from plants which confer health benefits to humans. The intake of vegetables and fruits with high phytochemical content results in health-promoting effects like lowering predominance of severe disorders, besides prolonging the shelf life and commercial value (Zhu et al. 2013; Ilahy et al. 2018). Antioxidants are the bioactive compounds that interact with reactive oxygen species (ROS) thereby checking oxidative damage to the components of the cell. This reduces the death rate of cells ultimately decreasing rapidity of ageing and related diseases. Foods biofortified with vitamins (e.g., vitamin C and E) also possess enhanced antioxidant properties thus enabling the consumers to combat different types of stress conditions (Amaya et al. 2015). High expression of vitamin C was obtained in transgenic stylo and tobacco plants and also improved the tolerance power of these plants against chilling and drought stress (Bao et al. 2016). The latest innovations in technologies resulted in an improvised insight into the biosynthesis of vitamin C and accelerated renewed interest in development of new functional foods by overcoming limitations associated with vitamin C biofortification (George et al. 2017).

An important class of antioxidants that could be utilized to biofortify the crops include flavonoids. The success rate of the crop biofortification program with flavonoids is dependent on the point that the increased production of antioxidants must not affect the plant's overall growth and fitness (Zhu et al. 2013). In a similar study, cherry tomato was biofortified with potassium, and higher accumulation of this micronutrient leads to improvement in storage of these fruits post harvesting. The fruits of cherry tomato showed better storage through antioxidant response with reduced peroxidation of lipids and efficient regeneration of ascorbic acid (Constán-Aguilar et al. 2014). Anthocyanins represent another class of phenolics that possess high antioxidant properties. The genetic biofortification of anthocyanins in staple crops, like rice, can promote improvisation of human health. A novel rice germplasm was generated by this genetic alteration of rice known as "Purple Endosperm Rice," which showed high contents of anthocyanin and subsequently high antioxidant activity (Zhu et al. 2017). A maize variety biofortified with provitamin A carotenoids was observed to be a good functional food to target vulnerable and at-risk populations in the developing countries. This variety revealed a rich mine of tocochromanols, vitamin E, phenolic compounds, etc. and thus showed heightened antioxidative nature (Muzhingi et al. 2017).

In the current scenario, foods are biofortified to improve their postharvest quality. The enhancement in pigment buildup in tomatoes boosted up availability of the phytonutrients in biofortified crops besides prolonging shelf life. All these factors provided a boost in the marketability of biofortified crops (Ilahy et al. 2018). In a similar study, tomatoes showed enhanced accumulation of minerals and natural antioxidants when they were grown on earthworm-grazed and *Trichoderma harzianum*-biofortified SMS (spent mushroom substrate). This substrate inhibited the peroxidation of lipids and protein oxidation along with considerable enhancement in content of polyphenols and flavonoids in tomato. Thus, it's a substantial example of an environment-friendly and practical methodology that comes up with biofortified tomatoes possessing high radical scavenging properties (Singh et al. 2018).

12.4 Public Acceptance and Concerns

The accomplishment of the biofortification approach to address the challenge of undernourishment universally depends upon many factors like cost efficiency, nutrition impact, acceptance by target consumers, sustainable implementation of products, etc. Numerous experimental findings have unfolded a comprehensive picture of approval of biofortified crops by consumers (Birol et al. 2015; Steur et al. 2017; Ricroch et al. 2018). These studies were founded on the interdisciplinary research methods involving sensory evaluation and consumer testing methods. Sensory evaluation represents a scientific method involving measurement and interpretation of man comebacks for various food products, identified through the sensory perceptions, viz., sight, odor, touch, odor, sound, and taste. Consumer or hedonic testing involves the measurement of personal response of consumer like preference, liking, or acceptance for concerned produce. A greater acceptance of the biofortified crops by the target consumers ensures better delivery and marketing of these crops (Birol et al. 2015). The approval of genetically altered biofortified crops totally depends on perception of this approach by the common people. The awareness of the public about the potential of the green biotechnology to combat the matter of undernourishment and calorie deficiency must be enhanced to ensure full acceptance of biofortified crops by the target consumer. As common people don't have thorough access to scientific technologies and practices used in agriculture, it is difficult to guarantee victory of biofortified crops in the market, and that considerably affects the perception of transgenics by the public. A number of ethical facts were proposed in different studies concerning utilization of transgenic biofortified crops, but, morally, biofortification advantages must reach one and all, especially the poor and vulnerable groups of populations (Ricroch et al. 2018).

To produce nutritionally valuable crops, much work must be done with involvement of different scientific disciplines like molecular biologists, plant biotechnologists, plant breeders, nutritionists, and even socialists. Biofortification of plants will gain success and will be considered worthwhile only if the people are prepared, knowledgeable, and ready to acknowledge this technology or modifications in the crop appearance. The newer nutritionally improved crop varieties must be evaluated at clinical alongside market level and must be provided to the populations who can benefit most from them. These target populations should be perceptive to the point that these crops with high-level nutrient enrichment will influence the overall health of the community. Cooperative action of governments, nonprofit organizations, and industries will promote true elimination of the hidden hunger especially from the poor and at-risk populations (Hefferon 2015; Ricroch et al. 2018).

12.5 Conclusion

A massive percentage of the Earth's population is encountering malnutrition; the foremost victims are people of developing countries. Although recent advances have contributed a lot to change the scenario, the question of undernutrition remains unanswered. The estimation of fundamental factors throws light on the still unsettled problem of malnutrition. Hidden hunger still poses as an obstacle in developing a healthy world. Studies underlined the grim situation that this crisis of malnutrition will turn out to be more prominent in the coming times. Unfortunately, our principal staple crops lack majority of the fundamental nutrients such as amino acids, vitamins, fatty acids, and minerals that are critical for usual growth and survival. Quite a few tactics were proposed to boost up the quality besides quantity of the staple food; among them, biofortification is an innovative and emerging practice to curb the risk of hidden hunger by increasing the bioaccessibility of key nutrients. To formulate the biofortification program as a successful strategy, effective implementation of biofortified products plays a major role, and it requires constant examination, quality guarantee, regulatory control, and remedial actions to ensure compliance of all factors.

References

Adarme-Vega TC, Thomas-Hall SR, Schenk PM (2014) Towards sustainable sources for omega-3 fatty acids production. Curr Opin Biotechnol 26:14–18

Amah D, Biljon A, Brown A et al (2018) Recent advances in banana (*Musa* spp.) biofortification to alleviate vitamin A deficiency. Crit Rev Food Sci Nutr. https://doi.org/10.1080/10408398.2018.1495175

Amaya I, Osorio S, Martinez-Ferri E et al (2015) Increased antioxidant capacity in tomato by ectopic expression of the strawberry D-galacturonate reductase gene. Biotechnol J 10: 490–500

Bachiega P, Salgado JM, de Carvalho JE et al (2016) Antioxidant and antiproliferative activities in different maturation stages of broccoli (*Brassica oleracea* Italica) biofortified with selenium. Food Chem 190:771–776. https://doi.org/10.1016/j.foodchem.2015.06.024

Bao G, Zhuo C, Qian C et al (2016) Co-expression of *NCED* and *ALO* improves vitamin C level and tolerance to drought and chilling in transgenic tobacco and stylo plants. Plant Biotechnol J 14:206–214

Barrameda-Medina Y, Blasco B, Lentini M et al (2017) Zinc biofortification improves phytochemicals and amino-acidic profile in *Brassica oleracea* cv. Bronco. Plant Sci 258:45–51. https://doi.org/10.1016/j.plantsci.2017.02.004

Beswa D, Dlamini NR, Siwela M et al (2016a) Effect of Amaranth addition on the nutritional composition and consumer acceptability of extruded provitamin A-biofortified maize snacks. Food Sci Technol 36:30–39

Beswa D, Dlamini NR, Amonsou EO et al (2016b) Effects of amaranth addition on the provitamin A content, and physical and antioxidant properties of extruded provitamin A biofortified maize snacks. J Sci Food Agric 96:287–294

Birol E, Meenakshi JV, Oparinde A et al (2015) Developing country consumers' acceptance of biofortified foods: a synthesis. Food Secur 7:555–568

Blancquaert D, De Steur H, Gellynck X et al (2017) Metabolic engineering of micronutrients in crop plants. Ann N Y Acad Sci 1390(1):59–73. https://doi.org/10.1111/nyas.13274

Bouis HE, Saltzman A (2017) Improving nutrition through biofortification: a review of evidence from HarvestPlus, 2003 through 2016. Glob Food Sec 12:49–58. https://doi.org/10.1016/j.gfs.2017.01.009

Cakmak I (2014) Agronomic biofortification. Conference brief #8, In: Proceedings of the 2nd global conference on biofortification: getting nutritious foods to people, Rwanda

Cakmak I, Kutman UB (2017) Agronomic biofortification of cereals with zinc: a review. Eur J Soil Sci 69:172–180. https://doi.org/10.1111/ejss.12437

Cashman KD (2015) Vitamin D: dietary requirements and food fortification as a means of helping achieve adequate vitamin D status. J Steroid Biochem Mol Biol 148:19–26

Chaudhary RC, Gandhe A, Sharma RK et al (2016) Biofortification to combat Vitamin A deficiency sustainably through promoting orange-fleshed sweet potato (*Ipomoea batatas*) in eastern Uttar Pradesh. Curr Adv Agric Sci 8:139–142. https://doi.org/10.5958/2394-4471.2016.00034.4

Constán-Aguilar C, Leyva R, Blasco B et al (2014) Biofortification with potassium: antioxidant responses during postharvest of cherry tomato fruits in cold storage. Acta Physiol Plant 36:283–293. https://doi.org/10.1007/s11738-013-1409-4

D'Imperio M, Brunetti G, Gigante I et al (2017) Integrated in vitro approaches to assess the bioaccessibility and bioavailability of silicon-biofortified leafy vegetables and preliminary effects on bone. In Vitro Cell Dev Biol Anim 53:217–224

De Moura FF, Miloff A, Boy E (2015) Retention of provitamin A carotenoids in staple crops targeted for biofortification in africa: cassava, maize and sweet potato. Crit Rev Food Sci Nutr 55:1246–1269. https://doi.org/10.1080/10408398.2012.724477

Dong W, Stockwell VO, Goyer A (2015) Enhancement of thiamin content in *Arabidopsis thaliana* by metabolic engineering. Plant Cell Physiol 56:2285–2296. https://doi.org/10.1093/pcp/pcv148

Duffner A, Hoffland E, Stomph TJ et al (2014) Eliminating zinc deficiency in rice-based systems, VFRC report 2014/2. Virtual Fertilizer Research Center, Washington, DC

FAO, IFAD, UNICEF, WHO (2017) The state of food security and nutrition in the world: building resilience for peace and food insecurity. FAO, Rome

Finkelstein JL, Haas JD, Mehta S (2017) Iron-biofortified staple food crops for improving iron status: a review of the current evidence. Curr Opin Biotechnol 44:138–145

Fudge J, Mangel N, Gruissem W et al (2017) Rationalising vitamin B6 biofortification in crop plants. Curr Opin Biotechnol 44:130–137. https://doi.org/10.1016/j.copbio.2016.12.004

Galili G, Amir R (2012) Fortifying plants with the essential amino acids lysine and methionine to improve nutritional quality. Plant Biotechnol J 11:211–222

Galili G, Amir R, Fernie AR (2016) The regulation of essential amino acid synthesis and accumulation in plants. Annu Rev Plant Biol 67:153–178

George GM, Ruckle ME, Abt MR et al (2017) Ascorbic acid biofortification in crops. In: Hossain M, Munné-Bosch S, Burritt D et al (eds) Ascorbic acid in plant growth, development and stress tolerance. Springer, Cham, pp 375–415. https://doi.org/10.1007/978-3-319-74057-7_15

Gödecke T, Stein AJ, Qaim M (2018) The global burden of chronic and hidden hunger: trends and determinants. Glob Food Sec 17:21–29

Haslam RP, Ruiz-Lopez N, Eastmond P et al (2013) The modification of plant oil composition via metabolic engineering—better nutrition by design. Plant Biotechnol J 11:157–168

Hefferon KL (2015) Nutritionally enhanced food crops; progress and perspectives. Int J Mol Sci 16:3895–3914. https://doi.org/10.3390/ijms16023895

Hu T, Liang Y, Zhao G et al (2018) Selenium biofortification and antioxidant activity in *Cordyceps militaris* supplied with selenate, selenite, or selenomethionine. Biol Trace Elem Res. https://doi.org/10.1007/s12011-018-1386-y

IHME (2018) The global burden of disease: a critical resource for informed policymaking. WWW document. http://www.healthdata.org/gbd/about

Ilahy R, Siddiqui MW, Tlili I et al (2018) Biofortified vegetables for improved postharvest quality: special reference to high-pigment tomatoes. In: Siddiqui MW (ed) Preharvest modulation of postharvest fruit and vegetable quality. Elsevier/Academic, London, pp 435–454

Kadam SS, Kumar A, Arif M (2018) Zinc mediated agronomic bio-fortification of wheat and rice for sustaining food and health security: a review. Int J Chem Stud 6:471–475

Kamotho SN, Kyallo FM, Sila DN (2017) Biofortification of maize flour with grain amaranth for improved nutrition. Afr J Food Agric Nutr Dev 17(4):12574–12588. https://doi.org/10.18697/ajfand.80.1594

Kaur G, Kalia A, Harpreet SS (2018) Selenium biofortification of *Pleurotus* species and its effect on yield, phytochemical profiles, and protein chemistry of fruiting bodies. Food Biochem 42:e12467. https://doi.org/10.1111/jfbc.12467

Kopec A, Piątkowska E, Bieżanowska-Kopec R et al (2015) Effect of lettuce biofortified with iodine by soil fertilization on iodine concentration in various tissues and selected biochemical parameters in serum of Wistar rats. J Funct Foods 14:479–486. https://doi.org/10.1016/j.jff.2015.02.027

Lawson PG, Daum D, Czauderna R et al (2015) Soil versus foliar iodine fertilization as a biofortification strategy for field-grown vegetables. Front Plant Sci 6:450–460

Li K, Hu G, Yu S et al (2018) Effect of the iron biofortification on enzymes activities and antioxidant properties in germinated brown rice. Food Meas 12:789–799. https://doi.org/10.1007/s11694-017-9693-0

Lividini K, Fiedler JL, De Moura FF et al (2018) Biofortification: a review of ex-ante models. Glob Food Sec 17:186–195

Melash AA, Mengistu DK, Aberra DA (2016) Linking agriculture with health through genetic and agronomic biofortification. Sci Res 7:295–307

Minhas AP, Tuli R, Puri S (2018) Pathway editing targets for thiamine biofortification in rice grains. Front Plant Sci. https://doi.org/10.3389/fpls.2018.00975

Murray CJL, Vos T, Lozano R et al (2012) Disability-adjusted life years (DALYs) for 291 diseases and injuries in 21 regions, 1990–2010: a systematic analysis for the global burden of disease study 2010. Lancet 380:2197–2223. https://doi.org/10.1016/S0140-6736(12)61689-4

Muzhingi T, Palacios Rojas N, Miranda A et al (2017) Genetic variation of carotenoids, vitamin E and phenolic compounds in Provitamin A biofortified maize. J Sci Food Agric 97:793–801. https://doi.org/10.1002/jsfa.7798

Oancea AO, Gaspar A, Seciu A-M et al (2015) Development of a new technology for protective biofortification with selenium of Brassica crops. AgroLife Sci J 4(2):80–85

Petry N, Olofin I, Boy E et al (2016) The effect of low dose iron and zinc intake on child micronutrient status and development during the first 1000 days of life: a systematic review and meta-analysis. Nutrients 8:1–22

Pourcel L, Moulin M, Fitzpatrick TB (2013) Examining strategies to facilitate vitamin B1 biofortification of plants by genetic engineering. Front Plant Sci. https://doi.org/10.3389/fpls.2013.00160

Rani A, Panwar A, Sathe M et al (2018) Biofortification of safflower: an oil seed crop engineered for ALA-targeting better sustainability and plant based omega-3 fatty acids. Transgenic Res 27:253–263. https://doi.org/10.1007/s11248-018-0070-5

Ricroch AE, Guillaume-Hofnung M, Kuntz M (2018) The ethical concerns about transgenic crops. Biochem J 475:803–811

Ruel MT, Alderman H (2013) Nutrition-sensitive interventions and programmes: how can they help to accelerate progress in improving maternal and child nutrition? Lancet 382:536–551

Ruiz-López N, Haslam RP, Venegas-Calerón M et al (2012) Enhancing the accumulation of omega-3 long chain polyunsaturated fatty acids in transgenic *Arabidopsis thaliana* via iterative metabolic engineering and genetic crossing. Transgenic Res 21:1233–1243

Ruiz-Lopez N, Haslam RP, Napier JA et al (2014) Successful high-level accumulation of fish oil omega-3 long-chain polyunsaturated fatty acids in a transgenic oilseed crop. Plant J 77:198–208

Saini RK, Keum Y-S (2018) Omega-3 and omega-6 polyunsaturated fatty acids: dietary sources, metabolism, and significance–a review. Life Sci 203:255–267. https://doi.org/10.1016/j.lfs.2018.04.049

Saltzman A, Birol E, Bouis HE et al (2013) Biofortification: progress toward a more nourishing future. Glob Food Sec 2:9–17

Sharma P, Aggarwal P, Kaur A (2017) Biofortification: a new approach to eradicate hidden hunger. Food Rev Intl 33:1–21. https://doi.org/10.1080/87559129.2015.1137309

Sida-Arreola JP, Sánchez-Chávez E, Ávila-Quezada GD et al (2015) Iron biofortification and its impact on antioxidant system, yield and biomass in common bean. Plant Soil Environ 61:573–576

Singh UB, Malviya D, Khan W et al (2018) Earthworm grazed-*Trichoderma harzianum* biofortified spent mushroom substrates modulate accumulation of natural antioxidants and bio-fortification of mineral nutrients in tomato. Front Plant Sci. https://doi.org/10.3389/fpls.2018.01017

Steur HD, Demont M, Gellynck X et al (2017) The social and economic impact of biofortification through genetic modification. Curr Opin Biotechnol 44:161–168

Taofiq O, Fernandes A, Barros L (2017) UV-irradiated mushrooms as a source of vitamin D2: a review. Trends Food Sci Technol 70:82–94

Valença AWD, Bake A, Brouwer ID et al (2017) Agronomic biofortification of crops to fight hidden hunger in sub-Saharan Africa. Glob Food Sec 12:8–14

Wang Y-H, Zou C-Q, Mirza Z et al (2016) Cost of agronomic biofortification of wheat with zinc in China. Agron Sustain Dev 36:44. https://doi.org/10.1007/s13593-016-0382-x

Wang G, Xu M, Wang W et al (2017) Fortifying horticultural crops with essential amino acids: a review. Int J Mol Sci 18:1306. https://doi.org/10.3390/ijms18061306

Witkowska Z, Michalak I, Korczyński M et al (2015) J Food Sci Technol 52:6484–6492. https://doi.org/10.1007/s13197-014-1696-9

Yang QQ, Zhang CQ, Chan ML et al (2016) Biofortification of rice with the essential amino acid lysine: molecular characterization, nutritional evaluation, and field performance. J Exp Bot 67:4285–4296. https://doi.org/10.1093/jxb/erw209

Zhu C, Sanahuja G, Yuan D et al (2013) Biofortification of plants with altered antioxidant content and composition: genetic engineering strategies. Plant Biotechnol J 11:129–141. https://doi.org/10.1111/j.1467-7652.2012.00740.x

Zhu Q, Yu S, Zeng D, Liu H (2017) Development of "purple endosperm Rice" by engineering anthocyanin biosynthesis in the endosperm with a high-efficiency transgene stacking system. Mol Plant 10:918–929. https://doi.org/10.1016/j.molp.2017.05.008

Microbes in Foods and Feed Sector 13

Rajni Singh, Prerna Gautam, Mahek Fatima, Sonali Dua, and Jyoti Misri

13.1 Introduction

Microorganisms such as fungi and bacteria are determined to be tremendously useful to individuals. Their application in the food and feed industry can be dated back to the Neolithic age. They help in the development of taste and texture and are also responsible for providing numerous metabolically important micronutrients. The ability of microbes to secrete inhibitory compounds also adds to their use in food industry by acting as a biological inhibitor and preservative.

In this era where extensive usage of chemicals and fertilizers has caused decline in the nutritional properties of crop, beneficial microorganisms have proven to be an apt substitute for nutrition in the food and feed industry. The rapidly increasing population of the world puts forward a great demand of protein-rich food, and the protein derived from animal and plant sources is not only expensive but also not able to fulfill this ever-growing protein demand. As microbes are easy to cultivate, have very low gestation period, and are a good source of protein, they are gaining lot of concentration as a source of good-quality protein for both animals and humans.

Lactic-acid-producing bacteria (LABs) have been found useful in the production of various food products that require fermentation such as dairy products, fermented meat and vegetables, and sourdough bread. Fungi, especially *Saccharomyces* spp., have found immense importance in brewing and baking industry particularly *S. cerevisiae*.

R. Singh (✉) · P. Gautam · M. Fatima · S. Dua
Amity Institute of Microbial Biotechnology, Amity University, Noida, Uttar Pradesh, India
e-mail: rsingh3@amity.edu

J. Misri
Division of Animal Science, Indian Council of Agricultural Research, New Delhi, India

D. P. Singh et al. (eds.), *Microbial Interventions in Agriculture and Environment*,
https://doi.org/10.1007/978-981-13-8391-5_13

13.2 Microbial Products for Direct Consumption

Microbial systems have been developed for use in the food industry because of various reasons. Firstly, microbial growth is very rapid than that of animals. Secondly, a wide range of substrates are suitable for consideration depending on the microorganism chosen (Adedayo et al. 2011). These proteins (macronutrient that is mostly obtained from animal products and other sources, such as nuts and legumes, and is vital for building muscle mass (Szalay 2015)) are easy to obtain as their growth requirements do not depend on seasonal or climatic factors unlike the proteins derived from plants and animals and also it only requires a small area of land as required in animal or plant protein production (Suman et al. 2015).

The two main approach related to substrate are low grade waste utilization or to use relatively simple carbohydrate as substrate to produce microbial material of very high-quality protein. Microorganisms also have the ability to produce high-protein food by upgrading low-protein organic material. Bacteria and yeast are the most widely used candidates for the production of SCP. Bacterial growth is much rapid and efficient than yeast on cheap substrates, which also provide a higher content of protein (Suman et al. 2015).

To encounter the request of enormously increasing population, nonconventional protein sources have become significant in our diet. Major nonconventional sources are fish protein concentrate (FPC), oil seed proteins, biomass protein (BMP) or single-cell protein (SCP), and leaf protein concentrate (Adedayo et al. 2011).

Microbial products for direct consumption from various microorganisms are as follows:

13.2.1 Yeast

Yeast can be consumed both directly and indirectly (Figs. 13.1 and 13.2). In both cases it offers various health benefits as it has probiotic and antioxidant properties and is also rich in nutrients. It improves the sensibility of insulin, boosts the immunity, provides good-quality protein, and helps in treating diarrhea and Crohn's disease (Curejoy 2018).

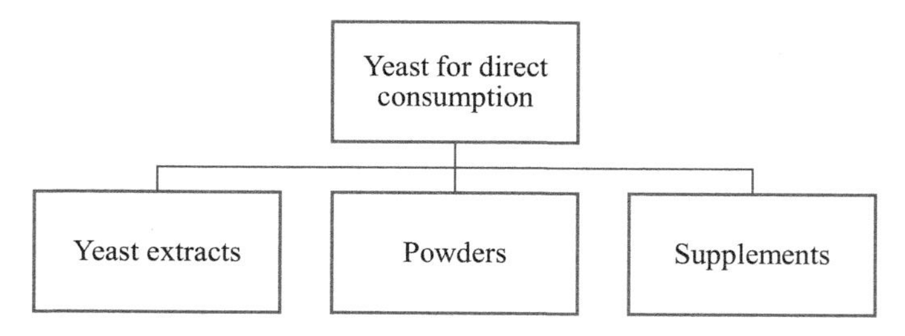

Fig. 13.1 Yeast for direct consumption

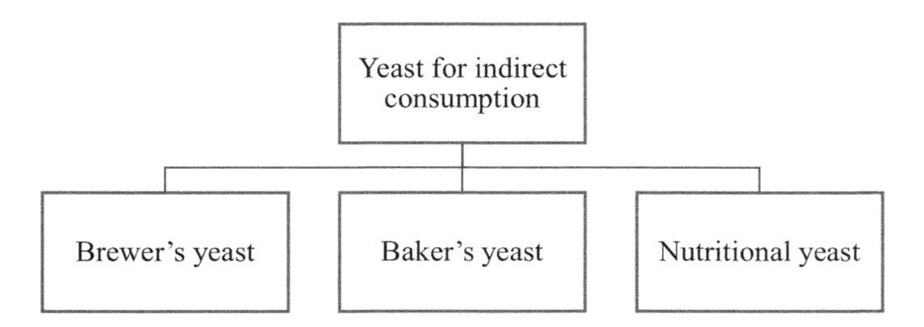

Fig. 13.2 Yeast for indirect consumption

Yeast Extract The popular name for various forms of processed yeast products that are used as edible flavors or food additives is yeast extract. They often contain free glutamic acid like MSG and are also utilized in a similar way as monosodium glutamate (MSG) (Differencebetween.net 2009).

Yeast Powder Yeast extract powder is the extract of a vacuum-dried concentrate of baker's yeast having excellent growth factors and providing high vitamin content for most microorganisms. It is a vacuum-dried autolyzed yeast extract obtained from 100% baker's yeast and is commonly added to culture media as a rich source of vitamins.

Yeast Supplement Yeast supplement plays a major part in flavor enhancement in fermented foods and release of antimicrobial compounds such as "mycocins" or antifungal killer toxins and various antibacterial compounds (Connolly 2017).

Brewer's Yeast Brewer's yeast is a single-celled fungus which contains a lot of proteins, minerals like chromium and selenium, and B vitamins. For direct consumption it is available in the form of powder, liquid, flakes, or tablet form.

Baker's Yeast Baker's yeast has proteins, carbohydrates, phosphorous, potassium, iron, magnesium, zinc, fibers, and B vitamins like folate and niacin. It is widely used in bread baking and various other bakery products.

Nutritional Yeast Nutritional yeast is rich in proteins; minerals like selenium, iron, potassium, and zinc; and B vitamins. It is used in various health products to make it healthy such as in pasta sauces and sprinkled over popcorn or chips.

Limitation of Yeast People with gluten sensitivity or celiac disease should avoid taking yeast as it is not gluten-free, and people with Crohn's disease should also avoid intake of yeast (Curejoy 2018).

13.2.2 Algae

Sea vegetables are the edible form of algae. Edible algae are popularly known as seaweed, and it comes in various shapes, textures, and tastes. Seaweeds offer great nutritional properties as they have various minerals which they absorb from the sea and also essential vitamins and are low in fat.

The three most common types of seaweed are nori, kombu, and brown algae.

Porphyra (Nori): The most widely consumed seaweed is nori. It is purplish to black in color. It has a very thin and flat texture. It has a mild sweet and slight meat flavor.

Laminaria (Kombu): Kombu is a very huge, bushy, brown alga with somewhat mushroom-like flavor. It is also rich in iodine.

Palmata (Dulse): It is another type of seaweed; it is a very thin, red alga with a smokish and nutty flavor (Renee 2017).

Other consumable algae are *Asparagopsis taxiformis* (limu), *Chondrus crispus* (Irish moss), *Gracilaria*, *Macrocystis*, *Undaria* (wakame), and green algae *Caulerpa racemosa*, *Ulva*, and *Codium* (Borowitzka 1998).

Spirulina Spirulina is a photosynthetic, multicellular, filamentous, helical-molded, blue-green microalga, and *Spirulina maxima* and *Spirulina platensis* are its two most popular species. They are procured from natural water, dried, and eaten as human food as it is a vital source of protein. It is used as protein supplement and as human health food specially for people trying to eliminate meat from their diet as it is 60–70% protein and 83–95% digestible (Sancbez et al. 2003).

Spirulina has been used as feed for poultry and majorly as a protein and vitamin supplement to aquafeeds. Spirulina has a fine body surface and without any covering, so it is easily digestible by simple enzymatic systems.

Spirulina is a miraculous food which is rich in the following nutrients:

(a) *B-12*: spirulina is a great food source of it. Vitamin B-12 is lacking in the diets of many vegetarians which is important for normal growth and neurological function and also provides energy.
(b) *Beta-carotene and mixed carotenoids*: help build the immune system nutritionally.
(c) *Full-spectrum antioxidants* from natural vitamins and pigments.
(d) *Phycocyanin*: a powerful, blue, immune-stimulating biliprotein found in excessively high concentration in spirulina.

(e) **Rhamnose:** a biologically active sugar, which speeds up the nutrient transport across the blood-brain barrier and the cell.
(f) **Glycogen:** stored energy form of glucose
(g) **EFAs:** omega 3 and 6 essential fatty acids.
(h) *Chlorophyll*: a deep-green blood builder that directly absorbs sunlight (Ahsan et al. 2008).

13.2.3 Mushrooms

Mushrooms have been devoured since soonest history and are also known as "Food of the Gods," and the Chinese culture has cherished mushrooms as sustenance for well-being, a "remedy of life." Out of more than 2000 mushroom species available in nature, those widely accepted as food are 25, and only a few of it are commercially cultivated (Valverde et al. 2015).

Nowadays, mushrooms are conventional commercial nourishments since they are low in fat, calories, starches, and sodium. Moreover, they provide us with essential supplements, including potassium, selenium, riboflavin, fiber, vitamin D, niacin, and proteins, and with such vast history as a potential nourishment source, they are imperative for their recuperating limits and properties in conventional pharmaceutical. It has detailed positive influence on well-being and is used in the treatment of a few illnesses. Innumerous nutraceutical properties are portrayed in mushrooms, for example, avoidance or treatment of Parkinson, Alzheimer, hypertension, and stroke.

Some popular edible mushroom species are *Agaricus bisporus*, *Lentinus edodes*, *Pleurotus sajor-caju*, *Pleurotus eryngii*, *Pleurotus ostreatus*, *Pleurotus giganteus*, *Huitlacoche* (*U. maydis*), and *Agaricus blazei.*

Huitlacoche (*U. maydis*) is a plant pathogenic fungus that infects only maize and its progenitor plant teosinte (*Zea mays*). It is also known as a nutraceutical food of high quality and an attractive ingredient to enhance other dishes, mainly for its exceptional quality and extraordinary flavor (Chatterjee and Patel 2016).

Mushrooms have more than a hundred medicinal values, and their prime medicinal uses are antioxidant, anticholesterolemic, anticancer, antidiabetic, antiallergic, antibacterial, cardiovascular protector, antiviral, antifungal, antiparasitic, detoxification, and hepatoprotective effects; they also safeguard against tumor formation and inflammatory processes. Macrofungi synthesize numerous bioactive molecules, and these bioactive compounds found in cultured mycelium, fruit bodies, and cultured broth include polysaccharides, proteins, fats, minerals, alkaloids, tocopherols, phenolics, glycosides, flavonoids, carotenoids, volatile oils, terpenoids, folates, lectins, enzymes, ascorbic, and organic acids, in general, and are found in dietary supplements (Valverde et al. 2015).

13.2.4 Single-Cell Protein (SCP)

Recently, the major protein that gained attention is SCP as it has the potential to increase the demand of protein supply. It is rich in protein with a broad amino acid spectrum. It contains less fat and high protein-carbohydrate ratio as compared to forages. SCPs are also environment-friendly as they help in recycling of waste by growing on it. For the production of SCP, different types of organic wastes like cellulose, hydrocarbon, hemicelluloses, and various kinds of agricultural wastes are also used (Fig. 13.3). SCP has been found to be an appropriate diet supplement especially in the developing country of Africa for both human and livestock (Meyer and Goldberg 1985).

The idea of eating microbes is hardly new as many microbes have high nutritional therapeutic values/properties and have been grown and eaten in many parts of the world for centuries. A wide range of microbes such as bacteria, yeast, algae, and fungi can be utilized to synthesize essential nutrients. Among the microorganisms; bacteria contains highest amount of proteins (in % dry weight) i.e. 50-65 followed by yeast (45-55), algae (40-60) and fungi (30-45) (Szalay 2015).

13.2.4.1 SCP Production by Bacteria

Bacteria is a potent microorganism for production of SCP as it has various characteristics which include short generation time, fast growth, and twofold increase in their cell mass (20 mins–2 h). They have a wide range of substrates on which they are capable of growing ranging from carbohydrates such as starch and sugars,

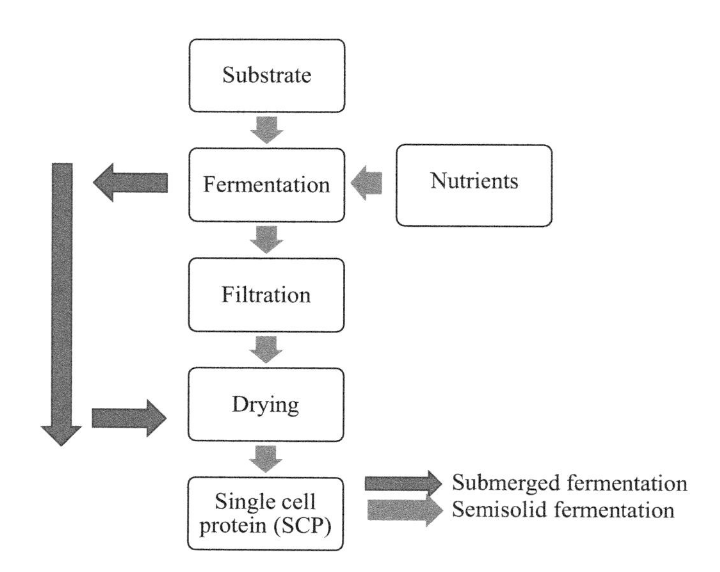

Fig. 13.3 Flow chart for SCP production

gaseous hydrocarbons, liquid hydrocarbons like urea, nitrates, ammonium salts, ammonia, and organic nitrogen present in waste, and to overcome any deficiency, it is always recommended to supplement the bacterial culture medium with nutrient minerals. Potential phototrophic and methanotrophic bacterial strains are the ones highly recommended for single-cell protein production.

Methylophilusis has the generation time of about 2 hours and is used in animal feed. This in general produces a more favorable protein composition than fungi or yeast. For animal feed, a huge amount of SCPs can be formed by bacterial species like *Brevibacterium*, *Methylophilus*, *Achromobacter delvaevate*, *Bacillus megaterium*, *Aeromonas hydrophila*, *Acinetobacter calcoaceticus*, *Bacillus subtilis*, *Lactobacillus* species, *Cellulomonas* species, *Rhodopseudomonas capsulate*, *Pseudomonas fluorescens*, *Methylomonas* species, *Flavobacterium* species, and *Thermomonospora fusca*.

13.2.4.2 SCP Production by Algae

Spirulina is the utmost, extensively used algae cultivated by African people and is dried and consumed as food species (*Chlorella* and *Scenedesmus*) by tribal communities in many parts of the world. Alga is commonly used as a food source, and its advantages include rapid growth, simple cultivation, high protein content, and effective utilization of solar energy. *Spirulina* has been acknowledged for use as an efficient additional protein.

13.2.4.3 SCP Production by Yeast

The topmost nutrient feed substitute is yeast single-cell protein (SCP). The most widespread yeast species are *Pichia*, *Candida*, *Hansenula*, *Saccharomyces*, and *Torulopsis*. *Saccharomyces cerevisiae* is grown on various fruit wastes and orange and cucumber peels as a substrate for the production of SCP.

13.2.4.4 SCP Production by Fungi

Various fungal species are sources of protein-rich food. *Actinomycetes* and filamentous fungi produced protein from various substrates. *Aspergillus oryzae* or *Rhizopus arrhizus* inoculums were preferred because of their nontoxic nature. There are a few very harmful species of molds, for example, *Aspergillus niger*, *Fusarium graminearum*, and *A. fumigates*, which are very dangerous to humans, as they are toxic in nature; therefore, such fungi must not be used. In recent times, SCP technology uses fungal species for bioconversion of lignocellulosic wastes (Suman et al. 2015).

Polysaccharides are vital for advanced medicine; β-glucan is the eminent and most versatile metabolite with a wide spectrum of biological activities (Valverde et al. 2015; Chatterjee and Patel 2016).

13.3 Microorganisms in Fermented Foods

Across the world, several species or genera of microbes have been described in relation to several beverages and fermented foods. Fermented foods are the center of a consortia of microbes, as they exist as native functional microbiota in raw flora or fauna substrates, containers, earth pots, utensils, environment, and starter culture which amend the substrates biochemically into comestible foods that are accepted on a social and traditional basis by the consumers. Microorganisms' main role is to alter the biochemical composition and organic compounds of uncooked constituents and convert them into desirable products like ethanol, lactic acid, and carbon dioxide during the fermentation process. This process improves the functionality, flavor, texture, and shelf life and increases the nutritive value of products, hence imparting varieties of health benefits to prevent diseases and improve the bioregulation of behavioral issues such as anxiety and stress.

Fermented food and beverages broadly consist of lactic acid bacteria (LAB). Several species of *Staphylococcus* (firmicutes), *Micrococcus* (actinobacteria), and *Kocuria* are associated with fermented milk, fish, sausages, and meat products. Different genera of yeast like *Cryptococcus*, *Candida*, *Brettanomyces*, *Debaryomyces*, *Pichia*, *Rhodosporidium*, and *Saccharomyces* and species of molds such as *Neurospora*, *Aspergillus*, *Amylomyces*, *Monascus*, *Rhizopus*, *Ustilago*, *Actinomucor*, *Paecilomyces*, and *Penicillium* are also tested in Asian nonfood amylolytic starter, alcoholic beverages, and fermented foods (Tamang et al. 2016a). The functional attributes of microbes in fermented food products include antimicrobial property, probiotic attributes, polyglutamic acid production, fibrinolytic activity, peptide production, and degradation of antinutritive compounds. Health welfares of several global fermented foods are prevention of gastrointestinal disorder, allergic reactions, cancer, diabetes, osteoporosis, synthesis of nutrients etc. (Tamang et al. 2016b). Global fermented foods can be categorized into nine key sets on the basis of raw materials (substrates) used from animal and plant sources (Tamang et al. 2016a).

13.3.1 Fermented Bamboo Shoots and Vegetables

Bamboo shoots are traditionally found in majority of Asian countries like Japan, China, Thailand, Bhutan, Indonesia, Nepal, Malaysia, and India. They are considered as food of importance because they are low in cholesterol and fats and very high in dietary fibers, potassium, vitamin B and C, and carbohydrates. They also contain antioxidants (phenols, steroids, and flavones) and nutritious and active minerals (amino acids and vitamins). Species like *Lactobacillus plantarum*, *L. mesenteroides*, *L. brevis*, *L. corniformis*, *Enterococcus durans*, *L. casei*, *L. fermentum*, *Streptococcus lactis*, *Leuconostoc fallax*, *Lactococcus lactis*, and *Tetragenococcus halophilus* are involved in bamboo shoots fermentation (Thakur et al. 2016).

Fermented vegetables are low-calorie food products as they contain less sugar in comparison with their raw counterparts. They are a source of nutritional fibers, which obstruct the acclimatization of fats and control peristalsis in the intestines, and are valuable sources of phenols, vitamin B and C, and various other nutrients Mir et al. 2018).

Seasonal and green leafy vegetables, cucumbers, radish, and edible shoots are habitually fermented into edible foods. *Leuconostoc mesenteroides* and associated LAB, including *Weissella* and other *Leuconostoc* spp., are significant for the beginning of fermentation of many vegetables. The most dominated species used for fermentation of vegetables are *Pediococcus* and *Lactobacillus* trailed by *Tetragenococcus, Weissella, Leuconostoc* and *Lactococcus* (Tamang et al. 2016a). Examples of more fermented vegetable products are given in Table 13.1.

Table 13.1 Some of the fermented vegetables and their substrates

S. no.	Fermented vegetable products	Substrate	Regions	Microbes involved during fermentation	Nature of use
1.	Sinki	Radish taproot	Nepal, Bhutan, and India	*L. brevis*, *L. fermentum*, and *L. plantarum*	Cure diarrhea, relieve stomach pain, and effective appetizer
2.	Inziangsang	Leafy vegetable	Manipur and Nagaland	*Pediococcus*, *L. plantarum*, and *L. brevis*	In soup
3.	Goyang	Leaves of magane-saag	Nepal and Sikkim	*Candida* sp., *Enterococcus faecium*, *L. lactis*, *Pediococcus pentosaceus*, *L. plantarum*, and *L. brevis*	
4.	Anishi	Yam leaves	Nagaland		Condiment
5.	Ekung	Bamboo tender shoot product	Arunachal Pradesh		
6.	Lung-Siej	Bamboo shoot	Meghalaya		Curry mix with meat and fish
7.	Soibum	Edible shoots of choya bans, bhalu bans, and karati bans	Sikkim and Darjeeling hills	*L. curvatus*, *Pediococcus pentosaceus*, *L. brevis*, *L. plantarum*, *Leuconostoc citreum*	Pickles
8.	Gundruk	Leafy vegetables	Nepal	*Pediococcus* sp., *Lactobacillus* sp. like *L. plantarum*, *L. cellobiosus*	Soup, pickle, and appetizer
9.	Khalpi	Fermented cucumber	Nepal	*Leuconostoc*, *L. brevis*, *L. plantarum*	

13.3.2 Fermented Legumes and Soybeans

Soybean meals (SBM) are commonly used as protein source for animal fodder containing good amino acid balance. The native components of SBM are altered through the fermentation process, predominately by bacterial and fungal strains (mainly *Lactobacillus subtilis* and *Aspergillus oryzae*) where the nutritive value is increased and allergic and antinutritive contents are decreased.

(a) **Fermentation by fungi**: Numerous fungal species of Aspergillus like *A. usamii, A. niger, A. oryzae,* and *A. awamori* eliminate phytates, stachyose, and raffinose and diminish trypsin inhibitors and large-size protein.
(b) **Fermentation by bacteria**: Usually *Bacillus* species are used in fermentation of SBM. For example, *B. subtilis* plays a major role in the fermentation of soybean natto (Mukherjee et al. 2016). Few strains of *Bacillus subtilis* yield gamma polyglutamic acid that exists in Asian fermented soybean foods, providing the adhesive surface to the products (Tamang et al. 2016a). *Lactobacillus plantarum* is another bacterial strain used for fermentation which results in increased hydrolysis and liberation of free amino acids, thus increasing the total free amino acid content (Mukherjee et al. 2016).

13.3.3 Fermented Milk Products

Milk encompasses crucial constituents mandatory for good nutrition of human beings. This nutritional value is further upgraded beneath the impact of metabolic activities of starter cultures during fermentation. The dietary value of Indian fermented milk products (IFMPs) is due to carbohydrates, minerals, proteins, vitamins, and several therapeutic activities. IFMPs utilize only 7% of the total milk produced, mainly including three products like *shrikhand* (sweet concentrated curd), *lassi* (stirred milk), and *dahi* (curd).

During manufacture of dahi, there is an increase in folic acid (165–331%), riboflavin (121–131%), niacin (160–201%), and thiamine. The protein value of dahi is specified to be 3–30% higher as compared to milk and comprises of essential amino acids (1470–2433 mg/100 ml). It possesses various antibacterial and antagonistic activities and activates the nonspecific immune system.

Lately, efforts have been made to embrace beneficial bacteria and probiotics with the objective of increasing the dietary properties of traditional dahi (Sarkar 2008). Probiotics are the viable preparation in dietary or food supplements to improvise the health of animals and humans. It is mainly composed of live microbes like lactic-acid-producing bacteria (*Streptococci, Lactobacillus, Lactococci*, and *Enterococci*), *Bifidobacteria*, and other *Bacillus* species as well as some yeast like *Saccharomyces* sp. They provide health benefits to hosts such as immune modulation, lactose intolerance, anticarcinogenic effect, maintenance of mucosal integrity, modulation of metabolic activities of colonic microbes, and reduction in allergic reactions and serum cholesterol (Shah et al. 2016).

Yogurt turns out to be a widespread transport for the assimilation of probiotic cultures such as *Lactobacillus acidophilus* and *Bifidobacterium bifidum* to upgrade nutrition and health benefits. It has been described that milk proposed for fermented milk products should be free from inhibitory substances (residues of preservatives, sanitizers, antibiotics, and natural inhibitors) and pathogens and should have low bacterial count (Sarkar 2008).

Fermented milk products are classified on the basis of microorganisms into two groups:

(a) Fungal lactic fermentations, in which yeast and lactic acid bacteria produce the final products like moldy milks and alcoholic milks (*kefir*, acidophilus yeast milk, *koumiss*).
(b) Lactic fermentation governed by species of lactic acid bacteria, containing mesophilic type (e.g., cultured milk, natural fermented milk, cultured buttermilk, cultured cream), probiotic type (e.g., bifidus milk, acidophilus milk), and thermophilic type (Bulgarian buttermilk, yogurt)

13.3.4 Fermented Cereals

Cereal grains are considered as one of the most significant sources of carbohydrates, dietary proteins, vitamins, minerals, and fibers for individuals all over the world (Blandino et al. 2003). Fermented cereal food is mainly characterized by lactic acid bacteria and yeast. The most common bacteria associated with fermented cereals are *Pediococcus, Lactobacillus, Lactococcus, Leuconostoc, Enterococcus, Weissella,* and *Streptococcus*. In continents of America, Europe, and Australia, most cereals (maize, rye, wheat, and barley) are fermented by addition of profitable baker's yeast into the batter of dough loaves/breads or via natural fermentation, whereas in Asia some countries ferment rice either by using food beverages or by using mixed cultures into alcoholic beverages (Tamang et al. 2016a).

In general, natural fermentation of cereals sometimes leads to poor nutritional quality in comparison with diary and dairy products. Basically, this is due to the absence of certain vital amino acids, lesser protein content, existence of certain antinutrients (polyphenols, tannins, and phytic acids), abrasive nature of grains, and low starch content (Blandino et al. 2003).

13.3.5 Fermented Roots and Tubers

Tuber crops and starchy roots hold second position after cereals as a universal source of carbohydrates. These plants originate from diversified botanical sources, which store comestible starch material in subterranean roots, rhizomes, stems, roots, tubers, and corms, where yams and potatoes are tubers; sweet potatoes and cassava are storage roots; taro and cocoyams are derived from swollen hypocotyls, underground stems, and corms; and arrowroots and canna are comestible rhizomes (Chandrasekara and Kumar 2016).

Cassava root is conventionally fermented into a sugary desert named "tape" in Indonesia (Tamang et al. 2016a). A major fraction of cassava produced in Latin America and Africa are involved in fermented fodder and food seasonings such as monosodium glutamate and organic acids. Lactic acid bacteria and yeasts are two chief groups of microorganisms that are utilized in cassava fermentation (Chandrasekara and Kumar 2016). These fermented food products comprise of *fufu* in Togo, Benin, Nigeria, and Burkina Faso; chikwangue in Zaire; *foo foo* in Benin, Togo, Ghana, and Nigeria; *gari* in Nigeria; lafun, agbelima, kivunde, and attieke in Africa; tape in Asia; and "coated peanut" and "cheese" bread in Latin America. Fermented roots and tubers have antioxidative, immunomodulatory, hypoglycemic, antimicrobial, and hypocholesterolemic properties (Chandrasekara and Kumar 2016).

13.3.6 Fermented, Dried, and Smoked Fish Products

People living in nearby lakes, rivers, and coastal regions preserve fishes through sun or smoke drying, salting, and fermentation (Tamang et al. 2016a,b). In Southeast Asian countries like in the case of northeast regions of India (NEI), fermentation of fishes is a well-acknowledged tactic of food preservation. Widespread fermented products of this area include lona ilish and shidal. NEI have the maximum rainfall in the world which does not provide a pleasant atmosphere for simple sun drying of fishes. Individuals preserve fishes for use in lean periods by drying beneath the sun. Henceforth, such drying used to be extended due to recurrent rainfall and high humid atmosphere mainly during the peak rainy seasons.

As in the case of fermentation, it encompasses the breakdown of proteins of raw fishes to simpler constituents which are steady at standard temperature for packaging and storage. This breakdown of proteins by native protease or microbes produces bioactive peptides leading to a significant increase in the biological properties of foods like organic compounds, crude proteins, and total lipids. Usually cured fishes are the main sources of nutritional proteins in various emerging countries out of which NEI has numerous fermented products such as*numsing, shidal, hentaak, ngari, tungtap* and *laonailish* etc. Besides fermented fish products, diverse smoked and dried fish products also exist in NEI, including maacha, sidra, karati, lashim, gnuchi, and shukti (Majumdar et al. 2016).

The storage and food safety stability of these outmoded fermented products can be seen as a descriptive hurdle principle, where not a single but an amalgamation of hurdles of microbial growth, such as salt concentration, pH, and *Lactobacillus*, make the ultimate product stable and safe by preventing the development of microbial pathogens and spoilage of vegetation (Skara et al. 2015).

13.3.7 Alcoholic Beverages

Traditionally, alcoholic beverages are socially and culturally accepted for drinking, customary practices, religious purpose, entertainment, and consumption. They are the principal example of "biotic ennoblement" (Tamang 2010), and some of them are given in Table 13.2.

13.3.8 Fermented, Preserved Meat Products

Meat and meat products play a vital role in the diet of established countries. Both of them can be altered by addition of ingredients or by reducing or eliminating components that are considered harmful for health. Their major constituents in addition to water are fats and proteins, with an extensive contribution of minerals and vitamins of a high degree of bioavailability (Fernandez-Gines et al. 2005).

Fermented meat products are bifurcated into two classes: those made by chopping the meat (sausages) and those made from meat slices or whole pieces (jerky and dried meat). The major group of microbes involved in meat fermentation are lactic acid bacteria trailed by *Enterobacteriaceae, Micrococci*, and coagulase-negative *Staphylococci*. Some species of molds and yeast also play a major function in ripening of meat (Tamang et al. 2016a).

13.3.9 Miscellaneous Fermented Products

(a) Some traditional Asian societies enjoy special fermented teas such as *fuzhuan* brick (*Eurotium*, *Penicillium*, and *Aspergillus* are major fungus), *puer* tea (*Aspergillus niger* is the predominant fungus), and *kombucha* (*Gluconacetobacter* is the predominant bacteria) of China and *miang* of Thailand.
(b) For making chocolates, bacterial species such as *Acetobacter pasteurianus* and *L. fermentum* are reported in cocoa bean fermentation.
(c) Vinegar is prepared from ethanol or sugar comprising substrates and starchy ingredients which are hydrolyzed by aerobic process to acetic acid via *Acetobacter malorum, Acetobacter pomorum, Acetobacter polyxygenes, Acetobacter pasteurianus,* etc. and yeast like *Candida stellate* and *Zygosaccharomyces lentus* (Tamang et al. 2016a).

13.4 Vitamins Derived from Microorganisms

Vitamins are essential micronutrients that perform specific biological functions. Unlike plants and microbes, mammals either produce these micronutrients in very small quantities or cannot synthesize them at all. Therefore, vitamins must be consumed through the diet.

Table 13.2 Alcoholic beverages of the world are classified into ten categories

S. no.	Alcoholic beverages	Produced by	Microorganisms involved	Examples
1.	Non-distilled and unfiltered alcoholic beverages	Amylolytic starters	*Rhizopus* spp., *S. fibuligera*, *S. cerevisiae*, *Amylomyces*, *Torulopsis*, and *Hansenula*	Bhaati jaanr (fermented rice) of Nepal and India; kodo ko jaanr (fermented finger millets), and makgeolli (fermented rice) of Korea
2.	Distilled alcoholic beverages	Amylolytic starters		Soju of Korea and shochu of Japan
3.	Non-distilled and filtered alcoholic beverages	Amylolytic starters	*Asp. oryzae*	Saké of Japan
4.	Alcoholic beverages	Amylase in human saliva	*S. cerevisiae*, *S. pastorianus*, *S. apiculata*, species of *Lactobacillus* and *Acetobacter, Leuconostoc*, *Bacillus*, *Escherichia*, *Enterococcus*, *Enterobacter, Streptomyces*, *Cronobacter, Acinetobacter*, *Klebsiella*, *Propionibacterium*, *and Bifidobacterium*	Chicha of Peru
5.	Alcoholic beverages	Honey	*S. cerevisiae*, *Debaromyces phaffi*, *Kl. veronae*, *Kluyveromyces bulgaricus*, *and* LAB species of *Lactobacillus*, *Pediococcus*, *Leuconostoc*, and *Streptococcus*	Tej of Ethiopia
6.	Alcoholic beverages	Germination or malting	*L. fermentum*	Pito of Nigeria and Ghana, sorghum beer of South Africa, and tchoukoutou of Benin
7.	Alcoholic beverages	Mono fermentation		Beer
8.	Distilled alcoholic beverages	Cereals and fruits		Brandy and whisky
9.	Non-distilled alcoholic beverages	Fruits	*Saccharomyces*, *Candida colliculosa*, *Hansenia sporauvarum*, *C. stellata*, *Torulaspora delbrueckii*, *Metschnikowia pulcherrima*, *Kloeckera apiculata*, *Kl. thermotolerans*	Cider and wine

(continued)

Table 13.2 (continued)

S. no.	Alcoholic beverages	Produced by	Microorganisms involved	Examples
10.	Alcoholic beverages	Plant parts	LAB (*L. lactis, Lactobacillus acidophilus, Lactobacillus acetotolerans, Lactobacillus hilgardii, Leuc. citreum, Leuc. mesenteroides, Lactobacillus kefir, Lactobacillus plantarum, Leuc. kimchi, Leuc. pseudomesenteroides, Microbacterium arborescens, Acetobacter malorum, S. cerevisiae, Clavispora lusitaniae*, etc.	Kanji of India, pulque of Mexico, and toddy of India

Microorganisms are a rich source of vitamins. Inside the human system the gut microbes produce vitamins that are actively absorbed. It is due to this reason that extra vitamin supplementation is required after administration of antibiotics by a diseased individual. In commercial sectors, microorganisms are widely used for the large-scale production of various vitamins such as vitamin B_{12}, riboflavin, ascorbic acid, pantothenic acid, biotin, folic acid, and thiamine.

13.4.1 Vitamin B_2 (Riboflavin)

Vitamin B_2 or riboflavin is a water-soluble essential human nutrient belonging to the vitamin B family. Commercially it is used as an animal food supplement. Another commercial use of riboflavin is in the form of a food colorant. Biochemically, it is the pioneer molecule for flavin adenine dinucleotide (FAD) and flavin mononucleotide (FMN). Riboflavin deficiency also known as ariboflavinosis results in cheilosis and angular stomatitis. It can also lead to anemia. This type of anemia is different from the one caused by vitamin B_9 and vitamin B_{12} deficiency. This is due to the fact that ariboflavinosis leads to interference in iron absorption that leads to anemia despite normal hemoglobin content and cell size.

13.4.1.1 Vitamin B_2 (Riboflavin): Microbial Production

Unlike higher animals, riboflavin is synthesized by all plants, fungi, and most bacteria. Therefore, it is essential for animals to obtain this micronutrient through their diet. Large-scale production of riboflavin has been done both chemically and through microbial fermentations, but the requirements for expensive chemicals and toxicity limit this method. Biosynthesis of riboflavin is seen in both gram-negative and gram-positive bacteria (Lin et al. 2014). The synthetic pathway of riboflavin is a seven-step process wherein one molecule of GTP and two molecules of ribulose-5-phosphate form one molecule of riboflavin. The process is the same in almost all bacterial species.

Among the microbes that are used in the commercial production of riboflavin are *Bacillus subtilis*, *Candida famata,* and *Ashbya gossypii*. Many lactic-acid-producing bacteria have also been screened for enhanced riboflavin production. Some of these bacteria include *Lactobacillus fermentum* MTCC8711, *Lactobacillus plantarum, Leuconostoc mesenteroides, L. fermentum, L. plantarum* CRL725, and *L. acidophilus* (Thakur et al. 2015).

13.4.2 Vitamin C (L-Ascorbic Acid)

Vitamin C or L-ascorbic acid is a strong reducing agent which contributes to its high antioxidant property. Vitamin C also acts as a cofactor for eight human enzymes. These enzymes are involved in the following biosynthetic procedures:

1. Carnitine biosynthesis
2. Collagen hydroxylation
3. Hormone and amino acid biosynthesis

The most well-known disorder that arises from vitamin C deficiency is scurvy. Scurvy is prevented and treated by administration of vitamin C. Vitamin C has also proved itself to be important in brain development. Deficiency of vitamin C has resulted in defects in cerebellum development in neonates (Kim et al. 2015).

13.4.2.1 Vitamin C: Microbial Production

Industrially, vitamin C is synthesized from glucose by two different methods: Reichstein process and two-step fermentation process. The first method involves six chemical stages and one bacterial fermentation stage for the conversion of D-glucose to vitamin C. During the chemical steps, d-sorbitol is converted to l-sorbose. The two-step fermentation process involves a mixed fermentation step that converts l-sorbose to 2-keto-l-gulonic acid (2-KLG) (Fig. 13.4). 2-KLG is the precursor molecule for vitamin C. The mixed fermentation step bypasses the chemical conversion involved in the Reichstein process (Zou et al. 2013).

Commercial production of vitamin C occurs through the modern two-step fermentation process accounting for about 80% of the worldwide vitamin C available in the market (Hancock 2009).

Microorganisms involved in the commercial production of vitamin C are *Glucuno melanogenus*, *Bacillus megaterium*, and *Ketogulonicigenium vulgare*. *G. melanogenus* is used in the first step of fermentation, and *B. megaterium* and *Ketogulonicigenium vulgare* are involved in the second step. In the second step use of other bacterial species such as *Pseudomonas striata, Xanthomonas maltophilia, Bacillus thuringiensis*, and *Bacillus cereus* has also been reported in place of *Bacillus megaterium* (Zou et al. 2013).

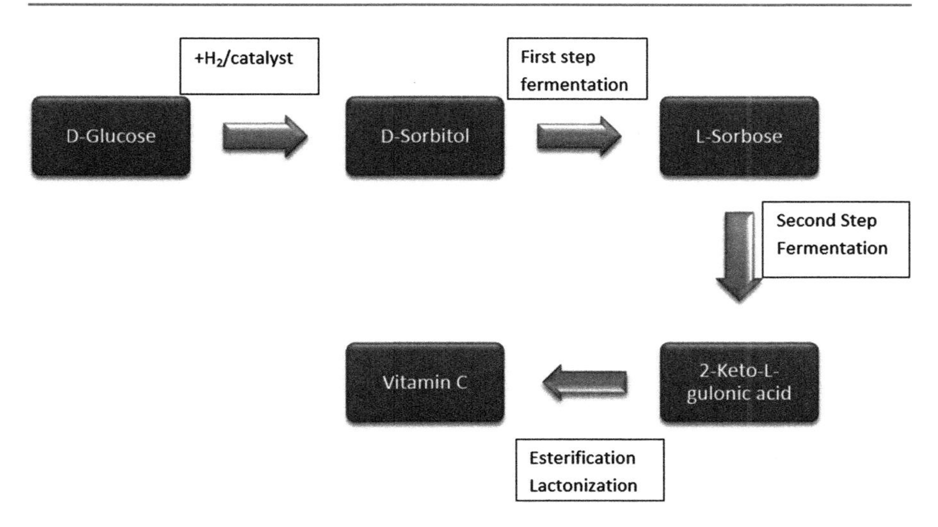

Fig. 13.4 The modern two-step fermentation process for vitamin C biosynthesis

13.4.3 Vitamin B_{12}: Cobalamin

Vitamin B_{12} also known as cyanocobalamin is an important organic compound in both medicine and food industry. It acts as a coenzyme for three types of enzymes, isomerases, methyltransferases, and dehalogenases, all three of which are involved in various biochemical pathways. Vitamin B12 deficiency can lead to specific myelin damage, development of pernicious anemia, coronary disease, stroke (Li et al. 2017), and myocardial infractions. All these facts state that indeed vitamin B_{12} is an important micronutrient.

13.4.3.1 Vitamin B_{12}: Chemical Structure

Chemically, the compound is a member of the cobalt corrinoid family. A vitamin B12 molecule consists of a central cobalt ion with four ligands: two lower ligands and two upper ligands. The former, also known as the alpha ligand, is DMBI, and the latter, also known as the beta ligand, is made up of either methyl group, adenosyl group, hydroxyl group, or cyano group (Fang et al. 2017).

13.4.3.2 Vitamin B_{12}: Microbial Production

Production of vitamin B12 can occur via both salvage and de novo pathways. In the de novo pathway usually, bacteria carry out the aerobic pathway, while the archaea take the anaerobic route. Various microorganisms that carry out the synthesis of vitamin B12 are listed in Table 13.3.

Table. 13.3 Microorganisms capable of producing vitamin B_{12}

S. no	Microorganism	De novo synthesis of vitamin B12		Salvage pathway of vitamin B12 synthesis
		Aerobic pathway	Anaerobic pathway	
1	*Pseudomonas denitrificans* SC510	Yes	No	Yes
2	*Propionibacterium shermanii*	No	Yes	–
3	*Salmonella enterica*	No	Yes	No
4	*Bacillus megaterium*	No	Yes	–
5	*Thermotoga lettingae*	No	No	Yes
6	*Lactobacillus reuteri*	No	Yes	Yes
7	*Halobacterium* sp. strain NRC-1	No	No	Yes

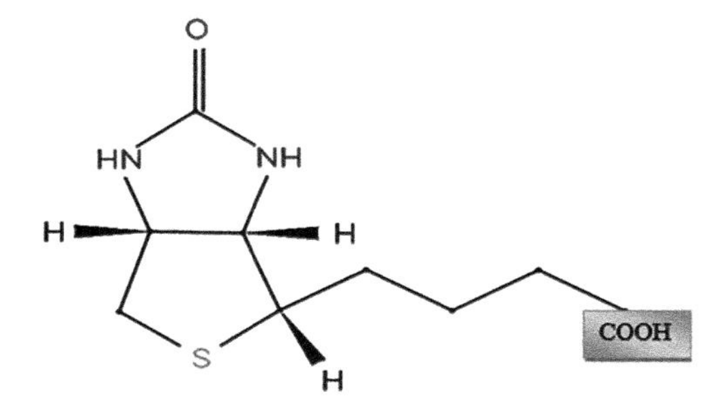

Fig. 13.5 Chemical structure of biotin

13.4.4 Vitamin B$_7$: Biotin

Biotin also known as vitamin B_7 or vitamin H is an essential micronutrient. Biochemically, it is found in the form of an enzyme-associated prosthetic group that acts as a carrier for carbon dioxide. The enzymes are involved in various carboxylation, decarboxylation, and transcarboxylation reactions. The ability of biotin to carry carbon dioxide is facilitated by its chemical structure. The compound consists of a bicyclic ring. This ring is attached to a valeryl side chain (Fig. 13.5).

Studies have shown that deficiency of biotin results in thinning of hair and skin rashes, increases inflammatory responses of dendritic cells, decreases the ability to reduce blood glucose, and decreases insulin sensitivity (Larrieta et al. 2012). As mammals are unable to synthesize this vital nutrient, it is essential for them to consume it through their diet. Plants, fungi, and microbes are a rich source of biotin. Commercially, biotin is majorly produced through chemical processes. But the increasing awareness toward restricting the use of hazardous chemicals has resulted in the exploitation of microbes that can be engineered for the large-scale production of biotin. Biosynthesis of biotin is a conserved two-stage process. Stage one is the synthesis of a pimelate moiety, and stage two is the assembly of a bicyclic ring. The

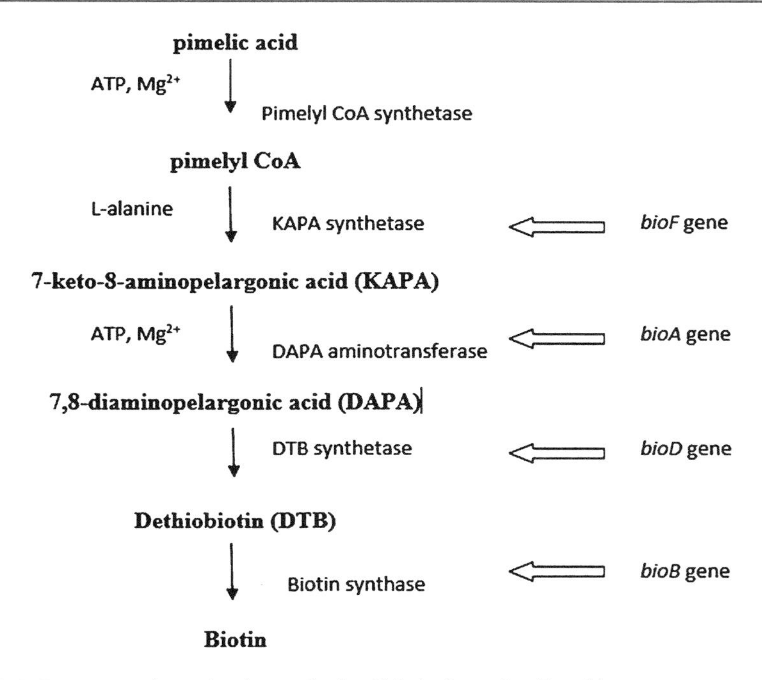

Fig. 13.6 De novo pathway for the synthesis of biotin from pimelic acid

pimelate moiety is synthesized via de novo pathway and is best understood in *E. coli* and *Bacillus subtilis* (Figs. 13.3 and 13.6).

Microorganisms that are able to excrete considerate quantities of biotin include *Bacillus sphaericus*, *E. coli* (recombinant strain), *Agrobacterium, Rhizobium* HK4, and *Sphingomonas* sp. Psp304. In the genetically engineered *E. coli*, the bio operon was altered to enhance the production of biotin. One of the modifications was that the bio promoter was substituted by tac which is a strong artificial promoter. This modified operon was introduced in many plasmids which resulted in enhanced biotin production (Survase et al. 2006).

13.5 Microbes in Feed Sector

Livestock nutritionists all over the world, in order to avoid the use of chemical growth stimulants, are driven by the idea to use alternative safer feed additives such as naturally occurring microbes for animal feed. The introduction of the formulation direct-fed microbial (DFM) has led to exploitation of microbes that are capable of improving the health and productivity of livestock. Many bacteria and yeast that have been screened for the same have resulted in enhanced productivity and digestion by the ruminants found in the gut of animals.

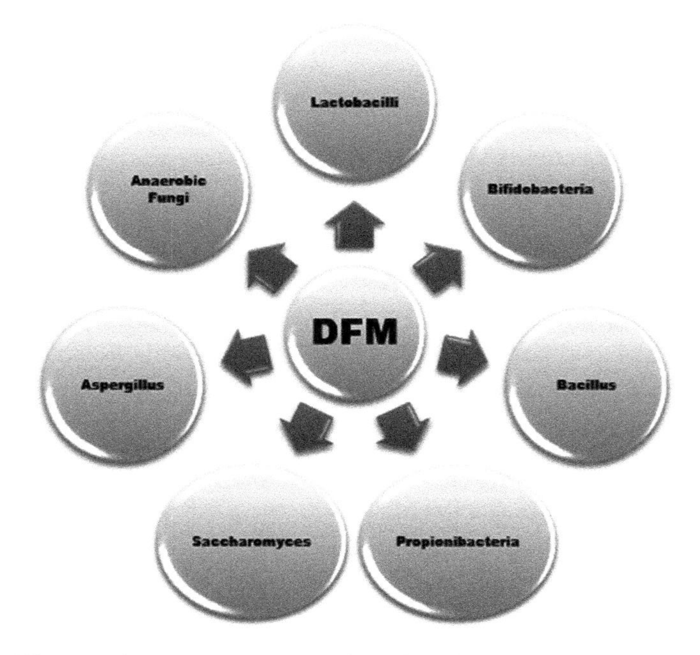

Fig. 13.7 Microorganisms commonly used as direct-fed microbial (DFM)

13.5.1 Direct-Fed Microbial (DFM)

DFM are nothing but probiotics that are administered in order to produce beneficial impacts in host animals. DFM include a variety of microorganisms such as bacteria, yeast, and fungi. The most frequently used microorganisms are exemplified in Fig. 13.7 (Nagpal et al. 2015).

13.5.1.1 Bacteria as DFMs

Most bacterial species used as DFM belong to the genus *Lactobacilli, Propionibacteria, Bifidobacterium*, and *Bacillus*. Examples of LAB species involved in growth promotion in animal livestock include *L. acidophilus, L. lactis, L. delbrueckii, L. plantarum, L. fermentum, L. salivarius*, and *L. bulgaricus* (Khan et al. 2016). These organisms are administered into calves in the form of bolus, while in beef and dairy animals, they are administered by incorporating them in the animal's diet. LABs are involved in prevention of acidosis inside the gut of animals by promoting growth of only those microbes that can withstand the concentrations of lactic acid. LABs are also involved in the synthesis of bacteriocin, benzoic acids, diacetyls, etc. (McAllister et al. 2011).

Another type of bacteria known as LUBs or lactate-utilizing bacteria keeps the pH of the gut at normal levels by utilizing lactic acid. One such bacterium is *Megasphaera elsdenii*. It is found in lactating animals. But when the animal is shifted to high-concentrate diet, it is unable to prevent acidosis in the organism (Seo et al. 2010). Species of *Bifidobacterium* such as *B. longum*, *B. infantis*, *B. animalis*, *B. lactis*, etc. are involved in starch digestion. Although these organisms are not rumen originated but are helpful when administered through diet.

Propionibacteria is also beneficial to the host animal when introduced into the rumen. *Propionibacteria* convert lactate and glucose to acetate and propionate. When in higher concentrations, propionate can be absorbed by the blood and converted to glucose inside the liver of the host. Though *Propionibacteria* has advantages, it is not used commercially for animal feed as it is slow growing and is unable to tolerate high lactic acid concentration inside the rumen. Other bacterial species that have been used as successful DFMs are *Bacillus subtilis*, *Bacillus coagulans*, *B. licheniformis*, *B. lentus*, etc. *Bacillus* species are involved in the degradation of cell wall (Nagpal et al. 2015).

13.5.1.2 Yeast as DFMs

Yeast in the form of feed additives has been used alone and in combination with fungal species such as those of *Aspergillus* as a growth promoter in lactating cows. They are useful in rumen fermentation and are generally classified as rumen modifiers. Addition of yeast in fodder has resulted in enhanced gut function and an increase in overall body weight in calves. As live yeast results in a huge increment in ethanol production inside the host, dead yeast cells are administered inside the livestock.

The most commonly used yeast in the form of DFM is *Saccharomyces cerevisiae*. It is used abundantly as a feed additive for daily cattle. The use of *Saccharomyces cerevisiae* has the following advantages:

(a) Enhances digestion
(b) Utilizes oxygen making the gut environment more suitable for anaerobic bacteria
(c) Prevents acidosis (Nagpal et al. 2015)

13.5.1.3 Fungi as DFMs

The role of fungi in the breakdown of complex plant tissues is no new knowledge. Anaerobic fungi are capable of penetrating these complex tissues that many rumen bacteria fail to do. Therefore, they enhance fiber digestion inside the host animal. Adding to these advantages is the fact that fungi are a rich source of proteolytic enzyme and thus are rich suppliers of protein. Fungi such as *Aspergillus niger*, *A. oryzae*, *Neocallimastix*, *Piromyces*, etc. have been used as direct-fed microbial in animal feed. The role of these fungi, as discussed, is limited to fibrinolysis (McAllister et al. 2011).

13.5.2 Benefits of DFMs

As discussed earlier, there are many microbes which are used extensively in the feed sector in order to nourish livestock. By using these DFMs which are nothing but probiotics, many benefits are provided to the animals. Some of the major benefits derived from them are:

(a) Enhancement in nutrition utilization
(b) Improved digestion
(c) Weight gain
(d) Improved quality of products derived from the livestock
(e) Stable rumen pH
(f) Lowered acidosis
(g) Immunomodulation: increased antibody production, effective cell-mediated response, enhanced dendritic cell-T cell interaction, etc.
(h) Enhanced milk production
(i) Production of antimicrobial compounds (Khan et al. 2016)

13.6 Conclusion

All recent investigations that contribute to the study of microbes in the feed sector have proved that administration of direct-fed microbial is the future for an environmentally safe growth-promoting option in animals. Further improvement in these formulations will result in better production performances in the ruminants.

References

Adedayo MR, Ajiboye EA, Akintunde JK, Odaibo A (2011) A single cell proteins: as nutritional enhancer. Adv Appl Sci Res 2:396–409

Ahsan M, Habib B, Parvin M (2008) A review on culture, production and use of Spirulina as food for humans and feed for domestic animals and fish. http://www.fao.org/3/a-i0424e.pdf. Accessed 13 Aug 2018

Blandino A, Al-Aseeri ME, Pandiella SS, Cantero D, Webb C (2003) Cereal-based fermented foods and beverages. Food Res Int 36:527–543

Borowitzka(1998)Algaeasfood.https://link.springer.com/chapter/10.1007/978-1-4613-0309-1_18. Accessed 13 Aug 2018

Chandrasekara A, Kumar TJ (2016) Roots and tuber crops as functional foods: a review on phytochemical constituents and their potential health benefits. Int J Food Sci:15

Chatterjee B, Patel T (2016) Edible mushroom – a nutritious food improving human health. Int J Clin Biomed Res 2:34–37

Connolly B (2017) How to avoid fermented foods. https://www.livestrong.com/article/252663-how-to-avoid-fermented-foods/. Accessed 12 Aug 2018

Curejoy (2018) Discover the health benefits of yeast in food. https://india.curejoy.com/content/benefits-of-yeast-in-food/#. Accessed 11 Aug 2018

Differncebetween.net (2009) Difference between yeast and yeast extract. http://www.differencebetween.net/object/difference-between-yeast-and-yeast-extract/. Accessed 11 Aug 2018

Fang H, Kang J, Zhang D (2017) Microbial production of vitamin B12: a review and future perspectives. Microb Cell Factories 16:15. https://doi.org/10.1186/s12934-017-0631-y

Fernandez-Gines JK, Fernandez-Lopez J, Sayas-Barbera E, Perez-Alvarez JA (2005) Meat products as functional foods: a review. J Food Sci 70(2):R37–R43

Hancock RD (2009) Recent patents on vitamin C: opportunities for crop improvement and single-step biological manufacture. Recent Pat Food Nutr Agric 1:39–49

Khan RU, Naz S, Dhama K, Karthik K, Tiwari R, Abdelrahman M, Alhidary IA, Zahoor A (2016) Direct-fed microbial: beneficial applications, modes of action and prospects as a safe tool for enhancing ruminant production and safeguarding health. Int J Pharmacol 12:220–231

Kim H, Kim Y, Bae S, Lim SH, Jang M, Choi J, Jeon J, Hwang YI, Kang JS, Lee WJ (2015) Vitamin C deficiency causes severe defects in the development of the neonatal cerebellum and in the motor behaviors of Gulo$^{-/-}$ mice. Antioxid Redox Signal 23:1270–1283

Larrieta E, de la Vega-Monroy MLL, Vital P et al (2012) Effects of biotin deficiency on pancreatic islet morphology, insulin sensitivity and glucose homeostasis. J Nutr Biochem 23:392–399

Li P, Gu Q, Yang L, Yu Y, Wang Y (2017) Characterization of extracellular vitamin B12 producing *Lactobacillus plantarum* strains and assessment of the probiotics potentials. Food Chem. https://doi.org/10.1016/j.foodchem.2017.05.037. Assessed on 20 Aug 2018

Lin Z, Xu Z, Li Y, Wang Z, Chen T, Zhao X (2014) Metabolic engineering of *Escherichia coli* for the production of riboflavin. Microb Cell Factories 13:104

Majumdar RK, Roy D, Bejjanki S, Bhaskar N (2016) An overview of some ethnic fermented fish products of the Eastern Himalayan region of India. J Ethn Foods 3:276–283

McAllister TA, Beauchemin KA, Alazzeh AY, Baah J, Teather RM, Stanford K (2011) Review: the use of direct fed microbials to mitigate pathogens and enhance production in cattle. Can J Anim Sci 91:193–211

Meyer J Goldberg J (1985) Using microbes as food source. http://articles.chicagotribune.com/1985-05-09/entertainment/8501290080_1_protein-micro-organisms-single-cell. Accessed 10 Aug 2018

Mir SA, Raja J, Masoodi FA (2018) Fermented vegetables, a rich repository of beneficial probiotics- a review. Ferment Technol 7:1. https://doi.org/10.4172/2167-7972.1000150

Mukherjee R, Chakraborty R, Dutta A (2016) Role of fermentation in improving nutritional quality of soybean meal – a review. Asian Australas J Anim Sci 29:1523–1529

Nagpal R, Shrivastava B, Kumar N, Dhewa T, Sahay H (2015) Microbial feed additives. In: Rumen microbiology from evolution to revolution Springer India, 2015, pp 161–175

Renee J (2017) Algae as a food source for humans. https://www.livestrong.com/article/458681-algae-as-a-food-source-for-humans/. Accessed 13 Aug 2018

Sancbez M, Bernal-Castillo J, Rozo C, Rodriguez I (2003) *Spirulina* (Arthrospira): an edible microorganism: a review. Univ Sci 8:7–24

Sarkar S (2008) Innovations in Indian fermented milk products— a review. Food Biotechnol 22:78–97

Seo JK, Kim SW, Kim MH, Upadhaya SD, Kam DK, Ha JK (2010) Direct-fed microbials for ruminant animals. Asian-Australas J Anim Sci 23:1657–1667

Shah C, Mokashe N, Mishra V (2016) Preparation, characterization and in vitro antioxidative potential of synbiotic fermented dairy products. J Food Sci Technol 53:1984–1992

Skara T, Axelsson L, Stefansson G, Bo E, Hagen H (2015) Fermented and ripened fish products in the northern European countries. J Ethnic Foods 2:18–24

Suman G, Nupur M, Anuradha S, Pradeep B (2015) Single cell protein production: a review. Int J Curr Microbiol App Sci 4:251–262

Survase SA, Bajaj IB, Singhal RS (2006) Biotechnological production of vitamins. Food Technol Biotechnol 44:381–396

Szalay J (2015) What is protein? https://www.livescience.com/53044-protein.html. Accessed 10 Aug 2018

Tamang JP (2010) Diversity of fermented beverages and alcoholic drinks. In: Tamang JP, Kailasapathy K (eds) Fermented foods and beverages of the world. CRC Press/Taylor and Francis group, Boca Raton, pp 85–125

Tamang JP, Watanabe K, Holzapfel WH (2016a) Review: diversity of microorganisms in global fermented foods and beverages. Front Microbiol 7:377

Tamang JP, Shin DH, Jung SJ, Chae SW (2016b) Functional properties of microorganisms in fermented foods. Front Microbiol 7:578

Thakur K, Tomar SK, De S (2015) Lactic acid bacteria as a cell factory for riboflavin production. Microb Biotechnol. https://doi.org/10.1111/1751-7915.12335

Thakur K, Rajani CS, Tomar SK, Panmei A (2016) Fermented bamboo shoots: a rich niche for bioprospecting lactic acid bacteria. J Bacteriol Mycol 3(4):00030. https://doi.org/10.15406/jbmoa.2016.02.00030

Valverde ME, Hernández-Pérez T, Paredes-López O (2015) Edible mushrooms: improving human health and promoting quality life. Int J Microbiol 2015:14

Zou W, Liu L, Chen J (2013) Structure, mechanism and regulation of an artificial microbial ecosystem for vitamin C production. Crit Rev Microbiol 39:247–255

New Age Agricultural Bioinputs

14

Bhavana V. Mohite, Sunil H. Koli, Hemant P. Borase, Jamatsing D. Rajput, Chandrakant P. Narkhede, Vikas S. Patil, and Satish V. Patil

14.1 Introduction

Nitrogen-based biofertilizers are significant bioinputs, but according to current environmental changes and ever-increasing food demand, it is the need of time to popularize more efficient bioinputs for soil. These bioinputs will help to fight against problems like an unpredictable monsoon, global warming, and decreasing soil fertility, and indiscriminate use of agrochemicals.

Besides chemical fertilizers, organic soil conditioners, the application of phosphate solubilizers, nitrogen fixers, and *Trichoderma*, *Verticillium*, *Metarhizium* like versatile biocontrolling agents are the common strategies of soil conditioning. In the past 50 years, there is tremendous work published on nitrogen fixers and phosphate solubilizers. The results of these findings directed to the exploitation of common biofertilizers like *Azotobacter* and *Rhizobium* as a nitrogen fixer and other organic inputs. In addition to above, phosphate, zinc, sulphur, potassium solubilizers are a

B. V. Mohite · S. H. Koli · J. D. Rajput · C. P. Narkhede
School of Life Sciences, Kavayitri Bahinabai Chaudhari North Maharashtra University, Jalgaon, Maharashtra, India

H. P. Borase
School of Life Sciences, Kavayitri Bahinabai Chaudhari North Maharashtra University, Jalgaon, Maharashtra, India

C. G. Bhakta Institute of Biotechnology, Uka Tarsadia University, Surat, Gujarat, India

V. S. Patil
University Institute of Chemical Technology, Kavayitri Bahinabai Chaudhari North Maharashtra University, Jalgaon, Maharashtra, India

S. V. Patil (✉)
School of Life Sciences, Kavayitri Bahinabai Chaudhari North Maharashtra University, Jalgaon, Maharashtra, India

North Maharashtra Microbial Culture Collection Centre (NMCC), Kavayitri Bahinabai Chaudhari North Maharashtra University, Jalgaon, Maharashtra, India

D. P. Singh et al. (eds.), *Microbial Interventions in Agriculture and Environment*,
https://doi.org/10.1007/978-981-13-8391-5_14

significant part of current agricultural practices. Although these practices proved beneficial to uphold soil fertility and other agronomical problems like pest attack and plant susceptibility to various infections, physiological problems due to the change in the atmosphere need some novel strategies or additional bioinputs.

There are various significant bioinputs like the application of 1-aminocyclopropane-1-carboxylic acid (ACC) enzyme and phytase producing microorganisms and bacterivorous flora. These are which were reported, but unfortunately remain as neglected practices by Indian farmers. The following three major bioinputs are need of time to use as new soil bioinputs in modern agricultural practices:

1. Use of ACC oxidase and deaminase producer bioinputs
2. Use of phytase producer
3. Use of bacterivorous soil microbes

The central idea of this chapter is presented in Fig. 14.1, which represents the ability of major modern agricultural bioinputs.

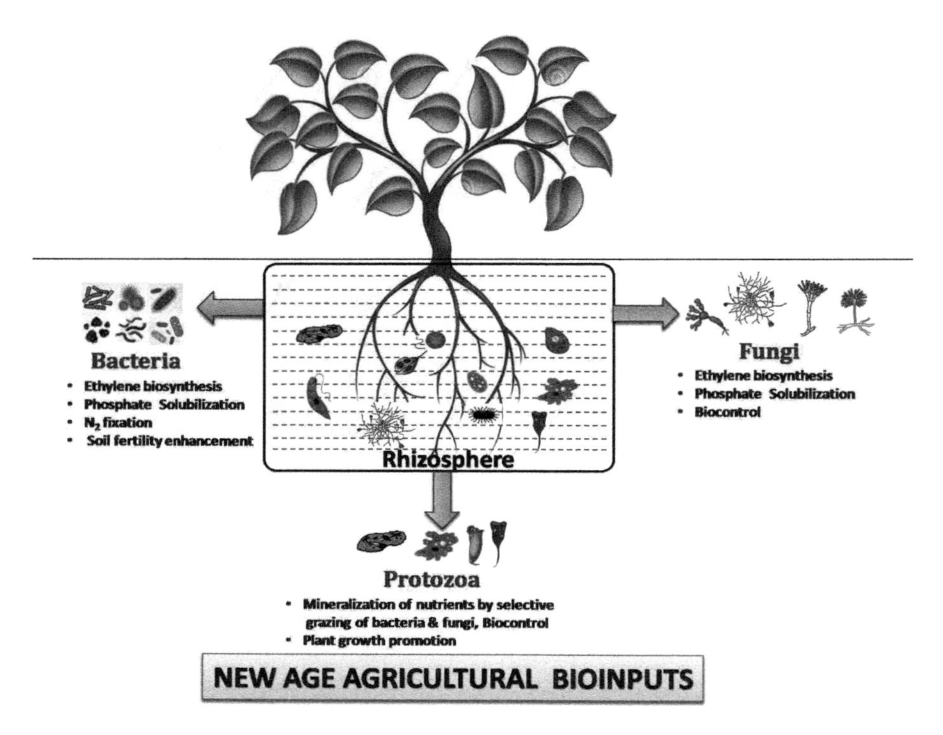

Fig. 14.1 Schematic representation for the new age agricultural bioinputs

14.2 Application of ACC Oxidase and Deaminase Producer Bioinputs

14.2.1 ACC and ACC-Degrading Enzymes

The Yang cycle produces 1-aminocyclopropane-1-carboxylic acid (ACC) and ACC oxidase and deaminase (ACCO and ACCD) (Yang and Hoffman 1984). Shang Yang unlocked the mystery of freshness of fruit, flowers, defoliation, and ripening of fruits by proposing a continuous biochemical cycle known as the Yang cycle. The Yang cycle biosynthesizes ethylene in plants. Ethylene is important in host–pathogen interactions, seed germination, flowering, and fruit ripening. It establishes the central role of methionine in ethylene synthesis. Yang's study proved the genesis of S-adenosylmethionine as a transitional compound which is further converted into ACC and then ethylene (Fig. 14.2).

ACC is the signaling molecule of a plant, easily transported through intra- and intracellular tissues over short and long distances.

ACC is a cyclic α-amino acid with a three-membered cyclopropane ring merged to an α-carbon atom of the amino acid (Fig. 14.3) and chemical formula $C_4H_7NO_2$ with a molar mass of 101.0 g/mol^{-1}. ACC is considered an essential intermediate that regulates ethylene biosynthesis. The enzyme ACCO is a member of the oxidoreductase class, which is responsible for the transformation of

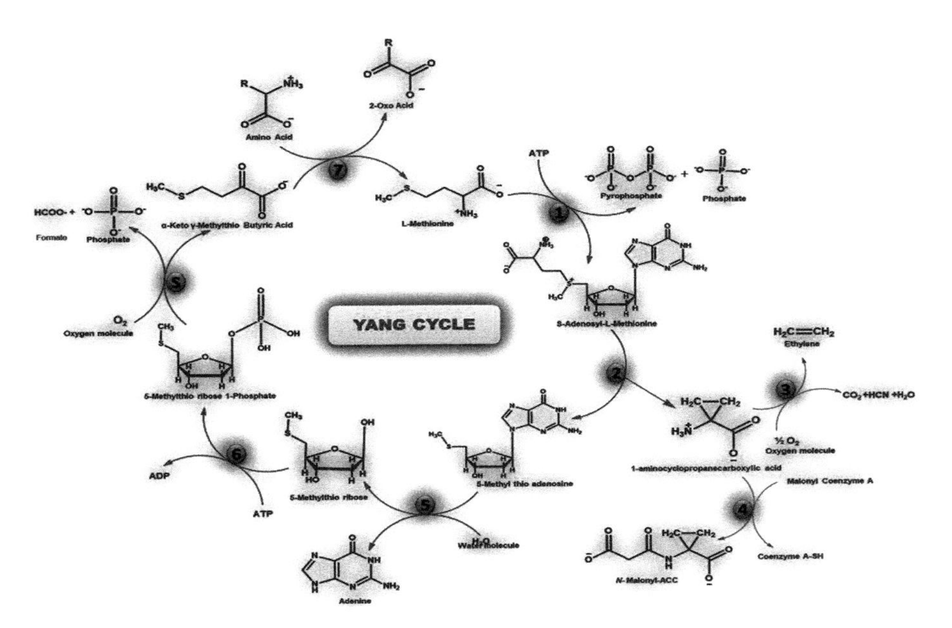

Fig. 14.2 Yang cycle for ethylene biosynthesis. Cycle path: (1) SAM synthetase, (2) ACC synthase, (3) ACC oxidase, (4) ACC N-malonyltransferase, (5) MTA nucleosidase, (6) MTR kinase, and (7) transaminase, (S) spontaneous reaction

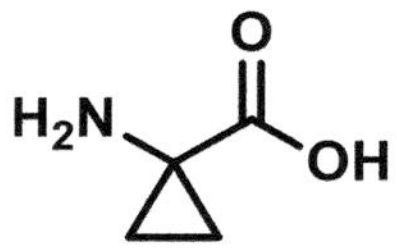

Fig. 14.3 Chemical structure of ACC

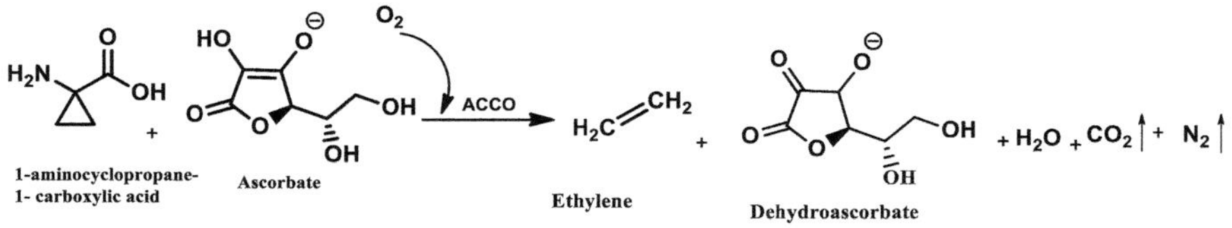

Fig. 14.4a Transformation of ACC to ethylene with ACCO

1-aminocyclopropane-1-carboxylate to ethylene with carbon dioxide, water, and other by-products (Fig. 14.4a).

In drought stress conditions, ethylene synthesis is rapidly increased (Morgan and Drew 1997). Ethylene is the one of the marker compounds of drought conditions and is also known as stress ethylene. Nitrogen fixation and nodulations are influenced by the various effects of high ethylene synthesis through water and temperature stress, like reduction of transpiration rate by closing stomata to regulate the abscisic acid pathway (Tanaka et al. 2005; Tamimi and Timko 2003; Penmetsa and Cook 1997; Guinel 2015). Hence, if the ACCO is regulated, then the natural synthesis of ethylene is regulated. Various researchers advocated that various rhizospheric microbes also control the ethylene level in a plant by deaminating ACC diffused through root cells and seeds (Finlayson et al. 1991; Penrose and Glick 2001; Penrose and Glick 2003).

14.2.2 Aminocyclopropane-1-Carboxylic Acid Oxidase (ACCO)

Aminocyclopropane-1-carboxylic acid oxidase is an enzyme recognized to fight against the consequences of drought in plants. It was well documented that drought affects various biochemical, morphological, and physiological activities of plants, e.g., turgor pressure, transport of soil nutrients, nutrient transport to root, nutrient diffusion through root mass, and a run of water-soluble nutrients such as silicon, manganese, and sulphate. Besides these, it leads to oxidative stress, which causes a decrease in chlorophyll synthesis, membrane deterioration, and protein degradation in plants (Hsiao 2000; Selvakumar et al. 2012; Sgherri et al. 2000; Rahdari et al. 2012).

14.2.3 Aminocyclopropane-1-Carboxylic Acid Deaminase (ACCD)

ACCD is the enzyme synthesized in the cytoplasm of bacteria. It is a multimeric sulfhydryl enzyme having a monomeric subunit with molecular weight of 35–42 KD (Glick et al. 2007). ACCD catalyses ACC conversion and produces α-ketoglutaric acid and ammonia (Fig. 14.4b). It was reported that D-serine and D-cysteine (D-amino acids) also act as a substrate for ACCD. Previously, the optimum temperature and pH for ACC deaminase were reported as 30–35 °C and 8.5 (Jacobson et al. 1994; Honma and Shimomura 1978; Jia et al. 1999). But currently, there is significant research going on to screen a versatile ACC deaminase producer who has a broad temperature and pH range (Xuguang et al. 2018). Various bacteria were reported for the production of ACCD, e.g., *Enterobacter cloacae*, *Pseudomonas putida*, *Pseudomonas* sp., *Alcaligenes*, *Hansenula*, *Rhizobium*, *Sinorhizobium* sp., *Pseudomonas chlororaphis*, *Rhizobium leguminosarum*, and *Bacillus subtilis* (Klee et al. 1991; Glick 1995; Belimov et al. 2007; Tittabutr et al. 2013; Ma et al. 2004; Duan et al. 2009). Similarly, some fungi and yeast were also reported for ACCD production, e.g., *Penicillium citrinum* (Minami et al. 1998; Jia et al. 1999).

Glick (1995) described the role and importance of some plant growth–enhancing *Rhizobacterium* in the management of drought pressure and various physiological activities of plants. Glick (1995) illustrated that ACC is produced in more quantity during drought stress and exudated outside of the root cells. The plant growth-inducing bacteria around the roots are recognized for its versatile activity and utilize the ACC exudate by ACC deaminase, and to keep the balance in internal and external ACC level, internal ACC is transported outside of the root. This process reduces the amount of ACC required for the biosynthesis of ethylene inside plant cells. Hence, if such ACCD-producing *Rhizobacterium* is present around the rhizospheric area of vegetation in a drought condition, ethylene production is suppressed, further leading to restrain inhibitory stress; ethylene causes defoliation, inhibition of root elongations, and nodulation transpiration (Glick et al. 2007). The presence of ACCD-producing microbes in soil proved their significance in a variety of plant growth–promoting activities, e.g., the existence of ACCD producer enhances the nitrogen fixations by inducing the normal process of root nodule organization in drought or temperature stress conditions.

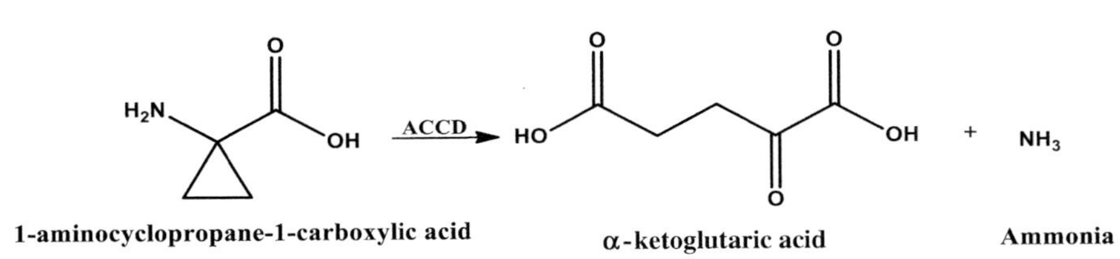

Fig. 14.4b Conversion of ACC to ethylene with ACCD

14.3 Application of Phytase Producer

14.3.1 Importance of Phosphorous

Phosphorous (P) is the next main macronutrient required for plant growth after nitrogen. It accounts for about 0.2% of dry weight of a plant. It makes vital biomolecules like nucleic acids, ATP, and phospholipids, and ultimately plant growth is inhibited without the supply of this nutrient. It also has a role in the regulation of the metabolic pathway and enzyme-catalyzed reactions. Phosphate affects germination and seed maturity and eventually plant development. Plant development comprising of root, stem, and stalk is dependent on phosphate. Phosphate has a role in the formation of seed and flower, which ultimately has an effect on crop development and yield (Khan et al. 2009). It has a remarkable function in N fixation in legumes, energy metabolism, membrane synthesis, photosynthesis, respiration, enzyme regulation, crop value, and abiotic and biotic stress resistance. No atmospheric source of phosphate could be made available to plants (Ezawa et al. 2002), and soils normally contain trace quantities of available phosphate (predominantly as HPO_4^{2-} and $H_2PO_4^-$) that is readily available for plant uptake. Phosphate addition in the soil in the form of fertilizers fulfills the plant requirement (Richardson et al. 2009). The unavailability of phosphate in soluble form is a vital factor (Xiao et al. 2011) that restricts the agricultural production worldwide (Ramaekers et al. 2010). Both organic and inorganic phosphate accumulate in soil and consequently not available for plant consumption. Inorganic phosphate is fused through chemical adsorption and precipitation, while immobilization of organic phosphate occurs in soil organic matter (Sharma et al. 2012).

Even phosphatic fertilizers fail due to their conversion to an insoluble form like calcium phosphate and aluminum phosphate (>70%) (Mittal et al. 2008). Phosphate is available in low quantity in soil (1.0 mg kg^{-1} soil); additionally, it becomes unavailable by reacting with reactive metals like Al^{3+} in acidic, calcareous, or normal soils (Gyaneshwar et al. 2002; Hao et al. 2002). Crop plants can, therefore, make use of only a little bit of phosphorus, which eventually results in reduced crop performance (Reddy et al. 2002). The high percentage of an insoluble type of phosphate leads to eutrophication, while frequent use of phosphate causes soil infertility and rapid depletion of nonrenewable phosphate reserves. The outcome of this event would be the lake's biological death i.e. cyanobacterial blooms, hypoxia, and death of aquatic animals due to depleted bioavailable oxygen and buildup of nitrous oxide. (Vats et al. 2005). In the plant, a range of morphological and physiological changes was observed due to deficiency in phosphate, which consecutively affects plant growth, productivity, and survival (Tran et al. 2010), and hence are a significant pin down for the agriculture industry worldwide.

Hence, effective phosphorous utilization is crucial for the sustainable expansion and prevention of undesirable environmental effects (Scholz et al. 2015). The translation of a phytate–phosphate compound in the soil in crop accessible orthophosphate would mitigate phosphate-related obstacles.

14.3.2 What Is Phytate?

Phytate is a significant storage compound of phosphorus in seeds. Eighty percent of the total seed phosphorus is made by phytate, which accounts for 1.5% of seed dry weight (Raboy and Dickinson 1987). The myo-inositol hexakisphosphate is a phosphate salt of myo-inositol having all six hydroxyl groups substituted by phosphate residues (Fig. 14.5). The myo-inositol 1,2,3,4,5,6-hexakis (dihydrogen) phosphate is commonly called myo-inositol hexakisphosphate, or phytate, which is a collection of the organic form of phosphorus compounds found widely in nature. The prefix "hexakis" designates that the phosphates are not internally connected and the compound is formed by a polydentate ligand, which binds with more than one metal atom coordination site. Each phosphate group is in ester form within an inositol ring and binds entirely with 12 protons (Bohn et al. 2008; Cao et al. 2007).

Phytate usually presents as a salt of monovalent and divalent cations (Fe^{2+}, Mn^{2+}, K^{+}, Mg^{2+}, and Ca^{2+}) and formed in seeds at the stage of ripening. In phytic acid, the negatively charged phosphate sturdily binds with positively charged metallic cations resulting in an insoluble complex and restricting the accessibility of nutrients. Phytic acid and its derivatives are accountable for various cellular events such as signaling, RNA export, endocytosis, DNA repair, and vesicular cell trafficking (Bohn et al. 2008; Frias et al. 2003). In plants, phytate is the prime storage type of inositol phosphate. The plant root has 30% phosphorus fractions, while seeds and cereal grains have 80% phosphorus (Lott et al. 2000; Turner et al. 2002; Haefner et al. 2005). Two pathways are considered for the biosynthesis of phytate: lipid-dependent and lipid-independent. The synthesis of phytic acid starts from myo-inositol via a series of phosphorylation steps. In the former route, phytate is attained by the successive phosphorylation of Ins(1,4,5)P3 (inositol 1,4,5-triphosphate) and Ins(1,3,4)P3 (inositol 1,3,4-triphosphate). The subsequent compound is released from PtdIns(4,5)P2 (phosphatidylinositol 4,5-biphosphate) by the effect of a specific phospholipase C. The intracellular location of the intermediates of phytic acid biosynthesis is not fully explored.

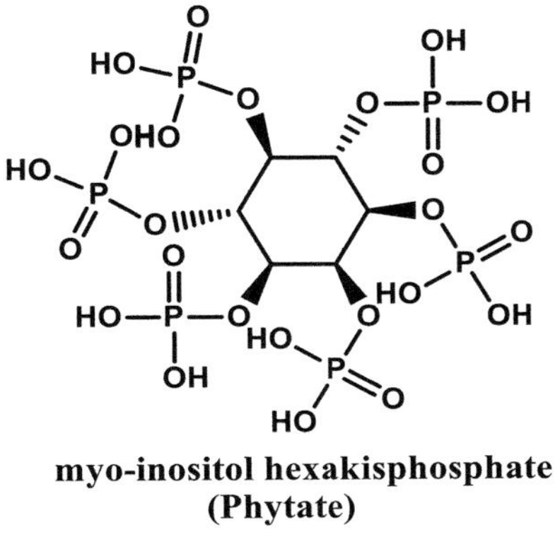

Fig. 14.5 Structure of phytate

myo-inositol hexakisphosphate (Phytate)

Organic phosphate in rhizosphere has a high affinity to soil particles by precipitation and adsorption and hence it creates deprived accessibility to the plant as it cannot be desorbed (Menezes-Blackburn et al. 2013). Phytic acid is degraded in seed germination by a precise assembly of enzymes called phytases.

14.3.3 Phytase Enzyme

Phosphorus deficiency results from the phytase secretion of a variety of plant roots (Minggang et al. 1997). The distinct phosphatases phytases (myo-inositol hexakisphosphate phosphohydrolase) sequentially hydrolyze the phosphomonoester bonds from phytic acid, thereby liberating lower inositol phosphates and inorganic phosphate (Singh et al. 2011). These catalysts commence phytic dephosphorylation at various positions on the inositol ring, and it produces diverse isomers of lower inositol phosphates (Turk et al. 2000).

14.3.4 Structure and Mechanism of Action of Phytase

Phytase (myo-inositol hexakisphosphate phosphohydrolase) is a homodimaeric enzyme (EC 3.1.3.26 and EC 3.1.3.8) (Hegeman and Grabau 2001; Guimarães et al. 2004). Phytases carried out the subsequent release of inorganic phosphorus from phytic acid. Phytases act hydrolytically to break the phosphate ester bond of phytate and release inositol phosphates and phosphorus with other essential nutrients, which are required for plant absorption (Angel et al. 2002) (Fig. 14.6). Phytases are involved in the dephosphorylation of inositol-6-phosphate and high-order inositol hexakisphosphate hydrolyze sequentially to form lower-order esters like inositol monoesters (Hayes et al. 1999; Vats and Banerjee 2004). The inositol penta- and hexakisphosphate (phytate) hydrolyzing enzymes are of interest because they constitute a high percentage of the whole organic phosphate (Turner et al. 2002).

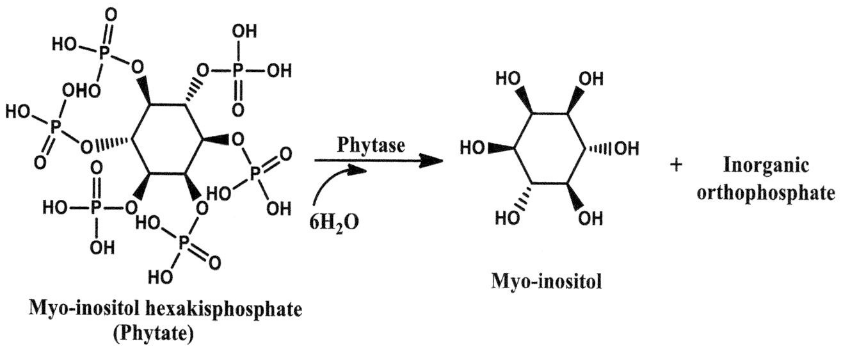

Fig. 14.6 Phytase action on phytate

The phytase protein has substrate binding and catalyzation conserved domains. The substrate binding domain is present at the N-terminal with RHGxRxP conserved sequence for substrate binding. The C-terminal catalyzation theme comprises of particular HD components. The "pocket" structure is framed by the connection of residues in the motif (Mullaney et al. 2000). The substrate restricting site with RHGxRxP arrangement responds with the substrate and frames the chemical substrate complex. The phosphate groups are then released from the substrate by the HD element (Li et al. 2010).

Phytate hydrolysis occurs in two stages: the nucleophilic attack and protonation. The histidine in the dynamic site of the catalyst caused a nucleophilic assault to the fragile phosphoester bond of phytate and caused the protonation by the aspartic acid of the leaving cluster (Li et al. 2010). The ß-propeller alkaline phytases lack the RHGXRXP sequence motif, and hence it needs calcium thermostability as well as enzyme activity to produce the IP3 (inositol triphosphate) (Kim et al. 1998a; Mullaney and Ullah 2003).

Phosphatases cause hydrolysis of 60% of the total organic phosphate. The highest quantity of phosphate was released by phytases from phytate (Bünemann 2008). The release of orthophosphate from soil natural phosphate is effective in microbes as well as in plants. Plant phytases have been distinguished in roots and root exudates during the early stage of seed germination; they frequently show a poor action, making them inefficient for hydrolyzing soil phytic acid as well as phosphorous usage (Hayes et al. 1999; Richardson et al. 2009) and thus suggest that the microbial catalyst demonstrates superior, effective liberation of phosphorous (Tarafdar et al. 2001).

14.3.5 Categorization of Phytases

Phytases are assembled by their enzyme action, pH action, and the initiation site of dephosphorylation of phytate. They are categorized into 3-phytases (EC 3.1.3.8), 5-phytases (EC 3.1.3.72), and 6-phytases (EC 3.1.3.26) on account of the initial hydrolysis position of phytate according to IUPAC-IUBMB (Bohn et al. 2008), which were subsequently alienated into alkaline and acid phytases (Jorquera et al. 2008). The three-dimensional structure and catalytic mechanism cause classification into four classes: histidine acid phytases (HAP) (EC 3.1.3.2), cysteine phytase or purple acid phosphatase (PAP) (EC 3.1.3.2), beta-propeller phytase (BPP) (EC 3.1.3.8), and protein tyrosine phosphatase (PTP)-like phytases (Li et al. 2010), which have recently been characterized (Lei et al. 2007). HAPs and BPPs are the most well-known and contemplated phytases. Various bacterial, fungal, and plant phytases have a place with the HAP family, while BPP has all the earmarks of being the prevalent phytase in *Bacillus* species (Greiner et al. 2007; Huang et al. 2009). These two most important categories have a different catalytic activity that results in distinct end products. While HAPs catalyze the hydrolysis of PA in myo-inositol and Pi, BPP activity results in the creation of the inositol-triphosphates – either Ins(1,3,5)P3 or Ins(2,4, 6)P3 (Greiner et al. 2007; Kerovuo et al. 2000).

As per the optimum pH, acid phytases, for the most part, incorporate HAP, PAP, and PTP-like phytases, though alkaline phytases include just BPPs from *Bacillus* species (Singh and Satyanarayana 2015; Tye et al. 2002). Alternatively, carbon position of dephosphorylation initiation resulted in phytases grouping into 3-phytase (myo-inositol hexakisphosphate 3-phosphohydrolase), 6-phytase (myo-inositol hexakisphosphate 6-phosphohydrolase), and 5-phytase (myo-inositol hexakisphosphate 5-phosphohydrolase).

The categorization of phytase into EC 3.1.3.8, EC 3.1.3.26, and EC 3.1.3.72 (myo-inositol-hexaphosphate phosphohydrolases) was organized on the background of protein sequencing, and successive dephosphorylation (George et al. 2007) of P occurs at three and six positions, correspondingly. The labeling basis is the three- and six-bond position of myo-inositol 6-phosphate. The 3-phytases (EC 3.1.3.8) are present in filamentous fungi like *Aspergillus* sp. and 6-phytases (EC 3.1.3.26) are found in plants, e.g., wheat.

14.3.6 Reserve of Phytase

Phytases can be formed by microorganisms, plants, and animals. Wheat, rice, soybeans, barley, peas, corn, and spinach are examples of plant sources. Microorganisms like bacteria, fungi, and yeast are the real source of phytase found in the blood of vertebrates such as fish and reptiles (Gupta et al. 2015; Bohn et al. 2008). Among the phytases from microorganisms, attention is focused on *Aspergillus* sp. because of its high production and extracellular activity (Gupta et al. 2015). To circumvent this obstacle the sole strategy is the application of phytases which hydrolyze the phytate and increase availability of P to plants. Commercially available phytase addition is costly and time-consuming, and hence the maintenance of rhizospheric phytase producer is important. Another engineering approach involves incorporation of genes behind phytase production from microbes into transgenic plants. However, there is a range of constraints for phytase engineered crop plants like loss of seed viability, yield, vulnerability for ecological pressure, and rejection of genetically modified organisms (GMOs) (Reddy et al. 2017).

14.3.7 Microorganisms Producing Phytase

Phytases of microbial origins are of rigorous significance among plants, animals, and microorganisms owing to the ease of genetic manipulation and large-scale production (Adhya et al. 2015). Microorganisms are the key drivers in the soil, which regulates phytate mineralization. The occurrence of microorganisms in soil rhizosphere may balance plants inability to procure P directly from phytate. In microorganisms, bacteria, yeast, and fungi have been effectively researched for extracellular phytase action (Pandey et al. 2001). A single phytase cannot address the issues of business and ecological applications (Bakthavatchalu et al. 2013). Microbial phytases are investigated mainly from fungi of a filamentous type such as *Aspergillus ficuum* (Gibson 1987), *Mucor piriformis* (Howson and Davis 1983), *Aspergillus fumigatus* (Pasamontes et al. 1997), *Cladosporium* sp. (Quan et al. 2004), and *Rhizopus oligosporus* (Casey and Walsh 2004). Phytase production by different bacteria has been described, viz., *Bacillus* sp. (Kim et al. 1998b; Choi et al. 2001), *Citrobacter braakii* (Kim et al. 2003), *Pseudomonas* sp. (Richardson & Hadobas 1997), *Escherichia coli* (Greiner et al. 1993), *Raoultella* sp. (Sajidan et al. 2004), and *Enterobacter* (Yoon et al. 1996). The anaerobic rumen bacteria, mainly *Selenomonas ruminantium*, *Prevotella* sp., *Megasphaera elsdenii*, and *Mitsuokella multiacidus* (Richardson et al. 2001b) and *Mitsuokella jalaludinii* (Lan et al. 2002), have also been investigated for phytases. The γ-proteobacteria group possesses the phytase production potential among the majority of soil bacteria. Fungi have extracellular phytases, while bacteria produce cell-linked phytases. *Bacillus* (Choi et al. 2001; Kerovuo et al. 1998; Kim et al. 1998a; Powar and Jagannathan 1982; Shimizu 1992) and *Enterobacter* (Yoon et al. 1996) are the only bacterial genera having extracellular phytase activity. The phytase activity of *Selenomonas ruminantium* and *Mitsuokella multiacidus* (D'Silva et al. 2000) is outer membrane linked, while *Escherichia coli* produces the periplasmic phytase enzyme (Greiner et al. 1993).

B. subtilis is as a competent of phytase producer owing to its nonpathogenic and safe nature for industrial-level phytase production. This microorganism has numerous additional advantageous properties like organic acid production and antibiosis for phosphate solubilization in the soil. Currently, *Aspergillus* and *E. coli* are the commercial phytase producers. Among the various organisms reported, the inhabitant *E. coli* enzyme demonstrates the maximum phytase activity.

Phytases from bacterial sources are a genuine option in contrast to fungal enzymes because of their specificity to the substrate, protection from proteolysis, and effective catalytic action (Konietzny and Greiner 2004). *Bacillus* phytases are exceptionally effective due to its higher thermal stability and neutral pH. The *Bacillus* phytase has stringent specificity for a substrate for the calcium–phytate complex effective for application in the environment (Farhat et al. 2008; Fu et al. 2008). Nevertheless, owing to inefficient enzyme production methods for *Bacillus* sp., it could not be produced at commercial scale as only a few strains have been significantly commercialized for phytase production (Zamudio et al. 2001). *Lactobacillus sanfranciscensis* is the main sourdough lactic acid bacteria that demonstrated a significant level of phytate degrading action (De Angelis et al. 2003). The HAP are specifically produced from *Aspergillus* sp. like *A. terreus*, *A. ficuum*, and *A. niger* (Wyss et al. 1999), while the alkaline phytases are produced from *Bacillus amyloliquefaciens* (Idriss et al. 2002) and *Bacillus subtilis* (Kerovuo et al. 2000). Escobin-Mopera et al. (2012) had purified phytase from *Klebsiella pneumoniae* 9–3B. Rhizobacteria can mineralize phytate and may enhance P uptake of plants in soils (Patel et al. 2010). A better and substitute resource of phytase is continuously searched by screening new organisms that may produce novel and effective phytases. The ultimate aim is to produce phytase cost-effectively with optimized conditions for industrial application.

14.3.8 Why Do Bacteria Produce Phytase?

Bacterial phytase production is an inducible complex regulatory mechanism. Phytase synthesis control is different in various bacteria. Phytase production is not a condition for balanced bacterial growth, but it is the response to an energy or nutrient constraint. Phytase formation takes place when bacterial cells face environmental variations prior to the commencement of growth or when actively growing culture faces a stressful condition. The metabolic regulation by signal transduction is also a mechanistic role (Zamudio et al. 2002).

14.3.9 Parameters Affecting the Activity of Phytases

The soil environment presents extreme difficulties like denaturation, degradation, adsorption, and dilution to extracellular chemicals (Wallenstein and Burns 2011). The constancy of extracellular and intracellular enzymes is variable. Stability is portrayed more in extracellular than intracellular proteins and is credited by glycosylating disulfide bonds that alter thermal soundness, an expansive pH scope of action, and some protection from proteases. Some are stabilized by binding with humic substances and clay minerals (Quiquampoix and Burns 2007). Biological and physicochemical procedures influence phytase action. The former causes changes in enzyme creation rates leading to isoenzyme generation and changes in microbial network synthesis, while the latter causes changes in absorption desorption responses, substrate dissemination rates, and enzyme degradation rates (Wallenstein et al. 2009). Essential elements influence the action of enzyme include the amount and kind of substrate (Fitriatin et al. 2008), type of solvent, pH, temperature, the existence of an inhibitor and activator, the quantity of the enzyme, and the reaction product (Sarapatka 2002).

14.3.9.1 Effect of Substrate on Phytase Action

Phytase action shifts with various substrates. The different substrates include 1-naphthyl phosphate, 2-glycerolphosphate, glucose-6-phosphate (Escobin-Mopera et al. 2012), 2-glycerolphosphate, fructose-6-phosphate, calcium phytate, sodium phytate, p-nitrophenyl phosphate, ß-glycerol phosphate, adenosine-5′-monophosphate (AMP), guanosine-5′-triphosphate (GTP), adenosine-5′-diphosphate (ADP), adenosine-5′-triphosphate (ATP), and nicotinamide adenine dinucleotide phosphate (NADP) (Farouk et al. 2012; Bakthavatchalu et al. 2013). Phytases are categorized as substrate particular and nonparticular acid phosphatases (Rossolini et al. 1998; Rodríguez and Fraga 1999).

14.3.9.2 Effect of pH on Phytase Action

The activity of phytases relies on the pH and temperature. Plant phytases have less pH and thermal stability than microbial phytases. The optimum pH for phytase activity is 5.0–8.0, hence classified as acid or alkaline phytases, respectively (Konietzny and Greiner 2002). The optimum pH for fungal phytases is 4.5–6.5 with

80% activity; for example, *Rhizoctonia* sp. and *F. verticillioides* have an optimum pH of 4.0 and 5.0, respectively (Marlida et al. 2010). The optimum pH for bacterial phytases is 6.0–8.0 (Kerovuo et al. 1998; Kim et al. 1998a). Acidic phytases have an optimum pH range from 4.5 to 6.0 (Konietzny and Greiner 2002), and pH 8.0 is the optimum for alkaline phytases in legume seeds (Scott 1991), lily pollen (Baldi et al. 1988), and cattail (Kara et al. 1985; Scott 1991).

14.3.9.3 Effect of Temperature on Phytase Action

Temperature is the most indispensable factor of enzyme action, influencing both enzyme generation and degradation rates by microorganism. The ideal temperature of phytate-degrading enzyme fluctuates from 35 to 77 °C. Predominantly plant phytases have the greatest action at lower temperature compared to microbial phytases (Konietzny and Greiner 2002). The ideal temperature for plant phytases ranges from 45 to 60 °C (Johnson et al. 2010). In general, metabolic rate of enzyme producing life forms increases with temperature over the range 5–40 °C. In this way, temperature supposes a more vital job in the rate of extracellular enzyme activity when contrasted with enzyme kinetics itself.

14.3.9.4 Effect of Soil Type on Phytase Action

The action of phytase in soil is additionally influenced by physicochemical properties of the soil, which incorporates soil compose, organic matter content, nitrogen content, C/N proportion, and aggregate P content (Djordjevic et al. 2003). The soil performance of phytase fluctuates with soil compose, and the movement of phytase lost expeditiously is dependent on three differentiating soil nature. The initial fate of phytase is confined by adsorption in the soil. The degradation and magnitude of phytase adsorbed continue as before for a wide range of soil arrangements. The highest adsorption was recorded at low pH, and it becomes nearly equivalent to zero when pH is adjusted to 7.5. The adsorption bestows defense to phytase degradation in the soil, but also limits loss of enzyme activity in the adsorbed state.

14.3.10 Mechanism of Phytase Activity

Microorganisms can enhance the capacity of a plant to acquire P through various mechanisms, and the important one is phytase like enzyme production (Richardson and Simpson 2011). The purified crystalline form of phytase has different catalytic properties with specific diverse mechanisms. The principal action of all portrayed phytases depends on the enzymatic hydrolysis of the bonds among inositol and phosphoric acid deposits. Enzymatic hydrolysis of bonds happens among inositol and phosphoric acid deposits whereupon the component of activity of all phytases is based. The results of this arrangement of responses are six-fold alcohol and phosphates (Mukhametzyanova et al. 2012). Microbial phytases decay fresh plant build-ups in the soil prompting the release of phosphorus from organic compounds. There are various arrangements alongside differing rates of responses by which the phosphoric acid deposits are discharged through microbial hydrolysis of phytate

(Mukhametzyanova et al. 2012). The histidine acidic phytases catalyze the release of phosphates in neighboring free hydroxyl group, after the dephosphorylation of a first phosphate group. For the most part, plant phytases display a difference in transitional myo-inositol pentaphosphate development among the first phase of the response. In the course of the first venture of hydrolysis, microbial 6-phytases frame a different set of intermediates. The acid phosphatases with phytate hydrolyzing properties hydrolyze glucose-1-phosphate in *Enterobacteriaceae* (Greiner and Sajidan 2008). Alkaline phosphatases in lily pollen, *B. subtilis*, and reed mace formed myo-inositol triphosphates as end products (Greiner et al. 2007; Greiner and Sajidan 2008; Mukhametzyanova et al. 2012).

14.3.11 Importance of Microbes for Phosphorous Mobility with Phytase

Soil microorganisms, particularly the higher plant rhizosphere, are exceptionally powerful in discharging P from natural pools of aggregate soil P by mineralization and inorganic complexes through solubilization (Hayat et al. 2010).

Mineralization results from the transformation of organic P, for example, phytate to plant-accessible inorganic P, by microorganisms through their expressed enzyme phytase (Ariza et al. 2013). Phytases have been recognized in roots and root exudates in plants (Li et al. 1997; Hayes et al. 2000; Richardson et al. 2000). Despite the fact that it is accounted for the enzymatic action in root exudates, it is not sufficient for efficient use of natural phosphorous (Brinch-Pedersen et al. 2002; Richardson et al. 2000). The addition of exogenous phytase into the media resulted in phytate availability for plant growth (Hayes et al. 2000; Idriss et al. 2002; Unno et al. 2005). The addition of exogenous phytase (Idriss et al. 2002; Richardson et al. 2001b; Singh and Satyanarayana 2010; Hayes et al.2000) or expression of phytase gene of microbial origin in plant (Richardson et al. 2001a; Li et al. 2007a, b, 2009) resulted in growth of plant with phytate as solitary source of phosphate. The current research is targeted on the genetic expression of phytase genes in the plant for organic P utilization from the soil. The graphic demonstration of the function of microorganisms in phosphate solubilization is described in Fig. 14.7.

The action of plant phytases comprises just a little extent of the aggregate phosphatase reaction and is viewed as insufficient for guaranteeing adequate phosphate securing (Richardson et al. 2000; Findenegg and Nelemans 1993; Hayes et al. 2000). Bacterial phytases are effective for growth and yield of the plant. The limitation of plants to extort P from soil phytate could be overcome by treatment with phytate-degrading bacteria, like biofertilizer. Microbial phytase plays a very important role for the availability and mobility of phosphorous in soil because of its agronomic and ecological value for the growth of the plant as suggested by the recent scientific research. The long-term phosphorous deprivation in plants could be met by phytase from microorganisms; hence, the use of microbial phytase on an industrial scale is very appealing nowadays (Jorquera et al. 2008). The fungal extracellular phytase-treated seeds support the plant phosphorus nutrition in high phytate

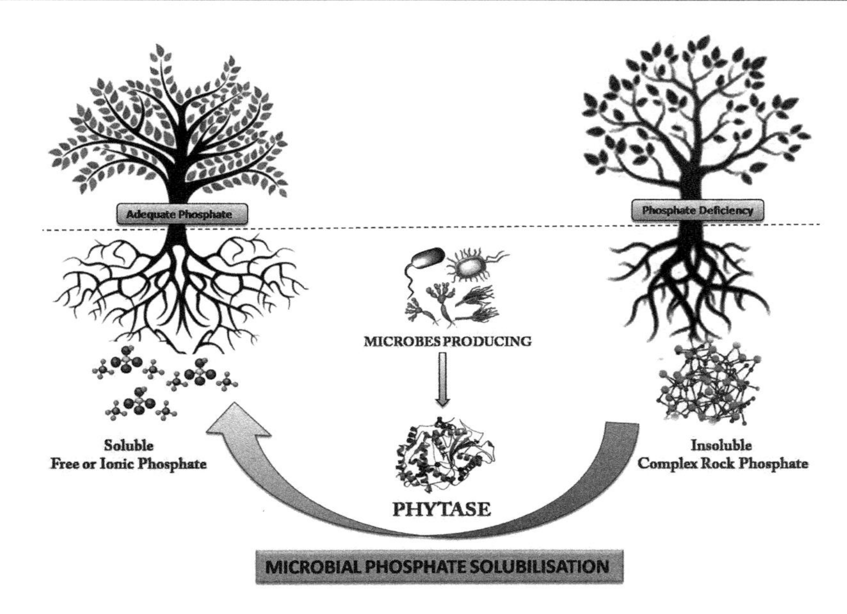

Fig. 14.7 Role of phytase from microorganisms in phosphate solubilization

content soil (Tarafdar 1995). The enrichment of soil with phytase from bacteria like *B. amyloliquefaciens* and *Bacillus mucilaginosus* advances the development of corn and tobacco, respectively (Li et al. 2007a, b; Idriss et al. 2002). Phytases from bacteria also release the vital soil micronutrients by phytate chelation and make it available to the plant. The purified microbial phytase or phytase-producing microbial strains could be functional as an effective and eco-friendly way to increase bioavailable soil phosphorus and limit the wide utilization of inorganic phosphate fertilizers.

14.3.11.1 Transgenic Plants for Phytase

Gene for phytase from a microorganism is integrated into plants like tobacco with a phyA gene from *A. niger* constituting phytase as soluble proteins in tobacco seeds. Genetically modified plants produce extracellular phytase from roots, which showed significant improvement in P nutrition in the soil, with higher phytate content or artificially modified for phytate (George et al. 2004, 2005). Thus the phytase from a microorganism is the critical element, and their existence in the rhizosphere helps the plant to recover from its inability to use the unavailable phytate.

Phytases have developed to be a valuable key to supportable agribusiness. It gives an approach to stop the revenue costs that turn out to be superfluously high because of the expansion of phosphorus manures. Broad research on phytase utilizing biotechnological applications will unquestionably give efficient arrangements towards practical agribusiness and ecological insurance in the coming years.

14.4 Use of Bacterivorous Microbes from Soil

14.4.1 Bacterivorous Protozoan

It was an accepted truth that soil microbes provide essential functions supporting soil fruitfulness and plant well-being. Recent evolution in molecular techniques like molecular sequencing resulted in a boom in studies of various microflora like an insect, animal gut, lakes, ponds, and terrestrial flora. However, all these studies cover bacteria and fungi only and neglect other trophic levels. But most attempts to use these bacteria and fungi as bioinputs in natural soil have been reported unsuccessful.

For the past 50 years the terms "biofertilizer" and "PGPR bacteria" only represent nitrogen fixer and phosphate and growth hormone producer. However, the truth is there is still no confirmation that these added bioinputs sustain soil fertility. The accepted truth is that these fungal and bacterial bioinputs have significant selective pressures of predation and not resource availability. These predators are bacterivorous and fungivorous protist. Protists massively consume bacteria as well as other soil microbes like fungi and yeast, and unicellular algae and release various micronutrients, growth-promoting substances, and different assimilable nitrogenous compounds and mineral (Ekelund and Rønn 1994).

Although various soil protozoans and nematodes are reported for their bacterivorous role, very few reports exist discussing the function of protozoans in the development of crop plant or soil richness (Bonkowski and Brandt 2002; Bonkowski 2004). The size of most soil protozoan ranges from 10 to 100 μm in diameter, but their weight is negligible. It was assumed that the biomass of total protozoan in soil is equal to the biomass of all other clusters of soil animals together except earthworm (Schaefer and Schauermann 1990; Schröteret al. 2003). In the biological energy coordination, the soil organic cycle plays an important role, which involves anabolic and catabolic steps of energy investment and energy escape or lost. Protozoans are major engineers which motion this organic energy cycle in the soil. Protozoa drive this cycle continuously where there is sufficient water available like moisture-containing intersoil capillaries, pore spaces, and fissures. Besides these, protozoans account for significant respiration of soil. It was noted that they contribute to 15–70% of the entire soil respiration. These indicate that protozoans are a vital component of the soil. The soil protozoans majorly include ciliates, flagellates, and naked and testate amoebae (Fig. 14.3). Although these protozoans have an extensive array of food assimilation and enzyme syntheses like a higher animal, they are not capable of synthesizing some vitamins and cofactors, and hence they depend on some microbial population for it.

Ciliates are one of the group including protozoan, which are identified for its extraordinary bacterivorous capacity (Sherr et al. 1987); owing to their large size. Algae, fungi, and small animals are foods for these ciliates (Bernard and Rassoulzadegan 1990; First et al. 2012). They have various habitats like freely swimming in the water, crawling on surfaces, and physically attached to surfaces by very flexible spring-like stalk, e.g., *Paramecium*, *Euplotes*, and *Vorticella* (James and Hall 1995). There are some ciliates, which have special cilia for swimming and hairs for predation known as membranelle, which help for catching massive bacteria or prey in food vacuoles. Ciliate feeding rates are very high; it was recorded that single ciliates can digest 1254 bacteria h^{-1} (Iriberri et al. 1995).

Flagellates are another member of protozoans bearing one or more flagella having a different size from 2 to 20 μm. They are versatile in nature like swimming freely or attaching to solid surfaces by trailing flagellum or stalks. Flagellates using these flagella either create feeding current or exploit it to put the water and prey in the oral furrow and at the base of the flagellum where the pseudopodia ingest the prey. Flagellates show selective grazing as per their size. They prefer smaller-size organisms as significant prey. It was reported that bacteria are more susceptible to flagellate grazing than other microbes having size >2.4 μm. Chrzanowski and Šimek (1990) reported that flagellate bacterial grazing rate varies from 2 to 300 bacteria h^{-1} (Davis and Sieburth 1984; Eccleston-Parry and Leadbeater 1994a).

Amoebas are widely occurring protozoans and are very normal in water, soil, and other habitats. They are abundant in the soil, i.e., 103–107 g^{-1} of dry soil, with varying size <10 μm. Amoebas play a very important function in the cycling of various minerals and minute supplements such as nitrogen and phosphorus, particularly in shallow levels of nutrient environments (Goldman et al. 1985; Eccleston-Parry and Leadbeater 1994b). Amoebae, ciliates, and flagellates together selectively nurture on bacteria and control bacterial soil population (Table 14.1). They act as an essential constituent of the "microbial loop" (Azam et al. 1983). They are well recognized as Rhizopoda amoebae because they use their cytoplasmic protrusions, i.e., pseudopodia, for locomotion and nourishment. Amoebae are of two types, naked amoebae and shelled amoebae (testate amoebae).

Naked amoebae have no perfect shape but show three major morphological forms, i.e., floating, active form with extended lobose; fan-shaped, slug-like pseudopodial form trophozoites; and smaller and dormant form called cyst, an unusual rounded form (Page 1988; Griffiths 1970). Typical examples of naked amoeba are *Amoeba*, *Acanthamoeba*, *Vannella*, and *Vampyrella.*

Testate amoebae secrete the siliceous shell around the body. These testate are species-specific architectures. The testate shell amoebae designate the nutritional category of the living environment. The aperture is at one side of a shell, which is used for feeding or catching of different preys (Jassey et al. 2012). The dominant victims of amoebae are bacteria; the intake rate of the amoebic cell was reported to be 0.2–1465 bacteria h^{-1} (Heaton et al. 2001; Huws et al. 2005).

14.4.2 Role of Protozoans as New Bioinputs

Various studies indicated that protozoans majorly preyed upon bacteria. Bacteria, unicellular fungi, yeast, algae, and cyanobacteria were assumed as a nutritional capsule. In addition to nitrogen and carbon sources, these nutritional capsules are enriched with micro- and macronutrients in addition to various growth factors (Table 14.2). It was formerly confirmed that the nitrogen and carbon content of a

Table 14.1 Bacterivorous capacity of various protozoans

Types	Example	Bacterivorous capacity (bacterial cell h^{-1})	References
Amoeba			
Naked	*Saccamoeba*	0.2–1465	Heaton et al. (2001) and Huws et al. (2005)
	Acanthamoeba		
	Euglypha cristata		
	Hartmannella		
	Cf. *Mayorella*		
	Cf. *Polychaos*		
	Vannella		
	Vampyrella		
Shelled	*Arcellinid testate*		
	Euglypha cristata		
	Arcella gibbosa		
	Difflugia		
	Foraminifera		
	Nebela		
Flagellates	*Giardia intestinalis*	2–300	Davis and Sieburth (1984) and Eccleston-Parry and Leadbeater (1994a)
	Peltomonas hanelisp. nov.		
	Apusomonas australiensis sp.		
	Cetcomonar crassicauda		
Ciliates	*Paramecium*	20–1254	Iriberri et al. (1995)
	Vorticella		
	Balantidium coli		
	Oxytricha trifallax		
	Stentor roeselii		

Table 14.2 Elemental composition of bacteria and fungi

Element	Bacteria (% dry weight)	Fungi (% dry weight)
Carbon	50–53	40–63
Hydrogen	7	–
Nitrogen	12–15	7–10
Phosphorus	2.0–3.0	0.4–4.5
Sulphur	0.2–1.0	0.1–0.5
Potassium	1.0–4.5	0.2–2.5
Sodium	0.5–1.0	0.02–0.5
Calcium	0.01–1.1	0.1–1.4
Magnesium	0.1–0.5	0.1–0.5
Chloride	0.5	–
Iron	0.02–0.2	0.1–0.2
References	Luria (1960)	Lilly (1965)
	Aiba et al. (1973)	Aiba et al. (1973)
	Herbert (1956)	

fungal and bacterial cell are 10–15% and 50–63% by dry weight of fungi and bacteria, respectively. Similarly, bacterial and fungal mass sufficiently contain valuable micronutrients such as phosphate, potassium, sulphur, calcium, and iron (Luria 1960; Herbert 1956; Aiba et al. 1973). All protozoans are well characterized for their enormous feeding habits on other microbes such as bacteria and other microbes. Different soil bacterial flora assimilated the atmospheric nitrogen with organic and inorganic matters from the soil and locked in their cells, which are not freely accessible for the plants. The enormous grazing activity remobilized this immobilized nitrogen and released ammonia, which is ultimately utilized by the plant (Goldman and Caron 1985). Griffith and Bardget (1997) proved that the nitrogen requirement of protozoans is comparatively less, and they make about 60% of ingested nitrogen available to plants in the form of ammonia. Hence after the ingestion of bacteria by a protozoan, nitrogen is not only released but also various nutrients like 50–63% carbon, 2.0–4.5% phosphorus, and 0.02–0.5% iron (Table 14.3). Bonkowski (2004) reported the essential function of protozoa in sustaining soil productiveness and plant health.

Protozoa provide all essential nutrients by mineralizing complex material in bacteria during feeding. They also control the structure and activity of bacterial loops of soil and root-associated communities (Sieburth and Davis 1982; Bonkowski and Brandt 2002). Krome et al. (2010) reported that selective predation of bacteria promotes the production of various plant growth hormones. Besides offering different mineralized nutrients, it was proved that protozoans also increased the nutrient assimilation rate by altering the root morphology. Bonkowski and Brandt (2002) reported that when the *Acanthamoeba castellanii* was inoculated in the rhizosphere, it induces the extensive fibrous and fine root, suggesting that protozoans play an important role like plant growth hormones (Krome et al. 2010). Jousset et al. (2010) also proved that protozoans not only stimulate growth but also play a noteworthy function in pathogen suppressions by encouraging other bacterial soil flora for antibiotics like chemicals. Similarly, it induces iron chelating organic molecule production, which makes iron unavailable for plant pathogen growth and multiplication (Levrat 1989; Mazzola et al. 2009; Müller et al. 2013; Mellano et al. 1970).

Nielsen et al. (2002) proved that bacteria such as *Pseudomonas* and *Bacillus* produce various antipathogenic compounds such as phenazines, DAPG (diacetyl phloroglucinol), and cyclic lipopeptides like tensin, amphisin, and viscosinamide, but Mazzola et al. (2009), Jousset and Bonkowski (2010), and Weidner et al. (2017) revealed that protozoan grazing pressure induced the making of such antipathogenic

Table 14.3 Performance of protozoans for phosphatases, ACCD, and tryptophan

Sr. no.	Bacterivorous organism	Phosphatase (IU/h)	ACC deaminase activity (μM of α-ketoglutarate/mg/h)	Tryptophan (μg/h)
1	*Acanthamoeba* sp.	16.20	0.161	15
2	*Paramecium* sp.	18.40	0.093	17
3	*Amoeba* sp.	11.20	0.218	11
4	*Tetrahymena* sp.	14.00	0.187	07

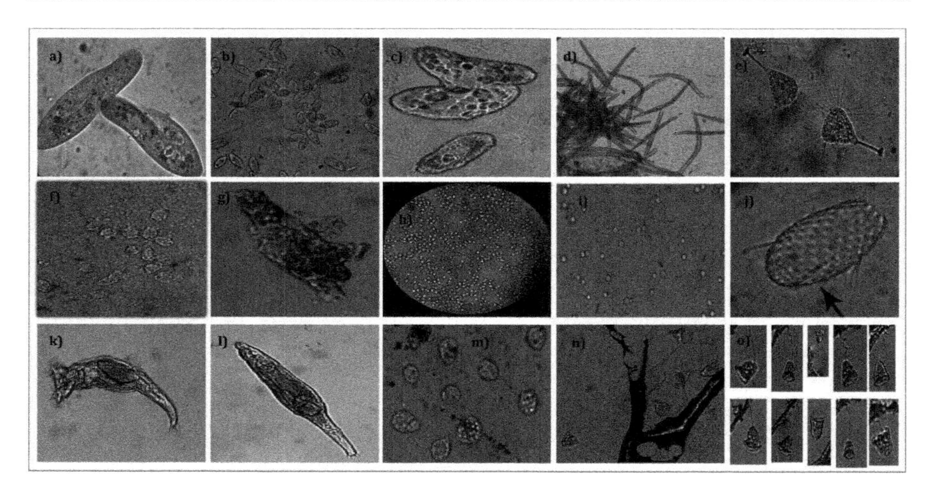

Fig. 14.8 Bacterivorous animals of soil cultured at School of Life Sciences, KBC NMU laboratory (**a–c**) *Paramecium* sp., (**d**) *Spirostomum* sp., (**e**) *Suctoria* sp., (**f, g**) *Acanthamoeba* sp., (**h, i**) cyst of amoebae, (**j**) testate amoebae, (**k, l**) Rotifer, (**m**) *Actinosphaerium* sp., (**n, o**) *Vorticella* sp.

fungal and bacterial compound. Recently in our laboratory studies at KBC North Maharashtra University (KBC NMU), Jalgaon, we have isolated and cultured various important agricultural bacterivorous animals, viz., *Paramecium*, *Amoeba*, Rotifer, and Vorticella (Fig. 14.8). It was revealed that *Acanthamoeba castellanii*, *Paramecium caudatum*, *Spirostomum*, and *Amoeba* spp. have the potential to produce various enzymes like phytase, phosphatase, and ACC deaminase. All these enzymes previously assumed the essential character of plant growth–promoting bacteria (Zahir et al. 2004). In laboratory-grown culture studies, it was discovered that *Paramecium* and *Acanthamoeba* efficiently utilized ACC and phytate and phosphate. Similarly, *Suctoria* sp. and *Spirostomum* were also investigated to use phosphate, phytic acid, and ACC like substrate at low concentrations (Table 14.3). *Amoeba* sp., *Acanthamoeba*, and *Paramecium* sp. were also found to be the producer of metabolic products such as amino acids like tryptophan, which was previously reported for a vital role in the stimulation of auxin production (Krome et al. 2010).

Sayre (1973) reported the potential of *Amoebae* as a future potent nematicidal agent. At KBC NMU laboratory, the cultured *Amoebae* sp. was also established to have an extraordinary potential of controlling invasive plant nematodes. Nematodes are the root-knot disease-causing agents of tomato and brinjal, i.e., *Meloidogyne incognita* and *Meloidogyne javanica*. It was observed that amoeba had 50–65 egg ingestion rate per amoeba per 24 h of both *Meloidogyne incognita* and *Meloidogyne javanica* and the 10–20 juvenile and 6–7 adult nematode ingestion per amoeba in 24 h.

14.5 Conclusion

Currently, nitrogen fixers, phosphate solubilizers, mycorrhiza, and biocontrolling agents like *Trichoderma* sp. are the most popular bioinputs throughout the world, even though it is necessary to recommend the utilization of other microbial bioinputs like ACCD, phytase producing microorganisms, Zn, K, S mobilizers. Besides that, latest studies proved the extraordinary potential of protozoa as the real new age bioinput, which proved their beneficial power for plant growth development, soil fertility augmentation, and biocontrol of soilborne pathogen. Recent advances in protozoans as bioinput will open a new avenue for plant–microorganism interaction research to solve current agricultural problems. The microbes present in the soil employ different strategies, and these beneficial belowground microbial interventions influence the plant beneficially. The character of these new age agricultural bioinputs is noteworthy for soil and plant well-being through nutrient fixation, solubilization, mineralization, and mobilization that are eventually accountable in the agroecological perspective. Such modern biological inputs in agriculture will help to achieve the future food demand of a growing world population and address the global problem of food security and malnutrition. So there is much more to do with nature's gift microorganisms which have tremendous metabolic flexibility and potential functionality.

Acknowledgment The corresponding author, SVP, is kindly acknowledging the Department of Biotechnology, New Delhi, for the Indo-US Foldscope Major Research Project grant (Grant No. BT/IN/Indo-US/Foldscope/39/2015).

References

Adhya TK, Kumar N, Reddy G et al (2015) Microbial mobilization of soil phosphorus and sustainable P management in agricultural soils. Curr Sci 108(7):1280–1287

Aiba S, Humphrey AE, Millis NF (1973) Scale-up. In: Biochemical engineering, 2nd edn. Academic, New York, pp 195–217

Angel R, Tamim NM, Applegate TJ, Dhandu AS, Ellestad LE (2002) Phytic acid chemistry: influence on phytin-phosphorus availability and phytase efficacy. J Appl Poult Res 11:471–480

Ariza A, Moroz OV, Blagova EV et al (2013) Degradation of phytate by the 6-phytase from *Hafnia alvei*: a combined structural and solution study. PLoS One 8(5):e65062

Azam F, Fenchel T, Field JG (1983) The ecological role of water-column microbes in the sea. Mar Ecol Prog Ser 20:257–263

Bakthavatchalu S, Thiam B, Lokanath CK (2013) Partial purification and characterization of phytases from newly isolated *Pseudomonas aeruginosa*. Asiat J Biotechnol Resour 4:7–12

Baldi BG, Scott JJ, Everard JD et al (1988) Localization of constitutive phytases in lily pollen and properties of the pH 8 form. Plant Sci 56:137–147

Belimov AA, Dodd IC, Safronova VI (2007) *Pseudomonas brassicacearum* strain Am3 containing 1-aminocyclopropane-1-carboxylate deaminase can show both pathogenic and growth-promoting properties in its interaction with tomato. J Exp Bot 24:1–11

Bernard C, Rassoulzadegan F (1990) Bacteria or microflagellates as a major food source for marine ciliates: possible implications for the microzooplankton. Mar Ecol Prog Ser 64(1):147–155

Bohn L, Meyer AS, Rasmussen SK (2008) Phytate: impact on environment and human nutrition, a challenge for molecular breeding. J Zhejiang Univ Sci B 9:165–191
Bonkowski M (2004) Protozoa and plant growth: the microbial loop in soil revisited. New Phytol 162(3):617–631
Bonkowski M, Brandt F (2002) Do soil protozoa enhance plant growth by hormonal effects? Soil Biol Biochem 34(11):1709–1715
Brinch-Pedersen H, Sørensen LD, Holm PB (2002) Engineering crop plants: getting a handle on phosphate. Trends Plant Sci 7:118–125
Bünemann EK (2008) Enzyme additions as a tool to assess the potential bioavailability of organically bound nutrients. Soil Biol Biochem 40:2116–2129
Cao L, Wang L, Yang W et al (2007) Application of microbial phytase in fish feed. Enzyme Microb Technol 40:497–507
Casey A, Walsh G (2004) Identification and characterization of a phytase of potential commercial interest. J Biotechnol 110:313–322
Choi YM, Suh HJ, Kim JM (2001) Purification and properties of extracellular phytase from Bacillus sp. KHU-10. J Protein Chem 20:287–292
Chrzanowski TH, Šimek K (1990) Prey-size selection by freshwater flagellated protozoa. Limnol Oceanogr 35(7):1429–136s
D'Silva CG, Bae HD, Yanke LJ et al (2000) Localization of phytase in *Selenomonas ruminantium* and *Mitsuokella multiacidus* by transmission electron microscopy. Can J Microbiol 46:391–395
Davis PG, Sieburth JM (1984) Estuarine and oceanic microflagellate predation of actively growing bacteria: estimation by frequency of dividing-divided bacteria. Mar Ecol Prog Ser 19(3):237–246
De Angelis M, Gallo G, Corbo MR et al (2003) Phytase activity in sourdough lactic acid bacteria: purification and characterization of a phytase from *Lactobacillus sanfranciscensis* CB1. Int J Food Microbiol 87:259–270
Djordjevic S, Djukic D, Govedarica M et al (2003) Effects of chemical and physical soil properties on activity phosphomonoesterase. Acta Agric Serbica 8:3–10
Duan J, Müller KM, Charles TC (2009) 1-aminocyclopropane-1-carboxylate (ACC) deaminase genes in rhizobia from southern Saskatchewan. Microbial Ecol 57:423–436
Eccleston-Parry JD, Leadbeater BS (1994a) A comparison of the growth kinetics of six marine heterotrophic nanoflagellates fed with one bacterial species. Mar Ecol Prog Ser 105:167–177
Eccleston-Parry JD, Leadbeater BS (1994b) The effect of long-term low bacterial density on the growth kinetics of three marine heterotrophic nanoflagellates. J Exp Mar Biol Ecol 177:219–233
Ekelund F, Rønn R (1994) Notes on protozoa in agricultural soil with emphasis on heterotrophic flagellates and naked amoebae and their ecology. FEMS Microbiol Rev 15(4):321–353
Escobin-Mopera L, Ohtani M, Sekiguchi S et al (2012) Purification and characterization of phytase from *Klebsiella pneumoniae* 9-3B. J Biosci Bioeng 113:562–567
Ezawa T, Smith SE, Smith FA (2002) P metabolism and transport in AM fungi. Plant Soil 244(1–2):221–230
Farhat A, Chouayekh H, Farhatben M et al (2008) Gene cloning and characterization of a thermostable phytase from *Bacillus subtilis* US417 and assessment of its potential as a feed additive in comparison with a commercial enzyme. Mol Biotechnol 64:1234–1245
Farouk AE, Greiner R, Hussain ASM (2012) Purification and properties of a phytate-degrading enzyme produced by *Enterobacter sakazakii* ASUIA279. J Biotechnol Biodivers 3:1–9
Findenegg GR, Nelemans JA (1993) The effect of phytase on the availability of P from myo-inositol hexaphosphate (phytate) for maize roots. Plant Soil 154:189–196
Finlayson SA, Foster KR, Reid DM (1991) Transport and metabolism of 1-aminocyclopropane-carboxylic acid in sunflower (*Helianthus annuus* L.) seedlings. Plant Physiol 96:1360–1367
First MR, Park NY, Berrang ME (2012) Ciliate ingestion and digestion: flow cytometric measurements and regrowth of a digestion-resistant *Campylobacter jejuni*. J Eukaryot Microbiol 59:12–19

Fitriatin BN, Joy B, Subroto T (2008) The influence of organic phosphorous substrate on phosphatase activity of soil microbes. In: Proceedings of international seminar on chemistry. 2008 Oct 30–31. Universitas Padjadjaran, Jatinangor
Frias J, Doblado R, Antezana JR et al (2003) Inositol phosphate degradation by the action of phytase enzyme in legume seeds. Food Chem 81:233–239
Fu S, Sun J, Qian L et al (2008) Bacillus phytases: present scenario and future perspectives. Appl Biochem Biotechnol 151:1–8
George TS, Richardson AE, Hadobas PA et al (2004) Characterization of transgenic *Trifolium subterraneum* L. which expresses phyA and releases extracellular phytase: growth and P nutrition in laboratory media and soil. Plant Cell Environ 27:1351–1361
George TS, Simpson RJ, Hadobas PA et al (2005) Expression of a fungal phytase gene in *Nicotiana tabacum* improves phosphorus nutrition of plants grown in amended soils. Plant Biotechnol J 3:129–140
George TS, Simpson RJ, Gregory PJ et al (2007) Differential interaction of *Aspergillus niger* and *Peniophora lycii* phytases with soil particles affects the hydrolysis of inositol phosphates. Soil Biol Biochem 39:793–803
Gibson DM (1987) Production of extracellular phytase from *Aspergillus ficuum* on starch media. Biotechnol Lett 9:305–310
Glick BR (1995) The enhancement of plant growth by free-living bacteria. Can J Microbiol 41:109–117
Glick BR, Cheng Z, Czarny J (2007) Promotion of plant growth by ACC deaminase-producing soil bacteria. Eur J Plant Pathol 119:329–339
Goldman JC, Caron DA (1985) Experimental studies on an omnivorous microflagellate: implications for grazing and nutrient regeneration in the marine microbial food chain. Deep-Sea Res 32:899–915
Goldman JC, Caron DA, Andersen OK (1985) Nutrient cycling in a microflagellate food chain. I. Nitrogen dynamics. Mar Ecol Prog Ser 24:231–242
Greiner R, Sajidan I (2008) Production of D-myo-inositol (1, 2, 4, 5, 6) pentakisphosphate using alginate-entrapped recombinant *Pantoea agglomerans* glucose-1-phosphatase. Braz Arch Biol Technol 51:235–246
Greiner R, Konietzny U, Jany KD (1993) Purification and characterization of two phytases from *Escherichia coli*. Arch Biochem Biophys 303:107–113
Greiner R, Lim BL, Cheng C (2007) Pathway of phytate dephosphorylation by β-propeller phytases of different origins. Can J Microbiol 53:488–495
Griffiths AJ (1970) Encystment in amoebae. Adv Microb Physiol 4:105–120
Griffiths BS, Bardgett RD (1997) Interactions between microbe-feeding invertebrates and Soil Microorganisms. In: van Elsas JD, Trevors JT, Wellington EMH (eds) Modern soil microbiology. Marcel Dekker, New York, pp 165–182
Guimarães LH, Terenzi HF, Jorge JA et al (2004) Characterization and properties of acid phosphatases with phytase activity produced by *Aspergillus caespitosus*. Biotech Appl Biochem 40:201–207
Guinel FC (2015) Ethylene, a hormone at the center-stage of nodulation. Front Plant Sci 6:1121
Gupta RK, Gangoliya SS, Singh NK (2015) Reduction of phytic acid and enhancement of bioavailable micronutrients in food grains. J Food Sci Technol 52:676–684
Gyaneshwar P, Kumar GN, Parekh LJ et al (2002) Role of soil microorganisms in improving P nutrition of plants. Plant Soil 245:83–93
Haefner S, Knietsch A, Scholten E et al (2005) Biotechnological production and applications of phytases. Appl Microbiol Biotechnol 68:588–597
Hao X, Cho CM, Racz GJ et al (2002) Chemical retardation of phosphate diffusion in an acid soil as affected by liming. Nutr Cycle Agroecosyst 64:213–224
Hayat R, Ali S, Amara U et al (2010) Soil beneficial bacteria and their role in plant growth promotion: a review. Ann Microbiol 60:579–598
Hayes JE, Richardson AE, Simpson RJ (1999) Phytase and acid phosphatase activities in extracts from roots of temperate pasture grass and legume species. Aust J Plant Physiol 26:801–809

Hayes J, Simpson R, Richardson A (2000) The growth and phosphorus utilisation of plants in sterile media when supplied with inositol hexaphosphate, glucose 1-phosphate or inorganic phosphate. Plant Soil 220:165–174

Heaton K, Drinkall J, Minett A et al (2001) Amoeboid grazing on surface associated prey. In: Gilbert P, Allison DG, Brading M et al (eds) Biofilm community interactions: chance or necessity? Bioline Press, Cardiff, pp 293–301

Hegeman CE, Grabau EA (2001) A novel phytase with sequence similarity to purple acid phosphatases is expressed in cotyledons of germinating soybean seedlings. Plant Physiol 126:1598–1608

Herbert D (1956) Stoichiometric aspects of microbial growth. In: Evans C, Melling J (eds) Continuous culture 6: applications and new field, vol 6. Ellis Horword, Chichester, pp 1–30

Honma M, Shimomura T (1978) Metabolism of 1- aminocyclopropane-1-carboxylic acid. Agric Biol Chem 42:1825–1831

Howson S, Davis R (1983) Production of phytate hydrolyzing enzymes by some fungi. Enzym Microb Technol 5:377–382

Hsiao A (2000) Effect of water deficit on morphological and physiological characterizes in rice (*Oryza sativa*). J Agric Res 3:93–97

Huang H, Shi P, Wang Y (2009) Diversity of beta-propeller phytase genes in the intestinal contents of grass carp provides insight into the release of major phosphorus from phytate in nature. Appl Environ Microbiol 75:1508–1516

Huws SA, McBain AJ, Gilbert P (2005) Protozoan grazing and its impact upon population dynamics in biofilm communities. J Appl Microbiol 98:238–244

Idriss EE, Makarewicz O, Farouk A et al (2002) Extracellular phytase activity of *Bacillus amyloliquefaciens* FZB45 contributes to its plant growth-promoting effect. Microbiology 148:2097–2109

Iriberri J, Ayo B, Santamaria E (1995) Influence of bacterial density and water temperature on the grazing activity of two freshwater ciliates. Freshw Biol 33:223–231

Jacobson CB, Pasternak JJ, Glick BR (1994) Partial purification and characterization of ACC deaminase from the plant growth-promoting rhizobacterium *Pseudomonas putida* GR12-2. Can J Microbiol 40:1019–1025

James MR, Hall JA (1995) Planktonic ciliated protozoa: their distribution and relationship to environmental variables in a marine coastal ecosystem. J Plankton Res 17:659–683

Jassey VE, Shimano S, Dupuy C et al (2012) Characterizing the feeding habits of the testate amoebae *Hyalosphenia papilio* and *Nebela tincta* along a narrow "fen-bog" gradient using digestive vacuole content and 13C and 15N isotopic analyses. Prosit 163:451–464

Jia YJ, Kakuta Y, Sugawara M (1999) Synthesis and degradation of 1-aminocyclopropane-1-carboxylic acid by *Penicillium citrinum*. Biosci Biotech Biochem 63:542–549

Johnson SC, Yang MP, Murthy PN (2010) Heterologous expression and functional characterization of a plant alkaline phytase in *Pichia pastoris*. Protein Express Purif 74:196–203

Jorquera M, Martinez O, Maruyama F (2008) Current and future biotechnological applications of bacterial phytases and phytase-producing bacteria. Microbes Environ 23:182–191

Jousset A, Bonkowski M (2010) The model predator Acanthamoeba castellanii induces the production of 2, 4, DAPG by the biocontrol strain *Pseudomonas fluorescens* Q2-87. Soil Biol Biochem 42:1647–1649

Jousset A, Rochat L, Scheu S et al (2010) Predator-prey chemical warfare determines the expression of biocontrol genes by rhizosphere-associated *Pseudomonas fluorescens*. Appl Environ Microbiol 76:5263–5268

Kara A, Ebina S, Kondo A et al (1985) A new type of phytase from pollen of *Typha latifolia* L. Agric Biol Chem 49:3539–3544

Kerovuo J, Lauraeus M, Nurminen P et al (1998) Isolation, characterization, molecular gene cloning and sequencing of a novel phytase from *Bacillus subtilis*. Appl Environ Microbiol 64:2079–2085

Kerovuo J, Rouvinen J, Hatzack F (2000) Analysis of myoinositol hexakisphosphate hydrolysis by Bacillus phytase, indication of a novel reaction mechanism. Biochem J 352:623–628

Khan AA, Jilani G, Akhtar MS et al (2009) Phosphorus solubilizing bacteria: occurrence, mechanisms and their role in crop production. J Agric Biol Sci 1:48–58

Kim Y-O, Lee J-K, Kim H-K et al (1998a) Cloning of thermostable phytase gene (phy) from Bacillus sp. DS11 and it's over expression in *Escherichia coli*. FEMS Microbiol Lett 162:185–191

Kim YO, Kim HK, Bae KS et al (1998b) Purification and properties of thermostable phytase from Bacillus sp. DS11. Enzym Microbiol Technol 22:2–7

Kim H-W, Kim Y-O, Lee J-H et al (2003) Isolation and characterization of a phytase with improved properties from *Citrobacter braakii*. Biotechnol Lett 25:1231–1234

Klee HJ, Hayford MB, Kretzmer KA (1991) Control of ethylene synthesis by expression of a bacterial enzyme in transgenic tomato plants. Plant Cell 3:1187–1193

Konietzny U, Greiner R (2002) Molecular and catalytic properties of phytase degrading enzymes (phytases). Int J Food Sci Technol 37:791–812

Konietzny U, Greiner R (2004) Bacterial phytase: potential application, in vivo function and regulation of its synthesis. Braz J Microbiol 35:12–18

Krome K, Rosenberg K, Dickler C (2010) Soil bacteria and protozoa affect root branching via effects on the auxin and cytokinin balance in plants. Plant Soil 328:191–201

Lan GQ, Abdullah N, Jalaludin S et al (2002) Culture conditions influencing phytase production of *Mitsuokella jalaludinii*, a new bacterial species from the rumen of cattle. J Appl Microbiol 93:668–674

Levrat P (1989) Actiond' *Acanthamoeba castellarni* (Protozoa: Amoebida) sur la production de siderophores par la bacterie *Pseudomonas putida*. C R Acad Sci Sér 3 Sci Vie 308:161–164

Li M, Osaki M, Madhusudana Rao I et al (1997) Secretion of phytase from the roots of several plant species under phosphorus-deficient conditions. Plant Soil 195:161–169

Li XG, Porres JM, Mullaney EJ et al (2007a) Phytase: source, structure and application. In: Industrial enzymes. Springer, Dordrecht, pp 505–529

Li X, Wu Z, Li W et al (2007b) Growth promoting effect of a transgenic *Bacillus mucilaginosus* on tobacco planting. Appl Microbiol Biotechnol 74:1120–1125

Li G, Yang S, Li M et al (2009) Functional analysis of an *Aspergillus ficuum* phytase gene in *Saccharomyces cerevisiae* and its root-specific, secretory expression in transgenic soybean plants. Biotechnol Lett 31:1297–1303

Li R, Zhao J, Sun C et al (2010) Biochemical properties, molecular characterizations, functions, and application perspectives of phytases. Front Agric China 4:195–209

Lilly VG (1965) The chemical environment for growth. 1. In: Ainsworth GC, Sussman AS (eds) The fungi, media, macro and micronutrients, vol 1. Academic, New York, pp 465–478

Lott JN, Ockenden I, Raboy V et al (2000) Phytic acid and phosphorus in crop seeds and fruits: a global estimate. Seed Sci Res 10(1):11–33

Luria SE (1960) The bacterial protoplasm: composition and organization. Bacteria 1:1–34

Ma W, Charles TC, Glick BR (2004) Expression of an exogenous 1-aminocyclopropane-1-carboxylate deaminase gene in *Sinorhizobium meliloti* increases its ability to nodulate alfalfa. Appl Environ Microbiol 70:5891–5897

Marlida Y, Delfita R, Adnadi P et al (2010) Isolation, characterization and production of phytase from endophytic fungus its application for feed. Pak J Nutr 9:471–474

Mazzola M, De Bruijn I, Cohen MF et al (2009) Protozoan-induced regulation of cyclic lipopeptide biosynthesis is an effective predation defense mechanism for *Pseudomonas fluorescens*. Appl Environ Microbiol 75:6804–6811

Mellano HM, Munnecke DE, Endo RM (1970) Relationship of seedling age to development of *Pythium ultimum* on roots of *Antirrhinum majus*. Phytopathology 60:935–942

Menezes-Blackburn D, Jorquera MA, Greiner R et al (2013) Phytases and phytase-labile organic phosphorus in manures and soils. Crit Rev Environ Sci Technol 43:916–954

Minami R, Uchiyama K, Murakami T (1998) Properties, sequence and synthesis in *Escherichia coli* of 1-aminocyclopropane-1-carboxylate deaminase from *Hansenula saturnus*. J Biochem 123:1112–1118

Minggang L, Mitsuru O, Idupulapati MR, Tadano T (1997) Secretion of phytase from the roots of several plant species under phosphorus-deficient conditions. Plant Soil 195:161–169
Mittal V, Singh O, Nayyar H et al (2008) Stimulatory effect of phosphate solubilizing fungal strains (*Aspergillus awamori* and *Penicillium citrinum*) on the yield of chickpea (*Cicer arietinum* L. cv.GPF2). Soil Biol Biochem 40:718–727
Morgan PW, Drew MC (1997) Ethylene and plant response to stress. Physiol Plant 100:620–630
Mukhametzyanova AD, Akhmetova AI, Sharipova MR (2012) Microorganisms as phytase producers. Microbiology 81:267–275
Mullaney EJ, Ullah AHJ (2003) Phytases: attributes, catalytic mechanisms and applications. Biochem Biophys Res Commun 312:179–184
Mullaney EJ, Daly CB, Ullah AH (2000) Advances in phytase research. Adv Appl Microbiol 47:157–199
Müller MS, Scheu S, Jousset A (2013) Protozoa drive the dynamics of culturable biocontrol bacterial communities. PLoS One 8:e66200
Nielsen TH, Sorensen D, Tobiasen C et al (2002) Antibiotic and biosurfactant properties of cyclic lipopeptides produced by fluorescent *Pseudomonas* spp. from the sugar beet rhizosphere. Appl Environ Microbiol 68:3416–3423
Page FC (1988) A new key to freshwater and soil Gymnamoebae: with instructions for culture. Freshwater Biological Association, Ambleside
Pandey A, Szakacs G, Soccol CR et al (2001) Production, purification and properties of microbial phytases. Bioresour Technol 77:203–214
Pasamontes L, Haiker M, Wyss M (1997) Gene cloning, purification, and characterization of a heat-stable phytase from the fungus *Aspergillus fumigatus*. Appl Environ Microbiol 63:1696–1700
Patel KJ, Singha AK, Nareshkumarb G (2010) Organic-acid-producing, phytate-mineralizing rhizobacteria and their effect on growth of pigeon pea (*Cajanus cajan*). Appl Soil Ecol 44:252–261
Penmetsa RV, Cook DR (1997) A legume ethylene-insensitive mutant hyperinfected by its rhizobial symbiont. Science 275:527–530
Penrose DM, Glick BR (2001) Levels of ACC and related compounds in exudate and extracts of canola seeds treated with ACC deaminase-containing plant growth-promoting bacteria. Can J Microbiol 47:368–372
Penrose DM, Glick BR (2003) Methods for isolating and characterizing ACC deaminase containing plant growth-promoting rhizobacteria. Physiol Plant 118:10–15
Powar VK, Jagannathan V (1982) Purification and properties of phytate-specific phosphatase from *Bacillus subtilis*. J Bacteriol 151:1102–1108
Quan C-S, Tian W-J, Fan S-D et al (2004) Purification and properties of a low-molecular weight phytase from Cladosporium sp. FP-1. J Biosci Bioeng 97:260–266
Quiquampoix H, Burns RG (2007) Interactions between proteins and soil mineral surfaces: environmental and health consequences. Elements 3:401–406
Raboy V, Dickinson DB (1987) The timing and rate of phytic acid accumulation in developing soybean seeds. Plant Physiol 85:841–844
Rahdari P, Hosseini SM, Tavakoli S (2012) The studying effect of drought stress on germination, proline, sugar, lipid, protein and chlorophyll content in purslane (*Portulaca oleracea* L.) leaves. J Med Plant Res 6:1539–1547
Ramaekers L, Remans R, Rao IM (2010) Strategies for improving phosphorus acquisition efficiency of crop plants. Field Crop Res 117:169–176
Reddy MS, Kumar S, Babita K (2002) Biosolubilization of poorly soluble rock phosphates by *Aspergillus tubingensis* and *Aspergillus niger*. Bioresour Technol 84:187–189
Reddy CS, Kim SC, Kaul T (2017) Genetically modified phytase crops role in sustainable plant and animal nutrition and ecological development: a review. 3 Biotech 7:195
Richardson AE, Hadobas PA (1997) Soil isolates of Pseudomonas spp. that utilize inositol phosphates. Can J Microbiol 43:509–516
Richardson AE, Simpson RJ (2011) Soil microorganisms mediating phosphorus availability. Plant Physiol 156:989–996

Richardson A, Hadobas P, Hayes J (2000) Acid phosphomonoesterase and phytase activities of wheat (*Triticum aestivum* L.). roots and utilization of organic phosphorus substrates by seedlings grown in sterile culture. Plant Cell Environ 23:397–405
Richardson AE, Hadobas PA, Hayes JE (2001a) Extracellular secretion of Aspergillus phytase from Arabidopsis roots enables plants to obtain phosphorus from phytate. Plant J 25:641–649
Richardson AE, Hadobas PA, Hayes JE (2001b) Utilization of phosphorus by pasture plants supplied with myo-inositol hexaphosphate is enhanced by the presence of soil micro-organisms. Plant Soil 229:47–56
Richardson AE, Barea J-M, McNeill AM (2009) Acquisition of phosphorus and nitrogen in the rhizosphere and plant growth promotion by microorganisms. Plant Soil 321:305–339
Rodríguez H, Fraga R (1999) Phosphate solubilizing bacteria and their role in plant growth promotion. Biotechnol Adv 17:319–339
Rossolini GM, Schippa S, Riccio ML et al (1998) Bacterial nonspecific acid phosphohydrolases: physiology, evolution and use as tools in microbial biotechnology. Cell Mol Life Sci 54:833–850
Sajidan A, Farouk A, Greiner R (2004) Molecular and physiological characterisation of a 3-phytase from soil bacterium Klebsiella sp. ASR1. Appl Microbiol Biotechnol 65:110–118
Sarapatka B (2002) Phosphatase activity of Eutric cambisols (Uppland, Sweden) in relation to soil properties and farming systems. Acta Agric Bohem 33:18–24
Sayre RM (1973) *Theratromyxa weberi*, an amoeba predatory on plant-parasitic nematodes. J Nematol 5:258
Schaefer M, Schauermann J (1990) The soil fauna of beech forests: comparison between a mull and a modern soil. Pedobiologia 34:299–314
Scholz RW, Hellums DT, Roy AA (2015) Global sustainable phosphorus management: a transdisciplinary venture. Curr Sci 108:3–12
Schröter D, Wolters V, De Ruiter PC (2003) C and N mineralisation in the decomposer food webs of a European forest transect. Oikos 102:294–308
Scott JJ (1991) Alkaline phytase activity in nonionic detergent extracts of legume seeds. Plant Physiol 95:1298–1301
Selvakumar G, Reetha S, Thamizhiniyan P (2012) Response of biofertilizers on growth, yield attributes and associated protein profiling changes of blackgram (*Vigna mungo* L. Hepper). WASJ 16:1368–1374
Sgherri C, Stevanovic B, Navari-Izzo F (2000) Role of phenolic acids during dehydration and rehydration of *Ramonda serbica*. Physiol Plant 122:478–485
Sharma A, Rawat US, Yadav BK (2012) Influence of phosphorus levels and phosphorus solubilizing fungi on yield and nutrient uptake by wheat under sub-humid region of Rajasthan, India. ISRN Agron 15:2012
Sherr BF, Sherr EB, Fallon RD (1987) Use of monodispersed, fluorescently labeled bacteria to estimate in situ protozoan bacterivory. Appl Environ Microbiol 53:958–965
Shimizu M (1992) Purification and characterization of a phytase from *Bacillus subtilis* (natto) N-77. Biosci Biotechnol Biochem 56:1266–1269
Sieburth JM, Davis PG (1982) The role of heterotrophic nanoplankton in the grazing and nurturing of planktonic bacteria in the Sargasso and Caribbean Seas. Ann Inst Oceanogr 58(S):285–296
Singh B, Satyanarayana T (2010) Plant growth promotion by an extracellular HAP-phytase of a thermophilic mold *Sporotrichum thermophile*. Appl Biochem Biotechnol 160:1267–1276
Singh B, Satyanarayana T (2015) Fungal phytases: characteristics and amelioration of nutritional quality and growth of non-ruminants. J Anim Physiol Anim Nutr 99:646–660
Singh B, Kunze G, Satyanarayana T (2011) Developments in biochemical aspects and biotechnological applications of microbial phytases. Biotechnol Mol Biol Rev 6:69–87
Tamimi SM, Timko MP (2003) Effects of ethylene and inhibitors of ethylene synthesis and action on nodulation in common bean (*Phaseolus vulgaris* L.). Plant Soil 257:125–131
Tanaka Y, Sano T, Tamaoki M (2005) Ethylene inhibits abscisic acid-induced stomatal closure in Arabidopsis. Plant Physiol 138:2337–2343

Tarafdar JC (1995) Dual inoculation with *Aspergillus fumigatus* and *Glomus mosseae* enhances biomass production and nutrient uptake in wheat (*Triticum aestivum* L.) supplied with organic phosphorus as Na-Phytate. Plant Soil 173:97–102

Tarafdar JC, Yadav RS, Meena SC (2001) Comparative efficiency of acid phosphatase originated from plant and fungal sources. J Plant Nutr Soil Sci 164:279–282

Tittabutr P, Piromyou P, Longtonglang A (2013) Alleviation of the effect of environmental stresses using co-inoculation of mungbean by *Bradyrhizobium* and *Rhizobacteria* containing stress-induced ACC deaminase enzyme. Soil Sci Plant Nutr 59:559–557

Tran HT, Hurley BA, Plaxton WC (2010) Feeding hungry plants: the role of purple acid phosphatases in phosphate nutrition. Plant Sci 179:14–27

Turk M, Sandberg AS, Carlsson N et al (2000) Inositol hexaphosphate hydrolysis by baker's yeast. Capacity, kinetics and degradation products. J Agric Food Chem 48:100–104

Turner BL, Papházy MJ, Haygarth PM et al (2002) Inositol phosphates in the environment. Philos Trans R Soc Lond B Biol Sci 357:449–469

Tye AJ, Siu FKY, Leung TYC et al (2002) Molecular cloning and the bio-chemical characterization of two novel phytases from *Bacillus subtilis* 168 and *Bacillus licheniformis*. Appl Microbiol Biotechnol 59:190–197

Unno Y, Okubo K, Wasaki J et al (2005) Plant growth promotion abilities and microscale bacterial dynamics in the rhizosphere of Lupin analysed by phytate utilization ability. Environ Microbiol 7:396–404

Vats P, Banerjee UC (2004) Production studies and catalytic properties of phytases (myo-inositol hexakisphosphate phosphohydrolases): an overview. Enzym Microb Technol 35:3–14

Vats P, Bhattacharyya MS, Banerjee UC (2005) Use of phytases (myo-inositolhexakis phosphate phosphohydrolases) for combating environmental pollution: a biological approach. Crit Rev Environ Sci Technol 35:469–486

Wallenstein MD, Burns RG (2011) Ecology of extracellular enzyme activities and organic matter degradation in soil: a complex community-driven process. In: Dick RP (ed) Methods of soil enzymology. Soil Sci Soc Am, Madison, pp 35–55

Wallenstein MD, McMahon SK, Schimel JP (2009) Seasonal variation in enzyme activities and temperature sensitivities in Arctic tundra soils. Glob Chang Biol 15:1631–1639

Weidner S, Latz E, Agaras B (2017) Protozoa stimulate the plant beneficial activity of rhizospheric pseudomonads. Plant Soil 410:509–515

Wyss M, Brugger R, Kronenberger A et al (1999) Biochemical characterization of fungal phytases (myo-inositol hexakisphosphate phosphohydrolases): catalytic properties. Appl Environ Microbiol 65:367–373

Xiao C, Chi R, Li X et al (2011) Biosolubilization of rock phosphate by three stress-tolerant fungal strains. Appl Biochem Biotechnol 165:719–727

Xuguang N, Lichao S, Yinong X et al (2018) Drought-tolerant plant growth-promoting Rhizobacteria associated with foxtail millet in a semi-arid agroecosystem and their potential in alleviating drought stress. Front Microbiol 8:2580

Yang SF, Hoffman NE (1984) Ethylene biosynthesis and its regulation in higher plants. Ann Rev Plant Physiol 35:155–189

Yoon SJ, Choi YJ, Min HK et al (1996) Isolation and identification of phytase-producing bacterium, *Enterobacter* sp. 4, and enzymatic properties of phytase enzyme. Enzym Microb Technol 18:449–454

Zahir ZA, Arshad M, Frankenberger WT (2004) Plant growth promoting rhizobacteria: applications and perspectives in agriculture. Adv Agron 81:98–169

Zamudio M, González A, Medina JA (2001) *Lactobacillus plantarum* phytase activity is due to nonspecific acid phosphatase. Lett Appl Microbiol 32:181–184

Zamudio M, González A, Bastarrachea F (2002) Regulation of *Raoultella terrigena* comb.nov. phytase expression. Can J Microbiol 48:71–81

Microbial Bio-production of Proteins and Valuable Metabolites

15

Abiya Johnson, Prajkata Deshmukh, Shubhangi Kaushik, and Vimal Sharma

15.1 Introduction

Microbes like bacteria, fungi, yeast, and microalgae are the prolific source of large number of valuable natural compounds of commercial and therapeutic interest. Microbes are capable of synthesizing structurally divergent compounds. According to a recent report by Business Communication Company (BCC), the global market of microbes and microbial products would reach to $302.4 billion by 2023 from $186.3 billion in 2018 (McWilliams 2012). The microbial products are comprised of either the whole microbial cells or the metabolites derived from the microbes. Various products including pharmaceuticals, bulk and fine chemicals, metabolites, proteins, nutraceuticals, biofuels, antibiotics, bioplastics, food supplements, and biofertilizers are produced using biocatalytic processes, microbial cell factories, or cell-free processes (Schmidt-Dannert 2017). The producer microbes are identified through various approaches followed by establishment of microbial growth as well as production parameters for target molecule under laboratory conditions and further optimizations for large-scale production of target molecule in fermenters. The microbial production platforms are successfully becoming an effective alternative to traditional chemical synthesis due to various advantages offered by them such that microbial biosynthesis does not require heavy metals, solvents, strong acids, or bases unlike chemical synthesis, enzymes exhibit broader substrate specificity resulting in lesser by-products, natural synthetic pathways are already available for some compound with complex structures, engineering of biosynthetic pathways can

A. Johnson · P. Deshmukh · S. Kaushik (✉)
Department of Biotechnology, National Institute of Pharmaceutical Education and Research-Guwahati, Guwahati, Assam, India
e-mail: shubhangi.kaushik@niperguwahati.ac.in

V. Sharma
Department of Biochemistry, Royal Global University, Guwahati, Assam, India

D. P. Singh et al. (eds.), *Microbial Interventions in Agriculture and Environment*,
https://doi.org/10.1007/978-981-13-8391-5_15

further improve yield, and productivity of the compound of interest or novel pathways can be constructed in host microbe (Du et al. 2011).

The advances in recombinant DNA technology have prompted the development of microbial systems for bio-manufacturing of various valuable chemicals and natural products (Chemier et al. 2009). Microbes with well-studied genetics, physiology, and biochemistry like *Escherichia coli* and *Saccharomyces cerevisiae* are commonly used as bio-production platform. *Pseudomonas putida*, *Bacillus*, *Cyanobacteria*, and *Streptomyces* species have also been used for biosynthesis of target compounds (Chemier et al. 2009). Metabolic engineering and synthetic biology approaches have significantly contributed in development of engineered microbes for production of various useful compounds from simple and cheap substrates not only at laboratory scale but also at the industrial scale (Jullesson et al. 2015). Computational softwares are tremendously used in metabolic engineering to extract the information from big datasets as well as to assist in designing and optimizing the novel pathways in microbes (Reed et al. 2011).

15.2 Microbial Enzymes

The enzymes are biological molecules, usually proteinaceous in nature with the exception of ribozymes (catalytic RNA molecules), and they play crucial role in different stages of metabolism or biochemical reactions as bio-catalysts (Cech and Bass 1986; Gurung et al. 2013). Enzymes possess several features that make them attractive candidates for various applications such that they enhance the rate of reaction under mild physico-chemical conditions without being consumed, they are non-toxic, and they exhibit remarkable chemoselectivity, enantioselectivity, regioselectivity, and substrate specificity.

15.2.1 Potential Sources of Enzymes

Nature contributes an extensive amount of enzyme resources. In the beginning of enzyme biotechnology era, the plant tissues and animals were the most important sources of enzymes. However, currently microbes represent the largest and useful sources of many enzymes (Demain and Adrio 2008; Volesky et al. 1984). Most of enzymes which are used commercially are obtained from aerobic strains. Majority of microbial enzymes are derived from *Aspergillus*, *Bacillus*, *Streptomyces*, and *Saccharomyces* species (Headon and Walsh 1994). Microbes are usually preferred over plants and animals as a source of enzymes because they represent amicable and economical way for enzyme production in short time, microbes have shorter generation time and genetic manipulation can be easily performed, microbial enzyme expression is controllable, microbial enzymes are more stable as well as active, and production in larger quantities can be achieved (Anbu et al. 2015; Gurung et al. 2013).

Microbial enzymes are obtained from different microorganisms. For example, proteases of commercial applicability are produced mainly by bacteria species such as *Pseudomonas*, *Clostridium*, and *Bacillus* and also by some fungal species (Nigam 2013). Studies on enzyme isolation, their characterization, and production on bench and pilot scale are continuously increasing. Owing to their commercial applications, the market for industrial enzymes is widespread (Sanchez and Demain 2017; Adrio and Demain 2014). The market for industrial enzymes will reach to nearly $6.2 billion by 2020 with annual growth rate (CAGR) of 7% (Singh et al. 2016b). In general, numerous microbial enzymes are already being exploited in many different industrial processes.

15.2.2 Microbial Enzyme Production

Microbes produce vast variety of enzymes but the absolute amount of produced enzyme differs markedly even between the strains of same microbial species. Thus, for the production of the desired enzyme for commercial applications, the strain that exhibit highest yield is ideally selected (Underkofler et al. 1958). The enzyme-based product, which is newly introduced in market, can become a commercial success if it has a large existing market share and if it is economically viable. For the successful development of a commercial enzyme process, various requirements should be fulfilled including the ability of producer microbe to grow at a rapid rate on an inexpensive medium, production of the enzymes in high yields as well as at high concentration, minimal generation of enzyme contaminants and other metabolites in the fermentation of broth, the possibility to grow the microbe on a concentrated medium in a dense culture which improves the enzyme productivity in fermenters, and easy as well as inexpensive recovery of the enzyme from the culture media (Headon and Walsh 1994; Volesky et al. 1984).

Generally, the production of the desired enzymes begins with the screening of the microbes present in the collected environmental samples to identify the producer strain using suitable selection procedures. It is followed by optimization of the culture conditions, physico-chemical properties, and process parameters to maximize the production of target enzyme. The screening processes on laboratory scale focuses on the search for a high titre enzyme-producing microorganism, and they are usually labour intensive, monotonous, and time consuming (Yoo et al. 2017). The advent of genetic engineering approaches facilitated the cloning of the gene encoding for the target enzyme in microbes with defined growth conditions, with controllable gene expression, and with GRAS status (generally recognized as safe), leading to impressive enzyme yields. The construction of metagenomic library by cloning of total isolated DNA from environmental samples in suitable vector system, and subsequent function-based screening is another powerful approach that allows to explore the potential of biological diversity in different ecosystems for the identification of target enzyme (Thies et al. 2016). This approach circumvents the need of culturing and isolation of individual microbe in laboratory (Guazzaroni et al. 2015). Another approach to obtain superior enzyme producer strain is mutagenesis where the microbial cultures are exposed to mutagenic agents like chemicals, heat, and radiations. The screening for survival of cells is then performed to select the strain that can overproduce the target enzyme (Ghazi et al. 2014).

For production of target enzyme, the producer microbes are cultivated by inoculation of the pure culture into the suitable sterile medium. Submerged fermentation and solid-state fermentation (SSF) are the methods used for the enzyme cultivation (Renge et al. 2012). In submerged fermentation, the microorganisms are cultivated in a closed vessel (fermenter) containing liquid nutrient media and a high concentration of oxygen. The growing microbes release the target enzyme in extracellular environment i.e. in fermentation broth. The biomass is then removed from fermentation broth by centrifugation and the enzymes in the broth are then concentrated by evaporation of media, membrane filtration, or crystallization. This approach was used traditionally to prepare the target enzymes due to easy handling and ability to control physico-chemical factors (Mrudula and Murugammal 2011). In solid-state fermentation, microbes are cultivated on a solid substrate like wheat bran, wheat straw, and rice straw. This method is used for the cultivation of fungi such as *Aspergillus and Penicillium* to obtain enzymes such as amylase, proteases, and pectinases (Volesky et al. 1984).

15.2.3 Applications of Microbial Enzymes

The demand of microbial enzymes in various industries is expanding rapidly. Their application in few sectors is summarized.

15.2.3.1 Industrial Application

Microbial enzymes are used in various industrial applications including production of pharmaceuticals or pharmaceutically important intermediates, leather processing, textile industry, and paper and pulp, detergents, and biofuel production. In laundry detergents, proteases are extensively used to remove the proteinaceous dirt from the fabric. Proteolytic enzymes in many commercially available detergents are derived from the *Bacillus* species (Kumar et al. 2008). Other enzymes are also used in combination with proteases to improve the cleaning performance of the detergent, which includes lipases, amylases, and cellulases to remove fats or oils, remove starch residues, and brighten colour, respectively (Hasan et al. 2010). Several active pharmaceutical ingredients are being generated using the enzymes because of their remarkable specificity and selectivity. Carbonyl reductases have been used to obtain an intermediate for synthesis of blockbuster drugs and statins, by reduction of ethyl 4-chloro-3-oxobutanoate (COBE) to ethyl (S)-4-chloro-3-hydroxybutanoate ((S)-CHBE) (Xu et al. 2016). Atorvastatin which is an important ingredient of Lipitor, a cholesterol-lowering drug, has also been shown to produce through enzymatic synthesis (Bornscheuer et al. 2012). The commercial manufacturing of telaprevir, boceprevir, and esomeprazole drugs against hepatitis C virus involved in the oxidase-catalysed desymmetrization (Li et al. 2012).

Tyrosine phenol lyase expressed in *Erwinia herbicola* cells has been used to produce L-3,4-dihydroxyphenylalanine (L-DOPA), a drug for treatment of Parkinson's disease (Patel 2008). Most of the enzymes used in textile industry are hydrolases like cellulases, pectinases, laccases, amylases, and catalases. These enzymes are being used as a substitute of stone wash, in bio-finishing, in bio-scouring, and in improving the look of material (Doshi and Shelke 2001). The involvement of lipases, cellulases, and xylanases has been reported for bioethanol production by decomposition of lignocellulosic material and also synthesis of fatty acid methyl esters (Liew et al. 2014). In leather industry, proteases and lipases are involved at different stages of leather processing. They are used in curing, soaking, dehairing, degreasing, tanning, and waste processing of leather (Choudhary et al. 2004).

15.2.3.2 Food

Microbial enzymes are significantly used in processing of food products such as cheese, beer, bread, and soft drinks, and the use of enzymes in manufacturing is increasing (Fernandes and Carvalho 2017). Amylases from the malted cereal, bacterial, or fungal sources are added to flour at the bakery and mill (Taylor and Richardson 1979). Another example of a microbial enzyme used in food industry is microbial transglutaminase which catalyses isopeptide bond formation between proteins. This property is widely used in manufacturing cheese and other dairy products, meat processing, manufacturing bakery products, and producing edible films (Kieliszek and Misiewicz 2014). Proteases are used in meat tenderization, ripening of cheese, and milk coagulation (Aruna et al. 2014). Lipases are also used in cheese flavour development and improving its texture. They are also in used in flavour development in butter and improving the shelf life of baking products (Aravindan et al. 2007). Galactosidases are used in lactose hydrolysis of milk-based products for lactose-intolerant people, in preparation of prebiotic food ingredient like galacto-oligosaccharides, and in lactose hydrolysis in whey (Rosenberg 2006).

15.2.3.3 Medicines

Therapeutic enzymes derived from microbial sources are used to treat various diseases. Nattokinase from *Bacillus subtilis* decreases the blood coagulation and removes existing thrombus. It is also used to decrease the lipids that can increase the chances of cardiovascular disease (Banerjee et al. 2004; Milner 2008). Streptokinase and urokinase are used for dissolving the blood clots in blocked blood vessels (Banerjee et al. 2004; Olson et al. 2011). Collagenases have been used to assist in healing skin burns and tumours in combination with antibiotics (Ostlie et al. 2012). In dental hygiene, enzymes like dextranase and cariogenanase from *Penicillium funiculosum* and *Bacillus* sp. are, respectively, used to reduce plaques and dental carries. Toothpastes containing a mixture of enzymes from *Aspergillus niger* and *Aspergillus oryzae* reduce calculus and soft accretions (Singh et al. 2016a). Tyrosine hydroxylase is responsible for catalysing the conversion of L-tyrosine to L-dopa, which is a useful agent in the treatment of Parkinson's disease (Taylor and Richardson 1979).

15.2.4 Strategies for Enhancing Applicability of Existing Microbial Enzymes

In spite of the significant advances in screening and selection approaches for identifying the novel enzymes to combat the ever-increasing industrial demands, there still remains the need of efficient ways to obtain enzymes with better catalytic performance for relevant industrial processes. In this connection, protein engineering strategies have been devised to improve the efficiency of existing enzymes (Kaushik et al. 2018). Protein engineering focuses on tailoring enzymes to overcome inherent shortcomings in existing enzymes like low activity, lack of specificity, and low stability or to introduce new functionalities. One of the protein engineering approaches is directed evolution or in vitro evolution, which mimics natural evolution process and does not require detailed knowledge on structure, function, and mechanistic aspects of target enzyme (Chen et al. 2012). It involves exposure of the gene encoding for the enzyme of interest to iterative rounds of random mutagenesis resulting in construction of library of gene variants; the resulting library is then screened for the variant that exhibits desired level of improvement (Chen and Arnold 1993). The process of directed evolution basically relies on effective mutagenesis method that generates significant genetic diversity and a robust screening method that leads to identification of the enzyme variant with desired catalytic characteristics as compared to wild-type enzyme.

The genetic diversity can be introduced by random mutagenesis methods like use of mutator strains, UV irradiation, chemical mutagenesis, error-prone PCR, and sequence saturation mutagenesis (SeSAM), or it can be introduced by gene recombination methods like DNA shuffling and oligonucleotide primer-based methods (Labrou 2010). Another protein engineering approach is rational redesign, which involves use of sequence and structure-based information with computational modelling to predict the hotspot residues which on mutagenesis are likely to result in improved enzyme functionalities. This approach dramatically reduces the library size and subsequently eliminates need of high-throughput screening methods. A semi-rational approach involving both the components of random and rational mutagenesis to design smart libraries with small size and high quality has shown to be practically more effective in generating tailor-made enzymes for specific needs (Lutz 2010). In recent years, engineering of access tunnel residues in enzymes with buried catalytic site has become an attractive approach to alter the enzyme properties. Access tunnels are the transport pathways that connect the buried active site of the enzyme to the exterior environment and allow the access or egress of substrates, reactive intermediates, solvents, ions, and products to the catalytic site (Damborsky et al. 2010; Timmis et al. 2010). Modification of the access tunnel lining residues doesn't affect the architecture of catalytic site and thus increases the chances of getting functional clones with tailored properties. This strategy has been applied on several enzymes with buried active site to improve their catalytic properties (Kaushik et al. 2018; Prokop et al. 2012; Sandström et al. 2012). De novo protein design is another strategy that can allow to introduce new catalytic functions in protein scaffold such that de novo enzymes have been successfully designed that can catalyse the Diels-Alder and Kemp elimination reactions (Blomberg et al. 2013; Siegel et al. 2010). Recently, possibility to introduce de novo functional tunnels in existing protein has been demonstrated to facilitate creation of better and efficient enzymes (Brezovsky et al. 2016).

15.3 Proteins

The word protein is derived from the Greek word 'protos' that means first or 'protieos' which means primary (Aronson 2012). Proteins are the primary constituent of living things and are part of the molecular machinery in living organisms. They form the fundamental basis of the structure and function of life. Peptides and proteins are polymers of amino acids. They are the products of translation of mRNA within the living cell (Berg et al. 2002; Nelson et al. 2008). Variety of proteins/peptides derived from microbial systems have direct implication in production of vaccines, as therapeutic agents and as food supplements (Akash et al. 2015).

15.3.1 Microbial Proteins and Their Utility in Vaccine Production

A vaccine is a biological substance that stimulates the active acquired immune system of the body to act against a particular germ, thereby preventing the disease caused by it. Microbial surface proteins are associated with pathogenesis and thus represents major target for vaccine development. The commercially available vaccines contain attenuated pathogenic microbes or the microbial antigenic protein (Table 15.1). Various in silico tools have been developed by researchers so as to rapidly identify the surface proteins which can possibly display antigenic properties (Giombini et al. 2010). Approaches like whole-genome sequencing, labeling of surface proteins by selective biotinylation of whole bacteria, identification of immunogenic proteins from pathogens on protein microarrays, and enzymatic shaving of surface of bacteria with proteases have made identification of the surface antigens easier for vaccine development (Grandi 2010).

Strategies used in the production of vaccines are attenuation of the live pathogenic microbe, structural vaccinology, reverse vaccinology, epitope mapping, recombinant protein synthesis, and microbial cell-surface display. Structural vaccinology or structure-based antigen design involves the use of high-resolution structural analysis in distinguishing structural components of the antigen that elicit protective and disease-enhancing immunity (Dormitzer et al. 2008). This strategy has effectively guided design of engineered RSV F subunit antigen against respiratory syncytial virus, GBS (group B *Streptococcus*) pilus-based fusion protein, and an improved MenB (serogroup B meningococcus) single-domain fHbp (factor H-binding protein) antigen against meningitis (Dormitzer et al. 2012). Reverse vaccinology uses whole genome sequencing and immunological information of the pathogen to identify the suitable candidate vaccine antigens (Sette and Rappuoli 2010). Bexsero™ is a meningococcal group B vaccine that was developed through

Table 15.1 Representative vaccines based on attenuated microbes and microbial proteins

Vaccine	Disease	Causative microbe	Components	Manufacturer	References
Tetravalent influenza vaccine (split virion) I.P. (TetIV)	Influenza	Influenza A and B viruses	Two A strains (H1N1 and H3N2) and two B strains (Yamagata and Victoria)	Cadila Healthcare Limited	Sharma et al. (2018)
Vaxigrip trivalent split-virion, inactivated influenza vaccine	Influenza	Influenza A and B viruses	Two A strains (H1N1 and H3N2) and either one of two B strains (Yamagata and Victoria)	Sanofi Pasteur	Haugh et al. (2017)
Zostavax	Herpes zoster (shingles)	Varicella-zoster virus	Live attenuated vaccine	Merck & Co.	Keating (2016) and Levin et al. (2018)
Rotavac oral	Rotaviral diarrhoea in children	Rotavirus	Live attenuated vaccine of human live rotavirus strain G9P11	Bharat Biotech International Limited	Chandola et al. (2017)
Dukora (monovalent oral cholera vaccine)	Cholera	*Vibrio cholerae*	Killed whole cell vaccine consisting of Inaba and Ogawa serotypes of *V. cholerae* O1 in conjunction with recombinant cholera toxin B subunit (WC-rBS)	Crucell Sweden AB, Stockholm, Sweden	Khan et al. (2017)
Shanchol (bivalent oral cholera vaccine)	Cholera	*Vibrio cholerae*	Killed whole cell cholera vaccine consisting of *V. cholerae* lacking cholera toxin B subunit	Shantha Biotechnics, Hyderabad, India	Ivers et al. (2015)
rVSVΔG-ZEBOV-GP vaccine	Ebola virus disease	Zaire Ebola virus	Recombinant vesicular stomatitis virus vaccine	Merck Phase 1 and phase 2 trial	Agnandji et al. (2017) and Kennedy et al. (2017)

(continued)

Table 15.1 (continued)

Vaccine	Disease	Causative microbe	Components	Manufacturer	References
ChinZIKV (recombinant chimeric ZIKV vaccine)	Zika virus disease	Zika virus	Replacement of the prM-E genes of Japanese encephalitis live attenuated vaccine JEV SA14-14-2 with the corresponding region of an Asian ZIKV strain FSS13025	Awaiting clinical development	Li et al. (2018)
Gardasil (recombinant human papillomavirus vaccine)	Cervical cancer	Human papillomavirus	Major capsid protein L1 of HPV types 6, 11, 16, and 18	Merck & Co.	Stanley (2007)

reverse vaccinology (Del Tordello et al. 2017). Recombinant protein subunit vaccines have been formulated with the help of protein antigens synthesized with heterologous host cells including *Escherichia coli*, *Saccharomyces cerevisiae*, and *Pichia pastoris* and mammalian cells. To design new recombinant protein production strategies, the gene sequence should be optimized to be stably expressed in the recombinant host cell. Optimizing culture conditions and induction protocols increases recombinant protein yields and it has been demonstrated in cultures of both *P. pastoris* and *E. coli* (Bill 2015). For posttranslational modifications such as glycosylation of the expressed protein, baculoviral system in insect cells is ideal (Demain and Vaishnav 2009). Microbial cell surface display is another strategy, which deals with expressing the protein of interest as a fusion to various anchoring motifs like surface proteins or their fragments. The host strain selected for display must be compatible with the protein of interest being displayed with minimal activity of proteases and should be able to cultivate without lysis (Lee et al. 2003). This approach has been used for the development of live vaccine where the heterologous epitopes were exposed on human commensal or attenuated pathogenic bacterial cells to evoke antibody responses specific to the antigen (Lee et al. 2000; Liljeqvist et al. 1997). More strategies for vaccine production are being developed for optimum yield in research laboratories globally in an effort to reduce the manufacturing costs, and microbes as hosts for the production of vaccine would be very advantageous in achieving this goal.

15.3.2 Toxins and Antimicrobial Peptides

15.3.2.1 Toxins

Microbial toxins are poisons produced biologically by either bacteria or fungi. They function as autonomous molecules, attacking specific cells in an organism by punching holes into the cell membranes or modifying intracellular components. Some bacteria secrete toxins into their surroundings to overcome host defence and are responsible for the symptoms of bacterial infections (de Wit 2013). Microbial toxins are typically soluble, stable, non-volatile, and highly bioactive compounds that may have cytotoxic, inflammatory, immunosuppressive, and carcinogenic effects (Korkalainen et al. 2017). Bacterial toxins are classified into two endotoxins and exotoxins while fungal toxins are classified into peptidic toxins and non-peptidic toxins. Despite of their detrimental effects, the toxins have been used as therapeutics, cosmetic agents, and adjuvants or drug delivery agents (Fabbri et al. 2008) (Table 15.2).

15.3.2.2 Antimicrobial Peptides and Proteins

Antimicrobial peptides and proteins (AMPs) or host defence peptides (HDP) are a diversified group of very small, normally positively charged molecules composed of varying number of amino acids. Multicellular organisms produce them as a first line of defence. They are used by unicellular organisms to compete for nutrients with other organisms. AMPs can be classified based on various parameters such as biological activity, 3D structure, and peptide family (Wang 2015). In 1939, Rene Dubos discovered and isolated the first microbial peptidic antibiotic Gramicidin from *Bacillus brevis* (renamed as *Brevibacillus brevis*) (Dubos and Cattaneo 1939). Since then, new AMPs are being discovered and their biochemical aspects were studied to shed light on their mechanism of action as well as their potential in clinical therapeutics. Some of these AMPs are listed in Table 15.3. Existing AMPs are being genetically engineered to create recombinant peptides with greater potency against infectious microorganisms. Thus, they represent attractive alternative to antibiotics in controlling pathogenic microbes and maintenance of human lifespan.

15.3.3 Microbial Proteins as a Food and Feed Source

Proteins are a dietary requirement for both humans and domesticated animals. Nutritious food is required in bulk quantities for livestock and pisciculture industry, both of which are among the major sources of proteins for humans. The concerns about future food security are raising due to rapidly increasing human population which is expected to reach ten billion in 2050 as per the United Nations report. Sustainable manufacturing of proteins in bulk will reduce the strain on the environment to provide sufficient nutritious food for the maintenance of these industries. Microbial proteins also known as single-cell proteins (SCP) can be a solution to this perplexing problem as bacteria already have high protein content and multiply exponentially using low-cost substrates under optimal conditions. SCP is a protein

Table 15.2 Microbial toxins, their mode of action, and applications

Toxin	Source microbe	Biochemistry	Deleterious effect	Potential applications	References
Viral toxin					
NSP4 viral		Viral replication and Ca^{2+} immobilization which may further induce signalling through other Ca^{2+}-sensitive cellular processes to potentiate fluid secretion while curtailing fluid absorption	Causes diarrhoea similar to rotaviral diarrhoea	Study of causes and possible vaccine for rotaviral diarrhoea	Morris and Estes (2001)
Rotavirus enterotoxin					
Bacterial exotoxins					
Botulinum neurotoxin	*Clostridium botulinum*	Single, inactive polypeptide (150 kDa). Acts by blocking the release of acetylcholine at neuromuscular junction	Causes botulism in humans	Treatment of dystonia, spasticity, hyperhidrosis, prophylactic migraine, strabismus, reduction of glabellar lines, and cosmetic surgery	Sharma (2016), Burroughs and Anderson (2015), and Foster et al. (2018)
Tetanus neurotoxin	*Clostridium tetani*	A heavy and a light chain linked via a disulphide bridge	Spastic paralysis in tetanus disease	Development of tetanus toxoid vaccine against tetanus	Masuyer et al. (2017) and Link et al. (1992)
		Inhibits neurotransmission of inhibitory interneurons			
Anthrax toxin	*Bacillus anthracis*	Consists of three proteins: protective antigen (83 kDa), lethal factor (90 kDa), and edema factor (89 kDa)	Anthrax disease	Tumour targeting and drug delivery	Bachran and Leppla (2016)
		Affects signalling pathways and modulates immunologic responses			
Cholera toxin B subunit	*Vibrio cholera*	Homopentameric protein, non-toxic in nature (55kD)	Subunit of virulence factor of *V. cholera* responsible for cholera	Part of a licensed oral cholera vaccine	Verier et al. (2006) and Baldauf et al. (2015)
		High affinity binding to the GM1 ganglioside present on mammalian cells		Adjuvant for mucosal vaccine	
				Immunomodulatory and anti-inflammatory agent	

(continued)

Table 15.2 (continued)

Toxin	Source microbe	Biochemistry	Deleterious effect	Potential applications	References
Fungal toxins					
α-Sarcin ribotoxin	*Aspergillus giganteus*	Composed of a rigid hydrophobic core and some exposed segments, mostly loops	Protein biosynthesis inhibition	Immunotoxin in antitumor therapy	Lacadena et al. (2007) and Olombrada et al. (2014)
		Cleaves one phosphodiester bond of the sarcin/ricin loop, impairing its essential function		Bioinsecticides	
Candidalysin	*Candida albicans*	Amphipathic protein containing an amino-terminal α-helical hydrophobic region	Mucosal infection	Novel therapeutic agents	Pasricha and Pearson (2016) and Moyes et al. (2016)
		Causes damage to epithelial membranes, activates a danger response signalling pathway and epithelial immunity			
Bassiacridin	*Beauveria bassiana*	Single polypeptide chain (60 kDa)	Cytotoxic effect on insect cells	Microbial control of locusts and grasshoppers	Quesada-Moraga and Alain (2004)
		Exhibits β-galactosidase, β-glucosidase, and N-acetylglucosaminidase enzyme activities			
Ribotoxin hirsutellin A	*Hirsutella thompsonii*	Single polypeptide chain (15 kDa) with at least two cysteine residues involved in the formation of an intrachain disulphide bridge	Cytotoxic effect on insect larvae	Model for bioinsecticide development	Mazet and Vey (1995)

Table 15.3 Representative antimicrobial peptides with their properties and applications

AMP	Source microbe	Type	Biochemistry	Mechanism of action	Target microbe	Activity	Applications	References
Nisin	*Lactococcus lactis* subsp. *Lactis*	Cationic lantibiotic	Exists in multimeric forms (3500 Da)	Interact with cytoplasmic membranes of bacteria and form transient pores, causing efflux of ions and small molecules	*Clostridium butyricurn, C. tyrobutyricum*, and *C. aporogenes*	Delayed toxigenesis of *C. botulinum* type A and type B	Increasing shelf life of processed cheese, dairy desserts, milk, fermented beverages, bacon, frankfurters, and fish	Vandenbergh (1993)
Linear gramicidin	*Bacillus brevis* (strain BG)	–	15-residue peptides with β-helix like structure	Forms transmembrane channels causing diffusion of intracellular component out of the cell leading to cell dysfunction	Gram-positive and Gram-negative bacteria	Antibacterial, antiviral, spermicidal	Ophthalmic use as solutions/drops (approved in 1940s)	Mishra et al. (2017), Yang and Yousef (2018), and Dubos (1939)
						Anti-HIV	Topical application as antibiotic cream	
Alamethicin	*Trichoderma viride*	Peptaibol	20-amino acid peptide with a backbone conformation resembling the flexible helix-bend-helix arrangement	Self-associates into hexameric barrel–stave transmembrane helices and permeabilizes membranes by forming channels through them	Gram-positive bacteria, such as *Enterococcus faecalis, Staphylococcus haemolyticus, Streptococcus viridans*, and *Staphylococcus aureus*	Inhibits methanogenesis and promotes acetogenesis in bioelectrochemical systems	Production of industrially relevant organic compounds by inhibition of methanogenesis	Bechinger (1997), Zhu et al. (2015), Ray et al. (2017), and Ageitos et al. (2017)
							Enhanced power generation in microbial fuel cells	

(continued)

Table 15.3 (continued)

AMP	Source microbe	Type	Biochemistry	Mechanism of action	Target microbe	Activity	Applications	References
Copsin	*Coprinopsis cinerea*	Defensin	CSαβ core fold interconnected by six disulphide bonds and N-terminal pyroglutamate	Binding to the peptidoglycan precursor lipid II and prevention of cell wall biosynthesis	Gram-positive bacteria, such as *Listeria*, *Enterococci*, and *Bacilli* including vancomycin-resistant *E. faecium*	Most potent against *L. monocytogenes*, a foodborne pathogen causing severe forms of listeriosis	Novel highly stabilized scaffold for antibiotics	Franzoi et al. (2017) and Essig et al. (2014)

source from microbial cultures such as bacteria, yeast, filamentous fungi, and algae with the potential to be animal feed as well as human protein supplements. They are either dehydrated microbial cell culture or purified proteins derived from microbial cell culture (Ugbogu and Ugbogu 2016).

Some SCPs that are available commercially or under study are indicated in Table 15.4. SCPs offer various advantages: they contain high protein content (60–82% of dry cell weight) along with other nutrients, they are good source of essential amino acids such as lysine and methionine which are limited in most plant- and animal-based foods (Suman et al. 2015), the microbes have rapid generation time, they are genetically modifiable (e.g. for composition of amino acids), and they require less space as compared to conventional agriculture. However, SCPs have some disadvantages like high nucleic acid content, accumulation of uric acid crystals caused by bacterial SCPs leading to gout, possibility of allergic reactions with fungal SCPs as mycotoxins are allergens, and slow digestibility due to rigid cell wall. Currently SCPs are produced using solid-state fermentation (Jaganmohan et al. 2013). Recent advances in fermentation, extraction, downstream processing techniques, and optimization of substrates/conditions resulted in large-scale production of protein biomass. Production and marketing of a wider range of SCPs could be a promising step to alleviate food shortage and malnutrition.

15.3.4 Microbial Factories for Production of Recombinant Proteins

Microbes represent convenient system for production of proteins which are difficult to obtain from their native sources (Ferrer-Miralles et al. 2009). The use of microbes for protein production has increased in recent times due to the low cost, high productivity, and rapid use (Terpe 2006). A range of microbes including bacteria such as *Escherichia coli and Bacillus megaterium*, filamentous fungi such as *Aspergillus niger* and *Trichoderma reesei*, and yeast such as *Saccharomyces cerevisiae* and *Pichia pastoris* are exploited as recombinant cell factories. The first licensed protein drug successfully produced by recombinant DNA technology was human insulin in *E. coli* by Genentech and was commercialized by Eli Lilly in 1982. At present, nearly 400 drugs out of approved 650 protein drugs are produced by recombinant technologies (Sanchez-Garcia et al. 2016).

Recombinant protein production involves manipulation of the gene expression system of microbes with the aim of producing large amounts of recombinant protein tailored for a specific function. For a microbe to express foreign protein, the gene encoding the protein of interest is cloned into an expression vector with a suitable promoter gene and then introduced into the microbe. If the gene contains introns, it is cloned from a cDNA library as bacteria cannot excise introns. The plasmid is then transformed into a suitable host that is able to produce the desired protein. The transformed strain is transferred to liquid media and cultured. At a specific stage of growth, a chemical inducer triggers the promoter of the expression vector and induces expression of recombinant gene. The polypeptide produced folds into the

Table 15.4 Representative antimicrobial peptides with their properties and applications

Type of microbe	Microbe	Substrate	Cultivation	Commercial name	Crude protein content (%)	Remarks/uses	References
Algae	*Euglena gracilis*	Cramer–Myers medium	Photo-bioreactor	Whole cell form marketed by Algaeon Inc.	47	Potential animal feed source	Chae et al. (2006) and Rodríguez-Zavala et al. (2010)
						Biotechnological production of metabolites like α-tocopherol and paramylon	
Bacteria	*Rhodococcus opacus*	Corn stover	Bioreactor	NA	47.0–56.9	Fortification to livestock and poultry feed	Mahan et al. (2018)
		Orange waste					
		Lemon waste				Food source in fish farming	
	Arthrospira platensis, Arthrospira maxima	Grows photosynthetically, molasses can be used to stimulate growth	Photo-bioreactor	Spirulina marketed by Cyanotech Corporation, DDW Inc., C.B.N. Spirulina Canada Co., Ltd	50–70	Supplement for both humans and animals	Andrade and Costa (2007)
	Methylococcus capsulatus	Methane	U-loop fermentation	*UniProtein*® marketed by UniBio	70–72.9	Animal feed supplement	Prado-Rubio et al. (2010)
Yeast	*Kluyveromyces marxianus*	Whey	Laboratory scale fermenter	NA	83.33	Food source	Nayeem et al. (2017)

Filamentous fungi	*Fusarium venenatum*	Wheat starch as carbon source	Batch culture,	Quorn® (mycoprotein) marketed by Marlow foods	56	Protein-rich food containing essential amino acids	Garodia et al. (2017)
		Ammonia as nitrogen source	Continuous culture			Fat or cereal replacer	
			Air-lift fermenters			Helps control blood lipids, blood glucose, and appetite due to dietary fibre content	
	Penicillium janthinellum	Sugarcane bagasse	Laboratory scale fermentation	NA	40–50	SCP as animal feed and xylanase by-product as prebleach agent in paper and pulp industry	Rao et al. (2010)

NA not available

recombinant protein of interest, which can be further purified by suitable purification approaches. The production of the target protein can also be scaled up from initial batch cultures to stirred tank bioreactors on fed-batch regimens to manufacture large protein biomass, which is then released and purified (Overton 2014).

E. coli has been one of the most commonly employed microbial cell factories for heterologous expression, and it has been used for the production of 30% of recombinant proteins approved by the FDA (Rosano and Ceccarelli 2014). It has been used for producing a range of biopharmaceuticals ranging from growth hormones (Goeddel et al. 1980; Olson et al. 1981), growth factors (Kwong et al. 2016), peptides (Zorko and Jerala 2010), and therapeutic proteins (Mane and Tale 2015). However, the major hurdles in exploiting *E. coli* as an expression host include inclusion body formation due to aggregation of overexpressed protein. Proteins derived from eukaryotes often undergo posttranslational modifications to achieve proper folding, but *E. coli* lacks such system and thus recombinant proteins expressed in *E. coli* microenvironment does not fold properly or misfolding occurs (Sharma and Chaudhuri 2017). The membrane proteins and the proteins with molecular weight more than 60 kDa are also difficult to express in *E. coli*. Toxic nature of heterologous protein and instability of the plasmid are other obstacles affecting successful expression in *E. coli*. *Saccharomyces cerevisiae* is another conventionally used host for recombinant protein production. Other non-conventional yeasts are *Hansenula polymorpha*, *Pichia pastoris*, and *Yarrowia lipolytica* (Kim et al. 2015).

The dominant role of yeast is seen in production of human blood proteins (Martinez et al. 2012), insulin analogues, and hepatitis vaccine (Wang et al. 2017). Efforts are being done to improve the titre, rate, and yield of the yeast cell factory through rational metabolic engineering in *Saccharomyces cerevisiae*. Multiple-genome integration was observed to be an ideal approach for generating stable strains with high copy numbers of heterologous genes. Strong glycolytic promoters (PGK1p, TPI1p, ADH1p) and inducible promoters have been developed to induce heterologous protein expression at various levels as the glycosylation capability of yeast is inappropriate for human proteins (Hou et al. 2012; Wang et al. 2017). One of the bottlenecks of protein production in yeasts is the protein secretory machinery, which may not be able to handle a high flux of proteins requiring specific posttranslational modification. This can result in missorting where the heterologous protein is targeted to the vacuole for degradation instead of being secreted. The use of systems biology integrates large-scale datasets (-omics) with mathematical modelling to direct metabolic engineering and site-directed mutagenesis towards overcoming the limitations of the protein secretion machinery (Martínez et al. 2012; Wang et al. 2017).

Information obtained using systems biology involving the study of the transcriptomics, proteomics, metabolomics, and metabolic flux analysis of *P. pastoris*, a methylotrophic yeast, is being utilized to enhance protein folding and secretion as well as engineer the recombinant protein process towards maximizing the yield and improving the yeast strain (Zahrl et al. 2017). High-throughput screening of improved strains with high protein yield in *S. cerevisiae* and *P. pastoris* specific to the target protein is the final step in the development of recombinant strains (Ahmad et al. 2014; Wang et al. 2017). Filamentous fungi are other candidates for recombinant protein production. They have mostly been used as robust cell factories for producing pharmaceutically relevant enzymes. Examples of recombinant enzymes are catalase, glucose oxidase, and phytase from *Aspergillus niger* and cellulose and xylanase from *Trichoderma reesei* (Archer 2000). Filamentous fungi have enormous potential in efficient large-scale production of recombinant proteins as they are cheap to cultivate and downstream processing is easier as the proteins are secreted through hyphae (Nevalainen and Peterson 2014). Numerous efforts have been made to develop filamentous fungi as a host for recombinant proteins, but further improvement is required for the expression of wider range of heterologous proteins. To achieve this, proteome profile of filamentous fungi like recombinant strains of *Aspergillus nidulans* is being performed to identify the bottlenecks in heterologous protein expression (Zubieta et al. 2018). These findings help us to understand the mechanisms underlying protein production and to rationally manipulate target genes for the improvement of fungal strains.

15.4 Secondary Metabolites

Secondary metabolites derived from microbes represent the important group of compounds with a wide range of applications. The term secondary metabolite has been introduced by Bu'LocK in 1961. Secondary metabolites are the low-molecular-weight products with no direct involvement in physiology and development of microbe but may render several benefits to the organism (Bu'Lock 1961). For instance, antibiotics are one of the well-known secondary metabolites, which confer selective growth advantage and better survival ability to the host microbe. Other examples of secondary metabolite from microbial origin with varied biological functions include antibiotics, alkaloids, pigments, antitumour agents, toxins, growth promoters, carotenoids, and enzyme inhibitors.

15.4.1 Microbial Source of Secondary Metabolites

Secondary metabolites or small molecule natural products are synthesized by prokaryotes like bacteria to eukaryotes like fungi, plants, and animals, although the secondary metabolite producing ability is unevenly distributed. Secondary metabolites are formed by the biosynthetic pathways which branch off from the primary metabolic pathways. Secondary metabolism in fungi occurs during stationary phase in the liquid cultures and is often linked to the onset of morphological developments in surface-grown cultures. Similarly, in bacteria the secondary metabolites are formed during the late growth phase. Nearly 20,000 so-called microbial secondary metabolites are known (Marinelli 2009). Among prokaryotes, the filamentous actinomycetes species has been reported to produce over 10,000 bioactive compounds, streptomyces produces 7600 compounds, and rare actinomycetes produces nearly

2500 bioactive compounds, and they produce 45% of known bioactive microbial metabolites, representing the largest producer group (Bérdy 2005). Streptomyces is the largest antibiotic-producing genus and it alone provides more than 60% of the antibiotics (Esnault et al. 2017).

The genome sequencing of model actinomycete *Streptomyces coelicolor* A3(2) led to identification of more than 20 gene clusters capable for coding the secondary metabolites (Bentley et al. 2002). The gene clusters (polyketide synthases type I and II, nonribosomal peptide synthetases) were found in its genome. This strain also produces metabolites like methylenomycin, prodigiosin, actinorhodin, and a calcium-dependent antibiotic. The microbes with lesser ability to produce secondary metabolites include mycoplasma, mycoplasmatales, and spirotheces. Among the eukaryotic fungi, ascomycetes and endophytic fungal species are frequent producers, while yeasts, phycomycetes, and slime moulds are less frequent producers. The fungal bioactive compounds constitute 38% of known microbial products (Bérdy 2005). It has been shown that a large number of microbial species that cannot grow under standard laboratory conditions, known as 'unculturable' strains, can also be potential source of novel secondary metabolites. Development of methods to culture such microbes would further allow the exploitation of microbial diversity to produce interesting metabolites (Lewis et al. 2010; Newman 2016).

15.4.2 Approaches for Isolation and Identification of Bioactive Secondary Metabolites

In 1929, the serendipitous discovery of antibiotic penicillin G from *Penicillium notatum* (Fleming 1929) established the therapeutic potential of this fungal secondary metabolite and further expedited the exploration of novel bioactive metabolites. Since then various microbial metabolites have been isolated including β-lactams, aminoglycosides, glycopeptides, tetracyclines, and cephalosporins. The classical approach leading to the antibiotic discovery was based on the growth inhibition of target microbes. However, in recent times the screening methods based on growth inhibition has turned out to be unsuccessful in identifying new antibiotics. This propelled the development of modern methodologies and techniques to accelerate the discovery process (Davies 2011).

15.4.2.1 Isolation of Secondary Metabolite Producing Microbes and Strain Improvement

The screening of microbial fermentation extracts to identify biologically active compound was practiced previously. For successful screening, the selection of growth conditions that can initiate the synthesis of secondary metabolites in microbes and the bioassays or analytical methods that allow detection of the secondary metabolite are the general requirements. Once the desired strain that can overproduce a particular compound is isolated, the next step involves improving the concentration of the compound. It may be achieved by optimization of the culture conditions like medium composition, pH, temperature, agitation, and aeration.

Various additives can also be tested in culture media as limiting precursors of desired compound; e.g. lysine is added to the culture media as a precursor and cofactor to enhance the production of cephamycin by *Streptomyces clavuligerus* (Demain 1998; Gonzalez et al. 2003; Khetan et al. 1999).

The advent of recombinant DNA techniques led to manipulation and improvement of microbial strain for enhanced production of target secondary metabolite. In classical genetics, mutations are introduced randomly or on rational basis followed by screening/selection to identify the mutants with desired improvements (Sharma et al. 2014). The random screening method requires the limited knowledge of genetics, biochemistry, and physiology of biosynthetic pathway. On the other hand, rational screening requires basic knowledge of pathway regulation and product metabolism. For example, *Streptomyces hygroscopicus* mutant strain producing higher titre of rapamycin was obtained after mutagenesis and screening of parent culture (Cheng et al. 2001).

15.4.2.2 Mining Microbial Genomes for New Natural Products

The whole genome sequencing enabled rapid identification of the producer strains. Only specific regions of genome, namely, biosynthetic gene clusters, are involved in formation of valuable bioactive molecules. These gene clusters encode for proteins, which participate in synthesis of bioactive molecule using building blocks derived from primary metabolism. The ribosomal peptide synthetases and polyketide synthases have particularly much attention in recent years as they account for majority of structurally diverse, clinically and commercially important molecules (Naughton et al. 2017). Recently, microbial genome sequencing analysis has revealed the presence of numerous cryptic or orphan gene clusters which are responsible for production of a number of unknown secondary metabolites (Chiang et al. 2009). Various strategies have been devised to identify the metabolic products of the microbial cryptic gene clusters. It includes isotopic tracer technique, in vitro reconstitution, sequence analysis to predict physico-chemical properties of product, gene knockout or comparative metabolic profiling, and heterologous expression of cryptic gene cluster (Bentley 1999; Challis 2008; Davati and Habibi Najafi 2013). Web-based platforms like antiSMASH 2.0 (Blin et al. 2013), ClustScan (Starcevic et al. 2008), and CLUSEAN (Weber et al. 2009) have also been developed to automate the identification and characterization of bioactive secondary metabolites.

15.4.2.3 Metabolic Engineering

Metabolic engineering is the approach to modify the existing metabolic pathway or combining the pathways or enzymes from different host to single microbe with an objective of improved production of target compound or to produce new compounds in host cells from simple, inexpensive starting material (Keasling 2010). The important design parameters in production of secondary metabolite are yield and productivity. Thus, in optimizing the production of microbial metabolite, the primary aim is to enhance the metabolic flux towards the compound of interest and to minimize the flux towards the by-products. Increasing the flux towards the product increases both the overall productivity and yield (Nielsen 1998). Metabolic engineering has been successfully applied for the efficient production of amino acids like L-threonine and L-valine, antimalarial drugs like artemisinin, anticancer drugs like taxol, antibiotics like β-lactams and cephalosporins, and benzylisoquinoline alkaloids (Davati and Habibi Najafi 2013; Minami et al. 2008).

15.4.3 Biosynthesis of Secondary Metabolites and Its Regulation

The secondary metabolite production is not only strain dependent but it is also influenced by diverse regulatory conditions like growth stage, optimum supply of nutrients, and the regulatory effects imparted by them (Liu et al. 2013). The production of particular secondary metabolite initiates due to the recognition of specific signal, transduction of this signal to generate the required regulators followed by regulator-mediated activation of biosynthetic gene cluster to produce the secondary metabolite, and then transport of the produced metabolite (Chang and Stewart 1998). The physiological regulation for production of secondary metabolites usually differs with the kind of microbe and metabolic pathway involved. It has been shown that when antibiotic-producing strain like streptomyces are cultivated under conditions that leads to nutritional stress, the stationary growth phase conforms to the onset of biosynthesis of secondary metabolite (Bibb 2005). Nutrients in culture media have been reported to be exerting their regulatory effects by activating or repressing the transcription factors and regulatory proteins.

Fine-tuning of optimal concentration of carbon source in medium is an important parameter to balance the qualitative production of the secondary metabolite and growth of the microbe. Presence of glucose as a carbon source usually improves the growth of the host but could interfere with the production of varied secondary metabolites like cephalosporin, alkaloids, and actinomycin. However, in some cases glucose acts as a good substrate for growth and differentiation as well as for the secondary metabolite production like aflatoxin (Luchese and Harrigan 1993). Glucose in high concentration of 100 g/L maximizes the production of the anticapsin by *Streptomyces griseoplanus*. The type of nitrogen sources employed in the medium affects the secondary metabolic pathways differently. Ammonium ions cause inhibition of novobiocin, cephamycin, and rifamycin production (Aharonowitz 1980). The biosynthesis of gibberellins by the fungus *Gibberella fujikuroi* was shown to be suppressed by the presence of ammonium ions and glucose as well (Brückner 1992). L-amino acids were found to positively influence the production of actinomycin D by *Streptomyces parvulus* (Bennett et al. 1977). The type of L-amino acids added to synthetic media strongly influenced the production of mycotoxins like emodin, catenarin, and islandicin by isolates of *Pyrenophora tritici-repentis* from wheat (Bouras et al. 2016). The concentration of inorganic phosphate that favours growth of the microbes generally exerts negative control on synthesis of secondary metabolites. However, in some cases high phosphate concentration is well tolerated for production of secondary metabolite as reported in the case of avermectin biosynthesis by *Streptomyces avermitilis* (Čurdová et al. 1989).

Secondary metabolite production also requires trace elements like manganese, iron, and zinc, although their required optimal concentration may vary depending on the metabolite to be produced.

15.4.4 Applications of Secondary Metabolites

Secondary metabolites are valuable compounds with a wide range of applications (Williams et al. 1989). The microbial secondary metabolites are now progressively used as drugs for the treatment of various diseases in place of synthetic drugs. They are widely used as uterocontractants, anti-inflammatory agent, anticancer drug, cholesterol-lowering agent, hypotensive agent, immunosuppressant, antibacterial/antifungal agent, and antiparasitic agent (Gonzalez et al. 2003). They are also being used for non-medical applications like weed management and plant growth regulation (Cutler 1995; Sadia et al. 2015).

Secondary metabolites in addition to their known activities have also shown alternative activities, and thus they have been unexpectedly used as possible solution to other diseases for which the effective treatment is not available. β-Lactams are known for their antibiotic action, and their derivatives have also displayed antitumour prodrug activity (Xing et al. 2008). Prodigines, pigmented antibiotics, display antifungal, antiprotozoal, antimalarial, anticancer, and immunosuppressive activities in addition to their antibiotic activity (Williamson et al. 2006). Squalestatin, a fungal metabolite known for lowering the cholesterol by inhibiting 3-hydroxy-3-methylglutaryl-CoA reductase enzyme of cholesterol biosynthesis pathway, has been identified as a potential drug against prion disease (Bate et al. 2004). Thus, exploring the new functions of existing secondary metabolites along with speeding up the process of identification of the novel secondary metabolites can allow the better targeting of the diseases for which currently no effective solutions are present (Vaishnav and Demain 2011).

15.5 Valuable Chemicals

Numerous chemicals are used in everyday life to serve various purposes such that they act as drugs for treating diseases, as fertilizers, as disinfectants, as industrial solvents, as pest control agents, and as health or hygiene products. These chemicals are produced using defined chemical synthesis reactions where simple chemicals are reacted to generate target products. The chemicals can be categorized into bulk chemicals, fine chemicals, and speciality chemicals. Bulk or commodity chemicals are produced on large scales and used as intermediates for production of other chemicals. Fine chemicals are produced as pure chemical substance in small quantities unlike bulk chemicals and are often used for production of speciality chemicals such as agrochemicals and pharmaceuticals.

15.5.1 Microbial Platform for Production of Bio-based Chemicals

The need of improved biotechnological processes for production of target chemicals is increasing with each passing year owing to the limited fossil resources and serious climate changes (Wu et al. 2018). The popularity of microbial systems as a tool for biological synthesis of chemicals is gaining momentum as they can produce a variety of complex molecules, and they require relatively less energy resources as compared to chemical synthetic techniques, thus making it a feasible option to produce fine chemicals. Many fine chemicals have been found to be ideal for microbial biosynthesis as they are intermediates or products of the natural metabolic pathways of various microbes. Industries that benefit from microbial biosynthesis include food, agriculture, chemical, pharmaceutical, and cosmetics (Gurung et al. 2013). The natural biosynthetic pathway in a microbial cell can also be modified by combining various approaches to produce target chemicals.

1. Enzymatic synthesis of fine chemicals where enzymes with or without coenzymes convert the substrate to the chemical of interest. The genes responsible for expressing the enzymes capable of catalysing the bio-based reaction are identified and isolated. The computational tools are used to mine genome and transcriptome data to identify novel biosynthetic pathways and enzymes (Lautru et al. 2005; Zhao et al. 2013). The identified enzymatic synthesis system is introduced into microbial cell to create a microbial cell factory.
2. Metabolic engineering is used to increase yield as well as productivity by redesigning the existing biosynthetic pathway to optimize the production of target compound. Various tools used in designing metabolic pathways are biochemical network integrated computational explorer (BNICE), RetroPath, GEM-Path, OptStrain, and DESHARKY (Chae et al. 2017). Flux balance analysis is a method that indicates how gene deletion and expression can be manipulated to distribute carbon towards chemicals of interest without blocking or reducing cell proliferation. This is a standard method to optimize metabolic pathways (Orth et al. 2010).
3. Genetic manipulation according to the redesigned pathway map obtained by computer simulation can be performed to give a relatively efficient recombinant strain of the selective microorganism. This involves heterologous expression, overexpression, downregulation, deletion, or mutation of the gene of interest.

Despite of various advantages offered by microbial systems for bio-based production of chemicals and other valuable materials, their potential could not be fully exploited as new alternative energy sources are coming to existence. Moreover, higher production cost of bio-products, lower yields, relatively decreased efficiency of bioprocesses as compared to chemical processes, and longer production periods due to slow microbial growth are other factors hindering the development of bio-based products at commercial scale (Chen 2012).

15.5.2 Bio-manufacturing of Bulk and Speciality Chemicals

The production of bulk chemicals is primarily driven by petrochemical feedstocks. However, as demand for bio-based chemicals is increasing, the chemical processes are being replaced with microbial catalysts and improved fermentation methods. Thus, the possibilities to utilize renewable resources for sustainable production of commodity chemicals are rapidly progressing in the current scenario (Hermann et al. 2007). Bio-based production of several commodity chemicals including alcohols, organic acids, amino acids, aromatic amines, diols, polyhydroxyalkanoates, and polysaccharides through fermentation has been successfully reported (Table 15.3). In parallel to fermentation approaches, system metabolic engineering has also been successfully used in production of commodity chemicals like amino acids (Ma et al. 2017). Such engineering strategies have been applied mainly in *Corynebacterium glutamicum* and *Escherichia coli* for amino acid production. Dedicated attempts are being made by researchers worldwide to construct novel pathways in microbes for bio-manufacturing of target bulk chemicals (Shin et al. 2013) (Table 15.5).

Like bulk chemicals, the fine chemicals were also conventionally produced by energy-intensive multistep chemical processes that resulted in high levels of wastes and by-products. However, the efforts are being made to exploit biological routes for chemical production on par with chemical synthetic techniques. The fine chemicals are synthesized by microbes, either as products of their natural metabolic pathways or by genetically engineering their metabolic pathways to produce the desired product (Hara et al. 2014). A range of speciality chemicals like isoprenoids, flavonoids, alkaloids, aromatic compounds, polyphenols, peptides, drugs, organic acids, and oligosaccharides has been reported to be produced by microbes using synthetic biology principles. The production strategy of few chemicals in microbial systems has been summarized.

15.5.2.1 Artemisinin

The antimalarial drug artemisinin is a sesquiterpene lactone with an endoperoxide bridge. It is naturally produced by *Artemisia annua* (sweet wormwood) (Liu et al. 2006; Rathod et al. 1997). However, the methods for extraction of artemisinin were not economical and resulted in insufficient production levels. This led to development of recombinant strains as microbial factories to produce artemisinic acid, which is a precursor of artemisinin. This precursor was then converted to artemisinin by following synthetic organic chemistry steps (Paddon and Keasling 2014). In one of the studies, *E. coli* strain was engineered to synthesize the precursor amorphadiene by introduction of heterologous, high-flux isoprenoid pathway from *S. cerevisiae* to *E. coli* (Martin et al. 2003). The pathway genes were coexpressed with a codon modified amorphadiene synthase (Martin et al. 2003) resulting in a recombinant strain that could produce amorphadiene up to 24 mg/L. In a follow-up study, a higher yield of amorphadiene was achieved by utilizing a two-phase partitioning bioreactor (TPPB) strategy that resulted in efficient separation of amorphadiene from the fermentation broth (Newman et al. 2006). Much later, production of

Table 15.5 Representative examples showing bio-production of commodity chemicals by fermentation methods

Commodity chemical	Involved microbe	Applications	References
Ethanol	*Saccharomyces cerevisiae*	Industrial chemical, motor fuel, gasoline additive, solvent, preparation of tonics and cough syrups, beverages	Stewart et al. (1983) and Dien et al. (2000)
	Zymomonas mobilis		
	Ethanologenic *Escherichia coli strains*		
Lactic acid	Lactic acid bacteria like *Lactobacillus fermentum*, *Lactobacillus bulgaricus*	Pharmaceutical, food, and cosmetic industry	Wee et al. (2006)
Polyhydroxyalkanoates	*Ralstonia eutropha*, *Alcaligenes latus*, *Chromobacterium violaceum*	Substitute of synthetic plastic, biomaterial for tissue engineering	Singh Saharan et al. (2014)
1,3 propanediol	*Clostridium pasteurianum*, *Klebsiella pneumoniae*, *Citrobacter freundii*	Production of polyesters, polyurethanes, polyethers	Biebl et al. (1999)
Butanol	*Clostridium acetobutylicum*	Fuel, paint thinner, solvent	Abou-Zeid et al. (1978)
Citric acid	*Aspergillus niger*, *Bacillus licheniformis*, *Saccharomycopsis lipolytica*	Food and pharmaceutical industry	Vandenberghe et al. (1999)
L-lysine	*Brevibacteria*, *Corynebacterium glutamicum*	Animal feed supplement, medicament, chemical agent, food industry	Hodgson (1994)
Acetic acid	*Acetic acid bacteria like Acetobacter aceti*	Food and beverage industry, production of industrial chemicals	Raspor and Goranovič (2008)
L-glutamic acid	*Corynebacterium glutamicum*	Food additive, flavour enhancer, therapeutic agent, infusion compound	Nakamura et al. (2007)
Acrylic acid	*Rhodococcus rhodochrous*	Chemical industry, manufacture of plastics, coatings, polymers	Nagasawa et al. (1990)
Glycolic acid	*Candida tropicalis*, *Gluconobacter*, *Pichia naganishii*	Leather industry, texture washing, cosmetics	Kataoka et al. (2001)
Succinic acid	*Basfia succiniciproducens*, *Actinobacillus succinogenes*, *Anaerobiospirillum succiniciproducens*	Food, surfactants, pharmaceutical products, detergents, plastics, precursor molecule	Tan et al. (2014) and Zheng et al. (2009)

artemisinic acid at gram scale (25 g/L) was achieved by optimizing the expression of CYP71AV1:CPR1 along with co-expression of cytochrome b5 and two dehydrogenases (Paddon et al. 2013).

15.5.2.2 γ-Aminobutyric Acid (GABA)

GABA, a non-protein amino acid, is synthesized by microbes, plants, and animals. It acts as an inhibitory neurotransmitter in the central nervous system of mammals and as a stimulant for immune cells (Dhakal et al. 2012). In microbes, it is involved in spore germination in the case of *B. megaterium* and *N. crassa* (Foerster and Foerster 1973; Schmit et al. 1975), while it provides resistance to acidic pH in *L. lactis, E. coli, and other microbes* (Castanie-Cornet et al. 1999; Sanders et al. 1998). It has a wide application in food, cosmetic, and pharmaceutical industry. The biosynthetic route of GABA involves a single-step reaction involving decarboxylation of glutamate to GABA, catalysed by glutamate decarboxylase (GAD) (Ueno 2000). The main GABA-producing microbes are lactic acid bacteria (LAB) (Dhakal et al. 2012). *Corynebacterium glutamicum* expressing Escherichia coli glutamate decarboxylase (GAD) has been engineered for production of GABA, and in order to further enhance its production, protein kinase G has been disrupted resulting in increased intracellular concentration of glutamate precursor and eventually improved yield of GABA (Okai et al. 2014).

15.5.2.3 Resveratrol

Resveratrol (trans-3,5,4′-trihydroxystilbene) is a plant-derived polyphenol that is present in red wine. It is used as an antioxidant, in cosmetic and food industry and as therapeutic agent due to its anticarcinogenic, anti-inflammatory, anti-diabetic, and anti-ageing properties (Beekwilder et al. 2006; Mei et al. 2015). As such, production of polyphenols in microbes is a challenging task due to antibacterial and antifungal activity of these compounds (Daglia 2012). However, still metabolic engineering principles have been utilized to produce such compounds via microbial systems. Engineered *E. coli* and *S. cerevisiae* strains expressing 4-coumarate:coenzyme A ligase from tobacco and stilbene synthase from grapes has been developed to achieve resveratrol accumulations in the culture medium by supplying *p*-coumaric acid as a precursor molecule. These engineered strains showed relatively low production titres (Beekwilder et al. 2006). Another research group investigated various constructs for resveratrol synthesis, different *E. coli* strains, promoters and gene expression combinations, sequence, and structure analysis to achieve high titres (g/l) of resveratrol from biotransformation of *p*-coumaric acid (Lim et al. 2011).

15.5.2.4 Cinnamic Acid

Cinnamic acid is a phenylpropanoid acid, which is used as a cinnamon flavouring agent, in high performance thermoplastics, as precursor for chemical compounds, and as nutraceutical and pharmaceutical products (Vargas-Tah and Gosset 2015). It can be obtained by either chemical synthesis or by extraction from source plant. It can also be produced by engineered microbes like *Escherichia coli*, *Streptomyces lividans, Saccharomyces cerevisiae*, and *Pseudomonas putida.* Genes encoding phenylalanine ammonia-lyase (PAL) and tyrosine ammonia-lyase (TAL) have been expressed in *E. coli* and *S. cerevisiae* to allow the conversion of L-phenylalanine and L-tyrosine to cinnamic acid and *p*-hydroxycinnamic acids (*p*-coumaric acid) (Vannelli et al. 2007). *Pseudomonas putida* S12 strain was engineered for conversion of p-hydroxycinnamic acid from glucose (Nijkamp et al. 2007). The heterologous expression of PAL encoding gene from *Streptomyces maritimus* in *Streptomyces lividans* resulted in production of cinnamic acid from glucose with maximum titre of 450 mg/L (Noda et al. 2011).

15.6 Conclusion

Microbes play a significant role in maintaining the ecological sustainability. They synthesize a wide range of products like antibiotics, toxins, antimicrobial peptides or proteins, and enzymes that help them to thrive in the varied environmental conditions and provide them an ability to compete with other species in their ecological niche. These products are valuable due to their application in industrial bioprocesses, as they are used in nutraceuticals, in agriculture for production of drugs or vaccine, and for generation of clean fuel and bioremediation. For instance, enzymes derived from microbial source have potential applicability in different fields as they are used in pharmaceutical industry, in processing of food products, as therapeutic agents, and in production of biofuels and bioplastics. In order to further enhance their usefulness, protein engineering methods are being employed to generate custom-made biocatalysts for the desired processes.

Microbial surface proteins with antigenic properties represent major target for generation of vaccines, and various strategies have been devised to utilize microbial systems for production of recombinant vaccines to decrease the production costs. Antimicrobial peptides derived from microbes are another group of interesting biomolecules with therapeutic applications owing to their utility as alternative to antibiotics. Similarly, microbial toxins produced by bacteria or fungi are utilized in cosmetic industry, as therapeutic agent, and for drug delivery. Microbial proteins or whole microbial cells are used as food source and feed supplement. A variety of bioactive secondary metabolites have been derived from microbes. The discovery of new bioactive compounds has been achieved by advent of modern techniques like genome sequencing, metabolic engineering, proteomics, and advance computational tools. Metabolic engineering and synthetic biology principles have been successfully employed to develop the engineered microbial strains as cell factories for heterologous expression of recombinant proteins and bio-based production of bulk and speciality chemicals. The natural biosynthetic pathways can be either fine-tuned or novel pathways can be assembled in host microbe to optimize the production of the target compounds. In conclusion, microbes share a major role in bio-production of valuable chemicals, toxins, metabolites, proteins, and peptides with broad scope of applications.

Acknowledgement SK is thankful to the Department of Biotechnology (DBT) (BT/P19588/BIC/101/425/2016) for their financial assistance and National Institute of Pharmaceutical Education and Research (NIPER), Guwahati, for providing the necessary support.

References

Abou-Zeid A, Fouad M, Yassein M (1978) Microbiological production of acetone-butanol by *Clostridium acetobutylicum*. Zentralbl Bakteriol Parasitenkunde Infektionskr Hyg Zweite Naturwiss Abt Mikrobiol Landwirtsch Technol Umweltschutzes 133:125–134

Adrio JL, Demain AL (2014) Microbial enzymes: tools for biotechnological processes. Biomol Ther 4:117–139

Ageitos J, Sánchez-Pérez A, Calo-Mata P, Villa T (2017) Antimicrobial peptides (AMPs): ancient compounds that represent novel weapons in the fight against bacteria. Biochem Pharmacol 133:117–138

Agnandji ST et al (2017) Safety and immunogenicity of rVSVΔG-ZEBOV-GP Ebola vaccine in adults and children in Lambaréné, Gabon: a phase I randomised trial. PLoS Med 14:e1002402

Aharonowitz Y (1980) Nitrogen metabolite regulation of antibiotic biosynthesis. Ann Rev Microbiol 34:209–233

Ahmad M, Hirz M, Pichler H, Schwab H (2014) Protein expression in *Pichia pastoris*: recent achievements and perspectives for heterologous protein production. Appl Microbiol Biotechnol 98:5301–5317

Akash MSH, Rehman K, Tariq M, Chen S (2015) Development of therapeutic proteins: advances and challenges. Turk J Biol 39:343–358

Anbu P, Gopinath SC, Chaulagain BP, Tang T-H, Citartan M (2015) Microbial enzymes and their applications in industries and medicine. Biomed Res Int 2015:816419

Andrade MR, Costa JA (2007) Mixotrophic cultivation of microalga *Spirulina platensis* using molasses as organic substrate. Aquaculture 264:130–134

Aravindan R, Anbumathi P, Viruthagiri T (2007) Lipase applications in food industry. Indian J Biotechnol 6:141–158

Archer DB (2000) Filamentous fungi as microbial cell factories for food use. Curr Opin Biotechnol 11:478–483

Aronson SM (2012) A proliferation of pro-words Rhode island. Med J 95:371

Aruna K, Shah J, Birmole R (2014) Production and partial characterization of alkaline protease from *Bacillus tequilensis* strains CSGAB0139 isolated from spoilt cottage cheese. Int J Appl Biol Pharm 5:201–221

Bachran C, Leppla SH (2016) Tumor targeting and drug delivery by anthrax toxin. Toxins 8:197

Baldauf KJ, Royal JM, Hamorsky KT, Matoba N (2015) Cholera toxin B: one subunit with many pharmaceutical applications. Toxins 7:974–996

Banerjee A, Chisti Y, Banerjee U (2004) Streptokinase—a clinically useful thrombolytic agent. Biotechnol Adv 22:287–307

Bate C, Salmona M, Diomede L, Williams A (2004) Squalestatin cures prion-infected neurons and protects against prion neurotoxicity. J Biol Chem 279:14983–14990

Bechinger B (1997) Structure and functions of channel-forming peptides: magainins, cecropins, melittin and alamethicin. J Membr Biol 156:197–211

Beekwilder J, Wolswinkel R, Jonker H, Hall R, de Vos CR, Bovy A (2006) Production of resveratrol in recombinant microorganisms. Appl Environ Microbiol 72:5670–5672

Bennett WM, Singer I, Golper T, Feig P, Coggins CJ (1977) Guidelines for drug therapy in renal failure. Ann Intern Med 86:754–783

Bentley R (1999) Secondary metabolite biosynthesis: the first century. Crit Rev Biotechnol 19:1–40

Bentley SD et al (2002) Complete genome sequence of the model actinomycete *Streptomyces coelicolor* A3(2). Nature 417:141

Bérdy J (2005) Bioactive microbial metabolites. J Antibiot 58:1

Berg JM, Tymoczko JL, Stryer L (2002) Protein structure and function. In: Biochemistry, 5th edn. W. H. Freeman & Co Ltd, New York

Bibb MJ (2005) Regulation of secondary metabolism in streptomycetes. Curr Opin Microbiol 8:208–215

Biebl H, Menzel K, Zeng AP, Deckwer WD (1999) Microbial production of 1, 3-propanediol. Appl Microbiol Biotechnol 52:289–297

Bill RM (2015) Recombinant protein subunit vaccine synthesis in microbes: a role for yeast? J Pharm Pharmacol 67:319–328

Blin K, Medema MH, Kazempour D, Fischbach MA, Breitling R, Takano E, Weber T (2013) antiSMASH 2.0—a versatile platform for genome mining of secondary metabolite producers. Nucleic Acids Res 41:W204–W212

Blomberg R et al (2013) Precision is essential for efficient catalysis in an evolved Kemp eliminase. Nature 503:418

Bornscheuer U, Huisman G, Kazlauskas R, Lutz S, Moore J, Robins K (2012) Engineering the third wave of biocatalysis. Nature 485:185

Bouras N, Holtz MD, Aboukhaddour R, Strelkov SE (2016) Influence of nitrogen sources on growth and mycotoxin production by isolates of *Pyrenophora tritici*-repentis from wheat. Crop J 4:119–128

Brezovsky J et al (2016) Engineering a de novo transport tunnel. ACS Catal 6:7597–7610

Brückner B (1992) Regulation of gibberellin formation by the fungus *Gibberella fujikuroi*. In: Secondary metabolites: their function and evolution. Wiley, Chichester, pp 129–143

Bu'Lock J (1961) Intermediary metabolism and antibiotic synthesis. In: Advances in applied microbiology, vol 3. Elsevier, pp 293–342

Burroughs JR, Anderson RL (2015) Cosmetic botulinum toxin applications: general considerations and dosing. In: Pearls and pitfalls in cosmetic oculoplastic surgery. Springer, New York, pp 393–394

Castanie-Cornet M-P, Penfound TA, Smith D, Elliott JF, Foster JW (1999) Control of acid resistance in *Escherichia coli*. J Bacteriol 181:3525–3535

Cech TR, Bass BL (1986) Biological catalysis by RNA. Annu Rev Biochem 55:599–629

Chae S, Hwang E, Shin H-S (2006) Single cell protein production of *Euglena gracilis* and carbon dioxide fixation in an innovative photo-bioreactor. Bioresour Technol 97:322–329

Chae TU, Choi SY, Kim JW, Ko Y-S, Lee SY (2017) Recent advances in systems metabolic engineering tools and strategies. Curr Opin Biotechnol 47:67–82

Challis GL (2008) Mining microbial genomes for new natural products and biosynthetic pathways. Microbiology 154:1555–1569

Chandola TR et al (2017) ROTAVAC® does not interfere with the immune response to childhood vaccines in Indian infants: a randomized placebo controlled trial. Heliyon 3:e00302

Chang C, Stewart RC (1998) The two-component system: regulation of diverse signaling pathways in prokaryotes and eukaryotes. Plant Physiol 117:723–731

Chemier JA, Fowler ZL, Koffas MA, Leonard E (2009) Trends in microbial synthesis of natural products and biofuels. Adv Enzymol Relat Area Mol Biol 76:151

Chen G-Q (2012) New challenges and opportunities for industrial biotechnology. Microb Cell Factories 11:111

Chen K, Arnold FH (1993) Tuning the activity of an enzyme for unusual environments: sequential random mutagenesis of subtilisin E for catalysis in dimethylformamide. Proc Natl Acad Sci 90:5618–5622

Chen MM, Snow CD, Vizcarra CL, Mayo SL, Arnold FH (2012) Comparison of random mutagenesis and semi-rational designed libraries for improved cytochrome P450 BM3-catalyzed hydroxylation of small alkanes. Protein Eng Des Sel 25:171–178

Cheng YR, Huang J, Qiang H, LIN WL, Demain AL (2001) Mutagenesis of the rapamycin producer *Streptomyces hygroscopicus* FC904. J Antibiot 54:967–972

Chiang Y-M, Lee K-H, Sanchez JF, Keller NP, Wang CC (2009) Unlocking fungal cryptic natural products. Nat Prod Commun 4:1505

Choudhary R, Jana A, Jha M (2004) Enzyme technology applications in leather processing. Indian J Chem Technol 11:659–671
Čurdová E, Jechová V, Zima J, Vaněk Z (1989) The effect of inorganic phosphate on the production of avermectin in *Streptomyces avermitilis*. J Basic Microbiol 29:341–346
Cutler HG (1995) Microbial natural products that affect plants, phytopathogens, and certain other microorganisms. Crit Rev Plant Sci 14:413–444
Daglia M (2012) Polyphenols as antimicrobial agents. Curr Opin Biotechnol 23:174–181
Damborsky J, Chaloupkova R, Pavlova M, Chovancova E, Brezovsky J (2010) Structure–function relationships and engineering of haloalkane dehalogenases. In: Handbook of hydrocarbon and lipid microbiology. Springer, Berlin/Heidelberg, pp 1081–1098
Davati N, Habibi Najafi MB (2013) Overproduction strategies for microbial secondary metabolites: a review. Int J Life Sci Pharma Res 3:23–27
Davies J (2011) How to discover new antibiotics: harvesting the parvome. Curr Opin Chem Biol 15:5–10
Del Tordello E, Rappuoli R, Delany I (2017) Reverse vaccinology: exploiting genomes for vaccine design. In: Human vaccines. Elsevier, pp 65–86
Demain AL (1998) Microbial natural products: alive and well in 1998. Nat Biotechnol 16:3
Demain AL, Adrio JL (2008) Contributions of microorganisms to industrial biology. Mol Biotechnol 38:41
Demain AL, Vaishnav P (2009) Production of recombinant proteins by microbes and higher organisms. Biotechnol Adv 27:297–306
Dhakal R, Bajpai VK, Baek K-H (2012) Production of GABA (γ-aminobutyric acid) by microorganisms: a review. Braz J Microbiol 43:1230–1241
Dien BS, Nichols NN, O'bryan PJ, Bothast RJ (2000) Development of new ethanologenic *Escherichia coli* strains for fermentation of lignocellulosic biomass. Appl Biochem Biotechnol 84:181–196
Dormitzer PR, Ulmer JB, Rappuoli R (2008) Structure-based antigen design: a strategy for next generation vaccines. Trends Biotechnol 26:659–667
Dormitzer PR, Grandi G, Rappuoli R (2012) Structural vaccinology starts to deliver. Nat Rev Microbiol 10:807
Doshi R, Shelke V (2001) Enzymes in textile industry-an environment-friendly approach. Indian J Fibre Text Res 26:202–205
Du J, Shao Z, Zhao H (2011) Engineering microbial factories for synthesis of value-added products. J Ind Microbiol Biotechnol 38:873–890
Dubos RJ (1939) Studies on a bactericidal agent extracted from a soil bacillus: I. Preparation of the agent. Its activity in vitro. J Exp Med 70:1
Dubos RJ, Cattaneo C (1939) Studies on a bactericidal agent extracted from a soil bacillus: III. Preparation and activity of a protein-free fraction. J Exp Med 70:249–256
Esnault C et al (2017) Strong antibiotic production is correlated with highly active oxidative metabolism in *Streptomyces coelicolor* M145. Sci Rep 7:200
Essig A et al (2014) Copsin, a novel peptide-based fungal antibiotic interfering with the peptidoglycan synthesis. J Biol Chem 289:34953–34964. M114. 599878
Fabbri A, Travaglione S, Falzano L, Fiorentini C (2008) Bacterial protein toxins: current and potential clinical use. Curr Med Chem 15:1116–1125
Fernandes P, Carvalho F (2017) Microbial enzymes for the food industry. In: Biotechnology of microbial enzymes. Elsevier, Amsterdam, pp 513–544
Ferrer-Miralles N, Domingo-Espín J, Corchero JL, Vázquez E, Villaverde A (2009) Microbial factories for recombinant pharmaceuticals. Microb Cell Factories 8:17
Fleming AG (1929) Responsibilities and opportunities of the private practitioner in preventive medicine. Can Med Assoc J 20:11
Foerster CW, Foerster HF (1973) Glutamic acid decarboxylase in spores of *Bacillus megaterium* and its possible involvement in spore germination. J Bacteriol 114:1090–1098
Foster JA, Wulc AE, Straka D, Cahill KV, Czyz C, Tan J (2018) Cosmetic uses of botulinum toxin. In: Manual of oculoplastic surgery. Springer, New York, pp 165–172

Franzoi M, van Heuvel Y, Thomann S, Schurch N, Kallio PT, Venier P, Essig A (2017) Structural insights into the mode of action of the peptide antibiotic copsin. Biochemistry 56:4992–5001
Garodia S, Naidu P, Nallanchakravarthula S (2017) QUORN: an anticipated novel protein source
Ghazi S, Sepahy AA, Azin M, Khaje K, Khavarinejad R (2014) UV mutagenesis for the overproduction of xylanase from *Bacillus mojavensis* PTCC 1723 and optimization of the production condition. Iran J Basic Med Sci 17:844
Giombini E, Orsini M, Carrabino D, Tramontano A (2010) An automatic method for identifying surface proteins in bacteria: SLEP. BMC Bioinformatics 11:39
Goeddel DV et al (1980) Human leukocyte interferon produced by *E. coli* is biologically active. Nature 287:411
Gonzalez JB, Fernandez F, Tomasini A (2003) Microbial secondary metabolites production and strain improvement. Indian J Biotechnol 2:322–333
Grandi G (2010) Bacterial surface proteins and vaccines. F1000 Biol Rep 2:36
Guazzaroni ME, Silva-Rocha R, Ward RJ (2015) Synthetic biology approaches to improve biocatalyst identification in metagenomic library screening. Microb Biotechnol 8:52–64
Gurung N, Ray S, Bose S, Rai V (2013) A broader view: microbial enzymes and their relevance in industries, medicine, and beyond. Biomed Res Int 2013:329121
Hara KY, Araki M, Okai N, Wakai S, Hasunuma T, Kondo A (2014) Development of bio-based fine chemical production through synthetic bioengineering. Microb Cell Factories 13:173
Hasan F, Shah AA, Javed S, Hameed A (2010) Enzymes used in detergents: lipases. Afr J Biotechnol 9:4836–4844
Haugh M, Gresset-Bourgeois V, Macabeo B, Woods A, Samson SI (2017) A trivalent, inactivated influenza vaccine (Vaxigrip®): summary of almost 50 years of experience and more than 1.8 billion doses distributed in over 120 countries. Expert Rev Vaccines 16:545–564
Headon D, Walsh G (1994) The industrial production of enzymes. Biotechnol Adv 12:635–646
Hermann B, Blok K, Patel MK (2007) Producing bio-based bulk chemicals using industrial biotechnology saves energy and combats climate change. Environ Sci Technol 41:7915–7921
Hodgson J (1994) Bulk amino–acid fermentation: technology and commodity trading. Nat Biotechnol 12:152
Hou J, Tyo KE, Liu Z, Petranovic D, Nielsen J (2012) Metabolic engineering of recombinant protein secretion by *Saccharomyces cerevisiae*. FEMS Yeast Res 12:491–510
Ivers LC et al (2015) Immunogenicity of the bivalent oral cholera vaccine Shanchol in Haitian adults with HIV infection. J Infect Dis 212:779–783
Jaganmohan P, Daas BP, Prasad S (2013) Production of single cell protein (SCP) with *Aspergillus terreus* using solid state fermentation. Eur J Biol Sci 5:38–43
Jullesson D, David F, Pfleger B, Nielsen J (2015) Impact of synthetic biology and metabolic engineering on industrial production of fine chemicals. Biotechnol Adv 33:1395–1402
Kataoka M, Sasaki M, Hidalgo A-RG, Nakano M, Shimizu S (2001) Glycolic acid production using ethylene glycol-oxidizing microorganisms. Biosci Biotechnol Biochem 65:2265–2270
Kaushik S et al (2018) Impact of the access tunnel engineering on catalysis is strictly ligand-specific. FEBS J 285:1456–1476
Keasling JD (2010) Manufacturing molecules through metabolic engineering. Science 330:1355–1358
Keating GM (2016) Shingles (herpes zoster) vaccine (zostavax®): a review in the prevention of herpes zoster and postherpetic neuralgia. BioDrugs 30:243–254
Kennedy SB et al (2017) Phase 2 placebo-controlled trial of two vaccines to prevent Ebola in Liberia. N Engl J Med 377:1438–1447
Khan AI, Islam MT, Qadri F (2017) Safety of oral cholera vaccines during pregnancy in developing countries. Hum Vaccin Immunother 13:2245–2246
Khetan A, Malmberg LH, Kyung YS, Sherman DH, Hu WS (1999) Precursor and cofactor as a check valve for cephamycin biosynthesis in *Streptomyces clavuligerus*. Biotechnol Prog 15:1020–1027
Kieliszek M, Misiewicz A (2014) Microbial transglutaminase and its application in the food industry. A review. Folia Microbiol 59:241–250

Kim H, Yoo SJ, Kang HA (2015) Yeast synthetic biology for the production of recombinant therapeutic proteins. FEMS Yeast Res 15:1–16

Korkalainen M et al (2017) Synergistic proinflammatory interactions of microbial toxins and structural components characteristic to moisture-damaged buildings. Indoor Air 27:13–23

Kumar D, Savitri TN, Verma R, Bhalla T (2008) Microbial proteases and application as laundry detergent additive. Res J Microbiol 3:661–672

Kwong KW, Sivakumar T, Wong W (2016) Intein mediated hyper-production of authentic human basic fibroblast growth factor in *Escherichia coli*. Sci Rep 6:33948

Labrou NE (2010) Random mutagenesis methods for in vitro directed enzyme evolution. Curr Protein Pept Sci 11:91–100

Lacadena J et al (2007) Fungal ribotoxins: molecular dissection of a family of natural killers. FEMS Microbiol Rev 31:212–237

Lautru S, Deeth RJ, Bailey LM, Challis GL (2005) Discovery of a new peptide natural product by *Streptomyces coelicolor* genome mining. Nat Chem Biol 1:265–269

Lee J-S, Shin K-S, Pan J-G, Kim C-J (2000) Surface-displayed viral antigens on *Salmonella* carrier vaccine. Nat Biotechnol 18:645

Lee SY, Choi JH, Xu Z (2003) Microbial cell-surface display. Trends Biotechnol 21:45–52

Levin MJ, Buchwald UK, Gardner J, Martin J, Stek JE, Brown E, Popmihajlov Z (2018) Immunogenicity and safety of zoster vaccine live administered with quadrivalent influenza virus vaccine. Vaccine 36:179–185

Lewis K, Epstein S, D'Onofrio A, Ling LL (2010) Uncultured microorganisms as a source of secondary metabolites. J Antibiot 63:468

Li T et al (2012) Efficient, chemoenzymatic process for manufacture of the boceprevir bicyclic [3.1. 0] proline intermediate based on amine oxidase-catalyzed desymmetrization. J Am Chem Soc 134:6467–6472

Li X-F et al (2018) Development of a chimeric Zika vaccine using a licensed live-attenuated flavivirus vaccine as backbone. Nat Commun 9:673

Liew WH, Hassim MH, Ng DK (2014) Review of evolution, technology and sustainability assessments of biofuel production. J Clean Prod 71:11–29

Liljeqvist S, Samuelson P, Hansson M, Nguyen TN, Binz H, Ståhl S (1997) Surface display of the cholera toxin B subunit on *Staphylococcus xylosus* and *Staphylococcus carnosus*. Appl Environ Microbiol 63:2481–2488

Lim CG, Fowler ZL, Hueller T, Schaffer S, Koffas MA (2011) High-yield resveratrol production in engineered *Escherichia coli*. Appl Environ Microbiol 77:3451–3460. https://doi.org/10.1128/AEM.02186-10

Link E et al (1992) Tetanus toxin action: inhibition of neurotransmitter release linked to synaptobrevin proteolysis. Biochem Biophys Res Commun 189:1017–1023

Liu C, Zhao Y, Wang Y (2006) Artemisinin: current state and perspectives for biotechnological production of an antimalarial drug. Appl Microbiol Biotechnol 72:11–20

Liu G, Chater KF, Chandra G, Niu G, Tan H (2013) Molecular regulation of antibiotic biosynthesis in Streptomyces. Microbiol Mol Biol Rev 77:112–143

Luchese RH, Harrigan W (1993) Biosynthesis of aflatoxin—the role of nutritional factors. J Appl Bacteriol 74:5–14

Lutz S (2010) Beyond directed evolution—semi-rational protein engineering and design. Curr Opin Biotechnol 21:734–743

Ma Q et al (2017) Systems metabolic engineering strategies for the production of amino acids. Synth Syst Biotechnol 2:87–96

Mahan KM et al (2018) Production of single cell protein from agro-waste using *Rhodococcus opacus*. J Ind Microbiol Biotechnol 45(9):795–801

Mane P, Tale V (2015) Overview of microbial therapeutic enzymes. Int J Curr Microbiol Appl Sci 4:17–26

Marinelli F (2009) From microbial products to novel drugs that target a multitude of disease indications. Methods Enzymol 458:29–58

Martin VJ, Pitera DJ, Withers ST, Newman JD, Keasling JD (2003) Engineering a mevalonate pathway in *Escherichia coli* for production of terpenoids. Nat Biotechnol 21:796

Martínez JL, Liu L, Petranovic D, Nielsen J (2012) Pharmaceutical protein production by yeast: towards production of human blood proteins by microbial fermentation. Curr Opin Biotechnol 23:965–971

Masuyer G, Conrad J, Stenmark P (2017) The structure of the tetanus toxin reveals pH-mediated domain dynamics. EMBO Rep 18:1306–1317

Mazet I, Vey A (1995) Hirsutellin A, a toxic protein produced in vitro by *Hirsutella thompsonii.* Microbiology 141:1343–1348

McWilliams A (2012) Microbial products: technologies, applications and global markets. BCC Research

Mei Y-Z, Liu R-X, Wang D-P, Wang X, Dai C-C (2015) Biocatalysis and biotransformation of resveratrol in microorganisms. Biotechnol Lett 37:9–18

Milner M (2008) Nattokinase: clinical updates from doctors support its safety and efficacy. FOCUS Allergy Res Group News: Lett

Minami H, Kim J-S, Ikezawa N, Takemura T, Katayama T, Kumagai H, Sato F (2008) Microbial production of plant benzylisoquinoline alkaloids. Proc Natl Acad Sci 105:7393–7398

Mishra B, Reiling S, Zarena D, Wang G (2017) Host defense antimicrobial peptides as antibiotics: design and application strategies. Curr Opin Chem Biol 38:87–96

Morris AP, Estes MK (2001) VIII. Pathological consequences of rotavirus infection and its enterotoxin. Am J Physiol Gastrointest Liver Physiol 281:G303–G310

Moyes DL et al (2016) Candidalysin is a fungal peptide toxin critical for mucosal infection. Nature 532:64

Mrudula S, Murugammal R (2011) Production of cellulase by *Aspergillus niger* under submerged and solid state fermentation using coir waste as a substrate. Braz J Microbiol 42:1119–1127

Nagasawa T, Nakamura T, Yamada H (1990) Production of acrylic acid and methacrylic acid using *Rhodococcus rhodochrous* J1 nitrilase. Appl Microbiol Biotechnol 34:322–324

Nakamura J, Hirano S, Ito H, Wachi M (2007) Mutations of the *Corynebacterium glutamicum* NCgl1221 gene, encoding a mechanosensitive channel homolog, induce L-glutamic acid production. Appl Environ Microbiol 73:4491–4498

Naughton LM, Romano S, O'Gara F, Dobson AD (2017) Identification of secondary metabolite gene clusters in the Pseudovibrio genus reveals encouraging biosynthetic potential toward the production of novel bioactive compounds. Front Microbiol 8:1494

Nayeem M, Chauhan K, Khan S, Rattu G, Dhaka RK, Sidduqui H (2017) Optimization of low-cost substrate for the production of single cell protein using *Kluyveromyces marxianus.* Pharma Innov J 6:22–25

Nelson DL, Lehninger AL, Cox MM (2008) Lehninger principles of biochemistry. Macmillan

Nevalainen H, Peterson R (2014) Making recombinant proteins in filamentous fungi-are we expecting too much? Front Microbiol 5:75

Newman DJ (2016) Predominately uncultured microbes as sources of bioactive agents. Front Microbiol 7:1832

Newman JD et al (2006) High-level production of amorpha-4, 11-diene in a two-phase partitioning bioreactor of metabolically engineered *Escherichia coli.* Biotechnol Bioeng 95:684–691

Nielsen J (1998) The role of metabolic engineering in the production of secondary metabolites. Curr Opin Microbiol 1:330–336

Nigam PS (2013) Microbial enzymes with special characteristics for biotechnological applications. Biomol Ther 3:597–611

Nijkamp K, Westerhof RM, Ballerstedt H, De Bont JA, Wery J (2007) Optimization of the solvent-tolerant *Pseudomonas putida* S12 as host for the production of p-coumarate from glucose. Appl Microbiol Biotechnol 74:617–624

Noda S et al (2011) Cinnamic acid production using *Streptomyces lividans* expressing phenylalanine ammonia lyase. J Ind Microbiol Biotechnol 38:643–648

Okai N, Takahashi C, Hatada K, Ogino C, Kondo A (2014) Disruption of pknG enhances production of gamma-aminobutyric acid by *Corynebacterium glutamicum* expressing glutamate decarboxylase. AMB Express 4:20
Olombrada M, Martínez-del-Pozo Á, Medina P, Budia F, Gavilanes JG, García-Ortega L (2014) Fungal ribotoxins: natural protein-based weapons against insects. Toxicon 83:69–74
Olson KC et al (1981) Purified human growth hormone from *E. coli* is biologically active. Nature 293:408
Olson DM et al (2011) A qualitative assessment of practices associated with shorter door-to-needle time for thrombolytic therapy in acute ischemic stroke. J Neurosci Nurs 43:329–336
Orth JD, Thiele I, Palsson BØ (2010) What is flux balance analysis? Nat Biotechnol 28:245
Ostlie DJ et al (2012) Topical silver sulfadiazine vs collagenase ointment for the treatment of partial thickness burns in children: a prospective randomized trial. J Pediatr Surg 47:1204–1207
Overton TW (2014) Recombinant protein production in bacterial hosts. Drug Discov Today 19:590–601
Paddon CJ, Keasling JD (2014) Semi-synthetic artemisinin: a model for the use of synthetic biology in pharmaceutical development. Nat Rev Microbiol 12:355
Paddon CJ et al (2013) High-level semi-synthetic production of the potent antimalarial artemisinin. Nature 496:528
Pasricha S, Pearson J (2016) Lifting the veil on fungal toxins. Nature Publishing Group
Patel RN (2008) Synthesis of chiral pharmaceutical intermediates by biocatalysis. Coord Chem Rev 252:659–701
Prado-Rubio OA, Jørgensen JB, Jørgensen SB (2010) Systematic model analysis for single cell protein (scp) production in a u-loop reactor. In: Computer aided chemical engineering, vol 28. Elsevier, Amsterdam, pp 319–324
Prokop Z, Gora A, Brezovsky J, Chaloupkova R, Stepankova V, Damborsky J (2012) Engineering of protein tunnels: keyhole-lock-key model for catalysis by the enzymes with buried active sites, vol 3. Wiley-VCH, Weinheim
Quesada-Moraga E, Alain V (2004) Bassiacridin, a protein toxic for locusts secreted by the entomopathogenic fungus *Beauveria bassiana*. Mycol Res 108:441–452
Rao MB, Varma A, Deshmukh SS (2010) Production of single cell protein, essential amino acids, and xylanase by *Penicillium janthinellum*. BioResources 5:2470–2477
Raspor P, Goranovič D (2008) Biotechnological applications of acetic acid bacteria. Crit Rev Biotechnol 28:101–124
Rathod PK, McErlean T, Lee P-C (1997) Variations in frequencies of drug resistance in *Plasmodium falciparum*. Proc Natl Acad Sci 94:9389–9393
Ray G, Noori MT, Ghangrekar M (2017) Novel application of peptaibiotics derived from Trichoderma sp. for methanogenic suppression and enhanced power generation in microbial fuel cells. RSC Adv 7:10707–10717
Reed JL, Senger RS, Antoniewicz MR, Young JD (2011) Computational approaches in metabolic engineering. J Biomed Res 2010
Renge V, Khedkar S, Nandurkar NR (2012) Enzyme synthesis by fermentation method: a review. Sci Rev Chem Comm 2:585e590
Rodríguez-Zavala J, Ortiz-Cruz M, Mendoza-Hernández G, Moreno-Sánchez R (2010) Increased synthesis of α-tocopherol, paramylon and tyrosine by *Euglena gracilis* under conditions of high biomass production. J Appl Microbiol 109:2160–2172
Rosano GL, Ceccarelli EA (2014) Recombinant protein expression in *Escherichia coli*: advances and challenges. Front Microbiol 5:172
Rosenberg ZM-M (2006) Current trends of β-galactosidase application in food technology. J Food Nutr Res 45:47–54
Sadia S, Qureshi R, Khalid S, Nayyar BG, Zhang JT (2015) Role of secondary metabolites of wild marigold in suppression of Johnson grass and Sun spurge. Asian Pac J Trop Biomed 5:733–737
Sanchez S, Demain AL (2017) Useful microbial enzymes—an introduction. In: Biotechnology of microbial enzymes. Elsevier, Amsterdam, pp 1–11

Sanchez-Garcia L, Martín L, Mangues R, Ferrer-Miralles N, Vázquez E, Villaverde A (2016) Recombinant pharmaceuticals from microbial cells: a 2015 update. Microb Cell Factories 15:33

Sanders JW, Leenhouts K, Burghoorn J, Brands JR, Venema G, Kok J (1998) A chloride-inducible acid resistance mechanism in *Lactococcus lactis* and its regulation. Mol Microbiol 27:299–310

Sandström AG, Wikmark Y, Engström K, Nyhlén J, Bäckvall JE (2012) Combinatorial reshaping of the *Candida antarctica* lipase A substrate pocket for enantioselectivity using an extremely condensed library. Proc Natl Acad Sci 109:78–83

Schmidt-Dannert C (2017) The future of biologically inspired next-generation factories for chemicals. Microb Biotechnol 10:1164–1166

Schmit J, Edson CM, Brody S (1975) Changes in glucosamine and galactosamine levels during conidial germination in *Neurospora crassa*. J Bacteriol 122:1062–1070

Sette A, Rappuoli R (2010) Reverse vaccinology: developing vaccines in the era of genomics. Immunity 33:530–541

Sharma N (2016) How does recent understanding of molecular mechanisms in botulinum toxin impact therapy? In: Botulinum toxin therapy manual for dystonia and spasticity. InTech open. doi:https://doi.org/10.5772/66696

Sharma A, Chaudhuri TK (2017) Revisiting *Escherichia coli* as microbial factory for enhanced production of human serum albumin. Microb Cell Factories 16:173

Sharma A, Kumari N, Menghani E (2014) Bioactive secondary metabolites: an overview. Int J Sci Eng Res 5:1395–1407

Sharma S et al (2018) Immunogenicity and safety of the first indigenously developed Indian tetravalent influenza vaccine (split virion) in healthy adults ≥18 years of age: a randomized, multicenter, phase II/III clinical trial. Hum Vaccin Immunother 14(6):1362–1369

Shin JH, Kim HU, Kim DI, Lee SY (2013) Production of bulk chemicals via novel metabolic pathways in microorganisms. Biotechnol Adv 31:925–935

Siegel JB et al (2010) Computational design of an enzyme catalyst for a stereoselective bimolecular Diels-Alder reaction. Science 329:309–313

Singh Saharan B, Grewal A, Kumar P (2014) Biotechnological production of polyhydroxyalkanoates: a review on trends and latest developments. Chin J Biol 2014:802984

Singh HB, Jha A, Keswani C (2016a) Intellectual property issues in biotechnology. CABI, Wallingford/Boston

Singh R, Kumar M, Mittal A, Mehta PK (2016b) Microbial enzymes: industrial progress in 21st century. 3 Biotech 6:174

Stanley M (2007) Prevention strategies against the human papillomavirus: the effectiveness of vaccination. Gynecol Oncol 107:S19–S23

Starcevic A, Zucko J, Simunkovic J, Long PF, Cullum J, Hranueli D (2008) ClustScan: an integrated program package for the semi-automatic annotation of modular biosynthetic gene clusters and in silico prediction of novel chemical structures. Nucleic Acids Res 36:6882–6892

Stewart G, Panchal CJ, Russell I, Sills AM (1983) Biology of ethanol-producing microorganisms. Crit Rev Biotechnol 1:161–188

Suman G, Nupur M, Anuradha S, Pradeep B (2015) Single cell protein production: a review. Int J Curr Microbiol App Sci 4(9): 251–262

Tan JP, Md. Jahim J, Wu TY, Harun S, Kim BH, Mohammad AW (2014) Insight into biomass as a renewable carbon source for the production of succinic acid and the factors affecting the metabolic flux toward higher succinate yield. Ind Eng Chem Res 53:16123–16134

Taylor MJ, Richardson T (1979) Applications of microbial enzymes in food systems and in biotechnology. In: Advances in applied microbiology, vol 25. Elsevier, Burlington, pp 7–35

Terpe K (2006) Overview of bacterial expression systems for heterologous protein production: from molecular and biochemical fundamentals to commercial systems. Appl Microbiol Biotechnol 72:211

Thies S et al (2016) Metagenomic discovery of novel enzymes and biosurfactants in a slaughterhouse biofilm microbial community. Sci Rep 6:27035

Timmis KN, McGenity T, Van Der Meer JR, de Lorenzo V (2010) Handbook of hydrocarbon and lipid microbiology, vol DLII. Springer, Berlin, p 4716
Ueno H (2000) Enzymatic and structural aspects on glutamate decarboxylase. J Mol Catal B Enzym 10:67–79
Ugbogu E, Ugbogu O (2016) A review of microbial protein production: prospects and challenges. Fuw Trends Sci Technol J 1:182–185
Underkofler L, Barton R, Rennert S (1958) Production of microbial enzymes and their applications. Appl Microbiol 6:212
Vaishnav P, Demain AL (2011) Unexpected applications of secondary metabolites. Biotechnol Adv 29:223–229
Vandenbergh PA (1993) Lactic acid bacteria, their metabolic products and interference with microbial growth. FEMS Microbiol Rev 12:221–237
Vandenberghe LP, Soccol CR, Pandey A, Lebeault J-M (1999) Microbial production of citric acid. Braz Arch Biol Technol 42:263–276
Vannelli T, Qi WW, Sweigard J, Gatenby AA, Sariaslani FS (2007) Production of p-hydroxycinnamic acid from glucose in *Saccharomyces cerevisiae* and *Escherichia coli* by expression of heterologous genes from plants and fungi. Metab Eng 9:142–151
Vargas-Tah A, Gosset G (2015) Production of cinnamic and p-hydroxycinnamic acids in engineered microbes. Front Bioeng Biotechnol 3:116
Verier A, Chenal A, Babon A, Ménez A, Gillet D (2006) Engineering of bacterial toxins for research and medicine. In: The comprehensive sourcebook of bacterial protein toxins, 3rd edn. Elsevier, Amsterdam/Boston, p 991–1007
Volesky B, Luong JH, Aunstrup K (1984) Microbial enzymes: production, purification, and isolation. Crit Rev Biotechnol 2:119–146
Wang G (2015) Improved methods for classification, prediction, and design of antimicrobial peptides. In: Computational peptidology. Springer, New York/Heidelberg, pp 43–66
Wang G, Huang M, Nielsen J (2017) Exploring the potential of *Saccharomyces cerevisiae* for biopharmaceutical protein production. Curr Opin Biotechnol 48:77–84
Weber T, Rausch C, Lopez P, Hoof I, Gaykova V, Huson D, Wohlleben W (2009) CLUSEAN: a computer-based framework for the automated analysis of bacterial secondary metabolite biosynthetic gene clusters. J Biotechnol 140:13–17
Wee Y-J, Kim J-N, Ryu H-W (2006) Biotechnological production of lactic acid and its recent applications. Food Technol Biotechnol 44:163–172
Williams DH, Stone MJ, Hauck PR, Rahman SK (1989) Why are secondary metabolites (natural products) biosynthesized? J Nat Prod 52:1189–1208
Williamson NR, Fineran PC, Leeper FJ, Salmond GP (2006) The biosynthesis and regulation of bacterial prodiginines. Nat Rev Microbiol 4:887
de Wit PJ (2013) Microbial toxins in the green world. FEMS Microbiol Rev 37:1–2
Wu F, Cao P, Song G, Chen W, Wang Q (2018) Expanding the repertoire of aromatic chemicals by microbial production. J Chem Technol Biotechnol 93:2804–2816
Xing B, Rao J, Liu R (2008) Novel beta-lactam antibiotics derivatives: their new applications as gene reporters, antitumor prodrugs and enzyme inhibitors. Mini Rev Med Chem 8:455–471
Xu Q, Tao W-Y, Huang H, Li S (2016) Highly efficient synthesis of ethyl (S)-4-chloro-3-hydroxybutanoate by a novel carbonyl reductase from *Yarrowia lipolytica* and using mannitol or sorbitol as cosubstrate. Biochem Eng J 106:61–67
Yang X, Yousef AE (2018) Antimicrobial peptides produced by *Brevibacillus* spp.: structure, classification and bioactivity: a mini review. World J Microbiol Biotechnol 34:57
Yoo YJ, Feng Y, Kim Y-H, Yagonia CFJ (2017) Fundamentals of enzyme engineering. Springer, p X, 209
Zahrl RJ, Peña DA, Mattanovich D, Gasser B (2017) Systems biotechnology for protein production in *Pichia pastoris*. FEMS Yeast Res 17:fox068
Zhao S et al (2013) Discovery of new enzymes and metabolic pathways by using structure and genome context. Nature 502:698

Zheng P, Dong J-J, Sun Z-H, Ni Y, Fang L (2009) Fermentative production of succinic acid from straw hydrolysate by *Actinobacillus succinogenes*. Bioresour Technol 100:2425–2429

Zhu X, Siegert M, Yates MD, Logan BE (2015) Alamethicin suppresses methanogenesis and promotes acetogenesis in bioelectrochemical systems. Appl Environ Microbiol 81:3863–3868

Zorko M, Jerala R (2010) Production of recombinant antimicrobial peptides in bacteria. In: Antimicrobial peptides. Springer, pp 61–76

Zubieta MP, Contesini FJ, Rubio MV, Gonçalves AESS, Gerhardt JA, Prade RA, Damasio ARL (2018) Protein profile in *Aspergillus nidulans* recombinant strains overproducing heterologous enzymes. Microb Biotechnol 11:346–358

2,4-Diacetylphloroglucinol: A Novel Biotech Bioactive Compound for Agriculture

16

Raksha Ajay Kankariya, Ambalal Babulal Chaudhari, Pavankumar M. Gavit, and Navin Dharmaji Dandi

16.1 Introduction

The annual loss of global food crop production is estimated to be one-third due to transboundary plant diseases and insects posing major threat to world economy and food security (FAO 2017a, b). A severe decline in crop productivity is accredited to phytopathogen intervention worldwide (FAO 2017b). As agricultural practices exaggerated over the past four decades, farmers eventually became dependent on stupendous use of recalcitrant synthetic agrochemicals. About 70,000 species of pests are known to affect agricultural crops and responsible for 40% reduction of world food production (Pimentel 1997). Of these, around 10,000 species of fungal phytopathogens are known to cause wide range of diseases that lead to an extent of 20% loss of agricultural crops (Agrios 2005; Strange and Scott 2005). Generally, fungal diseases may be overcome by the reduction of the inoculum, inhibition of its virulence mechanisms and promotion of genetic diversity in the crop (Strange and Scott 2005). However, the use of chemical fungicides in agriculture has benefited through reducing the fungal infections and post-harvest spoilages against myco-toxic fungi. Without the use of pesticides, an overall agricultural loss may range from 32% to 74% at global scale (Pimentel 1997). During 2010–2014, an average of 2.784 kg/ha pesticides were used globally, wherein Japan used 18.94 kg/ha as opposed to India with 0.261 kg/ha use. Fungicides and bactericides alone contribute ~12% of total pesticides and include use of inorganics, dithiocarbamates, diazoles, triazoles, benzimidazoles, morpholines, diazines, etc. to an estimate of ~0.2 million and likely to double before 2020 (Zhang 2018). The extensive use of chemical pesticides is known for various environmental damage and health problems causing acute toxicity, chronic toxicity and severe poisoning (Miller 2004). The search for

R. A. Kankariya · A. B. Chaudhari · P. M. Gavit · N. D. Dandi (✉)
School of Life Sciences, Kavayitri Bahinabai Chaudhari North Maharashtra University, Jalgaon, India
e-mail: nddandi@nmu.ac.in

D. P. Singh et al. (eds.), *Microbial Interventions in Agriculture and Environment*,
https://doi.org/10.1007/978-981-13-8391-5_16

alternative to chemical control of plant pathogens has gained momentum in the recent past due to the emergence of fungicide resistance in pathogens in addition to increased health concerns for the producer and the consumer (Hawkins and Fraaije 2018).

The most viable alternative to chemical pesticides appeared to be biological control which is quite feasible to crop pest control from environmental, economic and functional perspective to improve utilization of introduced or resident living organism and suppress activities of the phytopathogens (Lugtenberg and Kamilova 2009). This includes the use of microbial inoculants in a single cropping system that serve as antagonists to suppress single type or class of plant disease, pathogen or pest. Any organism that negatively affects pest or pathogen is called as a biocontrol agent (Whipps 2001; Haas and Défago 2005; Daguerre et al. 2014). Ecological and anthropological concerns regarding agrichemicals are motivating interest in more use of biocontrol agents for environment-friendly plant disease control. The potency of a biocontrol agent depends on (i) its successful colonization in the roots, (ii) induction of immune response in the plant, and (iii) secretion of diffusible antibiotics (Whipps 2001). Among the microorganisms, the greatest potential as commercially viable biocontrol products are live bacterial strains of the soil-borne fluorescent *Pseudomonas* spp. (Haas and Défago 2005; Mishra and Arora 2017). The key determinants for biocontrol are the ability to colonize in the rhizosphere, compete for nutrients and produce antagonistic compounds. This chapter critically examines the microbial production, application and regulation and the understanding of biosynthesis of antibiotic 2,4-diacetylphloroglucinol (DAPG).

16.2 Secondary Metabolites: Biotic Commodity Molecule

Secondary metabolites are organic composites produced from rhizospheric microbes that are not directly involved in the regular development, progress or reproduction of an organism and frequently play significant role in defence systems of diverse organisms (Stamp 2003; Mishra and Arora 2017). These molecules include antibiotics, pigments, toxins, effectors of ecological competition and symbiosis, pheromones, enzyme inhibitors, immune-modulating agents, receptor antagonists and agonists, pesticides, antitumor agents and growth promoters of animals and plants and are used in human practice in medicines, flavourings, and recreational drugs (Davati and Najafi 2013). They have a foremost outcome on the health, nutrition and economics of our society. Numerous rhizospheric bacterial species produce (i) fluorescent pigments (Stanier et al. 1966); (ii) siderophores (Neilands and Leong 1986); (iii) antibiotics such as DAPG, phenazines, pyoluteorin, pyrrolnitrin, rhamnolipids, oomycin, kanosamine, karalicin, pantocin, aerugine, azomycin, ecomycins, cepaciamide A, zwittermycin-A, pseudomonic acid, cepafungins, antitumor and antiviral antibiotics (Fernando et al. 2005), surface-active antibiotics, etc. (iv) biocides such as hydrogen cyanide (HCN) (Raaijmakers et al. 2002); (v) cell wall lytic enzymes (Haas and Défago 2005); (vi) 1-aminocyclopropane-1-carboxylic acid (ACC) deaminase (Penrose and Glick 2003); (vii) cyclic lipopeptides (Muller et al.

2016); (viii) phytohormones such as indole acetic acid (IAA); and (ix) biosurfactants (Saikia et al. 2011). The microbial release of bioactive molecules is controlled by (i) nutrients, (ii) growth rate, (iii) feedback control, (iv) enzyme inactivation, (v) enzyme induction and (vi) motivated by exclusive low molecular mass compounds, transfer RNA, sigma factors and gene products designed throughout post-exponential progress (Davati and Najafi 2013). Genes responsible for biosynthesis of bioactive are usually present on chromosomal DNA and rarely on plasmid DNA. Microbial response such as elicitors, quorum sensing, genetic engineering, metabolic engineering and ribosome engineering approaches are useful tools for overproduction of bioactives (Davati and Najafi 2013). Likewise, genetic modification strategies including amplification of biosynthetic genes, inactivation of competing metabolic paths, interruption or amplification of controlling genes, management of secretory mechanisms, appearance of a convenient heterologous protein and combinatorial biosynthesis are newer avenues for higher production of bioactive metabolites (Gonzalez et al. 2003). Among the secondary metabolites, DAPG has received attention for biocontrol because it exhibits broad-spectrum activity, viz. antiviral, antimicrobial, anti-peronosporomycetes, ichthyotoxic, insect and mammal antifeedant, anti-helminthic, phytotoxic, antioxidant, cytotoxic, antitumor and plant-growth-regulating activities. Besides these, anti-leukemic, anti-lung and anti-breast cancer properties are also reported (de Souza et al. 2003; Islam and Von Tiedemann 2011; Veena et al. 2016).

16.3 DAPG: A Multipotent Antibiotic

DAPG is a low-molecular weight, non-nitrogen containing, non-volatile phenolic polyketide secondary metabolite derived from phloroglucinol (PG) that has proven biocontrol activities (Shanahan et al. 1992; Weller 2007; Troppens et al. 2013a), frequently produced by plants, algae and bacteria (de Souza et al. 2003; Nagel et al. 2012) and prominently by plant-associated PGP *Pseudomonas* spp. (Gutiérrez-García et al. 2017). Of the rhizobacteria, *Pseudomonas* spp. is most explored for DAPG due to its biological and medicinal significance (Stolp and Gadkari 1981). DAPG-producing *Pseudomonas* strains may benefit the plant directly via (i) diazotrophic N fixation (Mirza et al. 2006), (ii) 1-aminocyclopropane-1-carboxylate (ethylene precursor) deamination (Hontzeas et al. 2004), (iii) production of auxin (Picard and Bosco 2005) and/or (iv) induced systemic resistance (ISR) to pathogens (Bakker et al. 2007). *Pseudomonas* spp. with DAPG-producing ability are prominently found in the rhizosphere of major dicot and monocot crops (such as banana, cotton, cucumber, maize, pea, tobacco, tomato, wheat, etc.) and protect from phytopathogen challenge (Primrose 1976; Schippers et al. 1987; Keel et al. 1992; Picard et al. 2000; Kuiper et al. 2001; Ramette et al. 2003; de Souza et al. 2003; Haas and Défago 2005; Saravanan and Muthusamy 2006; Chaubey et al. 2015). DAPG plays a major role in the biocontrol of plant neo-sporophyte and root diseases, for example, take-all of wheat (Fenton et al. 1992), soft rot and pest of potato (Cronin et al. 1997), crown and root rot of tomato (Duffy and Défago 1997), black root rot of tobacco (Ramette 2002), *Pythium* damping-off of sugar beet (Bottiglieri and Keel 2006), *Fusarium* wilt of banana and chickpea (Saravanan and Muthusamy 2006; Saikia et al. 2009), root rot of cucumber (Shirzad et al. 2012), bacterial wilt of tomato and banana (Zhou et al. 2012), wilt and rot disease of cucurbit (Shanthi and Vittal 2013) and red rot of sugar cane (Hassan et al. 2014) etc. (Table 16.1).

There are three authenticated indications for participation of DAPG in crop protection (i) mutation in DAPG-producing gene decreased biocontrol action of antagonistic bacteria, (ii) cell mass of DAPG producers and production responsible for disease destruction in diverse soils, and (iii) association of different DAPG producers in the rhizosphere accountable for disease suppression (Keel et al. 1992; Nowak-Thompson et al. 1994; Raaijmakers et al. 1999). DAPG produced on the roots of *Arabidopsis thaliana* was revealed to induce resistance to *Peronospora parasitica* (Iavicoli et al. 2003) and *Pseudomonas syringae* pv. tomato (Weller et al. 2004). DAPG producers are assembled into diverse phenotypes based on extracellular production of various metabolites such as antibiotics and HCN. The most important phenotypic assemblies of DAPG producers include (i) DAPG and HCN producers; (ii) DAPG, HCN and pyoluteorin co-producers (Keel et al. 1996); and (iii) DAPG, pyoluteorin and pyrrolnitrin co-producers (Nowak-Thompson et al. 1999; Sharifi-Tehrani et al. 1998). The DAPG-producing *phlACBD* synthetic genes are rare in the β-proteobacteria *Pseudomonas* genera and include species phylogenetically close to or outside the *P. fluorescens* group, namely, *P. brassicacearum, P. protegens, P. kilonensis, P. corrugate, P. thivervalensis* and *Pseudomonas* sp. OT69 (Frapolli et al. 2012; Almario et al. 2017). Although

Table 16.1 Various plant diseases caused by phytopathogens

Phytopathogen	Host	Disease	References
G. Graminis tritici	Wheat	Take-all decline	Fenton et al. (1992)
Fusarium oxysporum	Tomato	Crown and root rot	Duffy and Defago (1997)
Erwinia carotovora	Potato	Soft rot	Cronin et al. (1997)
Globodera rostochiensis	Potato	Pest	Cronin et al. (1997)
Thielaviopsis basicola	Tobacco	Black root rot	Ramette (2002)
Pythium ultimum	Sugar beet	Pythium damping-off	Bottiglieri and Keel (2006)
Fusarium oxysporum	Banana	*Fusarium* wilt	Saravanan and Muthusamy (2006)
Fusarium oxysporum f. sp. *ciceri*	Chickpea	*Fusarium* wilt	Saikia et al. (2009)
Phytophthora drechsleri	Cucumber	Root rot	Shirzad et al. (2012)
Ralstonia solanacearum	Tomato, banana	Bacterial wilt	Zhou et al. (2012)
Fusarium spp.	Cucurbit	Wilt and rot disease	Shanthi and Vittal (2013)
Colletotrichum falcatum	Sugar cane	Red rot	Hassan et al. (2014)
Xanthomonas oryzae	Rice	Bacterial blight	Velusamy et al. (2013)
Ralstonia solanacearum	Tomato	Bacterial wilt	Zhou et al. (2012)

Pseudomonas (*phl*$^+$) inhabitants form a polyphyletic group, DAPG biosynthesis is not restricted to this genus. Amphibian skin bacterium *Lysobacter gummosus* produces DAPG and share as the innate immune system by preventing pathogens from colonizing amphibia (Brucker et al. 2008). Another strain of *L. capsica* isolated from rhizospheric soil is also reported for the presence of *phl*$^+$ genotype (Park et al. 2008; Wang et al. 2015). Yet, among the DAPG producers, *P. fluorescens* is the most prominent, where DAPG functions as an intercellular signal, both within and among species, in culture and in the rhizosphere (Maurhofer et al. 2004; Combes-Meynet et al. 2011; Kidarsa et al. 2011).

16.4 Disease Suppressive Soils

In the suppressive soils, the pathogen either does not originate or, even if it does, the pathogen causes slight damage, and after some time, the phytopathogen is dormant in the soil (Weller 1988, 2007). Each regular soil has the capacity to defeat the development or movement of soil-borne pathogens to a limited amount generated by the total microbial activity in the soil, and this phenomenon is called general disease suppression c.a. conducive soils (Weller et al. 2002; Haas and Défago 2005; Berendsen et al. 2012). DAPG producers demonstrate a significant role in numerous natural disease suppressive soils. In such soils, plant roots motivate and provision the soil plant-growth promoting microorganisms for defence against the soil-borne pathogens (Weller et al. 2002). Certain specific microbes transform resident soil to be suppressive to specific plant diseases. In contrast to general suppression, specific suppression is movable by addition of 0.1–10% of the suppressive soil to a conducive soil (Weller et al. 2002; Mendes et al. 2011). Eradication of complete suppression can be performed by fumigation or pasteurization (60 °C, 30 min) of the soil (Weller et al. 2002; Weller 2007). DAPG producers are the key microbial community found in several suppressive soils effective against unlike plant diseases in the world, e.g. *Thielaviopsis basicola* and *Gaeumannomyces graminis* var. *tritici*, which mediated black root rot of tobacco and take-all disease of wheat, respectively (Weller et al. 2002; Weller 2007).

16.5 Induced Systemic Resistance (ISR)

Biocontrol mediators bring a continuous change in the plant and increase plant tolerance to pathogenic infection, a phenomenon identified as 'induced resistance'. Two distinct types of induced resistance – ISR and SAR (systemic acquired resistance) – are known. In the SAR, virulent/non-virulent/non-pathogenic microbes or synthetic chemicals such as benzo(1,2,3)thiadiazole-7-carbothioc acid S-methyl ester (BTH), 2,6-dichloro-isonicotinic acid (INA) and salicylic acid (SA) induce gene expression and accumulation of SA in the plants, while in the ISR, PGPR strains elicit jasmonic acid (JA) and ethylene (ET)-dependent signal transduction pathway to impart pathogen resistance (Ross 1961; Uknes et al. 1992; Ryals et al.

1996; Vallad and Goodman 2004). The SAR and ISR are responsible for overall elevated immunity established in plants that respond to exact biotic or chemical stimuli (van Loon et al. 1998, Bakker et al. 2007). Biocontrol agents prevent the growth of phytopathogens and increase soil disease suppression and/or induce systemic plant resistance (Heydari and Pessarakli 2010). Furthermore, biocontrol agents also amend some abiotic and physiological stresses and improve nutrient interest in plants (Shoresh et al. 2010; Bajsa et al. 2013). The PGP *Pseudomonas* biocontrol agents used in agriculture induce ISR through mixed path antagonism through (i) production of antibiotics (DAPG, phenazines, pyrrolnitrin, pyoluteorin, cyclic lipopeptides), (ii) lytic enzymes (chitinases, glucanases, proteases), (iii) extracellular metabolites (ammonia, HCN, siderophores, etc.) and (iv) physical and chemical interference from blockage of soil pores and confused molecular cross-talk thereby avoiding the dangers associated with synthetic pesticides (Pal and Gardener 2006). Phytopathogen resistance to such antibiotics produced by biocontrol agents are likely to develop slowly because (i) maximum biocontrol agents harvest supplementary antibiotic than one, and resistance to multiple antibiotics occur only at a very low occurrence, and (ii) total experience of the pathogen inhabitants to the antibiotics is low since, in general, the residents of biocontrol agents are localized on the root, therefore, minimizing selection pressures (Handelsman and Stabb 1996). Local suppression of root immune responses is a communal feature of ISR producing beneficial microbes. ISR triggered by beneficial soil-borne microbes is often regulated by a JA/ET, but cooperative microbes provoke the SA-dependent SAR pathway (Weller et al. 2012). Antibiotic-deficient mutants demonstrate significant reduction in ISR activity, for example, (i) *P. fluorescens* CHA0r mutant is reported to induce specific resistance to *Peronospora parasitica* in *Arabidopsis* sporophyte (Iavicoli et al. 2003), and (ii) ISR was found to be induced in *A. thaliana* against bacterial speck caused by *P. syringae* pv. *tomato* when its root was colonized by DAPG-producing *P. fluorescens* strains (Weller et al. 2012).

16.6 Biological Interactions of DAPG Producers

An enormous range of small-molecular-weight compounds (e.g. photosynthates) is formed by plant roots into the rhizosphere and responsible for triggering the metabolism of microbial population in the rhizosphere (Hirsch et al. 2003; Bais et al. 2006). The interactions include root-fungus, root-insect, root-microbe, microbe-microbe and microbe-fungus (Fig. 16.1).

Plants can actively secrete compounds by altering the composition of root exudates that selectively favour certain microorganisms in the rhizosphere, which in turn inhibit pathogenic microorganisms (Phillips et al. 2004; Mark et al. 2005; Doornbos et al. 2012). Noticeably, Phillips et al. (2004) found that plant roots secrete different types of photosynthates that impact on the production of DAPG by PGPR. DAPG increases the titre of plant root amino acid exudates and is scavenged by DAPG producers to synthesize additional metabolites as well as increase their movement.

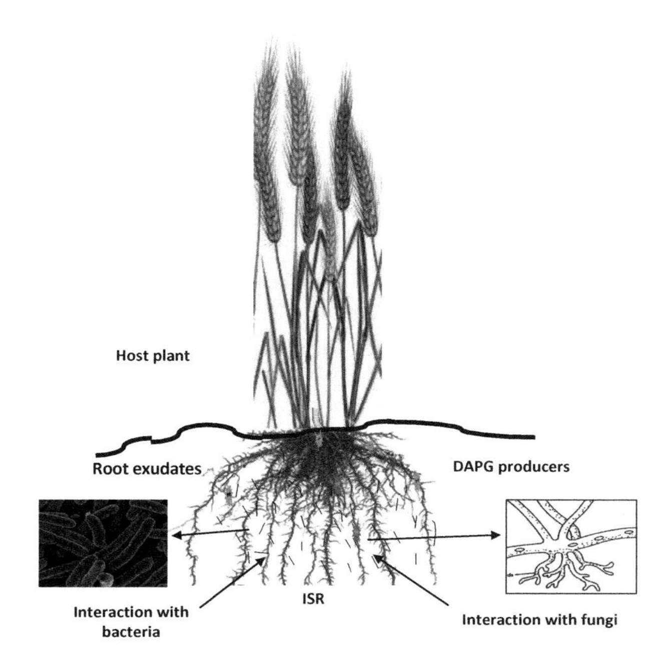

Fig. 16.1 Biological interactions between DAPG-producing fluorescent *Pseudomonas* spp. and host plant

In another strategy, efficacy of biocontrol treatments is improved with the use of DAPG-producing PGPR and fungal strain consortium. PGP fungi are combined with *P. fluorescens* which is attractive for (i) enhanced nutrition to plants (mycorrhizal fungi), (ii) superior disease control (e.g. *Trichoderma* spp.) or (iii) amended insect pest control (entomopathogenic *Beauveria* spp.) (Vega et al. 2009). DAPG-producing PGP pseudomonads positively affect plant metabolism by stimulating infection of *Beauveria* spp. to pest insects. Co-inoculation of *P. fluorescens* and *B. bassiana* simultaneously controlled the insect pests and pathogens on rice (Karthiba et al. 2010). Also, consortium of *Pseudomonads* and *Trichoderma* spp. is considered as potent biocontrol agents. *T. atroviride* secretes chitinases, viz. N-acetyl-β-D-glucosaminidase and endochitinase (encoded by *ECH42* and *NAG1*, respectively). Interaction of *T. atroviride* P1 with *P. fluorescens* strains CHA0 and Q2–87 was examined using *phlA'-'lacZ* translational fusion recombinant genes expressed in *P. fluorescens* CHA0 and *T. atroviride* P1 transformants with *ech42-goxA* or *nag1-goxA* fusions. The study showed suppression of both chitinases. However, accumulation of chitinases secreted by *T. atroviride* P1 in the culture filtrates enhanced *phlA* expression in *P. fluorescens,* suggesting both positive and negative regulatory effects on expression of biocontrol genes (Lutz et al. 2004). In another study, maize seeds were inoculated with *Pseudomonas*, *Azospirillum* or mycorrhizal genus *Glomus*. After 16 days inoculation, negligible influence on plant biomass as verified from maize root methanolic extracts (examined by RP-HPLC and secondary metabolites like phenolics, flavonoids, xanthones, benzoxazinoids, etc. identified by LC-MS) resulted in enhanced total root surface, total root volume and/or root number in certain inoculated treatments, with reduced fertilization (Walker et al. 2012).

DAPG synthesis in *P. fluorescens* is upregulated by amoebae grazing (Jousset and Bonkowski 2010) directly involving ISR in plants (Weller et al. 2012). *Pseudomonas* and bacterivorous nematodes in soil increase microbial biomass and stimulate PGPR activities (Jiang et al. 2012). *Burkholderia* and nematode combination showed higher proliferation of root tips than individual application (Jiang et al. 2012; Pedersen et al. 2009; Hol et al. 2013). In the rhizosphere, decomposers such as earthworms increase the nutrient accessibility for plants, inducing change in the root exudate composition leading to a secondary positive effect on *P. fluorescens* (Elmer 2009; Jin et al. 2010; Hol et al. 2013).

16.7 DAPG Biosynthesis and Regulation

DAPG is a complex natural secondary metabolite synthesized via polyketide pathway (Singh and Bharate 2006) through a series of decarboxylative condensation reaction of monomeric acetyl- and malonyl-coenzyme A (CoA) catalysed by polyketide synthases (PKSs) to form PG and then DAPG (Gao et al. 2010). The PKSs are classified as Type I, II or III based on structural and functional properties. The Type I and II PKSs are involved in fatty acid biosynthesis and aromatic polyketide synthetic pathway, respectively, whereas Type III PKSs are attributed for DAPG biosynthesis in *Pseudomonads* via PG implying iterative elongation of precursor acyl-CoA to malonyl-CoA, leading to formation of the PG core of DAPG (Song et al. 2006). In contrast to Type I (multifunctional non-iterative polypeptide modules) and Type II PKSs (discrete iterative catalytic complex), the Type III PKSs are single multifunctional enzyme complex that perform iterative catalytic condensation of acetate units derived from malonyl-CoA to a CoA-linked starter molecule up to a definite length and then cyclized in the ketoacyl synthase active site cavity. Thus, Type III PKSs display CoA bound substrate activation instead of an acyl-carrier protein (ACP) displayed by the Type I and II PKSs (Weissman 2009; Dairi et al. 2011). The biosynthetic genes (named *phl*) of the PKSs are arranged as clusters encoding PG and its derivatives. The *phl* nine-gene cluster (*phl*HGFOACBDE) from DAPG-producing *P. fluorescens* Q2-87 harbours genes for (i) biosynthesis (*phl*ACB-encoding acyltransferase and a highly conserved *phl*D-encoding Type III PKS), (ii) degradation (*phl*G-encoding hydrolase), (iii) regulation (*phl*H modulator, *phl*F repressor and *phl*F binding site – *phl*O), (iv) export/tolerance (*phl*E) and (v) *phl*I ORF encoding an uncharacterized protein (UniProt accession no. C0J9E0) spanning ~8 kb region (Figs. 16.2 and 16.3) (Bangera and Thomashow 1999; Moynihan et al. 2009; Hayashi et al. 2012; Gutiérrez-García et al. 2017). The genes *phl*GACBDE have a common transcriptional orientation with *phl*HGF located downstream to the promoter-proximal *phl*ACB. The *phl*H and *phl*F are oppositely oriented. The core *phl*ACBD operon is cumulatively responsible for DAPG synthesis, where (Step 1) *Phl*D is responsible for iterative condensation of malonyl-CoA (3 units) to form

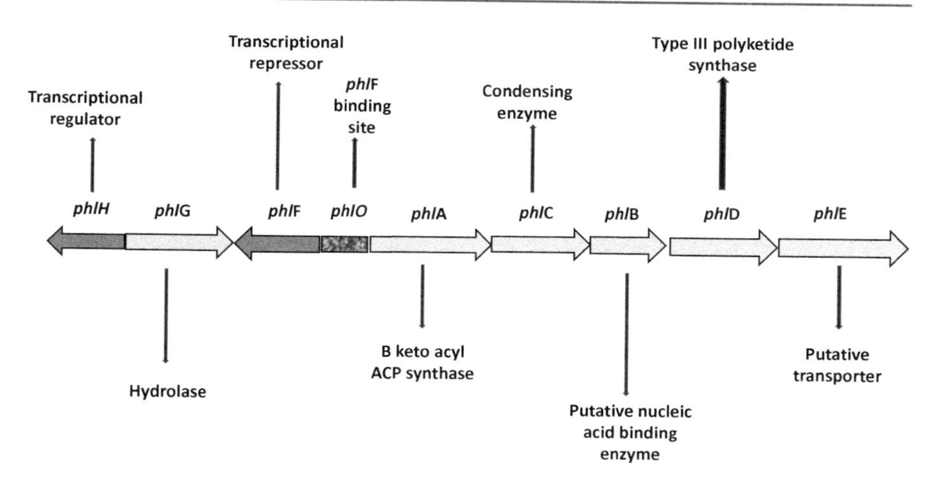

Fig. 16.2 Array of genes involved in biosynthesis and regulation of DAPG in *Pseudomonads*

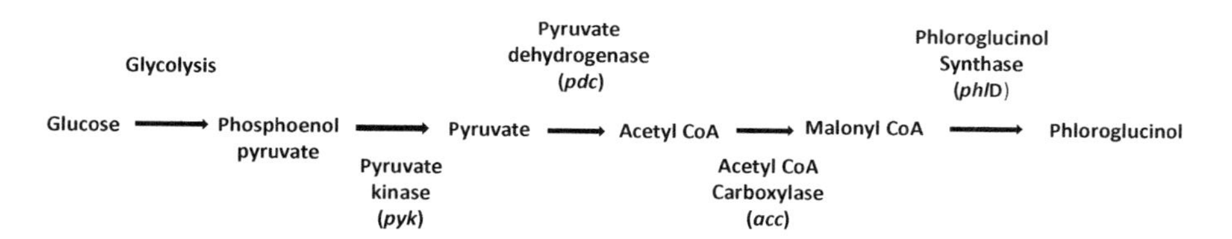

Fig. 16.3 Metabolic route for biosynthesis of PG via glycolytic pathway in microbes

3,5-diketoheptanedioate and later cyclize to form the PG (Zha et al. 2006) and (Step 2) *Phl*ACB jointly encodes a MAPG acyltransferase which catalyses acyl transfer at initial stage (acetyl-CoA to acetoacetyl-CoA) and at the final stage (the product DAPG from a MAPG precursor) (Achkar et al. 2005; Hayashi et al. 2012). The *phl*D is therefore the key biosynthetic gene and a genetic marker to identify PG producers and putative DAPG-producing strains (Picard and Bosco 2003). The *phl*D marker is relatively rare in phylum *Proteobacteria*, but commonly found in Gram-positive bacteria and plants that produce phenolic compounds containing PG core analogous to DAPG (Ramette et al. 2001; Gutiérrez-García et al. 2017).

The regulation of *phl* genes is central to overall DAPG production. The *phl*D and *phl*ACB gene expression for PG and DAPG biosynthesis is under transcriptional level control by PhlF and PhlH repressors (Schnider-Keel et al. 2000; Yan et al. 2017). PhlF contains a *phl*O-binding helix-turn-helix motif located upstream of the *phl*A transcriptional start site (Abbas et al. 2002). The *phl*ACB operon is also auto-induced by DAPG and probably mediated by *phl*F as revealed by a gene deletion study of *P. fluorescens* CHA0 (Schnider-Keel et al. 2000).

Recently, PhlH was found to repress *phl*G promoter (DAPG hydrolase) in *P. fluorescens* 2P24. Moreover, PhlH was also found to interact with MAPG as well as DAPG signalling molecules and activate expression of *phl*G. Thus, PhlH and PhlG were revealed to impose negative feedback regulation over DAPG biosynthesis (Yan et al. 2017). Besides gene regulatory elements, the *phl* operon is also repressed by extracellular metabolites, viz. salicylate, pyoluteorin and fusaric acid, which apparently interact with *phl*F or its product (Schnider-Keel et al. 2000).

The post-transcriptional regulation of DAPG apparently involves a quorum sensing signal transduction pathway of membrane bound sensor kinase GacS that complexes with small regulatory RNAs (srRNA) to activate response regulator GacA, which binds srRNA RsmX/Y/Z followed by RsmA/E binding to act as mRNA repressor in the trophophase, but neutralizes during idophase in *Pseudomonads* (Duffy and Défago 2000; Heeb and Haas 2001; Takeuchi et al. 2009). The expression of the stress σ^S (RpoS) is positively controlled by GacA and negatively by RsmA (Heeb et al. 2005). Alternatively, GidA and TrmE (tRNA-modifying enzyme and GTPase, respectively) positive regulatory system independent of the Gac/Rsm pathway was recently reported in *P. fluorescens* 2P24 (Zhang et al. 2014). Interestingly, the *gid*A and *trm*E mutants could convert PG to DAPG via MAPG, but could not produce PG. Besides, expression levels of RNA polymerase sigma factors, viz. housekeeping RpoD and stationary phase RpoS, upregulate and downregulate DAPG biosynthesis, respectively.

16.8 Mode of Action of DAPG

PG compounds are isolated from diverse natural sources (Singh and Bharate 2006). DAPG acts as a broad-spectrum polyketide antibiotic against soil-borne pathogens and displays antibacterial, antifungal, anthelmintic and phytotoxic activities (Dubuis et al. 2007; Brazelton et al. 2008) (Table. 16.2). The effect of DAPG on cell system is rather less specific and concentration dependent and impairs ionophore channels and the proton gradient across phospholipid bilayers. The action of DAPG is pinpointed to (i) mitochondrial dysfunctions, in which it interrupts membrane potential, changes cellular homeostasis and releases reactive oxygen in *S. cerevisiae,* (ii) inhibits photosynthesis in the chloroplast and (iii) alters bacterial cell membrane integrity (Terada 1981; Kwak et al. 2011; Troppens et al. 2013b) (Fig. 16.4). The inhibitory properties of DAPG are not constrained to phytopathogens, but also extend to non-pathogenic rhizosphere fungi and bacteria (Girlanda et al. 2001; Natsch et al. 1998). At high concentrations, DAPG has antimicrobial and phytotoxic action and at lower concentrations signal cascade effect on bacteria (Combes-Meynet et al. 2011) and plants (Weller et al. 2012).

16.8.1 Antibacterial Activity

DAPG sternly affects the cell membrane of a variety of bacteria ranging from phytopathogens to human pathogens in a dosage-dependent manner. In the rhizosphere, DAPG affects even N-fixing microbes such as *Azospirillum* spp. and *Rhizobium leguminosarum* and indirectly assists them to infect plant roots by increasing

Table 16.2 Bioactivity of DAPG on various microbes and cancer cell lines

Antagonism	Targets	Target characteristics	Effect of DAPG	References
Antibacterial	*Plant pathogens*			
	Erwinia carotovora subsp. *atroseptica*	Soft rot of potato	Inhibit growth at 15 μl	Cronin et al. (1997)
	Rhizobium leguminosarum	Nitrogen-fixing bacteria	Increase root cell permeability, enhance root exudation and nodulation of pea roots and C availability in the rhizosphere	de Leij et al. (2002)
	Xanthomonas oryzae pv. *oryzae*	Blight of rice	Inhibit growth of the devastating pathogen (59–64%)	Velusamy et al. (2006)
	Ralstonia solanacearum	Wilt pathogen	Reduce wilt incidence, targets one or more essential cellular processes in vascular pathogen causing wilt	Ramesh et al. (2009), Ramadasappa et al. (2012), and Zhou et al. (2012)
	Azospirillum brasilense	Nitrogen-fixing bacteria	Moderately damage cytoplasmic membrane, growth inhibition; however, co-inoculation with DAPG-producing *P. fluorescens* F113 induces accumulation of carotenoids and poly-hydroxybutyrate-like granules resulting in phytostimulation	Couillerot et al. (2011)
	Xanthomonas campestris DSM 3586	Black rot	–	Sekar and Prabavathy (2014)
	Erwinia persicina HMGU155	Soft rot		

(continued)

Table 16.2 (continued)

Antagonism	Targets	Target characteristics	Effect of DAPG	References
	Bacillus subtilis (DSM 347)	–	Growth inhibition (100%)	Nagel et al. (2012)
	Staphylococcus lentus (DSM 6672)		Growth inhibition (98%)	
	Human pathogens			
	Vibrio parahaemolyticus	Cholera	Sensitive at 24 μg/ml conc.	Kamei and Isnansetyo (2003)
	Staphylococcus aureus	Skin and systemic infections	Vancomycin-resistant strain inhibited at 4 mg/l conc.	Isnansetyo et al. (2003)
			Methicillin-resistant strain inhibited at 1 μg/ml conc.	Kamei and Isnansetyo (2003)
	Enterococcus spp. genotypes A, B	Urinary tract infections, endocarditis and meningitis	Vancomycin-resistant strain inhibited at 8 mg/l conc.	Isnansetyo et al. (2003)
	Escherichia coli	Intestinal infection	Repress cell progress at concentration 0.5 g/l conc.	Cao et al. (2011)
Antifungal	*Fusarium oxysporum* f. sp. *radicis-lycopersici*	Crown and root rot of tomato	Moderate control	Duffy and Défago (1997)
	Gaeumannomyces graminis var. *tritici*	Take-all decline	Effective at 5.04 μg/ml	Raaijmakers and Weller (1998), de Souza et al. (2003), Bakker et al. (2002), and Kwak et al. (2009)
	Pythium ultimum var. *sporangiiferum*	–	Disorganization in hyphal tips (disruption of the plasma membrane, proliferation, retraction, vacuolization and cell disintegration)	de Souza et al. (2003)
	F. oxysporum f. sp. *cubense*	Wilt in banana	Reduced vascular discoloration, bulb formation and lysis of fungal mycelia	Saravanan and Muthusamy (2006)

(continued)

Table 16.2 (continued)

Antagonism	Targets	Target characteristics	Effect of DAPG	References
	Thielaviopsis basicola	Black root rot of tobacco	–	Ramette et al. (2006)
	Rhizoctonia solani	Damping-off of cotton, bean and head rot	–	Reddy et al. (2007), Afsharmanesh et al. (2010), and Zhang et al. (2014)
	Magnaporthe grisea	Present on rice plant	Inhibition of mycelial growth	Reddy et al. (2007)
	Drechslera oryzae			
	Sarocladium oryzae			
	Botrytis cinerea	Necrotrophic to various plants	Resistance to non-degradative and degradative mechanisms of pathogen. Non-degradative mechanism involves efflux by the ABC transporter BcAtrB, whereas degradative mediated indirectly by the laccase BcLCC2	Schouten et al. (2008)
	Fusarium oxysporum f. sp. *ciceri*	Wilt of chickpea	–	Saikia et al. (2009)
	Saccharomyces cerevisiae	–	Impairs mitochondrial function by depolarization of the mitochondrial membrane. More toxic during the energy demanding early stages of exponential growth mimicking proton ionophore dissipating the proton gradient, respiration uncoupling and ATP synthesis	Gleeson et al. (2010) and Troppens et al. (2013a)

(continued)

Table 16.2 (continued)

Antagonism	Targets	Target characteristics	Effect of DAPG	References
	Aphanomyces cochlioides AC-5	Damping-off in sugar beet, spinach	Excessive branching and curling in the hyphae, inhibited mycelial growth and disrupted the organization in the cortical filamentous actin	Islam and Fukushi (2010)
	Plasmopara viticola	Downy mildew	Formation of round cytospores, zoospore germination with excessively branched germ tubes. Inhibition of zoosporogenesis and motility at 5 and 10 μg/ml, respectively. Substitution and extension of acyl group with H in the benzene ring corroborated to the level of bioactivity	Islam and Von Tiedemann (2011)
	Aphanomyces cochlioides	Damping-off		
	Phytophthora infestans (BASF)	Phytopathogenic fungi	Growth inhibition (86%)	Nagel et al. (2012)
	Septoria tritici (BASF)		Growth inhibition (100%)	
	Phytophthora drechsleri	Root and crown rot of cucumber	Strong inhibitory activity	Shirzad et al. (2012)
	Gaeumannomyces graminis var. *tritici*	Take-all decline	7.5 μg/ml reduced root growth; 10 μg/ml caused reduced root hair development, brown necrosis and tissue collapse of sporophytes	Okubara and Bonsall (2008), Kwak et al. (2012), Sekar and Prabavathy (2014)
	Colletotrichum falcatum	Red rot sugar cane	Mycelial growth inhibited from 14% to 52%	Hassan et al. (2014)
	Fusarium oxysporum f. sp. *cubense* FOC	Wilt and root necrosis	Suppressed the growth	Ayyadurai et al. (2006)

(continued)

Table 16.2 (continued)

Antagonism	Targets	Target characteristics	Effect of DAPG	References
	Pyricularia grisea TN 508	Rice blast	–	Sekar and Prabavathy (2014)
	Fusarium oxysporum DSM 62297	Banana wilt	–	
	Rhizoctonia bataticola	–	–	Chaubey et al. (2015)
	Fusarium culmorum and *Fusarium graminearum*	Head blight of wheat	Inhibit mycotoxin production	Muller et al. (2016)
	Alternaria alternata and *Alternaria tenuissima*	Black point of wheat leaves and ears		
	Batrachochytrium dendrobatidis	Amphibial pathogen	Effective at 136.13 μM (MIC)	Brucker et al. (2008)
	Trichophyton mentagrophytes	Dermatophyte	Growth inhibition (93%)	Nagel et al. (2012)
	Trichophyton rubrum		Growth inhibition (98%)	
Anthelmintic	*Globodera rostochiensis*	Pest of potato	Increase hatch ability and reduction in juvenile mobility	Cronin et al. (1997)
	Heterodera glycines	Plant-parasitic nematodes	Not effective	Meyer et al. (2009)
	Pratylenchus scribneri		Not effective	
	Meloidogyne incognita		Decreased egg hatch and induced mortality	
	Xiphinema americanum		Toxic to adults, decrease in viability at LD_{50} 8.3 μg/ml	
	Caenorhabditis elegans	Bacterial-feeding nematodes	Egg hatch in 0, 10 and 75 mg/ml DAPG was ca. 2.8%, 4.5% and 9.1%, respectively	
	Pristionchus pacificus		No effect	
	Rhabditis rainai			

(continued)

Table 16.2 (continued)

Antagonism	Targets	Target characteristics	Effect of DAPG	References
Antiprotozoal	*Vahlkampfia* spp. And12	Amoeba	Acute toxicity, which resulted in rapid cell lysis <1 h, with loss of motility within minutes, growth inhibition, encystation, paralysis and cell lysis	Jousset et al. (2006)
	Colpoda steinii Sp1	Ciliate		
	Neobodo designis And31	Flagellate	Inhibited at low concentrations	
Antiviral	Vesicular stomatitis virus (VSV)	–	Effective against RNA virus with envelope (VSV)	Tada et al. (1990)
	Herpes simplex virus type I (HSV-I)	–	Effective against DNA virus with envelope (HSV-I)	
	Polio virus type I		Ineffective against the RNA virus without envelope (polio-I)	
Herbicidal activity	*Linaceae, Cruciferae, Gramineae* and *Urticaceae*	Weed species	1000 μg/ml effected 100% inhibition of germination of *Linaceae, Cruciferae, Gramineae* and *Urticaceae* and 30% at 0.5–1000 μg/ml. *Urtica dioica* most sensitive at 0.5 μg/ml and most resistant species *Centaurea iberica*	Katar'yan and Torgashova (1976)
Algicidal	*Pseudoalteromonas elyakovii* and *Algicola bacteriolytica*	Baltic Sea	Beneficial effect on marine brown macroalga *Saccharina latissima*	Nagel et al. (2012)
Other	Breast, cervical, colon and lung cancer cell lines (MDA MB-23, HeLa, HCT-15 and A549, respectively)	–	Exhibit selective cytotoxicity and show anti-leukemic activity	Veena et al. (2016)

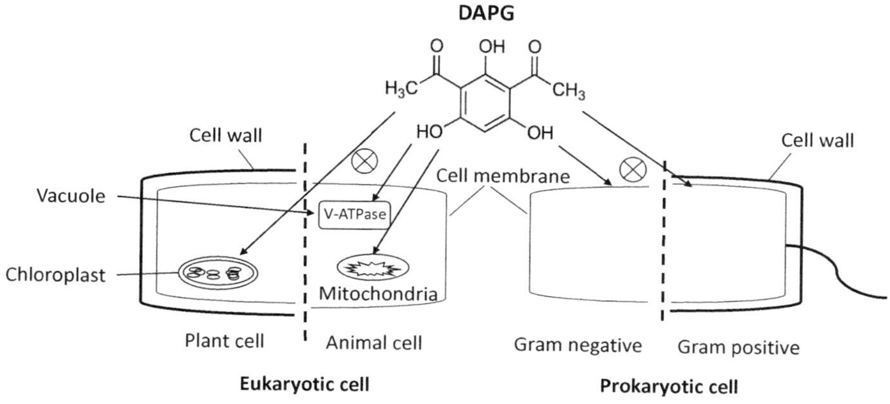

Fig. 16.4 Proposed cellular targets of DAPG

permeability of root cells, resulting in improved nodulation and alteration in root exudation pattern causing phytostimulation and increased plant nutrient uptake (Phillips et al. 2004; Couillerot et al. 2011). DAPG repressed growth of the phytopathogenic bacterium *Erwinia carotovora* subsp. *atroseptica* (Nowak-Thompson et al. 1994; Cronin et al. 1997) and *Erwinia persicina* HMGU155 that elicit soft rot in tuber crops. DAPG is also found to control wilt pathogen *Ralstonia solanacearum* (Ramesh et al. 2009; Ramadasappa et al. 2012; Zhou et al. 2012). DAPG (5 μg/ml) repressed some *Bacillus* spp. (128 μg/ml) toxic to dicotyledonous plants than monocotyledonous plants (256 μg/ml) (Keel et al. 1992). Stimulated root and shoot length along with grain yield in rice plants were observed against bacterial blight pathogen *Xanthomonas oryzae* pv. *oryzae* when IAA-producing *Pseudomonas* spp. PDY7 inoculations were examined (Velusamy et al. 2013). Similarly, phenolics salicylate and pyoluteorin from rhizobacteria strongly suppress biosynthesis of DAPG in *P. fluorescens* CHA0 (Schnider-Keel et al. 2000).

DAPG is also investigated as potential antibiotic against human pathogens. Despite its stability below temperature and pH conditions, DAPG does not produce acute toxicity in mice (Kamei and Isnansetyo 2003). DAPG caused lysis of methicillin-resistant Gram-positive *Staphylococcus aureus* within 2 h due to exposure of 5 μM DAPG, while Gram-negative *Vibrio parahaemolyticus* lysed more slowly at 114 μM concentration, demonstrating that the cell wall composition may impact DAPG uptake. Higher concentrations of DAPG exert lysis of *V. parahaemolyticus* more slowly compared to *S. aureus* (Kamei and Isnansetyo 2003). Amendment of DAPG at 0.5 g/L in culture media shows repression of *Escherichia coli* cell progression (Cao et al. 2011).

16.8.2 Antifungal Activity

The effect of DAPG on fungal and eukaryotic cells has been studied with *S. cerevisiae* as a cellular model (Gleeson et al. 2010). DAPG repressed growth of phytopathogenic fungi *Pythium ultimum* and *Rhizoctonia solani* (Nowak-Thompson et al. 1994). Rhizobial pH has a significant effect on the action of DAPG against mycelial growth of *Pythium ultimum* var. *sporangiiferum*: higher activity of DAPG at lower pH and disorganization in hyphal tips of *P. ultimum* var. *sporangiiferum* leading to cell disintegration and simultaneously blocking the maintenance of membrane integrity (de Souza et al. 2003). DAPG has impact on mycelial growth of wide host range of *R. solani* (a causative agent of wire stem, damping-off, root and collar rot) and defeats the disease by 33.34 and 14.29% through soil soaking and seed treatment, respectively (Afsharmanesh et al. 2010). Antagonistically, certain mycotoxins and fusaric acid produced from various *Fusarium* species, like *F. verticillioides*, decrease DAPG production and expression of *phl*D gene in *Pseudomonads* (Quecine et al. 2016).

16.8.3 Antiprotozoal Activity

DAPG exposure to the cells of amoeba *Vahlkampfia* spp. causes lysis within 1 h, with loss of motility within minutes. However, *Colpoda steinii* and *Neobodo designis* remain unaffected, while ingestion of DAPG was fatal to *N. designis* (Jousset et al. 2006). The talc formulation of DAPG-producing microbes also reduces vascular discoloration in banana plants when inoculated at 15 g/plant (Saravanan and Muthusamy 2006).

16.8.4 Anthelmintic Activity

The supernatant of DAPG is non-toxic when supplied exogenously to phagotrophic ciliate *Colpoda steinii* and flagellate *Neobodo designis*; however, rapid toxicity was seen upon ingestion of DAPG-producing microbes (Jousset et al. 2006; Brazelton et al. 2008). DAPG was found to reduce egg hatch in *Meloidogyne incognita,* but opposite effect was seen with microbivore *Caenorhabditis elegans* during the 1st hour of incubation (Meyer et al. 2009).

16.8.5 Phytotoxic Activity

DAPG amends root physiology (Brazelton et al. 2008) and showed more amino acid exudates (Phillips et al. 2004) that increase the root colonization in nodulating rhizobacteria. Increase in root mass and formation of lateral roots that were seen with inoculation of DAPG producers suggests that DAPG might act as a plant hormone-like substance (de Leij et al. 2002) and its specificity coincides with auxin-herbicide

2,4-dichlorophenonxyacetic acid (2,4-D) (Reddy et al. 1969). DAPG induces alteration in the root morphology and physiology leading to root tips browning and wrinkling with swollen roots. DAPG also stimulate host defence pathways promoting lateral root branching and inhibition of primary root growth (Meziane et al. 2005; Bakker et al. 2007). Microbes move toward root tips growth and form colonies depositing DAPG crystals inside and around the roots of tomato to inhibit primary root growth thereby stimulating secondary root (Brazelton et al. 2008).

High DAPG concentrations ranging from 32 to 1024 μg/ml were found to be phytotoxic to monocots (Keel et al. 1992); however, it is rare under in situ augmentation of PGPR *Pseudomonads* to monocots (Okubara and Bonsall 2008; Maurhofer et al. 1992, 1995). In dicots such as cucumber, phytotoxicity imparted by DAPG and allied compounds of *P. fluorescens* CHA0 was comparatively less than monocots, and the overall phytotoxicity is not specifically linked to DAPG, but the presence of hydrogen cyanide and other phytotoxic compounds (Notz et al. 2001). There is competition between pathogen and DAPG producers for organic nutrients or specific niches on the root (Nelson 2004; Heydari and Pessarakli 2010). As DAPG can cause ISR in the plant (Iavicoli et al. 2003), it affects pathogens indirectly (van Loon 2007; van Wees et al. 2008), but the significance is possibly negligible when other saprophytic bacteria advance root colonization.

16.9 Fermentative Production of DAPG

Both biotic and abiotic factors contribute to performance of fluorescent *Pseudomonas* for fermentative production of DAPG (Thomashow and Weller 1996; Duffy and Defago 1997; Notz et al. 2002) (Fig. 16.5). Biotic factors such as plant species, plant age, cultivar and pathogens alter the expression of *phl*A (Notz et al. 2001). In rhizosphere, production of DAPG depends on the (i) metabolic state of the bacteria, (ii) interaction with other organisms (Yang and Cao 2012), (iii) host factors such as root exudates (Kwak and Weller 2013) and (iv) cultivar (Okubara and Bonsall 2008), suggesting that plants as well as bacterial compounds has effect on the production of DAPG. For example, IAA stimulates PG gene expression (Dubuis et al. 2007). Presence of metabolites produced by other microbes also influence DAPG production; e.g. *F. oxysporum* f. sp. *radicis-lycopersici* produces fusaric acid which acts as a chemical signal to repress DAPG production by *P. protegens* CHA0 (Duffy and Défago 1997; Schnider-Keel et al. 2000).

Under gnotobiotic conditions, DAPG production by *Pseudomonads* is found to be influenced by a variety of carbon sources, inorganic phosphate and minerals. DAPG-producing organisms, yield and various culture media used for its production are enlisted in Table 16.3. Production of DAPG was improved in complex medium amended with organic and inorganic N sources at different phosphate concentrations (Saharan et al. 2011). Glucose stimulates DAPG production in a strain-specific manner. Various sugars such as sucrose, fructose and mannitol promoted high yields of DAPG in *Pseudomonas* spp. F113, whereas glucose and sorbose exert less amount of DAPG production (Shanahan et al. 1992), while in another study,

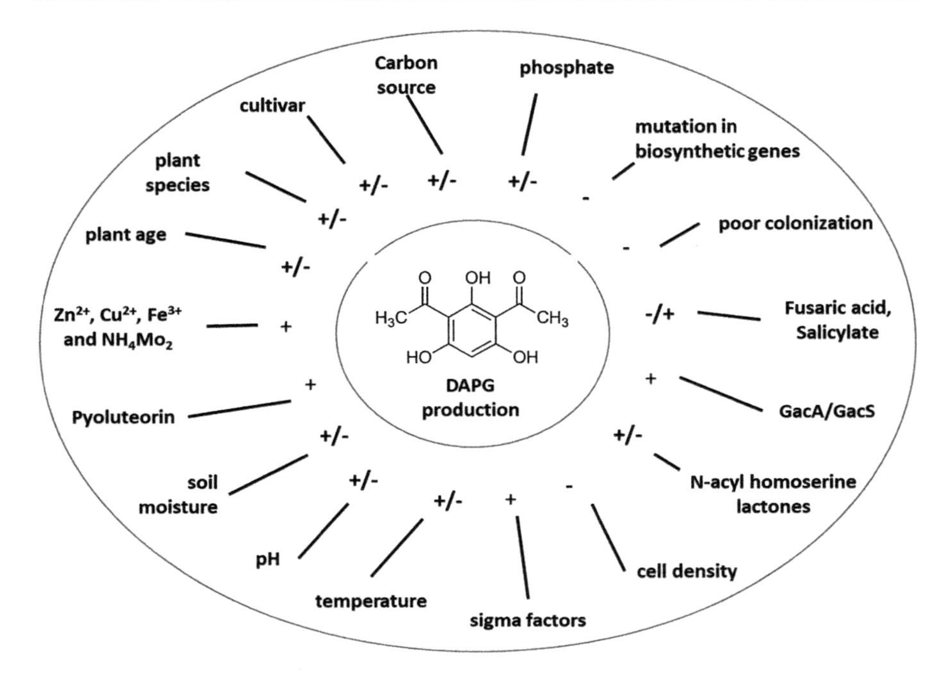

Fig. 16.5 Biotic and abiotic factors that modulate positively (+) or negatively (−) on DAPG production by *Pseudomonads*

glucose and fructose were seen to promote DAPG production (Standing et al. 2008). Approximately, 500 μg/ml of DAPG yield was obtained with ethanol as the sole carbon source under a high C/N ratio and limitation of inorganic phosphate in the medium (Yuan et al. 1998). Glycerol was found as the best carbon source for biomass and DAPG production, while NH_4Cl and urea had a steady effect on pH during batch cultivation (Shanahan et al. 1993; Duffy and Defago 1999; Hultberg and Alsanius 2008; Sarma et al. 2013).

The optimum temperature for DAPG production was found to be ≃12 °C, indicating that soil temperature is helpful to maximum antibiotic production by *Pseudomonas* sp. strain F113 (Shanahan et al. 1992) and promoted by exposure to stress, high ethanol and NaCl concentration, or heat shocking. About 870 mg/l DAPG concentration was reached with 1% NaCl within 3 days and a highest 1200 mg/l with a heat shock at 50 °C for 10 min (Nakata et al. 1999). Low pH had shown significant increase in DAPG. An optimized medium was proved to increase the DAPG production by 13-fold compared to unoptimized medium (De Souza et al. 2003). A pH-based fed-batch strategy achieved DAPG concentration of 298 and 342 mg/l for *Pseudomonas* spp. R62 and R81 strains, respectively (Sarma et al. 2013).

Abiotic factors like carbon sources and various minerals like Fe^{3+} and sucrose increase DAPG (Nowak-Thompson et al. 1994; Duffy and Defago 1999). Divalent

Table 16.3 Microbial production of DAPG in various media and its chromatographic detection

DAPG-producing organisms	Sources	Yield (μg/ml)	Production medium	Solvent system for TLC (v/v)	R*f*	HPLC RT(min)	Reference(s)
Pseudomonas sp. F113	Sugar beets	NR	SA	Dichloromethane: hexane: methanol (50:40:10)	NR	8.0	Shanahan et al. (1992)
P. fluorescens Pf-5	Rhizosphere	52.2 ± 3.2	523	NR	NR	6.4	Sarniguet et al. (1995)
P. fluorescens Q2–87	Rhizosphere	0.63	YMB	NR	NR	17.02	Bonsall et al. (1997)
P. fluorescens Q287(pPHL5122)	Rhizosphere	112					
P. fluorescens Q69c 80(pPHL5122)	Rhizosphere	1.5					
P. fluorescens	Tobacco roots	NR	PCG	NR	NR	11.6	Duffy and Défago (1997)
P. fluorescens S272+2% ethanol	NR	500	NR	NR	0.78	NR	Yuan et al. (1998)
P. fluorescens S272 with ethanol	Rhizosphere	670	YSE	$CHCl_2$: MeOH: H_2O (85:15:1)	0.75	NR	Nakata et al. (1999)
P. fluorescens S272 with a NaCl stress	Rhizosphere	870	YSG				
P. fluorescens S272 with a heat shock	Rhizosphere	1200	YSG				
P. fluorescens	Rhizosphere	1	S1	NR	NR	22.0	Picard et al. (2000)
Pseudomonas aeruginosa strain FP10	Banana	NR	PPM	NR	0.77	10.77	Ayyadurai et al. (2006)

(continued)

Table 16.3 (continued)

DAPG-producing organisms	Sources	Yield (μg/ml)	Production medium	Solvent system for TLC (v/v)	*Rf*	HPLC RT(min)	Reference(s)
P. fluorescens	Banana	NR	PPM	Acetonitrile: methanol: water (1:1:1)	0.88	20–21.30	Saravanan and Muthusamy (2006)
P. fluorescens (Pf4-92)	Chickpea	0.66	PCG	NR	NR	NR	Saikia et al. (2009)
P. fluorescens	Rice	NR	KBM	Benzene: acetic acid (95:5)	0.35	14.63	Reddy et al. (2007)
Pseudomonas spp. YGJ3	Rhizosphere	11	NM	$CHCl_3$: CH_3OH (30:1)	0.51	47.0	Matano et al. (2010)
Pseudomonas spp. R62	Rhizosphere	81	SM	NR	NR	NR	Saharan et al. (2010)
Pseudomonas spp. R81	Rhizobacteria	70	SM	NR	NR	NR	Sarma et al. (2010)
Fluorescent pseudomonad R62	Wheat	810	SM	NR	NR	NR	Saharan et al. (2010)
Fluorescent pseudomonad R62 (shake flask)	Wheat	125	SM	NR	NR	NR	Saharan et al. (2011)
Fluorescent pseudomonad R62 (bioreactor)	Wheat	135					
P. fluorescens	Rhizosphere	4.33	YME	NR	NR	NR	Shirzad et al. (2012)
P. brassicacearum J12	Tomato	40	LB	Chloroform:acetone (9:1)	0.70	5.9	Zhou et al. (2012)
P. aeruginosa MIC2	Cucurbit	NR	NR	Chloroform:methanol (9:1)	NR	0.77	Shanthi and Vittal (2013)
P. fluorescens PDY7	Rice	NR	NR	NR	0.32	NR	Velusamy et al. (2013)
P. protegens Pf-5	Soil	9.8 ± 0.9	NYBGly-Zn	NR	NR	NR	Quecine et al. (2016)
Pseudomonas gessardii	NR	NR	S1 agar	NR	NR	6.7	Muller et al. (2016)
P. protegens Pf-5	Soil	NR	NBGly	NR	NR	18.1	Yan et al. (2017)

Lysobacter gummosus	Salamander skin	NR	NB	NR	NR	13.46	Brucker et al. (2008)
Ochrobactrum intermedium NH-5	Sugar cane	0.4	LB	NR	NR	NR	Hassan et al. (2014)
E. coli (metabolically engineered)	NR	1280[a]	NR	NR	NR	NR	Zha et al. (2009)
E. coli JWF1 *phl*D$^+$ recombinant	NR	780[a]	MSM	NR	NR	NR	Yang and Cao (2012)

NR Not reported, *SA* sucrose asparagine medium, *YMB* yeast malt broth, *YSE* yeast extract soy flour ethanol, *YSG* yeast extract soy flour glucose (S1-NR), *PPM* pigment production medium, *PCG* NR, *NB* nutrient broth, *KBM* King's B medium, *NM* nutrient medium, *SM* Schlegel's medium, *YME* yeast malt extract medium, *LB* Luria Bertani, *NYBGly-Zn* nutrient broth fortified with glycerol and zinc, *MSM* minimal salt medium

[a]PG yield

cations Zn^{2+}, Cu^{2+} and NH_4Mo^{2+} stimulate the production of DAPG (Duffy and Defago 1999), and Zn^{2+}, Mn^{2+} and MoO_4^{2-} were the major components to attain 125 mg/l of DAPG, closely 13-fold more related to control and enough for possession of the bioinoculant viable in non-sterile talcum powder-based formulations which contained 25 µg DAPG/g carrier when stored at 28 °C for 6 months. A 14 L bioreactor experiment resulted in 135 mg/l of DAPG after 36 h (Saharan et al. 2011). Presence of Cl^- ions suppressed the production of DAPG in *Pseudomonas* spp. YGJ3. Without Cl^-, the cell-free supernatant exhibited more activity of the MAPG required for DAPG synthesis (Matano et al. 2010).

Apart from medium modification, genetic engineering has used PG encoded by a unigene *phl*D using acetyl CoA metabolic precursors to produce 1280 µg/ml PG (Zha et al. 2009). But conversion of PG to DAPG is possible only with *phl*DACB co-expression. Initial attempts to demonstrate increased DAPG production using a 6 kb fragment from *Pseudomonas* spp. F113 achieved 0.373 µM DAPG vis-a-vis 0.46 µM by native strain (Fenton et al. 1992). Increased DAPG production (3–12 nmol/g of roots with adhering rhizosphere soil) was observed with *P. fluorescens* CHA0 via the transfer of a recombinant cosmid pME3090 of 22 kb insert of CHA0 DNA (Maurhofer et al. 1995). In contrast, metabolically engineered *E. coli* produced moderate PG where *phl*D biosynthetic gene (Type III polyketide synthase) was cloned into a bacterial expression vector. In another study, PG resistance was enhanced due to more expression of *E. coli marA* (multiple antibiotic resistance) gene up to 0.27 g/g dry cell weight. DAPG production increased to around 0.27 g/g dry cell weight by increasing the level of malonyl coenzyme A through synchronized expression of four acetyl-CoA carboxylase subunits. Also, the co-expression of ACCase and *marA* produced improvement of PG production to 0.45 g/g dry cell weight, i.e. 3.3-fold to the original strain. Engineered strain produced PG up to 3.8 g/l under fed-batch conditions, after 12 h induction, consistent with a volumetric productivity of 0.32 g/l/h (Cao et al. 2011), and production reached 790 mg/l DAPG under fermenter-controlled conditions by recombinant strain in the presence of glucose (Achkar et al. 2005). A genetically modified strain of *P. fluorescens* F113 containing pCUP9 construct was developed for enhanced DAPG production containing the *phl*ACBDE genes to secrete up to 600 µg/ml from the log phase (6 h) of growth, but later decreased to 50 µg/ml (Delany et al. 2000). Recently, constitutive expression of DAPG biosynthetic gene cluster *phl*DACB sourced from *Pseudomonas* sp. G22 and *P. protegens* Pf-5 was achieved in endophytic *Pseudomonas* sp. WS5 that showed 12–14 µg/ml DAPG from <2 µg/ml by the native strains after 96 h. Non-*Pseudomonas* spp. like *Lysobacter gummous* (Brucker et al. 2008) isolated from salamander *Plethodon cinereus* and *Ochrobactrum intermedium* from sugarcane (Hassan et al. 2014) are also reported to produce DAPG, however, their biotechnological potential remains unexplored.

For detection of DAPG, a general solvent extraction strategy is adopted. Commonly, the broth culture is acidified to pH 2 with 5 N HCl (Shirzad et al. 2012) or 10% trifluoroacetic acid (TFA) (Bonsall et al. 1997) and extracted with equal volume of ethyl acetate for 30 min. Alternatively, extraction with 80% acetone at an acidic pH selective for non-polar compounds to eradicate polar contaminants such as soil humic and fulvic acids can be performed (Bonsall et al. 1997). Phase separation is accelerated by centrifugation (6000 rpm, 10 min) followed by flash evaporation of the organic phase in a round-bottom flask. The final residue is dissolved in 1 ml of HPLC-grade methanol before physico-chemical analysis. Further, characterization can be performed by TLC and HPLC, and the structure of the purified DAPG could be confirmed using ^{1}H NMR and mass spectrometry (Saharan et al. 2011).

16.10 Conclusion and Future Consideration

The review outlines the indispensable role of DAPG-producing *Pseudomonas* PGPR for overcoming various phytopathogen population, constituting the first line of defence by stimulation of induced resistance in plants, formation of disease suppressive soils and improving crop productivity. The fundamental aspects of biosynthesis of DAPG by *phl* genes could improve biocontrol activity and yield of DAPG-producing microbes. For the foregoing fact, more efforts are required on different aspect of DAPG-producing fluorescent *Pseudomonas* spp. to explain the mechanism for effective suppression of plant diseases and how the rhizobacteria affects DAPG production. These aspects may pave a new pathway to control phytopathogens to increase crop productivity. The endeavour would be useful for farmers and, in turn, society with improved food security and livelihood, reduced contamination from pesticides and increased biodiversity. It can be specified that, soon, industrial microbial processes will converge as the major source of PG compounds for medicine, cosmetics and agriculture.

Acknowledgement The authors RAK and NDD are grateful to the Department of Science and Technology, New Delhi, for providing WOS-A fellowship [SR/WOS-A/LS-1209/2014 (G)]. The authors also acknowledge DST-FIST and UGC-SAP for the infrastructural support to the host institute.

References

Abbas A, Morrissey JP, Marquez PC, Sheehan MM, Delany IR, O'Gara F (2002) Characterization of interactions between the transcriptional repressor PhlF and its binding site at the phlA promoter in *Pseudomonas fluorescens* F113. J Bacteriol 184:3008–3016

Achkar J, Xian M, Zhao H, Frost JW (2005) Biosynthesis of phloroglucinol. J Am Chem Soc 127:5332–5333

Afsharmanesh H, Ahmadzadeh M, Javan-Nikkhah M, Behboudi K (2010) Characterization of the antagonistic activity of a new indigenous strain of *Pseudomonas fluorescens* isolated from onion rhizosphere. J Plant Pathol 92:187–194

Agrios GN (2005) Plant pathology, 5th edn. Elsevier Academic Press, Burlington, pp 79–103

Almario J, Bruto M, Vacheron J, Prigent-Combaret C, Moënne-Loccoz Y, Muller D (2017) Distribution of 2, 4-diacetylphloroglucinol biosynthetic genes among the *Pseudomonas* spp. reveals unexpected polyphyletism. Front Microbiol 8:1218

Ayyadurai N, Naik RP, Rao SM, Kumar SR, Samrat SK, Manohar M, Sakthivel N (2006) Isolation and characterization of a novel banana rhizosphere bacterium as fungal antagonist and microbial adjuvant in micropropagation of banana. J Appl Microbiol 100:926–937

Bais HP, Weir TL, Perry LG, Gilroy S, Vivanco JM (2006) The role of root exudates in rhizosphere interactions with plants and other organisms. Annu Rev Plant Biol 57:233–266
Bajsa N, Morel MA, Braña V, Castro-Sowinski S (2013) The effect of agricultural practices on resident soil microbial communities: focus on biocontrol and biofertilization. Mol Microb Ecol Rhizosphere 2:687–700
Bakker PA, Glandorf DC, Viebahn M, Ouwens TW, Smit E, Leeflang P, Wernars K, Thomashow LS, Thomas-Oates JE, van Loon LC (2002) Effects of Pseudomonas putida modified to produce phenazine-1-carboxylic acid and 2, 4-diacetylphloroglucinol on the microflora of field grown wheat. Antonie Van Leeuwenhoek 81:617–624
Bakker PA, Pieterse CM, Van Loon L (2007) Induced systemic resistance by fluorescent *Pseudomonas* spp. Phytopathology 97:239–243
Bangera MG, Thomashow LS (1999) Identification and characterization of a gene cluster for synthesis of the polyketide antibiotic 2, 4-diacetylphloroglucinol from *Pseudomonas fluorescens* Q2-87. J Bacteriol 181:3155–3163
Berendsen RL, Pieterse CM, Bakker PA (2012) The rhizosphere microbiome and plant health. Trends Plant Sci 17:478–486
Bonsall RF, Weller DM, Thomashow LS (1997) Quantification of 2, 4-diacetylphloroglucinol produced by fluorescent *Pseudomonas* spp. *in vitro* and in the rhizosphere of wheat. Appl Environ Microbiol 63:951–955
Bottiglieri M, Keel C (2006) Characterization of PhlG, a hydrolase that specifically degrades the antifungal compound 2, 4-diacetylphloroglucinol in the biocontrol agent *Pseudomonas fluorescens* CHA0. Appl Environ Microbiol 72:418–427
Brazelton JN, Pfeufer EE, Sweat TA, Gardener BBM, Coenen C (2008) 2, 4-Diacetylphloroglucinol alters plant root development. Mol Plant-Microbe Interact 21:1349–1358
Brucker RM, Baylor CM, Walters RL, Lauer A, Harris RN, Minbiole KP (2008) The identification of 2, 4-diacetylphloroglucinol as an antifungal metabolite produced by cutaneous bacteria of the salamander *Plethodon cinereus*. J Chem Ecol 34:39–43
Cao Y, Jiang X, Zhang R, Xian M (2011) Improved phloroglucinol production by metabolically engineered *Escherichia coli*. Appl Microbiol Biotechnol 91:1545–1552
Chaubey S, Kotak M, Archana G (2015) New method for isolation of plant probiotic fluorescent Pseudomonad and characterization for 2, 4-Diacetylphluoroglucinol production under different carbon sources and phosphate levels. J Plant Pathol Microbiol 6:253
Combes-Meynet E, Pothier JF, Moënne-Loccoz Y, Prigent-Combaret C (2011) The Pseudomonas secondary metabolite 2, 4-diacetylphloroglucinol is a signal inducing rhizoplane expression of *Azospirillum* genes involved in plant-growth promotion. Mol Plant-Microbe Interact 24:271–284
Couillerot O, Combes-Meynet E, Pothier JF, Bellvert F, Challita E, Poirier MA, Rohr R, Comte G, Moënne-Loccoz Y, Prigent-Combaret C (2011) The role of the antimicrobial compound 2, 4-diacetylphloroglucinol in the impact of biocontrol *Pseudomonas fluorescens* F113 on *Azospirillum brasilense* phytostimulators. Microbiology 157:1694–1705
Cronin D, Moënne-Loccoz Y, Fenton A, Dunne C, Dowling DN, O'gara F (1997) Ecological interaction of a biocontrol *Pseudomonas fluorescens* strain producing 2, 4-diacetylphloroglucinol with the soft rot potato pathogen *Erwinia carotovora* subsp. atroseptica. FEMS Microbiol Ecol 23:95–106
Daguerre Y, Siegel K, Edel-Hermann V, Steinberg C (2014) Fungal proteins and genes associated with biocontrol mechanisms of soil-borne pathogens: a review. Fungal Biol Rev 28:97–125
Dairi T, Kuzuyama T, Nishiyama M, Fujii I (2011) Convergent strategies in biosynthesis. Nat Prod Rep 28:1054–1086
Davati N, Najafi MBH (2013) Overproduction strategies for microbial secondary metabolites: a review. Int J Life Sci Pharma Res 3:23–37
De Leij FA, Dixon-Hardy JE, Lynch JM (2002) Effect of 2, 4-diacetylphloroglucinol-producing and non-producing strains of *Pseudomonas fluorescens* on root development of pea seedlings in three different soil types and its effect on nodulation by rhizobium. Biol Fertil Soils 35:114–121

de Souza JT, Arnould C, Deulvot C, Lemanceau P, Gianinazzi-Pearson V, Raaijmakers JM (2003) Effect of 2, 4-diacetylphloroglucinol on Pythium: cellular responses and variation in sensitivity among propagules and species. Phytopathology 93:966–975

Delany I, Sheehan MM, Fenton A, Bardin S, Aarons S, O'Gara F (2000) Regulation of production of the antifungal metabolite 2, 4-diacetylphloroglucinol in *Pseudomonas fluorescens* F113: genetic analysis of phlF as a transcriptional repressor. Microbiology 146:537–546

Doornbos RF, van Loon LC, Bakker PA (2012) Impact of root exudates and plant defense signaling on bacterial communities in the rhizosphere. A review. Agron Sustain Dev 32:227–243

Dubuis C, Keel C, Haas (2007) Dialogues of root-colonizing biocontrol pseudomonads. Eur J Plant Pathol 119:311–328

Duffy BK, Défago G (1997) Zinc improves biocontrol of Fusarium crown and root rot of tomato by *Pseudomonas fluorescens* and represses the production of pathogen metabolites inhibitory to bacterial antibiotic biosynthesis. Phytopathology 87:1250–1257

Duffy BK, Défago G (1999) Environmental factors modulating antibiotic and siderophore biosynthesis by *Pseudomonas fluorescens* biocontrol strains. Appl Environ Microbiol 65:2429–2438

Duffy BK, Défago G (2000) Controlling instability in gacS-gacA regulatory genes during inoculant production of *Pseudomonas fluorescens* biocontrol strains. Appl Environ Microbiol 66:3142–3150

Elmer WH (2009) Influence of earthworm activity on soil microbes and soilborne diseases of vegetables. Plant Dis 93:175–179

Fenton AM, Stephens PM, Crowley J, O'callaghan M, O'gara F (1992) Exploitation of gene (s) involved in 2, 4-diacetylphloroglucinol biosynthesis to confer a new biocontrol capability to a *Pseudomonas* strain. Appl Environ Microbiol 58:3873–3878

Fernando WD, Nakkeeran S, Zhang Y (2005) Biosynthesis of antibiotics by PGPR and its relation in biocontrol of plant diseases. In: PGPR: biocontrol and biofertilization. Springer, Dordrecht, pp 67–109

Food and Agriculture Organization of The United Nations (2017a) Averting risks to the food chain: a compendium of proven emergency prevention methods and tools. ISBN:978-92-5-109539-3

Food and Agriculture Organization of the United Nations (2017b) The future of food and agriculture- trends and challenges. Food and Agriculture Organization of the United Nations, Rome

Frapolli M, Pothier JF, Défago G, Moënne-Loccoz Y (2012) Evolutionary history of synthesis pathway genes for phloroglucinol and cyanide antimicrobials in plant-associated fluorescent pseudomonads. Mol Phylogenet Evol 63:877–890

Gao X, Wang P, Tang Y (2010) Engineered polyketide biosynthesis and biocatalysis in *Escherichia coli*. Appl Microbiol Biotechnol 88:1233–1242

Girlanda M, Perotto S, Moenne-Loccoz Y, Bergero R, Lazzari A, Defago G, Bonfante P, Luppi AM (2001) Impact of biocontrol *Pseudomonas fluorescens* CHA0 and a genetically modified derivative on the diversity of culturable fungi in the cucumber rhizosphere. Appl Environ Microbiol 67:1851–1864

Gleeson O, O'Gara F, Morrissey JP (2010) The *Pseudomonas fluorescens* secondary metabolite 2, 4 diacetylphloroglucinol impairs mitochondrial function in *Saccharomyces cerevisiae*. Antonie van Leeuwenhoek 97:261–273

Gonzalez JB, Fernandez FJ, Tomasini A (2003) Microbial secondary metabolites production and strain improvement. Indian J Biotechnol 2:322–333

Gutiérrez-García K, Neira-González A, Pérez-Gutiérrez RM, Granados-Ramírez G, Zarraga R, Wrobel K, Barona-Gómez F, Flores-Cotera LB (2017) Phylogenomics of 2, 4-Diacetylphloroglucinol-producing pseudomonas and novel antiglycation endophytes from *Piper auritum*. J Nat Prod 80:1955–1963

Haas D, Défago G (2005) Biological control of soil-borne pathogens by fluorescent pseudomonads. Nat Rev Microbiol 3:307

Handelsman J, Stabb EV (1996) Biocontrol of soilborne plant pathogens. Plant Cell 8:1855

Hassan MN, Afghan S, ul Hassan Z, Hafeez FY (2014) Biopesticide activity of sugarcane associated rhizobacteria: *Ochrobactrum intermedium* strain NH-5 and *Stenotrophomonas maltophilia* strain NH-300 against red rot under field conditions. Phytopathol Mediterr 53:27–37

Hawkins NJ, Fraaije BA (2018) Fitness penalties in the evolution of fungicide resistance. Annu Rev Phytopathol 56:339–360

Hayashi A, Saitou H, Mori T, Matano I, Sugisaki H, Maruyama K (2012) Molecular and catalytic properties of monoacetylphloroglucinol acetyltransferase from *Pseudomonas* sp. YGJ3. Biosci Biotechnol Biochem 76:559–566

Heeb S, Haas D (2001) Regulatory roles of the GacS/GacA two-component system in plant-associated and other gram-negative bacteria. Mol Plant-Microbe Interact 14:1351–1363

Heeb S, Valverde C, Gigot-Bonnefoy C, Haas D (2005) Role of the stress sigma factor RpoS in GacA/RsmA-controlled secondary metabolism and resistance to oxidative stress in *Pseudomonas fluorescens* CHA0. FEMS Microbiol Lett 243:251–258

Heydari A, Pessarakli M (2010) A review on biological control of fungal plant pathogens using microbial antagonists. J Biol Sci 10:273–290

Hirsch AM, Bauer WD, Bird DM, Cullimore J, Tyler B, Yoder JI (2003) Molecular signals and receptors: controlling rhizosphere interactions between plants and other organisms. Ecology 84:858–868

Hol WH, Bezemer TM, Biere A (2013) Getting the ecology into interactions between plants and the plant growth-promoting bacterium *Pseudomonas fluorescens*. Front Plant Sci 4:81

Hontzeas N, Zoidakis J, Glick BR, Abu-Omar MM (2004) Expression and characterization of 1-aminocyclopropane-1-carboxylate deaminase from the rhizobacterium *Pseudomonas putida* UW4: a key enzyme in bacterial plant growth promotion. Biochim Biophys Acta (BBA)-Proteins Proteomics 1703:11–19

Hultberg M, Alsanius B (2008) Influence of nitrogen source on 2, 4-diacetylphloroglucinol production by the biocontrol strain Pf-5. Open Microbiol J 2:74

Iavicoli A, Boutet E, Buchala A, Métraux JP (2003) Induced systemic resistance in *Arabidopsis thaliana* in response to root inoculation with *Pseudomonas fluorescens* CHA0. Mol Plant-Microbe Interact 16:851–858

Islam MT, Fukushi Y (2010) Growth inhibition and excessive branching in Aphanomyces cochlioides induced by 2, 4-diacetylphloroglucinol is linked to disruption of filamentous actin cytoskeleton in the hyphae. World J Microbiol Biotechnol 26:1163–1170

Islam MT, Von Tiedemann A (2011) 2, 4-Diacetylphloroglucinol suppresses zoosporogenesis and impairs motility of Peronosporomycete zoospores. World J Microbiol Biotechnol 27:2071–2079

Isnansetyo A, Cui L, Hiramatsu K, Kamei Y (2003) Antibacterial activity of 2, 4-diacetylphloroglucinol produced by *Pseudomonas* sp. AMSN isolated from a marine alga, against vancomycin-resistant *Staphylococcus aureus*. Int J Antimicrob Agents 22:545–547

Jiang F, Chen L, Belimov AA, Shaposhnikov AI, Gong F, Meng X, Hartung W, Jeschke DW, Davies WJ, Dodd IC (2012) Multiple impacts of the plant growth-promoting rhizobacterium *Variovorax paradoxus* 5C-2 on nutrient and ABA relations of *Pisum sativum*. J Exp Bot 63:6421–6430

Jin CW, Li GX, Yu XH, Zheng SJ (2010) Plant Fe status affects the composition of siderophore-secreting microbes in the rhizosphere. Ann Bot 105:835–841

Jousset A, Bonkowski M (2010) The model predator *Acanthamoeba castellanii* induces the production of 2, 4, DAPG by the biocontrol strain *Pseudomonas fluorescens* Q2-87. Soil Biol Biochem 42:1647–1649

Jousset A, Lara E, Wall LG, Valverde C (2006) Secondary metabolites help biocontrol strain *Pseudomonas fluorescens* CHA0 to escape protozoan grazing. Appl Environ Microbiol 72:7083–7090

Kamei Y, Isnansetyo A (2003) Lysis of methicillin-resistant *Staphylococcus aureus* by 2, 4-diacetylphloroglucinol produced by *Pseudomonas* sp. AMSN isolated from a marine alga. Int J Antimicrob Agents 21:71–74

Karthiba L, Saveetha K, Suresh S, Raguchander T, Saravanakumar D, Samiyappan R (2010) PGPR and entomopathogenic fungus bioformulation for the synchronous management of leaffolder pest and sheath blight disease of rice. Pest Manag Sci Formerly Pestic Sci 66:555–564

Katar'yan BT, Torgashova GG (1976) Spectrum of herbicidal activity of 2, 4-diacetylphloroglucinol. Dokl Akad Nauk Armyanskoi SSR 63:109–112

Keel C, Schnider U, Maurhofer M, Voisard C, Laville J, Burger U, Wirthner T, Haas D, Defago G (1992) Suppression of root diseases by *Pseudomonas fluorescens* CHA0: importance of bacterial secondary metabolites 2, 4-diacetylphloroglucinol. Mol Plant-Microbe Interact 5:4–13

Keel C, Weller DM, Natsch A, Défago G, Cook RJ, Thomashow LS (1996) Conservation of the 2, 4-diacetylphloroglucinol biosynthesis locus among *fluorescent Pseudomonas* strains from diverse geographic locations. Appl Environ Microbiol 62:552–563

Kidarsa TA, Goebel NC, Zabriskie TM, Loper JE (2011) Phloroglucinol mediates cross-talk between the pyoluteorin and 2, 4-diacetylphloroglucinol biosynthetic pathways in *Pseudomonas fluorescens* Pf-5. Mol Microbiol 81:395–414

Kuiper I, Bloemberg GV, Noreen S, Thomas-Oates JE, Lugtenberg BJ (2001) Increased uptake of putrescine in the rhizosphere inhibits competitive root colonization by *Pseudomonas fluorescens* strain WCS365. Mol Plant-Microbe Interact 14:1096–1104

Kwak YS, Weller DM (2013) Take-all of wheat and natural disease suppression: a review. Plant Pathol J 29:125

Kwak YS, Bakker PA, Glandorf DC, Rice JT, Paulitz TC, Weller DM (2009) Diversity, virulence, and 2, 4-diacetylphloroglucinol sensitivity of *Gaeumannomyces graminis* var. *tritici* isolates from Washington State. Phytopathology 99:472–479

Kwak YS, Han S, Thomashow LS, Rice JT, Paulitz TC, Kim D, Weller DM (2011) *Saccharomyces cerevisiae* genome-wide mutant screen for sensitivity to 2, 4-diacetylphloroglucinol, an antibiotic produced by *Pseudomonas fluorescens*. Appl Environ Microbiol 77:1770–1776

Kwak YS, Bonsall RF, Okubara PA, Paulitz TC, Thomashow LS, Weller DM (2012) Factors impacting the activity of 2, 4-diacetylphloroglucinol-producing *Pseudomonas fluorescens* against take-all of wheat. Soil Biol Biochem 54:48–56

Lugtenberg B, Kamilova F (2009) Plant-growth-promoting rhizobacteria. Annu Rev Microbiol 63:541–556

Lutz MP, Wenger S, Maurhofer M, Défago G, Duffy B (2004) Signaling between bacterial and fungal biocontrol agents in a strain mixture. FEMS Microbiol Ecol 48:447–455

Mark GL, Dow JM, Kiely PD, Higgins H, Haynes J, Baysse C, Abbas A, Foley T, Franks A, Morrissey J, O'Gara F (2005) Transcriptome profiling of bacterial responses to root exudates identifies genes involved in microbe-plant interactions. Proc Natl Acad Sci 102:17454–17459

Matano I, Tsunekawa M, Shimizu S, Tanaka I, Mitsukura K, Maruyama K (2010) The chloride ion is an environmental factor affecting the biosynthesis of pyoluteorin and 2, 4-diacetylphloroglucinol in Pseudomonas sp. YGJ3. Biosci Biotechnol Biochem 74:427–429

Maurhofer M, Keel C, Schnider U, Voisard C, Haas D, Défago G (1992) Influence of enhanced antibiotic production in *Pseudomonas fluorescens* strain CHA0 on its disease suppressive capacity. Phytopathology 82:190–195

Maurhofer M, Keel C, Haas D, Défago G (1995) Influence of plant species on disease suppression by *Pseudomonas fluorescens* strain CHAO with enhanced antibiotic production. Plant Pathol 44:40–50

Maurhofer M, Baehler E, Notz R, Martinez V, Keel C (2004) Cross talk between 2, 4-diacetylp hloroglucinol-producing biocontrol pseudomonads on wheat roots. Appl Environ Microbiol 70:1990–1998

Mendes R, Kruijt M, De Bruijn I, Dekkers E, van der Voort M, Schneider JH, Piceno YM, DeSantis TZ, Andersen GL, Bakker PA, Raaijmakers JM (2011) Deciphering the rhizosphere microbiome for disease-suppressive bacteria. Science 332:1097–1100

Meyer SL, Halbrendt JM, Carta LK, Skantar AM, Liu T, Abdelnabby HM, Vinyard BT (2009) Toxicity of 2, 4-diacetylphloroglucinol (DAPG) to plant-parasitic and bacterial-feeding nematodes. J Nematol 41:274

Meziane H, Van Der Sluis I, Van Loon LC, Höfte M, Bakker PA (2005) Determinants of *Pseudomonas putida* WCS358 involved in inducing systemic resistance in plants. Mol Plant Pathol 6:177–185

Miller GT (2004) Sustaining the earth, vol 9. Thompson Learning, Pacific Grove, pp 211–216

Mirza MS, Mehnaz S, Normand P, Prigent-Combaret C, Moënne-Loccoz Y, Bally R, Malik KA (2006) Molecular characterization and PCR detection of a nitrogen-fixing Pseudomonas strain promoting rice growth. Biol Fertil Soils 43:163–170
Mishra J, Arora NK (2017) Secondary metabolites of fluorescent pseudomonads in biocontrol of phyto-pathogens for sustainable agriculture. Appl Soil Ecol 125:35–45
Moynihan JA, Morrissey JP, Coppoolse ER, Stiekema WJ, O'Gara F, Boyd EF (2009) Evolutionary history of the phl gene cluster in the plant-associated bacterium *Pseudomonas fluorescens*. Appl Environ Microbiol 75:2122–2131
Muller T, Behrendt U, Ruppel S, von der Waydbrink G, Müller ME (2016) Fluorescent pseudomonads in the phyllosphere of wheat: potential antagonists against fungal phytopathogens. Curr Microbiol 72:383–389
Nagel K, Schneemann I, Kajahn I, Labes A, Wiese J, Imhoff JF (2012) Beneficial effects of 2, 4-diacetylphloroglucinol-producing pseudomonads on the marine alga *Saccharina latissima*. Aquat Microb Ecol 67:239–249
Nakata K, Yoshimoto A, Yamada Y (1999) Promotion of antibiotic production by high ethanol, high NaCl concentration, or heat shock in *Pseudomonas fluorescens* S272. Biosci Biotechnol Biochem 63:293–297
Natsch A, Keel C, Hebecker N, Laasik E, Défago G (1998) Impact of *Pseudomonas fluorescens* strain CHA0 and a derivative with improved biocontrol activity on the culturable resident bacterial community on cucumber roots. FEMS Microbiol Ecol 27:365–380
Neilands JB, Leong SA (1986) Siderophores in relation to plant growth and disease. Annu Rev Plant Physiol 37:187–208
Nelson LM (2004) Plant growth promoting rhizobacteria (PGPR): prospects for new inoculants. Crop Management. https://doi.org/10.1094/CM-2004-0301-05-RV
Notz R, Maurhofer M, Schnider-Keel U, Duffy B, Haas D, Défago G (2001) Biotic factors affecting expression of the 2, 4-diacetylphloroglucinol biosynthesis gene *phlA* in *Pseudomonas fluorescens* biocontrol strain CHA0 in the rhizosphere. Phytopathology 91:873–881
Notz R, Maurhofer M, Dubach H, Haas D, Défago G (2002) Fusaric acid-producing strains of *Fusarium oxysporum* alter 2, 4-diacetylphloroglucinol biosynthetic gene expression in *Pseudomonas fluorescens* CHA0 in vitro and in the rhizosphere of wheat. Appl Environ Microbiol 68:2229–2235
Nowak-Thompson B, Gould SJ, Kraus J, Loper JE (1994) Production of 2, 4-diacetylphloroglucinol by the biocontrol agent *Pseudomonas fluorescens* Pf-5. Can J Microbiol 40:1064–1066
Nowak-Thompson B, Chaney N, Wing JS, Gould SJ, Loper JE (1999) Characterization of the pyoluteorin biosynthetic gene cluster of *Pseudomonas fluorescens* Pf-5. J Bacteriol 181(7):2166–2174
Okubara PA, Bonsall RF (2008) Accumulation of Pseudomonas-derived 2, 4-diacetylphloroglucinol on wheat seedling roots is influenced by host cultivar. Biol Control 46:322–331
Pal KK, Gardener BM (2006) Biological control of plant pathogens. Plant Health Instr 2:1117–1142
Park JH, Kim R, Aslam Z, Jeon CO, Chung YR (2008) *Lysobacter capsici* sp. nov., with antimicrobial activity, isolated from the rhizosphere of pepper, and emended description of the genus *Lysobacter*. Int J Syst Evol Microbiol 58:387–392
Pedersen AL, Nybroe O, Winding A, Ekelund F, Bjørnlund L (2009) Bacterial feeders, the nematode *Caenorhabditis elegans* and the flagellate *Cercomonas longicauda*, have different effects on outcome of competition among the Pseudomonas biocontrol strains CHA0 and DSS73. Microb Ecol 57:501–509
Penrose DM, Glick BR (2003) Methods for isolating and characterizing ACC deaminase-containing plant growth-promoting rhizobacteria. Physiol Plant 118:10–15
Phillips DA, Fox TC, King MD, Bhuvaneswari TV, Teuber LR (2004) Microbial products trigger amino acid exudation from plant roots. Plant Physiol 136:2887–2894
Picard C, Bosco M (2003) Genetic diversity of *phlD* gene from 2, 4-diacetylphloroglucinol-producing *Pseudomonas* spp. strains from the maize rhizosphere. FEMS Microbiol Lett 219:167–172

Picard C, Bosco M (2005) Maize heterosis affects the structure and dynamics of indigenous rhizospheric auxins-producing Pseudomonas populations. *FEMS Microbiol Ecol* 53:349–357
Picard C, Di Cello F, Ventura M, Fani R, Guckert A (2000) Frequency and biodiversity of 2, 4-diacetylphloroglucinol-producing bacteria isolated from the maize rhizosphere at different stages of plant growth. Appl Environ Microbiol 66:948–955
Pimentel D (1997) Techniques for reducing pesticide use: economic and environmental benefits. Wiley, New York
Primrose SB (1976) Formation of ethylene by *Escherichia coli*. Microbiology 95:159–165
Quecine MC, Kidarsa TA, Goebel NC, Shaffer BT, Henkels MD, Zabriskie TM, Loper JE (2016) An inter-species signaling system mediated by fusaric acid has parallel effects on antifungal metabolite production by *Pseudomonas protegens* Pf-5 and antibiosis of *Fusarium* spp. Appl Environ Microbiol. AEM-02574 82:1372–1382
Raaijmakers JM, Weller DM (1998) Natural plant protection by 2, 4-diacetylphloroglucinol-producing *Pseudomonas* spp. in take-all decline soils. Mol Plant-Microbe Interact 11:144–152
Raaijmakers JM, Bonsall RF, Weller DM (1999) Effect of population density of *Pseudomonas fluorescens* on production of 2, 4-diacetylphloroglucinol in the rhizosphere of wheat. Phytopathology 89:470–475
Raaijmakers JM, Vlami M, De Souza JT (2002) Antibiotic production by bacterial biocontrol agents. Antonie van Leeuwenhoek 81:537
Ramadasappa S, Rai AK, Jaat RS, Singh A, Rai R (2012) Isolation and screening of *phlD*+ plant growth promoting rhizobacteria antagonistic to *Ralstonia solanacearum*. World J Microbiol Biotechnol 28:1681–1690
Ramesh R, Joshi AA, Ghanekar MP (2009) Pseudomonads: major antagonistic endophytic bacteria to suppress bacterial wilt pathogen, *Ralstonia solanacearum* in the eggplant (*Solanum melongena* L.). World J Microbiol Biotechnol 25:47–55
Ramette A (2002) Diversity of biocontrol fluorescent pseudomonads producing 2, 4-diacetylphloroglucinol and hydrogen cyanide in disease suppressive soils. Doctoral dissertation, ETH Zurich
Ramette A, Moënne-Loccoz Y, Défago G (2001) Polymorphism of the polyketide synthase gene *phlD* in biocontrol fluorescent pseudomonads producing 2, 4-diacetylphloroglucinol and comparison of PhlD with plant polyketide synthases. Mol Plant-Microbe Interact 14:639–652
Ramette A, Moënne-Loccoz Y, Défago G (2003) Prevalence of fluorescent pseudomonads producing antifungal phloroglucinols and/or hydrogen cyanide in soils naturally suppressive or conducive to tobacco black root rot. FEMS Microbiol Ecol 44:35–43
Ramette A, Moënne-Loccoz Y, Défago G (2006) Genetic diversity and biocontrol potential of fluorescent pseudomonads producing phloroglucinols and hydrogen cyanide from Swiss soils naturally suppressive or conducive to Thielaviopsis basicola-mediated black root rot of tobacco. FEMS Microbiol Ecol 55:369–381
Reddy TK, Khudiakov I, Borovkov AV (1969) *Pseudomonas fluorescens* strain 26-o – producer of phytotoxic substances. Mikrobiologiia 38:909
Reddy KRN, Choudary KA, Reddy MS (2007) Antifungal metabolites of *Pseudomonas fluorescens* isolated from rhizosphere of rice crop. J Mycol Plant Pathol 37:280–284
Ross AF (1961) Systemic acquired resistance induced by localized virus infections in plants. Virology 14:340–358
Ryals JA, Neuenschwander UH, Willits MG, Molina A, Steiner HY, Hunt MD (1996) Systemic acquired resistance. Plant Cell 8:1809
Saharan K, Sarma MVRK, Roesti AS, Prakash A, Johri BN, Aragno M, Bisaria VS, Sahai V (2010) Cell growth and metabolites produced by fluorescent pseudomonad R62 in modified chemically defined medium. World Acad Sci Eng Technol 67:867–871
Saharan K, Sarma MVRK, Prakash A, Johri BN, Bisaria VS, Sahai V (2011) Shelf-life enhancement of bio-inoculant formulation by optimizing the trace metals ions in the culture medium for production of DAPG using fluorescent pseudomonad R62. Enzym Microb Technol 48:33–38

Saikia R, Varghese S, Singh BP, Arora DK (2009) Influence of mineral amendment on disease suppressive activity of *Pseudomonas fluorescens* to Fusarium wilt of chickpea. Microbiol Res 164:365–373

Saikia R, Sarma RK, Yadav A, Bora TC (2011) Genetic and functional diversity among the antagonistic potential fluorescent pseudomonads isolated from tea rhizosphere. Curr Microbiol 62:434–444

Saravanan T, Muthusamy M (2006) Influence of *Fusarium oxysporum* f. sp. *cubense* (ef smith) Snyder and Hansen on 2, 4-diacetylphloroglucinol production by *Pseudomonas fluorescens* migula in banana rhizosphere. J Plant Prot Res 46:241–254

Sarma MVRK, Saharan K, Kumar L, Gautam A, Kapoor A, Srivastava N, Sahai V, Bisaria VS (2010) Process optimization for enhanced production of cell biomass and metabolites of fluorescent pseudomonad R81. World Acad Sci Eng Technol 41:997–1001

Sarma MVRK, Gautam A, Kumar L, Saharan K, Kapoor A, Shrivastava N, Sahai V, Bisaria VS (2013) Bioprocess strategies for mass multiplication of and metabolite synthesis by plant growth promoting pseudomonads for agronomical applications. Process Biochem 48:1418–1424

Sarniguet A, Kraus J, Henkels MD, Muehlchen AM, Loper JE (1995) The sigma factor σ^s affects antibiotic production and biological control activity of *Pseudomonas fluorescens* Pf-5. Proc Natl Acad Sci 92:12255–12259

Schippers B, Bakker AW, Bakker PA (1987) Interactions of deleterious and beneficial rhizosphere microorganisms and the effect of cropping practices. Annu Rev Phytopathol 25:339–358

Schnider-Keel U, Seematter A, Maurhofer M, Blumer C, Duffy B, Gigot-Bonnefoy C, Reimmann C, Notz R, Défago G, Haas D, Keel C (2000) Autoinduction of 2, 4-diacetylphloroglucinol biosynthesis in the biocontrol agent *Pseudomonas fluorescens* CHA0 and repression by the bacterial metabolites salicylate and pyoluteorin. J Bacteriol 182:1215–1225

Schouten A, Maksimova O, Cuesta-Arenas Y, Van Den Berg G, Raaijmakers JM (2008) Involvement of the ABC transporter BcAtrB and the laccase BcLCC2 in defence of *Botrytis cinerea* against the broad-spectrum antibiotic 2, 4-diacetylphloroglucinol. Environ Microbiol 10:1145–1157

Sekar J, Prabavathy VR (2014) Novel Phl-producing genotypes of finger millet rhizosphere associated pseudomonads and assessment of their functional and genetic diversity. FEMS Microbiol Ecol 89:32–46

Shanahan P, O'Sullivan DJ, Simpson P, Glennon JD, O'Gara F (1992) Isolation of 2, 4-diacetylphloroglucinol from a fluorescent pseudomonad and investigation of physiological parameters influencing its production. Appl Environ Microbiol 58:353–358

Shanahan P, Glennon JD, Crowley JJ, Donnelly DF, O'Gara F (1993) Liquid chromatographic assay of microbially derived phloroglucinol antibiotics for establishing the biosynthetic route to production, and the factors affecting their regulation. Anal Chim Acta 272:271–277

Shanthi AT, Vittal RR (2013) Biocontrol potentials of plant growth promoting rhizobacteria against Fusarium wilt disease of cucurbit. Int J Phytopathol 2:155–161

Sharifi-Tehrani A, Zala M, Natsch A, Moënne-Loccoz Y, Défago G (1998) Biocontrol of soil-borne fungal plant diseases by 2, 4-diacetylphloroglucinol-producing fluorescent pseudomonads with different restriction profiles of amplified 16S rDNA. Eur J Plant Pathol 104:631–643

Shirzad A, Fallahzadeh-Mamaghani V, Pazhouhandeh M (2012) Antagonistic potential of fluorescent pseudomonads and control of crown and root rot of cucumber caused by *Phytophthora drechsleri*. Plant Pathol J 28:1–9

Shoresh M, Harman GE, Mastouri F (2010) Induced systemic resistance and plant responses to fungal biocontrol agents. Annu Rev Phytopathol 48:21–43

Singh IP, Bharate SB (2006) Phloroglucinol compounds of natural origin. Nat Prod Rep 23:558–591

Song L, Barona-Gomez F, Corre C, Xiang L, Udwary DW, Austin MB, Noel JP, Moore BS, Challis GLJ (2006) Type III polyketide synthase β-ketoacyl-ACP starter unit and ethylmalonyl-CoA extender unit selectivity discovered by *Streptomyces coelicolor* genome mining. J Am Chem Soc 128:14754–14755

Stamp N (2003) Out of the quagmire of plant defense hypotheses. Q Rev Biol 78:23–55

Standing D, Banerjee S, Ignacio Rangel-Castro J, Jaspars M, Prosser JI, Killham K (2008) Novel screen for investigating *in situ* rhizosphere production of the antibiotic 2, 4-diacetylphloroglucinol by bacterial inocula. Commun Soil Sci Plant Anal 39:1720–1732

Stanier RY, Palleroni NJ, Doudoroff M (1966) The aerobic pseudomonads a taxonomic study. Microbiology 43:159–271

Stolp H, Gadkari D (1981) Non-pathogenic members of the genus Pseudomonas. In: Starr MP, Stolp H, Truper HG, Ballows A, Shlegel HG (eds) The prokaryotes, A handbook on habitats, isolation and identification of bacteria, vol I. Springer, Berlin, pp 719–741

Strange RN, Scott PR (2005) Plant disease: a threat to global food security. Annu Rev Phytopathol 43:83–116

Tada M, Takakuwa T, Nagai M, Yoshii T (1990) Antiviral and antimicrobial activity of 2, 4-diacylphloroglucinols, 2-acylcyclohexane-1, 3-diones and 2-carboxamidocyclo-hexane-1, 3-diones. Agric Biol Chem 54:3061–3063

Takeuchi K, Kiefer P, Reimmann C, Keel C, Dubuis C, Rolli J, Vorholt JA, Haas D (2009) Small RNA-dependent expression of secondary metabolism is controlled by Krebs cycle function in *Pseudomonas fluorescens*. J Biol Chem 284:34976–34985

Terada H (1981) The interaction of highly active uncouplers with mitochondria. Biochim Biophys Acta (BBA)-Rev Bioenerg 639:225–242

Thomashow LS, Weller D (1996) Current concepts in the use of introduced bacteria for biological disease control: mechanisms and antifungal metabolites. Plant-Microbe Interact 1:187–235

Troppens DM, Dmitriev RI, Papkovsky DB, O'Gara F, Morrissey JP (2013a) Genome-wide investigation of cellular targets and mode of action of the antifungal bacterial metabolite 2, 4-diacetylphloroglucinol in *Saccharomyces cerevisiae*. FEMS Yeast Res 13:322–334

Troppens DM, Chu M, Holcombe LJ, Gleeson O, O'Gara F, Read ND, Morrissey JP (2013b) The bacterial secondary metabolite 2, 4-diacetylphloroglucinol impairs mitochondrial function and affects calcium homeostasis in *Neurospora crassa*. Fungal Genet Biol 56:135–146

Uknes S, Mauch-Mani B, Moyer M, Potte S, Williams S, Dincher S, Chandler D, Slusarenko A, Ward E, Ryals J (1992) Acquired resistance in Arabidopsis. Plant Cell 4:645–656

Vallad GE, Goodman RM (2004) Systemic acquired resistance and induced systemic resistance in conventional agriculture. Crop Sci 44:1920–1934

van Loon LC (2007) Plant responses to plant growth-promoting rhizobacteria. In: New perspectives and approaches in plant growth-promoting rhizobacteria research. Springer, Dordrecht, pp 243–254

van Loon LC, Bakker PAHM, Pieterse CMJ (1998) Systemic resistance induced by rhizosphere bacteria. Annu Rev Phytopathol 36:453–483

van Wees SC, Van der Ent S, Pieterse CM (2008) Plant immune responses triggered by beneficial microbes. Curr Opin Plant Biol 11:443–448

Veena VK, Kennedy K, Lakshmi P, Krishna R, Sakthivel N (2016) Anti-leukemic, anti-lung, and anti-breast cancer potential of the microbial polyketide 2, 4-diacetylphloroglucinol (DAPG) and its interaction with the metastatic proteins than the antiapoptotic Bcl-2 proteins. Mol Cell Biochem 414:47–56

Vega FE, Goettel MS, Blackwell M, Chandler D, Jackson MA, Keller S, Koike M, Maniania NK, Monzon A, Ownley BH, Pell JK (2009) Fungal entomopathogens: new insights on their ecology. Fungal Ecol 2:149–159

Velusamy P, Immanuel JE, Gnanamanickam SS, Thomashow L (2006) Biological control of rice bacterial blight by plant-associated bacteria producing 2, 4-diacetylphloroglucinol. Can J Microbiol 52:56–65

Velusamy P, Immanuel JE, Gnanamanickam SS (2013) Rhizosphere bacteria for biocontrol of bacterial blight and growth promotion of rice. Rice Sci 20:356–362

Walker V, Couillerot O, Von Felten A, Bellvert F, Jansa J, Maurhofer M, Bally R, Moënne-Loccoz Y, Comte G (2012) Variation of secondary metabolite levels in maize seedling roots induced by inoculation with *Azospirillum, Pseudomonas* and *Glomus* consortium under field conditions. Plant Soil 356:151–163

Wang X, Mavrodi DV, Ke L, Mavrodi OV, Yang M, Thomashow LS, Zheng N, Weller DM, Zhang J (2015) Biocontrol and plant growth-promoting activity of rhizobacteria from Chinese fields with contaminated soils. Microb Biotechnol 8:404–418

Weissman KJ (2009) Introduction to polyketide biosynthesis. In: Abelson JN, Simon MI (eds) Methods in Enzymology. Academic, London, pp 3–16

Weller DM (1988) Biological control of soilborne plant pathogens in the rhizosphere with bacteria. Annu Rev Phytopathol 26:379–407

Weller DM (2007) Pseudomonas biocontrol agents of soilborne pathogens: looking back over 30 years. Phytopathology 97:250–256

Weller DM, Raaijmakers JM, Gardener BBM, Thomashow LS (2002) Microbial populations responsible for specific soil suppressiveness to plant pathogens. Annu Rev Phytopathol 40:309–348

Weller DM, Van Pelt JA, Mavrodi DV, Pieterse CMJ, Bakker PAHM, Van Loon LC (2004) Induced systemic resistance (ISR) in *Arabidopsis* against *Pseudomonas syringae* pv. tomato by 2, 4-diacetylphloroglucinol (DAPG)-producing *Pseudomonas fluorescens*. Phytopathology 94:S108

Weller DM, Mavrodi DV, van Pelt JA, Pieterse CM, van Loon LC, Bakker PA (2012) Induced systemic resistance in *Arabidopsis thaliana* against *Pseudomonas syringae* pv. tomato by 2, 4-diacetylphloroglucinol-producing *Pseudomonas fluorescens*. Phytopathology 102:403–412

Whipps JM (2001) Microbial interactions and biocontrol in the rhizosphere. J Exp Bot 52:487–511

Yan Q, Philmus B, Chang JH, Loper JE (2017) Novel mechanism of metabolic co-regulation coordinates the biosynthesis of secondary metabolites in *Pseudomonas protegens*. elife 6:22835

Yang F, Cao Y (2012) Biosynthesis of phloroglucinol compounds in microorganisms. Appl Microbiol Biotechnol 93:487–495

Yuan Z, Cang S, Matsufuji M, Nakata K, Nagamatsu Y, Yoshimoto A (1998) High production of pyoluteorin and 2, 4-diacetylphloroglucinol by *Pseudomonas fluorescens* S272 grown on ethanol as a sole carbon source. J Ferment Bioeng 86:559–563

Zha W, Rubin-Pitel SB, Zhao H (2006) Characterization of the substrate specificity of PhlD, a type III polyketide synthase from *Pseudomonas fluorescens*. J Biol Chem 281:32036–32047

Zha W, Rubin-Pitel SB, Shao Z, Zhao H (2009) Improving cellular malonyl-CoA level in *Escherichia coli* via metabolic engineering. Metab Eng 11:92–198

Zhang W (2018) Global pesticide use: profile, trend, cost/benefit and more. Proc Int Acad Ecol Environ Sci 8:1

Zhang W, Zhao Z, Zhang B, Wu XG, Ren ZG, Zhang LQ (2014) Post-transcriptional regulation of 2, 4-diacetylphloroglucinol production by GidA and TrmE in *Pseudomonas fluorescens* 2P24. Appl Environ Microbiol. AEM-00455 80:3972–3981

Zhou T, Chen D, Li C, Sun Q, Li L, Liu F, Shen Q, Shen B (2012) Isolation and characterization of *Pseudomonas brassicacearum* J12 as an antagonist against *Ralstonia solanacearum* and identification of its antimicrobial components. Microbiol Res 167:388–394

Coral Reef Microbiota and Its Role in Marine Ecosystem Sustainability

17

Soumya Nair and Jayanthi Abraham

17.1 Introduction

Coral reefs are diversified marine ecosystem. They are responsible for secreting calcium carbonate assemblages into the environment which ultimately forms the hard exoskeleton. The carbonate exoskeleton supports and protects the coral polyps. The reefs are constructed by minute organisms (living in colonies), in environment with limited or scanty nutrients. Most of these reefs are erected from the excreted calcium carbonate forming the stony corals. These stony assemblages host a group of polyps (tiny microorganisms) in an association. Polyps belong to the class *Cnidaria.* Examples of polyps are sea anemones and jellyfishes. Most of the corals are capable to sustain in warm, trivial, clear and flustered water. Corals have been reported to produce the calcium carbonate assemblages throughout the year, and this is an important factor in shaping the reefs ecosystem and providing habitats to the associated flora and fauna. This in turn provides important ecosystem services to humans (Hallock et al. 2003). Humans are dependent on a number of services and applications delivered by the coral reef community.

Coral reefs are also called 'the rainforests of the ocean'. As mentioned earlier, the reefs are one of the most assorted and varied ecological communities on the planet, although they occupy <0.1% of the world's ocean surface. In spite of the space constraints, this particular marine ecosystem is a home for majority of all marine flora and fauna. Contradictorily, in spite of being surrounded by ocean water, the availability of nutrients is very scarce. As a result, they are predominantly present in shallow depths of the tropical water.

Reef ecosystem delivers services like tourism, fisheries, etc., to name a few. Statistical studies have reported that the annual economic value lies in between US$

S. Nair · J. Abraham (✉)
Microbial Biotechnology Laboratory, School of Biosciences and Technology, VIT, Vellore, Tamil Nadu, India

D. P. Singh et al. (eds.), *Microbial Interventions in Agriculture and Environment*, https://doi.org/10.1007/978-981-13-8391-5_17

30 and 375 billion. Despite its popularity, the reefs are very subtle and sensitive toward the water temperature. Moreover, they are under severe risk of global warming, blast fishing, ocean acidification, cyanide fishing, overuse of coral resources, dangerous land-use practices, water pollution, etc. (White and Vogt 2000; Jackson et al. 2001; Wilkinson 2004).

Corals form the unit of reef ecology. They have retracted growth rate ranging from 1 cm/year to 3–20 mm/year (Veron 2000; Lough et al. 2002). When these corals die, their carbonaceous exoskeletons remain, offering substratum for other corals and its associated microbiome to settle on and grow on it (Lough et al. 2002). The corals function in shielding the seashores from major waves. They also play a vital role in sustaining and maintaining the carbon–nitrogen cycle by fixing it in the ocean bed. Last, but not the least, the reefs play a vital role in nutrient recycling (Rädecker 2014).

17.2 The Organization of the Coral Microbiota

Reefs and its associated flora and fauna are known to possess a mutualistic association between the invertebrate tiny animals (*Symbiodinium*) and a varied group of bacteria, archaea and dinoflagellates. The symbiotic association helps in providing the corals with the energy required for its development via photosynthesis. Microbes play a critical role all through the development process of the corals (Sharp and Ritchie 2012). Larvae of several coral sp. settle on some crustose coralline algae (CCA), next to the planktonic dispersal phase. For example, *Pseudoalteromonas* sp. growing on CCA produces tetrabromopyrrole which helps in the generation or initiation of the larval settlement and its metamorphosis in numerous corals (Sneed et al. 2014; Thompson et al. 2015). After settlement, these corals develop a microbiome which helps in nutrition and developing secondary metabolites (e.g. antibiotics) (Ritchie 2006). Another example is *Exiguobacterium* sp. that produces a secondary metabolite of low molecular weight which decreases the coral pathogen *Serratia marcescens* (Krediet et al. 2013). Compounds of hydrophobic nature, present on the coral surfaces, help in inhibiting the biofilm formation of pathogenic bacteria (Alagely et al. 2011). Thus, the coral microbiota plays an important role in the form of senescence.

Coral reefs can undergo stressed condition due to ocean warming or competition from seaweeds. Consequently, the defensive microbiome may end in destabilization, leading to dysbiosis of defense mechanism (Barott and Rohwer 2012). Dissolved carbon released by the seaweeds does help in stimulating the development and evolution of microorganisms with virulence traits (Nelson et al. 2013).

17.3 Worldwide Locations of Coral Reef Ecosystem

Around the world, coral reefs are estimated to cover more than 2, 84,300 km^2 area (NEP 2001). The Indo-Pacific region (comprising the Indian Ocean and Red Sea), South-East Asia, and the Pacific account for a total 91.9% of the total area (Table 17.1) (Spalding et al. 2001; Vajed et al. 2013).

Some of the major reefs around the world are enlisted below:

- The Andros,
- The Bahamas Barrier Reef,
- The Great Barrier Reef,
- The Florida Reef Tract (NOAA CoRIS 2013),
- The Mesoamerican Barrier Reef System,
- The New Caledonia Barrier Reef,
- Northernmost coral reef, bay of Japan's Tsushima Island,
- The Philippines coral reef area,
- The Raja Ampat Islands (NGM 2007),
- The Red Sea,
- Southernmost coral reef, Lord Howe Island.

17.4 Coral Reef Zonation

Zonation in the reef ecosystem helps the coral reefs in supporting different species (Sheppard et al. 2005; Madin and Connolly 2006; Anthony and Kerswell 2007). Dome-shaped massive and columnar corals (*Diploria* sp. and *Dendrogyra* sp. respectively) are present on the transitional slopes of the reef front. Plate corals are predominant below this region, for example, *Pectinia* and *Agaricia*. Branched corals are predominantly found in the region where the ocean wave is felt at its peak, for example, Elkhorn coral (*Acropora palmata*). Staghorn corals (*Acropora cervicornis*), cluster (*Pocillopora*), finger (*Stylophora*), and lace corals (*Pocillopora damicornis*) are present past the reef front. In shallow regions of the reef, corals such as rose (*Meandrina, Manicina*), flower (*Mussa, Eusmilia*), and star (*Montastraea*) are found (Table 17.2).

Table 17.1 % Ocean surface area covered by the major reefs around the world

Region	% ocean surface area
South-east Asian reefs	32.3%
Pacific reefs	40.8%
Atlantic and Caribbean reefs	7.6%

Table 17.2 Microorganisms residing in the different zones of coral reefs

Zone	Microorganisms
Reef flat	*Acropora* sp., *Favites abdita, Grammia edwardsii, Goniastrea retiformis, Montipora* sp., *Platygyra* sp.
Reef crest	*Acropora cuneata, Acropora gemmifera, Acropora palifera, Goniastrea retiformis, Favia laxa, Favia stelligera, Platygyra daedalea*
Upper fore reef	*Acropora palifera, Favia palida, Montastrea annuligera, Goniastrea retiformis, Platygyra* sp., *Porites* sp.
Lower fore reef	*Diasens fragilis, Diploastrea heliopora, Favia matthai*

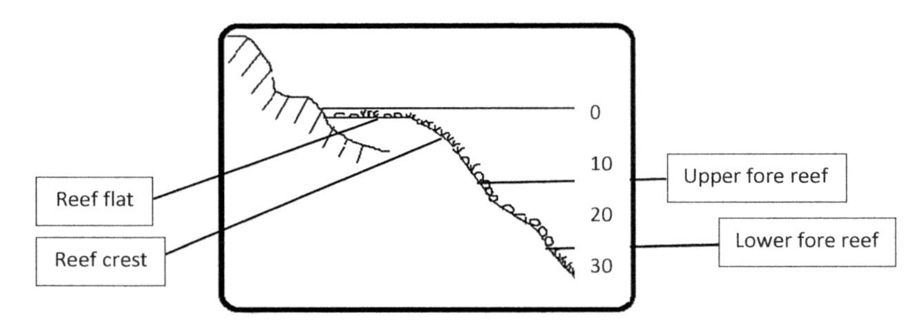

Fig. 17.1 Different zones of coral reef

Coral ecosystem is divided into major zones which represent different habitats. Usually, three distinct zones are accepted and documented (Montaggioni and Braithwaite 1988):

(a) The fore reef,
(b) The reef crest, and
(c) The back reef.

The reef zones are interconnected based on the ecology (Fig. 17.1). The major components are the nutrients, marine flora and fauna, seawater exchange, oceanic processes, etc., to name a few. Each component plays an important role in supporting the reefs' diversity and assemblages. Due to the lack of nutrient upwelling in the continental shelf, some of the corals are predominantly found in the tropics (for example, The Great Barrier Reef, Maldives, etc.).

Moyle and Cech have divided the coral reef ecosystem into six zones (Moyle and Cech 1988) (Fig. 17.2). They are as follows:

(a) The reef,
(b) The off-reef,
(c) The reef drop-off,
(d) The reef face,
(e) The reef flat, and
(f) The reef lagoon.

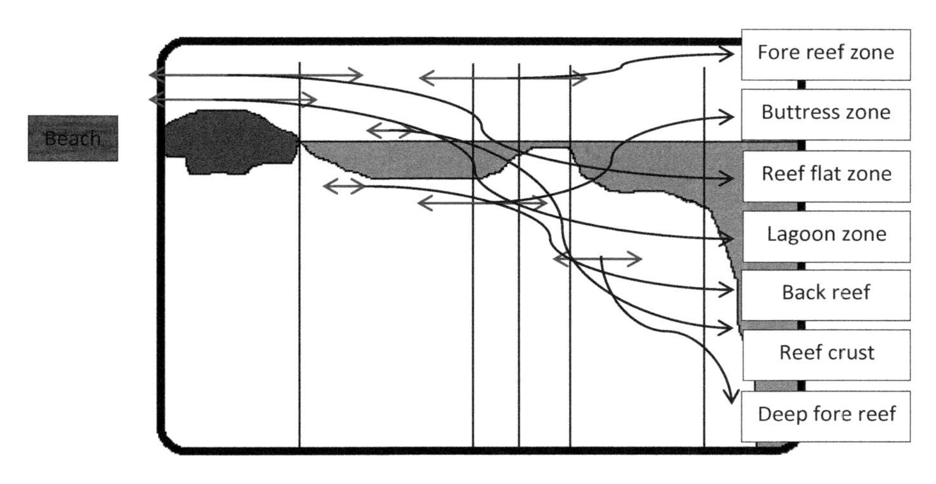

Fig. 17.2 Coral reef ecosystem zonation

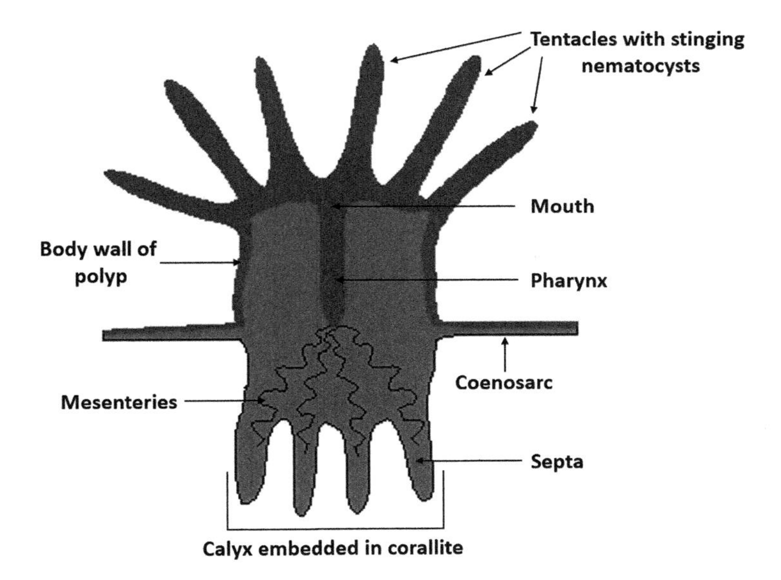

Fig. 17.3 Coral polyp

17.5 Reef Anatomy

The coral reef unit constitutes of the following components (Fig. 17.3):

1. Corals polyps,
2. Corallites, the exoskeleton of the corals and polyps,
3. Reef (calcium carbonate structure).

A solitary coral branch consists of many polyps grouped on it. Each polyp has a tentacle which helps the coral to seize food particle from the surrounding. These polyps are present in diverse shapes and sizes, ranging from a pinpoint to above 30 cm.

Reef-building corals live in the photic zone, allowing the absorption of sunlight by the residing algae (*Symbiodinium* sp.) followed by photosynthesis. The corals tend to grow faster in pollution-free water due to the mutualistic symbiosis. Reef-building corals get their maximum nutrients from their symbionts.

Fauna such as parrotfish, sea urchins and sponges act as bio-eroders. They function by breaking down the coral exoskeletons into smaller fragments which settle down in between the spaces of the carbonate structures.

17.6 Types of Coral Reefs

The most widely used method for differentiating coral reefs is by the following characteristics: morphology, size, shape and its relation to the nearby land.

As mentioned above, coral reefs are classified into different ways depending on their morphology and the location.

17.6.1 On the Basis of the Reef Nature, Shape and Mode of Occurrence

17.6.1.1 Fringing Reefs or Shore Reefs

It is one of the most important and common types of coral reefs. They propagate from a shore without any lagoon in between them. These reefs mostly grow near the shore around islands or continents. If at all a lagoon is present in between the reef, it is both narrow and shallow in nature. Since, majority of the time, a lagoon is not present to efficiently safeguard the freshwater runoff, pollution, or sedimentation, these reefs tend to be very delicate to the anthropological activities. Therefore, due to the rise in coastal population around the world, the fringing reef is deteriorating in number over the recent years (Fig. 17.4).

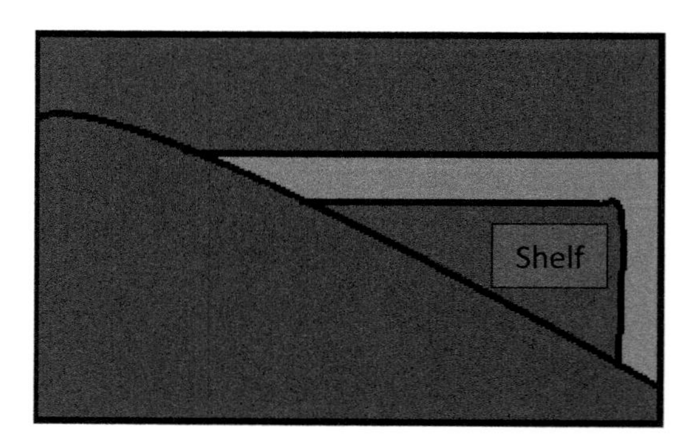

Fig. 17.4 Diagrammatic representation of fringing reefs or shore reefs

17.6.1.2 Barrier Reefs

These are reef complexes running parallel to the seashore and separated by a lagoon. These reefs are less in number and are mostly set up in the tropical area of the Atlantic or the Pacific region. The Great Barrier Reef (Australia) is the largest known barrier reef in the world. Other major reefs belonging to this category are The Belize and The New Caledonian Barrier Reefs. Some of these reef complexes are also present along the coast of Providencia, Mayotte, Kalimantan, Sulawesi and New Guinea, to name a few (Fig. 17.5).

17.6.1.3 Atolls

Atolls are categorized by a circular reef system adjacent to a deep central lagoon. This reef system forms when the level of the seawater rises due to the sinking of the islands surrounding the fringing reef. Atoll can be of varying shapes such as oval, circular, or horseshoe. These are commonly located throughout the Indian Ocean and the Pacific Ocean, for example, Maldives, The Chagos Islands, The Seychelles, Caroline Islands, Cook Island, French Polynesia and Micronesia. The Maldives consist of 26 atolls (Fig. 17.6).

17.6.1.4 Platform Reef

Platform reefs (bank or table reefs) tend to form mostly in the continental shelf and in the open ocean to some extent. This is because around these areas, the seabed rises to the level of the ocean surface, enabling the development of the reef-forming

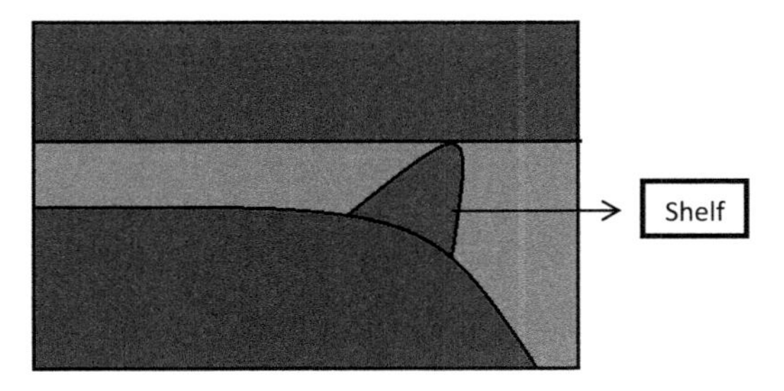

Fig. 17.5 Diagrammatic representation of barrier reefs

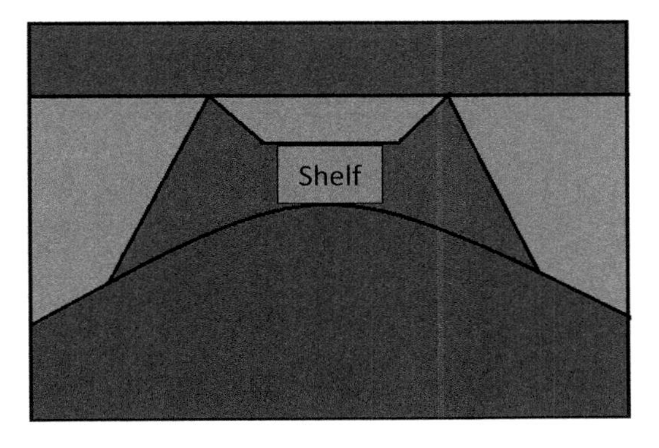

Fig. 17.6 Diagrammatic representation of Atoll

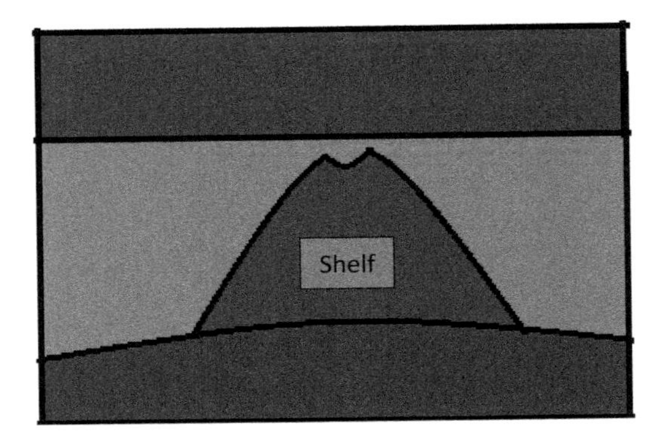

Fig. 17.7 Diagrammatic representation of platform reefs

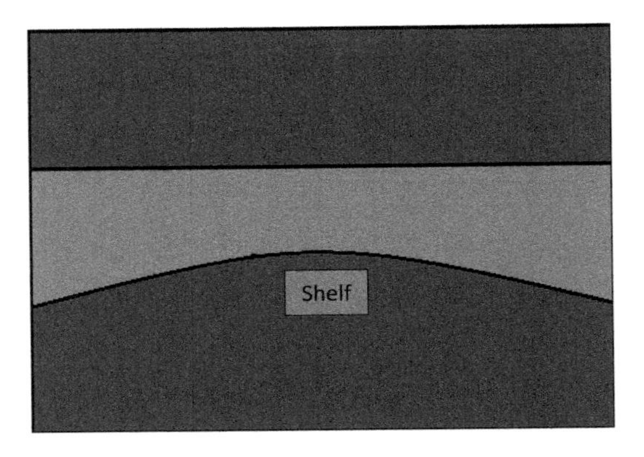

Fig. 17.8 Diagrammatic representation of tropical coral reefs

zooxanthemic corals. This kind of complex reef system is characterized by its growth in all directions and varying shapes and sizes. Sometimes, table reefs are located within certain atolls (east coast of the Red Sea). Pseudo-atolls are formed when the internal part of the ancient platform reefs gets heavily eroded. Some table reefs are U-shaped (located in Laccadives). This is essentially owing to the corrosion mediated by wind and water (Fig. 17.7).

17.6.2 On the Basis of Reef Location

17.6.2.1 Tropical Coral Reefs

Tropical coral reef ecosystem is related to clear and low latitude areas. These reefs prefer to grow in warm temperature conditions (Fig. 17.8).

17.6.2.2 Cold-Water Coral Reefs

This particular group of corals reefs is mostly located at depths greater than 2000 m. In deep ocean, the microorganisms depend on the organic debris and plankton for nutrition and survival in the cold environment (Fig. 17.9).

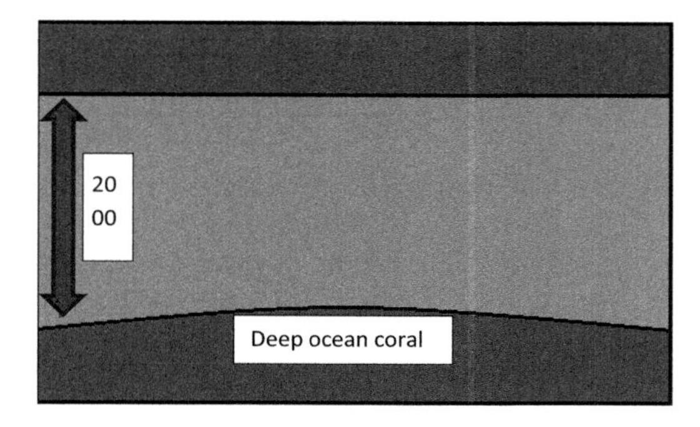

Fig. 17.9 Diagrammatic representation of cold-water coral reef

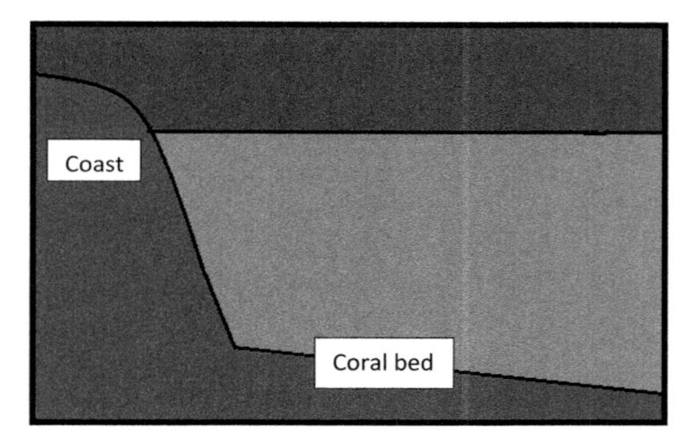

Fig. 17.10 Diagrammatic representation of marginal belt coral reefs

17.6.2.3 Marginal Belt Coral Reefs

Marginal reefs are characterized based on their juxtaposition to the environmental limits. They are further characterized by few volcanic islands surrounded by submarine banks of depths ranging from 10 to 60 miles in diameter (Fig. 17.10).

17.6.3 Other Reef Types or Variants

1. Apron reef: It is the preliminary stage of a fringing reef but more sloped.
2. Bank reef: It is larger than a patch reef, belonging to the subgroup of fringing reefs. It is usually located on the mid-shelf areas and diverges from a linear to a semicircular shape.
3. Patch reef: It belongs to the subgroup of platform reef which is isolated like that of the bank reef. It is mostly located within a lagoon and is often circular in shape.
4. Ribbon reef: These complex reef systems also referred to as a sill reef or a shelf edge reef. It is long, narrow and usually related to an atoll.
5. Habili: This kind of reef is specific only to the Red Sea.
6. Microatoll: They are a major community of coral species which are characterized by its vertical growth. This is because of the limitations bestowed upon them by the tidal height.

7. Cays: These are small, low-altitude islands formed on the superficial exoskeleton of the coral reefs. They are sandy in nature because of the eroded materials that pile up. As they reach above the sea level, plants become habitable. These reefs are usually located in the tropical environments.
8. Seamount: They are named as guyots (depending on the shape of the reef top). Seamounts are usually formed when a reef recedes on a volcanic island. They are usually round at the top. Flat top reefs are called table mounts.

17.7 Conditions Influencing the Growth of Corals

There are certain environmental parameters which are prerequisite and mandatory for the optimum growth of the corals and for the calcium carbonate secretion, because of which coral reefs and their associates are not found uniformly across the warm tropical ocean waters.

The major environmental parameters essential for the optimum growth of the corals are listed below:

1. Sunlight: It is an important factor for the survival and growth of the corals. Corals depend on zooxanthellae living inside them (symbiotic relationship) for oxygen and nutrients which in turn requires sunlight for its own survival. This is the main reason why corals rarely form in water depth greater than 50 m.
2. Sediment-free water: Coral reef and their associates are highly sensitive to sedimentation and water pollution. These two factors tend to cloud the water body, decreasing the amount of sun's rays reaching the algae zooxanthellae. Moreover, sediment tends to get deposited on the corals, thereby blocking the sun's rays. This results in the polyps getting damaged. Wastewater runoffs from the nearby industries contain many nutrients causing the seaweeds to overgrow. This condition results in the algal bloom. High concentration of organic and inorganic sediments inhibits the progression of the corals by clogging the mouth with cloudy water, thereby causing the corals to die of starvation.
3. Warmwater temperature: Corals are temperature sensitive. The reefs require optimum water temperature for survival. The reported water temperature ranges between 20 and 32 °C. Usually, the corals cannot thrive neither in very cold nor very warm sea temperature.
4. Salinity: Corals can tolerate salt concentration up to a certain level. This is the prime reason why corals do not survive in regions where rivers drain freshwater into the oceans, also termed as the estuaries. The ocean salinity value between 27 and 30 ppm is considered to be optimum for the survival of the reefs.
5. Depth of seawater: Corals have the ability to grow in ocean water with a depth lesser than 30 m below sea level. The maximum depth for survival is 80 m.
6. Freshwater influx: Influx of freshwater in high volume into the seas is toxic for coral growth and proper development. Henceforth, corals avoid growing in the surrounding area of mouths of major freshwater rivers but they thrive around islands (For example, around Lakshadweep islands, India).

7. Ocean currents and waves: Ocean currents and sea waves are favourable for the growth and survival of corals as they bring nutrients in high amount to the zooxanthellae algae (presiding in the outer tissue of the corals). These algae utilize the incoming nutrients to prepare food through photosynthesis.
8. Submarine platforms: Submarine platforms (not deeper than 90 m below sea level) are required for the formation of coral polyps.
9. Calcium carbonate saturation: This phenomenon is defined as the complete saturation of the total calcium and carbonate molecules in the ocean bed. Reef-building corals require calcium and carbonate molecules to build their exoskeletons. Low calcium carbonate concentration results in the inhibition of calcification by these coral polyps.

17.8 Different Colony Growth Forms

Corals have different colony growth forms (Fig. 17.11). They are as follows (Table 17.3):

1. Massive or lobate: These corals look like dome-shaped big boulders which help them in withstanding the ocean currents. They are categorized based on their slow and sturdy growth.
2. Columnar: These corals are cylindrical in shape (finger-like). They are categorized based on their upward growth and lack of secondary branches like the branching corals.
3. Ramose or branching: These corals are categorized based on the secondary branch which tends to break-off. These corals reproduce asexually via fragmentation. Branching corals grow faster than the other coral forms.
4. Foliaceous: Foliaceous corals form layered growth arrangements like that of petals in an open flower. The layered growth helps in increasing the surface area of

Fig. 17.11 Different coral growth forms

Table 17.3 Different coral growth forms with examples

Form	Description	Example
Branching	Presence of secondary branches	*Acropora cervicornis, Pocillopora* sp.
Columnar	Finger-like form	Pillar coral, *Dendrogyra cylindricus, Pavona clavus, Coscinaraea columna*
Foliaceous	Broad and thin leaf-like structure clustered together	Cabbage coral, *Astreopora randalli, Merulina ampliata, Turbinaria reniformis*
Massive	Dome-shaped boulders	Boulder brain coral, *Colpophyllia natans, Astreopora myriophthalma, Favites* sp., *Platygyra sinensis*
Encrusting	Grows as layers on substratum, Lichen-like appearance	Disc coral, *Turbinaria stellulata, Pavona varians, Cyphastrea* sp., *Acanthastrea echinata*
Plate-like	Flat upper surface giving table-like appearance	Brush coral, *Acropora hyacinthus, Leptoseris* sp., *Mycedium* sp., *Merulina ampliata*
Free-living	Solitary polyps that do not form colonies	Mushroom corals, *Fungia fungites, Herpolitha limax*
Soft corals	Belonging to the group *Octocorallia*	*Dendronephthya* sp., Toadstool coral, (*Sarcophyton glaucum*), *Sinularia notanda*
Phaceloid	Tubular in shape extending from a common base	*Caulastrea furca, Lobophyllia corymbosa, Lobophyllia hemprichii*

the corals which furthermore helps in absorbing maximum sunlight for photosynthesis.

5. Laminar or plate-like: These corals are categorized by their thin plate-like structure which helps them to procure food that is drifting down the slope. They grow horizontally. Such type of corals is usually located in the deeper regions of the fore reef.
6. Encrusting: This form typically sticks to the rocky substratum for support. They grow in the outward direction, thereby covering the entire rocky surface. Encrusting corals can tolerate strong ocean currents.
7. Free-living: Some corals live as solitaires and not in a colony. These coral forms are usually round, oval, or oblong in shape.
8. Soft corals: These corals belong to the group *Octocorallia*. This includes species of sea fans and sea whips. They are found in different shapes, colours and sizes.
9. Phaceloid: Each coral polyp is tubular in shape with individual wall extending from a common base.

17.9 Factors Influencing Colony Growth

There are many factors which influence the growth and the development of the corals (Fig. 17.12). Some are listed below.

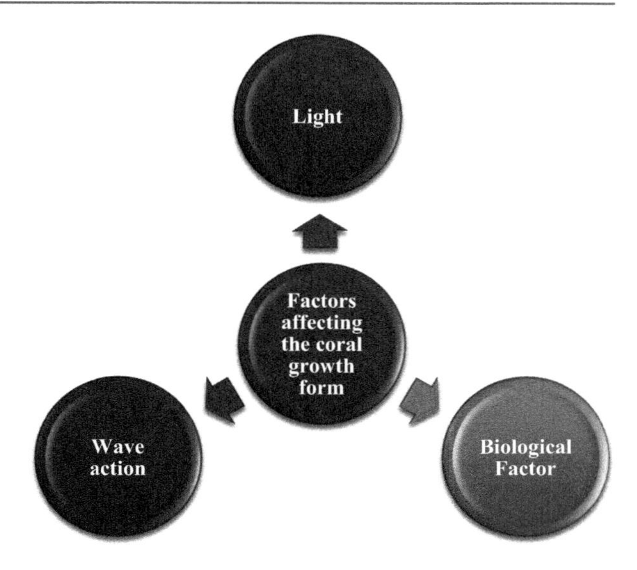

Fig. 17.12 Factors affecting the growth of coral forms

1. Water motion: Due to the wave motion, certain coral species tend to form dense and robust exoskeletons. Some corals may alter their development form depending on the wave motion and the surrounding environment. One such example of coral altering their structure is the lace coral, *Pocillopora damicornis*. The different growth forms are listed below as follows:
   (a) Short-branched and robust: they are slow growers found in areas of heavy wave motion and ocean current.
   (b) Delicate and fine-branched: they are fast growers found in calm water areas (protected bays).
   (c) Mushroom corals have a shape that stimulates self-righting by water motion.
2. Light: Corals alter their forms depending upon the extent of sun's rays falling on it. For example, Rice coral (*Montipora capitata*) forms hemispherical branching colonies in low depth water but it forms laminar, plate-like colonies in deeper or shaded water.
3. Biological factors: Factors such as genetics and coral symbionts are responsible for the alteration in the coral growth forms. For example, coral symbionts in gall crabs result in the gall-shaped branches.

17.10 Threats to Coral Reefs

Coral reefs are recognized for their huge biological, economical and striking values. It is reported that about 20% of the world's coral reef is under danger chiefly due to the numerous natural and anthropogenic activities (Miththapala 2008). Such combined exploitations result in the degradation of the reefs leading to loss in biodiversity, food provisions and economic revenue. It has been estimated that if such destruction persists worldwide, more than 70% of the coral population will be damaged by 2050 (Johnson et al. 2008). Anthropogenic activities such as human-induced marine pollution, industrialization, sweltering and excessive use of fossil fuels, overfishing, overexploitation, urbanization, or climate changes are few of the most significant dangers to the corals worldwide. As a result, the atmospheric CO_2 is continuously increasing (30% since the preindustrial period), which has headed to global warming. The increased use of chemicals in agricultural land management has resulted in the pollutants to contaminate the waterbody. The net result of such environmental stress triggers the degradation of the coral reefs in different parts of the globe (Sheppard and Loughland 2002; Sheppard 2003). The different natural and anthropological impacts on the coral reefs are listed below.

17.10.1 Natural Threats

1. Natural calamity

Natural calamities such as storms and earthquakes destroy large expanses of the reef ecosystem. Such disasters tend to be more severe if the reef-building communities are already deteriorated by other natural or anthropogenic impacts or if its recovery is inhibited by algal overgrowth.

2. Rise in ocean temperature

Climatic patterns or rise in the ocean temperature can often result in tension or stress to the reefs. For example, El Nino had severe impact on the sustainability of the corals. Rise in the seawater temperature can lead to coral bleaching where the corals expel their symbionts and turn bright white in colour.

3. Crown-of-Thorns (CoTs)

Sudden increase in the coral-eating Crown-of-Thorns sea star has started posing threat to reefs. In its presence, these tiny animals can destroy huge surface areas of the coral reefs. Recovery from such outbreaks may take more than 20–40 years if the damage is not severe.

4. Disease

Corals under severe environmental stress suffer from infections due to the production of excessive mucus resulting from natural and anthropogenic influences. It might also increase the number of blue-green algae (causative agent for black band disease among the coral population).

5. Coral bleaching

This phenomenon refers to the loss of symbiotic algae from the associated corals resulting in the white colouration (indicative of death of corals). One of the key factors responsible for bleaching is said to be global warming. This phenomenon (large-scale death of the corals) had occurred during 1997–1998. It is recorded as the most calamitous event, where corals in the tropical oceans of more than 50 countries and island nations were found dead.

Coral bleaching was first observed by Alfred Mayer in the year 1919 but not until the year 1998 was it known properly. Seventy percent of the corals died off the coasts of many tropical oceans and island nations such as Kenya, Maldives, Andaman and Lakshadweep islands. Three levels of coral bleaching have been observed in such places. They are as follows.

1. Catastrophic bleaching: Majority of the coral reef was adversely affected. Around 95% of corals in Bahrain, the Maldives, Sri Lanka and Tanzania were affected.
2. Severe bleaching: This phenomenon results in the loss of 50–70% of the reef in areas around Kenya, Seychelles, Japan, Thailand and Vietnam.
3. Moderate bleaching: This phenomenon results in the loss of 20–50% coral.
4. Insignificant bleaching: This phenomenon is also called as 'no-bleaching'.

17.10.2 Anthropogenic Threats

There are several causes due to which the anthropogenic threats lead to more than 75% loss of corals. Some of the factors are mentioned below.

1. Ocean Acidification: Coral helps in carbon sequestration. However, over the last few years, the levels of atmospheric CO_2 have been constantly increasing causing its absorption by the ocean water. This thereby increases the acidity level in the ocean. As ocean water's pH drops considerably, the coral exoskeletons tend to get dissolved by the acid, preventing new development of the corals.
2. Land pollution: This includes wastewater runoffs, sewage and sedimentation. As a result, the nutrient level increases which allows the overgrowth of macroalgae. This prevents them from receiving sunlight (essential for photosynthesis).
3. Overexploitation: Corals, fishes and other fauna such as turtles and sharks are being used for trade, food and for its aesthetic value in home décor. The above-mentioned fauna are involved in the food chain of the marine ecosystem, the loss of which can cause destructive and deleterious impacts on the ecosystem.
4. Vicious fishing techniques: Fishermen indulge in activities such as the blast and cyanide fishing which is dangerous for the wildlife. These processes involve either stunting or killing the fish in order to catch them easily. Blast fishing results in distortion of the reef structure, whereas the cyanide cloud interferes with the zooxanthellae photosynthesis.

5. Reef mining: It is the most devastating direct damage caused to the ecological structures of the corals. Mining provides calcium for the construction and manufacturing of cement.
6. Dynamiting: Dynamiting is a process similar to blast and cyanide fishing, where the fishes are killed and later collected. As a result, the corals suffer from serious structural damage caused due to the explosions. In certain cases, poisons are castoff to intoxicate fish so that they can be collected with ease manually.
7. Tourism: Tourism can be extremely dangerous causing physical damage by the boat anchors. Snorkelers may cause substantial damage by swimming or walking over the corals as less robust species can be easily broken. Some visitors like to collect corals as souvenirs, thereby contributing to the reef degradation.
8. Collection of coral for construction and use in the curio trade: Corals can be used as a construction tool or for the production of limekilns, house foundations and embankment of streets and canals, to name a few. There is an entire business that deals with collecting and selling corals as souvenirs or selling it in the aquarium trade.

17.11 Microeukaryotes and Coral Reef Habitats

Corals are the foundation of one of the rich marine ecosystems on earth. But the reefs are threatened on a large scale due to natural as well as anthropological effects (Hoegh-Guldberg et al. 2007; Casey et al. 2014; Graham et al. 2014; Peters 2015; Mora et al. 2016). Much research is being conducted to understand how corals respond to these environmental stresses. The coral microbiota is said to affect the marine ecosystem up to some extent (Gates and Ainsworth 2011; Zaneveld et al. 2016; Sweet and Bulling 2017). It includes a vast group of eukaryotes, bacteria, archaea and viruses. These organisms, in combination with the coral host, form the meta-organism. The research study on coral microbiome focusses on illustrating and distinguishing the marine microbial community associated with the coral host. Much research is being focused toward reviewing the symbiont dinoflagellate, *Symbiodinium* sp. (Knowlton and Rohwer 2003; Lesser et al. 2013; Baker et al. 2016).

17.11.1 Coral-Associated Bacteria

Coral-associated bacteria have been established but very little information is available about the operational activity of the coral holobiont. Reports suggest that coral microbiota is fundamental to host corals' biogeochemical cycling and pathogen resistivity. Nutrient cycling permits and facilitates the corals to prosper in oligotrophic waters. Recently, experiments with culture-dependent and culture-independent techniques have established that coral microbiota play a vital role in the reef biogeochemistry (Lesser et al. 2007; Chimetto et al. 2008; Raina et al. 2009; Williams et al. 2015). Nitrogen fixation within the coral exoskeleton by the nitrogen fixers has

Table 17.4 Bacterial species associated with the corals

Bacteria	Associated coral	References
Corynebacterium, *Acinetobacter*, *Parvularculaceae*, *Oscillatoria*	*Stephanocoenia intersepta*	Sweet et al. (2011)
Alteromonas, *Arhodomonas*, *Idiomarina*, *Pseudomonas*, *Spongiobacter*, *Roseobacter*	*Acropora millepora*, *Montipora aequituberculata*	Raina et al. (2009)
Roseobacter, *Marinobacter*, *Oceanospirillae*	*Porites astreoides*	Sharp et al. (2012)
Serratia marcescens	*Acropora palmata*	Sutherland et al. (2011)
Vibrio splendidus	Mediterranean coral	Reshef et al. (2006)

been documented using acetylene reduction assays (Lesser et al. 2007; Chimetto et al. 2008). Bacterial strains possessing genes responsible for nitrogen fixation has been identified (Lesser et al. 2004). In addition, experimental studies suggest that the bacterial strains associated with the corals are also involved in biogeochemical processes such as nitrification, denitrification, ammonium assimilation, carbon and sulphur cycling (Table 17.4).

17.11.2 Coral-Associated Fungi

Fungal strains associated with corals are suggested to have significant influences on the reef ecosystem (Priess et al. 2000; Ravindran et al. 2001; Blackall et al. 2015). They are extensively considered as primary pathogens. But research also mentions the existence of nonpathogenic invaders present in the unoccupied niches during environmental stresses. *Aspergillus sydowii,* a pathogenic fungus, is associated with the corals and it is extensively studied and characterized. It is the contributing pathogen responsible for the deterioration and degeneration of gorgonian sea fans in the Caribbean. In a study, Randall et al. (2016) concluded that the dark-spot syndrome is not a transmissible disease. In another study, Meyer et al. (2016) provided evidences stating that corals could be easily affected by dark spot and could fall prey for microbial colonization and overgrowth. Sweet (2013) in one such study concluded that the *Rhytisma acerinum* (marine fungal pathogen), associated with the tar spot disease, is present in high abundance. Endolithic coral fungi enhance the resistance and its survival caused due to the environmental stress (Table 17.5).

17.11.3 Endolithic Algae

Endolithic algae are the prevailing members of the coral microbiome. The best studied example of an endolithic microalga is the *Ostreobium* sp. The diversity and functional ecology of the abovementioned algae remains unknown (Marcelino and Verbruggen 2016). These microorganisms exist in the coral skeleton at low

Table 17.5 Marine fungi associated with the corals

Fungi	Associated coral	References
Ostracoblabe implexa	*Crassostrea cucullata*	Raghukumar and Lande (1988)
	Ostrea edulis	Alderman (1982)
	Crassostrea angulata	
Chama gryphoides	Corals	Durve and Bal (1960)
Scolecobasidium sp.	Corals from Andaman and Lakshadweep island	Raghukumar and Raghukumar (1991)
	Porites lutea	
	Porites lichen	
	Montipora tuberculosa	
	Goniopora sp.	
	Goniastrea sp.	
Curvularia lunata	*Porites lutea*	Ravindran et al. (2001)
Aspergillus sp.		
Cladosporium		
Mycelial yeast		
Rhytisma acerinum	Venezuela corals	Michael et al. (2013)
	Stephanocoenia intersepta	
	Siderastrea siderea	
	Montastraea annularis	
Mitosporic fungi		
Alternaria sp., *Aspergillus* sp., *Beauveria* sp., *Blodgettia* sp., *Camarosporium* sp., *Cladosporium* sp., *Curvularia* sp., *Cylindrocarpon* sp., *Diheterospora* sp., *Dreschlera* sp., *Exserohilum* sp., *Nigrospora* sp., *Oidiodendron* sp., *Papulaspora* sp., *Penicillium* sp., *Periconia* sp., *Scolecobasidium* sp., *Stachybotrys* sp., *Thysanophora* sp., *Torulomyces* sp., *Trichothecium* sp., *Tritirachium* sp., *Wardomyces* sp., *Zalerion* sp., *Zygosporium* sp.	Coral reefs in tropical Australian marine environments	Morrison-Gardiner (2002)
Ascomycota		
Aniptodera sp., *Carbosphaerella* sp., *Cochliobolus* sp., *Coronopapilla* sp., *Gaeumannomyces* sp., *Haloguignardia* sp., *Pseudeurotium* sp., *Sporormiella* sp., *Pestalotiopsis* sp.		
Oomycetes		
Pythium sp.		
Zygomycota		
Absidia sp., *Mucor* sp., *Rhizopus* sp., *Sterile mycelia*		

densities. During bleaching, corals tend to lose the primary source of carbon. As a result, the skeletal deposition of the exoskeleton is greatly reduced, causing bio-erosion.

Endosymbiotic algae are said to deliver benefit to its host coral during environmental stress condition (Fine et al. 2005; Gutiérrez-Isaza et al. 2015; del Campo et al. 2017). Fine and Loya (2002) in a particular study concluded that during the bleaching event, the algae undergo rapid growth. They were able to bleach the host tissues directly. The microeukaryotes have the potential to increase the survival rate of during severe stress conditions (Fine et al. 2006; del Campo et al. 2017). Endolithic microalgae have the potential to become secondary symbionts and offer the host protection during stress conditions (Fine and Loya 2002; Fine et al. 2005) (Table 17.6).

17.11.4 Coral-Associated Protists

Protists were the first microorganisms to be described as associated with the corals. The ecology of these organisms remains largely unexplored (Antonius and Lipscomb 2000; Palmer and Gates 2010; Sweet and Séré 2016; Page et al. 2017). Research proposes that protists may be opportunistic invaders of corals (Sweet and Bythell 2012; Sweet and Séré 2016; Verde et al. 2016). Members of the *alveolate superphylum* have been related with the corals worldwide.

A novel phylum of alveolates called the *Chromerida* has been discovered, and it is closely related to the apicomplexans, ciliates and dinoflagellates (Moore et al. 2008). Within the phylum Chromerida, *Chromera velia* and *Vitrella brassicaformis* have received much attention as they can photosynthesize. Both the species are closely related to the parasitic protist (Cumbo et al. 2013; Cumbo and Baird 2013; Linares et al. 2014; Oborník and Lukeš 2015). Close relationship between corals and the predatory protists has been witnessed in several corals worldwide. These protists are also associated with diseases in sea fans (Thompson et al. 2014).

17.12 Coral Reefs and Climate Change

Coral reefs form the major ecosystem engineers and are in charge of the diversity and abundance of marine life (Knowlton and Jackson 2001). However, this ecosystem in under severe threat of global decline due to reasons such as poor water quality, nutrients, sedimentation and presence of pollutants, and global climate shift. This leads to the elevation of ocean temperature and acidification (De'ath et al. 2012). The scleractinian corals, living close to their upper temperature limit, are prone to even slightest of the temperature changes. Temperature variation may lead to the large-scale coral bleaching (Hoegh-Guldberg et al. 2007). The generation time between 4 and 8 years is considered too long for the genetic adaptation to keep up with current change in the rate of climate change.

Table 17.6 Endolithic algae and its associated coral

Algae	Associated coral	References
Dictyota bartayresiana	*Montastrea annularis*	Barott et al. (2011)
Halimeda opuntia		
Acanthophora spicifera	*Montastraea faveolata*	Barott et al. (2012)
Dictyota sp.	*Porites astreoides*	
D.pulchella		
Lobophora variegate		
Lyngbya polychroa		
Lyngbya majuscula		
Dictyota sp.	*Montastraea faveolata*	Smith et al. (2006)
Halimeda opuntia		
Lobophora variegata		
Caulerpa, Cyanobacteria, Dictyosphaeria cavernosa	*Acropora* sp.	
	Favia sp.	
Halimeda	*Fungia* sp.	
Microdictyon	*Hydnophora* sp.	
Peyssonnelia	*Montastrea* sp.	
Turf mixed	*Montipora* sp.	
	Porites sp.	
	Stylophora sp.	
Caulerpa cupressoides	*Acropora muricata*	Sweet et al. (2013)
Caulerpa racemosa		
Chlorodesmis fastigiata	*Montastraea faveolata*	
Dictyota frabilis		
Halimeda macroloba		
Hincksia sp.		
Hydroclathrus clathrus		
Hypnea sp.		
Laurencia sp.		
Padina australis		
Sargassum polyceratium		
Dictyota menstrualis	*Porites astreoides*	Thurber et al. (2012)
Galaxaura obtusata		
Halimeda tuna		
Lobophora variegate		
Sargassum polyceratium		
Lobophora monticola	*Acropora muricata*	Vieira et al. (2016)
Lobophora rosacea		
Ulva fasciata	*Montipora capitata*	Vermeij et al. (2009)
Acanthophora spicifera		
Pterocladiella caerulescens		
Sargassum polyphyllum		

Scleractinian corals and zooxanthellae form the coral holobiont. Holobiont microbes influence reefs biogeochemical processes (carbon fixation and calcification). Calcification helps the reefs to maintain its biomass and diversity in low-nutrient waters.

Human-induced impacts are causing shifts in the tropical reef structure. This, in turn, causes the reef community to weaken over the time span. The science of microbial diversity is vital in understanding the connection between the corals and its associated microbial community. It also helps in establishing the relation between the climate shift and the coral microbiome. For instance, a transition to algae dominance and increase in oxygen consumption creates a hypoxic condition by the microbes.

17.13 Challenges and Opportunities

A major challenge for research in coral reef microbiome is that we are unaware of the functional roles played by the marine microbes within the complex reef communities, the biomechanics operating within them and the ecological. Change in the coral reef microbiome composition is known as dysbiosis. Egan and Gardiner (2016) in one study had provided an overview of challenges relating to dysbiosis.

Majority of the coral reef microbiota research is established on the dysbiosis, competition, or mutualism than factors such as predation or parasitism (Estes et al. 2011; Ohgushi et al. 2012; Parker et al. 2015). For example, *Halobacteriovorax,* a predatory bacterium, is present in most of the coral microbiome. They nurse on the *Vibrio coralliilyticus* and *Vibrio harveyi,* known pathogen (Webster et al. 2013). Eating pathogens are effective as poisoning them, as both the biotic interactions are chemically mediated. Similarly, many corals use biological warfare causing diseases in other corals. For example, *Campylobacteraceae* bacteria are mostly found on *Acropora* sp. and are often associated with the disease in *Montastrea* (Chu and Vollmer 2016).

Chemically mediated interactions and coral–host interactions are often limited by environmental conditions. It is highly possible that global change in the climate might destabilize chemical stability in the environment. For example, coral settlements from CCA-associated microbes are lost because of ocean acidification.

17.14 The Future of Coral Reef Ecosystem Sustainability

Anthropogenic stress results in the coral reef ecosystem decline. Owing to the climate shift, mass coral bleach occurs at an alarming rate each year. Reductions in carbon emissions must happen in order to fight the coral reef decline. There are new techniques in hand to understand the coral biology, even at molecular cellular level. The complex nature of the reefs makes the coral ecosystem conservation and its management challenging. Studies focussing on the structure and functions of the corals at a molecular level in a given environment setting will provide mechanistic insights in areas such as stress response, symbiosis and biomineralization. Genomic and metagenomic research on corals is still in its infancy, with most known about small bacterial genomes, a growing body of host coral studies, and limited information on large and complex Symbiodinium genomes. Genomic sequencing and analysis are keys to understanding the phenotypic responses and the adaptive capability of the corals. The complexity of the coral microbiome provides a vast array of evolutionary trails which is yet to be thoroughly explored.

References

Alagely A, Krediet CJ, Ritchie KB et al (2011) Signaling-mediated cross-talk modulates swarming and biofilm formation in a coral pathogen Serratia marcescens. ISME J 5:1609–1620

Alderman DJ (1982) Fungal disease of aquatic animals. Microbial diseases of fish, Spec. Publ.9. R. J. Roberts (Soc. Gen. Microbiol. London), 189–1242

Anthony KRN, Kerswell AP (2007) Coral mortality following extreme low tides and high solar radiation. Mar Biol 151:1623–1631

Antonius AA, Lipscomb D (2000) First protozoan coralkiller identified in the Indo-Pacific. Smithsonian Institution, National Museum of Natural History

Baker AC et al (2016) Diversity, distribution and stability of Symbiodinium in reef corals of the Eastern Tropical Pacific. In: Glynn PW (ed) Coral reefs of the Eastern Tropical Pacific. Springer, pp 405–420

Barott KL, Rohwer FL (2012) Unseen players shape benthic competition on coral reefs. Trends Microbiol 20:621–628

Barott KL, Rodriguez-Brito B, Janouškovec J, Marhaver KL, Smith JE, Keeling P et al (2011) Microbial diversity associated with four functional groups of benthic reef algae and the reef–building coral Montastraea annularis. Environ Microbiol 13:1192–1204

Barott KL, Rodriguez-Mueller B, Youle M, Marhaver KL, Vermeij MJ, Smith JE et al (2012) Microbial to reef scale interactions between the reef-building coral Montastraea annularis and benthic algae. Proc Biol Sci 279:1655–1664

Blackall LL et al (2015) Coral the world's most diverse symbiotic ecosystem. Mol Ecol 24:5330–5347

Casey JM et al (2014) Farming behaviour of reef fishes increases the prevalence of coral disease associated microbes and black band disease. Proc Biol Sci. 7 281(1788):20141032. https://doi.org/10.1098/rspb.2014.1032

Chimetto LA, Brocchi M, Thompson CC, Martins RCR, Ramos HR, Thompson FL (2008) Vibrios dominate as culturable nitrogen-fixing bacteria of the Brazilian coral Mussismilia hispida. Syst Appl Microbiol 31:312–319

Chu ND, Vollmer SV (2016) Caribbean corals house shared and host-specific microbial symbionts over time and space. Environ Microbiol Rep 8:493–500

Cumbo V, Baird A (2013) Chromera velia: coral symbiont or parasite? Galaxea J Coral Reef Stud 15:15–16

Cumbo VR et al (2013) Chromera velia is endosymbiotic in larvae of the reef corals Acropora digitifera and A. tenuis. Protist 164:237–244

De'ath G, Fabricius KE, Sweatman H, Puotinen M (2012) The 27-year decline of coral cover on the Great Barrier Reef and its causes. Proc Natl Acad Sci U S A 109:17995–17999

del Campo J et al (2017) The 'other'coral symbiont: Ostreobium diversity and distribution. ISME J 11:296–299

Durve VS, Bal DV (1960) Shell disease in Crossostrea gryphoides (Schlotheim). Curr Sci 29:489–490

Egan S, Gardiner M (2016) Microbial dysbiosis: rethinking disease in marine ecosystems. Front Microbiol 7:991

Estes JA, Terborgh J, Brashares JS, Power ME, Berger J, Bond WJ, Carpenter SR, Essington TE, Holt RD, Jackson JB, Marquis RJ, Oksanen L, Oksanen T, Paine RT, Pikitch EK, Ripple WJ, Sandin SA, Scheffer M, Schoener TW, Shurin JB, Sinclair AR, Soulé ME, Virtanen R, Wardle DA (2011) Trophic downgrading of planet Earth. Science 333:301–306

Fine M, Loya Y (2002) Endolithic algae: an alternative source of photoassimilates during coral bleaching. Proc R Soc Lond B Biol Sci 269:1205–1210

Fine M et al (2005) Tolerance of endolithic algae to elevated temperature and light in the coral Montipora monasteriata from the southern Great Barrier Reef. J Exp Biol 208:75–81

Fine M et al (2006) Phototrophic microendoliths bloom during coral ‘white syndrome. Coral Reefs 25:577–581

Gates RD, Ainsworth TD (2011) The nature and taxonomic composition of coral symbiomes as drivers of performancelimits in scleractinian corals. J Exp Mar Biol Ecol 408(1–2):94–101

Graham NA et al (2014) Coral reefs as novel ecosystems:embracing new futures. Curr Opin Environ Sustain 7:9–14

Gutiérrez-Isaza N et al (2015) Endolithic community composition of Orbicella faveolata (Scleractinia) underneath the interface between coral tissue and turf algae. Coral Reefs 34:625–630

Hallock P, Lidz BH, Cockey-Burkhard EM et al (2003) Environ Monit Assess 81:221. https://doi.org/10.1023/A:1021337310386

Hoegh-Guldberg O et al (2007) Coral reefs under rapid climate change and ocean acidification. Science 318:1737–1742

Jackson JBC, Kirby MX, Berger WH, Bjorndal KA, Botsford LW et al (2001) Historical overfishing and the recent collapse of coastal ecosystems. Science 293:629–638

Johnson RJ, Knap AH, Bates NR, White field JD, Kadko D, Lomas MW (2008) Coordinated change in the heat, salinity and CO 2 budgets of the mesopelagic zone at the Bermuda time-series sites. Abstract ASLO/AGU Ocean Sciences meeting, Orlando, March 2008

Knowlton N, Jackson JBC (2001) The ecology of coral reefs. In: Bertness MD, Gaines SD, Hay ME (eds) Marine community ecology. Sinaur Associates Incorporated, Sunderland, pp 395–422

Knowlton N, Rohwer F (2003) Multispecies microbial mutualisms on coral reefs: the host as a habitat. Am Nat 162:S51–S62

Krediet CJ, Ritchie KB, Paul VJ, Teplitski M (2013) Coral-associated micro-organisms and their roles in promoting coral health and thwarting diseases. Proc R Soc B Lond [Biol] 280:20122328

Lesser MP, Mazel CH, Gorbunov MY, Falkowski PG (2004) Discovery of symbiotic nitrogen-fixing cyanobacteria in corals. Science 305:997–1000

Lesser MP, Falcon LI, Rodriguez-Roman A, Enriquez S, Hoegh-Guldberg O, Iglesias-Prieto R (2007) Nitrogen fixation by symbiotic cyanobacteria provides a source of nitrogen for the scleractinian coral Montastraea cavernosa. Mar Ecol Prog Ser 346:143–152

Lesser M et al (2013) The endosymbiotic dinoflagellates (Symbiodinium sp.) of corals are parasites and mutualists. Coral Reefs 32:603–611

Linares M et al. (2014) Novel photosynthetic alveolates and Apicomplexan relatives. In: Löffelhardt W (ed) Endosymbiosis Springer, pp 183–196

Lough JM, Barnes DJ, McAllister FA (2002) Luminescent lines in corals from the GreatBarrier reef provide spatial and temporal records of reefs affected by land runoff. CoralReefs 21:333–343

Madin JS, Connolly SR (2006) Ecological consequences of major hydrodynamic disturbances on coral reefs. Nature 444:477–480

Marcelino VR, Verbruggen H (2016) Multi-marker metabarcoding of coral skeletons reveals a rich microbiome and diverse evolutionary origins of endolithic algae. Sci Rep 6:31508

Meyer JL et al (2016) Epimicrobiota associated with the decay and recovery of Orbicella corals exhibiting dark spot syndrome. Front Microbiol 7:893

Michael S, Bythell J, Nugues M (2013) Algae as reservoirs for coral pathogens. PLoS One. https://doi.org/10.1371/journal.pone.0069717

Miththapala S (2008) Coral reefs, Coastal Ecosystems Series, vol 1. Ecosystems and Livelihoods Group Asia, IUCN, Colombo, pp 1–36. + iii
Montaggioni L, Braithwaite CJR (1988) Quaternary coral reef systems: history, development processes and controlling factors, developments in Marine Geology 5. January 2009 Elsevier Science, Published Date: 13th August 2009
Moore RB et al (2008) A photosynthetic alveolate closely related to apicomplexan parasites. Nature 451:959–963
Mora C et al (2016) Ecological limitations to the resilience of coral reefs. Coral Reefs 35:1271–1280
Morrison-Gardiner S (2002) Dominant fungi from Australian coral reefs. Fungal Divers 9:105–121
Moyle, Cech JJ (1988) An introduction to Ichthyology (4th edition) fishes: introduction to Ichthyology
Nelson CE, Goldberg SJ, Wegley KL, Haas AF, Smith JE, Rohwer F, Carlson CA (2013) Coral and macroalgal exudates vary in neutral sugar composition and differentially enrich reef bacterioplankton lineages. ISME J 7:962–979
NEP (2001) UNEP-WCMC World Atlas of coral reefs coral reef unit
NGM nationalgeographic.com (September 2007) Ultra Marine: in far eastern Indonesia, the Raja Ampat islands embrace a phenomenal coral wilderness, by David Doubilet, National Geographic
NOAA CoRIS (2013) Regional Portal – Florida. Coris.noaa.gov (August 16, 2012). Retrieved on March 3, 2013
Oborník M, Lukeš J (2015) The organellar genomes of Chromera and Vitrella, the phototrophic relatives of apicomplexan parasites. Annu Rev Microbiol 69:129–144
Ohgushi T, Schmitz O, Holteds RD (2012) Trait-mediated indirect interactions: ecological and evolutionary perspectives. Cambridge University Press, New York
Page CA et al. (2017) Halofolliculina ciliate infections on corals (skeletal eroding disease). In: Woodley CM et al. (eds) Diseases of coral. Wiley, pp 361–375, 2016
Palmer CV, Gates RD (2010) Skeletal eroding band in Hawaiian corals. Coral Reefs 29:469–469
Parker IM, Saunders M, Bontrager M, Weitz AP, Hendricks R, Magarey KS, Gilbert GS (2015) Phylogenetic structure and host abundance drive disease pressure in communities. Nature 520:542–544
Peters EC (2015) Diseases of coral reef organisms. In: Birkeland C (ed) Coral reefs in the Anthropocene. Springer, pp 147–178
Priess K et al (2000) Fungi in corals: black bands and density banding of Porites lutea and P. lobata skeleton. Mar Biol 136:19–27
Rädecker et al (2014) Ocean acidification rapidly reduces dinitrogen fixation associated with the hermatypic coral Seriatopora hystrix. Mar Ecol Prog Ser 511:297–302
Raghukumar C, Lande V (1988) Shell disease of rock oyster *Crassostrea cucullata*. Dis Aquat Org 4:77–81
Raghukumar C, Raghukumar S (1991) Fungal invasion of massive corals. PSZNI Mar Ecol 12:251–260
Raina JB, Tapiolas D, Willis BL, Bourne DG (2009) Coral-associated bacteria and their role in the biogeochemical cycling of sulfur. Appl Environ Microbiol 75:3492–3501
Randall CJ et al (2016) Does dark-spot syndrome experimentally transmit among Caribbean corals? PLoS One 11:e0147493
Ravindran J, Raghukumar C, Raghukumar S (2001) Fungi in Porites lutea: association with healthy and diseased corals. Dis Aquat Org 47:219–228
Reshef L, Koren O, Loya Y, Zilber-Rosenberg I, Rosenberg E (2006) The coral probiotic hypothesis. Environ Microbiol 8:2068–2073
Ritchie KB (2006) Regulation of microbial populations by coral surface mucus and mucus-associated bacteria. Mar Ecol Prog Ser 322:1–14
Sharp K, Ritchie Kim B (2012) Multi-partner interactions in corals in the face of climate change. Biol Bull 223:66–77
Sharp KH, Distel D, Paul VJ (2012) Diversity and dynamics of bacterial communities in early life stages of the Caribbean coral Porites astreoides. ISME J 6:790–801

Sheppard CRC (2003) Predicted recurrences of mass coral mortality in the Indian Ocean. Nature 425:294–297
Sheppard CRC, Loughland R (2002) Coral mortality and recovery in response toincreasing temperature in the southern Arabian Gulf. Aquat Ecosyst Health Manag 5:395–402
Sheppard C, Dixon DJ, Gourlay M, Sheppard A, Payet R (2005) Coral mortality increases wave energy reaching shores protected by reef flats: examples from the Seychelles. Estuar Coast Shelf Sci 64:223–234
Smith JE, Shaw M, Edwards RA, Obura D et al (2006) Indirect effects of algae on coral: algae-mediated, microbe-induced coral mortality. Ecol Lett 9:835––845
Sneed JM, Sharp KH, Ritchie KB, Paul VJ (2014) The chemical cue tetrabromopyrrole from a biofilm bacterium induces settlement of multiple Caribbean corals. Proc R Soc B 281:20133086
Spalding M, Ravilious C, Edmund P (2001) Green. World Atlas of coral reefs. University of California, Berkeley, p 16
Sutherland KP, Shaban S, Joyner JL, Porter JW, Lipp EK (2011) Human pathogen shown to cause disease in the threatened elkhorn coral Acropora palmata. PLoS One 6:e23468
Sweet MJ, Bulling MT (2017) On the importance of the microbiome and pathobiome in coral health and disease. Front Mar Sci. Published online January 20, 2017. https://doi.org/10.3389/fmars.2017.00009
Sweet M, BythelL J (2012) Ciliate and bacterial communities associated with White syndrome and Brown Band disease in reef-building corals. Environ Microbiol 14:2184–2199
Sweet M, Séré MG (2016) Ciliate communities consistently associated with coral diseases. J Sea Res 113:119–131
Sweet MJ, Croquer A, Bythell JC (2011) Bacterial assemblages differ between compartments within the coral holobiont. Coral Reefs 30:39–52
Sweet M et al (2013) Characterisation of the bacterial and fungal communities associated with different lesion sizes of dark spot syndrome occurring in the coral Stephanocoenia intersepta. PLoS One 8:e62580
Thompson JR et al (2014) Microbes in the coral holobiont: partners through evolution, development, and ecological interactions. Front Cell Infect Microbiol 4:176
Thompson JR, Rivera HE, Closek CJ, Medina M (2015) Microbes in the coral holobiont: partners through evolution, development, and ecological interactions. Front Cell Infect Microbiol 4:176
Thurber Vega R, Burkepile DE, Correa AMS, Thurber AR et al (2012) Macroalgae decrease growth and alter microbial community structure of the reef-building coral, Porites astreoides. PLoS One 7:e44246
Vajed Samiei J, Dab K, Ghezellou P, Shirvani A (2013) Some Scleractinian corals (class: Anthozoa) of Larak Island, Persian Gulf. Zootaxa 3636(1):101–143. https://doi.org/10.11646/zootaxa.3636.1.5
Verde A et al (2016) Tissue mortality by Caribbean ciliate infection and white band disease in three reef-building coral species. Peer J 4:e2196
Vermeij M, Smith J, Smith C, Thurber RV, Sandin S (2009) Survival and settlement success of coral planulae: independent and synergistic effects of macroalgae and microbes. Oecologia 159:325–336. https://doi.org/10.1007/s00442-008-1223-7
Veron JEN (2000) In: Stafford-Smith M (ed) Corals of the world, vol 1–3. Australian Institute of Marine Science, Townsville. 1382 p
Vieira C, Engelen AH, Guentas L, Aires T, Houlbreque F, Gaubert PJ (2016) Species specificity of bacteria associated to the brown seaweeds *Lobophora* (Dictyotales, Phaeophyceae) and their potential for induction of rapid coral bleaching in *Acropora muricata*. Front Microbiol 7:316. https://doi.org/10.3389/fmicb.2016.00316
Webster NS, Uthicke S, Botte ES, Flores F, Negri AP (2013) Ocean acidification reduces induction of coral settlement by crustose coralline algae. Glob Chang Biol 19:303–315
White AT, Vogt HP (2000) Philippine coral reefs under threat: lessons learned after 25 years of community-based reef conservation. Mar Pollut Bull 40:537–550
Wilkinson C (ed) (2004) Status of coral reefs of the world: 2004. Australian Institute of Marine Science, Townsville

Williams AD et al (2015) Age-related shifts in bacterial diversity in a reef coral. PLoS One 10:e0144902

Zaneveld JR et al (2016) Overfishing and nutrient pollution interact with temperature to disrupt coral reefs down to microbial scales. Nat Commun. https://doi.org/10.1038/ncomms11833

Diversity and Ecology of Ectomycorrhizal Fungi in the Western Ghats

18

Kandikere R. Sridhar and Namera C. Karun

18.1 Introduction

Fungi are the largest community after insects that have evolved about 1800 mya possessing complex morphology (unicellular to filamentous to sporocarps) with different lifestyles and known to have only about 100,000 species (Kirk et al. 2008; Hawksworth 1991). The previous fungal species estimate from 1.5 million has been updated to a range of 2.2–3.8 million (Hawksworth 1991; Hawksworth and Lucking 2017). Hawkswroth (2019) has estimated the global existence of macrofungi as much as 220,000–380,000 species by considering 10% of all the fungi. Another global estimate based on plant–macrofungal ratio reveals existence of 53,000–110,000 macrofungi (Mueller et al. 2007). Among the mushrooms, the ectomycorrhizal (EM) fungi serve as one of the important components of the ecosystem and seedling establishment of tree species across the globe. The EM fungi encompass a broad variety of ascomycetes, basidiomycetes, and mucoromycetes associated with several tree species. A rough global estimate reveals the existence between 20,000 and 25,000 EM fungi that have relationship with 6000 plant species (Rinaldi et al. 2008; Tedersoo et al. 2010). Paleontological evidences suggest that members of the family Pinaceae (~156 mya) are symbiotically associated with EM fungi (~50 mya) (LePage et al. 1997; LePage 2003; Hibbett and Matheny 2009). Interestingly, the origin of EM fungal species (class *Agaricomycetes* and order *Pezizales*) as depicted by the molecular evidences dates back to ~200 and ~150 mya, respectively (Hibbett et al. 1997; Berbee and Taylor 2001, 2010). The phylogenetic diversity reveals that the EM fungal symbiotic association and lifestyle have evolved in multiple occasions (~66 times) independently from the saprophytes (Tedersoo et al. 2010, 2012; Floudas et al. 2012). The MGIC (Mycorrhizal Genomics Initiative Consortium) hypothesized that the EM fungal symbiosis with plant species evolved, owing to the loss of lignocellulose-degrading genes in comparison to saprotrophic relatives (Wolfe and Pringle

K. R. Sridhar (✉) · N. C. Karun
Department of Biosciences, Mangalore University, Mangalore, Karnataka, India

D. P. Singh et al. (eds.), *Microbial Interventions in Agriculture and Environment*,
https://doi.org/10.1007/978-981-13-8391-5_18

2012). According to Marcel et al. (2015), the EM fungi evolved after the appearance of brown-rot and white-rot fungi as common ancestor (~300 mya). Tedersoo et al. (2010) depict that majority of the EM fungi has evolutionary origin from the humus and wood-degrading ancestors. Two hundred twenty-five ITS sequences derived from 105 basidiomycetous mushrooms (59 genera, 29 families) in 10 locations across the State of Gujarat showed wide distribution of *Ganoderma* and *Schizophyllum* (Bhatt et al. 2018). This study revealed that from the *Agaricomycetes*, *Dacrymycetes* (wood-decomposing basidiomycetes) diverged during the Neoproterozoic era, while the *Hymenochaetales* (one of the orders of *Boletales*, *Hymenochaetales*, and *Russulales*) were evolved separately from *Agaricomycetes* during the Silurian period. According to Peintner et al. (2003), three *Cortinarius* spp. reported from India which belong to different clades were evolved independently. Being Gondwanan ancestors, these taxa served as typical examples that they are not geographically radiating clades owing to the endemic nature of their host tree species.

Studies on the diversity and distribution pattern of EM fungi are confined to the temperate and subarctic ecosystems (Smith and Read 2008). The largest numbers of EM fungi were known from the Holarctic regions ascribed to extensive exploration in comparison with Austral and tropical regions (Tedersoo et al. 2010). The Indian subcontinent consists of about 16,000 vascular plants; thus, at a 1:6 ratio, the fungal estimate will be around 96,000, but only 14,500 species have description till date (Hawkswroth 2019). The EM fungal research has been carried out in the Indian subcontinents such as Himalayas, Central India, Western Ghats, and Southern India (Sharma 2009, 2017; Kumar and Atri 2018). The Western Ghats of India has been classified as one of the important hotspots of biodiversity owing to the incidence of a wide variety of endemic species (Myers et al. 2000). Mountain range of the Western Ghats of India stretches about 1600 km (~160,000 km^2) in different states (Gujarat, Maharashtra, Goa, Karnataka, Tamil Nadu, and Kerala). The pattern of vegetation and ecosystems drastically differs from the Deccan plateau (plains) than the peak in Western Ghats (high-altitude) and thereafter mid-altitude, foothill, and coastal belt. It consists of a mixture of vegetation at different altitudinal ranges (~500 to 1200 m asl) (forests: shola, deciduous, moist-dry deciduous, evergreen, semi-evergreen, grasslands, scrub jungles). Diversity of EM fungi in the Western Ghats has been linked to the diversity of tree species by Riviere et al. (2007).

The main feature of EM fungi is to develop external sheath around the roots, and the penetrated hyphae establish Hartig net (in the intercellular gaps of the cortex and epidermis) (Smith and Read 2008). The main functions of EM fungi are to increase the absorptive surface, nutrient acquirement, and resistance to pathogenic organisms in the rhizosphere (Agerer 2006). Such interaction in turn facilitates the EM fungi to draw several organic compounds and energy sources from the host plant species (Bonfante and Genre 2008). In the Western Ghats of India, several pioneering studies have been undertaken on different issues of macrofungi (diversity and distribution; taxonomy and phylogeny; ecological aspects; nutritional and bioactive potential) (Table 18.1). The main objective of this review was to document scientific data on EM fungi of the Western Ghats with emphasis on their diversity, ecology, and future perspectives.

Table 18.1 Selected contributions on diversity, ecology, nutritional, and bioactive potential of macrofungi in the Western Ghats

Topic	Reference
Overview, diversity, distribution, and phylogeny	Sathe and Daniel (1980), Sathe and Deshpande (1980), Bhavanidevi (1995), Natarajan (1995), Leelavathy and Ganesh (2000), Bhagwat et al. (2005), Manoharachary et al. (2005), Brown et al. (2006), Leelavathy et al. (2006), Manimohan et al. (2007), Riviere et al. (2007), Swapna et al. (2008), Bhosle et al. (2010), Pradeep and Vrinda (2010), Mohanan (2011), Ranadive et al. (2011), Thiribhuvanamala et al. (2011), Farook et al. (2013), Karun and Sridhar (2013), Karun and Sridhar (2014a, 2015a, b), Karun et al. (2014), Mohanan (2014), Senthilarasu (2014), Aravindakshan and Manimohan (2015), Borkar et al. (2015), Pavithra et al. (2015), Greeshma et al. (2016), Karun and Sridhar (2016), Pavithra et al. (2016b), Senthilarasu and Kumaresan (2016), Karun and Sridhar (2017), Latha and Manimohan (2017), Pavithra et al. (2017a), Bhatt et al. (2018), and Karun et al. (2018b)
Ecological perspectives	Brown et al. (2006), Karun and Sridhar (2014a), Pavithra et al. (2015), Greeshma et al. (2016), Karun and Sridhar (2016), and Pavithra et al. (2016b)
Nutritional value	Sudheep and Sridhar (2014), Pavithra et al. (2017b), Karun et al. (2018a), and Greeshma et al. (2018a, b)
Bioactive potential	Karun et al. (2016), Pavithra et al. (2016a), Ghate and Sridhar (2017), and Karun et al. (2017)

18.2 Diversity and Distribution

The geographical setup and climatic environment of the Western Ghats offer ample scope to support perpetuation of macrofungi (Brown et al. 2006). In natural ecosystems, the EM fungal communities depend on the availability of suitable host tree species and competition among them to utilize the available resources (Kennedy 2010; Tedersoo et al. 2012). The diverse EM fungi of the Western Ghats are represented by a wide variety of host trees in different geographic locations. Table 18.2 summarizes the 148 species of EM fungi (34 genera) in 60 host tree species distributed in different locations of the Western Ghats. Sporocarps of ten representative EM fungi have been presented in Fig. 18.1. Considering all EM fungi on different host trees, the most dominant genus was *Inocybe* (36 spp.) followed by *Russula* (31 spp.), *Amanita* (13 spp.), and *Boletus* (8 spp.), 13 genera represented by 2–6 species, and 17 genera represented by 1 species (Fig. 18.2). A maximum number of EM fungi were found to be associated with the trees of Dipterocarpaceae (Fig. 18.3). In only eight hosts of Dipterocarpaceae, *Inocybe* is represented by 33 species followed by *Russula* (31 spp.), *Amanita* (10 spp.), and *Boletus* (7 spp.), eight genera represented by 2–4 species and 12 genera known by 1 species each. Interestingly, only eight Dipterocarpaceae members were the hosts for all 33 species of *Russula* obtained so far in the Western Ghats.

Among the trees (excluding the exotic species), *Vateria indica* possesses as high as 69 species of EM fungi, followed by *Hopea ponga* (50 spp.), *Hopea parviflora* (48 spp.), *Diospyros malabarica* (37 spp.), *Myristica malabarica* (10 spp.), and *Dipterocarpus indicus* (10 spp.) (Fig. 18.4). The rest of the hosts supported between one and seven species of EM fungi. Interestingly among 148 species, eight

Table 18.2 Ectomycorrhizal fungi recorded on the host tree species and geographic location in the Western Ghats

Ectomycorrhizal fungus	Host tree species	Geographic location	References
Amanita angustilamellata (Hohn.) Boedijn	*Terminalia bellirica* and *Vateria indica*	Kallar (Kerala)	Pradeep and Vrinda (2010)
	Hopea parviflora, H. ponga, and *Vateria indica*	Kuruva (Kerala)	Mohanan (2011)
Amanita antillana Dennis	*Vateria indica*	Uppangala (Karnataka)	Natarajan et al. (2005a)
Amanita aureofloccosa Bas	*Acacia auriculiformis*	Perayam (Kerala)	Pradeep and Vrinda (2010)
	Hopea ponga and *Vateria indica*	Iringolkav (Kerala)	Mohanan (2011)
Amanita bisporigera G.F. Atk.	*Diospyros malabarica, Hopea parviflora,* and *H. ponga*	Kuruva (Kerala)	Mohanan (2011)
Amanita cinerea Lam.	*Dipterocarpus indicus*	Uppangala (Karnataka)	Natarajan et al. (2005a)
Amanita elata (Massee) Corner & Bas	*Calophyllum calaba, Tectona grandis,* and *Terminalia paniculata*	Kallar (Kerala)	Pradeep and Vrinda (2010)
	Calophyllum calaba, Hopea parviflora, H. ponga, and *Vateria indica*	Sasthanada (Kerala)	Mohanan (2011)
Amanita griseofarinosa Hongo	*Garcinia morella* and *Hopea racophloea*	Kallar (Kerala)	Pradeep and Vrinda (2010)
	Hopea parviflora and *Vateria indica*	Arippa (Kerala)	Mohanan (2011)
Amanita hemibapha (Berk. & Broome) Sacc.	*Vateria indica*	Uppangala (Karnataka)	Natarajan et al. (2005a)
	Hopea parviflora, Myristica fragrans, and *Vateria indica*	Kallar (Kerala)	Pradeep and Vrinda (2010)
	Diospyros malabarica, Hopea parviflora, Myristica fragrans, and *Vateria indica*	Chandhakkunnu (Kerala)	Mohanan (2011)
Amanita magniverrucata Thiers & Ammirati	*Xanthophyllum arnottianum*	Kallar (Kerala)	Pradeep and Vrinda (2010)
Amanita muscaria (L.) Lam.	*Acacia mearnsii, Cupressus macrocarpa, Eucalyptus globulus,* and *Pinus patula*	Nilgiri Hills (Tamil Nadu)	Mohan (2008)
Amanita porphyria Alb. & Schwein	*Diospyros malabarica, Hopea parviflora, Terminalia paniculata,* and *Vateria indica*	Iringolkav (Kerala)	Mohanan (2011)

(continued)

Table 18.2 (continued)

Ectomycorrhizal fungus	Host tree species	Geographic location	References
Amanita vaginata (Bull.) Lam.	*Dipterocarpus indicus*	Uppangala (Karnataka)	Natarajan et al. (2005a)
Amanita volvata (Peck) Lloyd	*Hopea parviflora*	Kallar (Kerala)	Pradeep and Vrinda (2010)
Anamika indica K.A. Thomas, Peintner, M.M. Moser & Manim.	*Hopea parviflora*	Uppangala (Karnataka)	Natarajan et al. (2005a)
Astraeus hygrometricus (Peres.) Morgan	*Pinus roxburghii* and *Shorea robusta*	Chandhakkunnu (Kerala)	Mohanan (2011)
	Areca catechu	Konaje (Karnataka)	Karun and Sridhar (2014a)
	Anacardium occidentale, Artocarpus hirsutus, Holigarna arnottiana, Hopea parviflora, H. ponga, Phyllanthus emblica, and *Syzygium cumini*	Chinnibettu (Karnataka)	Pavithra et al. (2015)
Astraeus odoratus Phosri	*Hopea ponga*	Konaje (Karnataka)	Greeshma et al. (2015) and Pavithra et al. (2015)
Astrosporina amygdalina E. Horak	*Dipterocarpus indicus*	Uppangala (Karnataka)	Natarajan et al. (2005a)
Astrosporina avellana E. Horak	*Dipterocarpus indicus*	Uppangala (Karnataka)	Natarajan et al. (2005a)
Astrosporina calospora (Quél.) E. Horak	*Dipterocarpus indicus*	Uppangala (Karnataka)	Natarajan et al. (2005a)
Austroboletus gracilis (Peck) Wolfe	*Hopea ponga* and *Vateria indica*	Chandhakkunnu (Kerala)	Mohanan (2011)
Boletellus ananas (M.A. Curtis) Murrill	*Holigarna arnottiana*	Kallar (Kerala)	Pradeep and Vrinda (2010)
Boletinellus merulioides (Schwein.) Murrill	*Artocarpus heterophyllus, Cassine glauca,* and *Mangifera indica*	Kodagu (Karnataka)	Karun and Sridhar (2017)
Boletus alutaceus Morgan	*Vateria indica*	Brahmagiri (Kerala)	Mohanan (2011)
Boletus chrysenteron Bull.	*Holigarna arnottiana*	Kallar (Kerala)	Pradeep and Vrinda (2010)
Boletus edulis Bull.	*Dipterocarpus malabarica, Hopea parviflora,* and *Vateria indica*	Brahmagiri (Kerala)	Mohanan (2011)
	Canarium strictum, Holigarna nigra, and *Hydnocarpus pentandra*	Kodagu (Karnataka)	Karun and Sridhar (2017)

(continued)

Table 18.2 (continued)

Ectomycorrhizal fungus	Host tree species	Geographic location	References
Boletus hongoi T.N. Lakh. & Sagar	*Dipterocarpus malabarica, Hopea parviflora, Terminalia paniculata,* and *Vateria indica*	Mananthavdy (Kerala)	Mohanan (2011)
Boletus huronensis A.H. Sm. & Thiers	*Vateria indica*	Perumbavoor (Kerala)	Mohanan (2011)
Boletus pallidus Frost	*Hopea parviflora*	Chandhakkunnu (Kerala)	Mohanan (2011)
Boletus patriciae A.H. Sm. & Thiers	*Hopea parviflora*	Chandhakkunnu (Kerala)	Mohanan (2011)
Boletus reticulatus Schaeff.	*Dipterocarpus malabarica, Hopea parviflora,* and *Vateria indica*	Brahmagiri (Kerala)	Mohanan (2011)
Cantharellus cibarius Fr.	*Hopea parviflora* and *Myristica malabarica*	Kallar (Kerala)	Pradeep and Vrinda (2010)
	Diospyros malabarica, Hopea parviflora, Myristica malabarica, and *Vateria indica*	Kuruva, Mananthavady and Sasthanda (Kerala)	Mohanan (2011)
Cantharellus lateritius (Berk.) Singer	*Diospyros malabarica, Hopea parviflora,* and *Vateria indica*	Kuruva (Kerala)	Mohanan (2011)
Cantharellus minor Peck	*Hopea parviflora, Diospyros malabarica,* and *Vateria indica*	Sasthanda (Kerala)	Mohanan (2011)
Collybia fusipes (Bull.) Quél.	*Acacia auriculiformis* and *Casuarina equisetifolia*	Someshwara (Karnataka)	Ghate and Sridhar (2016a)
Cortinarius causticus Fr.	*Hopea parviflora*	Uppangala (Karnataka)	Natarajan et al. (2005a)
Cortinarius phlegmophorus K.A. Thomas, M.M. Moser, Peintner & Manim.	*Cansjera rheedii* and *Meiogyne pannosa*	Ponkuzhy (Kerala)	Peintner et al. (2003)
Geastrum triplex Jungh.	*Terminalia paniculata*	Konaje and Shankaraghatta (Karnataka)	Karun and Sridhar (2014b)
Gyroporus castaneus (Bull.) Quél	*Hopea ponga* and *Vateria indica*	Iringolkav (Kerala)	Mohanan (2011)
Hydnum rufescens Schaeff.	*Hopea parviflora*	Ammayambalan, Chandhakkunnu and Vadakancherry (Kerala)	Mohanan (2011)

(continued)

Table 18.2 (continued)

Ectomycorrhizal fungus	Host tree species	Geographic location	References
Inocybe antillana Pegler	*Diospyros malabarica, Hopea parviflora,* and *H. ponga*	Kuruva (Kerala)	Mohanan (2011)
Inocybe babruka K.P.D. Latha & Manim.	*Hopea ponga*	Kuruva (Kerala)	Latha and Manimohan (2017)
Inocybe brunneosquamulosa K.P.D. Latha & Manim.	*Vateria indica*	Ponnakkuam (Kerala)	Latha and Manimohan (2017)
Inocybe crassicystidiata Pegler	*Hopea parviflora* and *H. ponga*	Perumbavoor (Kerala)	Mohanan (2011)
Inocybe cutifracta Petch	*Aporosa lindleyana, Terminalia paniculata,* and *Vateria indica*	Kallar (Kerala)	Pradeep and Vrinda (2010)
	Aporosa lindleyana, Terminalia paniculata, and *Vateria indica*	Iringolkav, Kulathupuzha and Thamboormuzhi (Kerala)	Mohanan (2011)
Inocybe flavosquamulosa C.K. Pradeep & Matheny	*Hopea ponga*	Kuruva (Kerala)	Latha and Manimohan (2017)
Inocybe floccosistipitata K.P.D. Latha & Manim.	*Vateria indica*	Iringolkav (Kerala)	Latha and Manimohan (2017)
Inocybe gregaria K.P.D. Latha & Manim.	*Vateria indica*	Iringolkav (Kerala)	Latha and Manimohan (2017)
Inocybe griseorubida K.P.D. Latha & Manim.	*Vateria indica*	Iringolkav (Kerala)	Latha and Manimohan (2015)
Inocybe hydrocybiformis (Corner & E. Horak) Garrido	*Hopea ponga* and *Vateria indica*	Muthunga (Kerala)	Latha and Manimohan (2017)
Inocybe ianthinofolia Pegler	*Hopea parviflora*	Kallar (Kerala)	Pradeep and Vrinda (2010)
	Diospyros malabarica, Hopea parviflora, and *H. ponga*	Kuruva and Karadimale (Kerala)	Mohanan (2011)
Inocybe ingae Pegler	*Xanthophyllum arnottianum*	Kallar (Kerala)	Pradeep and Vrinda (2010)
Inocybe insulana K.P.D. Latha & Manim.	*Hopea ponga*	Kuruva islets (Kerala)	Latha and Manimohan (2017)
Inocybe iringolkavensis K.P.D. Latha & Manim.	*Hopea ponga and Vateria indica*	Iringolkav (Kerala)	Latha and Manimohan (2017)

(continued)

Table 18.2 (continued)

Ectomycorrhizal fungus	Host tree species	Geographic location	References
Inocybe keralensis K.P.D. Latha & Manim.	*Hopea ponga and Vateria indica*	Iringolkav (Kerala)	Latha and Manimohan (2017)
Inocybe kurkuriya K.P.D. Latha & Manim.	*Hopea ponga*	Kuruva (Kerala)	Latha and Manimohan (2017)
Inocybe kuruvensis K.P.D. Latha & Manim.	*Hopea ponga*	Kuruva (Kerala)	Latha and Manimohan (2017)
Inocybe lanuginosa (Bull.) P. Kumm.	*Pinus patula*	Sandynallah (Tamil Nadu)	Natarajan et al. (2005b)
Inocybe lasseri Dennis	*Diospyros malabarica, Hopea parviflora,* and *H. ponga*	Iringolkav and Kuruva (Kerala)	Mohanan (2011)
Inocybe muthangensis K.P.D. Latha & Manim.	*Hopea ponga*	Muthanga Wildlife Sanctuary (Kerala)	Latha and Manimohan (2017)
Inocybe papilliformis C.K. Pradeep & Matheny	*Hopea parviflora* and *Vateria indica*	Mathunga Wildlife Sanctuary (Kerala)	Latha and Manimohan (2017)
Inocybe petchii Boedijn	*Hopea parviflora*	Kallar (Kerala)	Pradeep and Vrinda (2010)
	Hopea parviflora and *H. ponga*	Chandhakkunnu, Iringolkav and Kuruva (Kerala)	Mohanan (2011)
	Rhizophora mucronata	Nethravathi (Karnataka)	Ghate and Sridhar (2016b)
Inocybe pileosulcata E. Horak, Mathney & Desjardin	*Hopea ponga, H. parviflora, Myristica malabarica,* and *Vateria indica*	Silent Valley (Kerala)	Latha and Manimohan (2017)
Inocybe purpureoflavida K.B. Vrinda & C.K. Pradeep	*Hopea parviflora*	Kallar (Kerala)	Pradeep and Vrinda (2010)
	Hopea parviflora and *H. ponga*	Iringolkav and Sasthanda (Kerala)	Mohanan (2011)
Inocybe rekhankitha K.P.D. Latha & Manim.	*Vateria indica*	Iringolkav (Kerala)	Latha and Manimohan (2017)
Inocybe rubrobrunnea K.P.D. Latha & Manim.	*Hopea ponga*	Muthanga Wildlife Sanctuary (Kerala)	Latha and Manimohan (2017)
Inocybe saraga K.P.D. Latha & Manim.	*Hopea ponga*	Muthanga Wildlife Sanctuary (Kerala)	Latha and Manimohan (2017)
Inocybe silvana K.P.D. Latha & Manim.	*Hopea ponga*	Kuruva islets (Kerala)	Latha and Manimohan (2017)

(continued)

Table 18.2 (continued)

Ectomycorrhizal fungus	Host tree species	Geographic location	References
Inocybe snigdha K.P.D. Latha & Manim.	*Hopea ponga*	Muthanga Wildlife Sanctuary (Kerala)	Latha and Manimohan (2017)
Inocybe squamata J.E. Lange	*Hopea parviflora*	Kallar (Kerala)	Pradeep and Vrinda (2010)
	Diospyros malabarica, Hopea parviflora, and *H. ponga*	Iringolkav (Kerala)	Mohanan (2011)
Inocybe stellata E. Horak, Mathney & Desjardin	*Hopea ponga*	Kuruva islets (Kerala)	Latha and Manimohan (2017)
Inocybe stuntzii Grund	*Acacia auriculiformis*	Perayam (Kerala)	Pradeep and Vrinda (2010)
Inocybe viraktha K.P.D. Latha & Manim.	*Hopea ponga*	Kuruva (Kerala)	Latha and Manimohan (2017)
Inocybe viridiumbonata Pegler	*Diospyros malabarica, Hopea parviflora, H. ponga,* and *Vateria indica*	Kuruva (Kerala)	Mohanan (2011)
Inocybe virosa K.B. Vrinda, C.K. Pradeep, A.V. Joseph & T.K. Abraham	*Aporosa lindleyana, Hopea parviflora, Terminalia paniculata,* and *Xylia xylocarpa*	Kallar (Kerala)	Pradeep and Vrinda (2010)
	Aporosa acuminata, Knema attenuata, and *Vateria indica*	Iringolkav, Palode and Pukayilanpara	Mohanan (2011)
	Vateria indica	Muthanga Wildlife Sanctuary (Kerala)	Latha and Manimohan (2017)
Inocybe wayanadensis K.P.D. Latha & Manim.	*Hopea ponga* and *H. parviflora*	Muthanga Wildlife Sanctuary (Kerala)	Latha and Manimohan (2017)
Laccaria amethystina Cooke	*Dipterocarpus malabarica and Vateria indica*	Sasthanda (Kerala)	Mohanan (2011)
Laccaria bicolor (Maire) P.D. Orton	*Vateria indica*	Uppangala (Karnataka)	Natarajan et al. (2005a)

(continued)

Table 18.2 (continued)

Ectomycorrhizal fungus	Host tree species	Geographic location	References
Laccaria fraterna (Sacc.) Pegler	*Acacia mearnsii, A. melanoxylon, Cupressus macrocarpa, Eucalyptus globulu,* and *Eucalyptus grandis*	Nilgiri Hills (Tamil Nadu)	Mohan (2008)
	Eucalyptus camaldulensis, E citriodora, E. deglupta, E. grandis, E. pellita, E. regnans, and *E. tereticornis*	Devikulam, Vattavanda and Methap (Kerala)	Mohanan (2011)
Laccaria laccata (Scop.) Cooke	*Acacia mearnsii, A. melanoxylon, Cupressus macrocarpa, Eucalyptus globulus, E. grandis,* and *Pinus patula*	Nilgiri Hills (Tamil Nadu)	Mohan (2008)
	Ficus beddomei, Hopea parviflora, H. racophloea, and *Vateria indica*	Kallar (Kerala)	Pradeep and Vrinda (2010)
	Eucalyptus deglupta, E. grandis, and *E. regnans*	Iringolkav, Nadukani, Kuruva and Sasthanda (Kerala)	Mohanan (2011)
Laccaria ohiensis (Mont.) Singer	*Eucalyptus deglupta, E. grandis,* and *E. regnans*	Suryanelly (Kerala)	Mohanan (2011)
Laccaria tetraspora Singer	*Vateria indica*	Uppangala (Karnataka)	Natarajan et al. (2005a)
Lacterius ignifluus K.B. Vrinda & C.K. Pradeep	*Hopea ponga* and *Vateria indica*	Iringolkav and Chandhakkunnu (Kerala)	Mohanan (2011, 2014)
Leccinum scabrum (Bull.) Gray	*Vateria indica*	Sasthanda and Kuruva (Kerala)	Mohanan (2011)
Lycoperdon decipiens Durieu & mont.	*Casuarina equisetifolia*	Someshwara (Karnataka)	Ghate and Sridhar (2016a)
	Acacia auriculiformis and *Rhizophora mucronata*	Nethravathi (Karnataka)	Ghate and Sridhar (2016b)
Lycoperdon perlatum Pers.	*Pinus patula*	Nilgiri Hills (Tamil Nadu)	Mohan (2008)
Lycoperdon utriforme Bull.	*Acacia mangium* and *Artocarpus heterophyllus*	Kodagu (Karnataka)	Karun and Sridhar (2017)
	Acacia auriculiformis and *Mangifera indica*	Derlakatte (Karnataka)	Karun et al. (2018)
Macrolepiota dolichaula (Berk. & Broome) Pegler & R.W. Rayner	*Acacia auriculiformis* and *Casuarina equisetifolia*	Someshwara (Karnataka)	Ghate and Sridhar (2016a)

(continued)

Table 18.2 (continued)

Ectomycorrhizal fungus	Host tree species	Geographic location	References
Macrolepiota rhacodes (Vittad.) Singer	*Acacia auriculiformis* and *Casuarina equisetifolia*	Someshwara (Karnataka)	Ghate and Sridhar (2016a)
Panus natarajanianus Senthil.	*Dipterocarpus indicus* and *Vateria indica*	Sirsi (Karnataka)	Senthilarasu (2015)
Phlebopus marginatus (Berk. & Broome) Boedijn	*Bambusa burmanica*	Kodagu (Karnataka)	Karun and Sridhar (2017)
	Bambusa burmanica	Derlakatte (Kanrnataka)	Karun et al. (2018)
Phlebopus portentosus (Berk. & Broome) Boedijn	*Coffea robusta*	Kodagu (Karnataka)	Karun and Sridhar (2017)
Phylloporus septocystidiatus C.K. Pradeep & K.B. Vrinda	*Hopea parviflora* and *Xanthophyllus arnottianum*	Palode (Kerala)	Pradeep et al. (2015)
Pisolithus albus (Cooke & Massee) Priest	*Acacia auriculiformis, A. mangium, Casuarina equisetifolia,* and *Eucalyptus tereticornis*	Arippa, Chandhakkunnu, Kadamkode, and Vazhikkadavu (Kerala)	Mohanan (2011)
	Acacia auriculiformis and *Casuarina equisetifolia*	Someshwara (Karnataka)	Ghate and Sridhar (2016a)
Pisolithus indicus Natarajan & Senthil.	*Vateria indica*	Uppamga;a (Karnataka)	Natarajan et al. (2005a)
	Vateria indica	Uppangala (Karnataka)	Reddy et al. (2005)
Pisolithus tinctorius (mont.) E. Fisch	*Acacia mangium*	Kokan (Maharashtra)	Borkar et al. (2015)
Rhizopogon luteolus Fr.	*Pinus patula*	Nilgiri Hills (Tamil Nadu)	Mohan (2008)
Rubinoboletus caespitosus T.H. Li & Watling	*Hopea parviflora*	Chandhakkunnu (Kerala)	Mohanan (2011)
Russula aciculocystis Kauffman ex Bills & O.K. Mill.	*Calophyllum apetalum, Myristica malabarica,* and *Vateria indica*	Kallar (Kerala)	Pradeep and Vrinda (2010)
	Calophyllum apetalum, Hopea parviflora, H. ponga, Myristica malabarica, and *Vateria indica*	Ammayambalam and Arippa (Kerala)	Mohanan (2014)

(continued)

Table 18.2 (continued)

Ectomycorrhizal fungus	Host tree species	Geographic location	References
Russula adusta (Pers.) Fr.	*Hopea parviflora and Myristica malabarica*	Kallar (Kerala)	Pradeep and Vrinda (2010)
	Diospyros malabarica, Hopea parviflora, H. ponga, Myristica malabarica, and *Vateria indica*	Chandhkkunnu (Kerala)	Mohanan (2011, 2014)
	Vateria indica	Konaje (Karnataka)	Pavithra et al. (2017a)
Russula albonigra (Pers.) Fr.	*Vateria indica*	Uppangala (Karnataka)	Natarajan et al. (2005a)
Russula amoena Auél.	*Vateria indica*	Uppangala (Karnataka)	Natarajan et al. (2005a)
Russula atropurpurea Peck	*Hopea parviflora* and *Vateria indica*	Chandhkkunnu (Kerala)	Mohanan (2011, 2014)
	Vateria indica	Konaje (Karnataka)	Pavithra et al. (2017a)
Russula azurea Bres.	*Vateria indica*	Uppangala (Karnataka)	Natarajan et al. (2005a); Riviere et al. (2007)
Russula cinerella Pat.	*Diospyros malabarica, Hopea parviflora,* and *Vateria indica*	Kuruva (Kerala)	Mohanan (2011, 2014)
Russula congoana Pat.	*Hopea parviflora* and *Pongamia pinnata*	Kallar (Kerala)	Pradeep and Vrinda (2010)
	Diospyros malabarica, Hopea parviflora, H. ponga, Myristica malabarica, and *Vateria indica*	Chandhakkunnu and Sanjeevani vanam (Kerala)	Mohanan (2011, 2014)
Russula delica Fr.	*Vateria indica*	Uppangala (Karnataka)	Natarajan et al. (2005a) and Riviere et al. (2007)
Russula delicula Romagn.	*Hopea parviflora* and *Vateria indica*	Kallar (Kerala)	Pradeep and Vrinda (2010)
	Diospyros malabarica, Hopea parviflora, H. ponga, Myristica malabarica, and *Vateria indica*	Kuruva (Kerala)	Mohanan (2011, 2014)
Russula emeticella (Singer) Romagn.	*Vateria indica*	Uppangala (Karnataka)	Natarajan et al. (2005a) and Riviere et al. (2007)

(continued)

Table 18.2 (continued)

Ectomycorrhizal fungus	Host tree species	Geographic location	References
Russula hygrophytica Pegler	*Hopea parviflora* and *Vateria indica.*	Arippa and Iringolkav (Kerala)	Mohanan (2011, 2014)
Russula koleggiensis K. Das, S.L. Mill., J.R. Sharma & J. Hemenway	Hopea ponga	Koleggi (Karnataka)	Das et al. (2008)
Russula laurocerasi Melzer	*Artocarpus hirsutus* and *Hopea parviflora*	Kallar (Kerala)	Pradeep and Vrinda (2010)
Russula leelavathyi K.B. Vrinda, C.K. Pradeep & T.K. Abraham	*Hopea parviflora*	Palode and Agasthyamala (Kerala)	Vrinda et al. (1997)
	Artocarpus hirsutus and *Hopea parviflora*	Kallar (Kerala)	Pradeep and Vrinda (2010)
	Diospyros malabarica, Hopea parviflora, H. ponga, Myristica malabarica, and *Vateria indica*	Chandhakkunnu, Kuruva and Palode (Kerala)	Mohanan (2011, 2014)
Russula luteotacta Rea	*Hopea parviflora* and *Vateria indica*	Kallar (Kerala)	Pradeep and Vrinda (2010)
	Diospyros malabarica, Hopea parviflora, H. ponga, Myristica malabarica, and *Vateria indica*	Iringolkav and Perumbavoor (Kerala)	Mohanan (2011, 2014)
Russula mariae Peck	*Hopea racophloea*	Kallar (Kerala)	Pradeep and Vrinda (2010)
	Diospyros malabarica, Hopea parviflora, H. ponga, and *Vateria indica*	Kuruva and Shenkily (Kerala)	Mohanan (2011, 2014)
Russula martinica Pegler	*Hopea parviflora*	Chandhakkunnu (Kerala)	Mohanan (2011, 2014)
Russula michiganensis Shaffer	*Diospyros malabarica, Hopea parviflora, H. ponga,* and *Vateria indica*	Iringolkav and Kuruva (Kerala)	Mohanan (2011, 2014)
Russula netrabaricus K. Das, S.L. Mill., J.R. Sharma & J. Hemenway	*Hopea ponga*	Netrabari (Goa)	Das et al. (2008)
Russula parazurea Jul. Schäff	*Pinus patula*	Kodaikanal (Tamil Nadu)	Natarajan and Raman (1983)
	Pinus patula	Nilgiri Hills (Tamil Nadu)	Mohan (2008)
Russula pectinata Fr.	*Vateria indica*	Uppangala (Karnataka)	Natarajan et al. (2005a)

(continued)

Table 18.2 (continued)

Ectomycorrhizal fungus	Host tree species	Geographic location	References
Russula pectinatoides Peck	*Vateria indica*	Uppangala (Karnataka)	Natarajan et al. (2005a) and Riviere et al. (2007)
Russula periglypta Berk. & Broome	*Vateria indica*	Kallar (Kerala)	Pradeep and Vrinda (2010)
	Vatica chinensis	Calicut University (Kerala)	Manimohan and Latha (2011)
	Diospyros malabarica, Hopea parviflora, and *Vateria indica*	Chandhakkunnu (Kerala)	Mohanan (2011, 2014)
Russula pseudodelica J.E. Lange	*Dipterocarpus indicus*	Uppangala (Karnataka)	Natarajan et al. (2005a) and Riviere et al. (2007)
Russula purpureonigra Petch	*Myristica malabarica*	Kallar (Kerala)	Pradeep and Vrinda (2010)
	Vatica chinensis	Calicut University (Kerala)	Manimohan and Latha (2011)
Russula romagnesiana Shaffer	*Hopea parviflora* and *Vateria indica*	Kallar (Kerala)	Pradeep and Vrinda (2010)
Russula rosea Pres.	*Dipterocarpus indicus*	Uppangala (Karnataka)	Natarajan et al. (2005a) and Riviere et al. (2007)
Russula senecis S. Imai	*Vateria indica*	Uppangala (Karnataka)	Natarajan et al. (2005a) and Riviere et al. (2007)
Russula subfoetens W.G. Sm.	*Vateria indica*	Uppangala (Karnataka)	Natarajan et al. (2005a) and Riviere et al. (2007)
Russula variegatula Romagn.	*Dipterocarpus indicus*	Uppangala (Karnataka)	Natarajan et al. (2005a);
Scleroderma areolatum Ehrenb.	*Eucalyptus grandis* and *E. tereticornis*	Devikulam (Kerala)	Mohanan (2011)
Scleroderma bovista Fr.	*Eucalyptus deglupta, E. grandis,* and *E. tereticornis*	Chandhakkunnu, Kuppadi and Nadukani (Kerala)	Mohanan (2011)

(continued)

Table 18.2 (continued)

Ectomycorrhizal fungus	Host tree species	Geographic location	References
Scleroderma citrinum Pers.	*Pinus patula*	Nilgiri Hills (Tamil Nadu)	Mohan (2008)
	Acacia auriculiformis, A. mangium, Eucalyptus deglupta, E grandis, E. tereticornis, and *Vateria indica*	Chandhakkunnu, Iringolkav and Wadakkanchary (Kerala)	Mohanan (2011)
	Acacia auriculiformis and *Casuarina equisetifolia*	Someshwara (Karnataka)	Ghate and Sridhar (2016a)
	Artocarpus heterophyllus, Dysoxylum malabaricum, and *Schefflera racemosa*	Kodagu (Karnataka)	Karun and Sridhar (2017)
Scleroderma polyrhizum (J.F. Gmel.) Pers.	*Eucalyptus grandis, E. tereticornis,* and *Vateria indica*	Brahmagiri and Pambadumshola (Kerala)	Mohanan (2011)
Scleroderma verrucosum (Bull.) Pers.	*Acacia auriculiformis, A. mangium, Eucalyptus deglupta, E. grandis, E. tereticornis,* and *Vateria indica*	Ingar, Periya and Vattavada (Kerala)	Mohanan (2011)
	Canarium strictum, Holigarna nigra, and *Vateria indica*	Bettoli and V'Badaga (Karnataka)	Karun et al. (2014)
Strobilomyces annulatus Corner	*Diospyros malabarica, Hopea ponga,* and *Vateria indica*	Iringolkav and Kuruva (Kerala)	Mohanan (2011)
Strobilomyces floccopus (Vahl) P. Karst.	*Holigarna arnottiana*	Kallar (Kerala)	Pradeep and Vrinda (2010)
Strobilomyces mollis Corner	*Dipterocarpus indicus*	Uppangala (Karnataka)	Natarajan et al. (2005a)
	Dipterocarpus malabarica, Hopea ponga, and *Vateria indica*	Kuruva, Nelliampathy and Pothumala (Kerala)	Mohanan (2011)
Strobilomyces strobilaceus (Scop.) Berk.	*Dipterocarpus malabarica, Holigarna arnottiana, Hopea ponga,* and *Vateria indica*	Iringolkav (Kerala)	Mohanan (2011)
Suillus brevipes (Scop.) Berk.	Pinus patula	Kodaikanal (Tamil Nadu)	Natarajan and Raman (1983)
	Acacia mearnsii, Cupressus macrocarpa, Eucalyptus globulus, and *Pinus patula*	Nilgiri Hills (Tamil Nadu)	Mohan (2008)

(continued)

Table 18.2 (continued)

Ectomycorrhizal fungus	Host tree species	Geographic location	References
Suillus pallidiceps A.H. Sm. & Thiers	Pinus patula	Kodaikanal (Tamil Nadu)	Natarajan and Raman (1983)
Suillus placidus (Bonord.) Singer	*Hopea ponga* and *Vateria indica*	Iringolkav and Manathvady (Kerala)	Mohanan (2011)
Suillus punctatipes (Snell & E.A. Dick) Singer	*Pinus patula*	Thalaikunda (Tamil Nadu)	Natarajan and Raman (1983)
	Pinus patula	Thalaikunda (Tamil Nadu)	Natarajan et al. (2005b)
Suillus subluteus (peck) Snell	*Pinus patula*	Katharikai Odai (Tamil Nadu)	Natarajan and Raman (1983)
	Pinus patula	Nilgiri Hills (Tamil Nadu)	Mohan (2008)
Suillus tomentosus (Kauffman) Singer	*Hopea parviflora*	Chandhakkunnu (Kerala)	Mohanan (2011)
Thelephora palmata (Scop.) Fr.	*Acacia auriculiformis* and *Casuarina equisetifolia*	Someshwara (Karnataka)	Ghate and Sridhar (2016a)
Thelephora terrestris Ehrh.	*Pinus patula*	Nilgiri Hills (Tamil Nadu)	Mohan (2008)
Tricholoma rimosoides Dennis	*Acacia auriculiformis*	Kadamkode (Kerala)	Mohanan (2011)
Tylopilus alboater (Schwein.) Murrill	*Dipterocarpus malabarica*	Kadalkandam (Kerala)	Mohanan (2011)
Xylaria nigripes (kltzsch) Cooke	*Bougainvillea spectabilis*	Konaje (Karnataka)	Karun and Sridhar (2015b)

Dipterocarpaceae members itself supported 77% (114 spp.) of EM fungi (see Fig. 18.3). Although 148 EM fungi colonized 60 host tree species, the number of hosts supported a specific species was fairly low. For instance, occurrence in tree hosts ranged from 1 to 12 with the highest hosts for *Scleroderma citrinum* and *Laccaria laccata* (12 hosts each) followed by *Laccaria fraterna* (11 hosts), *Astraeus hygrometricus* (10 hosts), and *Scleroderma verrucosum* (8 hosts). Further, in decreasing order 2, 3, 8, 14, 19, 23, and 74 EM fungi occurred on 7, 6, 5, 4, 3, 2, and 1 hosts, respectively. The number of times each EM fungus reported was ranged only between 1 and 4.

The host trees that supported the EM fungi in Western Ghats consist of 60 species (40 genera). There seem to be the Dipterocarpaceae members that are responsible for the propagation of EM fungi. More than 75% of EM fungi reported from the trees belong to the single family Dipterocarpaceae (eight tree species) in the Western Ghats. Lee (1998) reviewed the EM fungal association with Dipterocarpaceae in the Southeast Asia and found *Amanita*, *Boletus*, and *Russula* as the most dominant genera. Bâ et al. (2010) addressed the research performed on the EM fungi on the African Dipterocarpaceae. Our review supported the view that the members of Dipterocarpaceae possess three dominant genera like *Amanita*, *Boletus*, and *Russula* in Southeast Asia (Lee 1998; Natarajan et al. 2005b). Further, our study also

Fig. 18.1 Representative EM fungi of the Western Ghats: *Astraeus hygrometricus* (**a**), *Astraeus odoratus* (**b**), *Boletinellus merulioides* (**c**), *Lycoperdon utriforme* (**d**), *Phlebopus marginatus* (**e**), Yerava tribe lady of Kodagu holding the fruit body of *Phlebopus portentosus* (**f**), *Pisolithus albus* (**g**), *Russula atropurpurea* (**h**), *Scleroderma citrinum* (**i**), and *Thelephora palmata* (**j**)

endorsed the notion that *Vateria indica* supports the highest species of EM fungi (Natarajan et al. 2005b). In addition, recent studies in Kerala on Dipterocarpaceae raised the genus *Inocybe* to the top level (Latha and Manimohan 2017). It is predicted that the partnership between Dipterocarpaceae and EM fungi might have established prior to separation of the Gondwana (Brearley 2012). Species belonging to Dipterocarpaceae that is mainly known from the tropical rainforests seems to be an important repository of EM fungi in the Western Ghats (Appanah and Turnbull 1998). Being valuable trees in the timber trade, caution should be exercised for tangible exploitation without causing impairment to the perpetuation and ecosystem services of associated EM fungi.

Bâ et al. (2010) reviewed the management aspects of EM fungi linked with alien trees in tropical Africa. Literature on symbiotic status of EM fungi with four exotic

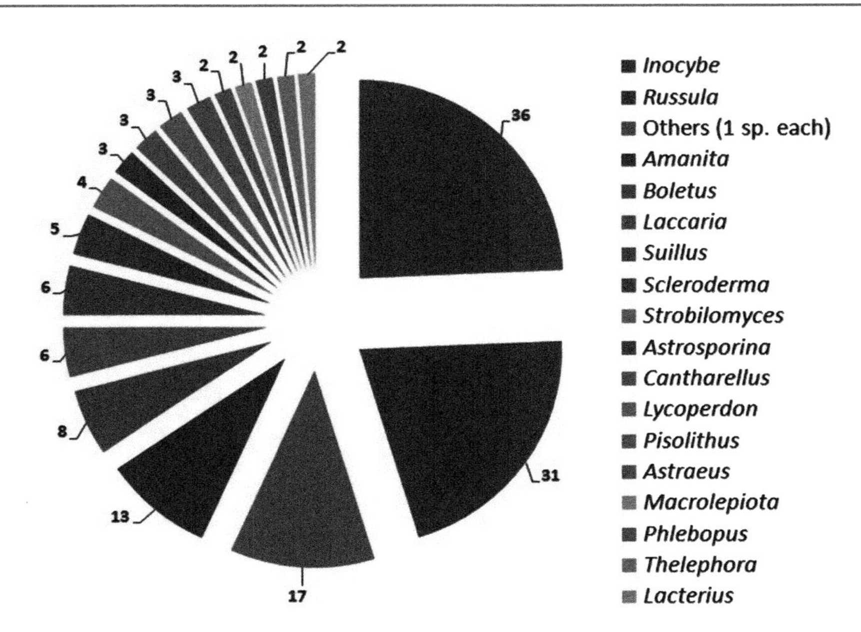

Fig. 18.2 Distribution pattern of EM fungi associated with tree species in the Western Ghats

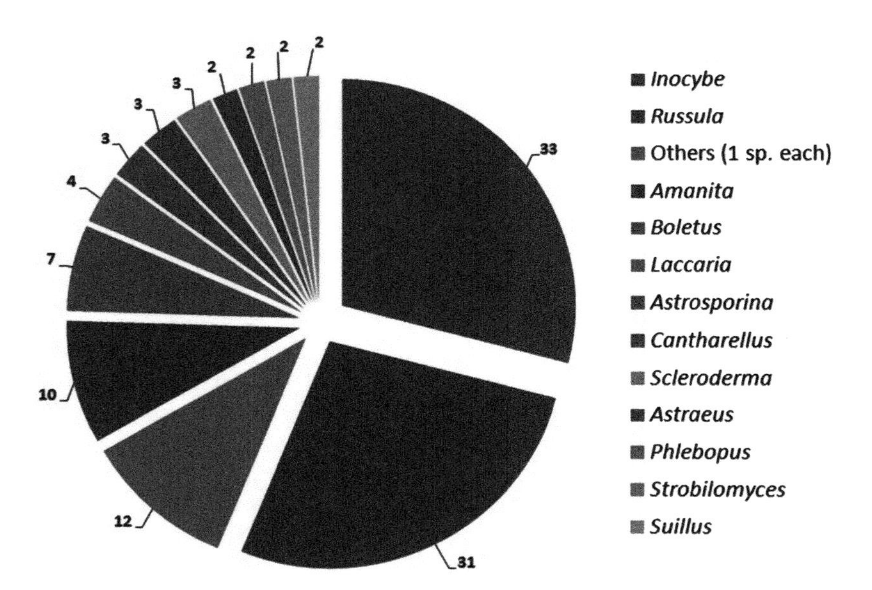

Fig. 18.3 Distribution pattern of EM fungi associated with Dipterocarpaceae in the Western Ghats

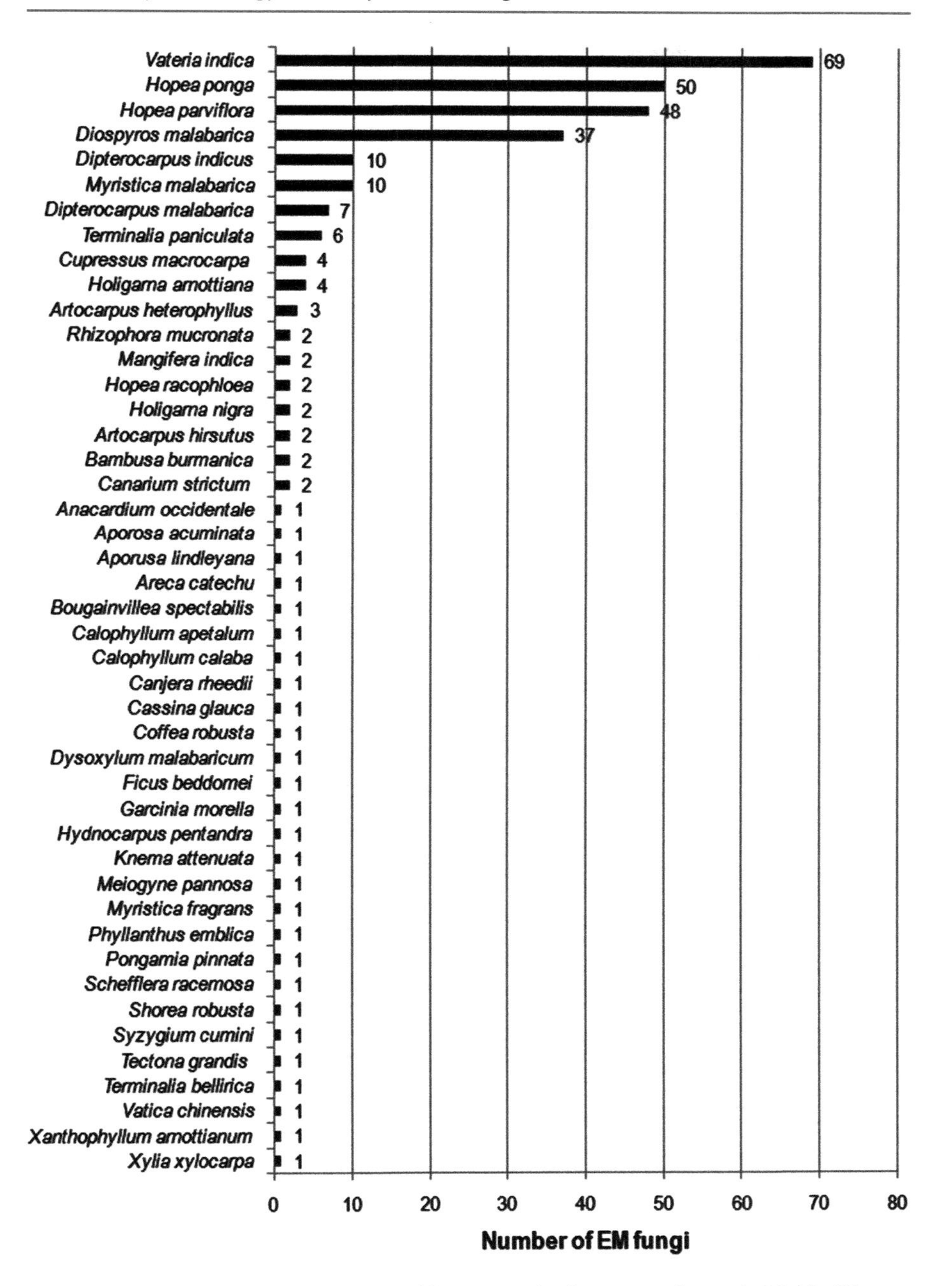

Fig. 18.4 Distribution pattern of EM fungi in tree species (except exotic species) in the Western Ghats

plant families (Casuarinaceae, Leguminosae, Myrtaceae, and Pinaceae) and strategies to be followed for improvement by inoculation for better management in sylviculture/reforestation have been addressed. Among the 60 tree species which serve as hosts for EM fungi in the Western Ghats, 15 were alien tree species belonging to four genera (*Acacia*, *Casuarina*, *Eucalyptus*, and *Pinus*). These 15 exotic species were the hosts of up to 29 EM fungi (20%) in the Western Ghats. The *Pinus patula* supported the highest species (13 spp.) of EM fungi, followed by *Acacia auriculiformis* (12 spp.), *Eucalyptus grandis* (9 spp.), and *Casuarina equisetifolia* and *Eucalyptus tereticornis* (7 spp. each) (Fig. 18.5). In the rest of species, EM fungal occurrence ranged from one to six species. In addition to the alien tree species (belonging to plant families Myrtaceae, Pinaceae, Casuarinaceae, and Leguminosae) in the Western Ghats, acacias, areca, bamboo, cashew, coffee, silver oak, and rubber are of special interest to study the EM fungal association to develop future eco-friendly plantations. Knowledge on symbiotic association of EM fungi with alien tree species is valuable in sylviculture/reforestation suitable to the geographic conditions of the Western Ghats. According to Smith and Read (2008), the EM fungal relationship with the family Pinaceae was as ancient as ~130 mya.

Although the EM fungi have been largely identified as useful macrofungi, some of them associated with alien tree hosts (e.g., alders, firs, pines, and willows) have

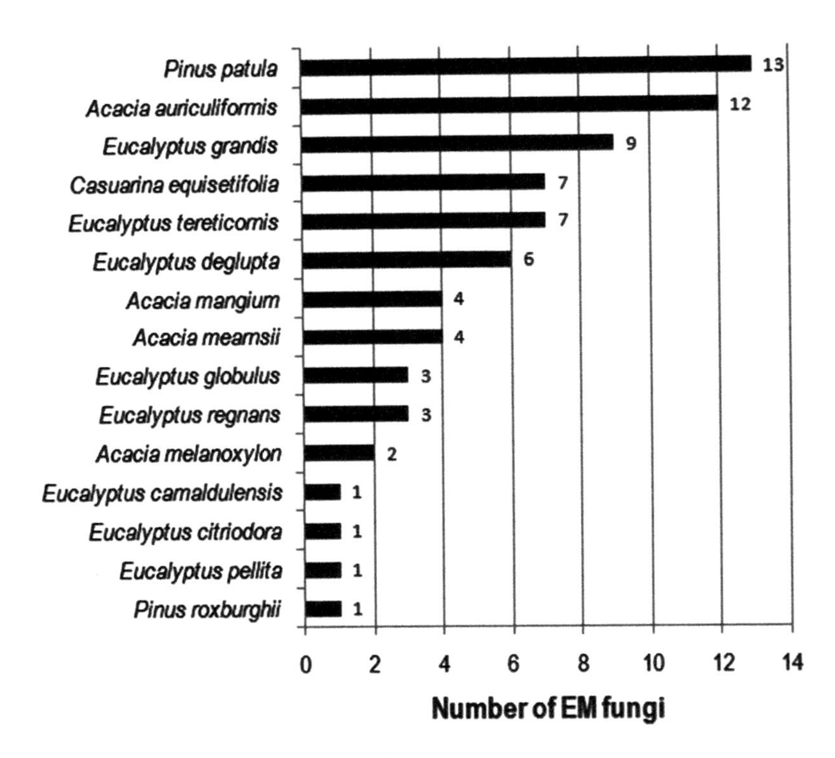

Fig. 18.5 Distribution pattern of EM fungi in exotic tree species in the Western Ghats

invasive qualities (see Dickie et al. 2016). Some examples of such EM fungi include *Amanita muscaria*, *Boletus edulis*, *Laccaria bicolor*, and *L. fraterna*. They usually invade the native forests in Australia, some regions of Europe, North America, South Africa, South America, and New Zealand. Caution needs to be exercised to balance the invasive nature of such EM fungi in alien tree species against the destruction of EM fungi of native tree species seems to be appropriate. Such invasions are challenging to the sylviculture practices and need to follow multicultural practices. However, *Pinus amamiana* being a timber-yielding tree is endemic to Japanese islands, which has been considered as an endangered and vulnerable tree species (Farjon 2005). Its abundance has been reduced owing to pine wilt disease, and now about 1600 trees exist in Japanese islands (http://ikilog.biodic.go.jp/Rdb/). The EM fungi play a major role in regenerating and conservation of such tree species; thus, Murata et al. (2017) demonstrated that *Rhizopogon* sp. bioassay was compatible in improving the status of *P. amamiana* especially in increasing growth and tolerance in transplantation in comparison with control.

In spite of the onset of sporocarp of EM fungi, their species richness depends on the age of the tree or plantation (Natarajan and Senthilarasu 2004; De Miguel et al. 2014). The present information on the EM fungi of the Western Ghats is the product of taxonomic and phylogenetic studies carried out mainly in the state of Kerala. About 160 collections from three forests of Kerala (evergreen, deciduous, and exotic forests) resulted in the highest EM fungi in evergreen forest (Pradeep and Vrinda 2010). Although several studies in the Western Ghats are not mainly intended to evaluate EM fungi, many new genera and species of EM fungi have been reported (see Table 18.2). Nearly 30 new species of EM fungi have been erected as new species from Kerala, which accounts to ~20% of known EM fungi in the Western Ghats. The dark-spored agaric fungus *Anamika*, a new genus of ectomycorrhizae of Cortinariaceae (new species, *Anamika indica*), is associated with *Hopea* sp. in Kerala (Thomas et al. 2002). Subsequently, two new species *Anamika angustilamellata* (associated with tree species belongins to plant families Fagaceae and Dipterocarpaceae in China and Thailand) and *Anamika lactariolens* (associated with *Quercus*-*Pinus* forest in Japan) have been described.

18.3 Ecological Perspectives

Regarding the EM fungi in the Western Ghats, the major contribution comes from the studies carried out in the state of Kerala, while reports from large areas of the rest of the Western Ghats are sporadic or neglected or underexplored. Under the human interference on the host tree species of the Western Ghats, an urgent need is to intensify research on the diversity, benefits, and conservation of EM fungi. Two important mechanisms are known to direct the distribution of the EM fungi which include dispersion and isolation. The disjunct distribution of *Inocybe* was seen between Australia and New Zealand among the tree species of Dipterocarpaceae and Fabaceae (see Vasco-Palacios et al. 2018). The long-distance dispersal is possible owing to the low host specificity, which increases the gene flow among geographically remote population of EM fungi. Such mechanisms are opposing to the dispersal of high host-specific EM fungi, leading to endemism and danger of extinction.

Recording associated EM fungi in a specific study is important since many EM fungi have bipartite or tripartite association. For example, spatial affinity has been seen in fruit bodies of *Chalciporus piperatus* with *Amanita muscaria* (Tedersoo et al. 2010) and also between *Amanita muscaria* and *Boletus edulis* (Wang and Hall 2004). Truffles are the most delicious and expensive EM fungi. In order to cultivate and harvest more truffles, truffle plantations or truffle orchards are established in some regions of Europe (e.g., France and Italy) (Lefevre and Hall 2001; Miguel et al. 2014). Such orchards require desired climatic/ecological conditions, specific trees, and prevention of invasive EM fungi. These orchards also the pave way for the expansion of other EM fungi like *Boletes*, *Hebeloma*, *Laccaria*, and *Russula*. However, no data are available on the truffles in the Western Ghats. Some of the recent studies in Western Ghats region revealed that the coffee agroforests and arboretum provide a scope for diverse macrofungi (Karun and Sridhar 2014a, 2016). Development of arboretum with native/endemic/endangered tree spices will pave the way to increase the production of macrofungi including EM fungi. Establishing such demonstration plots by the institutions will serve the purpose of extension education especially educating the public and students on the values of such ecosystems.

Although there are several studies that recorded the presence of EM fungi in the Western Ghats, their hosts are not authentically identified. During our forays in the Western Ghats, *Amanita*, *Amauroderma*, *Dictyophora*, *Geastrum*, *Phallus*, *Simblum*, and *Xylaria* have been associated with many tree species. Convergent evolution might have occurred among the genera *Astraeus* and *Geastrum* (Cannon and Kirk 2007). The diversity and evolutionary lineages especially ectomycorrhizal geasters have been studied by Tedersoo et al. (2010). Our unpublished observations reveal that *Geastrum fimbriatum* has been associated with *Artocarpus heterophyllus*, *Coffea robusta,* and *Mangifera indica* in coffee plantations in the Western Ghats, which needs further assessment. *Astraeus* is ectomycorrhizal in many tree species which belongs to *Dipterocarpaceae*, *Fagaceae*, and *Pinaceae*, but with a few exceptions in *Geatrum* (e.g., *Geastrum saccatum* and *G. triplex*) (Hibbett et al. 2000; Phosri et al. 2004; Fangfuk et al. 2010; Karun and Sridhar 2014b).

Boa (2004) presented an overview of global edible mushrooms. Several EM fungi in Western Ghats serve as edible and medicinal mushrooms. Many edible mushrooms could be recognized based on the traditional knowledge on the tree species and mushrooms by the tribals (Pavithra et al. 2015; Karun and Sridhar 2017). For instance, edible EM fungi like *Boletinellus merulioides*, *Boletus edulis*, *B. reticulatus*, *Lycoperdon utriforme*, *Phlebopus marginatus*, *P. portentosus*, *Rubroboletus caespitosus*, *Scleroderma citrinum*, *Suillus brevipes*, *S. placidus*, and *S. tomentosus* were recognized by the knowledge of tribals based on host tree species. Although many species of *Amanita* are toxic, one of its unknown species in scrub jungles is edible at tender stage (spherical, beak-like, and dumbbell-shape) prior to maturity (Greeshma et al. 2018a). As the matured ones are not eaten, the doubt remains as to whether the amatoxin accumulates when the mushroom matures. Similar to the edible EM fungi, many possess medicinal value and some produce metabolites of pharmaceutical interest (e.g., *Astraeus hygrometricus*) (Pavithra et al. 2015, 2016a).

The EM fungal association with tree species is governed by several features like geographic, climatic, soil, and host tree species. Usually, the EM fungal occurrence, association, and importance depend on the age of the forest, which determines ecosystem services (Fig. 18.6). Geographic and climatic conditions influence the soil edaphic features that in turn shape the host tree species. Depending on the climatic conditions, soil features, and host trees, the EM fungi spread in an ecosystem. Once they spread in a specific habitat, they are involved in the improvement of soil fertility, prevention of soil erosion, plant growth promotion, and prevention of plant diseases. In addition, many EM fungi produce CAZymes (carbohydrate-active enzymes: lignin-, cellulose-, hemicellulose-, and pectin-degrading enzymes) and metabolites of special interest. Besides, many EM fungi are not only edible, but they also have medicinal value. In this context, it is possible to recognize some keystone species among EM fungi, which are valuable in terms of association with hosts and associated EM fungi to enhance the ecosystem services. The complex ecosystem develops within a habit by mutualistic association of trees with EM fungi ultimately leading to increase the carrying capacity of the ecosystem. Several animal populations are involved to gear up such ecosystem processes. As these links are delicate, any anthropogenic intervention leads to the catastrophic collapse of the ecosystem leading to environmental degradation.

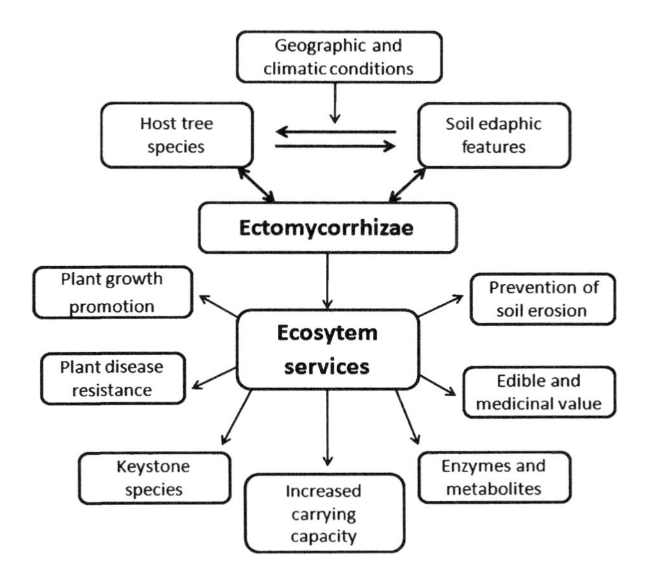

Fig. 18.6 Schematic outlook of EM fungal interactions and ecosystem services

18.4 Conclusion

The present study presents an overall view of research performed on EM fungi in the Western Ghats. It is not possible to decide the exact number of EM fungi occurring in the Western Ghats. However, it provides a scope to compare the EM fungal population of the Western Ghats with the Himalayas and Southeast Asia. According to Natarajan et al. (2005a), the generic composition of EM fungi of the Western Ghats (Nilgiri Biosphere Reserve) mirrors that of the Western Himalayas. Among the 148 species of EM fungi known from the Western Ghats, Dipterocarpaceae (8 species) offered up to 77%, while the alien tree species (15 species) offered about 20%. Several EM fungi have been listed in many taxonomic studies and checklists of the Western Ghats without much emphasis on their host trees (Natarajan et al. 2005a, b; Swapna et al. 2008; Farook et al. 2013; Senthilarasu 2014, 2015; Senthilarasu and Kumaresan 2016; Karun et al. 2014, 2016). If we add up the EM fungi reported (up to genus) in the Western Ghats without registering the host trees (up to species), then the total species will increase from 148 to ~250. Similarly, the number of species of host trees will rise from 60 to ~80 resulting in crossing the ratio of host/EM fungi from 1:2.5 to 1:3.

The quality not the quantity of the habitats of the Western Ghats appears to play a key role in the success of fungal conservation (Brown et al. 2006). According to Natarajan et al. (2005b), the EM fungi of Nilgiri Biosphere Reserve drastically differ from that of Uppangala forest (Karnataka); hence, these geographic regions could be considered as special localities (or outliers) to study and conserve the EM fungi in the Western Ghats. Such considerations will stimulate further research on the EM fungi in different forests and sylvicultures (e.g., young vs. old, monoculture vs. multiculture, artificial vs. natural, high altitude vs. low altitude). Similarly, the members of the family Dipterocarpaceae need special attention to follow the EM fungal diversity and distribution. Preliminary studies have been performed on the occurrence of EM fungi in forests of different age groups (3–17 years) by Natarajan and Senthilarasu (2004). Eucalypt plantations in Australia, Southeast Asia, India, and Portugal are other important areas to explore the diversity and allocation of EM fungi.

Comparing EM fungi of Dipterocarpaceae in the Western Ghats with those of the Himalayas and Southeast Asia is highly valuable in sylviculture. Similarly, comparison of EM fungi occurring in the lateritic scrub jungles of the Western Ghats with other regions of Indian subcontinent (Chhattisgarh, Orissa, West Bengal, Bihar, and Jharkhand) is another interesting approach. Special emphasis is necessary regarding the EM fungal association with *Acacia* and *Casuarina*, which are currently used to strengthen the coastal sand dunes of the west coast of India. Besides, there is a need to expand EM fungi with native tree species to prevent coastal erosion. Further interest on the association of EM fungi with endemic and endangered trees in the Western Ghats helps in conservation measures. In addition to EM fungi, other mutualistic associations (e.g., arbuscular mycorrhizae, nitrogen-fixing bacteria, and cyanobacteria) will strengthen the strategies necessary for ecosystem conservation. There are no information about the truffles in the Western Ghats. However, the ethnic population certainly consumes truffles thinking as tubers of plant species, further probing that such ethnic knowledge will provide a scope about the distribution and utility of truffles in the Western Ghats.

Acknowledgements The authors are grateful to Mangalore University for their support in carrying out macrofungal research in the Western Ghats in the Department of Biosciences. The award of UGC-BSR Faculty Fellowship by the University Grants Commission, New Delhi, is highly appreciated.

References

Agerer R (2006) Fungal relationships and structural identity of their ectomycorrhizae. Mycol Prog 5:67–107
Appanah S, Turnbull JM (1998) A review of dipterocarps: taxonomy, ecology and sylviculture. Centre for International Forestry Research, Indonesia
Aravindakshan D, Manimohan P (2015) Mycenas of Kerala. SporePrint Books, Calicut
Bâ AM, Diédhiou AH, Prin Y et al (2010) Management of ectomycorrhizal symbionts associated to useful exotic tree species to improve reforestation performances in tropical Africa. Ann For Sci 67:301. https://doi.org/10.1051/forest/2009108
Berbee ML, Taylor JW (2001) Fungal molecular evolution: gene trees and geologic time. Mycota 7B:229–245
Berbee ML, Taylor JW (2010) Dating the molecular clock in fungi – how close are we? Fungal Biol Rev 24:1–16
Bhagwat S, Kushalappa CG, Williams P, Brown N (2005) The role of informal protected areas in maintaining biodiversity in the Western Ghats of India. Ecol Soc 10:8. http://www.ecologyandsociety.org/vol10/iss1/art8/
Bhatt M, Mistri P, Joshi I et al (2018) Molecular survey of basidiomycetes and divergence time estimation: an Indian perspective. PLoS One 13:e0197306
Bhavanidevi S (1995) Mushroom flora of Kerala. In: Chadha KL, Sharma SR (eds) Advances in hortriculture – mushrooms, vol 13. Malhotra Publishing House, New Delhi, pp 277–316
Bhosle S, Ranadive K, Bapat G et al (2010) Taxonomy and diversity of *Ganoderma* from the Western parts of Maharashtra (India). Mycosphere 1:249–262
Boa ER (2004) Wild edible fungi: a global overview of their use and importance to people. Food and Agricultural Organization, Rome
Bonfante P, Genre A (2008) Plants and arbuscular mycorrhizal fungi: an evolutionary-developmental perspective. Tr Plant Sci 13:492–498
Borkar P, Doshi A, Navathe S (2015) Mushroom diversity of Konkan region of Maharashtra, India. J Threat Taxa 7:7625–7640
Brearley FQ (2012) Ectomycorrhizal associations of the Dipterocarpaceae. Biotropica 44:637–648
Brown N, Bhagwat S, Watkinson S (2006) Macrofungal diversity in fragmented and disturbed forests of the Western Ghats of India. J Appl Ecol 43:11–17
Cannon PF, Kirk PM (2007) Fungal families of the world. CAB International, Wallingford
Das K, Miller SL, Sharma JR, Hemenway J (2008) Two new species of Russula from Western Ghats in India. Ind J For 31:473–478
De Miguel AM, Águeda B, Sánchez S, Parladé J (2014) Ectomycorrhizal fungus diversity and community structure with natural and cultivated truffle hosts: applying lessons learned to future truffle culture. Mycorrhiza 24:5–18
Dickie IA, Nuñez MA, Pringle et al (2016) Towards management of invasive ectomycorrhizal fungi. Biol Invasions 18:3383–3395
Fangfuk W, Petchang R, To-Anun C et al (2010) Identification of Japanese Astraeus, based on morphological and phylogenetic analyses. Mycoscience 51:291–299

Farjon A (2005) Pines: drawings and descriptions of the genus *Pinus*, 2nd edn. Brill, Boston

Farook VA, Khan SS, Manimohan P (2013) A checklist of agarics (gilled mushrooms) of Kerala state, India. Mycosphere 4:97–131

Floudas D, Binder M, Riley R et al (2012) The Paleozoic origin of enzymatic lignin decomposition constructed from 31fungal genomes. Science 336:1715–1719

Ghate SD, Sridhar KR (2016a) Spatiotemporal diversity of macrofungi in the coastal sand dunes of southwestern India. Mycosphere 7:458–472

Ghate SD, Sridhar KR (2016b) Contribution to the knowledge on macrofungi in mangroves of the Southwest India. Pl Biosys 150:977–986

Ghate SD, Sridhar KR (2017) Bioactive potential of *Lentinus squarrosulus* and *Termitomyces clypeatus* from the southwestern region of India. Ind J Nat Prod Res 8:120–131

Greeshma AA, Sridhar KR, Pavithra M (2015) Macrofungi in the lateritic scrub jungles of southwestern India. J Threat Taxa 7:7812–7820

Greeshma AA, Sridhar KR, Pavithra M, Ghate SD (2016) Impact of fire on the macrofungal diversity of scrub jungles of Southwest India. Mycology 7:15–28

Greeshma AA, Sridhar KR, Pavithra M (2018a) Nutritional perspectives of an ectomycorrhizal edible mushroom *Amanita* of the southwestern India. Cur Res Environ Appl Mycol 8:54–68

Greeshma AA, Sridhar KR, Pavithra M (2018b) Functional attributes of ethnically edible ectomycorrhizal wild mushroom *Amanita* in India. Microb Biosys 3:34–44

Hawksworth DL (1991) The fungal dimension of biodiversity: magnitude, significance, and conservation. Mycol Res 95:641–655

Hawksworth DL, Lücking R (2017) Fungal diversity revisited: 2.2 to 3.8 million species. Microbiol Spectr 5:FUNK-0052-2016

Hawkswroth DL (2019) The macrofungal resource: extent, current utilization, future prospects and challenges. In: Sridhar KR, Deshmukh SK (eds) Advances of macrofungi: diversity, ecology and biotechnology. CRC Press, Boca Raton, pp 1–9

Hibbett DS, Matheny PB (2009) The relative ages of ectomycorrhizal mushrooms and their plant hosts estimated using Bayesian relaxed molecular clock analyses. BMC Biol 7:13. https://doi.org/10.1186/1741-7007-7-13

Hibbett DS, Grimaldi D, Donoghue MJ (1997) Fossil mushrooms from Miocene and cretaceous ambers and the evolution of homobasidiomycetes. Am J Bot 84:981–991

Hibbett DS, Gilbert L-B, Donoghue MJ (2000) Evolutionary instability of ectomycorrhizal symbiosis in basidiomycetes. Nature 407:506–508

Karun NC, Sridhar KR (2013) Occurrence and distribution of *Termitomyces* (Basidiomycota, Agaricales) in the Western Ghats and on the west coast of India. Czech Mycol 65:233–254

Karun NC, Sridhar KR (2014a) A preliminary study on macrofungal diversity in an arboretum and three plantations of the southwest coast of India. Cur Res Environ Appl Mycol 4:173–187

Karun NC, Sridhar KR (2014b) Geasters in the Western Ghats and west coast of India. Acta Mycol 49:207–219

Karun NC, Sridhar KR (2015a) Elephant dung-inhabiting macrofungi in the Western Ghats. Cur Res Environ Appl Mycol 5:60–69

Karun NC, Sridhar KR (2015b) *Xylaria* complex in the South Western India. Pl Pathol Q 5:83–96

Karun NC, Sridhar KR (2016) Spatial and temporal diversity of macrofungi in the Western Ghat forests of India. Appl Ecol Environ Res 14:1–21

Karun NC, Sridhar KR (2017) Edible wild mushrooms in the Western Ghats: data on the ethnic knowledge. Data Brief 14:320–328

Karun NC, Sridhar KR, Appaiah KAA (2014) Diversity and distribution of macrofungi in Kodagu region (Western Ghats) – a preliminary account. In: Pullaiah T, Karuppusamy S, Rani SS (eds) Biodiversity in India, vol 7. Regency Publications, New Delhi, pp 73–96

Karun NC, Sridhar KR, Niveditha VR, Ghate SD (2016) Bioactive potential of two wild edible mushrooms of the Western Ghats of India. In: Watson RR, Preedy VR (eds) Fruits, vegetables, and herbs: bioactive foods in health promotion. Elsevier, Oxford, pp 344–362

Karun NC, Sridhar KR, Ambarish CN, Pavithra M, Greeshma AA, Ghate SD (2017) Health perspectives of medicinal macrofungi of Southwestern India. In: Watson RR, Zibadi S (eds) Handbook of nutrition in heart health. Wageningen Academic Publishers, Netherlands, pp 533–548
Karun NC, Bhagya B, Sridhar KR (2018a) Biodiversity of macrofungi in Yenepoya campus, Southwest India. Microb Biosyst 3:1–11
Karun NC, Sridhar KR, Ambarish CN (2018b) Nutritional potential of *Auricularia auricula-judae* and *Termitomyces umkowaan* – the wild edible mushrooms of southwestern India. In: Gupta VK, Treichel H, Shapaval V et al (eds) Microbial functional foods and nutraceuticals. Wiley, Hoboken, pp 281–301
Kennedy P (2010) Ectomycorrhizal fungi and interspecific competition: species interactions, community structure, coexistence mechanisms and future directions. New Phytol 187:895–910
Kirk P, Cannon P, Minter D, Stalpers J (2008) Dictionary of the Fungi, 10th edn. CABI, Wallingford
Kumar J, Atri NS (2018) Studies on ectomycorrhiza: an appraisal. Bot Rev 84:108–155
Latha KPD, Manimohan P (2015) *Inocybe griseorubida*, a new species of *Pseudosperma* clade from tropical India. Phytotaxa 221:166–174
Latha KPD, Manimohan P (2017) Inocybes of Kerala. SporePrint Books, Calicut
Lee SS (1998) Root symbiosis and nutrition. In: Appanah S, Turnbull JM (eds) A review of dipterocarps taxonomy, ecology and Silviculture. Centre for International Forestry Research, Bogor, pp 99–114
Leelavathy KM, Ganesh PN (2000) Polypores of Kerala. Daya Publishing House, New Delhi
Leelavathy KM, Manimohan P, Arnolds EJM (2006) *Hygrocybe* in Kerala state, India. Persoonia 19:101–151
Lefevre C, Hall IR (2001) The global status of truffle cultivation. In: Mehlenbacher SA (ed) Fifth international congress on hazelnut, Corvallis. Acta Hort 556:513–520
LePage BA (2003) The evolution, biogeography and palaeoecology of the Pinaceae based on fossil and extant representatives. Acta Hortic 615:29–52
LePage BA, Currah RS, Stockey RA, Rothwell GW (1997) Fossil ectomycorrhizae from the middle Eocene. Am J Bot 84:410–412
Manimohan P, Latha KPD (2011) Observations on two rarely collected species of Russula. Mycotaxon 116:125–131
Manimohan P, Thomas KA, Nisha VS (2007) Agarics on elephant dung in Kerala state, India. Mycotaxon 99:147–157
Manoharachary C, Sridhar KR, Singh R et al (2005) Fungal biodiversity: distribution, conservation and prospecting of fungi from India. Cur Sci 89:58–71
Marcel G, van der Heijden A, Martin FM et al (2015) Mycorrhizal ecology and evolution: the past, the present, and the future. New Phytol 205:1406–1423
Miguel AMD, Águeda B, Sánchez S, Parladé J (2014) Ectomycorrhizal fungus diversity and community structure with natural and cultivated truffle hosts: applying lessons learned to future truffle culture. Mycorrhiza 24:5–18
Mohan V (2008) Diversity of ectomycorrhizal fungal flora in the Nilgiri Biosphere Reserve (NBR) area, Nilgiri Hills, Tamil Nadu. ENVIS Cent Newsl 6:1–6
Mohanan C (2011) Macrofungi of Kerala. Kerala Forest Research Institute, Peechi
Mohanan C (2014) Macrofungal diversity in the Western Ghats, Kerala, India; members of Russulaceae. J Threat Taxa 6:5636–5648
Mueller GM, Schmit JP, Leacock PR et al (2007) Global diversity and distribution of macrofungi. Biodivers Conserv 16:37–48
Murata M, Kanetani S, Nara K (2017) Ectomycorrhizal fungal communities in endangered *Pinus amamiana* forests. PLoS One 12:e0189957
Myers N, Mittermeier RA, Mittermeier CG et al (2000) Biodiversity hotspots for conservation priorities. Nature 403:853–858
Natarajan K (1995) Mushroom flora of South India (except Kerala). In: Chadha KL, Sharma SR (eds) Advances in horticulture. Malhotra Publishing House, New Delhi, pp 381–397
Natarajan K, Raman N (1983) South Indian Agaricales 20 – some mycorrhizal species. Kavaka 11:59–66

Natarajan K, Senthilarasu G (2004) Diversity of ectomycorrhizal fungi in Western Ghats of South India. In: Reddy MS, Khanna S (eds) Biotechnological approaches for sustainable development. Allied Publishers Pvt Ltd, New Delhi, pp 65–67

Natarajan K, Narayanan K, Ravindran C, Kumaresan V (2005a) Biodiversity of agarics from Nilgiri biosphere reserve, Western Ghats, India. Cur Sci 88:1890–1893

Natarajan K, Senthilarasu G, Kumaresan V, Riviere T (2005b) Diversity in ectomycorrhizal fungi of a dipterocarp forest in Western Ghats. Cur Sci 88:1893–1895

Pavithra M, Greeshma AA, Karun NC, Sridhar KR (2015) Observations on the *Astraeus* spp. of Southwestern India. Mycosphere 6:421–432

Pavithra M, Sridhar KR, Greeshma AA, Tomita-Yokotani K (2016a) Bioactive potential of the wild mushroom *Astraeus hygrometricus* in the Southwest India. Mycology 7:191–202

Pavithra M, Sridhar KR, Greeshma AA, Karun NC (2016b) Spatial and temporal heterogeneity of macrofungi in the protected forests of Southwestern India. Int J Agric Technol 12:105–124

Pavithra M, Sridhar KR, Greeshma AA (2017a) Macrofungi in botanical gardens of the southwestern India. J Threat Taxa 9:9962–9970

Pavithra M, Sridhar KR, Greeshma AA (2017b) Functional properties of edible mushroom *Astraeus hygrometricus*. Kavaka 49:22–27

Peintner U, Moser MM, Thomas KA, Manimohan P (2003) First records of ectomycorrhizal *Cortinarius* species (Agaricales, Basidiomycetes) from tropical India and their phylogenetic position based on rDNA ITS sequences. Mycol Res 107:485–494

Phosri C, Watling R, Martín MP, Whalley AJS (2004) The genus Astraeus in Thailand. Mycotaxon 89:453–463

Pradeep CK, Vrinda KB (2010) Ectomycorrhizal fungal diversity in three different forest types and their association with endemic, indigenous and exotic species in the Western Ghat forests of Thiruvananthapuram District, Kerala. J Mycopathol Res 48:279–289

Pradeep CK, Vrinda KB, Varghese SP, Kumar TKA (2015) A new species of Phylloporus (Agaricales, Boletaeae) from India. Phytotaxa 226:269–274

Ranadive KR, Vaidya JG, Jite PK et al (2011) Checklist of Aphyllophorales from the Western Ghats of Maharashtra state, India. Mycosphere 2:91–114

Reddy MS, Singla S, Natarajan K, Senthilarasu G (2005) *Pisolithus indicus*, a new species of ectomycorrhizal fungus associated with dipterocarps in India. Mycologia 97:838–843

Rinaldi AC, Comandini O, Kuyper TW (2008) Ectomycorrhizal fungal diversity: separating the wheat from the chaff. Fungal Ecol 33:1–45

Riviere TR, Diedhiou AG, Diabate M et al (2007) Genetic diversity of ectomycorrhizal basidiomycetes from African and Indian tropical rain forests. Mycorrhiza 17:145–248

Sathe AV, Daniel J (1980) Agaricales (mushrooms) of Kerala State. In: Sathe AV (ed) Agaricales (mushrooms) of South West India, Monograph # 1, Part # 3. Maharashtra Association of Cultivation of Science, Pune, pp 75–108

Sathe AV, Deshpande SD (1980) *Agaricales* (mushrooms) of Maharashtra State. In: Sathe AV (ed) Agaricales (mushrooms) of South West India, Monograph # 1, Part # 1. Maharashtra Association of Cultivation of Science, Pune, pp 1–66

Senthilarasu G (2014) Diversity of agarics (gilled mushrooms) of Maharashtra, India. Cur Res Environ Appl Mycol 4:58–78

Senthilarasu G (2015) The lentinoid fungi (*Lentinus* and *Panus*) from Western Ghats, India. IMA Fungus 6:119–128

Senthilarasu G, Kumaresan V (2016) Diversity of agaric mycota of Western Ghats of Karnataka, India. Curr Res Environ Appl Mycol 6(1):75–101

Sharma R (2009) Ectomycorrhizal mushrooms in Indian tropical forests. Biodiversity 10:25–30

Sharma R (2017) Ectomycorrhizal mushrooms: their diversity, ecology and practical applications. In: Verma A, Prasad R, Tuteja N (ed) Mycorrhiza – function, diversity, state of the art. Springer International Publishing, pp 99–131

Smith SE, Read DJ (2008) Mycorrhizal symbiosis, 3rd edn. Academic, London

Sudheep NM, Sridhar KR (2014) Nutritional composition of two wild mushrooms consumed by tribals of the Western Ghats of India. Mycology 5:64–72

Swapna S, Abrar S, Krishnappa M (2008) Diversity of macrofungi in semi-evergreen and moist deciduous forest of Shimoga District, Karnataka, India. J Mycol Pl Pathol 38:21–26
Tedersoo L, May TW, Smith ME (2010) Ectomycorrhizal lifestyle in fungi: global diversity, distribution, and evolution of phylogenetic lineages. Mycorrhiza 20:217–263
Tedersoo L, Bahram M, Toots M et al (2012) Towards global patterns in the diversity and community structure of ectomycorrhizal fungi. Mol Ecol 21:4160–4170
Thiribhuvanamala G, Prakasam V, Chandrasekar G et al (2011) Biodiversity, conservation and utilisation of mushroom flora from the Western Ghats region of India. In: Savoie J-M, Foulongne-Oriol M, Largeteau M, Barroso G (eds) Mushroom biology and mushroom products, vol 1. INRA, France, pp 155–164
Thomas KA, Peintner U, Moser MM, Manimohan P (2002) *Anamika*, a new mycorrhizal genus of Cortinariaceae from India and its phylogenetic position based on ITS and LSU sequences. Mycol Res 106:245–251
Vasco-Palacios AM, Hernandez J, Peñuela_Mora MC et al (2018) Ectomycorrhizal fungi diversity in a white sand forest in western Amazonia. Fungal Ecol 31:9–18
Vrinda KB, Pradeep CK, Abraham TK (1997) A new species of *Russula* from Kerala, India. Mycotaxon 52:389–393
Wang Y, Hall IR (2004) Edible ectomycorrhizal mushrooms: challenges and achievements. Can J Bot 82:1063–1073
Wolfe BE, Pringle A (2012) Geographically structured host specificity is caused by the range expansions and host shifts of a symbiotic fungus. ISME J 6:745–755

Halotolerant PGPR Bacteria: Amelioration for Salinity Stress

19

Brijendra Kumar Kashyap, Roshan Ara, Akanksha Singh, Megha Kastwar, Sabiha Aaysha, Jose Mathew, and Manoj Kumar Solanki

19.1 Introduction

Reading the chapter's title, the first question that pops up in our mind is what is stress and how does it develop in plants? So, to complement this query, the stress in plants may be defined as "Any external factor that negatively influences plant growth, productivity, reproductive capacity or survival". The stress can be broadly divided into two main categories: biotic (or biological stress) and abiotic (or environmental stress) factors. The biotic factors include pathogenic infections or other biological factors, and the abiotic factors include extremes of either high or low temperature, flooding and drought, deficiencies or excess of any of the micronutrients and macronutrients, extreme of soil pH, and high salinity. Among all the stresses, the decrease in yield of crop caused by soil salinity is not the only major problem of the world but also of developing countries like India which have a high growth rate of population in the range of 1.19%/annum (Etesami and Beattie 2018). This problem becomes more aggravated with continuously increasing population like India which has limited land area (seventh position in the world) among which most of these regions are jeopardized and competed for housing and industrial sectors. The entire land area on the Earth is 29.1% of which only 10.43% is arable land; the major land portion (20% of total cultivated land and 33% of agricultural land with irrigation) (Shrivastava and Kumar 2015) is oppressed by salinity, and this salinity stress makes plants water deficit resulting in curtailment of photosynthetic rate, nutrient and water deficiency, and mortification of plant cell. The salinity of a stressed soil means high salt concentration or sodium ions which destroy the soil structure and attracts water toward the soil after blocking its absorption through

B. K. Kashyap (✉) · R. Ara · A. Singh · M. Kastwar · S. Aaysha · J. Mathew
Department of Biotechnology, Bundelkhand University, Jhansi, Uttar Pradesh, India

M. K. Solanki
Department of Food Quality & Safety, Institute for Post-harvest and Food Sciences, The Volcani Center, Agricultural Research Organization, Rishon LeZion, Israel

D. P. Singh et al. (eds.), *Microbial Interventions in Agriculture and Environment*,
https://doi.org/10.1007/978-981-13-8391-5_19

plant roots, creating an environment of droughtiness even when the soil has abundant water. High soil salinity burns the leaves and stems and also shows its cruelty to earthworms and microorganisms (Waskom et al. 2012).

There are the bunch of problems that arise in the salt-affected soil including the destruction of healthy soil and microbial inhabitant. Therefore, to overcome these hurdles, halotolerant PGPR (plant growth-promoting rhizobacteria) (Etesami and Beattie 2018) are used to play a remedial role to wash out the salinity afflicts (Egamberdieva et al. 2017). The bioinoculant species may include various species of *Azospirillum*, *Agrobacterium*, *Pseudomonas*, and *Bacillus* (Wu et al. 2005). The bioinoculation of halotolerant PGPR approach is environment-friendly and economically suited for reclaiming salinity-affected land to attain maximal biomass production. The halotolerant bacteria can be inoculated to the rhizospheric zone of the various crops as these PGPR involve in nourishing plants through the various activities which may involve P solubilization, nitrogen fixation, siderophore reduction, phytohormones production, etc., thus, increasing the soil health followed by plant health and its yield.

19.2 Salinity Scales

The meaning of soil salinity is soil affected with saline condition. When the extract solution of soil has an electrical conductivity (EC) of 20 mM or more, the soil is considered as saline soil. As the soil EC increases (i.e., $EC_{1.5} \geq 50$ dS/m), there is a decrement in soil respiration leading to microbial shift, thus affecting soil health. But when salinity of soil goes beyond 4 dS/m (equivalent to 40 mM NaCl), a disrupted root growth of various crop had been observed (Gilliham 2015). Approximately, 7.0 million hectares land is covered by saline soil in India (Shrivastava and Kumar 2015). Salt-affected soil can be explained in two ways, i.e., salinity and sodicity. Salinity refers to salt concentration while sodicity is the salt composition. Broadly speaking, salinity means the salt concentration in the irrigation water or soil that adversely affects yield of crop and crop quality while sodicity refers to the sodium ion proportion in water that adsorbs to the soil surface, relative to Ca^{2+} and Mg^{2+}. Sodicity of soil may also be characterized by the exchangeable sodium percentage (ESP), and the sodicity of water is a measure of the sodium absorption ratio (SAR). SAR affects the physical properties of soil like soil texture, mineralogy, organic matter, pH, soil wetting, etc. (Läuchli and Grattan 2011; Shainberg et al. 2001).

Based on the aforementioned parameters, i.e., pH, ESP, and EC, soil can be further categorized into three types: (1) saline soil, (2) saline-sodic soil, and (3) sodic soil. Saline soils are those soils having EC in range of 6–16 ds/m and the pH value in between 1and 5, while sodic soils have their EC 0–4 ds/m, pH>8.5, and ESP 5–30. Saline-sodic soils are those soils having EC 4–6 ds/m and pH 6.5–8.50. The soil which has the pH value 5–6.50, considered as healthy soil, is described in Fig. 19.1(Yan et al. 2015).

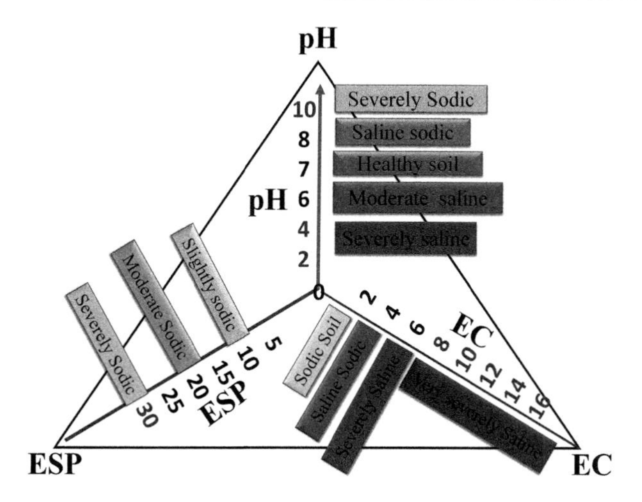

Fig. 19.1 Classification of saline and sodic soils based on different parameters (EC, electrical conductivity (dS/m)) on X-axis; pH on Y-axis; ESP, exchangeable sodium percentage) (Läuchli and Grattan 2011)

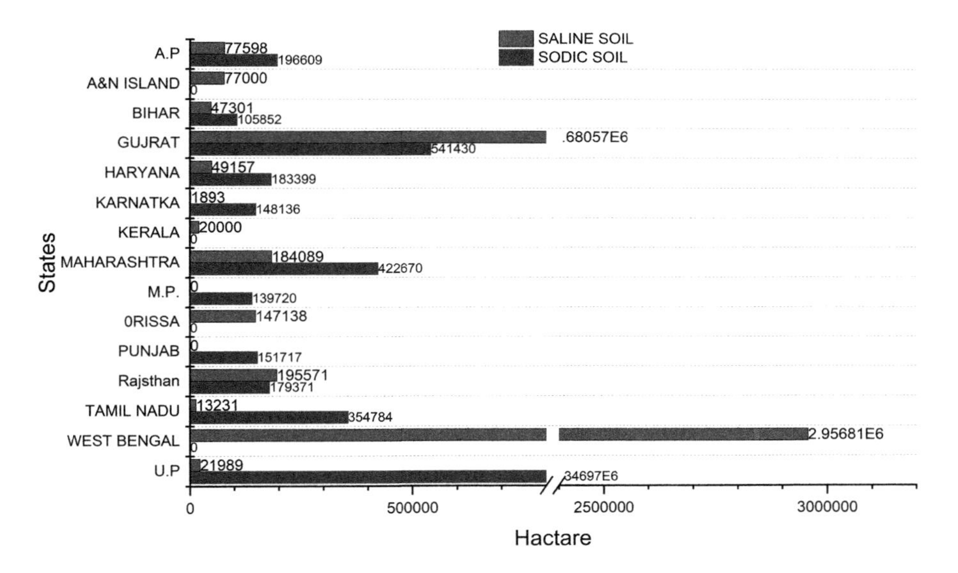

Fig. 19.2 Distribution of sodic and saline soils in India

19.2.1 Distribution of Salt-Affected Region in India

The report suggests that approximately 20% of the worldwide irrigated lands and 33% of cultivated lands are severely affected by salt (Nasher Mohamed et al. 2010). Figure 19.2 demonstrated the state-wise distribution of soil affected with salt in

India. The vast salt-affected area of soil occurs in Gujarat, preceded by Uttar Pradesh (UP) and Maharashtra which account for about 62.4%. 2.1% geographical area of soil in India is salt-affected. Out of 6.727 Mha of salt-affected soil, 2.956 Mha are saline and the rest 3.771 Mha are sodic.

19.2.2 Effects of Salinity

Salinity affects physiochemical, morphological, and biochemical processes of soil that includes germination, plant growth, and uptake of water and nutrient (Shrivastava and Kumar 2015). Increased soil salinity exposes the plant to the ionic form of sodium (Na^+) and chloride (Cl^-). Sodium ion can interact directly with components of the cell wall and can also modify their chemical properties. When the salt concentration increases up to a certain level, Na^+ starts accumulating in the apoplast leading to enhanced interaction between Na^+ and negatively (−vely) charged sites inside the cell wall polymers causing transient alkalinization in the apoplast which limit the growth of plant. Accumulation of Na^+ in the plant's tissue inhibits photosynthesis and has a drastic effect on electron transport chain (ETC) like deregulation, overflow, and even disruption of the ETC of chloroplast and mitochondria causing molecular oxygen as an electron acceptor leading to accumulation of reactive oxygen species (ROS) (Numan 2018).These radicals like OH^- (hydroxyl radical), a single oxygen, hydrogen peroxide, and superoxide are potentially harmful for the cell integrity as they are more potent oxidizing compounds, and these ROS adversely affect the plant, viz., by speeding up the toxic reaction, like mutation of DNA, degradation of protein, and membrane damage, causing programmed cell death (Filomeniet al. 2015; Zushi et al. 2009). When salinity increases, the ratio of Na^+ over Ca^{2+} becomes high causing replacement of Ca^{2+} from the binding site with Na^+, thus reducing pectin crosslinking, which results in slowing down of cell elongation (Proseus and Boyer 2012).

The changes in salinity of soil may cause modification in root's cell wall which effects transport of ions and water as reported in barley plants that salinity causes enhanced production of a different type of glucanase. This enzyme is involved in degrading callose, which somehow affects opening and closing of stomata, cell to cell communication and movement of nutrients (Byrt et al. 2018). Increased concentration of soluble salts affects plant in two ways, i.e., osmotic effect and specific ion effect (Fig. 19.3) (Zhang et al. 2008). During the starting/initial stage of salinity exposure to plant, the plant faces stress of water causing reduced leaf expansion, while long-term salinity exposure results in ionic stress causing premature senescence of adult leaves (Sultana et al. 1999). Soluble salt increases the osmotic potential (negative) causing withdrawal of water out of cells (i.e., plasmolysis) resulting in the death of beneficial soil microbes and plant roots. More salt concentration may reduce microbial activity, microbial biomass, and may cause microbial shifting (Yan et al. 2015). High salinity may also affect activities of soil enzyme like urease, alkaline phosphatase, and beta-glucosidase which get inhibited strongly by salinity. The phospholipid level in plant tissue is also damaged by NaCl salinity which is crucial as tolerance depends on the phospholipid level mainly on phosphatidyl choline (Valettiet al. 2018).

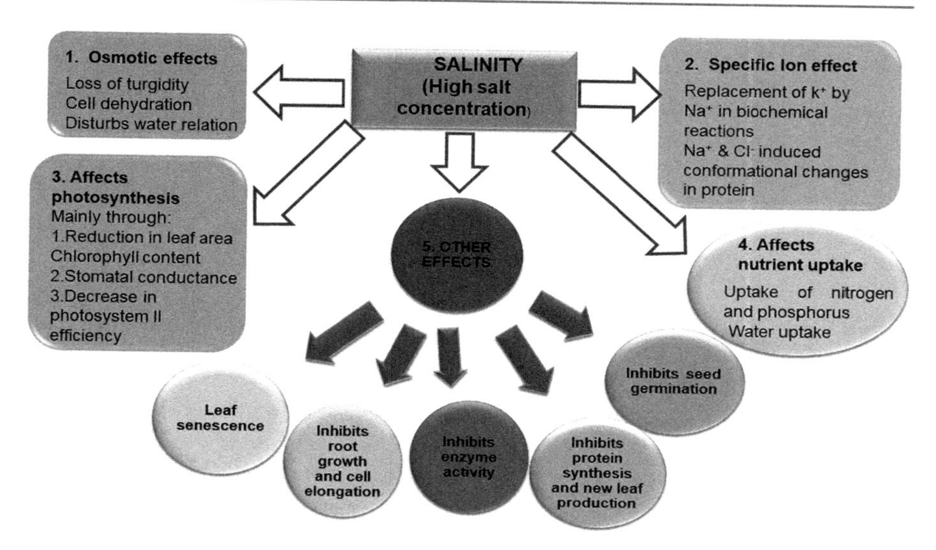

Fig. 19.3 Effects of salinity on plants

19.2.3 Response of Plant Under Saline Stress

Plants develop the certain mechanism to respond to various stresses including saline stress. These mechanisms are known as tolerance mechanisms, i.e., the plants' ability to grow in saline conditions without any adverse effect, involve production of various types of antioxidants to overcome the oxidative stress due to saline condition. These antioxidants may involve enzymatic and non-enzymatic antioxidants like peroxidase (glutathione peroxidase and ascorbate peroxidase), proline, catalase, superoxide dismutase, and glutathione reductase. These enzymes help in removal of overproduced ROS by scavenging (Asada 1999). On the other hand, those plants which can't tolerate salt stress, transportation of salt ions carried to the vacuole or transported to the older tissue and finally scarified with saline stress (Zhu 2003). Another important tolerance mechanism is salt overly sensitive stress signaling pathway, which consists of three proteins (SOS1, SOS2, SOS3), and these proteins are also regulate the transfer of membrane vesicles, pH homeostasis, and vacuole functions (Numan 2018).

19.3 Halotolerant Plant Growth-Promoting Rhizobacteria (PGPR)

The majority of PGPR colonizes around the plant root surface and thrives in the space between root hairs and rhizodermal layers. Among these PGPR some are even capable of surviving in high salt concentration called halotolerant. These halotolerant PGPR can be used as bioinoculant to recover the soil nutrients and may be promising agent for improving fertility of soil (Ahemad and Kibret 2014). These PGPR not only quash many plant pathogens but also produce various compounds including growth regulators, siderophores, and organic acids, fix atmospheric nitrogen, solubilize phosphorus, and produce antibiotics and lytic enzymes. They promote growth of plant by means of production of phytohormones, decomposition of organic matter, and improvement of the bioavailability of various mineral nutrients including iron and phosphorus (López-Bucio et al. 2007).

19.3.1 Role of Halotolerant PGPR

There are two mechanisms on which PGPR work (Kumar et al. 2018) – *direct mechanism*, (i) production of phytohormones, (ii) phosphate solubilization, and (iii) biological nitrogen fixation (Siddikee et al. 2010), and *indirect mechanism*, (i) induced systemic resistance and (ii) produc of antibiotic, siderophores, and lytic enzymes (Vacheron et al. 2013). PGPR induces systemic tolerance that further induces physical and chemical changes, which results in enhanced tolerance to abiotic stress. These facilitate indirect plant growth by reducing plant pathogen and enhancing the plant's innate immunity (Tabassum et al. 2017).

19.3.1.1 Production of Phytohormones

PGPR promotes growth of plant through production of various phytohormones like auxin (involved in cell elongation, division, and differentiation), gibberellin (involved in seed germination, flowering process, elongation of stem, and fruit setting), and cytokinin (involved in formation of shoot, development of root, and improved cell division) (Glick 2014).

19.3.1.2 Production of Exopolysaccharide (EPS)

Exopolysaccharides are polymeric metabolites secreted by bacteria, fungi, and microalgae. These EPS help the individual to attach themselves along with other bacteria to soil particles and root surfaces. They also help in stabilizing the soil structures and thus, enhance the water holding capacity of the soil (Ilangumaran and Smith 2017). These exopolysaccharides help in the synthesis of biofilm, where they get protection from environment anomalies and protect cell against toxic substances; EPS also serves as a carbon energy source. EPS also plays a crucial role in metal complexation and therefore reduces their bioaccessibility and bioavailability by filtration of heavy metals.

19.3.1.3 Production of Biosurfactant

Biosurfactant, a surface active agent that contains hydrophilic and hydrophobic groups, has application in metal reduction. In soil, biosurfactant agent weakens the strong bonds between metal and soil leading to acceleration in desorption of heavy metals from solid phases (Numan 2018).

19.3.1.4 Production of ACC (1-Aminocyclopropane-1-Carboxylate) Deaminase

The ethylene phytohormone production in plants is dependent on the endogenous level of ACC (1-aminocyclopropane-1-carboxylate). The enzyme ACC deaminase is present in many halotolerant rhizospheric bacteria. These bacteria can utilize ACC produced from roots of plants and convert it into ketobutyrate and ammonia causing reduction in ACC levels casing reduction in ethylene levels in the plants. This reduction in ethylene level causes alleviation in plant stress level (Egamberdieva and Lugtenberg 2014).

19.3.1.5 Phosphate Solubilization

Some halotolerant PGPR have been assayed as biofertilizers as they are capable of providing inorganic nutrients to plants. Phosphorus is one of the major and essential macronutrients of the plant. In soil, this element occurs in its organic and mineral forms and is absorbed by plants as phosphates. But the large part of this phosphate is immobilized and becomes unavailable for plants, even when its concentration is high. Therefore, phosphate solubilizing bacteria (PSB) play an important role in the transformation of phosphorus in the soil via solubilization of phosphate. Inoculation with halotolerant PSB results in higher crop yield, i.e., it can increase phosphorus availability by 15-folds in saline soil (Valetti et al. 2018).

19.3.1.6 Nitrogen Fixation

Many free-living halotolerant rhizobacteria like *Rhizobia*, *Azotobacter*, and *Azospirillum* spp. have the ability to fix nitrogen from environment which can be easily assimilated by plants (Bashan and Levanony 1990). These halotolerant bacteria also improve the fertility power of the saline soil by replenishing with nitrogen to saline soil.

19.3.1.7 Systemic Resistance Induction

Plants inoculated with PGPR induce systemic resistance which means induction of physical and chemical changes in plants that results in augmented plant tolerance to non-biological stresses (Yang et al. 2009). The possible causal determinants of induced systemic resistance (ISR) was found to have relations with certain structural components of bacteria such as flagella, lipopolysaccharides (LPS), siderophore, and antibiotic production (García-Gutiérrez et al. 2012; Pieterse et al. 2014; Hafeez et al. 2015). Apart from these, a few other biochemicals like N-alkylated benzylamine derivative produced by *P. putida*, dimethyl disulphide produced by *B. cereus* C1L (Meldau et al. 2013), DAPG (2,4-diacetylphloroglucinol) produced by *P. fluorescens* CHA0 (Hernández-León and Rojas-Solis 2015), volatile organic compounds (VOCs) produced by *B. amyloliquefaciens* and *B. subtilis* (Pérez-García et al. 2011; Yuan et al. 2012) are also responsible for ISR. This induced systemic tolerance had been reported for *Arabidopsis thaliana* against inoculated PGPR (Etesami and Maheshwari 2018) (Fig. 19.4).

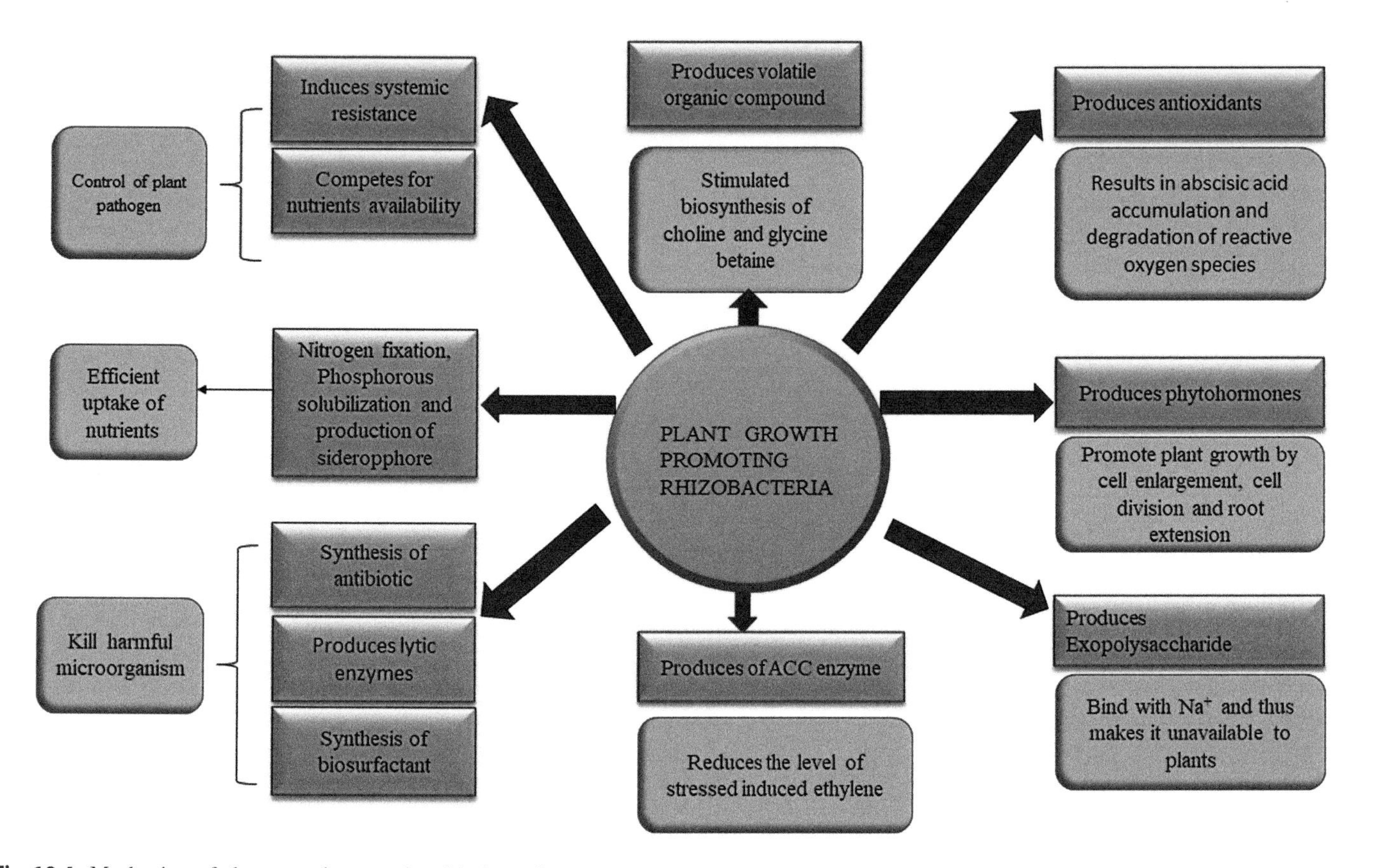

Fig. 19.4 Mechanism of plant growth-promoting rhizobacteria

19.4 Amelioration of Saline Stressed Crop Using Halotolerant PGPR

There are various crops grown worldwide in saline soil (Table 19.1), and their yield can be improvised by co-inoculating various spp. of halotolerant PGPR as mentioned below:

19.4.1 Tomato (*Lycopersicon esculentum*)

Mayak et al. (2004) isolated seven strains of halotolerant PGPR that have 1-aminocyclopropane-1-carboxylate (ACC) deaminase activity from rhizospheric soil sample of *Lycium shawii* plants grown in dry river-beds in Arava region of southern Israel where annual rainfall is below 50 mm. After 7 weeks of seedling growth on 43 mM NaCl, the halotolerant bacterium which promoted maximum growth of plant was genetically identified with 16S rDNA gene and was found to be *Achromobacter piechaudii* ARV8. Inoculating this *A. piechaudii* ARV8 halotolerant PGPR to tomato root plant with supplementation of 172 mM NaCl salt in plastic pot leads to increase in the dry and fresh weight of tomato seedling significantly. The increase in weight of seedling is due to reduction in ethylene production, increased uptake of phosphorus and potassium, and water utilization efficiency through bacterium (Mayak et al. 2004). Egamberdieva et al. (2017) isolated five halotolerant bacterial strains from rhizospheric region of wheat grown under saline soil from the National University, Uzbekistan. Among these five halotolerant PGPR, only two, i.e., *Pseudomonas chlororaphis* TSAU13 and *Pseudomonas extremorientalis* TSAU20, exhibited PGPR activity represented by increase in yield of plant height and fruit and acted as biocontrol agent against foot and root rot disease of tomato even in saline condition in pot experiment.

19.4.2 Cotton (*Gossypium hirsutum*)

Yao et al. (2010) isolated *Pseudomonas putida* Rs198 from the alkaline soil region of Xinjiang province, northwest of China. Inoculation with *P. putida* Rs198 in cotton seed was performed in pot experiment showing enhancement in rate of germination and healthy stand of *G. hirsutum*. Further, there was more than 10% improvement in height and weight (fresh and dry) of cotton seedling as compared to the control. In the field also, application of halotolerant Rs-198 to cotton seeds exhibited similar type of results as it was observed for pot experiment with slight variation. Further, under stress condition, Rs-198 increase the absorption of Mg^{2+}, K^{+}, and Ca^{2+} (Shahzad et al. 2010), enhance the endogenous level of IAA production, and reduce the abscisic acid (ABA) content of cotton seedling (Yao et al. 2010).

Wu et al. (2012) also screened out a halotolerant PGPR *Raoultella planticola* strain Rs-2 with 1-aminocyclopropane-1-carboxylate (ACC) deaminase activity (Jalili et al. 2009; Penrose and Glick 2003) from the saline soil of cotton

Table 19.1 Ameliorative effects of halotolerant PGPR on saline stressed crops

Halotolerant bacteria	Halotolerant PGPR			Impact on plants	References
	Isolation from plants/region	Growth attributes	Inoculation in plants		
Achromobacter piechaudii ARV8	Rhizosphere of *Lycium shawii* Arava, Israel	ACC deaminase activity	Tomato (*Lycopersicon esculentum*)	Increase in fresh and dry weight of tomato	Mayak et al. (2004)
Pseudomonas putida Rs 198	Xinjiang province, China	Improve the production of IAA	Cotton (*Gossypium hirsutum*)	Increased germination rate, healthy stand, growth parameters	Yao et al. (2010)
Dietzia natronolimnaea STR1	CSIR-CIMAP, Lucknow, India	Modulation of ABA signaling cascade, ion transporters and antioxidant machinery	Wheat (*Triticum asetivum*)	Higher biomass, shoot and root elongation	Bharti et al. (2016)
B. amyloliquefaciens SQR9	Rhizosphere soil of cucumber CGMCC, China	Phytohormone production, upregulation of RBCS, RBCL, HKT1, NHX1, 2, 3	Maize (*Zea mays*)	Promote maize seedling growth, enhance chlorophyll content	Chen et al. (2016)
Bacillus megaterium	Rhizosphere and plant root (Spain)	Increase proline and IAA content	Maize (*Zea mays*)	Root growth, necrotic leaf area, leaf relative water content	Marulanda et al. (2010)
Pseudomonas pseudoalcaligenes and *Bacillus pumilus*	Rhizosphere soil and roots of paddy	Induction of osmoprotectants and antioxidant proteins	Salt sensitive rice GJ-17 (*Oryza sativa*)	Enhance plant growth	Jha and Subramanian (2014)
P. extremorientalis TSAU20	Rhizosphere of wheat grown in salinated soil, National University of Uzbekistan	Enhanced proline concentration, enhanced antioxidant enzyme activities	Tomato (*Lycopersicon esculentum*)	Stimulate plant growth and biological controls of root rot disease of tomatoes	Egamberdieva et al. (2017)

Pseudomonas simiae AU	Noida, India	Production of IAA, siderophore, ACC deaminase, and phosphate solubilization	Soybean (*Glycine max*)	Promote seedling growth, increase chlorophyll content	Vaishnav et al. (2015)
Bacillus amyloliquefaciens NBRISN13	CSIR-National Botanical Research Institute, Lucknow, India	ACC deaminase, proline accumulation and enrichment of osmoprotectants	Rice (*Oryza sativa*)	Stimulation of root growth and effective root area for enhanced water and nutrient uptake	Nautiyal et al. (2013)
Azospirillum spp. *and Pseudomonas* spp.	Iran	Production of siderophores, ACC deaminase and antioxidant enzyme	*Canola* (*Brassica napus*)	Increase in plant growth, biomass and microelement uptake	Baniaghil et al. (2013)
Burkholdera cepcia SE4, *Acinetobacter calcoaceticus* SE370 *and Promicromonospora* spp.	Korea	Endogenous hormonal regulation (ABA, GA_4, SA)	Cucumber (*Cucumis sativas*)	Increased shoot and root growth, chlorophyll content stomatal closure to minimize water loss	Kang et al. (2014)
Halomonas desiderata, Exiguobacterium oxidotolerans and B. pumilus	Rhizospere of Poaceae family from Rae Bareilly, Uttar Pradesh, India	Production of siderophore and copious amounts of exopolysaccharides	Mint (*Mentha arvensis*)	Increased oil content and yield, fresh shoot weight and leaf-stem ratio	Bharti et al. (2014)
Enterobacter spp. UPMR18	University Putra, Malaysia	Increased antioxidant enzyme activities and upregulation of ROS pathway genes	Okra (*Abelmoschus esculentus*)	Enhance seed germination and growth of okra seedling under salinity stress	Habib et al. (2016)

(continued)

Table 19.1 (continued)

Halotolerant bacteria	Halotolerant PGPR			Impact on plants	References
	Isolation from plants/region	Growth attributes	Inoculation in plants		
Raoultella planticola RS-2	Salinized soil of cotton rhizosphere, Xinjiang province, China	Increase IAA, ACC deaminase activity	Cotton (*Gossypium hirsutum*)	Improve germination rate, average dry weight, fresh weight, and plant height	Wu et al. (2012)
Rhizobium and *Pseudomonas*	Chickpea, NARC Islamabad	Increased K uptake and proline accumulation	Maize (*Zea mays*)	Increased chlorophyll and carotenoid content, greater stem diameter	Bano and Fatima (2009)
Pseudomonas fluorescens	Rhizospheric soil of Tamil Nadu Agricultural University, Coimbatore, India	Enhanced ACC deaminase activity	Groundnut (*Arachis hypogea*)	Improvement in plant growth parameters and groundnut yield	Saravanakumar and Samiyappan (2007)

rhizospheric region of 5–25 cm depth in the Xinjiang province, China. After bio-inoculation with Rs-2 to cotton seeds, the rate of germination was increased by 29.5% in pot experiment compared with those in control saline treatment ($P < 0.05$). Further result showed that the quantities of phytohormone, i.e., ethylene and abscisic acid (ABA), get reduced, while the indole acetic acid (IAA) content gets increased in cotton seedlings under salinity stress with treated plant. In Rs-2-saline-treated cotton plant, uptake and accumulation of N, P, K^+, Ca^{2+}, and Fe^{2+} was increased significantly while that of Na^+ uptake gets decreased in cotton seedlings. This study suggests that *R. planticola* Rs-2 is a promising halotolerant PGPR for cotton growth under saline stress (Wu et al. 2012).

19.4.3 Wheat (*Triticum aestivum*)

Bharti et al. (2016) isolated bacterial strain *Dietzia natronolimnaea* STR1 (Accession no. KJ413139). Halotolerant PGPR (*D. natronolimnaea* STR1) were bio-inoculated with wheat (*Triticum aestivum*) (seeds of cv. HD 2285 from Indian Agricultural Research Institute, IARI, Pusa, New Delhi, India) in the pot under glasshouse conditions which suggest expression and involvement of abscisic acid cascade signaling and upregulation of salt stress genes like TaABARE, TaOPR1, TaMYB, TaWRKY, and TaST. Treated plant showed improved growth in terms of dry weight and plant height (higher biomass, shoot, and root elongation).

19.4.4 Maize (*Zea mays*)

Chen et al. (2016) reported *Bacillus amyloliquefaciens* SQR9 (China General Microbiology Culture Collection Center (CGMCC) accession no. 5808) from cucumber rhizospheric soil region, which had been used as an exogenous strain in a commercial bio-organic fertilizer (Cao et al. 2011; Qiu et al. 2012) for promotion of plant growth and soilborne disease suppression in the field. SQR9 produces phytohormones and antibiotics including indole-3-acetic acid (IAA) and bacillomycin D (Xu et al. 2012; Shao et al. 2015). In hydroponic system, the application of *B. amyloliquefaciens* SQR9 to maize after saline stress exposure of 20 days causes a significant growth promotion in maize seedlings and enhanced the chlorophyll content. Further, result showed that enhancement of osmolyte for reduction in cell destruction enhanced activity of peroxidase/catalase and glutathione content for scavenging ROS (Meyer et al. 2007) and decreased in sodium ion toxicity, upregulation of RBCS and RBCL (involved in photosynthesis), H^+-P Pase, HKT1, NHX1, NHX2, and NHX3 (ion transporters), as well as downregulation of 9-cis-epoxycarotenoid dioxygenase, NCED (biosynthesis of abscisic acid, ABA) (Chen et al.2016).

Marulanda et al. (2010) isolated *Bacillus megaterium* bacterial strain from the degraded soil of southern Spain. When maize plants were bioinoculated with the *B. megaterium* in pots. The results were showing higher root hydraulic conductance (L) value, ZmPIP1;1 protein amount, produces the auxin IAA which can up or downregulate plant aquaporin expression, necrotic leaf area, root growth, leaf relative water content.

Bano and Fatima (2009) bio-inoculated two isolates, i.e., *Rhizobium* spp. (strain THAL-8 chickpea nodulating) and P-solubilizing bacteria (*Pseudomonas* spp. 54RB), to two cultivars of maize (Agaiti 2002 and Av 4001) obtained from the National Agriculture Research Center, Islamabad (NARC) and grown in pots under natural condition. Microorganisms were inoculated during the seedling stage and induction of salt stress was done after 21 days of sowing. Co-inoculation resulted in some positive adaptive responses like decrease in electrolyte leakage (Lutts et al. 1999) and in osmotic potential and an increase in production of osmoregulant (like proline) (Jain et al. 2001), maintenance of relative water contents of leaves, and selective uptake of potassium ions increased chlorophyll content (Arnon 1949) and carotenoid content and greater stem diameter (Bano and Fatima 2009).

19.4.5 Rice (*Oryza sativa*)

Nautiyal et al. (2013) isolated halotolerant PGPR, i.e., *Bacillus amyloliquefaciens* NBRISN13 from the soil (alkaline) of Banthara research station, Council of Scientific & Industrial Research-National Botanical Research Institute, CSIR-NBRI, Lucknow, Uttar Pradesh, India. Application of this to rice plant in conditions of hydroponic and soil exposed to salinity causing enhancement in growth of plant with salt tolerance up to 200 mM NaCl concentration leads to expression of approximately 14 genes. Among these, four genes (SOS1, EREBP, SERK1, NADP-Me2, and BADH) (Zhang et al. 2010) were upregulated, while two (GIG and SAPK4) were repressed. This bio-inoculation of halotolerant PGPR (along with saline exposure) also stimulates concentration of betaine, sucrose, trehalose, and glutamine-utilizing bacteria, increases activity of ACC deaminase, and stimulates root growth and active root area for increased uptake of water and nutrient (Nautiyal et al. 2013).

Jha and Subramanian (2014) isolated *Pseudomonas pseudoalcaligenes* and *Bacillus pumilus* microorganism from the root tissue of paddy and rhizospheric soil. Salt-sensitive rice GJ-17 (obtained from the Main Rice Research Center, Nagawam, Anand, Gujarat) inoculated with *Pseudomonas pseudoalcaligenes* (endophytic bacterium) in pots showed significantly higher concentration of glycine betaine at higher salinity levels, while a combination of *Pseudomonas pseudoalcaligenes* and *Bacillus pumilus* gave much better response against the adverse effects of salinity such as enhanced antioxidant protein (reduced super oxide dismutase activity and lipid peroxidation) and increased growth of plant (increase in dry weight).

19.4.6 Soybean (*Glycine max*)

Vaishnav et al. (2015) screened out *Pseudomonas simiae* AU (NCBI accession no. LJ511869, MTCC No. 12057) and inoculated inside the magenta box containing King's B agar medium and sterilized seeds of soybean. It was placed on the box's bottom containing half Murashige and Skoog (MS) medium. *P. simiae* produces a putative volatile blend which can increase the growth of soybean seedlings and can elicit IST (induced systemic tolerance) against 100 mmol/l NaCl stress condition. Further expression studies with western blotting affirmed the upregulation of vegetative storage proteins (VSP), gamma-glutamyl hydrolase (GGH), and RuBisCo large chain protein and increased IAA production, siderophore production, P-solubilization, and ACC deaminase activities promote seedling growth, increase chlorophyll content, and maintain functioning of photosynthesis machinery.

19.4.7 Canola (*Brassica napus*)

Baniaghil et al. (2013) isolated three species of *Azospirillum* spp. and two strains of *Pseudomonas* spp. and co-inoculated a bacterial suspension of it to canola seeds (two cultivars, i.e., Hyola 401 and RGS 003) under 80 and 160mM NaCl under greenhouse conditions in a Leonard medium with Hoagland solution. *A. brasilense* effects on plant growth parameters while *A. lipoferum* showed maximum microelement uptake of Fe, Mn, and Zn (involves in ability to produce plants siderophores or microbial siderophores), increase in level of antioxidant enzymes (Baniaghil et al. 2013).

19.4.8 Cucumber (*Cucumis sativas*)

Kang et al. (2014) stated that *Burkholderia cepia* SE4, *Acinetobacter calcoaceticus* 370, and *Promicromonospora* spp. SE188 bacterial strain inoculated a cucumber plant with age of 1 week in the pot. The ameliorative effects of the halotolerant PGPR were reported to be increased water potential, decreased electrolyte leakage, decreased Na^+ concentration, decreased catalase activities, polyphenol oxidase, peroxidase, hormonal (endogenous) regulation (ABA, GA4, SA), oxidative damage prevention, and production of biologically active secondary metabolites including phytohormone, and it promotes closure of stomata to minimize loss of water and increase in growth of root and shoot along with content of chlorophyll.

19.4.9 Mint (*Mentha arvensis*)

Bharti et al. (2014) reported *Halomonas desiderata* STR8, *Exiguobacterium oxidotolerans* STR36, and *Bacillus pumilus* STR2 rhizobacteria from the rhizospheric region of the grass family (Poaceae) plants on the almost unprotective saline soils of Rae Bareilly, Uttar Pradesh, India. Inoculation of PGPR in plants conducted in pots containing field soil produces copious amounts of exo-polysaccharides (Siddikee et al. 2011). EPS does form a sheath of organo-mineral around cells leading to enhancement in micro-aggregates increasing aggregates stability. This provides peculiar water holding capacity and cementing properties which plays an important role in nutrients regulation and flow of water across roots of plants through biofilm formation. *Halomonas desiderata* treated plants (herb) showed the highest yield of herb at 100 and 300 mM NaCl salinity level, while *Exiguobacterium oxidotolerans* treated plants at 500 mM NaCl salinity level yielded maximum herb. The oil content of untreated, salt-stressed plants was 0.46%, 0.42%, and 0.35% at 100, 300, and 500 mM NaCl, respectively, while *Halomonas desiderata* treated plants showed an oil content of 0.71%, 0.60%, and 0.48% at 100, 300, and 500 mM NaCl, respectively (Bharti et al. 2014).

19.4.10 Okra (*Abelmoschus esculentus*)

Habib et al. (2016) isolated *Enterobacter* spp. UPMR18 bacteria from the crop field of the University Putra Malaysia, Malaysia, possessing N_2 fixation, P solubilization, IAA synthesis, and ACC deaminase activity. Inoculation of *Enterobacter* spp. UPMR18 in okra seed planted in pots of plastic for 15 days in saline condition enhances activity of antioxidant enzymes (SOD, CAT, and APX) and upregulation of genes related to reactive oxygen species pathway and growth parameter, that is, plant height, root length, fresh weight of leaf, stem and root, and increase in percentage of germination and chlorophyll content.

19.4.11 Groundnut (*Arachis hypogea*)

Saravankumar et al. (2007) isolated the four *Pseudomonas fluorescens* bacterial strain from rhizosphere region of soil from Tamil Nadu, India was isolated. Among the four plant growth-promoting rhizobacterial strains, *P. fluorescens* strain TDK1 exhibited better performance toward PGPR activity and yield of groundnut seedling in vitro. It showed high amount of ACC deaminase activity causing reduced ethylene synthesis (Saravanakumar and Samiyappan 2007).

19.5 Future Prospects and Challenges

The encroachment of biotechnology in the field of agriculture may lead to development of transgenic plants like cotton and brinjal using Bt strain leaving various challenging ethical issues before scientists. So, the researcher started working intensively on PGPR (halotolerant) bio-inoculation to the plant, but further, there is a requirement of biotechnological approach like genetic engineering to develop superior/superbug PGPR, but a few success research report is available in the literature. These halotolerant PGPR should be commercially launched by various companies (example of PGPR trade names: Bioboost, Bioplin, Bioyield, Compete, Kodiac) as startup program initiated by the current government of India, and for this farmer should be made aware. Apart from this, genetic modulation and biofertilizers, there is one more hypothesis to mobilize the nutrient and make it available for the plant growing under various stresses – this is through nanotechnology applied to the agriculture field, i.e., nano-fertilizers. Nano-fertilizers are nano-materials which can supply either of the nutrient to plant causing increased plant growth and yield without providing nutrient directly to crops (Benzon et al. 2015). Since application of studies of molecular regulators still need validation to be applied in the natural condition of stressed agricultural fields, the compensation of nano-particles is kept in consideration toward upcoming stress challenges. There are many various metal ions which are the essential prosthetic part of the metallo-enzymes taking part in the physiological functions of the plants. At increased pH, chlorosis due to iron deficiency (as the ferrous is oxidized to non-soluble ferric ions) is the major problem occurring in salt-stressed agricultural lands. So, the nano-particles of iron play the role of a trending solution of depression caused by salt stress and nutrient deficiency as these nano-particles are biocompatible and readily absorbed by the plants supplying the required nutrients. Apart from iron, there are the options of other nano-particles (Siddiqui and Al-whaibi 2014) such as zinc oxide (ZnO) (Faizan et al. 2018), titanium dioxide (TiO_2) (Haghighi et al. 2012), silica (SiO_2) (Siddiqui and Al-whaibi 2014), magnetite (Fe_3O_4), aluminum oxide (Al_2O_3), cupric oxide (CuO) (Siddiqui and Al-whaibi 2014), and carbon nano-tubes (CNTs) (Pandey et al. 2018; Jsarotia et al. 2018) which have been tested for their potent capability to ameliorate salinity stress. Although nano-fertilizer can be a good alternative to chemical fertilizers as the nano-particle plays a remedial role in resisting the salinity effect on the plants but their high concentration causes cyto- and geno-toxicity. This is the biggest challenge toward its virtue.

19.6 Conclusion

From the above-described details, it is obvious that there is enormous potential to increase crop yield with microbial co-inoculation to meet the demand of the increasing population of the world including developing country like India, where around 6.7 Mha of land is affected with salt. These microbes may be referred as halotolerant plant growth-promoting rhizobacteria (PGPR) and may involve various species of *Pseudomonas, Azospirillum, Bacillus, Rhizobium, Azotobacter, Halomonas*, etc. These PGPR can be applied to the rhizospheric region of the soil of various plant (like wheat, rice, cotton, tomato, maize, soybean, canola, cucumber, mint, okra, groundnut, etc.) and may involve in improved plant growth even under various stresses (biotic and abiotic) including salinity. These PGPR can be utilized as bio-fertilizers to enhance the uptake and mobilization of nutrient even under stress condition. If two or more halotolerant PGPR are used as bio-inoculants, their self-compatibility should be checked for the synergistic effects. If not, then further there is a need to use a biotechnological approach, i.e., genetic engineering tool to make superior/ superbug halotolerant PGPR by introducing various potent genes involved in promotion of growth and uptake of nutrient.

References

Ahemad M, Kibret M (2014) Mechanisms and applications of plant growth promoting rhizobacteria: current perspective. J King Saud Univ Sci King Saud Univ 26:1–20. https://doi.org/10.1016/j.jksus.2013.05.001

Arnon DI (1949) Copper enzymes in isolated chloroplasts. Polyphenoloxidase in *Beta vulgaris*. Plant Physiol 24:1–15

Asada K (1999) The water-water cycle in chloroplasts: scavenging of active oxygens and dissipation of excess photons. Annu Rev Plant Physiol Plant Mol Biol 50:601–639. https://doi.org/10.1146/annurev.arplant.50.1.601

Baniaghil N et al (2013) The effect of plant growth promoting rhizobacteria on growth parameters, antioxidant enzymes and microelements of canola under salt stress. J Appl Environ Biol Sci 3:17–27

Bano A, Fatima M (2009) Salt tolerance in *Zea mays* (L). Following inoculation with *Rhizobium* and *Pseudomonas*. Biol Fertil Soils 45:405–413. https://doi.org/10.1007/s00374-008-0344-9

Bashan Y, Levanony H (1990) Current status of *Azospirillum* inoculation technology: *Azospirillum* as a challenge for agriculture. Can J Microbiol 36:591–608. https://doi.org/10.1139/m90-105

Benzon HRL, Rubenecia MRU, Ultra VU Jr, Lee SC (2015) Nano-fertilizer affects the growth, development, and chemical properties of rice. Int JAgronAgric Res 7:2223–7054

Bharti N et al (2014) Plant growth promoting rhizobacteria alleviate salinity induced negative effects on growth, oil content and physiological status in *Mentha arvensis*. Acta Physiol Plant 36:45–60. https://doi.org/10.1007/s11738-013-1385-8

Bharti N, Pandey SS, Barnawal D, Patel VK, Kalra A (2016) Plant growth promoting rhizobacteria *Dietzia natronolimnaea* modulates the expression of stress responsive genes providing protection of wheat from salinity stress. Sci Rep 6:34768

Byrt CS, Munns R, Burton RA, Gilliham M, Wege S (2018) Plant science root cell wall solutions for crop plants in saline soils. Plant Sci 269:47–55. https://doi.org/10.1016/j.plantsci.2017.12.012

Cao Y, Zhang Z, Ling N, Yuan Y, Zheng X, Shen B, Shen Q (2011) *Bacillus subtilis* SQR 9 can control Fusarium wilt in cucumber by colonizing plant roots. Biol Fertil Soils 47:495–506. https://doi.org/10.1007/s00374-011-0556-2

Chen L, Liu Y, Wu G, Njeri KV, Shen Q, Zhang N, Zhang R (2016) Induced maize salt tolerance by rhizosphere inoculation of *Bacillus amyloliquefaciens* SQR9. Physiol Plant 158:34–44. https://doi.org/10.1111/ppl.12441

Egamberdieva D, Lugtenberg B (2014) Use of plant growth-promoting rhizobacteria to alleviate salinity stress. In: Miransari M (ed) Use of microbes for the alleviation of soil stresses. Springer, New York. https://doi.org/10.1007/978-1-4614-9466-9.

Egamberdieva D, Davranov K, Wirth S, Hashem A, Allah EFA (2017) Impact of soil salinity on the plant-growth – promoting and biological control abilities of root associated bacteria. Saudi J Biol Sci 24:1601–1608. https://doi.org/10.1016/j.sjbs.2017.07.004

El-Ramady H, El-Ghamry AM, Mosa A, Alshaal T (2018) Nanofertilizers vs. biofertilizers: new insights. Environ Biodivers Soil Secur 2:40–50. https://doi.org/10.21608/jenvbs.2018.3880.1029

Etesami H, Beattie GA (2018) Mining halophytes for plant growth-promoting halotolerant bacteria to enhance the salinity tolerance of non-halophytic crops. Front Microbiol 9:148. https://doi.org/10.3389/fmicb.2018.00148

Etesami H, Maheshwari DK (2018) Ecotoxicology and environmental safety use of plant growth promoting rhizobacteria (PGPRs) with multiple plant growth promoting traits in stress agriculture: action mechanisms and future prospects. Ecotoxicol Environ Saf 156:225–246. https://doi.org/10.1016/j.ecoenv.2018.03.013.

Faizan M, Faraz A, Yusuf M, Khan ST, Hayat S (2018) Zinc oxide nanoparticle-mediated changes in photosynthetic efficiency and antioxidant system of tomato plants. Photosynthetica 56:678–686. https://doi.org/10.1007/s11099-017-0717-0

Filomeni G, De Zio D, Cecconi F (2015) Oxidative stress and autophagy: the clash between damage and metabolic needs. Cell Death Differ 22:377–388. https://doi.org/10.1038/cdd.2014.150

García-Gutiérrez L, Romera D, Zeriouh H, Perez-Carcia A (2012) Isolation and selection of plant growth-promoting rhizobacteria as inducers of systemic resistance in melon. Plant Soil 358:201–212. https://doi.org/10.1007/s11104-012-1173-z

Gilliham M (2015) Salinity tolerance of crops – what is the cost? Tansley insight salinity tolerance of crops – what is the cost? New Phytol 208:668–673. https://doi.org/10.1111/nph.13519.

Glick BR (2014) Bacteria with ACC deaminase can promote plant growth and help to feed the world. Microbiol Res 169:30–39. https://doi.org/10.1016/j.micres.2013.09.009

Habib SH, Kausar H, Saud HM (2016) Plant growth-promoting rhizobacteria enhance salinity stress tolerance in okra through ROS-scavenging enzymes. BioMed Res Int 2016:6284547. https://doi.org/10.1155/2016/6284547

Hafeez FY, Al Harrasi A, Roberts MR (2015) Suppression of incidence of *Rhizoctonia Solani* in rice by siderophore producing rhizobacterial strains based on competition for iron. Eur Sci J 11:186–207

Haghighi M, Afifipour Z, Mozafarian M (2012) The effect of N-Si on tomato seed germination under salinity levels. J Biol Environ Sci 6:87–90

Hernández-León R, Rojas-Solis D (2015) Characterization of the antifungal and plant growth-promoting effects of diffusible and volatile organic compounds produced by *Pseudomonas fluorescens* strains. Biol Control 81:83–92. https://doi.org/10.1016/j.biocontrol.2014.11.011

Ilangumaran G, Smith DL (2017) Plant growth promoting rhizobacteria in amelioration of salinity stress: a systems biology perspective. Front Plant Sci 8:1–14. https://doi.org/10.3389/fpls.2017.01768

Ishitani M (2000) SOS3 function in plant salt tolerance requires N-myristoylation and calcium binding. Plant Cell 12:1667–1678. https://doi.org/10.1105/tpc.12.9.1667

Jain M, Mathur G, Koul S, Sarin NB (2001) Ameliorative effects of proline on salt stress-induced lipid peroxidation in cell lines of groundnut (*Arachis hypogaea* L.). Plant Cell Rep 20:463–468. https://doi.org/10.1007/s002990100353

Jalili F, Khavazi K, Pazira E, Nejati A, Rahmani HA, Sadaghiani HR, Miransari M (2009) Isolation and characterization of ACC deaminase- producing fluorescent pseudomonads, to alleviate salinity stress on canola (*Brassica napus* L.) growth. J Plant Physiol 166:667–674. https://doi.org/10.1016/j.jplph.2008.08.004

Jha Y, Subramanian RB (2014) PGPR regulate caspase-like activity, programmed cell death, and antioxidant enzyme activity in paddy under salinity. Physiol Mol Biol Plants 20:201–207. https://doi.org/10.1007/s12298-014-0224-8

Jsarotia P, Kashyap PL, Bhardwaj AK, Kumar S (2018) Nanotechnology scope and applications for wheat production and quality enhancement: a review of recent advances. Wheat Barley Res 10. https://doi.org/10.25174/2249-4065/2018/76672

Kang S, Khan AL, Waqas M, You Y-H, Kim J-H, Kim J-G, Hamaun M, Lee I-J (2014) Plant growth-promoting rhizobacteria reduce adverse effects of salinity and osmotic stress by regulating phytohormones and antioxidants in *Cucumis sativus*. J Plant Interact 9:673–682. https://doi.org/10.1080/17429145.2014.894587

Kumar V (2012) Phosphate solubilizing activity of some bacterial strains isolated from chemical pesticide exposed agriculture soil. Int J Eng Res Dev 3:1–6

Kumar A, Singh VP, Tripathi V, Singh PP, Singh AK (2018) Plant growth promoting rhizobacteria (PGPR): perspective in agriculture under biotic and abiotic stress. In: Agriculture under biotic and abiotic stress, crop improvement through microbial biotechnology. Elsevier B.V. https://doi.org/10.1016/B978-0-444-63987-5.00016-5

LäuchliA, GrattanSR (2011) Plant responses to saline and sodic conditions. In: Wallendar WW, Tanji KK (eds) Agricultural salinity assessment and management. p169–205. doi:https://doi.org/10.1061/9780784411698.ch06.

López-Bucio J, Campos-Cuevas JC, Hernández-Calderón E, Velásquez-Becerra C, Farías-Rodríguez R, Macías-Rodríguez LI, Valencia-Cantero E (2007) *Bacillus megaterium* rhizobacteria promote growth and alter root-system architecture through an auxin- and ethylene-independent signaling mechanism in *Arabidopsis thaliana*. Mol Plant-Microbe Interact 20:207–217. https://doi.org/10.1094/MPMI-20-2-0207

Lutts S, Majerus V, Kinet J (1999) NaCl effects on proline metabolism in rice (*Oryza sativa*) seedlings. Physiol Plant 105:450–458

Marulanda A, Azcon R, Chaumont F, Ruiz-Lozano JM, Aroca R (2010) Regulation of plasma membrane aquaporins by inoculation with a *Bacillus megaterium* strain in maize (*Zea mays* L.) plants under unstressed and salt-stressed conditions. Planta 232:533–543. https://doi.org/10.1007/s00425-010-1196-8

Mayak S, Tirosh T, Glick BR (2004) Plant growth-promoting bacteria confer resistance in tomato plants to salt stress. Plant Physiol Biochem 42:565–572. https://doi.org/10.1016/j.plaphy.2004.05.009

Meldau DG, Meldau S, Hoang LH, Underberg S, Wunsche H, Baldwin IT (2013) Dimethyl disulfide produced by the naturally associated bacterium *Bacillus* sp B55 promotes *Nicotiana attenuata* growth by enhancing sulfur nutrition. Plant Cell 25:2731–2747. https://doi.org/10.1105/tpc.113.114744

Meyer AJ, Brach T, Marty L, Kreye S, Rouhier N, Jacquot JP, Hell R (2007) Redox-sensitive GFP in *Arabidopsis thaliana* is a quantitative biosensor for the redox potential of the cellular glutathione redox buffer. Plant J 52:973–986. https://doi.org/10.1111/j.1365-313X.2007.03280.x

Nasher Mohamed A, Razi Ismail M, Hasan Rahman M (2010) In vitro response from cotyledon and hypocotyls explants in tomato by inducing 6-benzylaminopurine. Afr J Biotechnol 9:4802–4807. https://doi.org/10.5897/AJB09.1372

Nautiyal CS, Srivastava R, Chauhan PS, Seem K, Mishra A, Sopory SK (2013) Plant growth-promoting bacteria *Bacillus amyloliquefaciens* NBRISN13 modulates gene expression profile of leaf and rhizosphere community in rice during salt stress. Plant Physiol Biochem 66:1–9. https://doi.org/10.1016/j.plaphy.2013.01.020

Numan M (2018) Plant growth promoting bacteria as an alternative strategy for salt tolerance in plants: a review. Microbiol Res 209:21–32. https://doi.org/10.1016/j.micres.2018.02.003

Pandey K, Lahiani MH, Hicks VK, Hudson MK, Green MJ, Khodakovskaya M (2018) Effects of carbon-based nanomaterials on seed germination, biomass accumulation and salt stress response of bioenergy crops. PLOS One:1–17. https://doi.org/10.5061/dryad.h4r6h5n.Funding.

Penrose DM, Glick BR (2003) Methods for isolating and characterizing ACC deaminase-containing plant growth-promoting rhizobacteria. Physiol Plant 118:10–15

Pérez-García A, Romero D, de Vicente A (2011) Plant protection and growth stimulation by microorganisms: biotechnological applications of Bacilli in agriculture. Curr Opin Biotechnol 22:187–193. https://doi.org/10.1016/j.copbio.2010.12.003

Pieterse CMJ, Zamioudis C, Berendsen RL, Weller DM, Van Wees SCM, Bakker PAHM (2014) Induced systemic resistance by beneficial microbes. Annu Rev Phytopathol 52:347–375. https://doi.org/10.1146/annurev-phyto-082712-102340

Porcel R, Zamarreno AM, Carcia-Mina JM, Aroca R (2014) Involvement of plant endogenous ABA in *Bacillus megaterium* PGPR activity in tomato plants. BMC Plant Biol 14:36. https://doi.org/10.1186/1471-2229-14-36

Proseus TE, Boyer JS (2012) Pectate chemistry links cell expansion to wall deposition in *Chara corallina*. Plant Signal Behav 7:1490–1492. https://doi.org/10.4161/psb.21777

Qiu M, Zhang R, Xue C (2012) Application of bio-organic fertilizer can control Fusarium wilt of cucumber plants by regulating microbial community of rhizosphere soil. Biol Fertil Soils 48:807–816. https://doi.org/10.1007/s00374-012-0675-4

Saravanakumar D, Samiyappan R (2007) ACC deaminase from *Pseudomonas fluorescens* mediated saline resistance in groundnut (*Arachis hypogea*) plants. J Appl Microbiol 102:1283–1292. https://doi.org/10.1111/j.1365-2672.2006.03179.x

Shahzad SM, Khalid A, Arshad M (2010) Screening rhizobacteria containing ACC-deaminase for growth promotion of chickpea seedlings under axenic conditions. Soil Environ 29:38–46

Shainberg I, Levy GJ, Goldstein D, Mamedov AI (2001) Prewetting rate and sodicity effects on the hydraulic conductivity of soils. Aust J Soil Res 39:1279–1291. https://doi.org/10.1071/SR00052

Shao J, Li S, Zhang N, Cui X, Zhou X, Zhang G, Shen Q, Zhang R (2015) Analysis and cloning of the synthetic pathway of the phytohormone indole -3-acetic acid in the plant-beneficial

Bacillus amyloliquefaciens SQR9. Microb Cell Factories 14:130. https://doi.org/10.1186/s12934-015-0323-4

Shi H, Quintero FJ, Pardo JM, Zhu J-K (2002) The putative plasma membrane Na^+/H^+ antiporter SOS1 controls long-distance Na^+ transport in plants. PlantCell 14:465–477. https://doi.org/10.1105/tpc.010371.et

Shrivastava P, Kumar R (2015) Soil salinity: a serious environmental issue and plant growth promoting bacteria as one of the tools for its alleviation. Saudi J Biol Sci 22:123–131. https://doi.org/10.1016/j.sjbs.2014.12.001

Siddikee MA, Chauhan P, Anandham R, Han G-H (2010) Isolation, characterization, and use for plant growth promotion under salt stress, of ACC deaminase-producing halotolerant bacteria derived from coastal soil. J Microbiol Biotechnol 20:1577–1584. https://doi.org/10.4014/jmb.1007.07011

Siddikee A, Glick BR, Chauhan PS, Wj Y, Sa T (2011) Enhancement of growth and salt tolerance of red pepper seedlings (*Capsicum annuum* L.) by regulating stress ethylene synthesis with halotolerant bacteria containing 1-aminocyclopropane-1-carboxylic acid deaminase activity. Plant Physiol Biochem 49:427–434. https://doi.org/10.1016/j.plaphy.2011.01.015

Siddiqui MH, Al-whaibi MH (2014) Role of nano-SiO 2 in germination of tomato (*Lycopersicum esculentum* seeds Mill). Saudi J Biol Sci 21:13–17. https://doi.org/10.1016/j.sjbs.2013.04.005

Siddiqui MH, Al-Whaibi MH, Faisal M, Al Sahli AA (2014) Nano-silicon dioxide mitigates the adverse effects of salt stress on *Cucurbita pepo* L. Environ Toxicol Chem 33:2429–2437. https://doi.org/10.1002/etc.2697

Sultana N, Ikeda T, Itoh R (1999) Effect of NaCl salinity on photosynthesis and dry matter accumulation in developing rice grains. Environ Exp Bot 42:211–220. https://doi.org/10.1016/S0098-8472(99)00035-0

Tabassum B, Khan A, Tariq RM, Ramzan M (2017) Review- bottlenecks in commercialisation and future prospects of PGPR. Appl Soil Ecol 121:102–117. https://doi.org/10.1016/j.apsoil.2017.09.030

Vacheron J, Desbrosses G, Bouffaud M-L, Touraine B, Moënne-Loccoz Y, Muller D, Legendre L, Wisniewski-Dyé F, Prigent-Combaret C (2013) Plant growth-promoting rhizobacteria and root system functioning. Front Plant Sci 4:356

Vaishnav A, Kumari S, Jain S, Varma A, Choudhary DK (2015) Putative bacterial volatile-mediated growth in soybean (*Glycine max* L. Merrill) and expression of induced proteins under salt stress. J Appl Microbiol 119:539–551. https://doi.org/10.1111/jam.12866

Valetti L, Iriarte L, Fabra A (2018) Growth promotion of rapeseed (*Brassica napus*) associated with the inoculation of phosphate solubilizing bacteria. Appl Soil Ecol. https://doi.org/10.1016/j.apsoil.2018.08.017

Waskom RM, Bauder T, Davis JG, Andales AA (2012) Diagnosing saline and sodic soil problems, Crop Seris-Soil, Fact Sheet 0.521. Colarodo State University, Fort Collins, pp 1–2

Wu SC, Cao ZH, Li ZG, Cheung KC, Wong MH (2005) Effects of biofertilizer containing N-fixer, P and K solubilizers and AM fungi on maize growth: a greenhouse trial. Geoderma 125:155–166. https://doi.org/10.1016/j.geoderma.2004.07.003

Wu Z, Yue H, Lu J (2012) Characterization of rhizobacterial strain Rs-2 with ACC deaminase activity and its performance in promoting cotton growth under salinity stress. World J Microbiol Biotechnol 28:2383–2393. https://doi.org/10.1007/s11274-012-1047-9

Xu Z, Shao J, Li B, Yan X, Shan Q, Zhang R (2012) Contribution of bacillomycin D in *Bacillus amyloliquefaciens* SQR9 to antifungal activity and biofilm formation. Appl Environ Microbiol 79:808–815. https://doi.org/10.1128/AEM.02645-12

Yan N, Marschner P, Cao W, Zuo C, Qin W (2015) Influence of salinity and water content on soil microorganisms. Int Soil Water Conserv Res 3:316–323. https://doi.org/10.1016/j.iswcr.2015.11.003

Yang J, Kloepper JW, Ryu CM (2009) Rhizosphere bacteria help plants tolerate abiotic stress. Trends Plant Sci 14:1–4. https://doi.org/10.1016/j.tplants.2008.10.004

Yao L, Wu ZS, Zheng YY, Kaleem I, Li C (2010) Growth promotion and protection against salt stress by *Pseudomonas putida* Rs-198 on cotton. Eur J Soil Biol 46:49–54. https://doi.org/10.1016/j.ejsobi.2009.11.002

Yuan J, Raza W, Shen Q, Huang Q (2012) Antifungal activity of *Bacillus amyloliquefaciens* NJN-6 volatile compounds against *Fusarium oxysporum* f. sp. cubense. Appl Environ Microbiol 78:5942–5944. https://doi.org/10.1128/AEM.01357-12

Zhang L, Tian L-H, Zhao J-F, Song Y, Zhang C-J, Guo Y (2008) Identification of an apoplastic protein involved in the initial phase of salt stress response in rice root by two-dimensional electrophoresis. Plant Physiol 149:916–928. https://doi.org/10.1104/pp.108.131144

Zhang H, Liu W, Wan L (2010) Functional analyses of ethylene response factor JERF3 with the aim of improving tolerance to drought and osmotic stress in transgenic rice. Transgenic Res 19:809–818. https://doi.org/10.1007/s11248-009-9357-x

Zhu JK (2003) Regulation of ion homeostasis under salt stress. Curr Opin Plant Biol 6:441–445. https://doi.org/10.1016/S1369-5266(03)00085-2

Zushi K, Matsuzoe N, Kitano M (2009) Developmental and tissue-specific changes in oxidative parameters and antioxidant systems in tomato fruits grown under salt stress. Sci Horticult 122:362–368. https://doi.org/10.1016/j.scienta.2009.06.001

Microbial Degradation of Nitroaromatic Pesticide: Pendimethalin

20

Prasad Jape, Vijay Maheshwari, and Ambalal Chaudhari

20.1 Introduction

Agriculture is recognized as the main engine to drive the economy in Indian subcontinent, where 60–70% of the population has relied on agriculture for food. India's population is expected to reach approximately 1.3 billion by 2020 (Kanekar et al. 2003), while world population will increase to 12 billion before 2050 (Pimentel 1995). The weeds, insects, and microorganisms are the main competitors the moment humans settled to agriculture cropland, ravaging crops, food, and feed stores. Unfortunately, the worldwide crop losses have been estimated approx. 50% by pestilent, 13–16% by insect pest, 12–13% by phytopathogens, and 10–13% by weeds which cost to $ 244 billion loss of revenue per year (Pimentel 1997). For this purpose, intensive agricultural strategies are adopted to increase food grain production and prevent crop loss (Shroff 2000). For these tribulations, more emphasis is accorded to (i) use quality seeds, (ii) increased chemical fertilizer inputs for more crop productivity, and (iii) protection of crops against various plant pests that adversely affect crop productivity (Ahemad and Khan 2011). To ameliorate the enormous crop losses caused by pests, more use of chemically synthesized pesticides is promoted. Pesticide application to control plant pest was adopted as an effective regime to increase crop productivity, which promoted more production of pesticide widespread usage and spillage in the soil environment disposed or washed out in water, aquifers, etc. In recent years, a variety of pesticides of wide diversity chemical groups (>500) have been extensively employed for protection of the crop plants (Ahemad et al. 2009), large amount of which are lost in application process, and meager amount of pesticides reaches to the target pest (Pimentel 1995). The demand for pesticide in India is 3.75% of the total world consumption (Jogdand

P. Jape · V. Maheshwari · A. Chaudhari (✉)
School of Life Sciences, Kavayitri Bahinabai Chaudhari North Maharashtra University, Jalgaon, India

D. P. Singh et al. (eds.), *Microbial Interventions in Agriculture and Environment*, https://doi.org/10.1007/978-981-13-8391-5_20

2000). The most commonly used pesticides are categorized as (i) organophosphate, (ii) organochlorine, (iii) carbamate, (iv) pyrethroids, (v) neonicotinoids, (vi) nitroaromatics, and (vii) biopesticides. Among these, organophosphates and nitroaromatics are extensively used in the agriculture. Despite the benefits, the haphazard application of pesticides in the last two to three decades caused (i) serious environmental pollution; (ii) bulk of the residue (80–90%) was deposited on nontarget areas, such as soil, water, sediments; (iii) caused loss of vital plant pollinators; (v) threatened nontarget life forms; (vi) obligated public health issues; and (vii) damage loss to the tune of 100 billion every year (Sakata 2005; Parte et al. 2017). Of these, plethora of nitroaromatic compounds are manufactured for intended application, and tons of them finally come into water, retained in the soil, affect oil fertility, and impact various life forms in the ecosystem (Parte et al. 2017). Nitroaromatics are (i) stable to biotic and abiotic attack, (ii) persistent in the environment for prolonged time, (iii) synthesized in great volume and differ in chemical structure, (iv) used as chemical feedstock material for the synthesis of variety of pesticides, explosives, herbicides, dyes, etc. Indiscriminate application of nitroaromatics has caused inexorable amount of environmental pollution and was recognized as recalcitrant compound and priority hazardous type of pollutant by various regulatory systems. The recalcitrance nature of nitroaromatics is due to (i) unusual substitution, (ii) condensed aromatic ring, (iii) insolubility in aqueous phase, and (iv) resistance to abiotic and biotic degradation. Majority of them are identified as potent neurotoxin, endocrine disruptor, carcinogenic, mutagenic, teratogenic, toxic, and designated as a major priority pollutant by various regulatory systems. Moreover, the growing attention to public health hazards, environmental awareness, and legal requirement on the release of pesticides are becoming more complex, strict, and warranting for their removal. Hence, removal of nitroaromatic pesticide from contaminated environment is realized as peremptory art. Several conventional cleanup methods such as (i) incineration, (ii) volatilization, (iii) hydrolysis, (iv) photo-oxidation, (v) adsorption, (vi) percolator filters, (vi) advance oxidation, (vii) and photo-catalysis with TiO_2 are available for the removal of pollutant (Timmis et al. 1994). The cost-effective and eco-friendly biological system is also emerged out as effective alternative for removal of pesticides from contaminated areas either by (i) bioaugmentation, (ii) biostimulation, (iii) natural attenuation, (iv) biosparging, (v) in situ, (vi) ex situ, (vii) land farming, or (viii) composting. The physical cleanup methods (i) generate toxic (NO_x) nonintermediates that end up with enormous estimated cost of 3000–4000 USD per ton (Kanekar et al. 2003; Ortiz-Hernandez et al. 2011), (ii) cant handle complex chemistry of pesticides producing equally or even more toxic intermediates, and (iii) proved inefficient (Parte et al. 2017), while biological system follows biphasic mode of pesticide degradation but few of them may require prolonged time to recuperate the contaminated sites (Shaer et al. 2013; Ishag et al. 2017). Among the pesticides, a strategy to control weeds from crop area for more crop yield with herbicides has spurt the interest in world market as a profitable business, which is evidenced from steep rise in worldwide market demand for herbicide by 39% and projected to grow more by 11% (Gianessi 2013).

Presently, herbicides are grouped into 29 different classes on the basis of mechanism of action and generally applied either by (i) foliar spray, (ii) soil contact, (iii) broadcast, or (iv) spot contact (Singh and Singh 2014). The most commonly used herbicides in agriculture include (i) atrazine, (ii) metolachlor, (iii) glyphosphate (GP), (v) pendimethalin (PND), (vi) 2,4-dichlophenoxy acetic acid (2,4-D), (vii) clodinafop propargyl, and (viii) diuron (Singh and Singh 2014). Excess use of herbicides in the last two to three decades has caused (i) great concern to the environment, (ii) enormous water and soil pollution (Juhler et al. 2001), (iii) reduced biodiversity, (iv) lowered soil heterotrophic bacterial load, (v) and threat to nontarget life forms due to bioaccumulation risk in human, animal, and crop plants and disrupting the ecosystem through food chain, bioaccumulation, biomagnification, etc. (Singh and Singh 2014). Inadequate management and indiscriminate application of wide variety of chemical herbicides are the major root cause of contamination and irreparable damage to the ecosystem. Moreover, chemical properties, quantum of herbicide load, and its persistence determine the extent of impact. The necessity to remove the recalcitrant herbicides in an economical and eco-friendly manner constitutes the major objective (Singh and Singh 2014). The present review provides an overview of an attempt made for microbial removal of the third most frequently used herbicide in the world, pendimethalin (PND).

20.2 Pendimethalin: A Nitroaromatic Pesticide for Crop Protection

Pendimethalin (PND) (CAS registry number 40487-42–1); [N-(1-ethylpropyl)–2,6-dinitro-3,–4xylidine] is a dinitroaniline herbicide that has nitrated aromatic ring structure consisting of hydroxyl (-OH) and nitro (-NO_2) groups with molecular mass of 281.312 Da with empirical formula $C_{13}H_{19}N_3O_4$ and hydrophobic, sparingly soluble in 0.275 ppm water (Richardson and Gangolli 1992; Strandberg and Scott-Fordsmand 2004). PND is widely applied to soil as a selective preplant, preemergence, and sometimes postemergence herbicide in variety of crop plants including cotton, soybean, maize, wheat, rice, peas, and vegetable crops to control annual grasses, certain broad leaf weeds of dryland crops and non-crop areas, and also for plant growth promotion under tropical, subtropical, as well as temperate conditions (Ni et al. 2016a; 2018). PND is also recommended for use on fruit, grapes, vegetable, oil seeds, cereals, tobacco, and ornamental plants at 2 kg/ha in the European Union (EU) and at 6.7 kg/ha in the USA (European Community 2003).

Besides glyphosphate and parquet, PND is the third most frequently used selective herbicide throughout the world and has been on the market for almost 35 years (Ni et al. 2016b; Vighi et al. 2017). The demand for PND in crop protection raised from 9 to 114.3 tons, with more than 12-fold increase (Choudhury et al. 2016), and the northern part of India alone utilized almost 11.8 tons of PND a year for protection of cotton crop alone (Choudhury et al. 2016). PND is available in 30% EC or granule for manual application through spray method to pre- and postemergence or directs own crop plants.

20.3 Hazardous Implications of PND

PND has relatively (i) low volatility due to vapor pressure of 3×10^{-3} mmHg at 25 °C, and some meagre amount (10%) is lost through volatilization from surface soil; (ii) persist longer time in soil because of low leaching potential; (iii) hydrophobic nature assists to form strong physical bond with organic matter of soil and clay minerals (Walker and Bond 1977; Singh and Singh 2014); (iv) has high geometric mean (GM) with half-life of 76–98 days and 20 days in agriculturally relevant soils and sediment water under aerobic and anaerobic condition, respectively; (v) and there was a strong inhibitory action on mitotic cell division in developing root shoot system (Singh and Singh 2014). Although these attributes make PND a selective herbicide, it enters the surface water mainly as runoff from new application area due to heavy rains, which results in 2–134 μg l^{-1} residue to water sediments (Keese et al. 1994), and accumulates in onion up to 1 $mgkg^{-1}$ (Tsiropoulos and Miliadis 1998) making it a threat to the ecosystem. The major concern to herbicide use is that only meager amount of PND reaches the target and the remaining accumulates into the environment, where it adversely affects crop, animal, and public health (Pimentel 1995). PND contamination majorly occurred due to improper guidance on handling of herbicides on farm with moisture condition, temperature, and cultivation practices aiding the long-time persistence in the soil (Swarcewicz and Gregorczyk 2012). The widespread usage and contamination are an alarming environmental concern, and hence PND is listed as a persistent bioaccumulative toxin and a possible human carcinogen (group C) by US EPA (Ahmad et al. 2016). Excessive use of pendimethalin has further shown toxicity effects onto (i) onion and maize roots (Promkaew et al. 2010), (ii) the growth of funnel plants by inhibiting the tubulin production during mitosis (Engebretson et al. 2001; Fennell et al. 2006; El-Awadi and Hassan 2011), (iii) fish and other aquatic invertebrates on bioaccumulation, (iv) root knot nematode, (v) and humans through the food chain (Abd-Algadir 2011). Kidd and James (1991) observed oral LD_{50} of 1050–5000 $mgkg^{-1}$ in rats. Pendimethalin is (i) relatively nontoxic to humans by ingestion; (ii) slightly toxic by skin exposure, with dermal LD_{50} of $\geq$2000 $mgkg^{-1}$ in rats; (iii) and mildly irritant to the eye of rabbits.

20.4 Rationale Necessity for Removal of PND

PND is registered for herbicide use in several countries since two to three decades ago as the most effective, efficient, and economical entity to abate weed growth, but excessive application has raised these various concerns about potential environmental hazards. The environment fate of PND indicates that only 10% reaches to the target weed pest, 10–20% vaporizes in the first week after application, and the rest may dissipate via biological or chemical process with DT_{50} values between 30 and >200 days, suggesting (i) phytotoxicity to nontarget plant crops; (ii) enough time for physical adhesion to soil, organic fraction, sediment, and clay particles (Strandberg and Scott-Fordsmand 2004); (iii) more chances for entry into the food chain and lesser possibility for degradation; and (iv) decomposition to toxic NO_x. Consequently, PND (i) affects symbiosis between legume and *Rhizobium*; (ii) reduced nodulation by >25%; (iii) lowers VAM colonization by 36–69%; (iv) drops overall heterotrophic microbial activity for the initial 4–10 weeks; (v) suppresses rhizosphere nutrient cycling by microbes; (vi) exerts toxicity to plants, microbes, and also fish with LC_{50} (96 h) for rainbow trout and blue gill sunfish of 0.14 and 0.2 mgL^{-1} (Kidd and James 1991); (vii) reduces soil nematode by 35–36%; and (viii) inhibits roots and shoots in seedlings (Strandberg and Scott-Fordsmand 2004). Extensive use of PND as a preferred herbicide has now posed adverse toxicological impacts on flora and fauna through direct and indirect exposure. PND exposure in the agriculture health study had shown (i) increased incidences of lung, rectal, and pancreatic cancers (Ahmad et al. 2016), (ii) genotoxic effects on the fish species *Oreochromis niloticus* and aquatic invertebrate (El-Sharkawy et al. 2011), and (iii) mild hemotoxic effect in female rats after administration of dosage for 90 days (Ayub et al. 1997). As a result, PND is classified as a (i) persistent bioaccumulative toxic agent (Roca et al. 2009), (ii) possible human carcinogen (group C), and (iii) slight acute toxic compound (toxicity class III) (Ni et al. 2016b). Extensive exposure of PND for prolonged time can (i) cause cytotoxicity to living CHO cells (Patel et al. 2007); (ii) disrupt the endocrine, reproductive, and immune system; (iii) cause neurobehavioral disorders (Ritter et al. 1995); (iv) cause thyroid follicular cell adenoma; and (v) inhibit mitotic cell division in growing root system (Singh and Singh 2014). Overall, the forgoing discussion suggests the necessity for removal of PND from the contaminated environment.

20.5 Pendimethalin Degradation by Abiotic and Microbial Route

Until now, various abiotic avenues have been employed for removal of PND, but they have either lacked specificity or haven't proved to be reliable. Environmental parameters, such as reluctant species (complex structure, volatility, water solubility), pH, and dissolved oxygen matter (DOM), determine the PND degradation in the nature. DOM increase nitro group reduction in liquid solution of sulfide in anoxic black carbon-amended sediments (Gong et al. 2016). PND is sensitive to different wavelengths of UV light in water and soil-water suspension causing dealkylation of amino group (Scheunert et al. 1993), reduction into diamines by zero valent iron powder (Keum and Li 2004), and degradation which is achieved using TiO_2 (Pandit et al. 1995), nanoparticles of $BaTiO_3/TiO_2$ in the presence of peroxide, and per sulfate species by crystalline gel conversion methods (Gomathi Devi and Krishnamurthy 2008). Combination of ultraviolet light and sunlight had shown degradation of 99% PND (Dureja and Walia 1989; Moza et al. 1992), while electrolytic and electro-irradiated methods based on diamond anodes help to remove PND from soil washing effluents (Almazan-Sanchez et al. 2017). The abiotic mode of degradation (i) causes decomposition to toxic fumes of NO_x, (ii) separates unwanted compounds without destruction, (iii) generates toxic intermediates, and (iv) poses several issues for on-site or off-site treatment system. Hence, abiotic degradation route is obsolete and less preferred alternative in the present era.

In biotic degradation, microorganisms are the only tiny entities endowed with inherent abilities to transform complex compound to simple form and appeared as an effective strategy to decontaminate PND from the contaminated sites. Only microorganisms are empowered in the biosphere to bind, thrive, colonize, and metabolically utilize the compound as CorN and energy source for their growth and convert it into simple and nontoxic chemical structure of the target compound due to their involvement in nutrient cycling (Diez 2010; Pinto et al. 2012). This incredible versatility harbored by microbes can help to incorporate the recalcitrant PND into biogeochemical cycle. Bacterial and fungal entities are associated with significant role in transformation of nitroaromatic compounds (Pinto et al. 2012). Hence, applications of microorganisms are the most preferred strategy to degrade nitroaromatic compound, pendimethalin (More et al. 2015). At present, only few microbial systems for degradation of PND have been studied under both aerobic and anaerobic environments (Zheng and Cooper 1996). Collectively, three different mechanisms, namely, (i) oxidative N-dealkylation, (ii) cyclization, and (iii) nitroreduction, have been reported to initialize the PND degradation (Kole et al. 1994). Biodegradation of PND with *Azotobacter chroococcum* adopts N-dealkylation and reduction of more than one nitro group to form six metabolites (S_1, S_2, S_3, S_4, S_5, and S_6). S_1 metabolite is formed through oxidative complete N-dealkylation; further, it undergoes acetylation of the aniline nitrogen to S_3 and S_4 through elimination of nitro group at C-2 without substitution, S_2 by reduction of nitro group at C-6 position, and minor metabolite S_5 formed by aryl methyl group oxidation at C-3; oxidative cyclization reduced the 2-nitro group and N-dealkylation to S_6 (Kole et al. 1994). Likewise, several microbes (Table 20.1) including (i) fungus strain *Lecanicillium saksenae* had shown degradation of 250 ppm PND (Pinto et al. 2012); (ii) *Fusarium oxysporum* and *Paecilomyces variotii* converted PND into two metabolites, namely, N-(1-ethylpro-pyl)-3,4-dimethyl-2-nitrobenzene-l,6-diamine(II) and 3,4-dimethyl-2,6-dinitroaniline by nitroreduction and dealkylation (Singh and Kulshrestha 1991); (iii) *Bacillus circulans* degraded the PND and formed 6-amino pendimethalin and 3,4-dimethyl-2,6-dinitroaniline metabolites (Megadi et al. 2010; More et al. 2015); (iv) *Paracoccus* sp. P13 degrade 100 ppm PND within 2 days by ring cleavage through oxidation to yield 1,3-dinitro-2-(pentan-3-ylamino) butane-1,4-diol, an alkane organic compound (Ni et al. 2018); (v) *Bacillus subtilis* consumed 100 ppm PND within 2 days to form three metabolites, namely, 6-amino pendimethalin by nitroreduction using PND nitroreductase, 5-amino-2-methyl-3-nitroso-4-(pentan-3-ylamino) benzoic acid by nitroreduction at the nitro group connected to C-2, and 8-amino-2-ethyl-5-(hydroxymethyl)-1,2-dihydroquinoxaline-6-carboxylic acid by carboxylation of the aryl methyl group at C-4 (Ni et al. 2016b); (vi) six fungal species, *Aspergillus flavus*, *A. terreus*, *Fusarium solani*, *F. oxysporum*, *Penicillium citrinum*, and *P. simplicissimum*, have shown 66% of 500 ppm PND degradation in 15 days, and *Fusarium solani* alone displayed higher specificity to degrade 62% PND to form three metabolites through partial N-dealkylation to N-propyl-3-methyl-4-hydroxy-2,6-dinitroaniline, subsequent ring hydroxylation

Table 20.1 Summary of PND degrading microbial species isolated from various ecohabitats

Source	Medium	Condition	PND (mgL^{-1})	Degradation (%)	Metabolites	Reference
Bacteria						
Azotobacter chroococcum	N-free MS (pH 7.2)	31 °C in dark at steady state	500	55 (20 days)	2-Methyl-4,6-dinitro-5-[(lethylpropyl)amino] benzylalcohol,2,6-dinitro-3,4-xylidine,2-methyl4,6-dinitro-5-[(1-ethylpropy1) aminolbenzaldehyde,2-nitro-6-amino-(N-ethylpropyl)-3,-4xylidine,2,6-dinitro-3,4-xylidin	Kole et al. (1994)
Bacillus circulans	MS (pH 7)	30 °C at 150 rpm	1000	100	6-Amino pendimethalin by nitroreduction and form 3,4-dimethyl–2,6-dinitroaniline and pantane by oxidative dealkylation	Megadi et al. (2010)
Bacillus megaterium	MS (pH 7)	30 °C at 150 rpm	100	100 (5.6 days)	Undetected	Belal and Hassan (2013)
Bacillus lehensis XJU PUF cells	MS (pH 7)	(30 ± 2 °C) at 150 rpm	2000	100 (4 days)	Reduction to3,4-dimethyl 2,6-dinitroaniline by PND oxidative dealkylation	More et al. (2015)
Bacillus subtilis Y3	MS (pH 7)	30 °C 150 rpm	100	99.9 (2.5 days)	Nitroreductase catalyze nitroreduction of PND to 6-amino pendimethalin	Ni et al. (2016b)
Bacillus subtilis Y3	LB broth or MSM (pH 7.5)	PNR activity at 35 °C	7	100	Reduced the C-6 nitro group of the aromatic ring of PND to 2-nitro-6-amino-N-(1-ethylpropyl)-3,4-xylidine	Ni et al. (2016a)
Bacillus subtilis Y3	MS (pH 7)	30 °C at 150 rpm	100	99.5	Nitroreductase catalyze nitroreduction of PND to 6-amino pendimethalin	
B. safensis FO-36bT, *B. subtilis* subsp. *inaquosorum* KCTC *13429 T*, *B. cereus* ATCC 14579	MS (pH 7)	25 °C at stationary phase	28	90 (30 days)	N-(1-ethylpropyl)-3-methyl-2,6-diaminobenzine	Ishag et al. (2017)

(continued)

Table 20.1 (continued)

Source	Medium	Condition	PND (mgL^{-1})	Degradation (%)	Metabolites	Reference
Paracoccus sp. *P13*	MS (pH 7)	30 °C at150 rpmin (dark)	200	99.8 (5 days)	Oxidative ring cleavage converted PND to 1,3-dinitro-2-(pentan-3-ylamino)butane-1,4-diol	Ni et al. (2018)
			Fungi			
Aspergillus flavus, A. terreus, Fusarium solani, F. oxysporum, Penicillium citrinum, and P. simplicissimum	Czapek's broth (pH 6.5)	28 ± 2 °C in dark	500	60 (15 days)	N-(1-ethylpropyl)-2-amino-6-nitro-3,4-xylidin, by partial N-dealkylation by ring hydroxylation, N-propyl-3-methyl-4-hydroxy2,6 dinitroaniline by nitro group reduction and 2,6dinitro-3,4-xylidene by complete N-dealkylation	Barua et al. (1990)
Fusarium oxysporum and *Paecilomyces variotii*	Czapek's broth (pH 6.8)	31 ± 2°C (dark with intermittent shaking)	100	100 (6 days)	N-(1-ethylpro-pyl)-3,4-dimethyl-2-nitrobenzene-l,6-diamine (II) and 3,4-imethyl-2,6-dinitroaniline by nitroreduction or dealkylation of substituted amine	Singh and Kulshrestha (1991)
L. saksenae	MS (pH 7)	30 °C at 150 rpm	25	99.5 (10 days)	Metabolites undetected	Pinto et al. (2012)
Phanerochaete chrysosporium	MS (pH 7)	30°at 150 rpm	100	96	Metabolites undetected	Belal and Nagwa (2014)

MS, minimal salt medium

via C-dealkylation to N-(1-ethylpropyl)-2-amino-6-nitro-3,4-xylidine, and finally to 2,6-dinitro-3,4-xylidiene through complete N-dealkylation (Barua et al. 1990); and (vii) polyacrylamide and PUF-immobilized *Bacillus lehensis* XJU degraded the 100 ppm PND in 96 h and 6-amino pendimethalin through reduction reaction to form 3,4-dimethyl-2,6-dinitroaniline metabolites via oxidative dealkylation (More et al. 2015). In brief, the forgoing evidences suggest that more efforts were earlier focused on metabolites formed after nitroreduction reaction and more scope still exists to search for newer microbes with array of metabolic apparatus for effective remediation of PND using microbial system.

20.6 Pathway for Biodegradation of PND

The need for PND removal has been the focus of research more evidently in the recent years. Conventional methodologies used for pesticide treatment with (i) adsorption, (ii) photolysis, (iii) photolysis combined with oxidants, (iv) photo-fenton process, and (v) photocatalysis did not receive much commercial interest as these techniques are (i) just a segregation of pesticides rather than a treatment and (ii) often result into incomplete mineralization and (iii) more toxic residues that may even persist for longer duration in the ecosystem. These conventional physico-chemical approaches have proved to be (i) uneconomical, (ii) unreliable, and (iii) inconclusive due to incomplete conversion and (iv) failure with consequent unintentional damage to environment. On these evidences, microbial degradation to remediate polluted sites appeared as an emerging technology (Samanta et al. 2002). Biodegradation of herbicide using microbial system is (i) economic, (ii) is effective, and (iii) does not produce toxic products (Jiang and Li 2018), (iv) catalyzes either mineralization of compound to form inorganic end products, such as CO_2 and water, or (v) attempts co-dissimilatory nonspecific transformation with enzyme(s) specific for other substrates under aerobic or anaerobic conditions.

Several bacteria and fungi so far explored for degradation of PND have not deciphered the metabolic mechanism (Table 20.2). Ni et al. (2016b) reported the nitro-reduction is the first initial degradation and detoxification step for PND and recognized PND nitroreductase (PNR) encoded by *pnr* responsible for initial degradation step of PND from *Bacillus subtilis* Y3. PNR, a functional homodimer with a subunit molecular size of 23 kDa, showed reduction of C-6 nitro group of PND to yield 2-nitro-6-amino-N-(1-ethylpropyl)-3,4-xylidine which showed negligible inhibitory effect on *Saccharomyces cerevisiae* BY4741 during detoxification assay vis-a-vis parent PND, indicating potential role of PNR in detoxification of PND. More studies on such aspect are required to delineate the pathway for microbial mineralization of PND and, therefore, warrant search for robust microbes endowed with inherent capability to not only to degrade PND but also catabolize other toxic pesticides in the presence of metal ions in edaphic conditions to recoup the contaminated soil habitats.

More efforts to search the potent microbes which contain pesticide-degrading gene from the ecological habitat are highly essential for bioaugmentation,

Table 20.2 Metabolic pathway adopted for biodegradation of pendimethalin by various microbes

Microbes	Pathway		Reference
Bacteria			
A. chroococcum	PND	→2,6-Dinitro-3,4-xylidine	Kole et al. (1994)
		→6-Nitro-3,4-xylidine	
	PND	→2,6-Dinitro-3,4-xylidine (2,6-dinitro-3,4-dimethyl) phenyl cetamide	
	PND	→2-Methyl-4,6-dinitro-5-[(l-ethylpropyl) amino] benzyl alcohol	
		→2-Methyl 4,6-dinitro-5-[(1-ethylpropy1) aminol benzaldehyde	
	PND	→2-Methyl-4-nitro-5-N-(1- cyclopropyl)-6-nitrosobenzyl alcohol	
	PND	→N-2,6-dinitro-3,4- dimethyl) phenyl cetamide	
	PND	→2-nitro –6- amino-(N- ethylpropyl)-3, –4 xylidine	
Bacillus circulans	PND	→6-Aminopendimethalin	Megadi et al. (2010)
	PND	→3,4-dimethyl –2, 6- dinitroaniline	
		→Pantane	
Pseudomonas aeruginosa	PND	→*N*-(1-Ethylpropyl)-3-methyl-2, 6 diaminobenzin + CH_2O	Shaer et al. (2013)
Bacillus subtilis Y3	PND	→6-Aminopendimethalin	Ni et al. (2016a)
		→5-Amino-2-methyl-3-nitroso-4-(pentan-3-ylamino) benzoic acid	
		→8-Amino-2-ethyl-5-(hydroxymethyl)-1,2 dihydroquinoxaline-6-carboxylic acid	
Paracoccus sp. *P13*	PND	→1,3-Dinitro-2-(pentan-3- ylamino) butane-1,4-diol	Ni et al. (2018)
		→CO_2 + H_2O	
Fungi			
Fusarium solani	PND	→N-(1- ethylpropyl)-2-amino-6-nitro-3,4-xylidine	Barua et al. (1990)
	PND	→N-propyl- 3-methyl-4-hydroxy 2, 6 dinitroaniline	
	PND	→2,6 Dinitro-3,4-xylidene	
Fusarium oxysporum and *Paecilomyces variotii*	PND	→N-(1-Ethylpropyl)-3,4-dimethyl-2-nitrobenzene-l,6-diamine	Singh and Kulshrestha (1991)
	PND	→Isomeric diamine (N-(1-ethylpropyl)-3, 4-dimethyl-6-nitrobenzene-l, 2 diamine-	
		→3,4-Dimethyl-2,6-dinitroaniline	

biostimulation, or natural attenuation strategy. In bacteria, pesticide-degrading genes often reside on the plasmids (catabolic plasmid) and encode for the pollutant-degrading enzymes (Laemmli et al. 2000). The catabolic plasmids are now recognized from *Alcaligenes*, *Actinobacter*, *Arthrobacter*, *Cytophaga*, *Moraxella*, *Klebsiella*, and *Pseudomonas*. Plasmid-mediated augmentation method could possibly provide an effective solution to remove PND from the environment.

20.7 Conclusion

Ecosystems are under consistent threat due to the exposure to excess use of herbicide pollution. Microbes equipped with biodegradation pathways and response to biotic and abiotic system are providing the tool to design most suitable strategy for on-site or off-site removal of herbicides. Despite the availability of a gamut of microbes from the ecological habitat, the reach of bioremediation to degrade pesticide remains a great challenge. This review on environmentally relevant nitroaromatic pesticides reveals many limitations and future research scopes associated with the current body of knowledge. Effective and indigenous microbial strain or consortia with capability to tolerate and degrade pesticides in edaphic conditions as an effective biofertilizer could minimize chemical fertilizer application by 20–30%. Genetically engineered organism, with multiple nitroaromatic compound-degrading genes or enzyme systems, may play a crucial role in the biodegradation of these otherwise recalcitrant compounds. The development of a system-oriented understanding of natural pesticide attenuation with respect to pesticide degradation at low concentrations and in low-nutrient situations is urgently needed so as to ameliorate the toxicity of herbicide and safeguard the planet Earth.

References

Abd-Algadir M, Sabah Elkhier M, Idris O (2011) Changes of fish liver (*Tilapia nilotica*) made by herbicide (Pendimethalin). J Appl Biosci 43:2942–2946

Ahemad M, Khan MS (2011) Ecotoxicological assessment of pesticides towards the plant growth promoting activities of lentil (Lens esculentus)-specific *Rhizobium sp.* strain MRL3. Ecotoxicology 20:661–669

Ahemad M, Zaidi A, Khan MS, Oves M (2009) Factors affecting the variation of microbial communities in different agro-ecosystems. In: Microbial strategies for crop improvement. Springer, Berlin, p 301–324

Ahmad I, Ahmad A, Ahmad M (2016) Binding properties of pendimethalin herbicide to DNA: multispectroscopic and molecular docking approaches. Phys Chem Chem Phys 18:6476–6485

Almazán-Sánchez PT, Cotillas S, Saez C, Solache-Rios MJ, Martínez-Miranda V, Canizares P, Linares-Hernández I, Rodrigo MA (2017) Removal of pendimethalin from soil washing effluents using electrolytic and electro-irradiated technologies based on diamond anodes. Appl Catal B Environ 213:190–197

Ayub SM, Garg SK, Garg KM (1997) Sub-acute toxicity studies on pendimethalin in rats. Indian J Pharm 29:322

Barua AS, Saha J, Chaudhuri S, Chowdhury A, Adityachaudhury N (1990) Degradation of pendimethalin by soil fungi. Pestic Sci 29:419–425

Belal EB, Hassan NE (2013) Dissipation of pendimethalin by *Bacillus megaterium*. Mansoura J Plant Prot Pathol 5:463–472

Belal EB, Nagwa ME (2014) Biodegradation of pendimethalin residues by *P. chrysosporium* in aquatic system and soils. J Biol Chem Environ Sci 9:383–400

Choudhury PP, Singh R, Ghosh D, Sharma AR (2016) In herbicide recomandation for various crops. Herbicide use in Indian Agriculture ICAR – Directorate of Weed Research, Jabalpur, Bulletin No.22: 25–48

Diez MC (2010) Biological aspects involved in the degradation of organic pollutants. J Soil Sci Plant Nutr 10(3):244–267

Dureja P, Walia S (1989) Photodecomposition of pendimethalin. Pestic Sci 25:105–114

El-Awadi ME, Hassan EA (2011) Improving growth and productivity of fennel plant exposed to pendimethalin herbicide: stress–recovery treatments. Nat Sci 9:97–108

El-Sharkawy NI, Reda RM, El-Araby IE (2011) Assessment of stomp®(Pendimethalin) toxicity on *Oreochromis niloticus*. J Am Sci 7:568–576

Engebretson J, Hall G, Hengel M, Shibamoto T (2001) Analysis of pendimethalin residues in fruit, nuts, vegetables, grass, and mint by gas chromatography. J Agric Food Chem 49:2198–2206

European Community (2003) Review report for the active substance pendimethalin, Report 7477/VI/98-final. European Comission, Health and Consumer Protection Directorate General, Brussels, pp 1–43

Fennell BJ, Naughton JA, Dempsey E, Bell A (2006) Cellular and molecular actions of dinitroaniline and phosphorothioamidate herbicides on *Plasmodium falciparum*: tubulin as a specific antimalarial target. Mol Biochem Parasitol 145:226–238

Gianessi LP (2013) The increasing importance of herbicides in worldwide crop production. Pest Manag Sci 69:1099–1105

Gomathi Devi LN, Krishnamurthy G (2008) Photocatalytic degradation of the herbicide pendimethalin using nanoparticles of $BaTiO_3/TiO_2$ prepared by gel to crystalline conversion method: a kinetic approach. J Environ Sci Health B 43:553–561

Gong W, Liu X, Xia S, Liang B, Zhang W (2016) Abiotic reduction of trifluralin and pendimethalin by sulfides in black-carbon-amended coastal sediments. J Hazard Mater 310:125–134

Ishag AESA, Abdelbagi AO, Hammad AMA, Elsheikh EAE, Elsaid OE, Hur JH (2017) Biodegradation of endosulfan and pendimethalin by three strains of bacteria isolated from pesticides-polluted soils in the Sudan. Appl Biol Chem 60:287–297

Jiang J, Li S (2018) Microbial degradation of chemical pesticides and bioremediation of pesticide-contaminated sites in China. In: Twenty years of research and development on soil pollution and remediation in China. Springer, Singapore, p 655–670

Jogdand SN (2000) Biotechnology for hazardous waste management. In: Environmental biotechnology. Himalaya Publication House, New-Delhi, p 121–140

Juhler RK, Sorensen SR, Larsen L (2001) Analysing transformation products of herbicide residues in environmental samples. Water Res 35:1371–1378

Kanekar P, Daupure P, Sarnaik S (2003) Biodegradation of nitroexplosive Indian. J Exp Biol 41:991–1001

Keese RJ, Camper ND, Whitwell T, Riley MB, Wilson PC (1994) Herbicide runoff from ornamental container nurseries. J Environ Qual 23:320–324

Keum YS, Li QX (2004) Reduction of nitroaromatic pesticides with zero-valent iron. Chemosphere 54:255–263

Kidd H, James DRE (1991) The agrochemicals handbook. The Royal Society of Chemistry, Cambridge, UK

Kole RK, Saha J, Pal S, Chaudhuri S, Chowdhury A (1994) Bacterial degradation of the herbicide pendimethalin and activity evaluation of its metabolites. Bull Environ Contam Toxicol 52:779–786

Laemmli CM, Leveau JH, Zehnder AJ, van der Meer JR (2000) Characterization of a second tfd gene cluster for chlorophenol and chlorocatechol metabolism on plasmid pJP4 in *Ralstonia eutropha* JMP134 (pJP4). J Bacteriol 182:4165–4172

Megadi VB, Tallur PN, Hoskeri RS, Mulla SI, Ninnekar HZ (2010) Biodegradation of pendimethalin by Bacillus circulans. Indian J Biotechnol 9:173–177

More VS, Tallur PN, Niyonzima FN, More SS (2015) Enhanced degradation of pendimethalin by immobilized cells of *Bacillus lehensis* XJU. 3 Biotech 5:967–974

Moza PN, Hustert K, Pal S, Sukul P (1992) Photocatalytic decomposition of pendimethalin and alachlor. Chemosphere 25(11):1675–1682

Ni HY, Wang F, Li N, Yao L, Dai C, He Q, He J, Hong Q (2016a) The nitroreductase PNR is responsible for the initial step of pendimethalin degradation in *Bacillus subtilis* Y3. Appl Environ Microbiol 82:7052–7062

Ni H, Yao L, Li N, Cao Q, Dai C, Zhang J, He Q, He J (2016b) Biodegradation of pendimethalin by *Bacillus subtilis* Y3. J Environ Sci 4:121–127
Ni H, Li N, Qiu J, Chen Q, He J (2018) Biodegradation of pendimethalin by *Paracoccus sp.* P13. Curr Microbiol 75:1077–1083
Ortiz-Hernández ML, Sánchez-Salinas E, Olvera-Velona A, Folch-Mallol JL (2011) Pesticides in the environment: impacts and its biodegradation as a strategy for residues treatment. In: Pesticides-formulations, effects, fate. InTech, China, p 551–574
Pandit GK, Pal S, Das AK (1995) Photocatalytic degradation of pendimethalin in the presence of titanium dioxide. J Agric Food Chem 43:171–174
Parte SG, Mohekar AD, Kharat AS (2017) Microbial degradation of pesticide: a review. Afr J Microbiol Res 11:992–1012
Patel S, Bajpayee M, Pandey AK, Parmar D, Dhawan A (2007) In vitro induction of cytotoxicity and DNA strand breaks in CHO cells exposed to cypermethrin, pendimethalin and dichlorvos. Toxicol in Vitro 21:1409–1418
Pimentel D (1995) Amounts of pesticides reaching target pests: environmental impacts and ethics. J Agric Environ Ethics 8:17–29
Pimentel D, Wilson C, McCullum C, Huang R, Dwen P, Flack J, Cliff B (1997) Economic and environmental benefits of biodiversity. BioScience 47(11):747–757
Pinto AP, Serrano C, Pires T, Mestrinho E, Dias L, Teixeira DM, Caldeira AT (2012) Degradation of terbuthylazine, difenoconazole and pendimethalin pesticides by selected fungi cultures. Sci Total Environ 435:402–410
Promkaew N, Soontornchainaksaeng P, Jampatong S, Rojanavipart P (2010) Toxicity and genotoxicity of pendimethalin in maize and onion. Kasetsart J-Nat Sci 44:1010–1015
Richardson ML, Gangolli S (eds) (1992) The dictionary of substances and their effects. Royal Society of Chemistry, London. (1)
Ritter L, Solomon KR, Forget J, Stemeroff M, O'Leary C (1995) Persistent organic pollutants: an assessment report on: DDT-aldrin-dieldrin-endrin-chlordane-heptachlor-hexachlorobenzene-mirextoxaphene-polychlorinated biphenyls-dioxins and furans 1995. Inter-Organization Programme for the Sound Management of Chemicals (IOMC), Geneva
Roca E, D'Errico E, Izzo A, Strumia S, Esposito A, Fiorentino A (2009) In vitro saprotrophic basidiomycetes tolerance to pendimethalin. Int Biodeterior Biodegrad 63:182–186
Sakata M (2005) Organophosphorous pesticides. In: Suzuki O, Watanabe K (eds) Drugs and poisons in humans. Springer, Verlag, New York, pp 535–544
Samanta SK, Singh OV, Jain RK (2002) Polycyclic aromatic hydrocarbons: environmental pollution and bioremediation. Trends Biotechnol 20:243–248
Scheunert I, Mansour M, Doerfler U, Schroll R (1993) Fate of pendimethalin, carbofuran and diazinon under abiotic and biotic conditions. Sci Total Environ 132:361–369
Shaer IBS, Abdelbagi AO, Elmustafa EA, Ahmed SAI, Osama GE (2013) Biodegradation of pendimethalin by three strains of bacteria isolated from pesticides polluted soils. Univ Khartoum J Agric Sci 21:233–252
Shroff R (2000) Chairman address in pesticide information. Annual Issue. Pesticide Association of India Publication, New Delhi
Singh SB, Kulshrestha G (1991) Microbial degradation of pendimethalin. J Environ Sci Health B 26:309–321
Singh B, Singh K (2014) Microbial degradation of herbicides. Crit Rev Microbiol 42:245–261
Strandberg M, Scott-Fordsmand JJ (2004) Effects of pendimethalin at lower trophic levels-a review. Ecotoxicol Environ Saf 57:190–201
Swarcewicz MK, Gregorczyk A (2012) The effects of pesticide mixtures on degradation of pendimethalin in soils. Environ Monit Assess 184(5):3077–3084
Timmis KN, Steffan RJ, Unterman R (1994) Designing microorganisms for the treatment of toxic wastes. Annu Rev Microbiol 48:525–557
Tsiropoulos NG, Miliadis GE (1998) Field persistence studies on pendimethalin residues in onions and soil after herbicide postemergence application in onion cultivation. J Agric Food Chem 46:291–295

Vighi M, Matthies M, Solomon KR (2017) Critical assessment of pendimethalin in terms of persistence, bioaccumulation, toxicity, and potential for long-range transport. J Toxicol Environ Health B 20:1–21

Walker A, Bond W (1977) Persistence of the herbicide AC 92,553, N-(1-ethylpropyl) 2, 6-dinitro-3, 4-xylidine, in soils. Pestic Sci 8:359–365

Zheng SQ, Cooper JF (1996) Adsorption, desorption, and degradation of three pesticides in different soils. Arch Environ Contam Toxicol 30:15–20

Nisin Production with Aspects on Its Practical Quantification

21

Sunita Singh

21.1 Bacteriocin Nisin: A Gram-Positive Lactic Acid Bacterial Antibiotic

Bacteriocins are a diverse group of peptides (Hansen et al. 1991; Klaenhammer 1988) with bactericidal (antibacterial) action that varies in size (molecular weight), production, and biochemical properties (Gálvez et al. 2007). The narrow spectrum of action of these peptides (to kill other closely related bacteria of related species) or broad spectrum (toward bacterial species across genera) and their involvement in protecting host itself from its own antimicrobial activity (Bowdish et al. 2005) and the cell signaling mechanism are noteworthy. They can manipulate the food environment(s) by competitive exclusion, of the desirable strain(s), the particular bacteriocin can act on. The term "nisin" was coined (Mattick and Hirsch 1947) after it was initially discovered as "N inhibitory substance" produced by *Streptococcus* of the Lancefield serological group, now classified as *L. lactis* (Cleveland 2001).

Nisin lantibiotic, a bacteriocin produced by *Lactococcus lactis* subsp. *lactis*, a homolactic bacteria, produces (L+) lactic acid from glucose. The other homofermentative LABs are *L. lactis* subsp. *cremoris*, *Streptococcus thermophilus*, *Lactobacillus delbreuckii* biovar. *bulgaricus*, *L acidophilus*, *L. helveticus*, *L. casei*, and *L. plantarum*. This process of lactic fermentation with a glycolytic metabolism [$C_6H_{12}O_6 \rightarrow 2CH_3CHOHCOOH$+energy], of glucose to lactic acid, yields up to 98% lactic acid. When heterofermentative LABs metabolize glucose through a complex glucose-6-phosphate pathway, they produce acetic acid, ethyl alcohol, carbon dioxide, and other neutral by-products, such as diacetyl and acetoin, as major components. The lactates are metabolized by the phosphoenolpyruvate (PEP) pathway. During fermentative production of nisin, lactic acid is the other major end product

S. Singh (✉)
Division of Food Science and Post-harvest Technology, ICAR-Indian Agricultural Research Institute, New Delhi, India

D. P. Singh et al. (eds.), *Microbial Interventions in Agriculture and Environment*, https://doi.org/10.1007/978-981-13-8391-5_21

which is also responsible for the preservative action in foods. In the metabolism, the lactic acid produced has a negative feedback control on nisin production.

Nevertheless, LABs are the most exploited for preservation of traditional foods, with lactic acids in addition to nisin, which is a "GRAS" preservative (Food and Drug Administration 1988). It is most consumed for the preservation of fermented cheeses and milk. Such LABs can thus be used as multifunctional starters as per required. Use of nisin as an additive can not only curb use of harmful chemical preservatives in foods that are added at higher concentrations/levels but can also help to avoid use of various antibiotics in foods and feeds. In other words, they can be used to control pathogens that need antibiotics to target their removal. Such bio-preservatives can prevent chemicals and antibiotics in our foods. Nisin has wide medical applications too.

21.1.1 Studies with Nisin

Interest in nisin started when *Lactococcus lactis*, a natural flora found in dairy products, produced nisin (Rogers 1928; Chevalier et al. 1957). This led to use of nisin instead of nitrites to preserve canned foods (Rayman and Hurst 1984) in order to control clostridia and its spores. This could also reduce the excessive use of nitrites as a preservative, to deliver foods like meat-free and safe from spore growth. Studies on bacteriostatic effects on cell viability showed that nitrite inhibitions were possible only at greater levels than required to modify the sulfhydryl groups (Buchmann and Hansen 1987). Nitrite and nisin showed similar action because both target same specific sulfhydryl groups (Morris et al. 1984; Buchman and Hansen 1987). In nitrite-preserved foods, viable outgrowth of inhibited spores (Benedict 1980) normally develops again, when nitrite disappears from the inhibitory environment.

Due to multiple drug resistance developing all over the world, researchers are now constructing safer peptides from constructs of nisin, to overcome the antibacterial drug resistance to combat diseases (Yang et al. 2014). Replacement of such secondary metabolites (antibiotics) can play a big role in replacing vancomycin and oxacillin that have the same minimal inhibitory concentrations (MICs), as lantibiotic nisin and mutacin B-Ny266, and many bacteria cannot be treated, as they have developed resistance toward these antibiotics (Mota-Meira et al. 2000). So it is possible to increase/diversify its property, with newer peptides constructed from nisin residues, by chemical and genetic modifications (Sahl and Bierbaum 1998) in amino acid composition, their sequence, molecular characterizations, and antibiotic activity. These studies can be extended to other microbial peptides using molecular techniques too (Araya et al. 1992; Joerger and Klaenhammer 1990; Kemperman et al. 2003). Maganin is an antibacterial peptide of animal origin but contained lower specific activity than nisin. This is another reason to intrigue upon. The elucidation of such reasons may lead to answers, and creating newer and less harmful antibiotics (Breukink et al. 1997, 1999; Correia et al. 2015) was emphasized. Nisin can even act as an anticarcinogenic substance, at concentrations 40 and 80 μg/mL (Joo et al. 2012).

21.1.2 Biosynthesis of Nisin: A Small Peptide

The biosynthesis of nisin (Hurst 1966) is of great interest, as it serves as a model to study structure-function relationships in small proteins (Buchman et al. 1988). There is much scope further toward altered biosynthesis, in the majority of its small derivatives. The region altered in many small peptides of nisin, at its hinge, lies within a central, 3-amino acid stretch (Field et al. 2010). Such nisins include N20K and M21K, nisin M21V, nisin K22T (Field et al. 2010), and nisin N20P (Cotter 2012). The altered serine 29, at the "C" terminal end of nisin, helps to enhance the activity of nisin against Gram-positive and Gram-negative pathogens (Field et al. 2012). Nisins Z, F, and Q are variants produced by alteration of the structural gene that is flanked by open reading frames (ORFs) (Piper et al. 2011). These peptides can diffuse more profusely through complex matrices (Rouse et al. 2012).

Thus, to fully understand and produce nisin bacteriocin, to its full potential and in its active form (Schnell et al. 1988), it is also important to have an insight of its status classified under bacteriocins, microbial fermentation and conditions to produce it in medium, structure-function relationships along with schematic steps in its regulation, and immunity. The most important mode of action of the molecule as related to changes in peptide structure (Kleinnijenhuis et al. 2003) can give much of required information on its activity. Thus, research findings on its molecular mass, secretion after production, range/spectrum of activity, specific effects of heat, pH and proteolytic enzymes, salt, and assay of bacteriocin conditions can relate to appreciating nisin.

21.2 Classification of the Nisin Bacteriocin Produced by Gram-positive Bacteria (LABs)

Nisin is classified as a Class IA Gram-positive bacteriocin. It contains lanthionine and other dehydroamino residues post synthesis and is a lantibiotic. It acts against *Enterococcus* sp., *Lactobacillus* sp., *Lactococcus* sp., *Leuconostoc* sp., *Listeria* sp., *Staphylococcus* sp., *Micrococcus* sp., *Pediococcus* sp., and *Mycobacterium* sp. and is inclusive of spores of *Clostridium* sp. and *Bacillus* sp. If the Gram-negative bacteria yeasts are to be inhibited, synergistic substances like EDTA (chelators) are used with nisin (Delves-Broughton et al. 1996).

21.2.1 The Early Classification

The early classification (Klaenhammer 1993) of Gram-positive bacterial bacteriocins gave out four major classes:

Class I: Lanthionine-containing bacteriocins or lantibiotics. These lantibiotics have been resolved under electron microscope but are not separated by an ultracentrifuge.
Class II: Nonlanthionine-containing peptides are small and heat stable.
Class III: It comprises of large heat-labile protein murein hydrolases.

Class IV: It comprises of complex bacteriocins where the activity is due to chemical moieties with the protein.

21.2.2 The Revised Early Classification

The revised early classification (Moll et al. 1999; Nes et al. 1996) with heat-stable peptides is subdivided as:

Class I: Type "A" elongated lantibiotics have small peptides (<5 kDa); type "B" globular structured peptide lantibiotics are immunologically active and enzyme inhibitors (Jung 1991).
Class II: Class **IIA** have *Listeria*-active peptides, class **IIB** have the two peptide bacteriocins; class **IIC** have the secretion-dependent peptides, and class **IID** comprise of those that do not belong to other subgroups.
Class III: Class III are heat-labile peptides and large in size (>30 KDa).

21.2.3 The Modified Classification (Cotter et al. 2005)

Class I: Lantibiotics
Class II: Non-lantibiotics (class **IIA**, *Listeria*-active peptides; class **IIB**, bacteriocins with two peptides; class **IIC**, cyclic peptides; class **IID**, a repository for all other nonlanthionine linear peptide bacteriocins)
Class III: Bacteriolysins
Class IV: Non-bacteriocin lytic proteins

21.2.4 Most Recent Modified Classification Scheme

The classification scheme underwent modifications (Heng et al. 2007), before the most recent modified classification scheme (Karpinski and Szkaradkiewicz 2016) has considered the molecular weight, YGNGVXC motif (N-terminal sequences) of genes for bacteriocin located in DNA of plasmid (Holo et al. 1991) or chromosome for expression, disulfide bridges in primary structure of nisin, target organisms (wide or narrow spectrum), *Listeria* genus specific activity, and temperature sensitivity (thermostability of peptides):

(I) Lantibiotics, (II) non-lantibiotics (a, b, c, and d), (III) bacteriolysins, and (IV) non-bacteriocin (Fig. 21.1)

21.2.5 Variants of Nisin Bacteriocins

Nisin A lantibiotic protein is a cationic (net positively charged) protein with a linear/elongated flexible structure; the amphiphilic has a screw-shaped pentacyclic polypeptide (Jung 1991). Details of various other nisin variants can be downloaded from

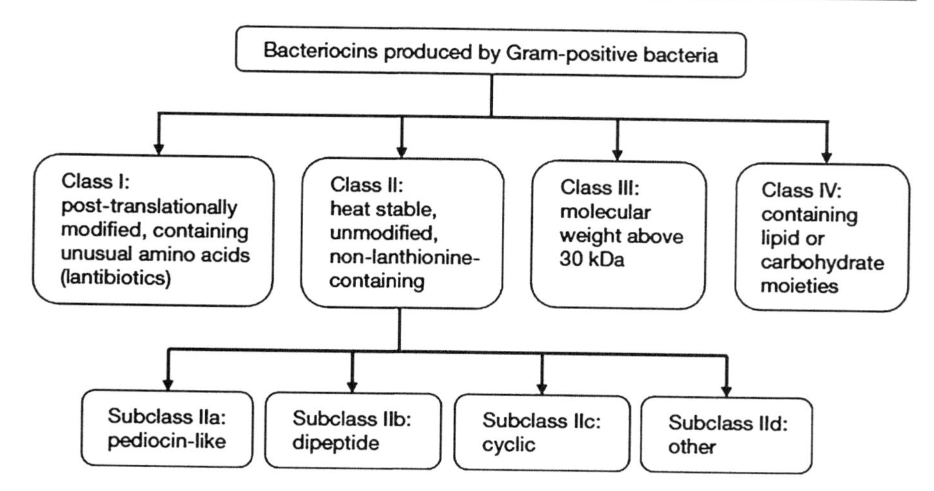

Fig. 21.1 Classification scheme for Gram-positive bacterial bacteriocins. (Source: Karpinski and Szkaradkiewicz 2016)

bactibase.hammamilab.org (Figs. 21.2, 21.3, 21.4, 21.5, and 21.6). *Lactococcus lactis* produces nisin A (MW 3352), produced by *L. lactis* subsp. *lactis* (Mattick and, Hirsch, 1947; Gross and Morell 1971; Field et al. 2012); nisin F, produced by *L. lactis* subsp. *lactis* (de Kwaadsteniet et al. 2008; Ustyugova et al. 2011); nisin H, produced by *Streptococcus hyointestinalis* (O'Connor et al. 2015); nisin Q, produced by *L. lactis* 61–14 (Zendo et al. 2003); nisin U, produced by *Streptococcus uberis* (Wirawan et al. 2006); nisin H, produced by *Streptococcus hyointestinalis* DPC6484 (O'Connor et al. 2015); and nisin Z, produced by *Lactococcus lactis* subsp. *lactis* biovar. *diacetylactis* UL 719 (Mulders et al. 1991; Meghrous et al. 1997; Graeffe et al. 1991; Zhang et al. 2014). Various other differences in variants of nisin are also reported (O'Connor et al. 2015 and others) (Table 21.1).

21.2.6 Other Gram-positive Bacterial Bacteriocins

Other bacteriocins that are produced by Gram-positive bacteria are, mersacidin [globular lantibiotic, acts by enzyme inhibition MW 1824 from *Bacillus* sp. HIL Y85, 54728] (Chatterjee et al. 1992), labrynthopeptin A2 [MW 1922 from *Actinomadura* sp.] (Meindl et al. 2010), subtilosin A [MW 3399 from *Bacillus subtilis* 168] (Babasaki et al. 1985; Gross et al. 1973), gallidermin (from Staphylococcus gallinarum) (Bierbaum et al. 1996), epidermin (Staphylococcus epidermidis) (Allgaier et al. 1986; Schnell et al. 1988), epilancin K7 (from *Staphylococcus epidermis* K7) (Van de Kamp et al. 1995a, b), Pep5 (from *Staphylococcus epidermidi* strain 5) (Kellner et al. 1991; Kaletta et al. 1989), lactocin S from *Lactobacillus sake* L45 (Mørtvedt et al. 1991), salivaricin A from *Streptococcus salivarius* 20P3 (Ross et al. 1993), lacticin 481 from *Lactococcus lactis* (Piard et al. 1993),

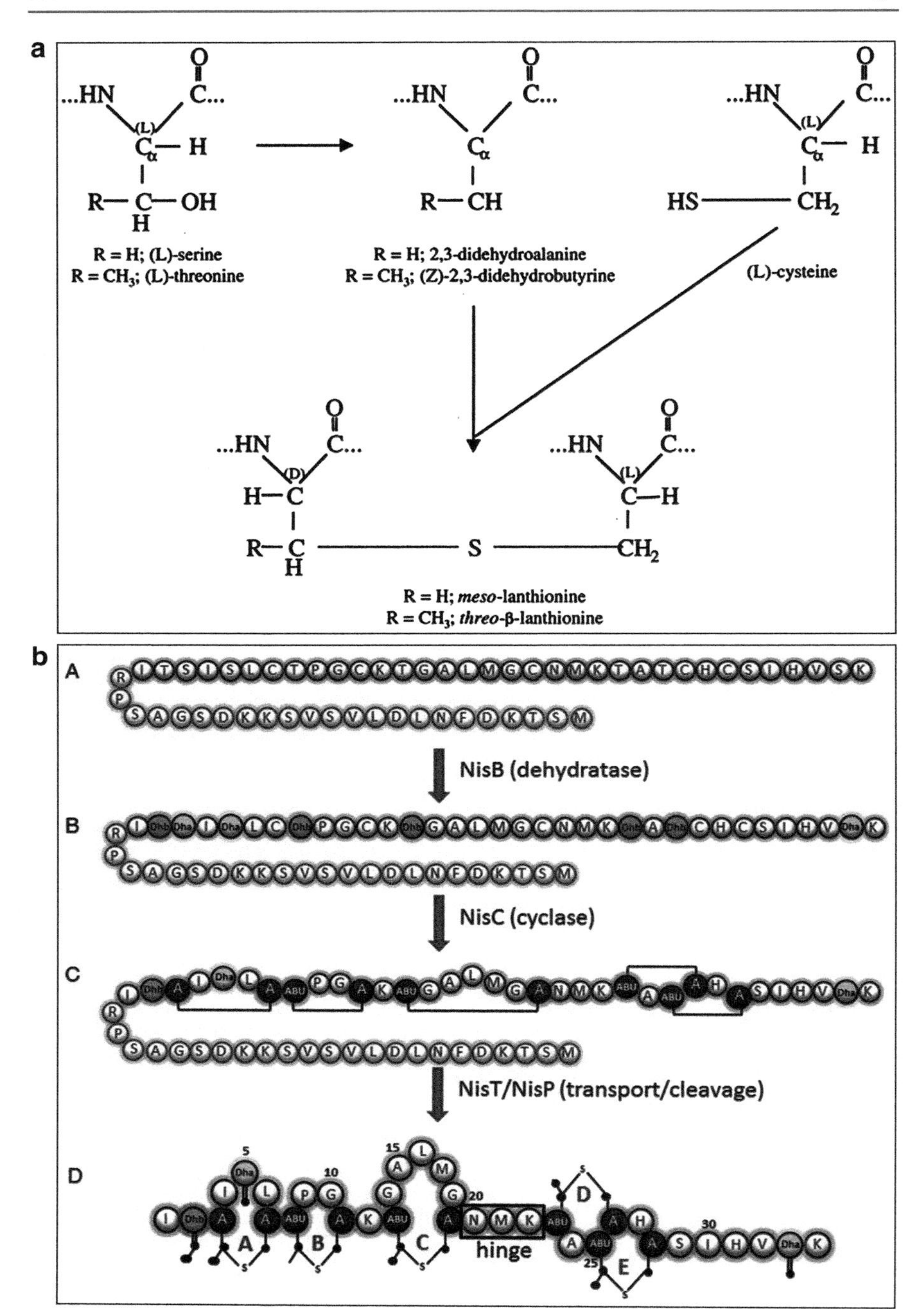

Fig. 21.2 (**a**) General mechanism for the formation of the thioether Lan during lantibiotic maturation; (**b**) posttranslational processing of nisin precursor pre-peptide to mature bioactive nisin peptide that is released. (Source: McAullife et al. 2001; Field et al. 2015)

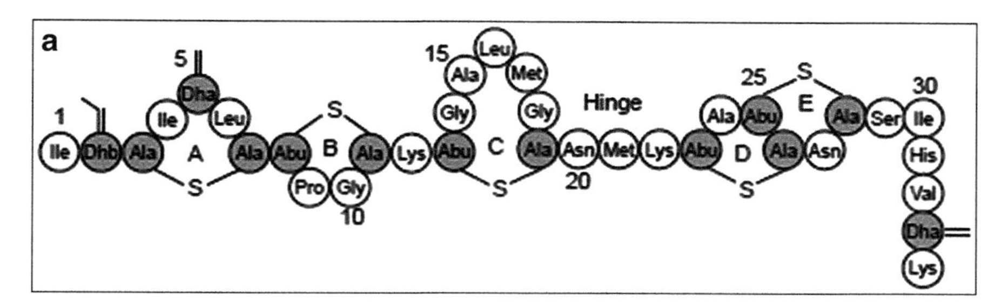

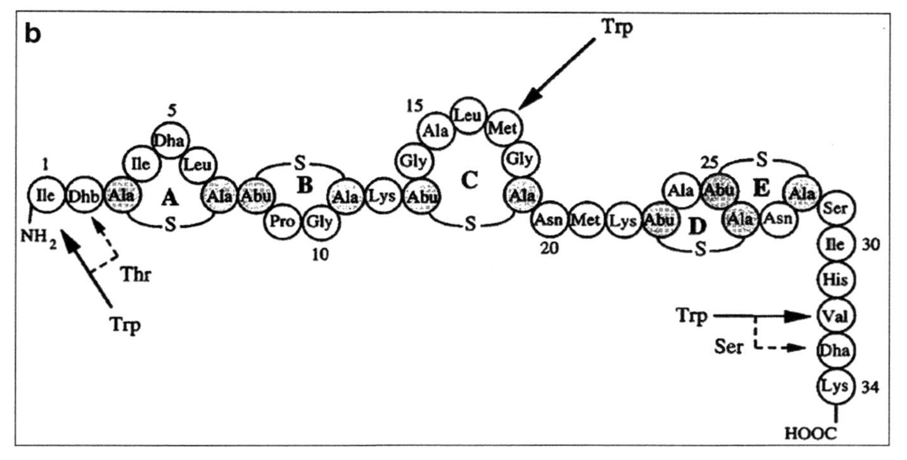

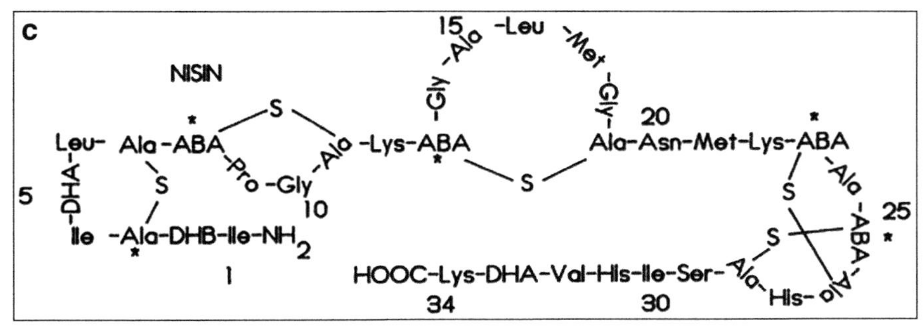

d

$CH_2=C(NH_2)-COOH$ dehydroalanine (DHA)

$CH_3CH=C(NH_2)-COOH$ dehydrobutyrine (DHB) (β-methyldehydroalanine)

$HOOC-CH(NH_2)-CH_2SCH_2-CH(NH_2)-COOH$ (D) (L) lanthionine

$HOOC-C(NH_2)(CH_3)-CH_2SCH_2-CH(NH_2)-COOH$ (D) (L) β-methyllanthionine

Fig. 21.3 (**a**) Primary structure of nisin: (**a**) nisin A, (**b**) nisin Z (a single tryptophan variant and histidine as in nisin A replaced by asparagine in nisin Z, (**c**) primary structure of nisin showing the D-stereo configuration (∗) for α-carbon, (**d**) structure of unusual amino acids of nisin (L to R): ABA, aminobutyric acid; Ala-S-Ala, lanthionine; β-methyllanthionine (Abu-S-Ala). (Sources: **a**, Hsu et al. 2004; **b**, Breukink et al. 1998; **c**, **d**, Wei and Norman 1990)

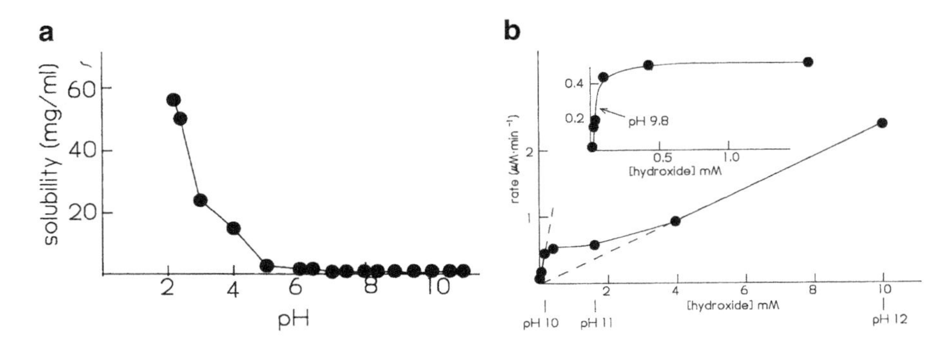

Fig. 21.4 (**a**) Solubility of nisin in the pH range 2.2 to 11.5 at 25 °C [in 0.1% TFA (pH 2.1), 100 mM sodium citrate (pH 2.2–6.0), 50 mM NaPi (pH 6.0–8.5 and pH 10.5–12.0), or 50 mM sodium carbonate (pH 8.5–10.5)]; (**b**) nisin reaction at high pH. The rate of disappearance of nisin at different hydroxide ion concentrations and on an expanded linear scale (inset). Experimental reaction times varied from more than 12 h at pH 6.5 to 10 h at pH 9.0 to 50 min at pH 12.0. (Source: Wei and Norman 1990)

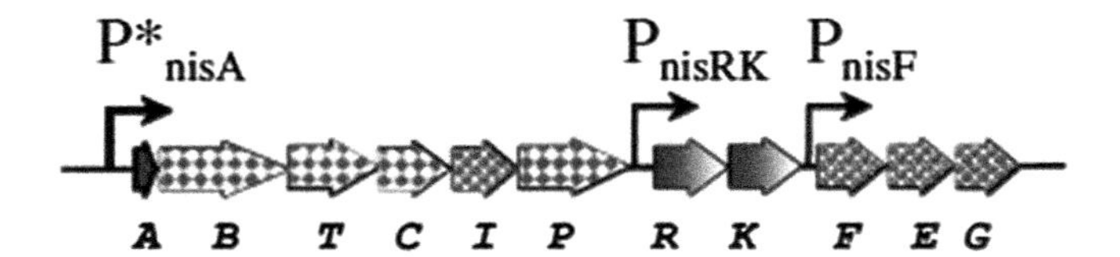

Fig. 21.5 Schematic representation of the nisin gene cluster. (Source: Mierau and Kleerebezem 2005)

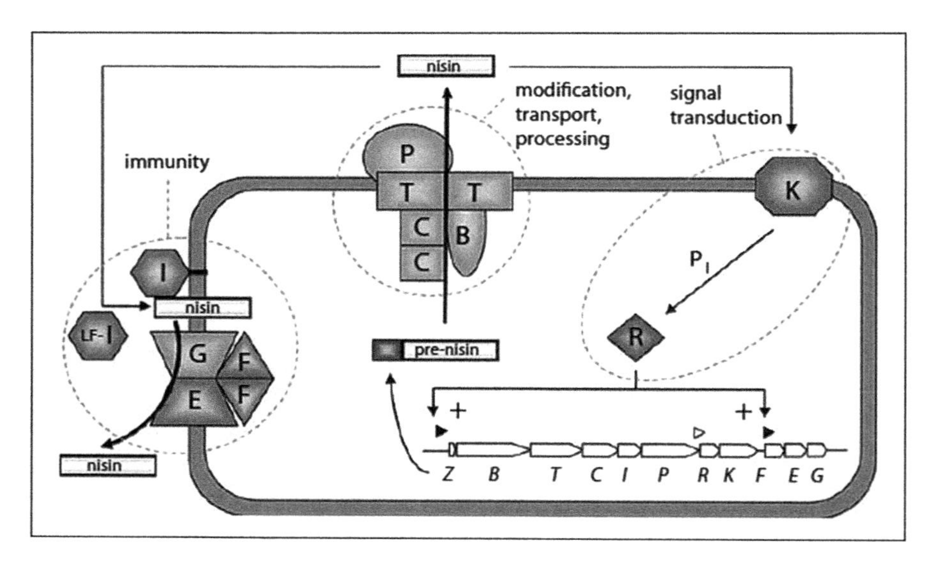

Fig. 21.6 The cellular model of nisin biosynthesis, regulation, and immunity in *L. lactis*. (Reproduced from Source: Cheng et al. 2007)

Table 21.1 Variations in nisin variants

S. No	Nisin peptides compared	Amino acids/other details after posttranslational modification to mature peptide nisin	References
1	Nisin A and nisin Z	Different by one amino acid [with His 27 substituted instead of Asn27 in mature nisin Z	Mulders et al. (1991)
		Nisin Z, a natural nisin variant, isolated from *L. lactis* subsp. *lactis* strain NIZO 22186	Abee et al. (1995)
2	Nisin A and nisin Q	Four amino acids are different in the mature peptide and two in the leader sequence	Zendo et al. (2003)
		His 27 replaced by Asn in nisin Q from *L. lactis* 61–14 like nisin Z	
3	Nisin Z and nisin Q	Three amino acids are substituted in the mature peptide; leucine is present instead of methionine in nisin Q	Zendo et al. (2003)
4	Nisin F and nisins A, Z, Q, and U	One or two bases in nisin F are different from nisins A, Z, and U; the position of two amino acids differs in nisin F compared to nisin A and nisin Z with His 27Asn; and nisin F with Ile30Val are obtained from gut-derived strain of *Streptococcus hyointestinalis*	de Kwaadsteniet et al. (2008) and O'Connor et al. (2015)

streptococcin A-FF22 from *Streptococcus pyogenes* strain FF22 (Jack et al. 1994a, b) cytolysin LL and cytolysin LS from *Enterococcus faecalis* (Gillmore et al. 1994), enterocin P from *Enterococcus faecium* P13 (Cintas et al. 1997), pedocin PA-1, from *Pediococcus acidilactici* strain PAC-1.0 (Henderson et al. 1992), acidocin B from *Lactobacillus acidophilus* (Leer et al. 1995; ten Brink et al. 1994), caseicin 80 from Lactobacillus casei strain B80 (Müller and Radler 1993).

21.3 Nisin: Structure, Composition, and Properties

21.3.1 Ribosomally Synthesized and Modified Structure

The peptide nisin is a polycyclic antibiotic or a lantibiotic, wherein it is biosynthesized in the cytoplasm on the ribosome as a precursor/pre-peptide protein (pre-nisin) with 57 amino acids in pre-peptide. That it was synthesized by a ribosomal mechanism was confirmed when this synthesis was blocked by protein synthesis inhibitors (Hurst and Peterson 1971). This pre-peptide protein formation was followed by posttranslational enzymatic modifications of precursor (Araya et al. 1992; Cheigh et al. 2002; De Vyust 1995; Mierau and Kleerebezem 2005).

The primary transcript pre-peptide of the linear lantibiotic has a leader sequence that is followed by the C-terminal pro-peptide from which the lantibiotic matures with a characteristic proteolytic processing site, having proline at position −2 (Engelke et al. 1992). The enzymatic modifications of pre-peptide include the formation of three unsaturated (or dehydro) residues of amino acids (i.e., one dehydroalanine {Dha} and two 3-methyl-dehydroalanines or two dehydrobutyrines

{Dhb}) formed from serine and threonine, respectively. A subsequent step further on adds cysteine sulfur to double bonding of di-dehydro amino acids and results in thioether bridge formation (De Vuyst 1995) (Fig. 21.2a). These then form thioether cross bridges with cysteine and create one lanthionine and four β-methyl-lanthionine S-bridges (i.e., in all five thioether amino acids) (Gross and Morell 1971; Jung and Sahl 1991; Moll et al. 1999; Schnell et al. 1988). The finally posttranslated protein peptide has a chain length of 34 amino acids (Figs. 21.2b and 21.3a, b). The final formed nisin modified after cleavage is an active lantibiotic (Schnell et al. 1988). It also has about 5–7% sulfur (Falconer 1949).

21.3.2 The Primary Structure of Nisin

The full primary and mature structure of nisin shows five regions (A, B, C, D, and E rings). The ring B connected to C and C to ring D lies in the hinge regions. The ring D interconnected to ring E with hinges that are flexible is also intertwined (Fig. 21.3a, b). The α-helical structure of a peptide is clear in the presence of trifluoroethanol. The domain regions with rings A, B, and C are hydrophobic, while the rings D and E are in hydrophilic domain of nisin (van den Hooven 1995), as analyzed by nuclear magnetic spin resonance (NMR) (Slijper et al. 1989; Chan et al. 1989).

The amphipathic nature of nisin A, due to both hydrophilic and hydrophobic regions, is due to (1) a single charged residue (Lys12) only, with the hydrophobicity dominant in N-terminal side, and (2) most of the charged and hydrophilic residues present on the carboxyl -terminal half. The 21–28 residues mainly represent amphipathic region of this molecule. The rings A, B, and C, with residues Ile 4, Leu 6, Pro 9, Leu 16, and Met 17 in the structure of nisin (Fig. 21.3a, b), are located opposite to the Lys12 and the thioether bonds. If these rings of nisin are modified, the activity of nisin is affected. However, still the N-terminal part of nisin has a major role in its activity (van Kraaij et al. 1997). When the nisin cleavage/modifications of nisin variants take place, the mutated nisin formed is mainly determined by the nature of the residue at amino acid preceding serine to be dehydrated. The activity of nisin Z reduces by negative charge at the C-terminal end of nisin Z. Since there are differences in the two ends of nisin, the properties of peptide environment around DHa5 different from DHa33 are responsible on the differences as determined in proton NMR spectrum (Wei and Norman 1990). Three other bacteriocins produced by *L. lactis* (lacticin 4811), *Lactobacillus sake* (lactocin *S*), and *Carnobactetium piscicola* (carnocin *UI49*) are similar to the molecular structure of mature nisin A (Klaenhammer 1993).

21.3.2.1 Nisin Molecule: Biochemical and Physicochemical Properties

The excess of lysine and arginine residues contributes to nisin being positively charged as a cationic peptide (Moll et al. 1999). The physicochemical characteristics (composition) of nisin variants are presented as compiled (Table 21.2). The mode of action of nisin as a function of nisin molecule is derived from its required physicochemical properties that help to depolarize the energised cytoplasmic membranes for pore formation across membranes (http://bactibase.hammamilab.org), as also detailed in latter section (Section 21.8). The residues and positions of the ribosomal modified peptide regions of the lantibiotic are reported as shown (Tables 21.3a, b). The details of nis genes in nisin variants are also available (bactibase.hammamilab.org).

Biochemical Properties, Stability, and Activity of Nisin

The application of nisin as an additive requires it to be a bioactive form with controlled stability. The brief composition (physicochemical) of its variants is compiled (Table 21.2) (Bactibase 2018). The gene sequences (Tables 21.3a, b) help in post translational modifications of nisin maturation, by enzymic conversion of Thr and Ser into dehydrated AA for the formation of thio-ether bonds with cysteine. The hydrophobicity in nisin is required for its separation and purification (Fremaux et al. 1993; Muriana and Klaenhammer 1991).

Structural Peptide Residues After Ribosomal Modifications of Nisin Peptide Variants

The details on its structure (Tables 21.2 and 21.3a, b) are still under continuous study by "Blast" analysis. Other details from protein sequencing studies are available (bactibase.hammamilab.org). Techniques are in use to study genes for nisin production, using specific probes of isolate *L. lactis* subsp. *lactis* in polymerase chain reaction (PCR) studies (Garde et al. 2001).

Also other detailed studies of various nisin variants are under continuous study along with their gene data (Tables 21.4 and 21.5) that are constantly being updated (http://bactibase.hammamilab.org).

Table 21.2 Nisin physicochemical composition/nature of the molecule (Bioinformatics)

Nisin	Nisin A	Nisin Z	Nisin Q	Nisin F	Nisin U
Formula	C_{143} H_{246} N_{42} O_{45} S_7	C_{141} H_{245} N_{41} O_{46} S_7	C_{143} H_{249} N_{41} O_{46} S_6	C_{140} H_{243} N_{41} O_{46} S_7	C_{134} H_{226} N_{36} O_{40} S_6
Absent amino acids	DEFQRWY	DEFQRWY	DEFQRWY	DEFQRWY	DENQRVWY
Mass (Da)	3516.78	3493.74	3489.73	3479.71	3192.37
Net charge	+5	+4	+4	+4	+4
Isoelectric point	8.52	8.51	8.51	8.51	8.51
Basic residues	5	4	4	4	4
Bowman index	–12.88	–14.86	–10.94	–15.74	–1.63
Instability index	27.52 (stable)	17.08 (stable)	13.45 (stable)	13.45 (stable)	30.65 (stable)
Extinction coefficient	250 M^{-1} cm^{-1}	250 M^{-1} cm^{-1}	0 M^{-1} cm^{-1}	0 M^{-1} cm^{-1}	0 M^{-1} cm^{-1}
Absorbance 280 nm	7.58	7.58	7.68	7.58	8.33

Source: http://bactibase.hammamilab.org

1. *Lantibiotic Nisin-A*

Table 21.3a Details of nisin A peptide

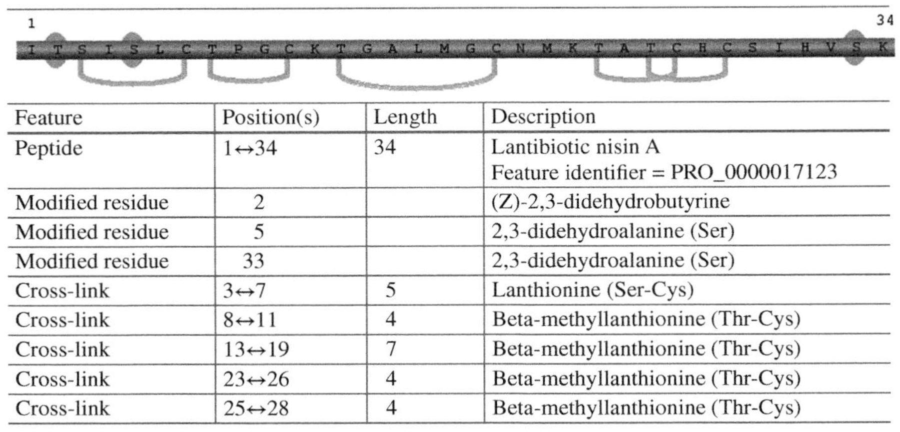

Feature	Position(s)	Length	Description
Peptide	1↔34	34	Lantibiotic nisin A Feature identifier = PRO_0000017123
Modified residue	2		(Z)-2,3-didehydrobutyrine
Modified residue	5		2,3-didehydroalanine (Ser)
Modified residue	33		2,3-didehydroalanine (Ser)
Cross-link	3↔7	5	Lanthionine (Ser-Cys)
Cross-link	8↔11	4	Beta-methyllanthionine (Thr-Cys)
Cross-link	13↔19	7	Beta-methyllanthionine (Thr-Cys)
Cross-link	23↔26	4	Beta-methyllanthionine (Thr-Cys)
Cross-link	25↔28	4	Beta-methyllanthionine (Thr-Cys)

Source: http://bactibase.hammamilab.org/BAC047

2. *Lantibiotic Nisin-Z*

Table 21.3b Details of nisin Z peptide

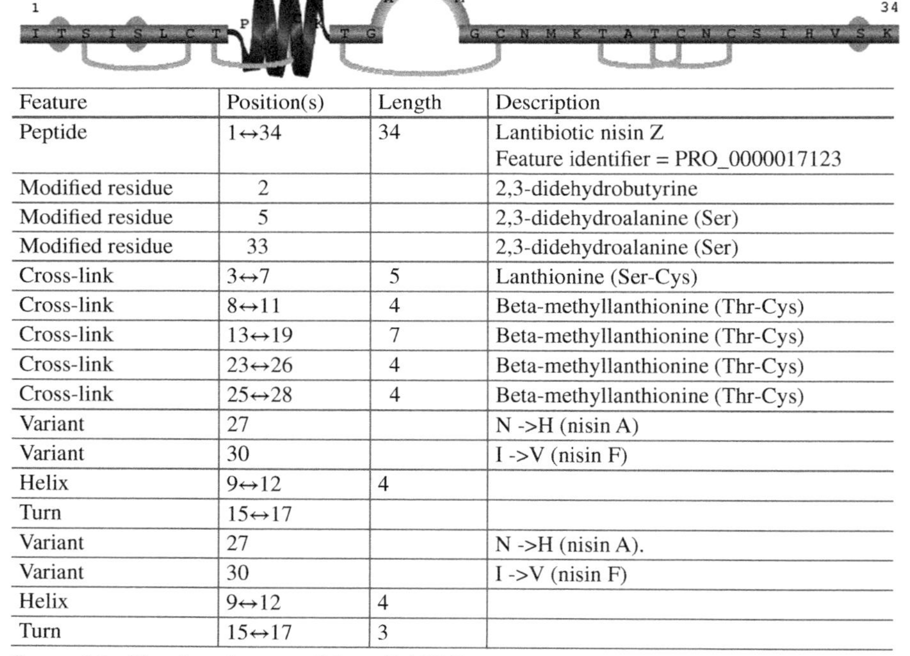

Feature	Position(s)	Length	Description
Peptide	1↔34	34	Lantibiotic nisin Z Feature identifier = PRO_0000017123
Modified residue	2		2,3-didehydrobutyrine
Modified residue	5		2,3-didehydroalanine (Ser)
Modified residue	33		2,3-didehydroalanine (Ser)
Cross-link	3↔7	5	Lanthionine (Ser-Cys)
Cross-link	8↔11	4	Beta-methyllanthionine (Thr-Cys)
Cross-link	13↔19	7	Beta-methyllanthionine (Thr-Cys)
Cross-link	23↔26	4	Beta-methyllanthionine (Thr-Cys)
Cross-link	25↔28	4	Beta-methyllanthionine (Thr-Cys)
Variant	27		N ->H (nisin A)
Variant	30		I ->V (nisin F)
Helix	9↔12	4	
Turn	15↔17		
Variant	27		N ->H (nisin A).
Variant	30		I ->V (nisin F)
Helix	9↔12	4	
Turn	15↔17	3	

Source: http://bactibase.hammamilab.org/BAC049

Table 21.4 Gene sequence of nisin A (LHS) and nisin Z (RHS) of *Lactococcus lactis* subsp. *lactis*

Gene id Nisin A	Name	Description	Gene id Nisin Z	Name Nisin Z	Description Nisin Z
BACGene166	nisX	Transposase	BACGene177	nisin	NisZ protein antibiotic
BACGene167	nisA	Nisin pre-peptide	BACGene178	nisB	NisB protein
BACGene168	nisB	Biosynthetic gene	BACGene179	nisT	NisT protein
BACGene169	nisT	Transport of nisin	BACGene180	nisC	NisC protein
BACGene170	nisC	Biosynthetic gene	BACGene181	nisI	NisI protein
			BACGene182	nisP	NisP protein

For loci of gene, see http://bactibase.hammamilab.org/ BAC047 nisin A http://bactibase.hammamilab.org/BAC145 nisin Q

Table 21.5 Gene sequence of nisin Q of *Lactococcus lactis* and nisin U of *Streptococcus uberis*

Gene id Nisin Q	Name	Description (Nisin Q)	Gene id (Nisin U)	Name	Description (Nisin U)
BACGene642	nisQ	Nisin Q precursor	BACGene654	nsuP	NsuP nisin U leader peptidase
BACGene643	niqB	Lantibiotic dehydratase	BACGene655	nsuR	NsuR response regulator
BACGene644	niqT	ATP-binding cassette transporter	BACGene656	nsuK	NsuK histidine kinase
BACGene645	niqC	Lantibiotic cyclase	BACGene657	nsuF	NsuF
BACGene646	niqI	Immunity protein self-protection	BACGene658	nsuE	NsuE
BACGene647	niqP	Leader peptidase	BACGene659	nsuG	NsuG
BACGene648	niqR	Response regulator	BACGene660	nsuA	NsuA nisin U lantibiotic
BACGene649	niqK	Histidine kinase	BACGene661	nsuB	NsuB lantibiotic biosynthesis protein
BACGene650	niqF	Immunity protein self-protection	BACGene662	nsuT	NsuT ABC transporter
BACGene651	niqE	Immunity protein self-protection	BACGene663	nsuC	NsuC
BACGene652	niqG	Immunity protein self-protection	BACGene664	nsuI	NsuI putative immunity peptide

http://bactibase.hammamilab.org/BAC145 nisin Q; http://bactibase.hammamilab.org/BAC147 nisin U

Structural Forms of Nisin

The ability of nisin to exist in monomer, dimer (7 KDa), and tetramer (14 KDa) forms (Adem et al. 2015) was observed while determining the molecular weight of bands formed, in SDS-PAGE, with Coomassie blue stain specific to nisin, a technique used to detect nisin.

When nisin protein was gamma-irradiated and then run on SDS-PAGE, one band of the monomeric protein was recovered on the gel. This nisin after irradiation and 2 weeks of storage gave a dimeric form (molecular weight 7 kDa), while several diffused bands of nisin were formed after even further storage, perhaps due to an irreversible degradation by Coomassie blue stain (Badr et al. 2005; Ivanova et al. 1998). The intermolecular reactions in nisin can change nisin to polycyclic structures, especially when dehydro groups of a molecule are combined with nucleophilic R-groups of other molecules (Wei and Norman 1990). Such intermolecular reactions and rearrangements were suggested to be important for membrane insertion property of nisin as a bacteriocin (McAuliffe et al. 2001).

Nisin Solubility

The application of purified nisin requires it to be in soluble form and is found to dissolve in dilute HCl (pH 2.5) (Wei and Norman 1990) (Fig. 21.4a). Nisin shows maximal solubility of 57 $mg.mL^{-1}$ (at pH 2.0) and a minimum of 0.25 $mg.mL^{-1}$ (at pH 8.0–12). It can be de-adsorbed from producer cells in a fermentative broth, by hot acidic extraction. For this, media with producer cells are adjusted to a lower pH ~2 to 3, using concentrated HCl, and placed in hot water bath (100 °C for 5 min) to release cell-bound nisin into solution (Badr et al. 2005). Such a heat treatment helps to destroy proteases produced/released in broth by the organism itself that cannot destroy nisin after hot extraction in the crude cell-free extract (Mitra et al. 2010).

The Antimicrobial Property of the Nisin Peptide

Nisin has an antimicrobial activity which is imparted by the rings: ring A of nisin A and ring C of nisin Z (Rollema et al. 1996; van Kraaij et al. 2000), the rare polycyclic thioether amino acids, lanthionine, and 3-methyl-lanthionines (Kuipers et al. 1993). In addition, the first three ring structures of nisin also have a profound effect on its bioactive property (Kuipers et al. 1992, 1996). The nisin activity can be reduced more by the breakdown of DHa5 of primary nisin than by a similar action on DHa33 (van Kraaij et al. 2000), since serine is unmodified at Ser_{33}. Such an arrangement could perhaps prevent steric hindrance of dehydrating enzymes of putative nisB (van Kraaij et al. 1997). Modified structures of nisin hamper the different steps in pore formation and affect overall influence depending on the organism it targets.

On the other hand, the C-terminal of nisin has a significant role in signaling potency (van Kraaij et al. 1997). The increased proteolytic resistance (Bierbaum et al. 1996; Rink et al. 2010) can be beneficial when nisin is extracted, whereas a greater tolerance to oxidation (Sahl et al. 1995) is imparted to nisin by the lanthionine and 3-methyl-lanthionines that are helpful during nisin production. The degrading enzymes (pancreatin, α-chymotrypsin, and ficin) can inactivate nisin by breaking its peptidic chain. Other gut enzymes, such as trypsin, pepsin, and carboxypeptidase, do not significantly affect antimicrobial power of nisin (Chollet et al. 2008).

Stability in pH

In assaying nisin, pH 5 and 6 are more appropriate pH levels for nisin activity than lower pH levels, when using *Staphylococcus aureus* and *Bacillus cereus* as indicator organisms. Still a lower pH shows more inhibitory effect, due to the higher solubility of nisin in acidic pH (Hurst and Hoover 1983). This solubility decreases in alkaline pH (Matsusaki et al. 1996;Yildirim and Johnson 1997). At higher pH, nisin is unstable and biologically inactive (Hurst 1981), due to individual or combined effect of denaturation and chemical modification of nisin (Wei and Norman 1990). Nisin completely loses its activity, at pH 11 and at 63 °C (after 30 min) (Wei and Norman 1990) (Fig. 21.4b). Its antibacterial activity measured at pH 6 and 7 was found much higher, ([10,000 AU/mL^{-1}] (Badr et al. 2005) against *Micrococcus luteus* than *Staphylococcus aureus* and *Bacillus cereus* (5040 and <5000 AU/mL^{-1}). This shows that antibacterial activity is defined under a set of comparable assay conditions, with the same target organism(s) used in assays. The variant nisin F is active as a bacteriocin over a broad pH range (from pH2 to 10) (de Kwaadsteniet et al. 2008). The nisin variant Z was found to be more stable at low pH, and also had a decreased solubility compared to nisin A (Rollema et al. 1995).

Stability of Nisin Under Different Temperatures

Nisin protein can tolerate high temperatures of 40–100 °C up to 30 min or more. Nisin produced by *L. lactis* WNC20, can tolerate high temperatures of 40–100 °C up to 30 min or more. At pH 7 it is inactivated at high temperature (121 °C for 15 min) (Todorov and Dicks 2005). However if suspended at pH 3.0 it could remain stable even at 121 °C for 15 min (Noonpakdee et al. 2003), being active in a wide pH range (2-10). Under long storage periods, it is stable under acidic pH (Badr et al. 2005; Jack et al. 1994a, b). The compact molecular structure is not disturbed with heat (Badr et al. 2005). However, when it is used in assay, it has higher activity (~9000 Au/mL) when incubation temperature for assay is 10 °C and not at 20 °C (Badr et al. 2005), as against *Micrococcus luteus.* Assay conditions (for bioactivity) and compact structure are thus two different entities of nisin protein that are affected by temperature.

Stability of Nisin in Food as a Matrix for Preservation

A low specific growth rate of nisin may help nisin to be established gradually in a food matrix, to be stable in it (Dykes and Hastings 1997). For this, it can be used as a starter culture (*L. lactis*) or a pure nisin formulation with low concentrations (<0.1%), in foods like milk. In fermenting milk, nisin can be added (0.05–5 mgL^{-1} nisin) which does not damage food. However, the subinhibitory amounts of nisin can induce nisin production from the adjunct starter strains already present in the food matrix. This can help nisin build-up gradually and preserve a food system (Kuipers et al. 1997).

21.4 Nisin Regulation

The non-conjugative plasmids are responsible for producing a majority of bacteriocins, and it is likely that long-term plasmid-host relationships led them to be evolved. With other structural genes transcribed from an operon, nisin gene is transcribed from an operon of size >8.5 kb (Steen et al. 1991) for nisin production. By conjugation, it is transferred to newer cells and gets inserted in chromosome on the transposon identified as "Tn5301," which is also responsible for sucrose fermentation along with nisin production too (Horn et al. 1991; Rauch and de Vos 1992).

The derivatives cured of the properties, to produce nisin or metabolize sucrose, in more extensive analysis can give more such evidence. The genetic traits to biosynthesize nisin and ability to ferment sucrose are linked and are co-transmissible during the bacterial multiplication and can together be irreversibly cured (Leblanc et al. 1980; Gasson 1984).

21.4.1 Biosynthesis of Nisin

Nisin is known to be self-regulated with a transport function to facilitate its release with an evolved mechanism (Riley 2009) for bacteriocin-specific regulation (Al Khatib et al. 2014). Its expression is self-induced (Kuipers et al. 1995). The 11 genes required for nisin nisABTCIPRKEFG are clustered (Fig. 21.5) (Buchman et al. 1988) on a large conjugally transmissible chromosomal gene block in *L. lactis*. They express biosynthesis and immunity (to not kill their own cells and other closely related spp.) and regulate nisin production (Buchman et al. 1988; Ra et al. 1999) and sucrose metabolism (Dodd et al. 1990; Williams and Delves-Broughton 2003). The genes to synthesize nisin are controlled by a conservative transposition event (Dodd et al. 1990). The gene has also been transferred to negative phenotype recipients of *L. lactis* (Williams and Delves-Broughton 2003).

21.4.2 Transcription of Nisin Pre-peptide

The structural gene nisA encodes the pre-peptide (Buchman et al. 1988) consisting of two components in signal transduction machinery (van der Meer et al. 1993), nisR and nisK, that activate its transcription to produce nisin (Kuipers et al. 1995; Abts et al. 2011). The three main genes, nisA, nisR, and nisP, produce nisin (van der Meer et al. 1993). Additionally, the genes nisR and nisK induce synthesis and regulate transport of nisin (Kuipers et al. 1995; Abts et al. 2011). The region upstream to nisA serves as a promoter sequence (Engelke et al. 1992). The nisin A protein thus belongs to the family of two-component regulatory system (van der Meer et al. 1993). Its biosynthesis involves the transport of precursor nisin out of cells by secreting it before it is cleaved from the leader peptide. The capacity required to then induce its expression depends on the interactions of specific amino acid residues with the N-terminal domain of the sensor protein (Kuipers et al. 1995).

A conjugative transposon "Tn5276" encodes the precursor peptide of nisin, located upstream, by encoding it on a 12 kb DNA region in *L. lactis*, along with a downstream region 10 kb long sequence of DNA for nisA gene and nisin production. The region downstream binds to ribosome and is transcribed by read through (Buchman et al. 1988). Two of the three promoters (PnisA and PnisF) in the nisin gene cluster are induced and transcribed by autophosphorylated and activated NisR (Fig. 21.5). When the nisR expression is driven by its promoter (Kuipers et al. 1995), nisin synthesis starts soon after. The nisR is essential for intermediate structural nisin precursor protein production (van der Meer et al. 1993). The putative regulatory protein is encoded by *nisR*, and a putative histidine kinase is encoded by *nisK* (Engelke et al. 1994; van der Meer et al. 1993). The nisR gene is essential for the intermediate structural nisin, precursor protein production (van der Meer et al. 1993) too.

The two reasons for the binding of nisin to nisK for transferring a phosphate group to NisR a regulator for nisin expresssion are as follows: (1) Nis R is autophosphorylated either by (i) histidine-protein kinase, a sensor protein on the cytoplasmic membrane which may be the receptor for the mature nisin molecule (Engelke et al. 1994; Mierau and Kleerebezem 2005) in *L. lactis*, or (ii) the protein kinase of the sucrose phosphotransferase uptake system encoded by Tn5276 (Thompson et al. 1991), and (2) Nis R becomes hyperexpressive due to being encoded from a multicopy plasmid. This promotes nisin expression that happens in the absence of a kinase gene, as in *B. subtilis* protease production (Tanaka and Kawata 1988). Thus, nisR can activate nisA and nisF promoters to transcribe the genes, under its control. The mRNA of precursor nisin is then synthesized (De Ruyter et al. 1996).

21.4.3 Pre-peptide to a Mature Protein Nisin Lantibiotic

Nisin needs to be secreted out of the cell to end up as an active molecule. For this, the enzymatic reactions due to genes nisB and nisC are required to form a mature nisin. The NisB helps to form mature nisin, once the pre-peptide reaches the membrane (Hess et al. 1988; Engelke et al. 1992). The nisB, nisC, and also nisT genes are ORFs (Engelke et al. 1992). The nisC gene overlaps nisT. The transport of nisin to the outside through the membrane is due to (Engelke et al. 1992) nisB and even nisC (Parada et al. 2007; Dodd et al. 1990). These genes together help in the formation of a mature nisin.

21.4.4 Immunity and Transport of Nisin for Secretion

A cell producing nisin is immune to get killed by its own bacteriocin(s). This is due to specific immunity proteins (Cotter et al. 2005). The immunity protein is a lipoprotein, encoded by nisI (Engelke et al. 1994). The nisFEG encodes a putative ATP-binding cassette exporter to help nisin extrusion (Siegers and Entian 1995; Stein et al. 2003) from cell and also protects producing cells from being killed.

The putative serine protease due to gene nisP processes the precursor nisin peptide (Engelke et al. 1994). It produces a leader sequence, upstream from nisR product. The leader sequence with a signal sequence at the N-terminal shares similarity to a subtilisin-like serine protease (responsible for modifications) and a putative C-terminal that anchors to the membrane. The N-terminal also helps in modifications and prevents toxicity (Schnell et al. 1988) of the final mature peptide, before the lantibiotic is cleaved off of its leader peptide, matures, and is then bioactive to be secreted from cell (Figs. 21.5 and 21.6).

21.4.5 Secretion of the Precursor Nisin

This process starts when the cell secretes the precursor, and later cleavage of N-terminal leads to the final step of secretion from membrane to allow it to gain immunity (Al Khatib et al. 2014). The formation of pore is destabilized by lipoprotein from nisI (Entian and de Vos 1996; Saris et al. 1996). The nisin is then transported by an assisted function of nisFEG gene (an ABC transporter) that hydrolyzes ATP and enables nisin with its necessary immunity property (Ra et al. 1999). The ATP-binding cassette transporter proteins, together with nisT, function (Hess et al. 1988) and help in bacteriocin secretion from the membrane of a cell (Karpinski and Szkaradkiewicz 2016). Lastly, an intermediate precursor nisin that was formed early during nisin synthesis is cleaved, to become bioactive. However, N-terminal leader sequence of nisin is cleaved in strains where nisP protease is not active (van der Meer et al. 1993). The overexpression of NisP later may cleave the precursor to even form a final and active modified peptide nisin. The proteolytic activation outside the cells releases an active nisin.

21.4.6 The Gene Sequences for Nisin Variants

The blast analysis and other techniques have identified gene loci for various genes and their functions that produce or regulate nisin (Tables 21.4 and 21.5).

Besides the various genes are identified with their identification numbers, gene function (Tables 21.4 and 21.5) (Fig. 21.7), and loci (bactibase.hammamilab.org). Nisin F is a structurally different protein as compared to other nisin variants (A, Z, Q, U) and is a fifth variant lantibiotic (de Kwaadsteniet et al. 2008). *L. lactis* F10 does not produce bacteriocin F when ethidium bromide (30 mgmL^{-1}) is present in a medium suggesting it to be encoded by plasmid-located genes. The similarities shared between Gram-positive bacterial bacteriocins including nisin Z, among others, in nisin F, being evolved genetically, like other variants of nisin, have been detailed in a previous work (Zouhir et al. 2010).

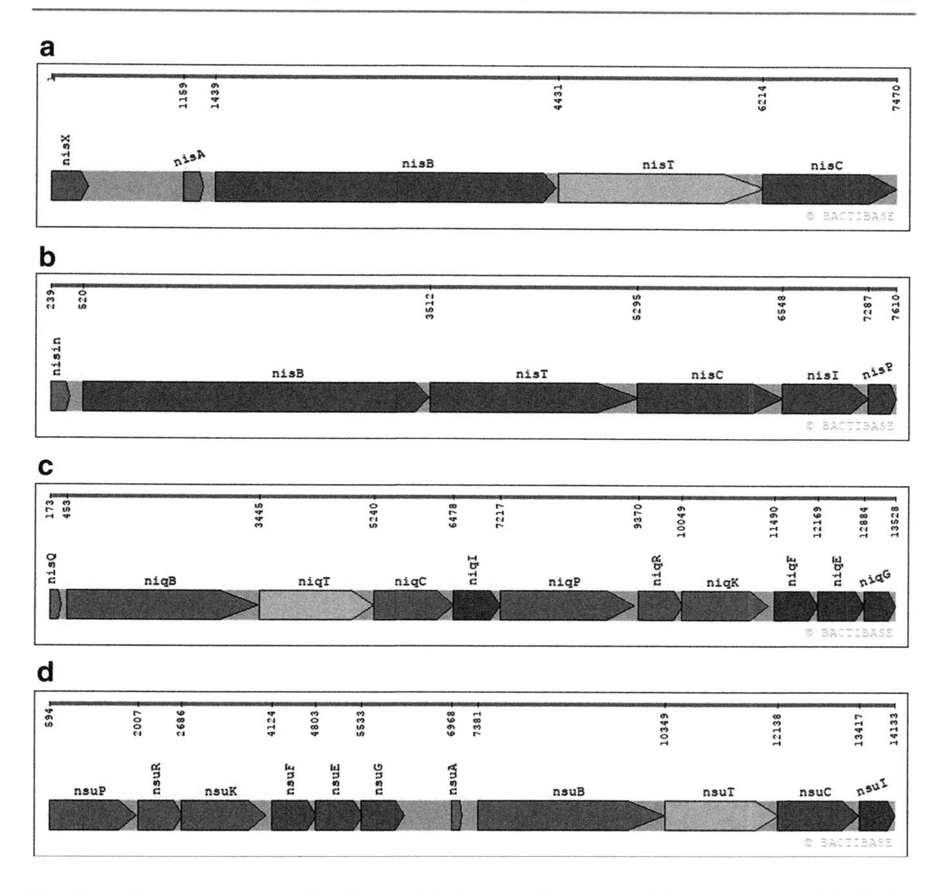

Fig. 21.7 The genes arranged in the lantibiotic gene clusters of nisin variants: (**a**) nisin A, (**b**) nisin Z, (**c**) nisin Q, and (**d**) nisin U. (Source: Bactibase: Bacteriocins)

21.5 Production of Nisin

21.5.1 Growth Media and Components on Biosynthesis and Release of Nisin

Nisin is a primary metabolite and formed at a rate that depends on its growth rate (Hirsch 1951; Luedeking and Piret 1959; De Vyust and Vandamme 1992; Matsusaki et al. 1996; Parente et al. 1994; Kaiser and Montville 1993; Kim et al. 1997; Egorov et al. 1971; Kozak and Dobrzanski 1977). The pre-nisin, precursor of nisin, is produced early in the exponential growth phase, and the rate at which it is produced is maximal at the end of the exponential growth. Its production stops in stationary phase (De Vuyst and Vandamme 1992). One can look for an increase in the biomass for an increased nisin metabolite production. With a high growth rate, a high biomass formation is preferred in a batch medium (Kim et al. 1997). However, this was not so in studies reported later with nisin produced in a 2 L fermenter (Singh et al. 2015). Here nisin production showed lower specific growth rates when aerated or under very low aerated conditions. Nisin produced adhered to cells and then are released with hot extraction, to purify it further. On the other hand, low nisin levels adhering to cells can elicit high nisin production rates, and high levels of nisin can possibly not be associated with nisin biosynthesis. In addition, other factors affecting nisin production like pH, magnesium sulfate, lactose concentrations, and nitrogen sources also interfere in nisin adhering to cells (Meghrous et al. 1992) that can affect nisin production.

Various nutrient sources have been used to optimize growth and nisin production. A few are highlighted here to show the variable conditions in which nisin production is affected. With glucose in solid complex M17 and minimal MS14 media (Cheigh et al. 2002), growth metabolism of *L. lactis* remained homolactic with lactic acid produced (Chandrapati and O'Sullvan 1998). The yield of lactic acid produced can reach 1.7 mol mol^{-1} (Novak et al. 1997). The amount and type of carbon source(s) supplied can also affect nisin levels produced by strains. Nisin production can also be affected by the ability of LAB to ferment carbon source "sucrose," since the same gene is involved in the control of nisin production/synthesis and also sucrose metabolism (De Vuyst and Vandamme 1992). As even sulfur "S" forms a part of a nisin structure, its biosynthesis also depends on a sulfur source. Inorganic salts (like magnesium sulfate or sodium thiosulfate) and amino acids containing "S" (like methionine, cysteine, or cystathionine) are S sources that can affect nisin production. Its production is highly stimulated by serine, threonine, and cysteine, but its final cell yield is not affected due to these amino acids which then confirm their role in precursor nisin biosynthesis (De Vuyst and Vandamme 1994).

The complex MRS medium, most suitable for lactobacilli cultivation, can be used for nisin production with a maximum specific productivity of 6.0 mg/mg dry cell weight (at 2.5 mg of pure nisin = 1.0×10^5 AU/mL specific activity). The MRS medium with specified C/N, C/P, and N/P ratios supplied (sucrose, asparagine, and phosphate as C, N, and P sources) with tween 80 in medium, when used for nisin production (Penna and Moraes 2002), promotes high nisin activities from cells that are released into the medium. This medium with specific contents of sucrose (12.5 g/L as C source), asparagine (75 g/L as N source), and phosphates (22 g/L from KH_2PO_4 to buffer) with tween (80 g/L) in the medium also helps to disperse and release nisin into culture medium. The MRS medium has been recommended due to reproducibility in the assay of nisin so produced (Penna and Moraes 2002).

Other workers have used complexly defined medium M17/MRS medium to produce/optimize nisin (Lan et al. 2006; Terzaghi and Sandine 1975; Zhang et al. 2009). The modified concentration of KH_2PO_4, carbohydrate (glucose/sucrose) (Zhang and Block 2009), and N contents (soy peptone) and other minor nutrients (Lv et al. 2005) can increase nisin production. After optimization (full CCD) among the medium (CM) nutrients, soy peptone and KH_2PO_4 were found to be the most important components of the media to allow accumulation of nisin (Li et al. 2002) to as high as 2150 $IU.mL^{-1}$ (Tramer and Fowler 1964). By using salts like KH_2PO_4 and NH_4PO_4 in M17-based synthetic media, nisin production was suppressed in *L.*

lactis ATCC11454 due to marked stimulatory effect on cell numbers that repress nisin production on per cell basis (Chandrapati and O'Sullvan 1998). In using complex media (CM, SM8, M17S including MRS) for growth, a high nisin accumulation with active nisin after purification is not always to be expected.

In M17 broth, batch fermentation (2.5 L, at 30 °C near neutral pH of 6.0) of *Lactococcus lactis* subsp. *lactis* A164 strain (isolate from kimchi), supplemented with lactose as the source of carbohydrate or added yeast extract (3%) as an optimal N source, a minimum fourfold increase in nisin production was observed. A maximal 131 × 10^3 AU $mL^{-1}$of active nisin was recovered, at an early stationary growth (20 h) in M17L broth (+ 3.0% lactose) (Cheigh et al. 2002; Terzaghi and Sandine 1975). Media formulations that contain 25% milk plus 25% M17 and 25% milk plus 25% MRS reportedly stimulate optimal bacteriocin production (Jozala et al. 2005). These media positively influence synthesis of nisin being released by *L. lactis*, where 885 mg/L nisin with 35,390 AU/ml activity in nisin was produced.

The MRS medium is suggested as a better medium for growth of LABs for cell growth and nisin production (Biswas et al. 1991; Daba et al. 1993; MacGroary and Reid 1988; ten Brink et al. 1994; Toba et al. 1991). Nisin production can increase consistently when the producing strain is transferred up to fifth time from its growth medium to a fresh medium that favored nisin expression and release into media at pH 4.6 and 4.8, respectively. Thus, batch growth and growth with transfers of the organism are two aspects of controlling nisin production in a medium.

A complex medium (CM medium) for nisin production, but not containing peptone in the inoculum medium, has the following ingredients:

Ingredients	($g.L^{-1}$)
Sucrose	10.0
Peptone (oxoid)	10–0
Yeast extract (oxoid)	10.0
KH_2PO_4	10.0
NaCl	2.0
$MgSO_4.7H_2O$	0.2
pH	6.8

Thus, in producing nisin, the source of carbohydrate and other nutrients supplied in batch fed or continuous system for a targeted high bioactivity in nisin is a much complex process, and thus optimization can achieve a control along with the controls the organism itself contributes to.

21.5.2 Producing Nisin Under Different pH and Metabolic Shift Conditions

To produce nisin by prolonged cultivation with a stepwise pH profile imposed during its growth, the productive period is prolonged, and it then acquires a secondary nature of metabolite (Cabo et al. 2001). This allowed a twofold increase in its production with an efficient nutrient consumption. In batch fermentative culture, fed with glucose at 15 g/L into the medium, maximal productivity increased 4.1- and 4.5-fold, respectively, giving 7.80×10^4 $U{\cdot}L^{-1}{\cdot}h^{-1}$ and 5.20×10^3 $U{\cdot}g^{-1}{\cdot}h^{-1}$, with respect to time and quantity of glucose consumed, respectively.

However, under a continuous mode of nisin production, in a medium similar to MRS, nisin productivity increased with enhanced glucose consumption (pH 6.8) and cultivation time, with a low lactic acid concentration that was maintained and coupled to microfiltration module (Taniguchi et al. 1994a). An online recovery system with silicic acid, as absorbent in reactor, for nisin Z production increased (7445 IU/ml) with a micro-filter module as compared to a batch fermentation without it (1897 IU/ml) (Pongtharangkul and Demirci 2007). Here a high cell density develops due to tangential flow in recovery that stimulates the production of nisin temporarily (Kuipers et al. 1995). The cell adsorption of nisin had reduced. Using a micro-filter module, nisin production can thus be enhanced (in contrast to a batch process), if cell adsorption of nisin can be reduced (Hao et al. 2017). In a fed batch fermentation with CM medium and a constant pH profile, the online recovery of nisin with high adsorption (67%) showed enhanced nisin (4.3×10^3 IU/mL), at pH 6.8 itself, as compared to recovery from pH 3.0 with only 54.0% adsorption. In the same biofilm reactor, with auto-acidification pH profiles, nisin production lowered significantly due to toxicity of lactic acid in the environment (Pongtharangkul and Demirci 2006). Thus, online recovery accompanied with production of nisin can help in active nisin recovery, in the presence of increased acidity formation.

In continuous cultures (lactose limited or non-lactose limited), nisin production can be maximized at intermediate growth rates ("μ" value of 0.2 h^{-1} and 0.3 h^{-1}) than at high μ of 0.6 h^{-1} (Meghrous et al. 1992) in MRS broth. The maximum active nisin (160 AU/mL) at specific growth rates ("μ" value) 0.25 h^{-1} can increase ninefold, compared to when dilution rates were very low 0.05 h^{-1} or much higher 0.4 h^{-1}. Under such dilution rates, the nisin titer increased from 12.5 to 164.2 AU/ml with increasing lactose consumption from 1 to 3.28 g of lactose/g (dry wt of cell mass) h^{-1}. At higher lactose values, nisin production declined. Thus, nisin biosynthesis can be repressed and derepressed with lactose utilization.

If metabolic pathway changes, nisin production can reduce/divert, under medium levels of specific growth rate, to avoid inhibitory effect(s) of lactate. In doing so, the inactivation of an arginine deaminase pathway helped to reduce consumption of amino acids. These amino acids like asparagine, serine, threonine, alanine, and cysteine reduced the conversion directed toward pyruvate synthesis and instead lead to a higher synthesis of specific protein, nisin by *Lactococcus lactis*, at intermediate specific growth rates of 0.35 h^{-1} (Adamberg et al. 2012). Such a strategy can also increase biomass yield of *Lactococcus lactis*. The strategy by carbon flux shift to alanine consumption provided to the cells at a concentration (150 nmM alanine or 15 g/L) reduced the impact of lactate when nisin was produced. Here a shift of a homolactate to homoalanine metabolism and a possible overexpression of alanine dehydrogenase took place. The reason for this shift was because L-alanine dehydrogenase competed more efficiently with L-lactate dehydrogenase, for an alternative pathway of pyruvate catabolism (Hols et al. 1999). In doing so, the alanine used did not inhibit growth and nisin production by *L. lactis* subsp. *lactis* ATCC 11454 (Wardani et al. 2006).

Nisin production can also increase by using disaccharides to create an apparent situation of nutrient limitation by use of disaccharides instead of monosaccharides that instead lower specific growth rates (De Vuyst and Vandamme 1992).

21.5.3 Effects of Aeration and Acidity (pH) on Growth and Nisin Production

Inoculum is the first requirement to produce nisin with an acidic level produced in a medium. A high nisin production is not always due to high inoculum size (Kim et al. 1997). With an initial Abs in bioreactor at Abs 1.0 (small), Abs 3.6 (medium), or Abs 6.1 (large), the growth Abs_{max} can reach 11, 13.2, and 16, respectively. The relative nisin concentration levels were the same (~11), and it closely correlated to growth and maximal levels reached when the inoculum's size served shorter lag times (Tramer and Fowler 1964).

When conditions of aeration are provided, the oxygen tolerance of LAB is associated with different substrates consumed that in turn affect nisin yields (Cabo et al. 2001). For nisin Z, 60% pO_2 is optimum (Amiali et al. 1998). Content of nisin can quadruple when oxygen saturation percentage increases from 50% to 100% (Cabo et al. 2001) and the biomass reaches its maxima point. However, the glucose fed and consumed in a medium was more important than re-alkalization of medium to produce nisin. With this, the pH control did control the production of nisin (Cabo et al. 2001). In an uncontrolled pH medium, a high biomass was not always associated with a very active nisin produced (Singh et al. 2016).

In a stationary batch (*S. lactis* in 100 mL MRS broth) with 40% air space (Erlenmeyer flask) and a starting pH of 6.5 (at 30 °C) without pH control, nisin of a high titer (9100 IU/mL) (Tramer and Fowler 1964; Wolf and Gibbons 1996) in crude cell-free extracts (CFE) in 17 h (Singh et al. 2013) was obtained. There was a high specific growth rate ($\mu = 0.77$), with a cell doubling time of 1.05 h that started 5.16 h after inoculation and pH of medium reached 4.4, when nisin was recovered. The final cell density was 0.35 Abs_{600}, in a 1–24 h growth in fermentative batch production (Fig. 21.8). Thus, the final pH did not damage the nisin produced in batch. Lowering of pH of a medium can promote nisin production, as long as the lowest possible pH is not reached (De Vuyst and Vandamme 1992).

For nisin production, strict anaerobiosis was recommended earlier (Hirsch 1951). No aeration and moderate shaking conditions significantly increased nisin production (De Vuyst and Vandamme 1992) under a pH drop gradient provided to *Lactococcus lactis* in batch. Its production lowered rapidly when the pH is below or above optimal (6.0–6.5) (Guerra and Pastrana 2002). With high levels of acidification in medium, cells collapse, not being able to make a balance against the high acidity in cytoplasm, which is a reason for the failure of *L. lactis* to grow further. On the other hand, when a medium is buffered at pH 6.0 than at 5.0 (Cabo et al. 2001), a higher nisin production (134.01 AU/mL]/substrate consumed [g/L]) ensued, as

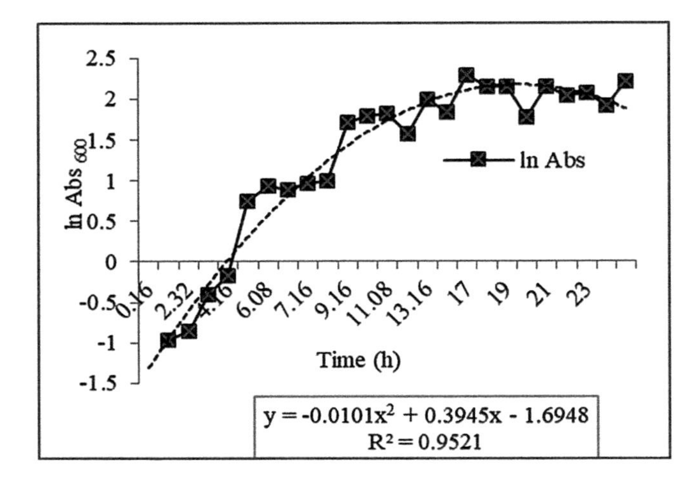

Fig. 21.8 Growth of *Lactococcus lactis* subsp. *lactis* for 24 h in MRS medium (100 mL batch). (Source: Singh et al. 2013)

compared to when this medium was re-alkalized. In the medium when pH was not controlled, bacteriocin units/protein unit increased when this medium was re-alkalized to pH 7 than to pH 6, with a concomitant increase in nutrient consumption efficiency. Thus, re-alkalization to pH 7 or pH 6 level(s) was controlled by substrate utilization, for activity units of nisin produced. If a pH drop in medium stops, pH is said to be stabilized. However, a gradual increase in medium acidity (stepwise pH drop profile) is always a phenomenon observed when lactic acid is produced by LAB and pH of medium reduced. The pH drop gradient (VpH) can help to increase nisin production with an increase in its biomass. In re-alkalizing the medium from pH 5.0 can also help to increase cell density Abs_{600} (Pedersen et al. 2002), but growth stopped when Abs_{600} reached 15 (with 15 g dry cell biomass/L) (Singh et al. 2015). Here the average specific growth rate was low at 0.26 h^{-1} with a lag period that extended to 0.82 h.

In a chemically defined medium, supplied with 75 g/L glucose to *L. lactis* ATCC 11454 growth, under pH control and moderate aeration, the titer of nisin was 3100 IU/mL after 8 h of growth (Papagianni and Avramidis 2012). In M17 broth (a complex defined medium) 30 g/L lactose to *L. lactis* growth, with pH control (pH 6.0), gave a high nisin activity (131 × 10^3 AU mL^{-1}) in its early stationary growth (20 h, 30 °C) (Cheigh et al. 2002). This growth increased eightfold with a fourfold higher nisin activity (16,384 AU/mL) in the presence of 5 g/L glucose as compared to MRS broth. In MRS broth, lactic acid is the only inhibitory by-product tolerated up to slightly >4 g/L lactic acid (Guerra and Pastrana 2002). However, this content of lactic acid was considered low to damage the strain from producing nisin (Matsusaki et al. 1996).

Nisin production can be enhanced by using acid-tolerant strains. In comparing a wild-type strain of *L. lactis* F44A producing 2884 and 3405 IU/mL nisin in batch and fed batch, it can increase to 3876 and 5346 IU/mL, when replaced with acid-tolerant strains to produce nisin (Hao et al. 2017).

In yet another batch fermentation (100 mL MRS broth) with *L. lactis* NCIM2114, lactic acid was monitored. Here the cell density Abs_{600} equivalent was 3.89, with a lactic acid content of 299.435 mg in 7 h, which increased to 653.48 mg % at a cell density 6.3 Abs_{600} growth and 694.92 mg % lactic acid at 6.88 cell density Abs_{600} growth, in 16 h (Singh 2013). High cell densities can inhibit nisin production (Kuipers et al. 1995), but not until a lowest possible achievable pH was reached (De Vuyst and Vandamme 1992).

The CFE from fermented broth, 17 h after growth of *L. lactis*, showed high nisin activity. This extract was preserved 6 months at 4 °C and assayed, to give 1360 IU/150 μl CFE (9068.52 IU/mL crude CFE) assayed against *Micrococcus luteus* as indicator organism (Tramer and Fowler 1964; Delgadoa et al. 2005). This nisin content in CFE equalled to a titer of 1.51 mg as against pure nisin powder (procured from HiMedia which contained 900 IU/mg in pure nisin powder) (Pongtharangkul and Demirci 2004; Singh et al. 2013). The same organism now in a 1.5 L MRS batch broth produced 3217 IU/g dry biomass of nisin, with an average productivity of 485.16 IU nisin/g.h^{-1} of cell dry biomass. The average biomass reached 9 g/L fresh biomass, when pH gradient fell 1.57 units from a starting pH of 5.8–6.0 (Singh 2014). Nisin from this LAB gave a high nisin concentration of 50,400 IU/mL (by UFLC), when harvested in early (4 h) exponential phase. However, nisin was bioactive in CFE only at 240 IU/mL (Singh et al. 2016). Thus, a CFE quantified (by UFLC) with high nisin content may not necessarily contain high bioactivity. There is thus a possible degradation of crude on purified nisin in CFE, by hydrolases that can take place, under uncontrolled pH conditions (Hao et al. 2017), or a possibility of an inactive pre-peptide if released before it matured (Wei and Norman 1990), early in exponential growth, cannot be ruled out (Singh et al. 2016). With the same inoculum size, growth levels differed in lag time in batch fermentative runs when the specific growth rates were between 0.23 and 0.30 h^{-1} in harvesting nisin at 4 h, 7 h, or 24 h of growth (Singh et al. 2015). This may be due to the self-inducing nature of nisin production and regulation.

Thus, the two main factors that determine nisin maximal levels (Kim et al. 1997):

1. After the utilization of available nutrients by a strain, premature ceiling of nisin production can take place. A high concentration of nisin produced can inhibit cells' own nisin production.
2. Even if a producing strain can grow well, nisin can switch off its own production when the specific concentration is reached by feedback inhibition.

21.5.4 Commercial Production Under Optimization of Factors

In a commercial production process (fed batch) for nisin, sucrose nutrient source (initial sucrose concentration (ISC fed at 10 g/L) can be one of the main factors in an initial stage along with time and rate of feeding (Wu et al. 2009) to produce nisin.

The titer and biomass, intermittently fed, in fed-batch culture were high (4490 $IUml^{-1}$ and 4.42 gl^{-1}) than in batch process (3375 IU ml/L, 4.12 g /L). Fed batch culture can eliminate inhibitions due to feeding high substrate levels (Lv et al. 2005) as was the case in a 3 h fermentation process, with an evenly fed substrate (1.9 g/L sucrose), to obtain a higher biomass and yield of nisin. This technology may be suitable for high-density culture and fermentation for final nisin concentration(s) (Wu et al. 2009). In order to exploit the specific growth rates, to manipulate growth and harvest cell mass with a consequent high nisin production, the growth of *L. lactis* can be modeled (Zwiettering et al. 1990; Singh et al. 2015). A Gompertz function with a stochastical approach was given out in growth of *L. lactis* in batch fermentation, up to a 24 h period. This can be used for scale-up to commercial production(s). The batch process in a 2 L bioreactor (5% inoculation) under different agitations and aeration levels helped to harvest nisin (at 4 h/7 h). The lag time and specific growth rate(s) could easily be determined for *L. lactis* (Mukhopadhyay 2007; Singh et al. 2015) and also other LABs (Zwiettering et al. 1990).

In large reactor systems with optimized fermentation conditions (pH 6.0 at 30 °C temperature) (Simsek and Saris 2009), *L. lactis* cells generally produce high nisin activity at 8–10 h of fermentation. It reduced after this point, probably due to degradation by proteolysis (Parente and Ricciardi 1999; Pongtharangkul and Demirci 2007) or nisin adsorbed onto producer cell surface, if produced in excess (Simsek and Saris 2009). A repeated cycle to change the medium after fixed periods (Simsek and Saris 2009) can help increase nisin production. A simultaneous extraction/recovery of nisin can reduce the chances to degrade/adsorb nisin and increases the nisin recovery (Pongtharangkul and Demirci 2007).

In commercial preparations (Özel et al. 2018), standard values of nisin preparation are 10^6 IUg^{-1} nisin with denatured milk proteins and NaCl. It is prepared with nisin content not exceeding 2.5 wt % pure nisin. A gram of pure nisin equals to 40 $\times$ 10^6 IU biological activity. In commercial preparations, 40 IU equals to 1 μg pure nisin, regardless of its purity. International units thus correspond to amount of nisin, that can inhibit a single cell of *Streptococcus agalactiae* in 1 ml of broth (Tramer and Fowler 1964). Thus, a nisin purity targeted to 2.5% and 40 $\times$ 10^6 IU/g is standard for commercial nisin target preparations (Patent 1960).

Various other innovative approaches to produce high levels of nisin by using genes in recombinants are reviewed (Özel et al. 2018) with newer approaches to fermentation optimization under pH, temperature, substrate, and DO controls which are necessary to regulate nisin, cell biomass, or the expression of nisin.

21.6 Purification of Nisin

21.6.1 Strategies Adopted in Nisin Purification

To improve process for nisin purification, the biochemical nature of bacteriocins can help make it easier to understand. A large-scale nisin produced in broth may require a downstream process to yield >50% yield with a 90% purity in recovery

(Schöbitz et al. 2006). Many purification techniques for antimicrobial peptides have been used that include precipitation with salt, various adsorption-desorption (AD) combinations, ion-exchange chromatography (IEC) and reversed-phase C-18 solid-phase extraction, reversed-phase high-performance liquid chromatography (RP-HPLC), and sodium dodecyl sulfate polyacrilamide gel electrophoresis (SDS PAGE) (PingitoreVera et al. 2007). A cost-effective method (Jozala et al. 2008) was suggested where an aqueous two-phase micellar system with a single nonionic surfactant like Triton X-114 was used to extract nisin.

Nisin loses its active concentration at higher pH. Thus, a CFE of nisin, after adsorption at high pH, is suspended in dilute HCl to extract (desorb) it at low pH. This is then followed by precipitation of nisin using solvent, for high potency (93.3%) (Mattick and Hirsch 1947; Yang et al. 1992). Ammonium sulfate precipitation and dialysis can give semi-purified nisin F (Sambrook et al. 1989) from its crude extract (de Kwaadsteniet et al. 2008). As high as 47%, nisin can be recovered using silicic acid adsorbent from a 1.0 L CM medium (pH 6.8 at 30 °C and 100 rpm agitation) by adsorption/desorption of nisin Z, produced by *L. lactis* subsp. *lactis* (NIZO 22186) (Pongtharangkul and Demirci 2007). This online nisin recovery system for nisin extraction along with production was most stable, and nisin was found to be soluble at low pH 3.0 along with the lactic acid that was produced during fermentation. The degradation by proteases and product inhibitions of nisin after its production and secretion into extracellular medium can be prevented by adsorption/desorption on an adsorbent, and simultaneous recovery of nisin was an added advantage. The requirement of pH adjustment in fermentation broth can help to recover the nisin henceforth.

Another single-step method of expanded bed ion-exchange chromatography (EB-IEC), to purify nisin from broth, is done by using pH-mediated producer cell adsorption in pH 3–4 and its subsequent elution with salt (0.15 M) that can achieve a 31-fold purification level and a 90% nisin yield from unclarified *Lactococcus lactis* subsp. *lactis* A164 culture broth (Cheigh et al. 2004). Under IEC separation, nisin cannot be monitored in IEC column run (2 mL/min) at 280 nm, as nisin does not contain any aromatic amino acids. Thus, nisin protein elution under IEC is monitored at Abs 215 nm. Another protocol optimized wherein 5 NaCl concentrations were used in five steps was 200 mM (step I), 400 mM (step II), 600 mM (step III), 800 mM (step IV), and 1 M (step V) NaCl by flow at 1 mL/min, separated nisin, in different fractions. The nisin protein can then be precipitated with Trichloroacetic acid (TCA) (20% (v/v)) at 4 °C. TCA is later removed by an ice-cold acetone wash. The NaCl helps to remove contaminants of low molecular weights from pure nisin. The nisin protein pellet is resuspended in 50 mM lactic acid pH 3 diluent that contains centrifuged supernatant to feed IEC again. In this way, pure nisin can be estimated spectroscopically at absorbance 584 nm (BCA protein assay) (Abts et al. 2011).

Immunoaffinity chromatography is another specific method that uses monoclonal antibodies to give 30-fold higher final yields of 33,690 ng nisin at A/20 mL in total eluted volume from a single 10 mL sample (Suárez et al. 1997). Other modified methods are elaborated (Jamaluddin et al. 2018). Partial purification of nisin by foaming of fermentates in large-scale reactors can be used to concentrate nisin

(Özel et al. 2018; Zheng et al. 2015). Extraction using ammonium sulfate for precipitation and desalting to partially purify nisin in the CFE can purify it further. A partially purified CFE can finally be purified (Boris et al. 2001; Pingitore Vera et al. 2007; Pongtharangkul and Demirci 2007; Abts et al. 2011) and assayed for inhibitory bioactive potential as nisin.

Nisin from food matrix can be isolated and partially purified using a food-grade process. It can give 100% desorption of nisin, using 1% sodium dodecyl sulfate (SDS) after being adsorbed into a food grade (calcium silicate), Micro-Cel, an anti-caking agent. Thus, subsequent adsorption/desorption by repeated elutions, or with use of surfactant in increasing concentration, can be used as a useful method for nisin extractions, from food matrix (Coventry et al. 1996).

After pure nisin is obtained, the purified nisin is characterized initially for (1) protease-sensitive degradation (with 1 mg.ml^{-1} proteinase K, trypsin, a-chymotrypsin, or pronase in 1 mol.L-1 Tris-HCl, pH 7.6, or pepsin in 100 mmol. L^{-1} citrate-phosphate buffer, pH2.8, at 37 °C for 2 h), (2) thermostability (at 90 °C or 100 °C for 15, 30, and 60 min and at 121 °C for 5 and 10 min) (Boris et al. 2001; Ivanova et al. 1998), and (3) nisin in solubility in different pH (2–12). Such characterizations are done to overcome added costs of further purifications and ensure nisin as a mature (secretion) protein, before it gets degraded in actual use (Wei and Norman 1990). The agar well diffusion assay for residual bioactivity characterizes the bactericidal property of purified nisin.

The final purified protein can be stored for months in 0.5% trifluoroacetic acid (pH 2.2) without detectable chemical or biological changes (Wei and Norman 1990). Storage at 80 °C in 30% 2-propanol (with 0.1% trifluoroacetic acid) until use was also suggested (Boris et al. 2001). Purified nisin can even be stored in 50–60% 2-propanol and/or ethanol containing 0.1% TFA at −20 °C (Rodríguez et al. 1995).

21.6.2 HPLC Technique

This is a very common method to study the release of nisin from a matrix (Table 21.6). It can detect/quantify nisin along with the understanding of its extents of modification in structure due to its solubility characteristics (Wei and Norman 1990) or modified peptides formed from nisin. Reversed-phase liquid chromatography (RP-LC) to quantify nisin has been used as summarized (Table 21.6).

21.7 Assay of Antimicrobial Activity, Spectra with Respect to Nisin, and Structural Variants

21.7.1 Quantification of Bioactivity

Inhibitory assay to quantify nisin bioactivity toward indicator microorganism can determine the antimicrobial activity of bacteriocins (Cabo et al. 1999) (Table 21.7). These assays use the simple measurement of zones of inhibitory activity on plates

Table 21.6 High-performance liquid chromatography (HPLC) conditions used to quantify nisin in sample

S. No	HPLC (for gradient elutions)	Column	Gradient mobile phase used, elution time (RT) for nisin	Detector at Abs λ	References
1	Beckman System Gold HPLC system (with a variable wavelength UV/VIS detector)	C18 Ultrasphere analytical column	Gradient of 0–100%: 27 min (RT) in acetonitrile with over 60 min run (see Fig. 21.9)	220–254 nm	Wei and Norman (1990)
			Gradient of 50–100%: 5.7 min (RT) in acetonitrile with over 30 min		
2	HPLC Waters 2695, (with a photodiode array UV detector 2998)	[a]C5 column (for small peptides), 5 μm size particles	Gradient of 0–100%: 16 min (RT) in acetonitrile with 30 min run	238 nm	Singh (2014)
3	Shimadzu LC at Prominence HPLC (with a SPD 20 A/SPD-20UV VIS detector) using windows based on EZStart software	[a]C5 column (for small peptides), 5 μm size particles	Gradient of 0–100%: ~3 min (RT) in acetonitrile with 16 min run (see Fig. 21.9)	238 nm	Singh et al. (2016)
4	C18 reversed-phase HPLC (Pharmacia LKB, Uppsala, Sweden)	C18 reversed-phase column (Ultrasphere ODS, 5 mm dia.)	A discontinuous gradient for preparative HPLC, followed by linear gradient of acetonitrile (0–60%): 34 min (RT) with 50 min run	220 nm	Meghrous et al. (1997)
5	HPLC (Pharmacia LKB Biotechnology)	PepRPC HR 5/5 C2/C18 reversed-phase chromatography column	Pre-extracted fraction diluted (5X in 0.1% TFA: in linear gradient ranging from 10% to 60% 2-propanol containing 0.1% TFA	254 nm	Rodríguez et al. (1995)

[a]The column was equilibrated with acetonitrile containing aqueous 0.1% TFA for 60 min and also checked for *column performance* as per specified elution characteristics of the compounds like toluene. A dilute HCl (0.02 N) solution or diluent of sample is used as zero control check. The peak total area at RT was quantitated by integration using

$$\text{Conc of Nisin in crude sample} = \frac{\{\text{Area of peak sample}\}}{\text{Area of peak standard}} \times\times \text{Conc of Std} \times \text{Dil factor}(\text{if any})$$

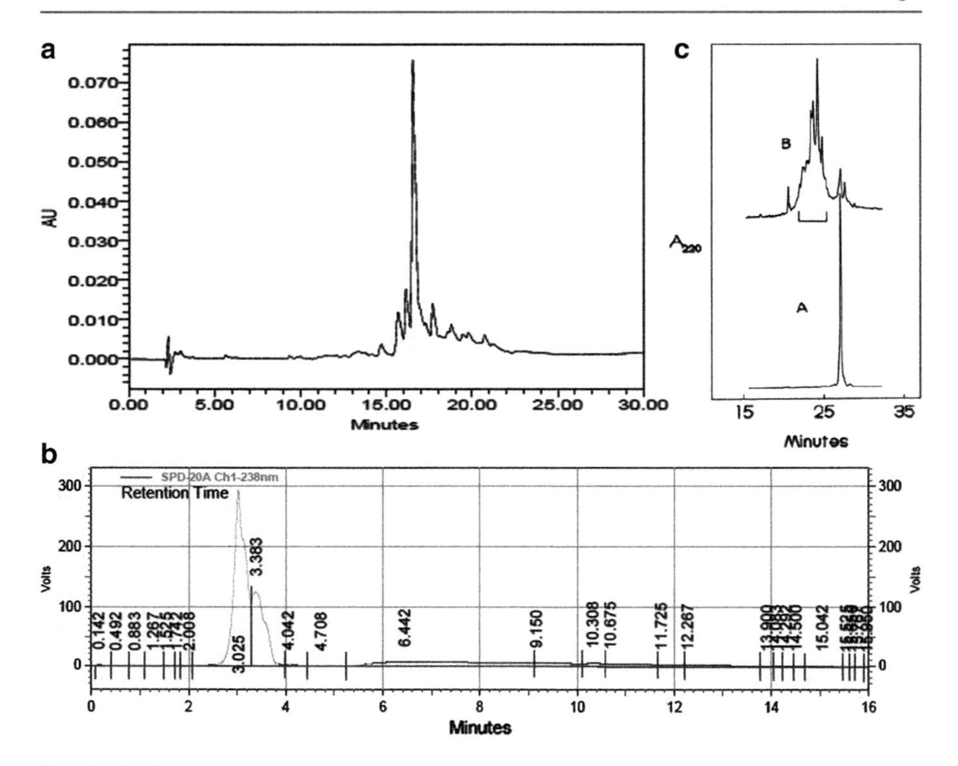

Fig. 21.9 Chromatogram profiles of unreacted nisin showing retention time (RT), using (**a**) HPLC (238 nm λ) and (**b**) FPLC (238 nm λ). (**c**) HPLC (220 nm λ). (Source: Singh et al. 2013 unpub; Wei and Norman 1990)

(Mocquot and Lefebvre 1956; Tramer and Fowler 1964) that are similar to and typical of antibiograms, as in checking antibiotic resistance. Turbidometry is another method used for assay (Berridge and Barrett 1952; Reeves 1965; Mortvedt and Nes 1990; Wu and Li 2007).

Different indicator strains used can give variations in determining the efficacy of the same bacteriocin (Yoneyama et al. 2008). Also, a single method used may not always be a proportionate representation on the content of nisin measured (Ripoche et al. 2006). The minimal inhibitory concentration of nisin (MIC = 3.3 Nm) is enough to kill indicator strains (*M. flavus* cells), used in nisin bioassay. Similarly, *L. lactis* and *S. thermophillus* are sensitive to be killed by nisin bacteriocin at 4.5 nM and 1.8 nM MIC (Breukink et al. 1999). Some of the useful assays for nisin are summarized (Table 21.7).

Assaying nisin activity, by a producer organism, cannot give the same assay as a pure form of nisin which has defined properties as a compound. There may be unfounded errors in the results on the activity in cellular preparations being different, from the purified inhibitor lantibiotic. When nisin is assayed from food, by diffusion assays, nisin cannot be distinguished from other interferences present in

Table 21.7 Various assay/quantification methods for bacteriocin (nisin) bioactivity

S.No	Assay medium/method	Indicator organism	Target measurement	References
1	Agar diffusion assays with modifications	*Micrococcus luteus*, *Lactobacillus agalactiae*	Inhibitory zones formed by diffusion through agar media are proportional to the concentration of nisin. This is valid over a limited linear range to check nisin in this range with accurate assessment. Detection limit at 12.5 ng/ml in pure solution	Tramer and Fowler (1964) and Wolf and Gibbons (1996)
2	Agar spot method[a] or microtitration method	*Lactobacillus* spp., *Listeria* spp., etc.	Semi-purified bacteriocin activity as arbitrary units (AU/ml). One AU is the reciprocal of the highest serial twofold dilution that shows a clear zone against the indicator strain	Schillinger and Lücke (1989), Tagg and McGiven (1971), Ivanova et al. 1998, and de Kwaadsteniet et al. (2008)
3	Critical dilution micromethod	*Listeria ivanovii*, etc.	Activity expressed as arbitrary unit (AU/mL)	Daba et al. (1991) and Cheighet al. (2002)
4	ID_{50}[b] using dose-response model	*Leuconostoc mesenteroides* subsp. *lysis* serially diluted nisin containing fractions (sample) with sensitive cells in fixed volumes were used in assay	$I R = I - (OD_m/OD_o)$	Cabo et al. (1999, 2000) and Abts et al. (2011)
			The concentration (ID_{50}) or dose that inhibits 50% of the active principle on indicator strain	
			Plot curve for normalized optical density (OD) against log of nisin concentration for ID_{50}	
			$Y = \frac{OD\min + (OD\max - OD\min)}{1 + 10^{((\log(IC50) - X) * slope)}}$.	
			See [c] below	

(continued)

Table 21.7 (continued)

S.No	Assay medium/method	Indicator organism	Target measurement	References
5	Dose-response[d] relationship to test bacteriocin solution	*Micrococcus luteus NCIM B287*	Measure of sensitivity of the indicator bacterium to the bacteriocin	Delgado et al. (2005) and Singh et al. (2013)
6	Rapid and sensitive dot immunoblot assay[e] using chemiluminescence	Using polyclonal antibodies directed against nisin Z&*P. acidilactici* UL 5 (by MIC)	Detection limit was the highest dilution with a detectable signal significantly different from the background	Bouksaim et al. (1998)
			The limit of detection is 375 ng/ml in pure solution and 155 ng/ml in milk and whey	
7	Bioluminescence-based bioassays	Using bacterial luciferase operon *luxABCDE* construct transformed into *Lactococcus lactis* strains NZ9800 and NZ9000	A linear dose-response relationship (nisin dilutions vs induction factor (IF lux))	Immonen and Karp (2007) and Wahlstrom and Saris (1999)
			A sensitivity of bioassay for nisin is 0.1 pg/ml in pure solution and 3 pg/ml in milk	
			0.0005–0.3 IU/ml (corresponding to 0.0125–7.5 ng/ml in pure solution) and 0.003 to 1 IU/ml (0.075–25 ng/ml) in milk	
8	Capillary zonal electrophoresis	By electropherograms with a peak at constant migration time of 11.6 min	Linear response to nisin concentration was observed in the range from 10 to 100 μg/ml in milk after removal of caseins and lipids from milk	Rossano et al. (1998)

[a]Based on a calibration curve of arbitrary units: to compare AU ml^{-1} vs IU ml^{-1}; 1.92 AU corresponds to 1 IU (40 IU = 1 μg of pure nisin A/nisin Z). AU = (1000/125)×(1/D)] where D is the highest dilution to allow growth of the indicator organism

[b]With OD_m being the optical density (700 nm) of the sample and ODo, with that of the control, the inhib ratio is calculated. The concentration or dose (ID_{50}) that inhibits 50% of the active principle (nisin) on indicator strain is a 1 unit of activity

[c]ODmax value = normalized OD_{600} value in the starting plateau; ODmin value = normalized OD_{600} of the end plateau; value *Y* stands for normalized optical density value, and *X* represents the logarithmic concentration of the peptide. The IC50=value of the peptide concentration used where the growth inhibition (or OD600) is 50%

[d]R = a + b ln C, where R = area of inhibition zone (mm^2); C = bacteriocin concentration ($mg.L^{-1}$); for crude bacteriocin, solution C is substituted by the dilution factor, fd; a = constant that corresponds to the y intercept or to the expected response from test solution (mm^2); b = constant that corresponds to the slope and gives a measure of sensitivity of the indicator bacterium to the bacteriocin (mm^2.ln C)

[e]Sensitivity is read from a standard curve, having serial twofold dilutions of pure nisin Z, starting at 0·120 $\mu g.mL^{-1}$

food matrix which may lead to false-positive results (Tramer and Fowler 1964). The assay medium was later modified (Wolf and Gibbons 1996) to remove such false-positive effects from a medium in assaying nisin. Still, the variations present may be due to other reasons while using different assay methods in place of agar diffusion. Nisin can in active form result in the loss of membrane potential of targeted cells and by the outward flow of small metabolites from these sensitive cells and also by its effect through enzyme inhibition (Jung 1991). Thus, under the inhibitory tests, there could be marked variations in nisin characteristics, when either (1) a strain that is producing nisin or (2) when pure nisin is under use in the test. Hence, nisin may not always inhibit the test strain(s) like *Bacillus cereus*, in an antagonism test (by using producer or pure nisin) (Cintas et al. 1998; de Vos et al. 1993; Ming and Daeschel 1993; Mota-Meira et al. 2000; Ray 1992). Various assays are listed (Table 21.7).

Production of other inhibitory by-products by the producer organism (Parrot et al. 1989) can be the main factors for variations in such tests. Also, a wider inhibition zone itself was reasoned on the substance having more activity or if it diffused faster or was produced in higher quantity when under the inhibition tests, which are other reasons. Even if nisin A and nisin Z differ in structure (primarily by only one amino acid), the similarity in percentage activity spectra was only 83% (Morency et al. 2001). This is attributed to increased diffusion of nisin Z in solid agar medium, which can form large inhibitory zones in tests of deferred antagonism, compared with nisin A (de Vos et al. 1993). Such characteristic variations associated with LABs, with direct application, make it all the more important in need to have more important newer lantibiotics.

The biological activity imparted to peptide nisin is by its characteristic ring structures, the rare polycyclic thioether amino acids, lanthionine, and 3-methyllanthionines (Kuipers et al. 1993), in nisin secreted after maturity. This helps nisin to be resistant from proteolysis (Bierbaum et al. 1996; Rink et al. 2010) with greater oxidation tolerance (Sahl et al. 1995).To help purify the nisin further, it is necessary to overcome any oxidative damage (Wilson-Stanford et al. 2009) during production.

21.8 Mode of Action of Lantibiotic Nisin

In action, producer cells of nisin cannot be killed by its own bacteriocin (Cotter et al. 2005) (Fig. 21.6 above). Bacteriocins of LABs differ importantly in their active mode (Karpinski and Szkaradkiewicz 2016), by the type of bacteriocin, genetic origin, and regulatory role (Riley 2009) they play. The nisin bacteriocin combines with high affinity to wall lipids II (which transports the peptidoglycan units from inner cytoplasm of the wall of cell (Fig. 21.10a), to anchor the receptor for nisin) and initiate the process of insertion onto sensitive membranes of target/indicator bacteria with its pore-forming ability (Fig. 21.10b). The primary target in sensitive cells is permeabilized by the cationic ends of nisin peptide. It allows pores to be formed across the bacterial membranes (Moll et al. 1999) (Fig. 21.11). Nisin

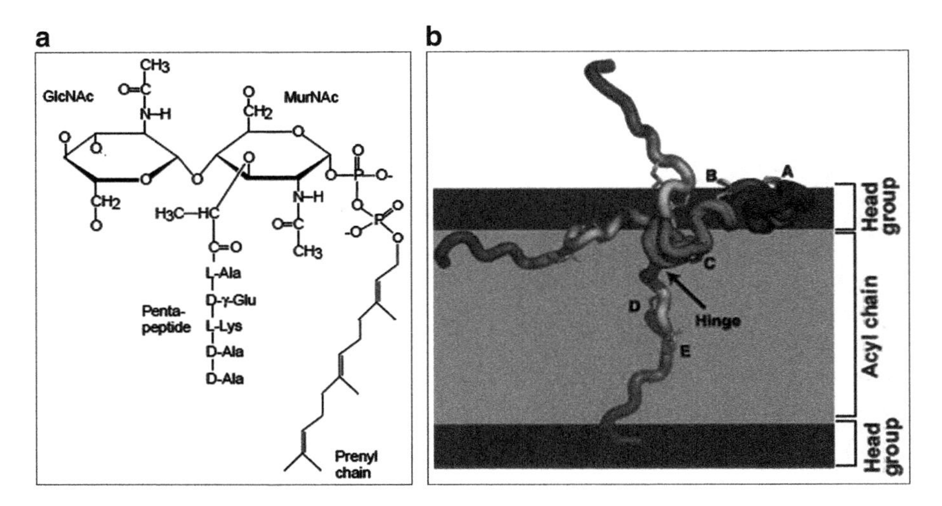

Fig. 21.10 (a) Chemical structure of a lipid II variant (3LII) which differs from the natural occurring full-length lipid II, with a shortened prenyl chain consisting of 3 isoprene repeats instead of 11. (b) Model of the nisin-lipid II complex in a membrane bilayer. The backbone of three representative structures of nisin is shown and colored from blue to red, from N to C termini. Each conformation, taken from the ensemble of structures of the nisin-3LII complex, corresponds to a possible orientation of the C-terminal part of nisin outside on the surface or inserted into the membrane bilayer, respectively. They are positioned so that the pyrophosphate group of 3LII (spheres) is at the same depth as head groups of lipid molecules (dark gray) in a membrane with a thickness of ~40 Å, which is based on a molecular dynamics simulation of free lipid II in the explicit membrane bilayer. (Source: Hsu et al. 2004)

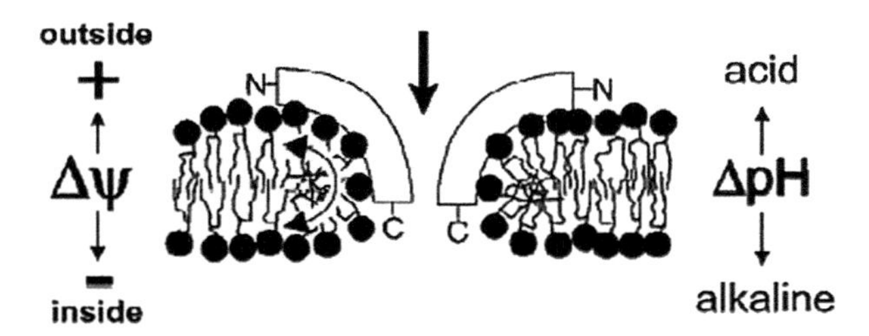

Fig. 21.11 Model of pore formation by nisin. (Source: Moll et al. 1999)

can also inhibit murein synthesis wherein the lipid II itself is not synthesized and it can be detrimental to vegetative cells formation, that are to develop from the spores of harmful pathogens in foods (Reisinger et al. 1980).

In the process of its activity, the amphipathic nisin being cationic with hinges across the ring structures (Moll et al. 1999) helps the molecule to insert into the membrane. It is the N-terminal as in nisin A with its lanthionine rings A and B that react with the pyrophosphate, the peptidoglycan MurNAc, and the first isoprene of lipid II (Hsu et al. 2004) on the membrane. In this, the carboxyl-terminal region remains bound at the membrane surface (Demel et al. 1996; Breukink et al. 1997, 1998; Van Kraaij et al. 1997) (Fig. 21.10b).

There is likeliness that a pore is made up of interaction of eight nisin and four lipid II molecules (Hasper et al. 2004; Martin et al. 2004). The action of nisin biological activity is mainly due to elongated nature of nisin seen by the third-order structures (Karpinski and Szkaradkiewicz 2016) among other bacteriocins. The peptide is highly active at nanomolar concentrations that can make the membrane permeable (Ruhr and Sahl 1985; van Heusden et al. 2002). In vitro studies show nisin as a flexible structure when it binds to the zwitterionic dodecylphosphocholine (DPC) micelles in solution (van den Hooven et al. 1993). This micelle DPC gives an impression of the bilayer environment of the membrane (Beswick et al. 1999). Nonselective pores in membranes are formed by reacting nisin molecules with anionic phospholipids and not with anionic solutes (Driessen et al. 1995) to undergo structural transition(s).

After insertion of nisin into membrane and increase in membrane permeability, a corresponding leakage of adenosine triphosphate and other ions (amino acids) leads to the loss of the ability of bacterial membrane potential (Penna and Moraes 2002) to remain intact and negatively charged in the inner side (Driessen et al. 1995), and its disruption takes place in a voltage-dependent fashion. It allows the small molecules to flow out leaving behind a depolarized membrane potential (Ruhr and Sahl 1985; Kordel and Sahl 1986; Sahl et al. 1987). The efflux of ions/essential compounds from the cytoplasm then inhibits the formation of macromolecules like DNA, RNA, protein, and polysaccharide in the cell (Cotter et al. 2005; Gillor et al. 2008; Karpinski and Szkaradkiewicz 2016; Wiedemann et al. 2001). An immediate collapse of the membrane potential leads to cell death (Kordel and Sahl 1986; Morris et al. 1984). Membrane disruption is the ultimate of nisin action (García Garcerá et al. 1993; Ruhr and Sahl 1985).

In germinated spore coverings, the dehydro residues of nisin (α- and β-unsaturated amino acids, e.g., DHA and DHB), also called the "Michael acceptors" (Gross and Morell 1967), can bind on to sulfhydryl groups. The nisin mimicry to "nitrite inhibition" (Buchmann and Hansen 1987) can help its dehydroamino residues to bind and inactivate the membrane sulfhydryl group(s) (Morris et al. 1984). Such "Michael acceptors" are placed on the peptide double bonds which conjugate to the carbonyl group (Fig. 21.3d) and react readily with nucleophiles (like mercaptans). This helps for a mechanical rupture of spore coats (Morris et al. 1984). Nisin is thus an inactivator of spore outgrowths, as in *Bacillus cereus*. The rupture of spore coat was easy to demonstrate, like nitrite inhibition, where it induces bacteriostasis in an aerobe (Morris et al. 1984). This inhibition was associated with inactivation of the membrane sulfhydryl group(s) and depletion of cell energy with the outflow of essential compounds from pores formed. This leads to inhibition in the formation of macromolecules that results in an ultimate cell death.

In the case of Gram-negative bacteria, their outer membranes protect them from bacteriocins and most of the other compounds of molecular weight >600 Da, to reach their cytoplasmic membrane (Abee et al. 1995). Thus, the inhibitory effect of nisin on the Gram-negative bacteria can be possible, only if chelators, such as EDTA, are also used alongwith, to destabilize the outer membrane, and only then can nisin use its own mode to kill cells (Stevens et al. 1991).

The magnesium ions in outer membranes function to keep the lipopolysaccharide layer of these membranes intact and stable. However, use of EDTA can destabilize it by its chelating mechanism (Stevens et al. 1991) that removes the magnesium ions from the membrane. This increases the susceptibility of cells to even antibiotics and detergents (Nikaido and Vaara 1987), and in this way nisin can exert inhibitory function via pore-forming action in Gram-negative cell membranes (Stevens et al. 1991).

21.9 Nisin Bactericidal Property as a Preservative

Nisin helps various fermented foods safe by its antimicrobial property. It is gaining attention in active packaging and MAP system of foods too (Chen and Hoover 2003). Since this product is still a very costly additive for extensive use, it is pertinent and important to study the conditions to use it and requirements to produce nisin in large scale at affordable costs. Nisin is now a commercial bacteriocin for biopreservation in various European countries in a big way. It is presently marketed as Nisaplin® (trade name). It is approved by the US Food and Drug Administration (Federal Register 1988) and licensed as a food additive in over 45 countries (Settanni and Corsetti 2008) for various food uses. During its wider use, aspects of its inactivation by specific enzymes like "nisinase" from *Streptococcus thermophilus* (Alifax and Chevalier 1962); "subtilopeptidase, an extracellular proteinase" of *Bacillus subtilis* (Jarvis 1967); and α-chymotrypsin (Hurst 1981) also need more attention.

The manufacturing or processing phase of foods needs the removal of any contamination in foods. Thus, if foods get contaminated in post-process stages, it is a major cause of contamination and outbreaks. Such contaminations lead to foodborne diseases, in which identifying the sources of contamination is very difficult, and hence go unnoticed. A comprehensive list of outbreaks due to post-process contamination in foods by various pathogens is available (Reij and Den Aantrekker 2004). The defective foods can be prevented from such contaminations using nisin as an additive (Zacharof and Lovitt 2012). The various aspects in its use are mainly related to its mode of action, singly or synergistically with other chelators or inhibitory conditions. The developing newer processes also need to be recorded.

This bacteriocin nisin protein complex is bioactive, antimicrobial/bactericidal, toward closely related Gram-positive species (broad spectrum) including *Lactobacillus* sp., *Lactococcus* sp., *Leuconostoc* sp., *Staphylococci*, *Streptococci*, *Bacilli*, *Micrococcus* sp., *Pediococcus* sp., *Enterococcus* sp., and even *Clostridia* and *Mycobacterium* sp. (Stoyanova et al. 2012; Karpinski and Szkaradkiewicz 2016), due to the typical action. Nisin is less effective on antibiotic-resistant pathogens like Mycobacteria (De Vuyst and Vandamme 1994; Field et al. 2015) (Mota-Meira et al. 2000). Nevertheless, its potency has been positively tested against some of the closely related Gram-positive bacteria and also antibiotic-resistant strains like *Neisseria gonorrhoeae*, *Campylobacter jejuni*, and *Helicobacter pylori* (Morency et al. 2001). Nisin controls *Listeria monocytogenes* sp. in seafoods, *Clostridium* spp. in soups and canned foods, and unwanted lactic acid bacteria (LABs) in some sauces, salads, and fermented drinks (beer and wine) (Karpinski and Szkaradkiewicz 2016). Since nisin producer cells are immune to being killed by themselves (Cotter et al. 2005), it can be used as a starter in foods. The best broad spectrum of activity for nisin use can be as much as 0.25–4 μg ml^{-1} (Moll et al. 1999).

As a biopreservative in sardine fish, nisin in combination with the lactoperoxidase system (LP system) inhibits spoilage flora. The inhibitory extent is strain dependent. The primary cellular target for both inhibitors is the cytoplasmic membrane, which explains their synergistic action. The growth of Gram-negative spoilage organisms *Pseudomonas fluorescens* and *Shigella putrefaciens* reduced significantly due to such an inhibition system (Elotmani and Assobhei 2003) during ice storage of fish.

A combination of carbon dioxide followed by nisin can work on pathogenic *Listeria monocytogenes* cells in fresh or minimally processed foods. A two-log reduction of the wild-type *Listeria monocytogenes* with nisin at 4 °C can be increased to four-log reduction (Nilsson et al. 2000) using nisin synergistically with carbon dioxide. Here, carbon dioxide is instrumental in increasing lag phase of *L. monocytogenes* by 6 days in wild-type cells, during which period the nisin can effectively reduce it further.

In the upcoming process technologies like pulsed electric fields (PEF), a combination with nisin can inactivate the pathogen *Listeria innocua* on liquid whole egg and skim milk foods (Calderon-Miranda et al. 1999a, b). Even though nisin is far better than antibiotics, still factors like low temperature, acidic pH, and presence of sodium chloride in foods can affect its inhibitory actions (Gravesen et al. 2002; Moll et al. 1997).

21.9.1 Use of Nisin in Packagings

To supply fresh fruits and vegetables, nisin can actively control contaminations in edible films and coatings used in packagings (Gandhi and Chikindas 2007). Biodegradable package films are also suitable to incorporate nisin with stearic acid that make an antimicrobial agent with a moisture barrier, respectively (Sebti et al. 2002). This film (hydroxylpropyl methylcellulose, HPMC) when adjusted to pH 3 can prevent nisin to interact with stearic acid in the package and also to induce inhibitions against *L. monocytogenes* and *Staphycoccus aureus* (Sebti et al. 2002). Other such packaging material films (poly lactic acid [PLA]) that are used to incorporate nisin can control microbial contaminations like *Escherichia coli* O157:H7 and other aerobic natural background microflorae as in strawberry puree (Jin et al. 2010).

Typical application of nisin for various food applications has been listed (D'Amato Daniela and Sinigaglia Milena 2010) including newer strategies (O'Shea et al. 2013). It is recommended to use nisin as in most applications between 250 and 500 IU/g (Coultate 2002). Nisaplin contained in cellophane-based coating (Guerra et al. 2005), corn zein with nisin and/or nisin + sodium diacetate and sodium lactate on films (Hoffman et al. 2001; Lungu and Johnson 2005), and soy-based films impregnated with nisin and lauric acid (Dawson et al. 2002) are various options to use as active packagings with nisin as bactericide. Recently, nanoparticle packaging materials (Behzadi et al. 2018) have been developed for nisin. The small pore diameters of nisin (average 4.1 nm) allow it to be incorporated on such nanoparticle carrier molecules (of 92 ± 10 nm diameter). Its application on such carriers helps to minimize the loss of nisin's antibacterial function in comparison to its direct application. The solubilization of nisin in an acidic buffer (sodium acetate buffer, 50 mM; pH, 5.5; isoelectric point, 8.5) ensures its incorporation on such surfaces, keeping its function intact. Nisin thus adsorbed onto such surfaces can retain its antimicrobial activities at pH as high as 9.5. This is unlikely when nisin is used alone. Thus, physical adsorption onto such nanoparticles can increase stability of films in active packagings with nisin (Behzadi et al. 2018). Based upon the charge on such surfaces of mesoporous silica nanopaticles (MSNs), unfunctionalized BMSNs (bare MSN) with~ −28 mV zeta (ζ) potential and CMSNs (carboxylated MSNs) with −40 mV potential values can help as barriers, by allowing controlled release of nisin in such nano-packaging materials, as required. In using nisin as a valuable microbial product, it must have a known active purified concentration as emphasized again.

More reviews can be read on the strategies to enhance nisin expressions in exploiting its potential (Sahl and Bierbaum 1998). As such it can also be used in foods via its in situ production (starter culture) (Etchells et al. 1964), or by adding purified or semi-purified preparations (e.g., nisin containing powders, Nisaplin), along with adjunct strains that may already be present (native) in fermented foods, to turn on its antimicrobial efficacy (Deegan et al. 2006) in food preservation. The highly stable nisin in a food matrix can be achieved by manipulating use of starters/adjunct nisin producers having low specific growth rates (Dykes and Hastings 1997). The major challenge in using nisin for preservation is to prevent it from rapid depletion after its initial application. These factors (heat treatment, storage at alkaline pH, long processing times, and interaction of peptide with food matrix itself) that reduce stability and antimicrobial effect of nisin can be taken care of at various stages to allow maximal potential of nisin as a preservative.

21.10 Conclusion

From existing developments on food preservation with nisin, a wholesome of commercial processes on nisin production are identified, in using potential LAB strain(s). To further improve the desired requirement(s), in harnessing capabilities of these microorganisms, traits can be molded into it with a sufficient knowledge on the regulatory aspects of nisin production. Moreover, higher nisin yields and purified preparations with suitable techniques are the main targets in such commercial ventures.

The feedback inhibitions and various other factors that control nisin production and synthesis by LAB strains need more attention to produce even newer lantibiotics. Since nisin can itself regulate its production, nisin resistance and mode of action of nisin or its newer lantibiotics are the two major ends to deal in producing nisin commercially. At present, the LAB strains used are plentiful, with several directions and protocols. Online systems are also available for recovery during production itself. Major attention is also needed to upscale the active nisin with high bactericidal activity and purification protocols in place.

The various food applications (processed foods/minimal processed fresh foods or packagings on foods) are an ever growing area of interest. The small peptide (with its several variants) is yet to be exploited with more understanding. The aspects pertaining to use in medical applications also can give insights into ventures that can tackle excess use of antibiotics to remove drug resistance, in food chains. Various reports of small-peptide engineering and use of nisin in cancer therapy are drawing more attention in the ongoing efforts for its medical applications by molecular studies. Nisin with its properties and applications can not only help keep our foods safer from pathogenic organisms but also can mitigate diseases due to ever-evolving newer pathogens in foods.

In a true sense, "nisin" is a benefactor for mankind as a biopreservative for foods and health.

References

Abee T, Krockel L, Hill C (1995) Bacteriocins: modes of action and potentials in food preservation and control of food poisoning. Int J Food Microbiol 28:169–185

Abts A, Mavaro A, Stindt J, Bakkes PJ, Metzger S, Driessen AJM, Smits SHJ, Schmitt L (2011) Easy and rapid purification of highly active nisin. Int J Peptides. Article ID 175145, 9 pages. https://doi.org/10.1155/2011/175145

Adamberg K, Seiman A, Vilu R (2012) Increased biomass yield of *Lactococcus lactis* by reduced overconsumption of amino acids and increased catalytic activities of enzymes. PLoS One 7:e48223. https://doi.org/10.1371/journal.pone.0048223

Adem G, Nadia O, Catherine J, Pascal D (2015) Nisin as a food preservative: part 1: physicochemical properties, antimicrobial activity, and main uses. Crit Rev Food Sci Nutr. https://doi.org/10.1080/10408398.2013.763765

Alifax R, Chevalier R (1962) Studies on nisinase produced by *Streptococcus thermophiles*. J Dairy Res 29:233–240. https://doi.org/10.1017/S0022029900011043

Allgaier H, Jung G, Werner RG, Schneider U, Zähner H (1986) Epidermin: sequencing a heterodetic tetracyclic 21-peptide amide antibiotic. Eur J Biochem 160:9–22

AlKhatib Z, Lagedroste M, Zaschke J, Wagner M, Abts A, Fey I, Kleinschrodt D, Smits SHJ (2014) The C-terminus of nisin is important for the ABC transporter NisFEG to confer immunity in *Lactococcus lactis*. Microbiologyopen 3:752–763. https://doi.org/10.1002/mbo3.205

Amiali MN, Lacroix C, Simard RE (1998) High nisin Z production by Lactococcus lactis UL719 in whey permeate with aeration. World J Microbiol Biotechnol 14(6):887–894. https://doi.org/10.1023/a:1008863111274

Araya T, Ishibashi N, Shimamura S (1992) Genetic Evidence that *Lactococcus lactis* JCM 638 produces a mutated form of nisin. J Gen Appl Microbiol 38:271–278

Babasaki K, Takao T, Shimonishi Y, Kurahashi K (1985) Subtilosin A, a new antibiotic peptide produced by *Bacillus subtilis* 168: isolation, structural analysis, and biogenesis. J Bacteriol 98:585–603. PMID: 3936839

Bactibase (2018) A data base dedicated to bacteriocins. http://bactibase.hammamilab.org/bacteriocinslist.php?q=Nisin. Accessed 9 Sept 2018

Badr S, Abdel Karem H, Hussein H, El-Hadedy D (2005) Characterization of nisin produced by *Lactococcus lactis*. Int J Agric Biol 7(3):499–503

Behzadi F, Darouie S, Alavi SM, Shariati P, Singh G, Dolatshahi-Pirouz A, Arpanaei A (2018) Stability and antimicrobial activity of nisin-loaded mesoporous silica nanoparticles: a game-changer in the war against maleficent microbes. J Agric Food Chem 66:4233–4243. https://doi.org/10.1021/acs.jafc.7b05492

Benedict RC (1980) Biochemical basis for nitrite inhibition of *Clostridium botulinum* in cured meat. J Food Prot 43:877–891. https://doi.org/10.4315/0362-028X-43.11.877

Berridge NJ, Barrett J (1952) A rapid method for the turbidimetric assay of antibiotics. J Gen Microbiol 6:14–20

Beswick V, Guerois R, Cordier-Ochsenbein F, Coïc YM, Tam HD, Tostain J, Noël JP, Sanson A, Neumann JM (1999) Dodecylphosphocholine micelles as a membrane-like environment: new results from NMR relaxation and paramagnetic relaxation enhancement analysis. Eur Biophys J 28:48–58. PMID: 9933923

Bierbaum G, Szekat C, Josten M, Heidrich C, Kempter C, Jung G, Sahl HG (1996) Engineering of a novel thioether bridge and role of modified residues in the lantibiotic Pep5. Appl Environ Microbiol 62:385–392. PMC167809

Biswas SR, Ray P, Johnson MC, Ray B (1991) Influence of Growth Conditions on the Production of a Bacteriocin, Pediocin AcH, by *Pediococcus acidilactici* H. Appl Environ Microbiol 57:1265–1267

Boris S, Jimenez-Diaz R, Caso JL, Barbes C (2001) Partial characterization of a bacteriocin produced by *Lactobacillus delbrueckii* subsp. *lactis* UO004, an intestinal isolate with probiotic potential. J Appl Microbiol 91:328–333

Bouksaim M, Fliss I, Meghrous J, Simard R, Lacroix C (1998) Immunodot detection of nisin Z in milk and whey using enhanced chemiluminescence. J Appl Microbiol 84:176–184. https://doi.org/10.1046/j.1365-2672.1998.00315.x

Bowdish DM, Davidson DJ, Hancock RE (2005) A re-evaluation of the role of host defence peptides in mammalian immunity. Curr Protein Pept Sci 6:35–51. PMID:15638767

Breukink E, van Kraaaij C, Demel RA, Siezen RJ, Kuipers OP, de Kruijff B (1997) The C-terminal region of nisin is responsible for the initial interaction of nisin with the target membrane. Biochemist 36:6968–6976. https://doi.org/10.1021/bi970008u

Breukink E, van Kraaij C, van Dalen A, Demel RA, Siezen RJ, de Kruijff B, Kuipers OP (1998) Biochemist 37:8153–8162

Breukink E, Wiedemann I, van Kraaij C, Kuipers OP, Sahi HG, de Kruijff B (1999) Use of the cell wall precursor lipid II by a pore-forming peptide antibiotic. Science 286:2361–2364

Buchman GW, Banerjee S, Hansen JN (1988) Structure, expression, and evolution of a gene encoding the precursor of nisin, a small protein antibiotic. J Biol Chem 263:16260–16266

Buchmann GW III, Hansen NJ (1987) Modification of membrane sulflhydryl groups in bacteriostatic action of nitrite. Appl Environ Microbiol 53:79–82

Cabo ML, Murado MA, González MP, Pastoriza L (1999) A method for bacteriocin quantification. J Appl Microbiol 87:907–914

Cabo ML, Murado MA, GonzaÂ lez MP, Pastoriza L (2000) Dose response relationships. A model for describing interactions, and its application to the combined effect of nisin and lactic acid on *Leuconostoc mesenteroides*. J Appl Microbiol 88:756–763

Cabo ML, Murado MA, González MP, Pastoriza L (2001) Effects of aeration and pH gradient on nisin production. A mathematical model. Enz Microb Technol 29:264–273. https://doi.org/10.1016/S0141-0229(01)00378-7

Calderon-Miranda ML, Barbosa-Canovas GV, Swanson BG (1999a) Inactivation of *Listeria innocua* in liquid whole egg by pulsed electric fields and nisin. Int J Food Microbiol 51:7–17

Calderon-Miranda ML, Barbosa-Canovas GV, Swanson BG (1999b) Inactivation of *Listeria innocua* in skim milk by pulsed electric fields and nisin. Int J Food Microbiol 51:19–30

Chan WC, Lian LY, Bycroft BW, Roberts GCK (1989) Confirmation of the structure of nisin by complete 1H NMR resonance assignment in aqueous and dimethyl sulphoxide solution. J Chem Sot Perkin Trans I: 235967

Chandrapati S, O'Sullvan DJ (1998) Procedure for quantifiable assessment of nutritional parameters influencing nisin production by *Lactococcus lactis* subsp. *lactis*. J Biotechnol 63:229–233. PMID:9803535

Chatterjee S, Chatterjee S, Lad SJ, Phansalkar MS, Rupp RH, Ganguli BN, Fehlhaber HW, Kogler H (1992) Mersacidin, a new antibiotic from *Bacillus*. Fermentation, isolation, purification and chemical characterization. J Antibiot (Tokyo) 45:832–838

Cheigh C-I, Choi H-J, Park H, Kim S-B, Kook M-C, Kim T-S, Hwang J-K, Pyun Y-R (2002) Influence of growth conditions on the production of a nisin-like bacteriocin by *Lactococcus lactis* subsp. *lactis*A164 isolated from kimchi. J Biotechnol 95:225–235. https://doi.org/10.1016/S0168-1656(02)00010-X

Cheigh C-I, Moo-Chang K, Seong-Bo K, Young-Ho H, YuRyang P (2004) Simple one-step purification of nisin Z from unclarified culture broth of *Lactococcuslactis* subsp. lactis A164 using expanded bed ion exchange chromatography. Biotechnol Lett 26:1341–1345. https://doi.org/10.1023/B:BILE.0000

Chen H, Hoover DG (2003) Bacteriocins and their food applications. Compr Rev Food Sci Food Saf 2:82–100

Cheng F, Takala TM, Saris PEJ (2007) Nisin biosynthesis in vitro. Mol Microbiol Biotechnol 13:248–254. https://doi.org/10.1159/000104754

Chevalier RJ, Fournaud J, Levebre E, Mocquot G (1957) Mise en evidence des streptocoques lactiques inhibiteurs et stimulants dans le lait et les fromages. Ann Technol Agric 2:117–137

Chollet E, Sebti I, Martial-Gros A, Degraeve P (2008) Nisin preliminary study as a potential preservative for sliced ripened cheese: NaCl, fat and enzymes influence on nisin concentration and its antimicrobial activity. Food Control 19:982–989

Cintas LM, Casaus LS, Håverstein LS, Hernandez PE, Nes IF (1997) Biochemical and genetic characterization of enterocin P, a novel Sec-dependent bacteriocin from Enterococcus faecium P13 with a broad antimicrobial spectrum. Appl Environ Microbiol 63:4321–4330

Cintas LM, Casaus P, Fernandez MF, Hernandez PE (1998) Comparative antimicrobial activity of enterocin L50, pediocin PA-1, nisin A and lactocin S against spoilage and food pathogenic bacteria. Food Microbiol 15:289–298

Cleveland J, Montville TJ, Nes IF, Chikindas ML (2001) Bacteriocins: safe, natural antimicrobials for food preservation. Int J Food Microbiol 71:1–20

Correia SS, Manuela O, Semedo LT (2015) Bacteriocins. In: Manuela O, Serrano Isa D (eds) The challenges of antibiotic resistance in the development of new therapeutics. Bentham Science Publishers Ltd, Sharjah, pp 178–302

Cotter PD (2012) Bioengineering: a bacteriocin perspective. Bioengineered 3:313–319

Cotter PD, Hill C, Ross RP (2005) Bacteriocins: developing innate immunity for food. Nat Rev Microbiol 3(10):777–788

Coultate TP (2002) Food: the chemistry of its components, 4th edn. Royal Society of Chemistry, RSC, Cambridge, pp 304–324

Coventry MJ, Gordon JB, Alexander M, Hickey MW, Wan J (1996) A food-grade process for isolation and partial purification of bacteriocins of lactic acid bacteria that used diatomite calcium silicate. Appl Environ Microbiol 62(5):1764–1769

Daba H, Pandian S, Gosselin JF, Simard RE, Huang J, Lacroix C (1991) Detection and activity of a bacteriocin by *Leuconostoc mesenteroides*. Appl Environ Microbiol 57:3450–3455

Daba H, Lacroix C, Huang J, Simard RE (1993) Influence of growth conditions on production and activity of mesenterocin 5 by a strain of *Leuconostoc mesenteroides*. Appl Microbiol Biotechnol 39:166–173

D'Amato, Daniela, Sinigaglia Milena (2010) Antimicrobial agents of microbial origin: nisin. Bevilacqua Antonio, Corbo Maria Rosaria and Sinigaglia MilenaApplication of alternative food preservation technologies to enhance food safety and stability,Sharjah:Bentham Science, 83–91

Dawson PL, Carl GD, Acton JC, Han IY (2002) Effect of lauric acid and nisin-impregnated soy-based films on the growth of *Listeria monocytogenes* on turkey bologna. Poult Sci 81:721–726. https://doi.org/10.1093/ps/81.5.721

Deegan LH, Cotter PD, Hill C, Ross P (2006) Bacteriocins: biological tools for bio-preservation and shelf-life extension. Int Dairy J 16:1058–1071. https://doi.org/10.1016/j.idairyj.2005.10.026

de Kwaadsteniet M, Doeschate K, Dicks LMT (2008) Characterization of the structural gene encoding nisin f, a new lantibiotic produced by a *Lactococcus lactis* subsp. *lactis* isolate from freshwater catfish (*Clarias gariepinus*). Appl Environ Microbiol 74:547–549

Delgado A, Brito D, Fevereiro P, Tenreiro R, Peres C (2005) Bioactivity quantification of crude bacteriocin solutions. J Microbiol Methods 62:121–124

Delves-Broughton J, Blackburn P, Evans RJ, Hugenholtz J (1996) Applications of the bacteriocin, nisin. Antonie Van Leeuwenhoek 69:193–202

Demel RA, Peelen T, Siezen RJ, de Kruijff B, Kuipers OP (1996) Nisin Z, mutant nisin Z and lacticin 481 interactions with anionic lipids correlate with antimicrobial activity. A monolayer study. Eur J Biochem 235:267–274

De Ruyter Pascalle G, Kuipers Oscar P, Beerthuyzen Marke M, van A-BI, de Vos Willem M (1996) Functional analysis of promoters in the Nisin gene cluster of *Lactococcus lactis*. J Bacteriol 178:3434–3439

de Vos WM, Mulders JWM, Siezen RJ, Hugenholtz J, Kuipers OP (1993) Properties of nisin Z and distribution of its gene, nisZ, in *Lactococcus lactis*. Appl Environ Microbiol 59:213–218

De Vuyst L (1995) Nutritional factors affecting nisin production by *Lactococcus lactis* subsp. lactis NIZO 22186 in a synthetic medium. J Appl Bacteriol 70:28–33

De Vuyst L, Vandamme EJ (1992) Influence of the carbon source on nisin production in *Lactococcus lactis* subsp.*lactis* batch fermentation. J Gen Microbiol 138:571–578

De Vuyst L, Vandamme EJ (1994) Nisin, a lantibiotic produced by *Lactococcus lactis* subsp. *lactis*: properties, biosynthesis, fermentation and applications. In: de Vuyst L, Vandamme EJ (eds) Bacteriocins of lactic acid bacteria, microbiology, genetics and applications. Blackie Academic and Professional, London, pp 151–121

Dodd HM, Horn N, Gasson MJ (1990) Analysis of the genetic determinant for production of the peptide antibiotic nisin. J Gen Microbiol 136:555–556. https://doi.org/10.1099/00221287-136-3-555

Driessen AJM, van den Hooven Henno W, Wieny K, van de Kamp M, Sahl H-G, Konings Ruud NH, Konings Wil N (1995) Mechanistic studies of lantibiotic-induced permeabilization of phospholipid vesicles. Biochemist 34:1606–1614. https://doi.org/10.1021/bi00005a017

Dykes GA, Hastings JW (1997) Selection and Fitness in Bacteriocin-Producing Bacteria. Proc Biol Sci R Soc Lond 264:683–687

Egorov NS, Baranova IP, Kozlova YI (1971) Optimization of nutrient medium composition for the production of the anti-biotic nisin by *Streptococcus lactis*. Mikrobiologiya 40:993–998

Elotmani F, Assobhei O (2003) *In vitro* inhibition of microbial flora of fish by nisin and lactoperoxidase system. Lett Appl Microbiol 38:60–65. https://doi.org/10.1046/j.1472-765X.2003.01441.x

Engelke G, Gutowski-Eckel Z, Hammelmann M, Entian KD (1992) Biosynthesis of the lantibiotic nisin: genomic organization and membrane localization of the NisB protein. Appl Environ Microbiol 58:3730–3743

Engelke G, Gutowski-Eckel Z, Kiesau P, Siegers K, Hammelmann M, Entian KD (1994) Regulation of nisin biosynthesis and immunity in Lactococcus lactis 6F3. Appl Environ Microbiol 60:814–825

Entian KD, de Vos WM (1996) Genetics of subtilin and nisin biosynthesis. Antonie Van Leeuwenhoek 69:109–177. PMID: 8775971

Etchells JL, Costilow RN, Anderson TE, Bell TA (1964) Pure culture fermentation of brined cucumbers. Appl Microbiol 12:523–535

Falconer R (1949) Private communication to Hirsch, In: Hirsch A (1951) Growth and Nisin Production of a Strain of *Streptococcus lactis*. J Gen Microbiol 5:208–221

Federal Register (1988) Nisin preparation: affirmation of GRAS status as a direct human food ingredient. Fed Regist 54:11247–11251

Field D, Quigley L, O'Connor PM, Rea MC, Daly KM, Cotter PD, Hill C, Ross RP (2010) Studies with bioengineered Nisin peptides highlight the broad-spectrum potency of Nisin V. Microb Biotechnol 3:479–496. https://doi.org/10.1111/j.1751-7915.2010.00184.x

Field D, Begley M, O'Connor PM, Daly KM, Hugenholtz F, Cotter PD (2012) Bioengineered nisin A derivatives with enhanced activity against both Gram positive and Gram negative pathogens. PLoS One 7:e46884. https://doi.org/10.1371/journal.pone.0046884

Field D, Cotter PD, Ross RP, Hill C (2015) Bioengineering of the model lantibiotic nisin. Bioengineered 6:187–192. https://doi.org/10.1080/21655979.2015.1049781
Food and Drug Administration (1988) Nisin preparation: affirmation of GRAS status as direct human food ingredient. Fed Reg 53:11247
Fremaux C, Ahn C, Klaenhammer TR (1993) Molecular analysis of the lactacin F operon. Appl Environ Microbiol 59:3906–3915
Gálvez A, Abriouel H, López RL, Ben Omar N (2007) Bacteriocin-based strategies for food biopreservation. Int J Food Microbiol 120:51–70. https://doi.org/10.1016/j.ijfoodmicro.2007.06.001
Gandhi M, Chikindas ML (2007) Listeria: a foodborne pathogen that knows how to survive A Review. Int J Food Microbiol 113:1–15
García Garcerá MJ, Elferink MGL, Driessen AJM, Konings WN (1993) In vitro pore-forming activity of the lantibiotic nisin. Eur J Biochem 212:417–422
Garde S, Rodriguez E, Gaya P, Medina M, Nunez M (2001) PCR detection of the structural genes of nisin Z and lacticin 481 in *Lactococcus lactis* subsp. *lactis* INIA 415, a strain isolated from raw milk Manchego cheese. Biotechnol Lett 23:85–89
Gasson MJ (1984) Transfer of sucrose fermenting ability, nisin resistance and nisin production into *Streptococcus lactis* 712. FEMS Microbiol Lett 21:7–10
Gillor O, Etzion A, Riley MA (2008) The dual role of bacteriocins as anti- and probiotics. Appl Microbiol Biotechnol 81:591–606. https://doi.org/10.1007/s00253-008-1726-5
Gillmore MS, Segarra RA, BoothMC BCP, Hall LR, Clewell DB (1994) Genetic structure of the *Enterococcus faecalis* plasmid pAS1-encoded cytolitic toxin system and its relationship to lantibiotic determinants. J Bacteriol 176:7355–7344
Graeffe T, Rintala H, Paulin L, Saris P (1991) A natural nisin variant. In: Jung G, Sahl HG (eds) Nisin and novel lantibiotics. ESCOM Science, Leiden, pp 260–268
Gravesen A, Jydegaard Axelsen AM, Mendes DS, Hansen TB, Knochel S (2002) Frequency of bacteriocin resistance development and associated fitness costs in *Listeria monocytogenes*. Appl Environ Microbiol 68:756–764. https://doi.org/10.1128/AEM.68.2.756-764.2002
Gross E, Morell JL (1967) The presence of dehydroalanine in the antibiotic nisin and its relationship to activity. J Am Chem Soc 89:2791–2792
Gross E, Morell J (1971) The structure of nisin. J Am Chem Soc 93:4634–4635. https://doi.org/10.1021/ja00747a073
Gross E, Kiltz HH, Nebelin E (1973) Die structur des subtilins. *Hoppe-Seyler ZPhysiol Chem* 354:810–812
Guerra NP, Pastrana L (2002) Modelling the influence of pH on the kinetics of both nisin and pediocin production and characterization of their functional properties. Process Biochem 37(9):1005–1015. https://doi.org/10.1016/S0032-9592(01)00312-0
Guerra NP, Macias CL, Agrasar AT, Castro LP (2005) Development of a bioactive packaging cellophane using Nisaplin as biopreservative agent. Lett Appl Microbiol 40:106–110. https://doi.org/10.1111/j.1472-765X.2004.01649.x
Hansen JN, Chung Y, Liu W (1991) Biosynthesis and mechanism of the action of nisin and subtilin. In: Jung G, Sahl HG (eds) Nisin and novel lantibiotics. ESCOM, Leiden, pp 287–302
Hao P, Liang D, Cao L, Qiao B, Wu H, Caiyin Q, Zhu H, Qiao J (2017) Promoting acid resistance and nisin yield of *Lactococcus lactis* F44 by genetically increasing D-asp amidation level inside cell wall. Appl Microbiol Biotechnol 101(15):6137–6153. https://doi.org/10.1007/s00253-017-8365-7
Hasper HE, De Kruijff B, Breukink E (2004) Assembly and stability of nisin-Lipid II pores. Biochemist 43:11567–11575
Henderson JT, Chopko AL, van Wassenaar PD (1992) Purification and primary structure of pediocin PA-1 produced by *Pediococcus acidilactici* PAC-1.0. Arch Biochem Biophys 295:5–12
Heng NCK, Wescombe PA, Burton JP, Jack RW, Tagg JR (2007) The diversity of bacteriocins in Gram positive bacteria. In: Riley MA, Chavan MA (eds) Bacteriocins: ecology evolution. Springer, Berlin/New York, pp 45–92
Hess JF, Bourret RB, Simon MI (1988) Histidine phosphorylation and phosphoryl group transfer in bacterial chemotaxis. Nature 336:139–143

Hirsch A (1951) Growth and nisin production of a strain of *Streptococcus lactis*. J Gen Microbiol 5:208–221
Hoffman KL, Han IY, Dawson PL (2001) Antimicrobial effects of corn zein films impregnated with nisin, lauric acid and EDTA. J Food Prot 64(6):885–889
Holo H, Nilssen O, Nes IF (1991) Lactococcin A, a new bacteriocin from *Lactococcus lactis* subsp. *cremoris*: isolation and characterization of the protein and its gene. J Bacteriol 173:3879–3887
Hols P, Kleerebezem M, Schanck AN, Ferain T, Hugenholtz J, Delcour J, de Vos WM (1999) Conversion of *Lactococcus lactis* from homolactic to homoalanine fermentation through metabolic engineering. Nat Biotechnol 17:588–592. https://doi.org/10.1038/9902
Horn N, Swindell S, Dodd H, Gasson M (1991) Nisin biosynthesis genes are encoded by a novel conjugative transposon. Mol Gen Genet 228:129–135
Hsu S-TD, Breukink E, Tischenko E, Lutters MAG, de Kruijff B, Kaptein R, Bonvin AMJJ, van Nuland NAJ (2004) The nisin–lipid II complex reveals a pyrophosphate cage that provides a blueprint for novel antibiotics. Nat Struct Mol Biol 11(10):963–967. https://doi.org/10.1038/nsmb830
Hurst (1966) Biosynthesis of antibiotic Nisin by whole *Streptococcus lactis* organisms. J Gen Microbiol 44:209–220
Hurst A (1981) Nisin. Adv Appl Microbiol 27:85–123
Hurst A, Hoover DG (1983) Nisin. In: Davidson PM, Branen AL (eds) Amimicrobials in foods, 2nd edn. Marcel Dekker Pub, New York, pp 369–394
Hurst A, Peterson GM (1971) Observations on the conversion of an inactive precursor protein to the antibiotic nisin. Can J Microbiol 17:1379–1384
Immonen N, Karp M (2007) Bioluminescence-based bioassays for rapid detection of nisin in food. Biosens Bioelectron 22:1982–1987. https://doi.org/10.1016/j.bios.2006.08.029
Ivanova I, Miteva V, Stefanova T, Pantev A, Budakov I, Danova S, Moncheva P, Nikolova I, Dousset X, Boyaval P (1998) Characterization of a bacteriocin produced by *Streptococcus thermophilus* 81. Int J Food Microbiol 42:147–158
Jack RW, Tagg JR, Ray B (1994a) Bacteriocins of Gram-positive bacteria. Microbiol Rev 59:171–200. PMC239359
Jack RW, Carne A, Metzger J, Stefanovitc S, Sahl H-G, Jung G, Tagg JR (1994b) Elucidation of the structure of SA-FF22, a lanthionine-containing antibacterial peptide produced by *Streptococcus pyogenes* strain FF22. Eur J Biochem 220:455–462
Jamaluddin N, Stuckey David C, Ariff Aarbakaritya B, Fadzlie W (2018) Novel approaches to purifying Bacteriocin.: a review. Crit Rev Food Sci and Nutr 58(14):1–13. https://doi.org/10.1080/10408398.2017.1328658
Jarvis B (1967) Resistance to nisin and production of nisin inactivating enzymes by several *Bacillus* species. J Gen Microbiol 47:33–48
Jin T, Zhang H, Boyd G (2010) Incorporation of preservatives in polylactic acid films for inactivating *Escherichia coli* o157:h7 and extending Microbiological shelf life of strawberry puree. J Food Protect 73:812–818
Joerger MC, Klaenhammer TR (1990) Cloning, expression, and nucleotide sequence of the *Lactobacillus helveticus* 481 gene encoding the bacteriocin helveticin J. J Bacteriol 172:6339–6347
Joo NE, Ritchie K, Kamarajan P, Miao D, Kapila YL (2012) Nisin, an apoptogenic bacteriocin and food preservative, attenuates HNSCC tumorigenesis via CHAC1. Cancer Med 1:295–305
Jozala AF, Novaes De Lencastre LC, Cholewa O, Moraes D, Vessoni Penna TC (2005) Increase of nisin production by *Lactococcus lactis* in different media. Afr J Biotechnol 4:262–265
Jozala AF, Lopes AM, Mazzola PG, Magalhães PO, Vessoni Penna TC, Pessoa A Jr (2008) Liquid–liquid extraction of commercial and biosynthesized nisin by aqueous two-phase micellar systems. Enz Microb Technol 42:107–112
Jung G (1991) Lantibiotics—ribosomally synthesized biologically active polypeptides containing sulfide bridges and α, β- di dehydroamino acids. Angew Chem Int Ed Engl 30:1051–1192
Jung G, Sahl HG (1991) Nisin and novel lantibiotics. Escom, Leiden

Kaletta C, Entian KD, Kellner R, Jung G, Reis M, Sahl HG (1989) Pep5, a new lantibiotic: structural gene isolation and prepeptide sequence. Arch Microbiol 152:16–19
Karpinski TM, Szkaradkiewicz AK (2016) Bacteriocins. In: Encyclopedia of food and health. Elsevier Pub. Ltd., Amsterdam, pp 312–319. https://doi.org/10.1016/B978-0-12-384947-2.00053-2
Kaiser AL, Montville TJ (1993) The influence of pH and growth rate on production of the bacteriocin, bavaricin MN, in batch and continuous fermentations. J Appl Bacteriol 75:536–540
Kemperman R, Jonker M, Nauta A, Kuipers OP, Kok J (2003) Functional analysis of the gene cluster involved in production of the bacteriocin circularin a by *Clostridium beijerinckii* ATCC 25752. Appl Environ Microbiol 69(10):5839–5848. https://doi.org/10.1128/AEM.69.10.5839-5848.2003
Kellner R, Jung G, Sahl H-G (1991) Structure elucidation of the tricyclic lantibiotic Pep5 containing eight positively charged amino acids. In: Jung G, Sahl H-G (eds) Nisin and novel lantibiotics. Escom Publishers, Leiden, The Netherlands, pp 141–1581
Kim WS, Hall RJ, Dunn NW (1997) Host specificity of nisin production by *Lactococcus lactis*. Biotechnol Lett 19(12):1235–1238
Klaenhammer TR (1988) Bacteriocins of lactic acid bacteria. Biochimie 70:337–349
Klaenhammer TR (1993) Genetics of bacteriocins produced by lactic acid bacteria. FEMS Microbiol Rev 12:39–86
Kleinnijenhuis AJ, Duursma MC, Breukink E, Heeren RMA, Heck A Jr (2003) Localization of intramolecular monosulfide bridges in lantibiotics determined with electron capture induced dissociation. Anal Chem 75:3219–3225
Kordel M, Sahl HG (1986) Susceptibility of bacterial, eukaryotic and artificial membranes to the disruptive action of the cationic peptides Pep 5 and nisin. FEMS Microbiol Lett 34:139–144
Kozak W, Dobrzanski WT (1977) Growth requirements and the effect of organic components of the synthetic medium on the biosynthesis of the antibiotic nisin in Streptococcus lactis strain. Acta Microbiol Pol 26:361–368
Kuipers OP, Rollema HS, WMGJ Y, Boot HJ, Siezen RJ, de Vos WM (1992) Engineering dehydrated amino acid residues in the antimicrobial peptide nisin. J Biol Chem 267:24340–24346
Kuipers OP, Rollema HS, de Vos Willem M, Siezen RJ (1993) Biosynthesis and secretion of a precursor of nisin Z by Lactococcus lactis, directed by the leader peptide of the homologous lantibiotic subtilin from Bacillus subtilis. FEBS Lett 330:23–27
Kuipers OP, Beerhuyzen MM, deRuyter PG, Luesink EJ, de Vos WM (1995) Autoregulation of nisin biosynthesis in *Lactococcus lactis* by signal transduction. J Biol Chem 270:281–291
Kuipers OP, Bierbaum G, Ottenwalder G, Dodd HM, Horn N, Metzger J, Kupke T, Gnau V, Bongers R, van den Boogaard P, Kosters H, Rollema HS, de Vos WM, Siezen RJ, Jung C, Gdtz F, Sahl HG, Gasson MJ (1996) Protein engineering of lantibiotics. Anthonie van Leeuwenhoek 69:161–169
Kuipers OP, de Ruyter PGGA, Kleerebezem M, de Vos Willem M (1997) Controlled overproduction of proteins by lactic acid bacteria. TIBTECH 15:135–140
Lan CQ, Oddone G, Mills DA, Block DE (2006) Kinetics of *Lactococcus lactis* growth and metabolite formation under aerobic and anaerobic conditions in the presence or absence of hemin. Biotechnol Bioeng 95:1070–1080
Leblanc DJ, Crow VL, Lee LN (1980) Plasmid mediated carbohydrate carbohydrate catabolic enzymes among strain of *Streptococcus lactis*. In: Stuttard C, Rozee KR (eds) Plasmids and transposons: environmental effects and maintenance mechanisms. Academic, New York, pp 31–41
Leer RJ, van der Vossen JMBM, van der Giezen M, van Noort JM, Pouwels PH (1995) Genetic analysis of acidocin B, a novel bacteriocin bacteriocin produced by *Lactobacillus acidophilus*. Microbiology 141:1629–1635
Li C, Bai J, Cai Z, Ouyang F (2002) Optimization of a cultural medium for bacteriocin production by *Lactococcus lactis* using response surface methodology. J Biotechnol 93(1):27–34. https://doi.org/10.1016/S0168-1656(01)00377-7

Luedeking R, Piret EL (1959) A kinetic study of the lactic acid fermentation. Batch process at controlled pH. J Biochem Microbiol Technol Engg 1:393–412. https://doi.org/10.1002/jbmte.390010406

Lungu B, Johnson MG (2005) Fate of *Listeria monocytogenes* inoculated onto the surface of model Turkey frankfurter pieces treated with zein coatings containing nisin, sodium diacetate, and sodium lactate at 4 °C. J Food Prot 68(4):855–859. https://doi.org/10.4315/0362-028X-68.4.855

Lv W, Zhang X, Cong W (2005) Modelling the production of nisin by *Lactococcus lactis* in fed-batch culture. Appl Microbiol Biotechnol 68(3):322–326. https://doi.org/10.1007/s00253-005-1892-7

MacGroary JA, Reid G (1988) Detection of a lactobacillus substance that inhibits *Escherichia coli*. Can J Microbiol 39:974–978

Martin N, Sprules T, Carpenter M, Cotter P, Hill C, Ross R, Vederas J (2004) Structural characterization of lacticin 3147, a two peptide lantibiotic with synergistic activity. Biochemist 43:3049–3056

Matsusaki H, End N, Sonomato A, Ishizaki A (1996) Lantibiotic nisin Z fermentative production by Lactococcus lactis 10–1: relationship between productions of the lantibiotic, lactate and cell growth. ApplMicrobiol Biotechnol 45:36–40

Mattick ATR, Hirsch A (1947) Further observations on an inhibitory substance (nisin) from lactic *streptococci*. Lancet 2:5–8

McAuliffe O, Ross RP, Hill C (2001) Lantibiotics: structure, biosynthesis and mode of action. FEMS Microb Rev 25:285–308

Meghrous J, Huot E, Quittelier M, Petitdemange H (1992) Regulation of nisin biosynthesis by continuous cultures and by resting cells of *Lactococcus lactis* subsp. *lactis*. Res Microbiol 143:879–890

Meghrous J, Lacroix C, Bouksaim M, LaPointe G, Simard RE (1997) Note: Genetic and biochemical characterization of nisin Z produced by*Lactococcus lactis ssp. lactis biovar. diacetylactis* UL 719. J Appl Microbiol 83:133–138

Meindl K, Schmiederer T, Schneider K, Reicke A, Butz D, Keller S, Gühring H, Vértesy L, Wink J, Hoffmann H, Brönstrup M, Sheldrick GM, Süssmuth RD (2010) Labyrinthopeptins: a new class of carbacyclic lantibiotics. Angew Chem Int Ed Engl 49:1151–1154. https://doi.org/10.1002/anie.200905773

Mierau I, Kleerebezem M (2005) 10 years of the nisin-controlled gene expression system (nice) in *Lactococcus lactis* mini-review. Appl Microbiol Biotechnol 68:705–717. https://doi.org/10.1007/s00253-005-0107-6

Ming X, Daeschel MA (1993) Nisin resistance of foodborne bacteria and the specific resistance responses of *Listeria monocytogenes* Scott A. J Food Prot 56:944–948

Mitra D, Pometto AL III, Khanal SK, Karki B, Brehm-Stecher BF, van Leeuwen HJ (2010) Value-added production of nisin from Soy Whey. Appl Biochem Biotechnol 162:1819–1833

Mocquot G, Lefebvre E (1956) A simple procedure to detect nisin in cheese. J Appl Bacteriol 19:322–323

Moll GN, Clark J, Chan WC, Bycroft BW, Roberts GC, Konings WN, Driessen AJ (1997) Role of transmembrane pH gradient and membrane binding in nisin pore formation. J Bacteriol 179:135–140

Moll GN, Konings WN, Driessen AJM (1999) Bacteriocins: mechanism of membrane insertion and pore formation. Antonie Van Leeuwenhoek 76:185–198

Morency H, Mota-Meira M, LaPointe G, Lacroix C, Lavoie MC (2001) Comparison of the activity spectra against pathogens of bacterial strains producing a mutacin or a lantibiotic. Can J Microbiol 47:322–331

Morris SL, Walsh RC, Hansen JN (1984) Identification and characterization of some bacterial membrane sulfhydryl groups which are targets of bacteriostatic and antibiotic action. J Biol Chem 259:13590–13594

Mortvedt CI, Nes IF (1990) Plasmid-associated bacteriocin production by a *Lactobacillus sake* strain. J Gen Microbiol 136:1601–1607

Mørtvedt CI, Nissen-Meyer J, Sletten K, Nes IF (1991) Purification and amino acid sequence of lactocin S, a bacteriocin produced by *Lactobacillus sake* L45. Appl Environ Microbiol 57:1829–1834

Mota-Meira M, Lapointe GL, Lacroix C, Lavoie MC (2000) MICs of mutacin B-Ny266, Nisin A, Vancomycin, and Oxacillin against bacterial pathogens. Antimicrob Agents Chemother 44:24–29

Mukhopadhyay SN (2007) Experimental process biotechnology protocols, 1st edn. Viva Books, New Delhi, pp 27–31

Mulders JW, Boerrigter IJ, Rollema HS, Siezen RJ, de Vos WM (1991) Identification and characterization of the lantibiotic nisin Z, a natural nisin variant. Eur J Biochem 201:581–584

Müller E, Radler F (1993) Caseicin, a bacteriocin from *Lactobacillus casei*. Folia Microbiol (Praha) 38:441–446

Muriana PM, Klaenhammer T (1991) Purification and partial characterization of Lactacin F, a bacteriocin produced by *Lactobacillus acidophilus* 11088. Appl Environ Microbiol 57:114–121. PMC182671

Nes IF, Diep DB, Håvarstein LS, Brurberg MB, Eijsink V, Holo H (1996) Biosynthesis of bacteriocins in lactic acid bacteria. Antonie Van Leeuwenhoek 70:113–128

Nikaido H, Vaara M (1987) Outer membrane. In: Neidhardt FC (ed) *Escherichia coli* and *Salmonella typhimurium*: cellular and molecular biology, vol 1. Am Soc Microbiol, Washington, DC, pp 7–22

Nilsson L, Chen Y, Chikindas ML, Huss HH, Gram L, Montville TJ (2000) Carbon dioxide and nisin act synergistically on *Listeria monocytogenes*. Appl Environ Microbiol 66:769–774

Noonpakdee W, Santivarangkna C, Jumriangrit P, Sonomoto K, Panyim S (2003) Isolation of nisin-producing *Lactococcus lactis* WNC20 strain from nham, a traditional Thai fermented sausage. Int J Food Microbiol 81:137–145

Novak L, Cocaign-Bousquet M, Lindley ND, Loubiere P (1997) Metabolism and energetics of *Lactococcus lactis* during growth in complex or synthetic media. Appl Environ Microbiol 63:2665–2670

O'Connor PM, O'Shea EF, Guinane CM, O'Sullivan O, Cotter PD, Ross RP, Hill C (2015) Nisin H is a new nisin variant produced by the gut-derived strain *Streptococcus hyointestinalis* DPC6484. Appl Environ Microbiol 81:3953–3960. https://doi.org/10.1128/AEM.00212-15

O'Shea EF, Cotter PD, Ross RP, Hill C (2013) Strategies to improve the bacteriocin protection provided by lactic acid bacteria. Curr Opin Biotechnol 24:130–134. https://doi.org/10.1016/j.copbio.2012.12.003

Özel B, Şimşek Ö, Akçelik M, Saris PEJ (2018) Innovative approaches to nisin production (Mini review). Appl Microbiol Biotechnol. Pub online. https://doi.org/10.1007/s00253-018-9098-y

Papagianni M, Avramidis N (2012) Engineering the central pathways in – *Lactococcus lactis*: functional expression of the phosphofructokinase (pfk) and alternative oxidase (aox1) genes from *Aspergillus niger* in *Lactococcus lactis* facilitates improved carbon conversion rates under oxidizing conditions. Enz Microb Technol 51:125–130

Parada JL, Caron CR, Medeiros ABP, Soccol CR (2007) Bacteriocins from lactic acid bacteria: purification, properties and use as biopreservatives. Braz Arch Biol Technol 50:521–542

Parente E, Ricciardi A (1999) Production, recovery and purification of bacteriocins from lactic acid bacteria. Appl Microbiol Biotechnol 52:628–638

Parente E, Ricciardi A, Addario G (1994) Influence of pH on growth bacteriocin production by*Lactococcus latis* subsp. *lactis* 140 NWC during batch fermentation. Appl Microbiol Biotechnol41:388–394

Parrot M, Charest M, Lavoie MC (1989) Production of mutacin-like substances by *Streptococcus mutans*. Can J Microbiol 35:366–372

Patent (1960) Production of nisin. Patented by United States Patent Office with number US2935503

Pedersen MB, Koebmann Brian J, Jensen Peter R, Dan N (2002) Increasing Acidification of Nonreplicating *Lactococcus lactis* Δ*thyA* Mutants by Incorporating ATPase Activity. Appl Environ Microbiol 68:5249–5257

CV, Moraes DA (2002) Optimization of Nisin Production by *Lactococcus lactis*. Appl hem Biotechnol 98(100):775–789
Pingitore Vera E, Salvucci E, Sesma F, Nader-Macías ME (2007) Different strategies for purification of antimicrobial peptides from Lactic Acid Bacteria (LAB). In: Méndez-Vilas A (ed) Communicating current research and educational Topics and trends in applied microbiology. FORMATEX, Argentina, pp 557–568
Piper C, Hill C, Cotter PD, Ross RP (2011) Bioengineering of a Nisin A producing *Lactococcus lactis* to create isogenic strains producing the natural variants Nisin F, Q and Z. Microb Biotechnol 4:375–382
Piard J-C, Kuipers OP, Rollema HS, Desmazeaud MJ, de VWM (1993) Structure, organization and expression of the lct gene for lacticin 481, a novel lantibiotic produced by *Lactococcus lactis*. J Biol Chem 268:16361–16368
Pongtharangkul T, Demirci A (2004) Evaluation of agar diffusion bioassay for nisin quantification. Appl Microbiol Biotechnol 65:268–272
Pongtharangkul T, Demirci A (2006) Effects of fed-batch fermentation and pH profiles on nisin production in suspended-cell and biofilm reactors. Appl Microbiol Biotechnol 73:73–79. https://doi.org/10.1007/s00253-006-0697-7
Pongtharangkul T, Demirci A (2007) Online recovery of nisin during fermentation and its effect on nisin production in biofilm reactor. Appl Microbiol Biotechnol 74:555–562
Ra R, Beerthuyzenf Marke M, de Vos Willem M, Saris Per EJ, Kuipers Oscar P (1999) Effects of gene disruptions in the nisin gene cluster of *Lactococcus lactis* on nisin production and producer immunity. Microbiology 145:1227–1233
Rauch PJG, de Vos WM (1992) Characterization of the novel nisin-sucrose conjugative transposon Tn5276 and its insertion in *Lactococcus lactis*. J Bacteriol 174:1280–1287
Ray B (1992) Nisin of *Lactococcus lactis* subsp. *lactis* as a food biopreservative. In: Ray B, Daeschel M (eds) Food biopreservatives of microbial origin. CRC Press, Boca Raton, pp 207–264
Rayman KN, Hurst A (1984) Nisin: properties, biosynthesis and fermentation. In: Vandamme EJ (ed) Biotechnology of industrial antibiotics. Marcel Dekker, New York, pp 607–628
Reeves P (1965) The bacteriocins. Bacteriol Rev 29:24–45
Reij MW, Den Aantrekker ED (2004) Recontamination as a source of pathogens in processed foods. Int J Food Microbiol 91:1–11. https://doi.org/10.1016/S0168-1605(03)00295-2
Reisinger P, Seidel H, Tschesche H, Hammes WP (1980) The effect of nisin on murein synthesis. Arch Microbiol 127:187–193
Riley MA (2009) Bacteriocins, biology, ecology, and evolution. Elsevier Univ Massach, Amherst, pp 32–44
Rink R, Arkema-Meter A, Baudoin I, Post E, Kuipers A, Nelemans SA, Akanbi MH, Moll GN (2010) To protect peptide pharmaceuticals against peptidases. J Pharmacol Toxicol Methods 61:210–218. https://doi.org/10.1016/j.vascn.2010.02.010
Ripoche A, Chollet E, Peyrol E, Sebti I (2006) Evaluation of nisin diffusion in a polysaccharide gel: Influence of agarose and fatty content. Innov Food Sci Emerg Technol 7:107–111
Rodríguez JM, Cintas LM, Casaus P, Horn N, Dodd HM, Hernández PE, Gasson MJ (1995) Isolation of nisin-producing *Lactococcus lactis* strains from dry fermented sausages. J Appl Bacteriol 78:109–115
Rogers LA (1928) The inhibiting effect of *Streptococcus lactis* on *Lactobacillus bulgaricus*. J Bacteriol 16:321–325. PMCID: PMC375033
Rollema HS, Kuipers OP, Both P, de Vos WM, Siezen RJ (1995) Improvement of solubility and stability of the antimicrobial peptide nisin by protein engineering. Appl Environ Microb 61:2873–2878. PMID: 7487019
Rollema HS, Metzger JW, Both P, Kuipers OP, Siezen RJ (1996) Structure and biological activity of chemically modified nisin A species. Eur J Biochem 241:716–722
Ross KF, Ronson CW, Tagg JR (1993) Isolation and characterization of the lantibiotic salivaricin a and its structural gene salA from *Streptococcus salivarius* 20P3. Appl Environ Microbiol 59:2014–2021

Rossano R, Del Fiore A, D'Elia A, Pesole G, Parente E, Riccio P (1998) New procedure for the determination of nisin in milk. Biotechnol Tech 12:783–786. https://doi.org/10.1023/A:1008820803

Rouse S, Field D, Daly KM, O'Connor PM, Cotter PD, Hill C, Ross RP (2012) Bioengineered nisin derivatives with enhanced activity in complex matrices. Microb Biotechnol 5:501–508

Ruhr E, Sahl HG (1985) Mode of action of the peptide antibiotic nisin and influence on the membrane potential of whole cells and on cytoplasmic and artificial membrane vesicles. Antimicrob Agents Chemother 27:841–845. PMID: 4015074

Sahl HG, Bierbaum G (1998) Lantibiotics: biosynthesis and biological activities of uniquely modified peptides from Gram-positive bacteria. Ann Rev Microbiol 52:41–79. https://doi.org/10.1146/annurev.micro.52.1.41

Sahl HG, Kordel M, Benz R (1987) Voltage dependent depolarization of bacterial membranes and artificial lipid bilayers by the peptide antibiotic nisin. Arch Microbiol 149:120–124. https://doi.org/10.1007/BF00425076

Sahl H, Jack R, Bierbaum G (1995) Biosynthesis and biological activities of lantibiotics with unique post-translational modifications. Eur J Biochem 230:827–853

Sambrook J, Fritsch EF, Maniatis T (1989) Molecular cloning: a laboratory manual, 2nd edn. Cold Spring Harbor Lab, Cold Spring Harbor

Saris P, Immonen EJ, Reis TM, Sahl HG (1996) Immunity to lantibiotics. Antonie Leeuwenhoek 69:151–159

Schillinger U, Lücke FK (1989) Antibacterial activity of *Lactobacillus sake* isolated from meat. Appl Environ Microbiol 55(8):1901–1906

Schnell N, Entian KD, Schneider U, Gotz F, Zahner H, Kellner R, Jung G (1988) Prepeptide sequence of epidermin, a ribosomally synthesized antibiotic with four sulphide-rings. Nature 333:276–278

Schöbitz RP, Bórquez PA, Costa ME, Ciampi LR, Brito CS (2006) Bacteriocin like substance production by *Carnobacterium piscicola* in a continuous system with three culture brooths Study of antagonism against *Listeria monocytogees* in vacuum packaged salmon. Braz J Microbiol 37:52–57

Sebti I, Ham-Pichavant F, Coma V (2002) Edible bioactive fatty acid-cellulosic derivative composites used in food-packaging applications. J Agric Food Chem 50:4290–4294. https://doi.org/10.1021/jf0115488

Settanni L, Corsetti A (2008) Application of bacteriocins in vegetable food biopreservation. Int J Food Microbiol 121:123–138. https://doi.org/10.1016/j.ijfoodmicro.2007.09.001

Siegers K, Entian KD (1995) Genes involved in immunity to the lantibiotic nisin produced by *Lactococcus lactis* 6F3. Appl Environ Microbiol 61:1082–1089

Simsek O, Saris PEJ (2009) Cycle changing the medium results in increased nisin productivity per cell in *Lactococcus lactis*. Biotechnol Lett 31:415–421. https://doi.org/10.1007/s10529-008-9891-2

Singh S (2013) Lactic acid production. In: RPF II- 2012-2013, In-house Project on 'Development of functional foods through valorization of horticultural produce' (Project No: IARI/ PHT/09/02), Division of Food Science and Post-harvest Technology, ICAR- IARI, New Delhi 110012 (For official use only)

Singh S (2014) Estimation of the recovery of nisin from medium (yield of the product in process). In 'Ann Rep 2013–14, Division of Food Science and Post-harvest Technology, ICAR- IARI, New Delhi 110012, p 14

Singh S, Gupta S, Pal RK, Kaur C (2013) Production of a biopreservative 'nisin': assay and quantification from a cell free extract. In: Abstracts of 7th Asian Conference on Lactic Acid Bacteria, India Habitat Centre, New Delhi, Sept 6–8th 2013

Singh S, Singh KNMS, Holmes L (2015) Modelling the growth of *Lactococcus lactis* NCIM 2114 under differently aerated and agitated conditions in broth medium. Fermentation 1:86–97. https://doi.org/10.3390/fermentation1010086

Singh S, Sukanta D, Kooliyottil R, Saha S, Gupta S, Mandjiny S, Upadhyay D, Holmes L (2016) Nisin production in a two liters bioreactor using *Lactococcus lactis* NCIM 2114. J Med Biol Sci Res 2:21–26

., Hilbers CW, Konigs ARN, van de Ven FJM (1989) NMR studies of antibiotics. ,nment of the H-NMR spectrum of nisin and identification of interresidual contacts. FEBS Lett 252:22–28

Steen MT, Joon CY, And Norman Hansen J (1991) Characterization of the nisin gene as part of a polycistronic operon in the chromosome of Lactococcus lactis ATCC 11454. Appl Environ Microbiol 57:1181–1188

Stein T, Heinzmann S, Solovieva I, Entian KD (2003) Function of *Lactococcus lactis* nisin immunity genes nisI and nisFEG after coordinated expression in the surrogate host *Bacillus subtilis*. J Biol Chem 278:89–94

Stevens KA, Sheldon BW, Klapes NA, Klaenhammer TR (1991) Nisin treatment for inactivation of *Salmonella* species and other gram-negative bacteria. J Food Prot 55:7763–7766

Stoyanova LG, Ustyugova EA, Netrusov AI (2012) Antibacterial metabolites of lactic acid bacteria: their diversity and properties. Prikladnaya Biokhimiya i Mikrobiologiya 48: 259–275. Appl Biochem Microbiol 48(3):229–243

Suárez AM, Azcona JI, Rodríguez JM, Sanz B, Hernández PE (1997) One-step purification of nisin A by immunoaffinity chromatography. Appl Environ Microbiol 63:4990–4992

Tagg JR, McGiven AR (1971) Assay system for bacteriocins. Appl Microbiol 21:943

Tanaka T, Kawata M (1988) Cloning and characterization of *Bacillus subtilis* iep, which has positive and negative effects on production of extracellular proteases. J Bacteriol 170:3593–3600

Taniguchi M, Hoshino K, Urasaki H, Fujii M (1994a) Continuous production of an antibiotic polypeptide (nisin) by *Lactococcus lactis* using a bioreactor coupled to a microfiltration module. J Ferment Bioeng 77:704–708. https://doi.org/10.1016/0922-338X(94)90159-7

ten Brink B, Minekus M, van der Vossen JM, Leer RJ, Huisin'tVeld JH (1994) Antimicrobial activity of lactobacilli: preliminary characterization and optimization of production of acidocin B, a novel bacteriocin produced by *Lactobacillus acidophilus* M46. J Appl Bacteriol 77:140–148

Taniguchi M, Hoshino K, Urasaki H, Fujii M (1994b) Continuous production of an antibiotic polypeptide (nisin) by *Lactococcus lactis* using a bioreactor coupled to a microfiltration module. J Ferment Bioeng 77:704–708. https://doi.org/10.1016/0922-338X(94)90159-7

Terzaghi BE, Sandine WE (1975) Improved medium for lactic streptococci and their bacteriophages. Appl Microbiol 29:807–813. PMID: 16350018

Thompson J, Sackett DL, Donkersloot JA (1991) Purification and properties of fructokinase I from *Lactococcus lactis*. J Biol Chem 266:22626–22633

Toba T, Samant SK, Yoshioka E, Itoh T (1991) Reutericin 6, a new bacteriocin produced by *Lactobacillus reuteri* LA 6. Lett Appl Microbiol 13:281–286. https://doi.org/10.1111/j.1472-765X.1991.tb00629.x

Todorov SD, Dicks LMT (2005) Characterization of bacteriocins produced by lactic acid bacteria isolated from spoiled black olives. J Basic Microbiol 45:312–322

Tramer J, Fowler GG (1964) Estimation of nisin in foods. J Sci Fd Agric 15:522–528

Ustyugova EA, Fedorova GB, Katrukha GS, Stoyanova LG (2011) Investigation of the antibiotic complex produced by *Lactococcus lactis* subsp. *lactis*194, variant K. Mikrobiologiya 80:644–650 (In Russian)

Van de Kamp M, van den Hooven HW, Konings RNH, Bierbaum G, Sahl H-G, Kuipers OP, Siezen RJ, de Vos WM, Hilbers CW, van de Ven FJM (1995a) Elucidation of the primary structure of the lantibiotic epilancin K7 from *Staphylococcus epidermis* K7. Cloning and characterization of the epilancin-K7-encoding gene and NMR analysis. Eur J Biochem 230:587–600

Van de Kamp M, Horstink LM, van den Hooven HW, Konings RNH, Hilbers CW, Frey A, Sahl H-G, Metzger JW, van de Ven FJM (1995b) Sequence analysis by NMR spectroscopy of the peptide lantibiotic epilancin K7 from *Staphylococcus epidermis* K7. Eur J Biochem 227:757–771

van den Hooven (1995) Structure elucidation of the lantibiotic nisin in aqueous solution and in membrane-like environments. Ph.D. Thesis, University of Nijmegen, The Netherlands

van den Hooven HW, Fogotari F, Rollema HS, Konings RNH, Hilbers CW, van de Ven FJM (1993) NMR and circular dichroism studies of the lantibiotic nisin in non-aqueous environments. FEBS Lett 319(1,2):189–194 van der Meer JR, Polman J, Beerthuyzen MM, Siezen RJ, Kuipers OP, De Vos WM (1993) Characterization of the *Lactococcus lactis* nisin A operon genes nisP, encoding a subtilisin-like serine protease involved in precursor processing, and nisR, encoding a regulatory protein involved in nisin biosynthesis. J Bacteriol 175:2578–2588

van Heusden HE, De Kruijff B, Breukink E (2002) Lipid II induces a transmembrane orientation of the pore-forming peptide lantibiotic nisin. Biochemist 41(40):12171–12178

van Kraaij C, Breukink E, Rollema HS, Siezen R, Demel, de Kruijff B, Kuipers OP (1997) Influence of charge differences in the C-terminal part of nisin on antimicrobial activity and signalling capacity. Eur J Biochem 247:114–120

van Kraaij C, Breukink E, Rollema HS, Bongers RS, Kosters HA, de Kruij¡ B, Kuipers OP (2000) Engineering a disulphide bond and free thiols in the lantibiotic nisin Z. Eur J Biochem 267:901–909

Wahlstrom G, Saris PEJ (1999) A nisin bioassay based on bioluminescence. Appl Environ Microbiol 65(8):3742–3745

Wei L, Norman HJ (1990) Some chemical and physical properties of nisin, a small protein antibiotic produced by *Lactococcus lactis*. Appl Environ Microbiol 56:2551–2558

Wiedemann I, Breukink E, van Kraaij C, Kuipers OP, Bierbaum G, de Kruijff B, Sahl HG (2001) Specific binding of nisin to the peptidoglycan precursor lipid II combines pore formation and inhibition of cell wall biosynthesis for potent antibiotic activity. J Biol Chem 276:1772–1779

Wardani AK, Egawa S, Nagahisa K, Shimizu H, Shioya S (2006) Computational prediction of impact of rerouting the carbon flux in metabolic pathway on cell growth and nisin production by *Lactococcus lactis*. Biochem Eng J 28:220–230

Williams GC, Delves-Broughton J (2003) Nisin – structure and biosynthesis. In: Encyclopedia of food sciences and nutrition, 2nd edn. Academic, Amsterdam

Wilson-Stanford S, Kalli A, Håkansson K, Kastrantas J, Orugunty RS, Smith L (2009) Oxidation of lanthionines renders the lantibiotic nisin inactive. Appl Environ Microbiol 75(5):1381–1387

Wirawan RE, Klesse NA, Jack RW, Tagg JR (2006) Molecular and genetic characterization of a novel nisin variant produced by *Streptococcus uberis*. Appl Environ Microbiol 72:1148–1156

Wolf CE, Gibbons WR (1996) Improved method for the quantification of the bacteriocin nisin. J Appl Microbiol 80:453–457

Wu Z, Li X (2007) Modification of the data-processing method for the turbidimetric bioassay of nisin. Appl Microbiol Biotechnol 74:511–516. https://doi.org/10.1007/s00253-006-0670-5

Wu Z, Wang L, Jing Y, Li X, Zhao Y (2009) Variable Volume Fed-Batch Fermentation for Nisin Production by *Lactococcus lactis* subsp. *lactis* W28. Appl Biochem Biotechnol 152:372–382. https://doi.org/10.1007/s12010-008-8335-8

Yang R, Johnson MC, Bibek R (1992) Novel method to extract large amounts of bacteriocins from lactic acid bacteria. Appl Environ Microbiol 58:3355–3359. PMID: 1444369

Yang S-C, Lin C-H, Sung CT, Fang J-Y (2014) Antibacterial activities of bacteriocins: application in foods and pharmaceuticals. Front. Microbiol 5: 241. https://doi.org/10.3389/fmicb.2014.00241. www.frontiersin.org

Yildirim Z, Johnson MG (1997) Detection and characterization of bacteriocin produced by *Lactococcus lactis* subsp *cremois* R isolated from radish. Lett Appl Microbiol 26:297–304. PMID: 9633097

Yoneyama F, Fukao M, Zendo T, Nakayama J, Sonomoto K (2008) Biosynthetic characterization and biochemical features of the third natural nisin variant, nisin Q, produced by *Lactococcus lactis* 61-14. J Appl Microbiol 105:1982–1990

Zacharof MP, Lovitt RW (2012) Bacteriocins produced by lactic acid bacteria – a review Article. APCBEE Procedia 2:50–56

Zendo T, Fukao M, Ueda K, Higuchi T, Nakayama J, Sonomoto K (2003) Identification of the lantibiotic nisin Q, a new natural nisin variant produced by *Lactococcus lactis* 61-14 isolated from a river in Japan. Biosci Biotechnol Biochem 67:1616–1619

Zhang YF, Liu SY, Du YH, Feng WJ, Liu JH, Qiao JJ (2014) Genome shuffling of *Lactococcus lactis* subspecies lactis YF11 for improving nisin Z production and comparative analysis. J Dairy Sci 97:2528–2541. https://doi.org/10.3168/jds.2013-7238

Block DE (2009) Using highly efficient nonlinear experimental design methods for optimization of *Lactococcus lactis* fermentation in chemically defined media. Biotechnol Prog 25:1587–1597. https://doi.org/10.1002/btpr.277

Zhang G, Mills DA, Block DE (2009) Development and optimization of chemically-defined media supporting high cell density growth of lactococci, enterococci, and streptococci. Appl Environ Microbiol 75:1080–1087

Zheng H, Zhang D, Guo K, Dong K, Xu D, Wu Z (2015) Online recovery of nisin during fermentation coupling with foam fractionation. J Food Eng 162:25–30. https://doi.org/10.1016/j.jfoodeng.2015.04.006

Zouhir A, Riadh H, Ismail F, Jeannette BH (2010) A new structure-based classification of Gram-positive bacteriocins. Protein J 29:432–439. https://doi.org/10.1007/s10930-010-9270-4

Zwiettering MH, Jongenburger I, Rombouts FM, Van 'T Riet K (1990) Modeling of the bacterial growth curve. Appl Environ Microbiol 56:1875–1881

9789811383939